SYNERGY MATTERS

Working with Systems
in the 21st Century

SYNERGY MATTERS
Working with Systems
in the 21st Century

Edited by

Adrian M. Castell

Nottingham Business School
Teent University, Nottingham, United Kingdom

Amanda J. Gregory
Giles A. Hindle
Mathew E. James
Gillian Ragsdell

Lincoln School of Management
University of Lincolnshire and Humberside
Lincoln, United Kingdom

Springer Science+Business Media, LLC

Proceedings of the Sixth International Conference of the United Kingdom Systems Society on Synergy Matters; Working with Systems in the 21st Century, held July 5–9, 1999, in Lincoln, United Kingdom

ISBN 978-1-4757-7182-4   ISBN 978-0-306-47467-5 (eBook)
DOI 10.1007/978-0-306-47467-5

© 1999 Springer Science+Business Media New York

Originally published by Kluwer Academic / Plenum Publishers, New York in 1999.

Softcover reprint of the hardcover 1st edition 1999

10 9 8 7 6 5 4 3 2 1

A C.I.P. record for this book is available from the Library of Congress.

PREFACE

The 21st Century is now almost upon us and whilst this represent a somewhat artificial boundary it provides an opportunity for reflection upon the changes, and the accelerating pace of change, in our social, economic and natural environments. These changes and their effects are profound, not least in terms of access to information and communication technologies, at once global in affect and manifest in the local.

Recognition of these changes and their consequent demands are reflected in the theme of this Conference, *Synergy Matters: Working with Systems in the 21st Century*. The word synergy has entered the everyday lexicon but it is useful to reflect on the etymology of the word: synergy comes from the Greek *synergos* which means working together. This demands a platform for participation through the development of dialogues between disciplines and people, the very stuff of systems thinking. Synergy matters, referring both to its critical importance as well as to issues associated with its realisation, represent(s) a *liet motif* for the explication of systems ideas into the next century. Achieving clarity, methodological transparency and participation in problem definition and resolution will remain essential canons in the best of systems thinking and practice. The contributions in this volume, from methodology development through to the problems of practical application, clearly illustrate these concerns and the tensions that exist between them.

Working together does not only mean establishing links with other disciplines but also with those at different stages of their careers as well as across different cultures. We are pleased to include in this collection of papers contributions from systems thinkers from nineteen countries. The high number of papers included within the student stream demonstrates the level of activity of researchers recently entering the domain and that systems is a 'live' discipline with the ideas of systems thinking resonating with a new generation.

The quality of the papers included within the Proceedings has been a concern of the Programme Committee and, in consideration of the forthcoming Research Assessment Exercise, rigorous review procedures were employed. In the first instance, each abstract submitted was reviewed 'blind' by at least two members of the Programme Committee and as a result of this process some 16% of the abstracts were rejected. Full papers were again reviewed by the Programme Committee and approximately 20% of the papers were inspected by the Executive Committee of the UKSS to ensure consistency of application of quality standards.

Quality in terms of academic standards and clarity of thought must be defining characteristics if the systems discipline is to fulfil its potential role in helping to address the problems of the new century. The future of the discipline will be shaped through the thoughts and actions of a new generation of systems thinkers who recognise the importance of synergy in providing both the tools and methodologies for enabling change and supporting creativity in environments bound ever more closely to emerging technologies. In determining the nature of synergy, and the ways that synergy matters, we cannot be prescriptive, rather we open the debate here in the expectation that it will be carried forwards by those actively engaged in its realisation in practice.

Thanks to all those who have and will contribute to this debate.

Amanda J. Gregory
Adrian M. Castell
Lincoln School of Management
University of Lincolnshire & Humberside
December 1998

ACKNOWLEDGEMENTS

The Programme Committee would like to express their thanks, and the thanks of the United Kingdom Systems Society, to a variety of people and institutions for their help in the organisation of this Conference.

We are grateful to Professor Roger King, Vice-Chancellor of the University of Lincolnshire and Humberside, and Professor Mike Jackson, Dean of the Faculty of Business and Management, University of Lincolnshire and Humberside, for their encouragement and support.

Our thanks also go to the members of the International Advisory Committee who have promoted the Conference to systems thinkers throughout the World. In addition, we would not have been able to complete the organisation of this Conference without the help of the UKSS Executive Committee and in particular the advice and encouragement of the Chair of the previous Programme Committee, Professor Frank Stowell.

Special thanks and a debt of gratitude is expressed to our Managing Editor, Joanna Lawrence of Plenum Publishing, for her assistance, patience and unstinting support in the preparation of these Proceedings.

The Committee and the Society would also like to thank the organisations who have kindly agreed to sponsor this event as well as Doreen Gibbs for all her administrative assistance.

PROGRAMME COMMITTEE

Amanda J. Gregory (Chair)   Giles A. Hindle   Gillian Ragsdell
Adrian M. Castell   Mathew E. James

Lincoln School of Management
University of Lincolnshire and Humberside

INTERNATIONAL ADVISORY COMMITTEE

Australia	William Hutchinson
Austria	Robert Trappl
China	Gu Jifa
Colombia	Alfonso Reyes
Czech Republic	Igor Vajda
Germany	Werner Schuhmann
Hong Kong	Joseph K.K. Ho
India	P.N. Murthy
Italy	Gianfranco Minati
Japan	Kyoichi Kijima
Korea	Yong Pil Rhee
Mexico	M. Adrian Flores
Netherlands	Gerard de Zeeuw
New Zealand	John Brocklesby
Poland	Wojciech Gasparski
Romania	Horatiu Dragomirescu
Slovak Republic	Ken Udas
South Africa	Tom Ryan
Sweden	Stig Holmberg
Switzerland	Werner Ulrich
U.S.A.	Peter A. Corning
U.S.A.	G.A. Swanson
Venezuela	Ramses Fuenmayor

CONTENTS

Change Management

Community & Social Systems Research

Critical Systems Thinking

Developments & Applications of Systems Methodologies

Information Systems

Organisational Learning

SEEING SYSTEMS: OVERCOMING ORGANISATIONAL FRAGMENTATION

Raul Espejo

Lincoln School of Management
Bayford Pool
Lincoln LN6 7TS
UK

INTRODUCTION

This paper deals with the issue of structural fragmentation in organisations. Very often we do not see how to relate resources in order to produce effective organisations. We fail to see these resources as parts of possible systems. I want to illustrate in this paper the value of *seeing systems*. Failing in this increases the difficulty of relating the thinking and doing of organisations, thus hindering their learning.

I first offer an introduction to the idea of seeing systems, second elaborate the idea of complexity unfolding as a natural process to deal holistically with task complexity, third discuss, supported by empirical work at the National Westminster Bank, a common case of fragmentation in task implementation and finally, offer the experience of designing a process to overcome this fragmentation.

ABOUT SYSTEMS

The challenge is seeing the systems implied by people's tasks. *A system is an observer's construct of interrelated resources constituting a whole.* In order for us to see an autonomous system, we must recognise those resources *creating, regulating* and *producing* a task. What is a common observation in all kinds of organisations, in both the public and private sectors, in large and small enterprises, is that we create fragmented structures. For instance, it is not uncommon to see in the public sector institutions producing policies whose regulation is the responsibility of other institutions and whose implementation is done by yet other institutions. This diagnosis would not be such a problem if all these institutions were effectively communicated, that is, if the concerned people were able to see and produce the system. But unfortunately, often this is not the case.

Synergy Matters: Working with Systems in the 21st *Century,*
Edited by Castell *et al.*, Kluwer Academic / Plenum Publishers, New York, 1999.

There are a number of implications emerging from this view of systems. First, working out an organisation's task, is working out its meaning as a social activity and this is often the outcome of hazy conversational processes. Who are the ones responsible for creating this meaning? If it is not created from within the organisation is not a system but only of a fragment. How can it be autonomous if it does not create its own task? Second, since we may be unclear or unsure about an organisation's task we may find it difficult to distinguish which resources are producing the system and which are regulating it. From particular viewpoints, it is not unusual to *see* in organisations, regulatory resources striving for their own autonomy at the expense of servicing the global interests. Third, to develop this capacity to observe systems requires practices. We need capacity to observe a total system, that is the resources and their relations creating, regulating and producing a social meaning. This requires capacity to see deeper structures in the huge number of contingencies in daily life.

UNFOLDING OF COMPLEXITY

People responsible for large complex tasks, beyond their action possibilities, must rely for their implementation in the autonomous actions of other people. Indeed, a natural strategy to cope with complexity is the emergence of autonomous units within autonomous units. This kind of embedding was recognised by Beer in his Viable System Model (Beer 1979), and if seen from the perspective of a total organisation, is what I have referred to as its *unfolding of complexity*. This is shown in Figure 1 (Espejo 1989).

Of course we can also think the other way round; a small autonomous unit as it becomes more functionally differentiated, to maintain its cohesion, may require the emergence of a larger autonomous unit subsuming the differentiated resources and this process may continue as the larger units become more and more functionally differentiated (Kelly, 1994). This is the strategy for nature's evolution.

Autonomous units within autonomous units is above all the outcome of self-regulating and self-organising processes.

From the viewpoint of management, it should be apparent the value of understanding how to enable *self-organisation* in a direction consistent with the organisation's self-constructed tasks. This is a critical strategy to enhance the performance of an organisation in its environment. Hence the relevance of seeing that an organisation consists of both, autonomous units, or as I call them, *primary activities*, producing its purposes and *regulatory functions* maintaining its cohesion. This applies to each autonomous unit within the organisation, and to each autonomous unit within each autonomous unit, until the complexity of the organisation's task is fully absorbed. Each autonomous unit depends on its own regulatory functions to construct its task and maintain cohesion. Autonomy emerges throughout the organisation in the form of smaller units so that each takes responsibility for co-evolving in their medium (i.e. in their environment, from the viewpoint of an observer). In managerial terms each has to pay attention to its short, medium and long-term. This is a requirement to enhance the organisation's performance and to manage effectively the complexity of environmental issues. Autonomy is limited only by the need for each unit to respect the cohesion of the larger unit within which it is embedded. It is not difficult to anticipate that cohesion requires communications between the autonomous *parts* and the regulating resources in the context of an autonomous *whole* (Figure 1). The structural requirements for these communications are given by the Viable System Model (Beer, 1979). We call this kind of organisation, based on autonomous units within autonomous units, within autonomous units, and so forth, *recursive organisations* (Espejo, 1996)

For us, whether managers, researchers or participants, as observers, the challenge is seeing how far are particular institutions/resoureces from these recursive organisations. A key aspect of recursive organisations is that each primary activity constructs its own task. If this is not the case, we are seeing a hierarchical and not a recursive organisation. It is clear that in today's societies the rule is hierarchical and not recursive organisations, however we ought to be aware of the implications of this fact.

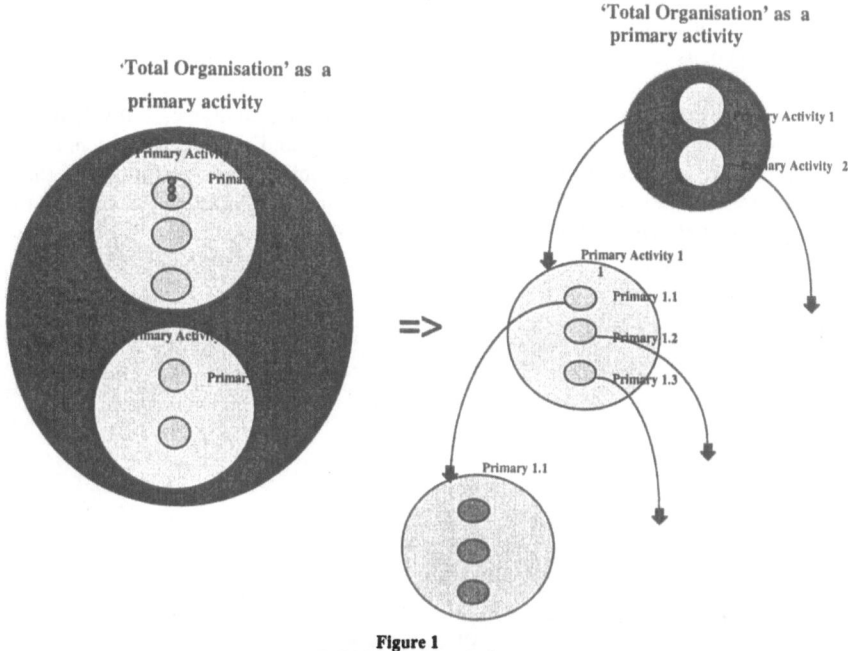

Figure 1
Unfolding of Complexity

A recursive organisation emerges from multiple communications, producing the alignment of self-defined tasks by each and every one of the primary activities, at all structural levels, and not by the unilateral imposition of corporate purposes. In this sense a recursive organisation is inherently pluralistic. An aspect which limits the construction of a recursive organisation is the fragmentation of resources. This is the case when the resources needed for creating, regulating and implementing a task are not integrated. Fragmentation may take place in the processes creating tasks and policies and also in the processes for regulating and implementing them. Here we will only focus on the implemetation of tasks.

FRAGMENTATION IN TASK IMPLEMENTATION

Fragmentation in the implementation of tasks is particularly clear in large functional structures, where this implementation is done by units operating largely in isolatation, while the management of their operational interdependencies is left to overviewing resources lacking in capacity for this purpose. This may make impossible their integration. During the 90's firms have carried out different forms of business and organisational processes re-engineering to overcome this fragmentation. This was one of the concerns of the SYCOMT project at the National Westminster Bank (Espejo and Gill, 1996)[1]. Its focus was with the retail operations of NatWest UK, encompassing over 2,000 branches structured into seventeen regions. As in most banks, they had to adjust their structures as a result of new information technology. Shortly before we started our study, a new service delivery strategy had been implemented which aimed to improve services to customers by centralising specialised resources for some common operations, thereby enabling higher qualified staff to be employed for these activities. The new strategy created Lending and Service Centres to carry out particular work that was previously undertaken in branch bank offices. Figure 2 shows our view of the bank's unfolding of complexity after this restructuring. In this unfolding we hypothesise areas and service and lending centres, as primary activities of a region; these are its autonomous units. Within the areas we find the high street branches as primary activities.

Areas often contained both new branches, which operated in collaboration with the new centres, and old branches still responsible for the total operations. Lending Centres could operate as autonomous units in managing and delivering loan products. These functions, below certain discretionary powers, could also be performed by the branches with which they were collaborating. Service centres also acted autonomously, essentially as computer centres processing the bank's transactions, including those of branches and Lending Centres.

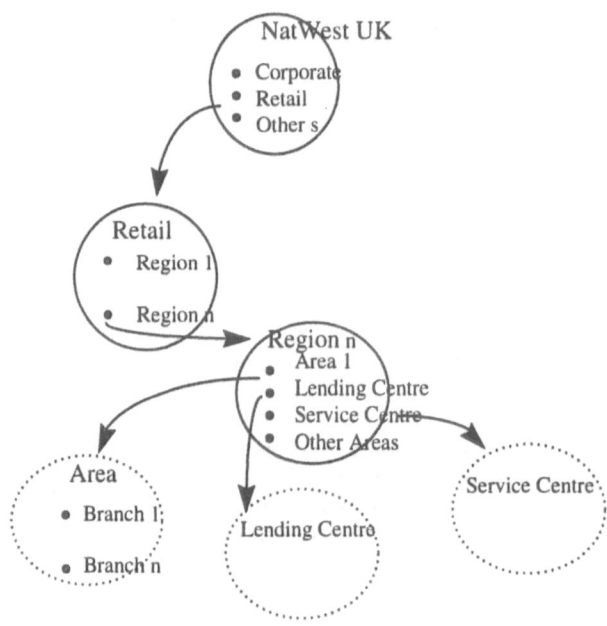

Figure 2: NatWest UK Delivery Strategy

This arrangement was creating fragmentation because branches, Lending Centres and Service Centres were operationally interdependent, but were part of different, independent business reporting structures. Branches and Lending Centres also competed

---

[1] The SYCOMT research project was part of the Computer Supported Co-operative Work (CSCW) research programme sponsored by the UK Department of Trade and Industry and the Engineering and Physical Sciences Research Council.

to a degree for a particular customer's business. Incentives were based on the loans secured within a branch or Lending Centre, so branch staff might be reluctant to pass on a customer contact to a Lending Centre and vice-versa.

OVERCOMING FRAGMENTATION: IMPROVING TEAM WORK

The bank agreed SYCOMT could establish a prototype to experiment with ways of overcoming this fragmentation. We proposed three options. One was to make the branch the lending primary activity within the Region, with Lending Centres offering expert advice to support these activities (i.e. operating as regulatory functions rather than as businesses in their own right). Another would have made the Lending Centres the primary activity and owner of the customer, with branches acting strictly as functional support to facilitate channels of communication with customers. The bank decided to proceed with a third 'hybrid' option, which was a mixed perspective combining roles. This was implemented initially for loans to small businesses within the Chester and Wirral Area of NatWest's Merseyside Region and, later on, for all the branches' operations.

The SYCOMT prototype established 'lending teams'. These were 'virtual' teams composed of people in branches and Lending Centres who could work together interactively using IT-based communication services, like email and bulletin boards accessed through intelligent desktop PCs. Figure 3 shows the unfolding of complexity that emerged along with the implementation of this strategy. The circle at the bottom makes apparent that the prototyped produced a *virtual area* before the intended virtual lending teams handling loans.

The reporting system was changed to ensure that everyone worked together in the same direction. Assistance was also provided to help staff from different locations to understand different roles and to get to know each other. Extensive job swaps, including ones at managerial levels, proved to be the most successful and popular way of achieving this. Effective team working from different locations depended on all staff having access to an appropriate PC-based network. Only dumb terminals were available previously, so a new network supporting co-operative team working had to be developed first. The implementation of the intended virtual teams was facilitated by the use of process flowcharting techniques, such as 'deployment flowcharting' (Howard 1995). This enabled the braiding of business and organisational processes by modelling the details of the lending business process in relation to the unfolding of complexity model, as well as linking all processes with their sub-processes.

However, what most likely triggered the emergence of the *virtual area* was the establishement of a Meta-level Area Team (MAT), consisting of the managers of the Area, Lending Centre and Service Centre, to address the need to provide co-ordination and a common framework across existing functional divides. The MAT was the mechanism integrating lending and service activities with branch activities. It was responsible for aspects like generating a shared vision and agreeing joint aims, performance measures and targets. The MAT was a 'steering' regulatory team only regulating and doing some policy creation, but depending on virtual teams to make things happen. The MAT together with these teams constituted the Virtual Area Primary Activity which embed lending teams. After a six months trial, this Area outperformed all others in the Region by at least 25%, that is, it became clear the experimental hybrid approach could boost performance substantially by forging a virtual organisation that operated effectively as an integrated autonomous unit. In addition to the performance outcomes, the success of the approach was indicated by the degree to which the technological drive was soon pulled by team members once they could see the tangible benefits, such as real-time access to operational information for which they previously had to wait one or more days as it was distributed through couriers.

Some ingredients of this success related to the specific environment, such as the bank's project manager who gained deep knowledge of the principles and practice of the unfolding of complexity method, as well as being an influential motivator and resolver of potential resistance to change. Job swaps also proved to be important.

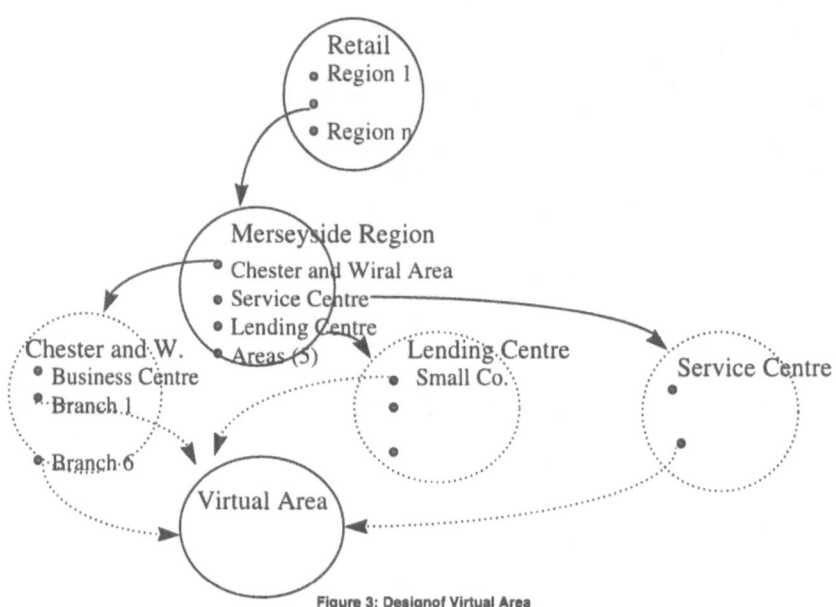

Figure 3: Designof Virtual Area

However, the enduring broader significance of this project is its demonstration of how structural changes can help to improve performance and replace fragmentation by cohesion. Once the structure was right, including the new communication links, all people in the virtual teams began to flow together rather than compete. The prototype also showed that introducing technology is much easier if attention is paid first to designing an appropriate structure.

REFERENCES

Beer, S. (1979). The Heart of Enterprise. Chichester: Wiley.

Espejo, R. (1989). The Viable System Model Revisited. In: Espejo, R & Harnden, R, eds. The Viable System Model: Interpretations and applications of Stafford Beer's VSM. Chichester: Wiley.

Espejo, R. and Gill, A. (1996). SYCOMT Handbook (eds.), Syncho Ltd. Aston Science Park, Birmingham, UK.

Howard, D. (1995) Deployment Flowcharting of Business Processes, Chislehurst, Kent (PO Box 281): Management-NewStyle.

Kelly, K. (1994). Out of Control: The New Biology of Machines. London: Fourth Estate Ltd.

FIGHTING FAILURE

Joyce Fortune

Technology and Manufacturing Management
Centre for Complexity and Change
The Open University
Walton Hall
Milton Keynes, MK7 6AA

INTRODUCTION

The preface to the volume of papers presented to the 1995 UKSS conference (Ellis, Gregory, Mears-Young and Ragsdell, 1995) began by pointing out that the systems movement was 40 years old before arguing that 'the future of the systems movement depends on our ability to address the question: "How can the various disciplines of the systems movement work together in addressing the problems which face the human race?" It then went on to assert that 'the systems movement is facing a crisis' but did not specify, or even speculate upon, the nature of this crisis. Instead, it followed the assertion with the very positive statement: 'the role of the trans-disciplines, such as the systems movement, as the means by which communication across the disciplines may be facilitated, is becoming increasingly important.'

This paper looks again at some of those 'critical issues' that were raised in 1995. In effect, it is a personal exploration by the author of the purpose of systems thinking and of some of the challenges the systems discipline faces and the sorts of questions it should be able to answer if it is to give people a chance of fighting failure and assuring success to an extent that makes a discernible difference to the total sums of failure and success that people experience.

THE PURPOSE OF SYSTEMS THINKING

Many sophisticated and intellectually impressive arguments have been put forward to explain why systems thinking is needed. In recent years these have often become bound up in political positions concerned with, for example, emancipation and sustainability. However, it will be argued here that most attempts to apply systems thinking are in effect motivated by a simpler concern. This concern was expressed by Jenkins, the founding Professor of Systems

Synergy Matters: Working with Systems in the 21st *Century*,
Edited by Castell *et al.*, Kluwer Academic / Plenum Publishers, New York, 1999.

7

at Lancaster University, in the first paper published in the first Journal of Systems Engineering. In that paper Jenkins (1969) used 'disasters that could have been avoided with systems engineering' and 'success stories that went with good systems engineering' to highlight 'the urgent need to apply a systems approach'. Although 'traditionalist' systems thinking of the type employed by Jenkins and his contemporaries has since been described as possessing 'obvious failings' (Jackson, 1991) and being 'too naive in its questions' (Checkland, 1995) the view that the purpose of a systems approach is to prevent failure and deliver success has not been bettered.

This notion, that the purpose of systems thinking is to prevent failure and deliver success, is stated explicitly in some areas of the systems literature (for example, Bignell and Fortune (1984), Fortune and Peters (1995), Bowonder, Arvind and Miyake (1991) and van Gigch (1988)), but it can be seen almost as clearly in the practice of many, many others who use systems approaches but do not ostensibly link their work to the study of failures. For example, a selection, taken at random, of papers published in Systems Practice in the first half of the nineties that had titles indicating an emphasis on application, revealed that their work was motivated by concerns similar to those of the authors referenced above. Amongst them were:

- Barstow (1990) - A study of the application of Interactive Planning at the Tennessee plant of the Aluminium Company of America was said to be triggered by 'many interrelated problems' but the problems listed can all be summarized as failures to control costs and quality and to develop new products.

-

- Fuenmayor, Bonucci and Lopez-Garay (1991) - Their interpretive-systemic study of the University of Los Andes in Venezuela was undertaken because the university was being perceived, in the 'opinions of most actors in the institution and some of the general public', to fail to accomplish its mission.

- Ledington (1992) - This application of Soft Systems Methodology at the Xenon Corporation followed identification of 'inadequacies in the management information system' (thus suggesting a pattern of partial failures) and by the potential failure of the computer system which was described as 'unreliable and prone to error'.

- Green (1992) - In this account of the application of Total Systems Intervention in the North Yorkshire Police Force failures as perceived at various organizational levels were used to describe the 'problem situation' in terms of a series of 'inabilities'. For example, the executive were said to be concerned by an 'inability to pass information down through the organizational hierarchy in a manner which allowed members of the organization to understand not only what was required of them but also why' while the junior ranks were experiencing an 'inability to pass information upward and have it acted upon'.

- Lartin-Drake and Curran (1996) - In this study a version of Ackoff's circular organization was used at University Hospitals, the Milton S. Hershey Medical Center, Penn State, US, to look for 'a better way to structure patient care'. The study began because the director of the Department of Nursing believed that 'the lack of co-ordinating mechanisms between nurses and physicians was detrimental to patient care'.

The appropriateness of systems approaches when looking at problem situations such as those described on the previous page has been argued very widely. (See, for example, Flood and Jackson (1991), Flood and Carson (1993), Fortune and Peters (1995) and Wilson (1990).) In fact, setting aside the 'pessimism about adopting any systemic or systematic approach [associated with postmodernism]' described by Jackson (1995), agreement over the benefits of systems approaches is so great amongst the work of those who acknowledge the existence of such approaches that it is difficult to find detractors. And where negative views do exist they tend to be focussed upon the weaknesses of systems approaches rather than claiming that the approaches are fundamentally inappropriate or useless. For instance, Collins (1998), 'drawing on historical and sociological perspectives' describes systems thinking as representing a 'unitary modelling of organizations' and criticises it on the grounds that it fails to consider conflict properly, because it regards conflict as 'dysfunctional' and 'something to be avoided' and because it asserts harmony above all else. Within the systems literature Herrscher (1996) is also critical, arguing that the view that the systems approach is 'devoid of absolute values' cannot be set aside. However, critics such as these stop far short of suggesting that systems thinking cannot make a contribution to understanding and, after all, as Churchman (1985) points out, 'Given the limited scope of our capabilities to solve the social problems we face, we have every right to question whether any approach ... is the correct approach to the understanding of our society.'

But despite the absence of criticism there is a major worry. Although advocates of systems thinking point to many successful interventions across a wide range of endeavours it is clear that systems thinking has become embedded in the way enterprises operate to only a limited extent. Furthermore, there is a vast body of work in the areas that systems thinking seeks to address that ignores such thinking entirely or presents it as a travesty of itself. As Warren, Ellis and Adman (1998) say: 'The lack of transfer of contemporary theory into application by practising managers and consultants must be a concern to all those engaged in the future development of Systems Thinking.'

Warren, Ellis and Adman recommend a way of addressing this concern; they suggest that 'academic Systems Thinkers' should 'promote' (i.e. sell) the benefits of systems thinking to 'enlightened managers and organizations' (i.e. potential customers). It can be argued, however, that this suggestion may only be tackling part of the problem because it is predicated upon the view that systems thinking, in its present state of development, can meet 'customer requirements'. This paper will now consider the extent to which that is the case.

MEETING CUSTOMER REQUIREMENTS

Failure is extremely widespread in all spheres of activity. For example, the UK usually faces at least two disasters a year in which up to 100 people are killed and every three or four years there is a major incident in which between 100 and 1000 die (Keller, Wilson and Kara-Zaitri, 1990). Bearing in mind the limited extent to which systems thinking is used, the high prevalence of failure probably says nothing about the effectiveness of systems thinking. However, it could be expected that there are pockets of endeavour where systems thinking is having an impact. One promising area where much work has been done to incorporate systems thinking into practice is information systems development. This has been the focus of much research and methodological development with many accounts of successful applications (see, for example, Avison and Wood-Harper (1990), Stowell (1995) and Bell and Wood-Harper (1998)) but unfortunately failure is still commonplace. High-profile examples that have appeared in the press recently include the Department of Social

Security's £25 million statistical information system which was aborted, British Gas's new billing system which initiated disconnection procedures against customers who had not yet been billed, the system designed to open the domestic electricity market, and the Swanwick air traffic control centre which was due to open in 1996 but which will not be on-line this century. The are not isolated examples; one of the major studies into the success of similar projects (OASIG, 1996) found 80 to 90 per cent of IT investments did not meet their performance objectives and around 80 per cent of new systems were delivered late and overbudget. Furthermore, the study showed that many of the problems encountered were precisely those systems approaches are designed to address. Those reported included the following:

- Most investments in IT are technology-led, addressing too narrow an agenda, and reflecting too technical an emphasis.
- Organizations are not successful at attending to the non-technical ... Most organizations lack an integrated approach to organizational and technical change.
- The majority of companies fail to consider how work should be organized and jobs designed to make the technologies effective.
- In most cases users do not have a substantial influence on system development.
- Organizational and managerial practice are characterized by high levels of fragmentation and differentiation, any by associated political concerns.

(OASIG, Section 1.4)

So what is going wrong? One possible answer is that although the knowledge to be found in the systems movement had become broader it has not become deeper and the consequence has been that some of the valuable original concepts have not been developed to the point where they become tools for action as well as tools for thought. The example that will be considered here as possibly exemplifying this neglect of basic concepts is emergence.

EMERGENCE

Although Ackoff (1971) did not include the concept in his 'system of system concepts it did feature in earlier systems literature (for example, Young (1964)) and its importance was acknowledged by Checkland (1979) when he identified emergence and hierarchy and communication and control as the two pairs of concepts upon which systems thinking was founded.

Checkland (1981) also provided a useful insight into emergence when he wrote: 'the concept of organized complexity .. became the subject matter of the new discipline "systems"; and the general model of organized complexity is that there exists a hierarchy of levels of organization, each more complex than the one below, a level being characterised by emergent properties which do not exist at the lower level.' However, in effect, little seems to have been done to develop understanding of the concept further or, perhaps even more importantly, to bring it into practical use. Some obscure papers such as that by Bahm (1989), who coins the word 'emergentism', have tackled the topic but for the most part, especially within the common parlance of systems teaching, this important concept has often been glossed over as being entirely represented by the phrase 'the whole is more than the sum of the parts'. Worse still, emergent properties have often been exemplified by trivial observations of the type 'wetness is a property of water that could not be predicted from the properties of hydrogen and oxygen'.

The treatment the concept has received begs a lot of important questions. For instance, is emergence anything more than the product of interaction? Is it possible to predict emergent properties? If so, how could the 'whole' be manipulated to maximise emergent properties that are desirable and eliminate those that are undesirable? Experience suggests that in complex situations the whole is often less than the sum of the parts; how does this fit the notion of emergence? When the purpose of using systems approaches is to fight failure all of these are very important questions.

CONCLUSION

This paper has argued that the purpose of systems thinking is not thinking per se but devising strategies and plans for action that will fight failure and deliver success. It has accepted that encouraging more widespread use of systems approaches is important but it has also argued that there is a failure to make full use of some of the concepts that derive from the early days of systems engineering.

It is such a shame that the phrase has been so discredited, otherwise my cry would be 'back to basics'!

REFERENCES

Ackoff, R.L., 1971, Towards a system of systems concepts, *Man Sci*, 17:11:661-71.

Avison, D. and Wood-Harper, T., 1990, "Multiview: An Exploration in Information Systems Development", Blackwell, Oxford.

Bahm, A.J., 1989, Methodologies of five types of philosophy, *General Systems: Yearbook of the Society for General Systems Research*, 32:197-200.

Barstow, A.M., 1990, On creating opportunity out of conflict: two case studies, *Systems Practice*, 3:4:339-55.

Bell, S. and Wood-Harper, T., 1998, "Rapid Information Systems Development", McGraw Hill, London.

Bignell, V. And Fortune, J., 1984 "Understanding Systems Failures", Manchester University Press, Manchester.

Bowonder, B., Arvind, S.S. and Miyake, T., 1991, Low probability—high consequence accidents: application of systems theory for preventing hazardous failures, *Systems Research*, 9:2:5-58.

Checkland, P.B., 1979, The shape of the systems movement, *J of Applied Systems Analysis*, 6:129-35.

Checkland, P.B., 1981, "Systems Thinking, Systems Practice", Wiley, Chichester.

Checkland, P., 1995, Systems theory and management thinking, in "Critical Issues *in:* Systems Theory and Practice", Ellis, K., Gregory, A., Mears-Young, B.R. and Ragsdell, G., eds., Plenum, New York.

Churchman, C.W., 1985, quoted in Schoderbek, P.P., Schoderbek, C.G. and Kefalas, A.G., "Management Systems", Business Publications, Texas.

Collins, D., 1998, "Organizational Change", Routledge, London.

Ellis, K., Gregory, A., Mears-Young, B.R. and Ragsdell, G., 1995, "Critical Issues in Systems Theory and Practice", Plenum, New York.

Flood, R.L. and Carson, E.R., 1993, "Dealing With Complexity: An Introduction to the Theory and Application of Systems Science", Plenum, New York.

Flood, R.L. and Jackson, M.C., 1991, "Creative Problem Solving", Wiley, Chichester.

Fortune, J. and Peters, G., 1995, "Learning from Failure", Wiley, Chichester.

Fuenmayor, R., Bonucci, M. and Lopez-Garay, H., 1991, An interpretive-systemic study of the University of Los Andes, *Systems Practice*, 4:5:507-25.

Green, S.M., 1992, Total systems intervention: organizational communication in North Yorkshire Police, *Systems Practice*, 5:6:585-600.

Herrscher, E.G., 1996, An agenda for enhancing systemic thinking in society, *Systems Research*, 13:2:159-64.

Jackson, M.C., 1991, "Systems Methodology for the Management Sciences", Plenum, New York.

Jackson, M.C., 1995, Beyond the fads: systems thinking for managers, *Systems Res*, 12:1:25-42.

Jenkins, G.M., 1969, The systems approach, *J. of Systems Eng*, 1:3-49.

Keller, A.Z., Wilson, H. and Kara-Zaitri, C., 1990, The Bradford disaster scale, *Disaster Man*, 2:207-13.

Lartin-Drake, J.M. and Curran, C.R., 1996, All together now: the circular organization in a University Hospital Part 1 planning and design, *Systems Practice*, 9:5:391-402.

Ledington, P., 1992, Intervention and the management process: an action-based research study, *Systems Practice*, 5:1:17-36.

OASIG, 1996, "The Performance of Information Technology and the Role of Human and Organizational Factors", Report to the ESRC, UK.

Stowell, F. (Ed), 1995, "Information Systems Provision", McGraw Hill, London.

van Gigch, J.P., 1988, Diagnosis and metamodelling of systems failures, *Systems Practice*, 1:31-35.

Warren, L., Ellis, K. and Adman, P., 1998, Reflections on systems theory and systems practice, *OR Insight*, 11:2:14-19.

Wilson, B., 1990, "Systems: Concepts, Methodologies, and Applications", Wiley, Chichester.

Young, O.R., 1964, *General Systems: Yearbook of the Society for General Systems Research*, 9:61.

CRITICAL MANAGEMENT EDUCATION: A CRITICAL ISSUE?

John Mingers[1]

[1]Warwick Business School
Coventry CV4 7AL, UK

INTRODUCTION

I will argue that the whole area of management education is indeed a *critical* issue for all of us. And that it deserves much more attention than it receives relative to, say, the development and application of systems methodologies (looking in the proceedings of previous UKSS conferences, I can find barely a handful of papers discussing the courses that we academics produce). I will illustrate this with an evaluative discussion of a innovative undergraduate course that aimed to encourage a genuinely *critical* approach in its students. In part, this will involve a discussion of what it is to be critical, and why this is vital within management education.

CRITICAL EDUCATION IS CRITICAL

My first argument concerns what is good for the systems discipline itself. Much of the research that goes on in the systems and critical systems areas is concerned with more effective ways of managing our institutions and organisations. Generally this involves the development and application of particular approaches or methodologies that we ourselves use and that, more importantly, we would like managers to use. One way of doing this is to try and persuade those already involved in managing, but a potentially much more effective way is to capture those who are still participants in management education. Increasingly, a management qualification is seen as vital and so those that have not got a relevant undergraduate degree end up taking some sort of post-graduate course, often an MBA. This is the point at which we really have the potential to gain some leverage. If we can convince such students, especially those who already have experience, of the value and practical relevance of systems thinking and systems methodologies then we should that they will carry on using them in their later work. Put rather more negatively, if we cannot win them over at that stage then we are hardly likely to be able to later.

The second argument concerns the wider question of the growing power and significance of managing organisations and the importance of a critical approach. Clearly, there is a fast developing trend for the creation of trans-national and global corporations. By their nature such institutions are not bounded or controlled by the laws of any one nation and in a very real sense have more power, both politically and economically, than most nations. In a list produced by the *Economist* in 1988 of the world's top ten economic entities, three Japanese organisations (Mitsubishi, Mitsui, and Itochu) were 24th, 25th, and 26th, ahead of Turkey, Denmark, and Saudi Arabia for example. Nine companies were in the top 40. Indeed, it has been argued (Angell, 1997) that we are moving into a future radically

Synergy Matters: Working with Systems in the 21st Century,
Edited by Castell *et al.*, Kluwer Academic / Plenum Publishers, New York, 1999.

13

different from the past. One in which nation states become irrelevant; a few transnationals hold power through their control of the information networks; and the world is stratified into a small, wealthy technological elite, and a massive, very poor underclass. Given such trends, it seems clear that the values, beliefs and actions of managers, rather than politicians, will increasing determine the social, economic and ecological state of our world. This makes it all the more imperative that in the field of management education we move away from the positivistic and reductionist approach that sees management as simply a technical application of value-free "knowledge", towards one that is *holistic*, that is *committed* in recognising the inherent involvement of the values beliefs and emotions of the individual, and that is *critical* in not simply accepting the prevailing dogmas, orthodoxes and established interests.

WHAT IS CRITICAL MANAGEMENT EDUCATION?

The prevailing view within business schools and management departments is the utilitarian one that management education is primarily concerned with enhancing managerial effectiveness. Grey and French (1996) contrast this managerialist view with a critical view that we should decouple management education from management activity in order that the claims and practices of management can be called into question. One should be able to study management as a practice without it being training for management, in the same way that one can study politics without being trained to be a politician. On this view, higher education should withdraw from management training all together. A variant of this might be that both should exist within academe but develop in separate ways with particular institutions should clearly specialise in one or the other (Thomas, 1997). Must it be assumed, however, that one either adopts the presuppositions of management as conventionally defined (and thus supports the status quo), or be antagonistic toward all management as an activity. Should we not instead move beyond a critique of management toward developing a critical practice of managing - "a qualitatively different form of management: one that is more democratically accountable to those whose lives are affected in so many ways by management decisions" (Alvesson and Willmott, 1996, p. 40). This might involve focusing attention away from *management* as a class-based hierarchy towards *managing* as an activity that we all do and that is done to us.

What might this mean for the practicalities of individual courses? Grey *et al* (1996) suggest that, in the main, current courses embody a positivist stance, assuming that there is a given and unquestionable body of valid knowledge that must be presented and then mastered by the student. The teaching approach is largely didactic and, although this is now changing towards more student-centred methods that stress practicality, role-playing, participation and so on, this is still within a context that does not problematise knowledge itself. In contrast, Grey *et al* suggest that a critical approach should start with the students' own lived experiences, not to make the transmission of knowledge more effective, but to "provide a basis for a critical reflection on experience as a means of subverting such knowledge" (Grey, Knights et al., 1996, p 100). The point here is to raise very fundamental questions about the status and validity of management theory and the extent to which it privileges only one, primarily functionalist, view of knowledge.

There are clearly practical problems with such an approach. Practically, there is the inevitable tension of teaching a critical course within a context and degree programmes that are largely positivist in the above sense. This is exacerbated by the current political and economic climate (in the UK anyway) of reduced funding, students incurring more and more debt, greater emphasis on relevance (to industry), practicality and skills, and increasing measurement of university performance by crude indicators. Theoretically, there is the almost inherent contradiction that we are encouraging students to question the validity of knowledge and authority and yet, by that very choice, imposing our visions upon them.

THE *CRITICAL ISSUES IN MANAGEMENT* COURSE

This section will briefly describe the development of our *Critical Issues in Management* course. Further details including a review and evaluation are given in Mingers (1999). At the undergraduate level Warwick Business School (WBS) has three major programmes - Management Sciences, Accounting and Finance, and International Business. Traditionally the major programmes had emphasised disciplinary-based core courses in the first two years while in the final year students choose a range of electives. This meant that in the final year there was no common core course for all students on a programme, nor was there much inter-disciplinary teaching throughout the three years. A new course was required to be a core across all three programmes (200 students). The specification of the new course was ambitious in that it should involve all the disciplines in an integrated manner; be academically rigorous; at the same time be participative and based on student-centred learning; and should develop the students' practical skills in presentations, report-writing and group work. A group of academics from across the School produced an initial proposal specifying a framework for how the course would operate, and the core idea that it should be about developing in the students a critical approach to management. A title was agreed - *Critical Issues in Management* - but it was open to many different interpretations as to what "being critical" actually meant.

The framework of the course was agreed in principle but was very complex in its logistical details. There were to be no lectures on the course but only fortnightly, two-hour seminars at which the students would give individual and group presentations on a particular topic or case study, and participate in discussions led by the tutor. There would be roughly 20 students to a seminar group and these would be split into five subgroups, two of which would present each week. Each student would also review a book from a list of books that the staff felt were important for management students. The review would be presented orally and in writing. The assessment for each student would consist of four written-up case studies, the book review, and a mark allocated by the tutor for classroom performance. The content was also agreed reasonably easily. Different staff members would contribute their own case studies that would then be taught by all members of the group. The cases were not to be too tied in to a particular discipline but should explore the different facets of the course.

What was not so clear, however, was the core intellectual foundation of the course - what was meant by a critical approach? Here there were a variety of views. Certainly, as the title implied, there was the idea of critical, as in crucial or vital, issues facing management and organisations in the future. There was also a commitment to critical thinking as in the ability to evaluate the validity and strength of arguments and proposals. But beyond this there was also the idea of adopting a critical stance towards the accepted, managerialist, assumptions underpinning most management education. It was felt important to raise issues such as the nature and effects of power in organisations; the relationship of organisations to local communities and to the environment; issues of race, culture and gender; and ethics and responsibility. It is to this question - what is the nature of a critical approach? - that I will now turn.

WHAT IS BEING CRITICAL - A FRAMEWORK?

There are many strands of thought in both the social and philosophical literature that can be labelled "critical" (Mingers, 1999) - critical thinking, critical social theory as in the Frankfurt School, critical management studies, critical systems thinking, and finally the work of Foucault, especially on power and its relationship to knowledge. These all related to or exemplified the different aspects of being critical that were expressed within the group. The problem from a pedagogic point of view was how present these different, and quite sophisticated, notions in a way that would be meaningful to our particular students in the context of this course. The main constraints were first, that many of the students, especially those from the Accounting and Finance course, would have no background in social science or organizational behaviour at all. Second, being undergraduates almost none would have, experience of real-world organizational work and many would have been taught a very rationalistic and abstract view of decision-making. Third, there were to be no lectures and so any material assigned to the students would have to be intelligible in its own right. And fourth, that this was to be the first session of the course and so could assume no prior material. The main conclusions were that a fairly simplistic framework would need to be developed to relate these different aspects together; that the reading material would need to be both straightforward and interesting; that some practical activity to allow the students to apply the material would be necessary; and that the students would have to become aware of the messy nature of real-world decision making to motivate their participation in the course. The rest of this section explains the response to these concerns.

In everyday language, "being critical" means finding fault and being negative about something. It can often be quite destructive rather than constructive, and is often done with a particular antagonistic motive or attitude. In developing a critical approach in the course, we were concerned that it should not be purely destructive; that it should be rigorous and structured; and that it should generate insights that are valuable in taking practical action. What all of the different aspects of being critical mentioned above seemed to share was not taking things for granted, not just accepting how the situation seemed or was portrayed but questioning or evaluating such claims before deciding or acting. This may seem quite simple or straightforward but, if done seriously, rigorously, and radically it can lead to far-reaching and unsettling conclusions.

Four different dimensions of questioning or scepticism were identified. The rationale for these was based, by analogy, on Habermas's (1984) theory of communicative action and in particular his theory of the validity claims of speech acts. Habermas argues that any communicative utterance aimed at generating understanding and agreement implicitly raises four validity claims - that it is comprehensible, that it is factually correct or in principle possible (truth), that it is acceptable normatively (rightness), and that it is meant sincerely (truthfulness). In our situation we are concerned with a wider range than simply speech acts - for example, plans, proposals, actions, and designs; and they may well not be communicative (i.e., oriented towards understanding) but may well be strategic (oriented towards getting one's way). In an analogous manner, we can say that proposals for action involve implicit assumptions or validity claims that should be questioned. First, the logical soundness of the argument and its manner of expression (*rhetoric*); second, the taken-for-granted assumptions about factual matters and acceptable social practices and values (*tradition*); third, assumptions made about legitimacy and whose views should be privileged (*authority*); and fourth assumptions concerning the validity of knowledge and information (*objectivity*). These four aspects of a critical approach are further developed below.

i) Critical Thinking - the Critique of Rhetoric: The first sense that is considered is that known as critical thinking. At the simplest level this concerns being able to evaluate whether peoples' arguments and propositions are sound in a logical sense (Hughes, 1996). Do the conclusions follow from the premises? Are the premises themselves justifiable? Is language being used in a fair way, or is it deliberately emotive or misleading? This might appear to be a simple technical skill concerned with the logical analysis of language, but in real situations it can become extremely difficult to fully understand what is meant or claimed by some assertion, or to discover whether particular claims are or are not valid. Critical thinking can be defined more widely (McPeck, 1981) to involve a scepticism or suspension of belief towards particular statements, information, or norms. To think critically is not purely abstract but is always about some particular problem or domain. It therefore requires knowledge and skills specific to the problem or disciplinary domain although Paul (1990) argues that critical thinking is a general skill rather than being domain specific. It should also be reflective scepticism - being aware of its purpose (why am I adopting this particular attitude?) and being capable of offering alternatives. This aspect of being critical could be called the *critique of rhetoric* as it is particularly concerned with the use of language.

ii) Being Sceptical of Conventional Wisdom - the Critique of Tradition: The other senses of the term critical that we will consider are really developments of this sceptical attitude, taking less for granted and questioning deeper and more fundamental assumptions that we usually make. One of the most common assumptions we meet in organisations (and society more generally) is that of tradition or custom - the taken-for-granted "way we do things around here". Organisations and parts of organisations develop particular cultures and particular practices. These may have originated for good reasons, or simply by chance, but they tend to become accepted and, indeed, unseen. However, they may well not be the most appropriate way of doing things either because the situation has changed, or because in fact they never were, or because they deny or contradict moral values such as sexism, racism or environmentalism. It is often not so much the long-standing practices or traditions of an organization, but assumptions that relate to a particular project or plan. These can be seen as boundary judgements (Ulrich, 1991), often set by technical experts or powerful groups, that limit (perhaps implicitly) what may be debated or challenged. Questioning such practices or judgements can often provoke strong reaction and the weight of tradition and authority may well be used to support them. Trying to change them can be extremely difficult as it will inevitably change the status quo and upset established patterns of power and authority. This can be called the *critique of tradition*.

iii) Being Sceptical of One Dominant View - the Critique of Authority: Another, deeper, assumption is that there should be just one right or dominant view as opposed to a plurality of different but valid perspectives. For students this is particularly difficult to accept since much of their education so far will have been aimed at teaching the "correct" answer, on the assumption that there is one. They will not have been encouraged to question the validity of their teachers. However, by this stage, in their final year, they should be appreciating that there are genuine disagreements and unresolved issues even within academic disciplines. The situation in the organizational world, which does not split itself into well-defined disciplines and problems, can be highly complex with many different stakeholders involved. These interest groups will all have different experiences of the situation, different relationships to it, and stand to benefit or lose in different ways. Recognising that there is a multiplicity of perspectives, questioning the dominant view or privileged position, and trying to "see the world through another's eyes" (Checkland and Scholes, 1990; Churchman, 1968) could be called the *critique of authority*.

iv) **Being Sceptical of Information and Knowledge - the Critique of Objectivity**: The final level to be considered is questioning the validity of the knowledge and information that is available, and recognising that it is never value-free and objective. At the simplest level students have to see that even seemingly objective "facts" such as quantitative data do not simply occur but are the result of particular processes involving a whole variety of people, operations, and decisions/choices. Which factors are recorded and which are not? How are they recorded or measured - there are usually several possibilities? Can important factors be measured at all or do we have to use some surrogate? Do the non-quantifiable judgemental factors get given their due weight (Mingers, 1989)? Even when some data has been produced, it only becomes usable as information when someone interprets it from their point of view and for their particular purposes. A simple table of data embodies many assumptions and has as many interpretations as there are readers. At a broader level it can be argued that information and knowledge always reflect or are shaped by the structures of power and interest within a situation (Foucault, 1980; Foucault, 1988). Which problems are raised and which are not? Which decisions get taken and which are always put off? To what extent are particular interest groups able to promote or suppress certain information, or shape the agenda's of discussion and meetings? This aspect of critical thinking can be called the *critique of objectivity* as it calls into question the whole idea of there being objective, value-free knowledge.

DISCUSSION

Where does our *Critical Issues in Management* course stand with respect to critical management education in general? First, it is clear that it actually embodies within it the central dichotomy between utilitarian and critical management education. Is it primarily concerned with problematising management knowledge or with improving the effectiveness of our students in their management careers? The answer is that it tries, perhaps unsuccessfully, to be both. This is partly because of its institutional context. Warwick Business School, like most university management departments, is both a business school concerned with effective management and a centre for research into management. Individual members of faculty may be more committed to one than another, but often have to embrace both - for instance, teaching on an MA in Critical Management and an MBA - simply because of the demands of the job.

To what extent does this tension and ambiguity undermine the claim that the course is an example of critical pedagogy? There are several possible responses. The most obvious is perhaps that smuggling in critical ideas is the best that can be managed in the current circumstances. Any attempt at raising the students' critical awareness is better than none. A more deliberative response is to see it as a "Trojan horse" strategy, disguising a subversive critical intent within a course apparently concerned with management effectiveness. A third, and in some ways attractive, approach is to argue against the supposed contradiction between the utilitarian and critical models. Does the course have to be either one or the other? Seen in this light, the course could be a first step toward synthesising the often competing demands of morality (our duties and responsibilities towards others), ethics (our concern with our own worth and self-identity), and pragmatics (the need to be effective in our activities) (Habermas, 1992; Habermas, 1993; Mingers, 1997).

The second issue is the extent to which CIM can claim to be critical in problematising the status of knowledge. Here, I would argue that the course certainly aims at this within the inevitable practical constraints outlined above. It is clearly addressed within the four aspects of being critical in terms of the critiques of authority and objectivity. The critique of authority denies the hegemony of a single legitimate viewpoint or interest promoting instead the acceptance of a plurality of positions. The critique of objectivity denies the assumption of pure, value-free knowledge and introduces the Foucauldian notion of power/knowledge. Both of these theoretical ideas are developed practically in the various case studies.

The third issue concerns questions of contradictions in the teaching approach. For example, we are essentially forcing the students to participate in seminars by using assessment rather than allowing their participation to be given freely. We have to accept that this is less than ideal, a justification being that it supports the wider benefit of having a non-lecture, student-based course. Hopefully the material will be interesting enough for the students to want to participate but this may well be an area for future developments. In a way we are placing the students in a very contradictory position - expecting them to become critical but at the same time to adhere to our rules. A more generalised problem is the extent to which our version of being critical is itself biased, representing a particular rationalistic, universalistic and gendered view. This is an important debate concerning the nature of rationality itself that cannot be pursued in detail here except to note that such criticisms have already been registered within Habermasian critical theory itself (Benhabib, 1992; Habermas, 1994; Young, 1990), sparked in part by clashes with the Foucauldian perspective.

REFERENCES

Alvesson, M. and Willmott, H., 1996, "Making Sense of Management: a Critical Introduction", Sage, London.

Angell, I., 1997, Welcome the the 'brave new world', *in*:"Information Systems: an Emerging Discipline?", (J. Mingers and F. Stowell, eds.), pp. 363-383, McGraw Hill, London.

Benhabib, S., 1992, "Situating the Self: Gender, Community and Postmodernism in Contemporary Ethics", Polity Press, Cambridge.

Checkland, P. and Scholes, J., 1990, " Soft Systems Methodology in Action", Wiley, Chichester.

Churchman, C. W., 1968, "The Systems Approach", Dell Publishing, New York.

Foucault, M., 1980, "Power/Knowledge: Selected Interviews and Other Writings 1972-1977", Harvester Press, Brighton.

Foucault, M., 1988, Truth, Power, Self: an Interview with Michel Foucault, *in*:"Technologies of the Self: An Interview with Michel Foucault", (L. Martin, H. Gutman and P. Hutton, eds.), pp. 9-15, University of Massachusetts Press, Amherst.

Grey, C. and French, R., 1996, Rethinking management education: an introduction, *in*:"Rethinking Management Education", (R. French and C. Grey, eds.), pp. , SAGE Publications, London.

Grey, C., Knights, D. and Willmott, H., 1996, Is a critical pedagogy of management possible?, *in*:"Rethinking Management Education", (R. French and C. Grey, eds.), pp. , SAGE Publications, London.

Habermas, J., 1984, " The Theory of Communicative Action Vol. 1: Reason and the Rationalization of Society", Heinemann, London.

Habermas, J., 1992, Discourse ethics: notes on a programme of philosophical justification, *in*:"Moral Consciousness and Communicative Action", (J. Habermas, ed.), pp. 43-115, Polity Press, Cambridge.

Habermas, J., 1993, On the pragmatic, the ethical, and the moral employments of practical reason, *in*:"Justification and Application", (J. Habermas, ed.), pp. 1-17, Polity Press, Cambridge.

Habermas, J., 1994, What theories can accomplish - and what they can't, *in*:"The Past as Future: Jurgen Habermas Interviewed by Michael Haller", (M. Haller, ed.), pp. 99-120, Polity Press, Cambridge.

Hughes, W., 1996, "Critical Thinking", Broadview Press, Ontario.

McPeck, J., 1981, "Critical Thinking and Education", Martin Robertson, Oxford.

Mingers, J., 1989, Problems of measurement, *in*:"Operational Research and the Social Sciences", (M. Jackson, P. Keys and S. Cropper, eds.), pp. 471-477, Plenum Press, New York.

Mingers, J., 1997, Towards critical pluralism, *in*:"Multimethodology: Theory and Practice of Combining Management Science Methodologies", (J. Mingers and A. Gill, eds.), pp. 407-440, Wiley, Chichester.

Mingers, J. , 1999, What is it to be critical? Teaching a critical approach to management undergraduates., *Management Learning* (forthcoming).

Paul, R., 1990, "Critical Thinking", Foundation for Critical Thinking, Santa Rosa.

Thomas, A. , 1997, The coming crisis of Western management education, *Systems Practice* **10**(6):681-701.

Ulrich, W., 1991, Critical heuristics of social systems design, *in*:"Critical Systems Thinking: Directed Readings", (R. Flood and M. Jackson, eds.), pp. 103-115, Wiley, Chichester.

Young, I., 1990, "Justice and the Politics of Difference", Princeton University Press, Princeton.

PUTTING SYSTEMS THEORY TO WORK - THE SYNERGY BETWEEN THEORY AND PRACTICE

Richard J. Ormerod

Warwick Business School
Warwick University
Coventry CV4 7AL

INTRODUCTION

Since joining academia 8 years ago I have been engaged in a series of consultancy interventions employing hard and soft systems methods. I use these interventions as field trials. At the outset I determined to place these field trials at the centre of my research programme. My area of expertise was in the area of strategy in general and information systems strategy in particular. My purpose was (i) to demonstrate that participative methods for developing information systems strategy are more effective than conventional methods (interview-analyse-propose), and (ii) to evaluate the role and effectiveness of soft methods in supporting such a participative strategy development exercise. Thus, the research approach I adopted was to conduct field trials of participative strategy development in large corporations. As a paid consultant I was under contract to do all I could to meet the aims of the consulting project; these aims being determined mainly by the client. Generally the aim of each project was to help senior managers develop an effective strategy which they were committed to implement.

Into each intervention I carried with me some explicit theories derived from the literature (mainly theories about the process of involving groups of people in structuring problems and making choices) and from my experience to date. I also deployed tacit knowledge based on many years in strategy and consulting. During each intervention I had to adapt my current understanding of the theory in the light of the internal and external context of the client organisation and the practical constraints of the particular project. After each intervention I reflected on what had occurred and wrote the experience up as a case study. For each of the major cases the result was a journal article. Thus at least one objective of the field studies has been achieved: a number of major interventions have been recorded and discussed in the literature. In parallel I have also written some papers about the theory of intervention. The overall research objective was, of course, to add to the body of knowledge related to the process of intervening in organisations to create change. It is a moot point whether the research method I adopted was an effective way of achieving the research objectives: this issue is addressed in this paper. I will refer to the research method adopted as reflection-on-action.

The purpose of the paper is to describe the programme as a whole and to critically assess its strengths and weaknesses. First, I will briefly outline the cases and point to where detailed accounts can be found in the literature. Second, I will summarise some of the theoretical lines of argument that I have been developing in parallel. Third, I will examine the essential steps involved in the research and compare them to more conventional theoretical and empirical approaches. Fourth, I will draw some conclusions about the limitations of the approach taken.

Synergy Matters: Working with Systems in the 21st *Century,*
Edited by Castell *et al.*, Kluwer Academic / Plenum Publishers, New York, 1999.

THE FIELD TRIAL CASES

The cycle of conducting a field trial, reflecting on it, and then conducting another trial has to start somewhere. My story starts in 1989 with the development of an information systems strategy at Sainsbury's, the UK retail supermarket chain.

1989 Sainsbury's

This intervention was conducted on a purely commercial basis with no prior research objectives. Nevertheless, soft systems methodology and other ideas had been employed and the opportunity to reflect on the design, conduct and outcome was there. The paper that resulted was thus reflection-on-action rather than action research (Ormerod, 1995b). The conclusion from Sainsbury's was that the participative approach and the methods used had been successful in the, albeit rather limited, sense that a strategy had been developed and both the client and the participants were pleased with the process and outcome. Further they were committed to implementing the strategy. The outcome of the 5-year strategy was reviewed five years later by examining what had been done and the benefits obtained (Ormerod, 1996b). This confirmed that the strategy had indeed been carried out and that the outcome, despite some pluses and minuses, was meeting the expectations of the authors of the original strategy. Thus, in a somewhat stronger sense, the strategy development process can be claimed to have been successful. However, in the intervening period some competitors had also moved forward and it was by no means clear that Sainsbury's had gained competitive advantage over its near competitors.

1991 Palabora

Trials, which had research as well as commercial objectives, commenced with the development of an information systems strategy at Palabora, a South African copper mine operated by RTZ (now Rio Tinto). Here the approach developed for Sainsbury's was tested under severe limitations of time and location in a very different cultural and organisational context. The outcome was perceived by those involved to have been successful but in the process I had adapted the methods used. For instance, the use of soft systems methodology was more structured and less interpretive than is usual. It could reasonably be argued that the adaptation had blunted some of the key attributes of the approach. While the importance of the formal methodologies was thus lessened, the importance of the management of the cognitive and political processes was apparent (Morley and Ormerod, 1996). The value of using a participative approach had been reinforced by its application in a different context. The paper describing the case sought to address some of the questions raised by the conduct of the project (Ormerod, 1998a).

1992 Richards Bay

Following the success at Palabora, the information strategy development at Richards Bay was intended by RTZ to confirm that the approach was more generally applicable. Richards Bay is a mineral dredging and processing operation located in South Africa. The application of the same methods at a different mine site, also in South Africa, represents only a modest contrast in contexts. The interest of the case lay in whether the success at Palabora was repeatable or whether it simply reflected a fortunate set of circumstances. The exercise at Richards Bay could draw on the Palabora experience, both good and bad, but inevitably a different set of problems was faced. The Richards Bay project also helped to build experience in the use of participative methods allowing more general reflection on applicability and good practice. In the paper describing the case (Ormerod, 1996d) conclusions were drawn about (i) the practical value of soft OR and systems approaches and the nature of their contribution to theory and practice, (ii) the diffusion of such approaches in an international context, and (iii) the strategy itself. One of the issues that was highlighted by the project was the importance of ensuring that senior managers were asked to fill roles appropriate to their level in the hierarchy. Some senior managers felt that they should be part of the steering committee rather than the task force. Another issue was the concentration on IT as opposed to other solutions to issues. The next field trial was not thus constrained.

1994 PowerGen

At PowerGen, a privatised UK electricity generator, there was an opportunity to diversify away from information systems strategy development. The project, initially aimed at developing performance benchmarks and an IS strategy, was widened to include business process redesign. Here, organisational questions were more directly addressed. This proved more satisfactory for those involved as there was no feeling that the wrong question was being addressed. However, because the impact of process and structural change on jobs and empires was clear, agreement was more difficult to reach. Conclusions about major organisational change were not implemented until 2 years after the end of the project. Although the project as a process worked well and it undoubtedly resulted in major cost savings, the results were not unequivocally accepted as successful. Nevertheless, from an academic perspective further evidence was gained in the use of the methods and their application to a wider set of issues. The resulting paper also attempted to set the project in the context of recent theoretical developments in the nature of OR and systems interventions (Ormerod, 1998b).

1995 Rio Tinto (previously RTZ)

In the previous four cases I had been central to the interventions as lead consultant. In Rio Tinto I now had the opportunity to act in an advisory capacity to a project conducted by internal Rio Tinto consultants. The aim was to improve operational performance globally at RTZ operated mine sites. Over an extended period a series of projects were conducted in Australia, South Africa and the USA. The challenge was to reconcile the desire of the corporats centre for comparative benchmarks with the mine managers' insistence that the performance of each mine should be determined in its own terms. The centre resolved to develop an approach that the mines could apply to themselves. The story of the resulting field trials is told in Pauley and Ormerod (1998a). It is a story of trials and tribulations until finally a successful method of intervention was developed. It had proved impossible to develop a method that mines could implement without outside assistance and the emphasis had moved from performance measurement to performance improvement. One finding was that it was easier to obtain improvement when the whole mine is considered rather than just the open pit operations. This echoes the finding in the earlier trials that restricting oneself to part of the system (to IT in the earlier case), provided an unnecessary and unhelpful constraint.

THEORETICAL DEVELOPMENTS

Reflection on the trials above, together with a number of smaller cases (Ormerod, 1995a, 1997a) has given support to a number of propositions which are now discussed.

Participative Approaches to Strategy Development

The advantages of participative processes has long been a theme in systems development (Mumford, 1981) and consulting firms have increasingly recognised the power and effectiveness of participation in strategy and change projects. The series of field trials described above support the proposition that participation in strategy making motivates line managers, makes strategy a lively and accessible subject, and results in ownership and commitment to implement.

Effectiveness of Soft Methods

Soft systems and other methods have been demonstrated to be effective in supporting participative processes. The methods support enquiry, debate, analysis and agreements to act. They involve small groups working together in an iterative fashion, they can be used to encourage participants to take a systemic view, and they enable uncertainties and different stakeholder viewpoints to be taken into account. The methods are designed to be discursive, iterative, exploratory and somewhat open ended. To meet the time constraints imposed on consultancy assignments I have found it necessary to adapt the methods. This goes against the grain of the original conception of the methods and can weaken some of their powerful features. Striking the right balance in the design of the intervention process poses a dilemma.

Management of Political and Cognitive Processes

In addressing the needs of the project, the methods used have to be embedded in a participative processes that manages the political and cognitive dilemmas that participants face. The process must encourage participants to (i) explore creative possibilities that go beyond normal every day linear thinking, (ii) analyse the possibilities in business terms, and (iii) build increasingly specific commitment to act (Morley and Ormerod, 1996). The achievement of these aims must be addressed in both the design and the conduct of the intervention whether or not formal methods are deployed. In considering why participative interventions using soft systems and other methods are generally successful it is difficult to separate out the contribution of the general approach of managing the political and cognitive processes and the specific use of the methods chosen.

Mixing Different Methods

The experience of the field trials was that more than one method (or parts of more than one method) was required to support the type of process envisaged. The issue of mixing methods therefore comes to the fore. There was no evidence from the trials that mixing methods, whether hard or soft, created problems in practice. Problems of incommensurability are more apparent than real. In practice managers have no difficulty in utilising both interpretive and empirical evidence; fact and opinion can together inform the process of decision making. The transformation-competence model can be used to guide the choice of methods and the process design (Ormerod, 1997a, 1997b; Pauley and Ormerod, 1998).

The Evolution of Practice

The fact that the field trials formed a series of interventions over an extended period of time provided an example of the evolution of a "strand of practice" (Corbett et al, 1995). Each intervention drew on the experience of previous projects, adjusting the approach to the new circumstances. Over time both the set of methods used and the issues addressed evolved resulting in a distinctive strand of practice. The collection of cases provides a detailed history of one particular strand (Ormerod, 1998b).

THE RESEARCH METHOD ADOPTED

Traditional academic methods of conducting research into organisational change involve observation of the flow of organisational action either directly or indirectly through interviews. The observations made and the interview content are informed by some theoretical propositions which are to be tested by the evidence obtained. The academic then engages in knowledge production, writing up the results, setting them in the context of the existing body of knowledge, and conducting some analysis. The evidence, analysis and findings are then published, adding to the body of knowledge of the subject area. The learning cycle is closed when some notice is taken of the findings by individuals embedded in the world of organisational action, and behaviour is changed as a result. Problems arise at each stage in the process. First, it is often difficult to gain access to organisations, particularly to people at or near the top. Second, easy to observe factual data may dominate the more difficult to access data on the tacit understanding and behaviour of individual actors. Third, the language and rigour of publication in academic journals may provide an effective barrier to access by organisational actors. Fourth, organisational actors may never become aware that the research findings exist. Fifth, even if an organisational actor becomes aware of the findings, the difficulty of translating theory into action should not be underestimated. Despite these shortcomings the traditional approach may be the best we have got.

Alternative research approaches should be seen in the context of the above potential shortcomings of the traditional approach. If the academic approach forms one extreme of organisational learning (theory-based approach), at the other extreme learning takes place as those engaged in the flow of action reflect on their experience, usually in an informal manner (practice-based approach). This reflection leads to some thoughts and theories about the behaviour of the system in which they are engaged and informs their future actions. The reflection is likely to be more practically orientated and the link between reflection and action is more directly in the hands of those able to do something about it. Of course, many managers are not particularly reflective and the learning that does occur is neither subjected to scrutiny nor usually is it made accessible to others through publication.

The approach taken in the field trials described here attempts to obtain synergy between theory-based and practice-based approaches by combining the two extremes described above in a research process that I refer to as reflection-on-action. Access is gained to the flow of organisational action by engaging in it directly, in this case as a consultant. Tacit knowledge is more accessible because it is experienced. The reflections are written up and published in the academic literature for scrutiny and this opens up the possibility of access by organisational actors. The findings are much more directly routed back into the world of organisational action by being utilised in the next field trial. An academic consultant is thus in a good position to translate theory into action. There is also the possibility that other organisational actors engaged in the field trials will apply their experience to other issues in their organisation (Pauley and Ormerod, 1998).

LIMITATIONS OF THE APPROACH ADOPTED

According to some definitions the approach I have described would be termed action research. Reflection-on-action is subject to the same criticisms levelled at action research and perhaps some others as well.

The Dominance of the Consulting Objectives

In taking on the role of consultant, the academic must meet the objectives of the project agreed with the client. Generally, the consulting objectives will take precedence over academic objectives during the project. For this reason a tightly prescribed set of theories to be tested might not be appropriate. In addition, the flow of action might provide insight into some unanticipated aspect.

The Inevitable Bias in the Account of Academic Consultants

The ability of academic consultants to give an unbiased account is prejudiced by their intellectual and emotional engagement in the events studied. In these circumstances it is difficult, if not impossible for the consultant to triangulate the findings through accessing the multiple viewpoints of different actors in a balanced way. It is difficult for a consultant heavily engaged in the flow of action to gauge the success of their own decisions and choices. At times consultants draw deep on their tacit knowledge of what works in different circumstances. Such decisions remain difficult to explain and justify.

Alignment with Senior Management Objectives

Ethnographic researchers, who observe activities without engaging in them, often sympathise with those lower in the hierarchy as they observe the impact of the decisions of senior managers remote from the action. Major change interventions are usually commissioned by senior mangers to meet organisational goals. When taking on a consulting assignment consultants are expected to (temporarily) align their goals with those of the top management. Alignment with the goals of top management gains their confidence and access to their thinking but will inevitable bias the researcher's viewpoint.

Control over Publication Content

It is common practice for consultants engaged in strategically important assignments to be required to sign confidentially agreements. At the outset, when success is far from sure and organisational reputations are at risk, clients seldom wants to contemplate public exposure of their organisational life, warts and all. Trust between client and consultant needs to build. In the event, given the pace of life in business, what was commercial important and sensitive today rapidly becomes openly discussed history. It is then possible to gain permission to publish despite earlier undertakings. Nevertheless, the client will want to look at what is being said and suspicion may remain that the academic will hold back criticism if that were to put the permission to publish at risk. In the extreme case of things going badly wrong, the academic consultant and the client might not wish to see their efforts exposed to public view.

CONCLUSIONS

The paper describes a series of field trials that have been conducted with the aim of accessing the effectiveness of the participative approach to strategy making supported by soft systems and other methods. The trials themselves are described together with some of the theoretical findings. The research approach adopted attempts to find a synergy between theory-based and practice-based approaches. The reflection-on-action approach shares many of the attributes and shortcomings of action research. Nevertheless, the approach offers an alternative source of insight to traditional research into organisational change. The research laid out in the paper illustrates one way in which synergy can be and has been developed between systems theory and practice.

REFERENCES

Corbett, C.J., Overmeer, J.A.M. and van Wassenhove, L.N., 1995, Strands of practice in OR (the practitioner's dilemma), *Eur. J. Opl Res.* 87: 484.

Morley, I.E. and Ormerod, R.J., 1996, Managing cognitive and political processes: a language - action perspective, *J. Opl Res. Soc.* 47: 731.

Mumford, E., 1981, Participative systems design: structure and method, *Systems, Objectives, Solutions.* 1: 5.

Ormerod, R.J., 1995a, Information systems strategy development and the practical use of SSM. *In* "Information Systems Provision," Stowell, F.A., ed., McGraw-Hill, Maidenhead.

Ormerod, R.J., 1995b, Putting soft OR methods to work - information systems strategy development at Sainsbury's, *J. Opl Res. Soc.* 46: 277.

Ormerod, R.J., 1996a, Combining management consultancy and research, *Omega, Int. J. Mgmt Sci.* 24: 1.

Ormerod, R. J., 1996b, Information systems strategy development at Sainsbury's Supermarkets using 'Soft' OR, *Interfaces.* 26: 102.

Ormerod. R.J., 1996c, On the nature of OR - entering the fray, *J. Opl Res. Soc.* 47: 1.

Ormerod, R.J., 1996d, Putting soft OR to work: information systems strategy development at Richards Bay, *J. Opl Res. Soc.* 47: 1083.

Ormerod, R. J., 1997a, Mixing methods in practice: a transformation-competence perspective, *In* "Multimethodology: The Theory and Practice of Integrating Methodologies," Mingers, J. and Gill, A., eds., John Wiley and Son, Chichester.

Ormerod, R. J., 1997b, The design of organisational intervention: choosing the approach, *Omega, Int. J. Mgmt Sci.* 25: 415.

Ormerod, R.J., 1998a, Putting soft OR to work: Information systems strategy development at Palabora, *Omega, Int. J. Mgmt Sci.* 26: 78.

Ormerod, R.J., 1998b, Putting soft OR to work: the business improvement project at PowerGen. *Eur. J. Opl Res.* forthcoming.

Pauley, G. S. and Ormerod, R.J., 1998, The evolution of a performance measurement project at RTZ . *Interfaces.* 28: 94.

MAKING SENSE OF CHAOS - SYSTEMS AND THE INFORMATION SOCIETY

Frank Stowell

Department of Computing and Information Sciences
De Montfort University
Milton Keynes Campus
Buckinghamshire
England
fstowell@dmu.ac.uk

INTRODUCTION

The social world is changing and we are told that we have become what is popularly referred to as the Information Society. Communications and cheap, powerful computers have provided an information infrastructure for a new world. For example, the world wide web (www) has changed the way that we learn, the way that we interrelate and the way that we communicate with each other forever. Rapid and easy access to information provides a route to new and hitherto unimagined alliances. Communication systems enable global control systems with large global corporations which have more influence upon world markets than some sovereign states, (e.g. General Motors) this is exemplified further by Logan who cites the example that 70% of market capitalisation in Holland is controlled by 10 top 10 companies (Logan, 1998). In many respects the Chief Executives of these enterprises have more power than democratically elected heads of state. It is difficult to anticipate what the long term effects upon society will be arising from the synergy created by the uncontrollable manipulation and processing of information. The speed of change is also a significant factor which will add to the uncertainty about the future felt by many people. The discussion in this paper may well be commonplace by the time this is published, which illustrates not just the pace of change but the way in which we have traditionally communicated. Discussion of some changes initiated by IT and the dramatic effect of these changes may have upon us are provided as a means of raising the question about how well we are equipped to deal with them. The validity of the claim that Systems Thinking and Practice help to appreciate and address messy situations is raised.

Synergy Matters: Working with Systems in the 21st *Century*,
Edited by Castell *et al.*, Kluwer Academic / Plenum Publishers, New York, 1999.

THE INFORMATION SOCIETY

What is an information society? The Information Society Commission define it as "A Society in which economic and cultural life is critically dependent on information and communication technologies and where people get the full benefit of that technology at work home and at play" (Information Society Commission 1998). Of course this begs the question what is information but that is an argument for another paper and will be left to one side on this occasion. The inference in the Commissions definition is upon access to information which can be used to enrich each of us in a variety of ways. In what way then will this affect our lives, will the effect be any different from other major changes, such as the railway, and will the effects all be beneficial? The information that is available to us will enable us to explain more, store more data about a huge variety of things and make this accessible, almost instantaneously, to the better part of the population of the world.

It is predicted that by 2000 one in five Europeans and 50% of Britons (Parsley, 1998) will be using a mobile communications device which will be followed with equal success by the internet, electronic commerce and digital TV (Bangemann, 1997). At an individual level it means that each of us can find alternative routes for the information that we need. If you are dissatisfied with what your GP, your Solicitor or your Professor has told you, you can quickly look up your subject on the web or put out a call for more information. It is possible for the individual to research an area quickly and access more information than the expert. Students with more time to concentrate upon a specific assignment can gain access to material that their tutor may not even have seen.

Businesses cannot afford to be without effective IT systems to enable control of the processes and activities of the enterprise. Advances in communications have made global operations a reality and the home office commonplace. The familiar locally based company where people went to work at regular hours, had a job for life and had a degree of company loyalty has virtually disappeared. Home shopping is set to become a major business market (Whittell, 1998) with predictions that by 2002 electronic commerce will reach over 300 billion ECU's (Telecommunications Council, 1998). With these changes will come the attendant changes to social habits and the design and development of towns and cities.

The Information Society has to address the problems of taxation, public protection, data protection. How will we be protected from intrusion from unwanted data and the use of our addresses? It is estimated that 81 million people will be users of the Internet by the end of the century (Smalley, 1998). Already those with access to the www are bombarded with unwanted mail which is quickly refined against the material rejected and apparent interests of the user analysed from the data derived from the users accesses to the www.

A FORCE FOR CHANGE

There is every reason to accept Bangemann's (1997) predictions that in a short time school children in different places in the world will be able to undertake joint projects on shared topics using video conferencing and electronic mail. It makes possible local, National and International projects between schools and at the same time creates an environment for shaping opinion. Of course the reverse can be true too as IT can be used to enforce ideas. The development of communication systems have helped to enhance the culture and language of western states, most especially Anglo-American. The availability of so much material on the www may be a valuable source of material but there is a substantive issue about the role of the education process and the ability of the user to critically evaluate what is there. Ironically the "modernisation" of the UK education system has in some respects reduced the opportunity to provide a rounded education for the students and in turn may affect their ability to critically evaluate the material that they uncover. For example, many students only attend University for their time-tabled classes, many have more powerful computing systems at home and can access the www easily. What will be the result of an education systems that does not include in its curriculum the aim of developing students ability to critically evaluate material and no provide facilities that provide an informal setting for the student to expose their thinking and have their ideas debated by their peers. Is there is a danger that the indiscriminate use of material from the www may be more harmful than beneficial in the short term?

Whilst there are many benefits to be gained by a free and ready access to information should we not also consider helping people to develop the skills which will enable them to get those benefits. It is an immediate problem. For example, a recent survey showed that the 75% of residents of one new town in Southern England has access to a PC and almost half have access to e-mail with 1 in 6 access to the Internet in their home. Within the UK and Europe there is the growth of the Learning City. These cities undertake to support life long learning and promote social and economic regeneration. A key feature is the IT infrastructure which will enable the distribution of information to be readily available to its citizens. Even in a relatively rich towns a proportion of the population will not have home access to the city intranet and will have to reply upon public kiosks. There is also a disparity between towns and between nations. This means that a select part of our population will have at their disposal a significant source of power and influence which the remainder do not.

Global communications make possible the sharing of knowledge and the use of expertise where it is needed sometimes without the expert moving from their office. For example in health care specialists can help to treat patients any where in the world, provided the communications infrastructure is in place. It makes possible round the clock medical aid and moreover the potential for an international data base of patient record and diseases (Telecommunications Council, 1998). Potentially an improving communication system can provide a basis for better health care for all but the proviso is that there is the opportunity and ready access to the communication system, and this is a long way off for a large percentage of the world population.

It may seem to be an irrelevant fact but there are more people members subscribing to the Royal Society for the Protection of Birds than for all the Political parties in the UK put together (Logan, 1998). Communication systems enable them to become a cohesive force that can quickly assemble powerful arguments and, as a body, pressurise the government to push through legislation which may affect the well being of the birds that the society pledges to protect. The sheer size of the group makes their argument a powerful one for any political party. It does not take too much imagination to provide many other examples. What effect will this have upon our politicians and upon our society as a whole? Perhaps the anti bird lovers will form another cohesive force, and so on. There are already examples where IT has been used to communicate the means to reject the establishment view of events and co-ordinate social unrest. The www provides the would-be terrorist with ways of manufacturing the means of terror and information about the way it can be done (Whittell, 1998). Doubtless the same infrastructure can be used to recruit like minded people. Indeed the terrorism itself may be the coordination (and maybe manipulation) of opinion which is used as a means of pressure. For example, a world boycott of a given "something" would not take too long before the owner of it wanted to negotiate. Present developments in warfare tactics concentrate upon the use of IT and IS as invisible weapons as strategies develop to control a variety of vital systems of other nations.

SYSTEMS AND THE INFORMATION SOCIETY

Clearly the above has been categorised according to the way that the author wishes to present the case for the development and potential effects of information system developments. These categorisations are in many ways sensible and provide a basis for argument. This categorisation is also divisive as it presents the points as if this is how the world is organised. The world as our grandfathers knew it no longer exists but it is the same world. What has happened is that each succeeding generation has redefined it to help make sense of the changes that have been taking place. It is arguable that because significant changes took place over at least one generation the problem was an evolutionary one. What this generation is facing are revolutionary changes.

There is an opportunity for Systems Thinkers to consider the what the effects the changes are having in Systems terms. If Systems is a meta discipline then can we explain the changes at a meta level?

The question might be asked about the effect that the www will have upon the individual and the related effects upon citizenship? Certainly there are indications of the spread of a particular kind of cultural experience. For example, it seems that modern western citizens demand instant gratification. The mobile telephone is one example. The explosion of mobile telephones is a modern phenomenon almost without parallel (Parsley,1998). But one could ask what is it that makes it necessary for so many people to have the means to make instant contact with anyone they chose? Will this develop similar characteristics to the way in which electronic mail is used? (I refer here to mail rage and flame mail) .

If we take the simple notions of Systems, say, Emergence, Hierarchy Communication and Control, (Checkland, 1981) will these help us to make sense of the situation? At the very least it may help us to ask questions.

What effect will the availability of information in birth/death rates, the male/female ratio, and the population distribution have upon decision takers? The way that food is grown and distributed may be influenced by the changes in population growth. This suggests further intervention of mankind with the natural control systems that act upon the provision of food, consumption of energy and the net cultural impact? Is it likely that the population system (as a label) begin to act as a positive feedback system?

What effect will the www have upon social order. Will the state and its officers remain the highest authority over its citizens? For example, being a member of a global enterprise may provide more benefits than being a citizen of a nation. Many multinational companies provide private health care (and access to the world finest clinics), education and training (and access to the worlds finest providers), freedom of movement, and status (an employees salary and personality profile is available in personnel files) and many more advantages over the ordinary citizen. The denial of these benefits provides the employer with more control over its employees than any police force.

How will the boundaries between nations be drawn? We read more and more about the importance of national sovereignty but does this become more intangible as IT redefines the boundaries? National frontiers become irrelevant as satellite communications provide a different kind of boundary. Angel argues that global communications will see the re-emergence of the City State (Angel, 1997) and the resultant fragmentation of National identities. Seen by some as a little extreme Angel's view non-the-less has some credibility. London for example operates as much as an international centre as the UK Capital.

EVOLUTION, REVOLUTION OR CHAOS?

There is little doubt over the past decade there have seen significant changes to world order. The rise and influence of the multinationals, the development of a more federal like European community, the collapse of the Tiger economies and the end of the cold war. The traditional enemies and wars of the 20th C are replaced by terrorism and the violent outcomes of the break-up of former Iron curtain countries. At the end of the century IT has made the was of the future more sinister as the silent power of computer control slowly replaces the traditional tank and gun, e.g. control of power generation, communications and propaganda.

Genetic engineering offers major benefits for health and with the developments the fear of what this may bring to society. Will it bring us closer to Huxley's "Brave New World" and its nightmarish representation of a people without emotion (Huxley, 1932,1959) or offer the prospect of better health? Communication Systems will enable medical knowledge to be used to help where such knowledge has hitherto be unattainable. What effect will these technological advances have upon the population patterns and food provision and what effect will such medical advances have upon faith and moral teaching?

IT has created the virtual enterprise, that is a company that exists in cyberspace where business transactions take place according to the "best deals" that are available in a virtual trading system. For example, financial transactions may be conducted in which a virtual enterprise emerges around a project base which in turn provides the basis of the corporate whole. A conglomerate might emerge around the notion of the development of the power source for the new personal transport system or the adoption of a single currency. Recent examples include the motor industry in which Jaguar is taken over by global Ford and, politically, England may be integrated into a federal Europe. But computer supported IS can enable these confederations to alter as easily as they began. Collaboration over significant distances are possible at all levels. Ford can install Honda engines or Honda cars be powered by Ford engines, Japanese cars built by British workers to Japanese specifications, British car manufacture can be controlled by a German company, British companies, chameleon like, become whatever country they happen to be in at the time. On the political map IS could make possible the re-emergence of the British Commonwealth as an economic force and a reduce the importance of the European federation for Britain. The communication infrastructure is in place to enable the formation and reformation of collaborative systems that suit particular business needs. Local communities that hitherto were devoted to a particular manufacturing activity, such as ship building or motor manufacture, can no longer have the security of continued business presence.

For the ordinary citizen the patterns of work characteristic of the 20th C have changed out of all recognition. Jobs for life, company loyalty and security have disappeared. Those that work have little time for leisure and those with leisure have little financial power to enjoy their leisure. Numerical Flexibility is the modern term for the old stevedores "hire and fire" system of years gone by. The modern version of labour market selection means that the work force has to have a greater functional flexibility of skills and greater flexibility in the location of work (Adnett, 1998). There is every indication that a percentage of every developed nations population will be unemployed and serious concerns about the cost of supporting them and those that retire. Paradoxically this follows a time of making so called lean and mean companies by shedding of 45plus year old staff and at the same time opportunities for individuals to continue working in some capacity until they decide to stop increases. Moreover, "ageism" is outlawed and opportunities for short term employment increases for those with the right education and skills (it is difficult to know the age of someone behind a computer terminal). Many people in their late 40's and early 50's are leaving their prime place of work with perhaps 30 years or more life expectancy. Many of these begin new interests and in some cases new careers. It is difficult to predict what effect this will have upon society and the way that opinion is shaped since there has never been such a significant group of healthy, experienced and financially secure individuals on the look out for new interests. Information technology enables a new set of alliances and allegiances unfettered by national boundaries and it is difficult to estimate what will bind them together and what systems will emerge.

TRYING TO MAKE SENSE OF CHAOS

Systems ideas are said to be at their most powerful in attempting to understand messy situations. It can be argued that the technology in the 20th C is creating one such messy situation and one which we have not any models at our disposal to mimic. This is a true opportunity for Systems thinking to play an important role. But will it? My guess is that systems will be used to analyse the aftermath and advise on what should have been or what can be done to fix the problem. There is little evidence that Systems Thinkers are at the heart of decision making and when decisions are made they seems to be made without regard to wider implications.

It seems important that such a conference as our that we should feel able to address such issues. If we cannot then the oft quoted strength of systems thinking and practice will be damaged.

REFERENCES

Adnett, N., 1998, Labour Flexibility, Business Review, Vol. 5, No.2, p12.

Angel, I., 1997, Welcome To The Brave New World, in, Information Systems, an Emerging Discipline? (ed. J.Mingers and F.A. Stowell), McGraw-Hill, Maidenhead.

Bangemann, M., 1997, A New World Order for Global Communications, Telecom Inter@ctive '97, http://www.ispo.cec.be/infosoc/speech/geneva.html

Checkland, P.B., 1981, Systems Thinking, Systems Practice, Wiley, Chichester.

Huxley, A., 1959, A Brave New World, Chatto and Windus, London

Information Society, 1998, http://www.ispo.cec.be/welcome.htm, and infoconc.htm

Logan, D., 1998, Corporate Citizenship in a Global Age, in, RSA journal, vol. Cxlvi, no. 5486, 3 / 4, pp65 - 71.

Parsley, D., 1998, Mobile Phone Firms Step-up Price War, "The Sunday Times", 8th November, 1998, p10.

Smalley, M., 1998, The World Wide Web - a Brave New World of Marketing, Business Review, Vol. 5, No.2, pp20-21

Telecommunications Council, 1998, Council Conclusions on Globalisation and the Information Society, 1998, http://www.ispo.cec.be/welcome.htm, and infoconc.htm

Whittell, G., 1998, Hoodlums can Boost Firepower on the Internet, "Times", 22 December, 1998, p13.

TRANSLATING PRODUCTION QUALITY DATA FOR USE IN THE IMPROVEMENT PROCESS

Alan J. Beckett, Dr Charles E.R. Wainwright, and David Bance

C.I.M,
Cranfield University,
Cranfield,
Bedford,
MK43 OAL

INTRODUCTION

The growth of common information technology platforms in the workplace has facilitated the development of integrated information systems. These systems are able to unite data held by different departments, that were previously 'ring-fenced' by the specialised nature of their information technology resources. Much of the data generated within these departments is used directly to measure and control the local processes. However, much of it also goes unused, either because the data itself is a by-product of another process, with no local relevance, or because it needs to be combined with other data to be of use. Further, the data used locally could also have a company-wide significance not obvious to the local users.

The evolution of common platforms offers the potential for rectifying this problem, gaining synergistic effects by integrating this data, and integrating the knowledge derived from it into the product (or service) creation process.

This paper describes the actions necessary to create and control such an integrative system, with particular emphasis upon data created during inspection processes. The theoretical processes are illustrated with practical experience gained during a series of pilot studies and system implementations within a medium sized manufacturing organisation.

DATA COLLECTION AND QUALITY MANAGEMENT

Clearly, if expensive data gathering techniques are not resulting in some payback for an organisation, then they will quickly be discontinued. As particular example, inspection data yields quality information that is applied to prevent sub-standard product from leaving the factory. This implies that the company must have calculated that it costs more to rectify faults after delivery than it costs to identify and rectify them before delivery. Therefore there is an immediate use that this data is put to.

Synergy Matters: Working with Systems in the 21st *Century,*
Edited by Castell *et al.*, Kluwer Académic / Plenum Publishers, New York, 1999.

The emphasis of contemporary Quality Management[1,2,3,4] is on process controls rather than simple product conformance. This has meant that more intelligence has to be extracted from the inspection data to enable process analysis, away from simple "go/no go" criteria to actual quantified product variables. In parallel, the smaller amount of pure conformance measurement that takes place has also become more accessible, because to minimise their cost many conformance tests have been automated and measured variables may be recorded on data logging equipment[5].

It is the contention of this research that this inspection data is also useful outside of the purpose for which it was generated. Particularly, by combining and integrating data from different sources, synergistic effects can be taken advantage of that contribute to value adding activities of the company. In addition to direct production process application, the data may be used to aid future designs, inform sales and marketing processes and support strategic business decisions generally.

RESEARCH APPROACH

Examining how quality data can be best applied 'out of context', there are two strategies that could be adopted. A data centred approach would take all the data available from inspection routines, subject them to a set of analytical methods, and present the findings of those analyses in the hope that useful knowledge is created. Whilst comprehensive this would be difficult to justify economically. Instead, a utility centred approach has been adopted, which concentrates upon the value adding activities that the system generates. Essentially this is a change management process which draws upon a knowledge generation process. To explain in more detail, this means that the organisation using the system will wish to use it to add value to its processes, for example to reduce costs, streamline its product mix, or to develop products with improved performance. To do this will require changes to the processes of the organisation. To enable these changes, an opportunity to gain advantage by changing must have been identified. This appears obvious, but in practice a structured method for applying this principle has been found to be necessary to make the advantages explicit to all participants.

DEFINITIONS

In order to describing how inspection data can be used to prompt process and non-process related changes, it is necessary to define a number of terms.

Continuous Improvement

The principle of Continuous Improvement centres upon the idea that products and processes may be made better by monitoring them, in order to identify areas that need to be addressed, and addressing those areas in turn. This assumes that with increasing knowledge of how a product or process operates, it may be made to operate better. It is a closed loop system, as shown in Figure 1 below.

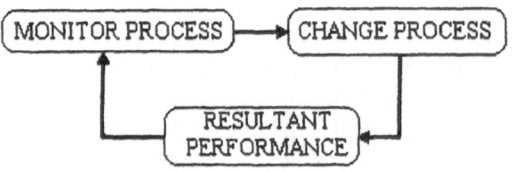

Figure 1. The Continuous Improvement closed loop system

Knowledge Generation and Application

One definition of knowledge[6,7] is the identification, within information, of an opportunity that may be exploited for competitive advantage. Knowledge is thus a nebulous commodity, distinguishable from information only by its context and the extent that it may be made use of. What constitutes knowledge at one instant may degrade into information when the opportunity for exploiting that knowledge passes. Information, on the other hand, is much more clearly definable as the identification of patterns within data. The knowledge obtained is finally exploited by a technology, such as a design improvement. This process of knowledge creation is shown in Figure 2.

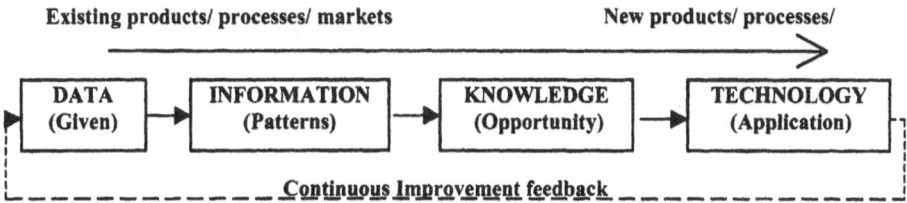

Figure 2. The DIKT model from reference 7.

If the principle of Continuous Improvement is applied to the DIKT process, it can be seen that the process becomes circular, in that the resultant technology change produces a changed data set itself and the DIKT process is free to repeat.

APPLYING THE UTILITY CENTRED APPROACH

Because the utility-centred approach was adopted, it can now be said that the system to extract benefits form inspection data has effectively to work backwards from the "technology application" benefits to product or process performance.

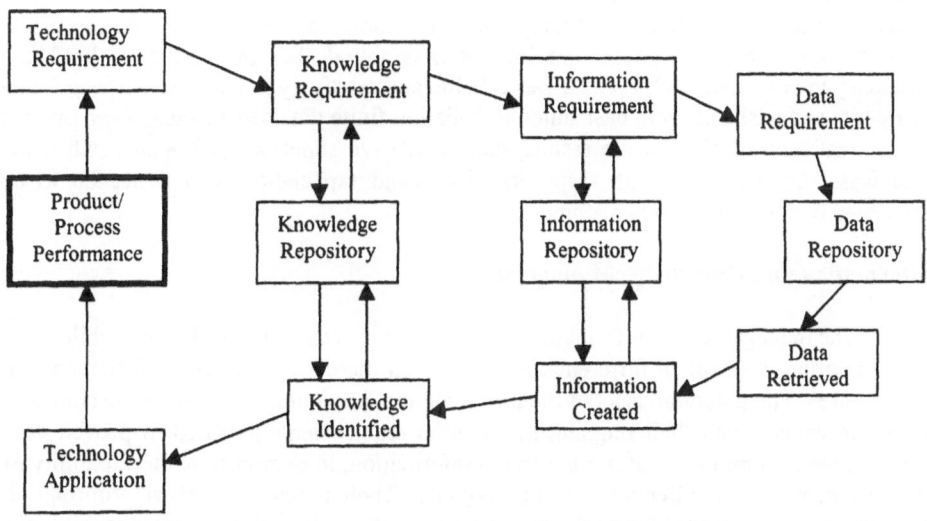

Figure 3. The Utility Centred Management Framework.

Figure 3 above shows the framework that was developed, to manage the creation, transmission and storage of data, information and knowledge. The product/process performance extends to include the development, research, or marketing processes. It can be seen that this leads to a technology requirement (for example an improved process or new product), influenced by the strategic objectives of the organisation. This in turn generates a requirement for the knowledge to create that technology, which could be satisfied by the existing knowledge base, or informed by it to create an information requirement. This in turn might be immediately satisfied from existing information, or partially satisfied, creating a follow-on data requirement. The data thus retrieved is analysed to create information, which is then evaluated to identify the knowledge it contains. This knowledge is then used to create the technology .

The next step of the research was to implement this framework as a system suitable for use by the industrial sponsor.

IMPLEMENTING THE SYSTEM

The sponsoring organisation is a manufacturer of industrial electrical appliances with of the order of 500 employees. The four major aspects of the framework were considered during a series of pilot studies, intended to identify the major tasks required and the associated issues.

Data Management

Data requirements were compared with a process map, which showed the data 'nodes' in the organisation where data was generated and collected. It could therefore be seen if the data required was already available, potentially available (ie: that it could be measured processes) or if a one-off experiment would be required.

Collecting the data was subject to data integrity issues. The format of the output of the automatic test equipment often required modification for use, while manually input data or data that was affected by manually input commands (such as download commands) were subject to quite extensive operator errors. These errors were addressed through training sessions and presentations to encourage buy-in amongst the operators.

Transmission was either digital, across the PC computer network or input manually from test result sheets. Storage in digital form was found to be the only useful method away from the production line. It was found that the computer hardware necessary to store and back-up data was reliable and relatively inexpensive. It was found that while spreadsheets were best suited to collating finite data sets (such as for a one-off experiment) and displaying the results, databases gave superior performance when the data was continuous (as with inspection data), and especially when it needed to be combined with data from other sources.

Information and Knowledge Management

The management of information and knowledge were combined, since while there is an important conceptual difference between them there are no physical differences in their natures. The potential users of the information at particular sites within the company were interviewed, and their ongoing information requirements assessed. It proved very useful to have an initial set of pre-generated information, to demonstrate the capability of the system, which the clients could then criticise. Their potential 'one-off' information requirements were catered for by the use of a query system linked to those responsible for managing the system.

Starting to make the information accessible was a complex task, given the variety of I.T. platforms (mostly MS Windows PCs and UNIX servers), and the nature of the information (published technical reports, continuously updated quality reports, weekly performance reports). Also, access via a PC was required by the sponsor. Therefore, Intranet technology was adopted to display the reports. As a separate development, the library catalogue was also put on-line. Clearly, passive accessibility was not enough, since even with a staffed company library, work in the technical library had been repeated by different departments over the years with near identical results. This was partly because of training issues, and partly because the relevance of the contents of one project based report to another piece of project work was not immediately apparent.

This was addressed by creating reference guides to highly relevant subjects, such as core technology areas, which described past and current research, design work, and management decisions and their effects. The personnel involved were also detailed so that the user could locate the subject experts within the company. This effectively created a 'road-map', which pointed the user at the research relevant to them. This was seen as the most important part of the system, because it explicitly detailed the organisation's experience in the core areas, and provided a level of continuity as personnel moved around or away from the organisation.

To manage the reference guides, a research panel was created, so that research findings throughout the company could be evaluated and their existence advertised to other researchers in the relevant fields, before being incorporated into the reference guide itself. The panel was also able to decide whether further research prompted by previous findings was worth pursuing. This was combined with a lecture programme to further transmit the combined experience to the workforce as a whole.

Technology Management

This was one of the most challenging aspects of the pilot study, since it involved disseminating new knowledge to workers for application. Historically the sponsor had performed this via new project developments or as a rapid response to serious process issues. There were no structures to enable incremental design development.

The change management process adopted was to introduce development and production panels in addition to the research panel, so that each of the three groups were prompted to absorb new knowledge into their own activities and pass actions along the product development process. The three panels were thus linked. In the case of one pilot study, the research was a combination of pure research supported by production evidence. After further work suggested by the research panel was carried out, the findings were presented to development, which initiated a design change programme, while production carried out a range of short-term measures.

CONCLUSIONS AND FURTHER WORK

This paper has described the actions necessary to create and control an integrative system, which analyses data created during inspection processes, and feeds the findings into the value creation process. The system centres around the management of knowledge generation combined with a change management programme. The work has been supported by a series of pilot studies with the industrial partner, before starting to implement an integrated system.

REFERENCES

1) W.E.Deming. "Out of the Crisis". Cambridge University Press. (1986).
    ISBN 0-58-130553-5
2) J.M.Juran. "Quality Control Handbook". 4th Ed. McGraw-Hill. (1989).
    ISBN 0-07-033176-6
3) W.A. Shewhart. "Economic Control of Quality of Manufactured Product". Van Nostrand. (1931).
4) G.Taguchi, E.A.Elsayed, T.C.Hsiang, "Quality Engineering in Production Systems".
    McGraw-Hill. (1989). ISBN 0-07-100358-4
5) J.D.T.Tannock. "Automating Quality Systems". Chapman & Hall. (1992).
    ISBN 0412-40910-0
6) I.Nonaka, The Knowledge Creating Company, *Harvard Business Review*, Nov-Dec Issue,
    p96-104. (1991).
7) V.Newman, Redefining Knowledge Management to Deliver Competitive Advantage. *Journal of
    Knowledge Management*. Vol.1, No.2, p123-128. (1997).

USING SYSTEMS CONCEPTS TO RESEARCH THE IMPLICATIONS OF EMERGING INFORMATION TECHNOLOGIES (ITs) FOR PEDAGOGY IN HIGHER EDUCATION (HE)

W. F. Ian Beggs

Department of Computing & Information Systems
University of Paisley
Paisley, PA1 2BE, United Kingdom

INTRODUCTION

Systems concepts have been used extensively in the context of research in areas of recognised complexity where established disciplinary research methodologies offer no ready answers. The issues surrounding the implementation of emerging information technologies in support of pedagogy in higher education would appear to give rise to such an area of complexity. Within the project being undertaken by the author, systems concepts such as *holism* and *synergy* have been used to help provide a working definition of the problem area, whilst concepts such as *hierarchy* and *transformation* have been used in the management of the research process. The research framework is itself offered as an emergent construction of the process.

USING SYSTEMS CONCEPTS TO DEFINE THE QUESTION

Amongst the first problems to confront a researcher whose interest lies in the implications of emerging ITs for pedagogy in HE are: (1) the potential breadth of the field; and (2) the associated diversity of perspectives which inform contributions to the ongoing discourse. Literary comment is offered from such apparently distinct knowledge domains as, for example, instructional design (Laurillard, 1993), educational theory (Barnett, 1990, 1997), history of the university (Walker, 1994), sociology (Birch, 1988), and policy analysis (Neave and Van Vught, 1991). The potential for bewilderment is significant. Indeed, Ball's acknowledgement of this state of affairs appears to carry with it the additional caveat that the attempt to offer an overview of HE as a whole, as opposed to analysis based on fragmentation, may be viewed by some as being 'almost unscholarly' (1989, p. 1).

Taking Ball's warning into account, this paper argues in favour of an holistic approach to the research question; thereby offering the prospect of accommodating differing standpoints and reconciling them within a whole which is greater and more robust than the sum of its parts.

Using Systems Concepts to Define the Field

The drawing of boundaries has been used within the context of the project to balance the desirability of adopting an holistic approach on the one hand whilst addressing the practical requirement for research focus on the other. Systems concepts have therefore been used to provide a context for discourse in the research project as illustrated by the following examples taken from the domain of implementing ITs in support of pedagogy in HE.

Higher education (Fig. 1) has thus been delineated in terms of the university sector as defined by the 1992 Further and Higher Education Act, whilst accepting that activities amounting to HE may take place beyond the universities, and may impact upon the activities of the universities themselves.

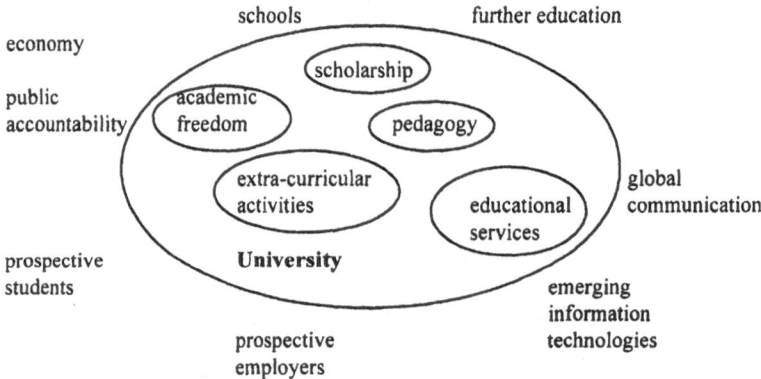

Figure 1. Systems map showing the University as a system.

Pedagogy in HE (Fig. 2.) has similarly been defined specifically in terms of undergraduate and postgraduate teaching and learning, whilst recognising, nonetheless, the potential influence of research for example.

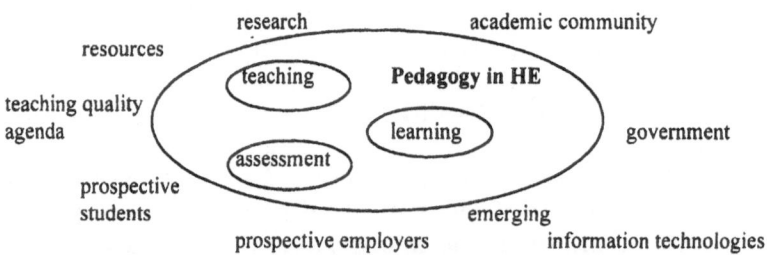

Figure 2. Systems map showing Pedagogy in HE as a system.

Emerging information technologies (Fig. 3) have been defined to include computer-mediated delivery, whilst acknowledging the potential impact of media such as conventional video.

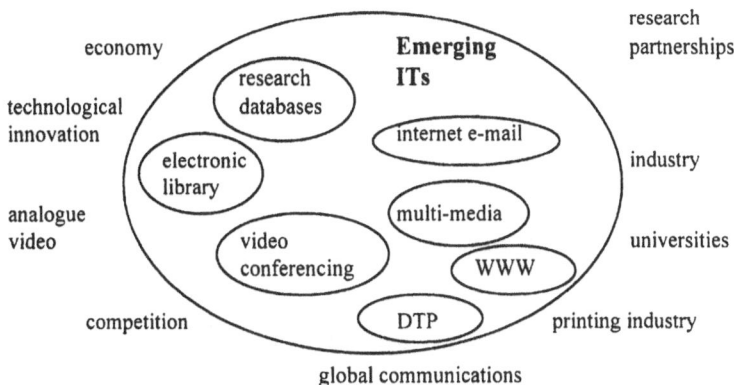

Figure 3. Systems map showing Emerging Information Technologies as a system.

Using Systems Concepts to Bridge the Gap

A review of the literature demonstrates the extent to which such an apparently fundamental concept as 'higher education' can be understood to connote such diverse meanings as academic achievement and social acceptance (Robinson, 1997). This research project is predicated upon a need to address the question as to the balance to be achieved between all those seeking to contribute to the debate as to how emerging ITs should be implemented in support of pedagogy in universities.

Duke (1992) is an example of an author who has sought to offer an account of the university from a number of perspectives and has suggested that it should be understood in terms of the following factors: (1) the essential character of the university; (2) the needs of modern society; and (3) the present concerns and activities of universities. It would seem that these factors provide an appropriate starting point for the research project.

Figure 4 provides an example of the way in which the author has brought together some of the factors emerging from the literature using an influence diagram based on a systems map. These factors are depicted either as components of systems and sub-systems or as environmental elements. Whilst it is accepted by the author that it is unlikely that such a diagram could represent the totality of relationships concerned, Figure 4 is nonetheless offered as a construction which allows further exploration of and discourse on the relationships suggested. It would seem likely that the nature of the real-world relationships may vary considerably from one institution to another. It may be, for example, that some institutions place a higher premium than would others on historical conceptions of the university or on industrial liaison.

USING SYSTEMS CONCEPTS TO MANAGE THE RESEARCH PROCESS

Having used the literature to construct an initial view of the potential relationships involving pedagogy in HE and emerging ITs, systems concepts have been employed in the subsequent management of the research project. It would seem that the area being addressed

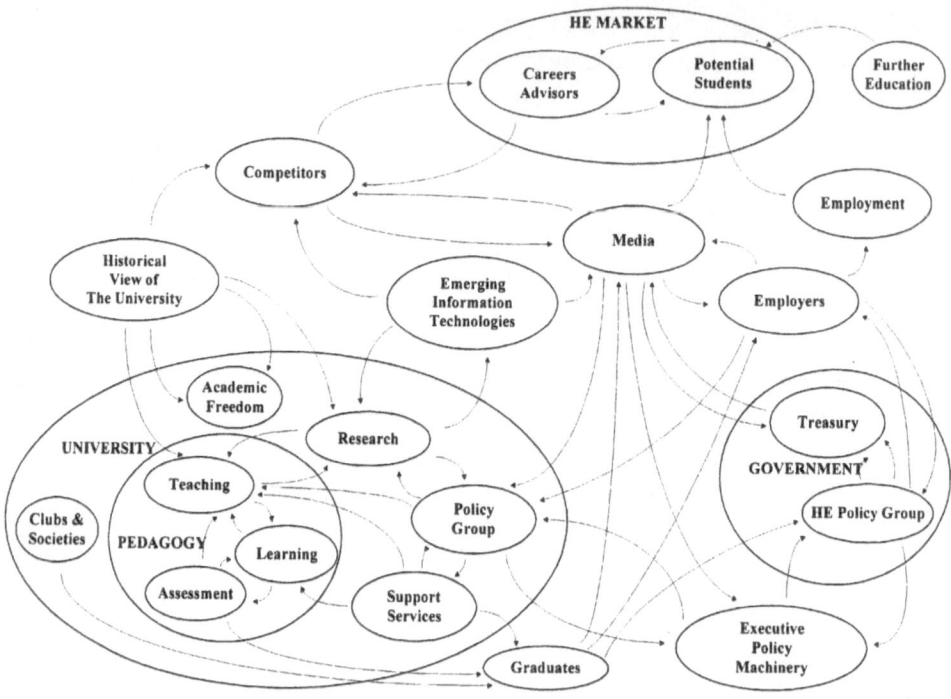

Figure 4. Influence diagram suggesting links between pedagogy in HE (within a university) and emerging information technologies.

in this research project falls within the world of living things defined by Bateson (1972, 1979) and Reason (1994, p. 12) in which "... order arises from the patterns of information flow rather than from physical relationships of cause and effect, and where differences in quality are more profoundly important than differences in quantity." It is on the basis of such an ontology that the author has adopted a qualitative approach to the research. Within this context systems concepts have been employed to construct a defensible inquiry framework taking into account 'worldview', 'methodology', and 'contribution to knowledge'.

Using hierarchy for planning, implementation and review

One of the uses to which systems concepts have been put in the project has been that of managing the research process. This aspect of the project has involved planning, implementing, and reviewing the research. At the planning level the idea of hierarchy has been particularly useful in terms of allowing for the process to be viewed at various degrees of resolution. An example of this is the way in which the research has been planned with a view to providing an account of the concerns and activities of a representative set of UK universities. The planning of the field research has been predicated upon a need to explore these concerns at both strategic and operational levels within each institution. At each of these levels it has been necessary to devise a means of implementing the research objectives. In this way the idea of hierarchy has been used as a means of managing: (1) a research process itself operating on different levels, from conceptual to implementation and review; and (2) the activities, such as fieldwork, associated with each level of the process.

Using Transformation to maintain Methodological Consistency

It is the concept of transformation which drives the research project, and it is argued in this paper that the nature and quality of the transformation depend upon the methodology by which the research is progressed. It has been argued that one of the characteristics of the qualitative approach is its tendency to be 'multimethod' in focus (Denzin and Lincoln, 1994). The availability of a range of methods arising from opposing ontological and epistemological approaches has provided the basis for criticism of researchers who have sought to mix and match methods derived from paradigms with opposing epistemic roots (Leininger, 1994).

It is argued in this paper that methodological consistency is crucial in terms of achieving a meaningful transformation. In practice this has meant that the researcher has had to plan and review each aspect of the process with the nature of the transformation in mind. An example is the way in which historical data might inform the outcome of the research. A positivistic approach might seek to identify those aspects of the history of the university that would be expected to determine the approach of a university to technological change. Such an approach, however, would be inconsistent with a subjectivist approach which emphasises voluntarism over determinism. Similarly, documents are to be viewed as adding breadth to the field research, rather than as substantiating an oral account. The maintenance of methodological consistency, as illustrated in Figure 5, through an explicit statement of the transformation of each aspect of the process has therefore been a constant theme of the research.

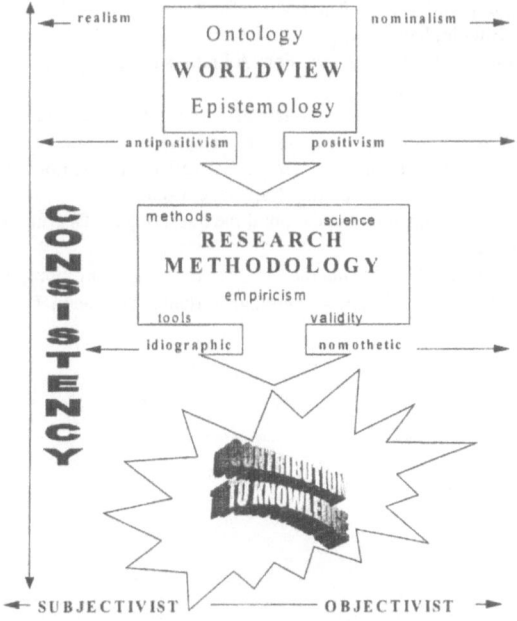

Figure 5. After Burrell and Morgan (1979).

CONCLUSION

This paper offers an account of the author's reflection of a research process in which systems concepts have been used explicitly to define the context of the research, provide focus at a variety of levels, and to address questions as to outcomes. It is the author's experience that systems thinking has become progressively pervasive as the process has developed, with concepts such as 'hierarchy', and 'transformation' having become meaningful in contexts other than those in which they were originally employed. Hierarchy, for example, became a means to understanding the management of the process as well as a means to conceptualising systems and subsystems. Similarly, transformation became a means to questioning the nature and quality of the influences depicted in the domain as well as a means to understanding the implications of employing particular methods. It seems to the author that one of the implications to arise from a systems approach to the research process is that the research framework itself may be presented as an 'emergent construction' (Weinstein and Weinstein, 1991).

REFERENCES

Barnett, R., 1990, "The Idea of Higher Education," The Society for Research into Higher Education & Open University Press, Buckingham.

Barnett, R., 1997, "Higher Education: A Critical Business," The Society for Research into Higher Education & Open University Press, Buckingham.

Ball, C., 1989, "Higher Education into the 1990s," The Society for Research into Higher Education and Open University Press, Milton Keynes.

Bateson, G., 1972, "Steps to an Ecology of the Mind," Ballantine, New York.

Birch, W., 1988, "The Challenge to Higher Education: Reconciling Responsibilities to Scholarship and to Society," The Society for Research into Higher Education & Open University Press.

Burrell, G., & Morgan, G., 1979, "Sociological Paradigms and Organisational Analysis," Arena, Aldershot.

Denzin, N.K., Lincoln, Y., 1994, eds., "Handbook of Qualitative Research," Sage, London.

Duke, C., 1992, "The Learning University," The Society for Research into Higher Education & Open University Press, Buckingham.

Laurillard, D., 1993, "Rethinking University Teaching: A Framework for the Effective Use of Educational Technology," Routledge, London.

Leininger, M., Evaluation criteria and critique of qualitative research studies, *in*: "Critical Issues in Qualitative Research Methodology," Morse, J.M., ed., Sage, London.

Neave, G. & Van Vught F.A., 1991, "Prometheus Bound," Pergamon Press, London.

Resaon, P., 1994, ed., "Participation in Human Inquiry," Sage, London.

Robinson, E., 1997, Mass, continuing adult education: three questions for Dearing, *Higher Education Review Vol. 29 No. 2.*

Walker, A.L., 1994, "The Revival of the Democratic Intellect," Polygon, Edinburgh.

Weinstein, D., Weinstein, M.A., Georg Simmel: sociological flaneur bricoleur, *Theory, Culture and Society Vol. 8.*

RICH PICTURES, METAPHORS AND STORIES AS MECHANISMS TO IMPROVE COLLECTIVE ACTIONS

Alvaro Carrizosa

Lincoln School of Management
Lincoln University Campus
Brayford Pool
Lincoln, LN67TS

INTRODUCTION

This paper presents a series of findings from an ongoing Action Research project in a manufacturing company in the UK. The organisation is undergoing an accelerated process of re-structuring, partially motivated by the need to increase its learning capacity. The previous structure split the engineering department and created small interdisciplinary teams that were spread across the company. Consequently, the technical knowledge of the members of this department was progressively diluted as a result of the limited interaction that engineers had between colleagues of the same speciality. The capacity of the company to transmit knowledge and learn from individual experiences was also progressively reduced.

Due to the fast pace of change in the business environment both internally and externally, managers currently spend most of their time fire-fighting. Before the re-structuring, they used to transmit information through bullet-point reports which were followed by brief explanations. Currently, the time they can dedicate to writing and explaining the information contained in these reports has been considerably reduced. This situation has impacted on the process of communication between managers and employees, principally with respect to a very important issue: transmitting the new vision and objectives of the organisation.

Observing the 'vision-gap' between the managers involved in the restructuring process and the rest of the company, indicates that novel ways of communicating are needed. The usual bullet point list appears to be an inadequate tool to involve people in the organisational change that is proposed. The impoverished process that is taking place between senders and recipients only increases the frustration and anxiety in both parties.

Synergy Matters: Working with Systems in the 21st *Century,*
Edited by Castell *et al.*, Kluwer Academic / Plenum Publishers, New York, 1999.

The situation is rather more complex than a 'vision-gap' between the two parties. As soon as an inquiry process started the weaknesses of the vision of the new structure emerged. The apparent good ideas and intentions that the group of middle managers presented, resulted in little understanding and coherence on the 'how' and 'what' to do. Moreover, since the company does not appear to have a workable strategy, only targets to be achieved, the reasons for adopting the new structure started to become blurred as the conversations advanced. This new structure was approved by the top management and now is in the process of being implemented. At the moment the middle managers are not only trying to implement the new structure, but also trying to create a collective view of the future of the company, making sense of the new structure for them and for the rest of the company.

From the research point of view this situation has been approached by promoting spaces for reflection and inquiry. New ways of improving communication have been considered and, rather than giving solutions, a self-discovery and a learning process have been encouraged. The use of systems tools is also encouraged, but it is not imposed, and before adopting any of these tools a critical appraisal is carried out. The researcher also documents experiences that have taken place in the organisation using alternative methods for communication and making them visible for the organisation as a whole.

The action-research process has become concerned with facilitating the managers of the Engineering Department towards developing ways of improving their own observations and common understanding of the objectives and vision of the company. Naturally this is a difficult and slow process that aims to improve collective action. Furthermore, it is through the participants' interactions continuously being maintained that they jointly agree that a given purpose has been provided, or better that a purpose has emerged. It is only then that it could be said 'closure over observations' (De Zeeuw, 1996) has been reached.

Therefore, the communication processes and the language used by the participants (including the researcher) in their interactions play an important part in the possibility of constituting this collective (Vahl, 1997). It is here where new language tools such as metaphors, stories and rich pictures have appeared as alternative methods of steering the collective and improving their actions.

EPISTEMOLOGICAL DEBATE

The theory and applications of rich pictures (Avison, D., Golder, P. & Shah, H.,1992; Bentley, 1993; Checkland, 1981,1990,1995; Finegan, 1995), metaphors (Morgan, 1986,1993; Palmer & Dulford, 1996; Tsoukas, 1991; Van de Ven & Poole, 1995), and stories (Boyse, 1995; Dyer & Wilkins, 1991; Smith, 1981; Vahl, 1995,1996) have been widely discussed and elaborated in the management science. This study has, however, focussed attention on epistemological evidence where these tools could help to steer collectives and improve their actions.

There is reported evidence that all these three 'tools' have a symbolic form by which groups and members of an organisation construct shared meaning and new understandings about a particular situation, stimulating new actions. From this emergent property of the group a competent collective could arise.

Ortony (1979) suggests two epistemological positions, which can be characterised as 'non-constructivist' and 'constructivist'. The first one suggests that this is an objective world, which through a systematic observation we can objectively discover. The 'constructivist' approach suggests that, even though, this is an objective world, we perceive this through our senses which build, construct and structure our ideas of reality. Morgan (1986,1993) and Checkland (1981,1990, 1995), the fathers of the metaphors and rich picture techniques, respectively, seem to perceive the world in a 'constructivist' manner. They argue that reality takes many different shapes according to the individual in a particular moment in time. Nevertheless, a group of people has the potential to agree on a view of the world and then act according to it.

This process of converging the different views into a rich picture, a metaphor or even a story requires an effort from the participants that constitutes in itself the most important outcome of the process. The processes of communication that emerges, being reinforced by these tools, results in a common framework or context in which to think and transfer thoughts to others. It is important to note that by no means, is it only a matter of consensus, it can also be a situation where participants' observations converge, allowing the observer to achieve observational closure. Naturally the different tools make it possible to observe different instances and properties of the 'object'. These tools are considered separately in the following section.

THE 'TOOLS' IN PRACTICE

The previous epistemological debate underlies the issue of the production of collective knowledge, which is inherent to its transferability. This section presents three cases where the 'tools' facilitated this process, in relation to the improvement of the understanding of the new vision and objectives of the organisation in question.

Rich Pictures

One of the key factors in the success of the new organisational structure for this company is the installation of a new information system. This information system centralises the data of the whole organisation, offering more accurate and updated information. It also gives the possibility of additional analysis and linking of the data generated in different departments.

The need for expressing and transmitting the implications of the additional relations available in the new system led the manager of the IT department to present the situation 'before' and 'after' using a rich picture. "It was the best way to communicate the implications of the new system", then he added, "the use of words or technical drawings only add to the confusion and misunderstanding ... only when people have grasped the concepts, then formal procedures can be introduced." The users' comments were also very favourable. One of the participants mentioned after this presentation: "the cartoons help me to visualise my relation with the system and other members of the company".

Since this presentation took place many of the conversations around the new information system changed from general to specific issues related to its applicability, and from a negative to a positive disposition towards the system. This shift could provide evidence that, as a result of the use of this 'tool', this collective is improving its understanding in relation to the introduction of this change, and consequently are aligning their actions towards this desirable future.

Metaphors

A workshop that took place in the Lincoln University Campus in August, 1997 with a group of 15 members of the Proposals Department. The purpose of the workshop was to think and generate new ideas on how to improve the structure of the department. The activity was organised in two main sections; the first part applied the Creativity phase of Total Systems Intervention (see, Flood and Jackson, 1991) to facilitate conversations and guide the choice of a dominant metaphor (that which best expressed the current situation), and the second part consisted of elaborating on metaphors which best represented the desired future departmental structure for the participants.

The mechanistic metaphor emerged as the one that best represented the present, and the organism metaphor, as the one that expressed the 'ought to be' of the department. The change in the language used by the participants was evident and influenced new insights, as well as, new more active dynamics within the group.

Towards the end of the activity, some people started to express their experiences through all sorts of metaphors as a result of the methodological freedom of the activity. The new metaphorical language gave to the participants the possibility to communicate beyond the constraints of their day to day language. Metaphors also promoted a mental shift which could be defined as turning from a 'unitary best solution' to a 'pluralistic way of improving'. This new way of thinking allowed different points of view about the 'reality' and people started to talk about improving the situation throughout the different interpretations presented.

The recommendations of the group were reported and some of the important changes in the new structure could be attributed to the new understanding and the ideas put forward by the group. An example of this is that the participants visualised themselves as assisting sales people more closely; being more aware about the environment where sales engineers were based. They also recognised that the previous departmental structure was an obstacle to achieving this contact with the customer. As a result, in the new structure the Proposals Department was divided, and now the functions of this department are assumed by the Sales Department and the Projects Department, allowing the people that constituted the Proposals Department to develop more awareness about the environment.

Stories

A group of around 15 middle managers held a series of meetings for two months in order to address issues related to interdepartmental communication and the layout of the offices. The emergent result was a new organisational structure, which required the approval of the top management to be implemented. After a long discussion on the form of presenting the conclusions of the group, they decided to write three stories.

Once the stories were created the idea of presenting them through a performance was approved by the participants. Initially, many were very sceptical about the effects, but soon agreed to take part in a dramatisation. The stories attempted to show the impact of the major changes proposed. In order to do so, the first play brought into scene the present situation, exaggerating the problems and difficulties that were being faced as a result of the present structure. The next two plays showed the new structure in action and how things would be improved through the new organisation. After a break, a formal presentation recapitulated the ideas that were dramatised before.

The impact of the representations was enormous. They managed to show the contrast between the present situation and the possibilities generated by the new structure in such a way that there was an acclaimed general approval of the proposed structure even before the formal presentation took place. The top management had grasped the concept and captured the enormous potential of the new structure.

CONCLUSIONS

As demonstrated by the examples provided, the use of metaphors, rich pictures and stories can help to centre thinking, and focus in particular aspects of reality which are highlighted by the specific 'tool' in use. This approach could also help in approximating reality from a holistic point of view. Moreover, according to the 'tool' in use, a language develops which provides all sorts of mental representations, codes, conventions and connections aiding the process of constructing complex reality.

In terms of the individuals involved in a particular reality, the 'tools' can pull and steer their experiences into an enactment of what is possible. This 'reality' that becomes expressible according to personal experiences, can create a shift of thinking in the group and in the individuals who share conversations. This mental shift could be the result of a new understanding about the organisation, which in fact induces creative action (Morgan, 1986, 1993).

Another shift could be in terms of looking for 'truths' rather than a 'truth'. This shift could emerge as a result of the contradictory and even paradoxical interpretations of the same 'reality'. This shift could be observed in terms of making sense of actions in order to improve situations, rather than holding objective solutions (Weick, 1989).

From the researcher's point of view, the examples presented, constitute cases in themselves. All these cases, with their constraints, represent observations of the organisational reality. From these observations it is possible to promote conversations about the fundamental assumptions on which the members of the organisation are basing their actions. By bringing to the surface those fundamental assumptions, the organisational learning process could begin. This organisational learning can be the result of the new expressible understandings facilitated and promoted by the 'tools' analysed. Consequently, once these tools are established as normal practice, the collective action might be considerably improved .

REFERENCES

Avison, D., Golder, P. & Shah, H., 1992, A tool kit for soft systems methodology, in "The Impact of Computer Supported Technologies on Information Systems Developments," Kendall, K., Lyytinen, K. & DrGross, J., Amsterdam, Elesevier

Axelrod, R., 1984, "The Evolution of Co-operation," Basic Books, New York.

Bentley, T., 1993, Soft systems methodology, *Management Accounting*, London 71, 7, 22-36.

Boyse, M., 1995, Collective centring and collective sense-ma king in the stories and storytelling of one organization, *Organization Studies*, 16, 1, 107-137.

Checkland, P., 1981, "Systems Thinking, Systems Practice," Wiley, Chichester.

Checkland, P., 1995, Model validation in soft systems practice, *Systems Research*, 12, 1, 47-54.

Checkland, P. & Scholes, J, 1990, "Soft Systems Methodology in Action," Wiley, New York.

De Zeeuw, G., 1996, Action and social order research. In: "Theory and Practice of Action Research," Boog, B., Coenen, H., Keune, L., Lammerts, R., Tilburg University Press, pp. 129-155.

De Zeeuw, G. & Koppelaar, H., 1995, Objects, systems and invariants, *Systemica*, 10, 2, 213-227.

Dyer, W.G. & Wilkins, A.L. 1991, Better stories, not better constructs, to generate better theory: a rejoinder to Eisenhardt. *Academy of Management Review*, 16, 3, 613-619.

Finegan A.D., 1995, Fuzzy logic and soft systems methodology-a complex connection. "The Australian Systems Conference-Systems for the Future," Australia.

Flood, R. L & Jackson, M. C., 1991, "Creative Problem Solving: Total Systems Intervention," Wiley, Chichester.

Morgan, G., 1993, "Imaginization: The art of creative management," Newbury Park, CA: Sage.

Morgan, G., 1986, "Images of organization." Beverly Hills, CA: Sage.

Morgan, G., 1983, More on metaphor: Why we cannot control tropes in administrative science. *Administrative Science Quarterly*, 28: 601-607.

Morgan, G., 1983, "Beyond Method: Strategies for Social Research," Beverly Hills, CA: Sage.

Ortony, A., 1979, Metaphor: a multidimensional problem, In "Methaphor and Thought," Ortony, A. (Ed.),. 1-16, Cambridge University Press.

Palmer, I. & Dulford R., 1996, Conflicting use of metaphors: reconceptualizing their use in the field of organizational change, *Academy of Management Review*, 21, 3, 691-717.

Popper, K. R., 1959, "The Logic of Scientific Discovery," Harper & Row, New York.

Schregenberger, J., 1982, The development of Lancaster soft systems methodologies: A review and some remarks from a sympathetic critic, *Journal of Applied Systems Analysis*, 9, 87-98.

Smith, B. H., 1981, Narrative versions, narrative themes, in "American Criticism in the Poststructuralist age". Pondy, L., Morgan, G. and Dandridge, T. (eds.) Ann Arbor: Univeristy of Michigan, 162-186.

Tsoukas, H., 1991, The missing link: a transformational view of metaphors in organizational science, *Academy of Management Review*, 16, 3, 566-585.

Vahl, M. 1997, A challenge to research: the improvement of communities, Working Paper, Nijenrode University, Centre for Corporate Renewal, Netherlands.

Vahl, M., 1996, Managing Stories: a Community Research Perspective on Water. In: "Cybernetics and Systems' 96," Trappl, R., (eds.), Vol. 2. Austrian Society for Cybernetics and Systems, Vienna, pp. 798-800.

Vahl, M., 1995, Theories that create coherence and other practical stories, in "Critical Issues in Systems Theory and Practice," Ellis, K., Gregory, A., Mears-Young, B., &Ragsdell, G., (eds.), , Plenum, London.

Van de Ven, A. & Poole, M., 1995, Exploring development and change in organizations, *Academy of Management Review*, 20, 510-540.

Weick, K., 1989, Theory construction as disciplined imagination, *Academy of Management Review*, 14: 516-531.

A NOVEL APPROACH TO QUANTIFY AGRICULTURAL SUSTAINABILITY USING FUZZY SET THEORY

A.M.G. Cornelissen

Animal Production Systems Group
Wageningen Institute of Animal Sciences
Wageningen Agricultural University
P.O. Box 338
NL-6700 AH Wageningen
The Netherlands

INTRODUCTION

The concept "agricultural sustainability" signifies a generally accepted concern for the future viability of agricultural production systems (e.g., Olson, 1992; Hansen, 1996). This concern is expressed from economic, ecological and societal perspectives (Shearman, 1990; Becker, 1997; Giampietro, 1997).

The different perspectives manifest themselves through the different demands that are made on agricultural production systems. These demands range from, for example, earning a proper income out of agricultural activities, or meeting a need for inexpensive food products, to annulling the undesirable consequences of agricultural practice, such as, soil erosion, or nutrient surpluses (Ikerd, 1993; Steinfeld et al., 1997). The economic, ecological and societal demands usually result in the formulation of aims and strategies to guide the sustainability development of agricultural production systems (Hansen, 1996).

To implement agricultural sustainability as a yardstick in practical situations, a quantitative measure is needed (Heinen, 1994; Becker, 1997). This paper presents a method to quantify agricultural sustainability.

AGRICULTURAL SUSTAINABILITY

Starting Point

A quantitative measure of agricultural sustainability has to consider the joint economic, ecological and societal perspectives on agricultural production systems. These perspectives are site-specific and they will change over time (Shearman, 1990; Fresco and Kroonenberg,

Synergy Matters: Working with Systems in the 21st Century,
Edited by Castell *et al.*, Kluwer Academic / Plenum Publishers, New York, 1999.

49

1992; Clayton and Radcliffe, 1996). Agricultural sustainability should, therefore, not aim at designing agricultural production systems that last forever in a definite form, but monitor the continuous process of adapting agricultural production systems to the specific economic, ecological and societal systems they are embedded in.

Ambiguity and Conflict

The supposition that agricultural production systems need to be considered as part of larger systems emphasizes the complexity of the problem at hand. This supposition makes agricultural sustainability an ambiguous concept. There are mainly two reasons for this ambiguity.

First, there is virtually an unlimited number of ways in which an agricultural production system may interact with the larger systems mentioned. A complete and unambiguous description is practically impossible. To reduce the complexity to a manageable level the process of sustainability development must, inevitably, be monitored by a selected group of sustainability variables (Klir and Folger, 1988; Zimmermann, 1996).

Second, different demands may present agricultural production systems with conflicting aims (De Wit et al., 1995). For example, keeping chickens in ground floor systems can be considered a societal aim in The Netherlands, considering that an increasing number of Dutch consumers object to cage housing systems for chickens that limit the natural behavior of the animals. Ground floor systems, however, also produce a higher ammonia volatilization compared to conventional housing systems. This is in conflict with ecological aims (Groot Koerkamp, 1994). Individual strategies can, consequently, lead to ambiguous and conflicting conclusions (Silvert, 1997). In this paper such ambiguous and conflicting conclusions, in relation to individual strategies, are referred to as trade-offs.

QUANTIFYING AGRICULTURAL SUSTAINABILITY

Fuzziness and Acceptability

Considering the necessary selection of a limited number of sustainability variables, and as a result of the mutually emerging trade-offs, it is impossible to determine indisputably whether an agricultural production system is sustainable or unsustainable. Applying conventional, two-valued logic (e.g., sustainable-or-unsustainable type decisions), therefore, comes to an unsatisfactory conclusion (Klir and Folger, 1988; Pedrycz, 1993; Zimmermann, 1996).

Fuzziness describes *event ambiguity*: it measures the degree to which an event occurs (Kosko, 1992). Fuzziness, therefore, relates to multi-valued logic (Klir and Folger, 1988): e.g., all intermediate situations between sustainable and unsustainable are possible. This means that agricultural production systems can be assessed as partially sustainable.

In this paper, agricultural sustainability is interpreted as to what extent agricultural production systems are able to meet the joint demands. Agricultural sustainability is interpreted, therefore, as an assessment of the degree to which agricultural production systems achieve acceptable results with respect to the demands made on them. Since agricultural sustainability is approximated by a selection of sustainability variables, the *acceptability* of achievements should be determined for every selected sustainability variable. Such a *degree of acceptability* can be determined using fuzzy set theory (Bockstaller et al., 1997; Silvert, 1997).

Hypothesis

A method that wishes to quantify agricultural sustainability satisfies the following conditions. First, the method should assess to what extent agricultural production systems are able to meet the joint economic, ecological and societal demands. Second, it should present a common denominator in which the selected sustainability variables can express the suggested assessment. Third, the method has to provide a framework that enables the assessment of important trade-offs.

It is hypothesized that fuzzy set theory offers a mathematical framework to develop a method that satisfies the formulated conditions.

FUZZY SET THEORY

Classical set theory is based on two-valued logic. Assuming that the symbol U defines a complete set of elements x (i.e., the Universe of discourse), and given a subset A of U, in classical set theory each element x of U ($x \in$ U) either belongs to A ($x \in$ A) or does not belong to A ($x \notin$ A). The process that determines whether $x \in$ A or $x \notin$ A can be defined by a *characteristic function* $\mu_A(x)$. Thus, $\mu_A(x)$ assign a value to every $x \in$ U such that $\mu_A(x) = 1$ iff (if and only if) $x \in$ A, and $\mu_A(x) = 0$ iff $x \notin$ A. There is an unambiguous distinction between members and nonmembers of the subset A (Klir and Folger, 1988; Pedrycz, 1993; Zimmermann, 1996).

Fuzzy set theory is based on multi-valued logic. Fuzzy set theory describes the situation in which partial membership in a set occurs, i.e., where transition between membership and nonmembership is gradual rather than abrupt. If A denotes a classical subset, then Ã denotes a fuzzy subset. In fuzzy set theory the analogous function to the characteristic function $\mu_A(x)$ for A is the *membership function* $\mu_{\tilde{A}}(x)$ for Ã, where $\mu_{\tilde{A}}(x)$ can take on values from 0 to 1, i.e., $\mu_{\tilde{A}}:U \rightarrow [0,1]$, depending on the degree of membership $\mu_{\tilde{A}}(x)$ of x in Ã (Klir and Folger, 1988; Pedrycz, 1993; Zimmermann, 1996).

The theory of graded concepts can be pursued to the determination of a degree of acceptability for specific sustainability variables. For example, as a result of the ammonia (NH_3) volatilization problem in intensive animal production systems in The Netherlands a dominant ecological demand made on these agricultural production systems is to reduce ammonia volatilization (Groot Koerkamp, 1994). In this case, ammonia volatilization (x) is considered a sustainability variable (Figure 1). A membership function $\mu_{\tilde{A}}(x)$ determines to what degree the observed level of ammonia volatilization (x) is acceptable. Thus, acceptability is interpreted as a subset Ã of U, i.e., the entire set of all possible x. For example, when the observed level of ammonia volatilization is x_i then $\mu_{\tilde{A}}(x_i) = 0.75$, or the degree of acceptability for x_i is 0.75 (Figure 1). Figure 1 also shows that for this hypothetical sustainability variable the degree of acceptability decreases as the level of ammonia volatilization increases.

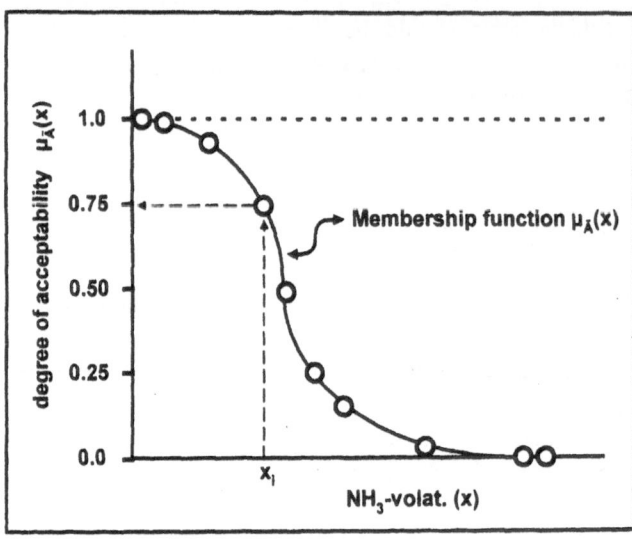

Figure 1. A membership function $\mu_{\tilde{A}}(x)$ determines the degree of acceptability (acceptability is defined as a fuzzy set $\tilde{A}$) of specific observations (x) ($x \in U$, the set of all possible observations of x) of the hypothetical sustainability variable Ammonia Volatilization (NH_3-volat.).

AN OUTLINE OF THE METHOD

Step 1: Sustainability Variables

The first step in the method that seeks to quantify agricultural sustainability is to determine which site-specific sustainability variables are taken into account. These sustainability variables can be roughly classified into three clusters, corresponding to the three perspectives on agricultural production systems (Table 1). The determination of suitable sustainability variables, however, is beyond the objectives of this paper.

Step 2: Membership functions

For each sustainability variable in each cluster a membership function $\mu_{\tilde{A}}(x)$ determines the degree of acceptability (acceptability is defined as a fuzzy set $\tilde{A}$) of specific observations x of that sustainability variable ($x \in U$, the set of all possible observations of x) (Table 1).

Step 3: Combining degrees of acceptability

Degrees of acceptability for observations of individual sustainability variables can be combined into a "partial degree of acceptability" for each cluster: a degree of economic acceptability ($\mu_{\tilde{A}\text{-en}}$), a degree of ecological acceptability ($\mu_{\tilde{A}\text{-el}}$), and a degree of societal acceptability ($\mu_{\tilde{A}\text{-so}}$) (Table 1). Thus, partial economic, ecological and societal degrees of acceptability are determined for individual or groups of agricultural production systems. The partial degrees of acceptability can, finally, be combined into an "overall degree of acceptability": $\mu_{\tilde{A}\text{-overall}}$ (Table 1). The resulting $\mu_{\tilde{A}\text{-overall}}$ is a site-specific and time-dependent judgement of the sustainability of individual or groups of agricultural production systems.

Table 1. A schematic outline of a method that quantifies agricultural sustainability, using Fuzzy Set Theory, on the basis of selected sustainability variables in three clusters.

Three Clusters of Sustainability Variables		
Economic Cluster Degrees of Acceptability for variables $x_1..x_k$	**Ecological Cluster** Degrees of Acceptability for variables $x_1..x_m$	**Societal Cluster** Degrees of Acceptability for variables $x_1..x_n$
$\mu_{\tilde{A}\text{-en}}(x_1)$ . . . . $\mu_{\tilde{A}\text{-en}}(x_k)$	$\mu_{\tilde{A}\text{-el}}(x_1)$ . . . . $\mu_{\tilde{A}\text{-el}}(x_m)$	$\mu_{\tilde{A}\text{-so}}(x_1)$ . . . . $\mu_{\tilde{A}\text{-so}}(x_n)$
Combined $\mu_{\tilde{A}\text{-en}}$	Combined $\mu_{\tilde{A}\text{-el}}$	Combined $\mu_{\tilde{A}\text{-so}}$
Combined $\mu_{\tilde{A}\text{-overall}}$		

CONCLUDING REMARKS

This paper proposes a method to quantify agricultural sustainability. By expressing each sustainability variable in a degree of acceptability, using fuzzy set theory, a common denominator is created, which enables a combination of degrees of acceptability of various selected sustainability variables into a site-specific and time-dependent judgement of the sustainability of individual or groups of agricultural production systems.

Although the use of selected sustainability variables may seem an analytical approach to describe a complex problem, the approach is holistic in a sense that through the choice of sustainability variables processes and effects at different hierarchical levels can be monitored (Bockstaller et al., 1997).

Membership functions constitute the essential basis on which fuzzy set theory is built. Paradoxically, the membership function is at the same time the strongest and the weakest point of the theory. Where on the one hand membership functions are seen as a very useful generalization of classical set theory, they, on the other hand, are sometimes considered to be too subjective (Munda et al., 1992). Future research will especially address the possibilities and problems in relation to defining membership functions.

Silvert (1997) gives a detailed discussion on the various possibilities of combining individual memberships into partial memberships. Future research will address possibilities and problems in relation to combination rules.

REFERENCES

Becker, B., 1997, "Sustainability Assessment: A Review of Values, Concepts, and Methodological Approaches," Worldbank-CGIAR, Washington D.C., Issues in Agriculture 10.

Bockstaller, C., Girardin, P., and Van der Werf, H.M.G., 1997, Use of agro-ecological indicators for the evaluation of farming systems, *Eur. J. Agron.* 7:261-270.

Brklacich, M., Bryant, C.R., and Smit, B., 1991, Review and appraisal of concept of sustainable food production systems, *Environ. Man.* 15(1):1-14.

Clayton, M.H., and Radcliffe, N.J., 1996, "Sustainability. A Systems Approach," Earth Scan Publications Ltd., London.

Dalsgaard, J.P.T., Lightfoot, C., and Christensen, V., 1995, Towards quantification of ecological sustainability in farming systems analysis, *Ecol. Engng.* 4:181-189.

De Wit, J., Oldenbroek, J.K., Van Keulen, H., and Zwart, D., 1995, Criteria for sustainable livestock production: a proposal for implementation, *Agric. Ecosyst. Environ.* 53:219-229.

Fresco, L.O., and Kroonenberg, S.B., 1992, Time and spatial scales in ecological sustainability, *Land Use Policy* July 1992.

Giampietro, M., 1997, Socioeconomic constraints to farming with biodiversity, *Agric. Ecosyst. Environ.* 62:145-167.

Groot Koerkamp, P.W.G., 1994, Review on emissions of ammonia from housing systems for laying hens in relation to sources, processes, building design and manure handling, *J. Agric. Engng. Res.* 59:73-87.

Hansen, J.W., 1996, Is agricultural sustainability a useful concept?, *Agric. Syst.* 50:117-143.

Heinen, J.T., 1994, Emerging, diverging, and converging paradigms on sustainable development, *Int. J. Sustain. Dev. World Ecol.* 1:22-33.

Ikerd, J.E., 1993, The need for a systems approach to sustainable agriculture, *Agric. Ecosyst. Environ.* 46:147-160.

Klir, G.J., and Folger, T.A., 1988, "Fuzzy Sets, Uncertainty, and Information," Prentice-Hall International Inc., New Jersey.

Kosko, B., 1992, "Neural Networks and Fuzzy Systems," Prentice-Hall International Inc., New Jersey.

Munda, G., Nijkamp, P., and Rietveld, P., 1992, "Multicriteria Evaluation and Fuzzy Set Theory: Applications in Planning for Sustainability," Free University, Amsterdam, Research-Memorandum 1992-68.

Neher, D., 1992, Ecological sustainability in agricultural systems: definition and measurement, *in*: "Integrating Sustainable Agriculture, Ecology, and Environmental Policy," R.K. Olson, ed., Haworth Press Inc., Binghampton.

Olson, R.K., 1992, The future context of sustainable agriculture: planning for uncertainty, *in*: "Integrating Sustainable Agriculture, Ecology, and Environmental Policy," R.K. Olson, ed., Haworth Press Inc., Binghampton.

Pedrycz, W., 1993, "Fuzzy Control and Fuzzy Systems," Research Studies Press Ltd., Taunton.

Shearman, R., 1990, The meaning and ethics of sustainability, *Environ. Man.* 14(1):1-8.

Silvert, W., 1997, Ecological impact classification with fuzzy sets, *Ecol. Modell.* 96:1-10.

Steinfeld, H., De Haan, C., and Blackburn, H., 1997, "Livestock-Environment Interactions, Issues and Options," FAO-Worldbank, Rome.

Zimmermann, H.-J., 1996, "Fuzzy Set Theory and its Applications," Kluwer Academic Publishers, Boston.

APPLIED ARTIFICIAL INTELLIGENCE AND THE MANAGEMENT OF KNOWLEDGE

Colquhoun-John Ferguson & Scott Goldie

Department of Management & Marketing
University of Paisley
High Street
PAISLEY
PA1 2BE
U.K.

INTRODUCTION

For any organisation, the single most important asset must be its individual members and their accumulated knowledge. It has been suggested that the sum of individual parts of knowledge may be less than the whole, and so methods of accessing, utilising, sharing and storing this knowledge are important factors to be addressed in organisational studies.

Knowledge management is a relatively new and developing area which has introduced a methodology for the planned capture and re-use of organisational knowledge. Successful application of knowledge management practices involves the understanding and constructive use of organisational learning and information flows within the organisation.

The research described in this paper focuses on the application of information technology as a means to manage information and link this to the development of a knowledge 'sink' within the organisation. Specifically, we use techniques from the field of artificial intelligence - in particular case-based reasoning and data mining - and apply these techniques to the functions of a manufacturing organisation. Our analyses have concentrated on how these techniques can enable a more efficient access, sharing and usage of accumulated knowledge as a means of enabling different functions within the organisation to perform their tasks more effectively.

Synergy Matters: Working with Systems in the 21st *Century,*
Edited by Castell *et al.*, Kluwer Academic / Plenum Publishers, New York, 1999.

RESOURCE-BASED VIEW OF ORGANISATIONS

Strategists have long sought to explain the reason for superior market performances, or above average profitability amongst organisations. However, strategists such as Porter (1985), have largely focused on external environmental factors in relation to the organisation and have concentrated on issues such as barriers to competition etc. Such an approach has acted as a catalyst for an alternative viewpoint in strategy which sees organisation's resources and competencies as critical factors for achieving superior performance and therefore competitive advantages. This resource-based view of firms originates in work carried out in the late 1950s, where researchers such as Selznick (1957) and Penrose (1959) postulated that organisations were collections of resources which had distinctive competencies which enabled them to *earn* competitive advantage. This approach has been substantiated recently and received much attention from researchers such as Barney (1991) and Kay (1995), who proposed that the analysis of an organisation's skills and capabilities is of greater strategic value than the analysis of its competitive environment. One such capability is innovation, and this paper focuses on an internal architectural aspect of the use of an organisation's capabilities and tools which may enable this. The value of architecture rests in the capacity of organisations to create organisational knowledge, and routines which achieve an open exchange of information (Kay, 1995).

There are two essential implications of the resource-based view of the organisation which make it relevant to the knowledge management concept:
- recognition that resources contained within the organisation comprise of both tangible (material) assets as well as intangible (tacit) assets,
- recognition of the relationship between the resource-based view of organisations and knowledge and learning i.e. the acquisition and use of resources to assist learning.

KNOWLEDGE MANAGEMENT

As has been stated previously, every organisation is composed of individuals, and the knowledge they possess. The overall knowledge of an organisation however may be less than that of its individuals (Argyris & Schön, 1978). A definition of what knowledge is in this, or any, context difficult. Any definition may be either brief and unhelpful to the reader (e.g. "knowledge is information which can be put to effective use"), or philosophically verbose and therefore possibly equally unhelpful. Metes & Gundry (1997) define certain operational characteristics of knowledge - firstly, knowledge is a human capability i.e. a capability to do or judge something now or in the future. Knowledge is the transformation of information by a person. Secondly, knowledge acquisition is a dynamic process. Knowledge management tools do not have an inherent capacity to manage knowledge - rather they help capture, organise, store and transmit source material from which an individual may acquire knowledge. Whether an individual does acquire knowledge from a source depends on a dynamic interaction in which two factors are important:
- the similarity between the person's context (their situation, history and assumptions) and the context described,

- the degree of congruence between how the material is structured and how the structure of the domain appears to the reader.

A third characteristic of knowledge is that it is generative and multi-dimensional. This means that it can be explored and have abstractions applied to it; to have knowledge enables the possessor to generate new statements about a subject rather than merely reproduce the statements that were received. The multi-dimensional nature of knowledge is linked to a fourth aspect i.e. that knowledge is elaborate. Knowledge is a complex body of organised information, delivered in large packets. The acquisition of knowledge by an individual then may be hindered by information overload without the aid of some filtering tool or map.

A definition of knowledge *management* in an organisational context is no less difficult. Of help though is reference to the resource-based view of the organisation, and from this viewpoint knowledge management can be thought of as a set of practices to identify and organise the intellectual assets (both explicit and tacit) of an organisation. This knowledge should then be accessible to members of the organisation to enable them to do their jobs more effectively.

PRODUCT DESIGN

Management of knowledge studies mostly concentrate on the facilitation of concurrent team applications. Our research however focuses on the acquisition of knowledge and interpretation by individuals working within a team within a particular domain i.e. the function of product design. This domain displays the criteria of the information being complex and multi-dimensional and prone to being delivered to the user resulting in 'information overload'.

The problems faced by designers of new product have been highlighted in a previous paper (Ferguson et. al., 1998) describing research being undertaken here at the University of Paisley in collaboration with a German manufacturing organisation. In this paper we described the organisational structure and information needs of the design team, and identified problem areas. This paper discussed the information needs of product design teams and focused on information feedback from customer complaints, standard parts and form features from existing product. We concentrated on the problems that the data/information received by the product development department were *restricted* in the sense that it was not always in a format immediately understandable, and therefore of instant use to designers. Another problem existed in that the sources from which these data came did not allow designers a complete view of the system - in particular, feedback from the customer base. This paper focused on the learning aspect of knowledge, and highlighted several areas as being important:
- knowledge acquisition - learning occurs when an organisation acquires knowledge, not only from the external environment, but from rearrangement of existing knowledge (Dodgson, 1993)
- information distribution - sharing of information between different units within the organisation increases learning (Brown and Duguid, 1991)
- information interpretation - the importance of information or data becoming knowledge before it can affect learning (Huber, 1991)
- organisational memory - the importance of corporate knowledge and learning histories to aid learning (Prahalad and Hamel, 1994)

Garvin (1993) highlights the organisation's ability to learn from experience in his definition of a learning organisation as, *"an organisation skilled at creating, acquiring, and transferring knowledge, and at modifying its behaviour to reflect new knowledge and insights."* The research described in this paper attempts to provide a link between knowledge management and organisational learning. Many practitioners see the significance of this concept as a mark of progress for organisations from a focus on material resources, to that of human potential (Senge, 1992; Prahalad & Hamel, 1994).

ARTIFICIAL INTELLIGENCE

Our current research addresses the question of how artificial intelligence (AI) tools may facilitate the organising, disseminating, storing and interpreting of knowledge. Two (AI) techniques have formed the focus for our research with the aim of combining these techniques into a hybrid system to manage the interpretation of corporate data and information for use by product designers: case -based reasoning and data mining. A full description of these techniques does not fall within the scope of this paper. However a brief explanation as to their functioning will act as a guide to why we consider them viable for this particular problem.

Case-based reasoning (CBR)

CBR has been heralded as one of the most promising recent technologies for building intelligence into computers (Watson, 1997) with the aim of supporting the management decision-making process. CBR is a problem-solving approach that takes advantage of the knowledge gained from previous attempts to solve a particular problem. A record of each past attempt is stored as a case, and it is the collection of historical cases which forms our model. When a CBR system solves a problem, rather than starting from scratch, it searches its case base for similar cases whose attributes are similar to the problem that it is being asked to solve. The CBR system then creates a solution by synthesising the similar cases and adjusting the final answer for differences between the current situation and the ones described in the cases. As the case base grows, the accuracy of the system should improve. A case is simply then a collection of attributes. In the scenario described in this paper, attributes consist of, firstly the form features associated with individual products (i.e. design features such as dimensions, strength, positioning and assembly), and secondly, fault reports from both the company's testing department and customer complaint forms. Cases have the ability to transform abstract concepts into real images. A case also provides inherently useful information because of how it organises information and accesses it. (Dhar & Stein, 1996).

Data mining

The other technique currently under investigation in our research is that of data mining. The limitations of traditional relational databases and data-server technology as a means to manage ever increasing masses of data has been described elsewhere. (Ferguson, 1997). Applications for data mining are varied, and include such areas as forecasting for retail product demand, promotion response, stock allocation, production planning, scoring for customer prospecting and profiling, anomaly detection used for fraud detection, and data integrity validation. Our previous research investigated the use of data mining as an approach to enable product designers access to corporate information and data sources. Briefly, data mining is the technique by which relationships and patterns in data are identified in large databases (Fayyad and Uthurusamy, 1995). Data mining differs from traditional on-line analytical processing tools (OLAP), which provide data access to end-users, thus allowing the users to view their data at whatever level of detail they need. Data mining tools take analysis a step further by automatically sifting through a large amount of data to find useful information. In general, data mining and OLAP work well together (Parsayae, 1997) - data mining finds something interesting, and OLAP tools allow the user to investigate further. Our research aims to replace OLAP with a case-based reasoning mechanism and investigate the viability of such a hybrid system.

Why use data mining? One of the key attributes of a data mining problem is that there are typically many contributing factors in the problem. This is no less true for the example given in this paper where product designers must contend with varying factors such as features and tolerances of materials and finished products and diverse error and failure reports from a range of sources relating to one or a combination of factors. Data mining attempts to locate the complex and subtle relationships among these variables.

CONCLUSION

Previously we suggested that knowledge can only be gleaned from data/information if it is presented in a meaningful way to its users (Ferguson, 1997). The approach described here builds on this previous research which agrees with Galliers' (1995) statement that the important question in information systems strategy is determination of the key information requirements to meet individual needs. This paper discusses the information needs of product design teams and focuses on customer feedback, standard parts and form features from existing product as being essential to their function. We propose that a hybrid approach using the techniques of data mining and case-based reasoning may provide an increased problem solving ability for product designers. Both data mining and CBR are used to solve a problem - not purely answer a query, and it is further proposed that enabling designers the ability to solve problems from disparate corporate data sources is a major step in the implementation of effective knowledge management within the organisation.

REFERENCES

Argyris, C., Schön, D.A., 1978, *Organizational Learning: A Theory of Action Perspective*, Addison-Wesley, Reading, MA

Barney, J.B., 1991, Firm resources and sustained competitive advantage, *Journal of Management,* **17,** 99-120

Brown, J.S., Duguid, P., 1991, Organizational learning and communities of practice: Toward a unified view of working, learning and innovation, *Organization Science*, **2**/1, 40-57

Dhar, V. & Stein, R., 1997, *Intelligent Decision Support Methods. The science of knowledge work*. Prentice Hall, Upper Saddle River, NJ

Dodgson, M., 1993, Organizational learning: A review of some literatures, *Organization Studies*, **14**/3: 375-394

Dosi, G., 1988, *The nature of the innovative process* in Technical Change and Economic Theory, (G. Dosi, ed), Pinter, London: pp.221-238

Drucker, P.F., 1994, The theory of the business, *Harvard Business Review*, Sept-Oct. 1994

Fayyad, U.M., 1996, Data mining and knowledge discovery: Making sense out of data, *IEEE Expert Systems*, **Oct. 96**: 20-25

Fayyad, U.M., Uthurusamy, R., 1995, *First International Conference on Knowledge Discovery and Data Mining*, AAAI Press, 1995

Ferguson, C-J. & Lees, B., 1998, An application of data mining for product design, *Colloquium on Data Mining and Knowledge Discovery*, IEE, London, May 1998

Ferguson, C-J., 1997, *Application of a data mining system as an aid to organisational learning*, in Systems for sustainability: people organizations and environments, (F.A. Stowell ed.) Plenum Press, New York: pp211-215

Galliers, B., 1995, *Re-orienting information systems strategy: Integrating information systems into business* in Information Systems Provision: The Contribution of Soft Systems Methodology (F.A. Stowell, ed), McGraw-Hill, London: pp.51-74

Huber, G.P., 1991, Organizational learning: The contributing processes and the literatures, *Organization Science*, **2**/1: 88-115

Katzenbach, J., Smith, D., 1993, *The Wisdom of Teams: Creating the High Performance Organization*, Harvard Business School Press, Boston, Mass.

Kay, J., 1995, *Foundations of corporate success*, Oxford University Press

Metes, G., Gundry, J. & Bradish, P., 1997, *Agile networking: competing for the future through the internet and intranets*, Prentice Hall PTR, Upper Saddle River, NJ

Nevis, E.C., DiBella, A.J. and Gould, J.M., 1995, Understanding organizations as learning systems, *Sloan Management Review*, Winter 1995: 73-85

Nonaka, I., 1991, The knowledge creating company, *Harvard Business Review*, **69**(6): 96-104

Parsayae, K., 1997, The bridge between OLAP and data mining, *2nd Annual Data Mining Summit*, Feb. 19, San Francisco

Penrose, E.T., 1959, *The theory of the growth of firms*, M.E. Sharpe

Porter, M., 1985, *Competitive advantage*, Free Press

Prahalad, C.K., Hamel, G., 1994, *Competing for the Future*, Harvard Business School Press, MA

Richter, F-J., 1994, Industrial organizations as knowledge systems, *Systems Practice*, 7(2): 205-216

Selznik, P., 1957, *Leadership in administration: a sociological interpretation*, Harper Row

Senge, P.M., 1992, *The Fifth Discipline: The Art & Practice of The Learning Organization*, Century Business, London

Shonk, J.H., 1992, *Team Based Organisations*, Business One Irwin Homewood, Illinois

Watson, I., 1997, *Applying case-based reasoning: techniques for enterprise systems*, Morgan Kaufman

Weston, D.M., 1994, *Organizational Learning as Strategy*, SRI International, Menlo Park, CA

DIVERSITY MANAGEMENT AND ORGANISATIONAL CHANGE

Guangming Cao, Steve Clarke and Brian Lehaney

Department of Finance, Systems and Operations
Luton Business School
University of Luton
Park Square
Luton
LU1 3JU

INTRODUCTION

Whilst there is considerable empirical evidence suggesting that the failure rate of organisational change programmes is unacceptably high, there are examples of approaches to change management derived from contemporary systems theory that have proved successful in addressing complex, changing organisational problem contexts (Checkland, 1981; Flood and Jackson, 1991; Flood, 1995). The suggestion to be drawn from this is that theory and practice of the management of change (MOC) could be improved by an investigation from a perspective based on contemporary systems theory. The research on which this paper is based is undertaking such an investigation, the key objectives of which are to summarise the characteristics of problem contexts in organisational change; to critique current MOC approaches; to briefly review systemic perspectives on MOC; to classify organisational change and the approaches taken to it; and finally to suggest a systemic framework for MOC.

THE COMPLEXITY OF ORGANISATIONAL CHANGE

To explore the diverse nature and complexity of change management, this paper focuses on the four dimensions of an organisation: process, design, culture and politics (for an overview see Flood, 1995). Using this classification, different types of complexity in organisational change can be related to particular dimensions of an organisation. However, in systemic terms, this does not mean that organisational change can be *reduced to* changes of process, design, culture or politics, since organisations are not susceptible to simple classification under one of these headings, but must be viewed as a complex, interacting system, with, for example, change in one dimension resulting in change in other dimensions (DeLisi, 1990). Organisational change is a dynamic process of many stages (Pettigrew, 1985; Dawson, 1994), not a set of discrete stages.

A CRITIQUE OF CURRENT APPROACHES TO THE MANAGEMENT OF CHANGE

In recent years, the management of change has become dominated by methods which are largely pragmatically informed. To address these, we will raise a critique of arguably the three

Synergy Matters: Working with Systems in the 21st *Century,*
Edited by Castell *et al.*, Kluwer Academic / Plenum Publishers, New York, 1999.

61

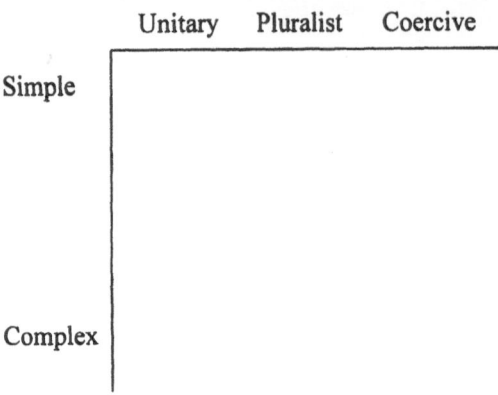

Figure 1. A Classification of Problem Contexts
[Jackson, 1995 #58]

This framework provides a unified approach that can be used to guide problem solvers to understand the problem contexts and to choose appropriate systems intervention methodology(ies) for tackling the perceived problem context.

A complementarist approach to diversity management recognises the value of a mix of methods in a given change management context, whilst valuing each method for the contribution it can make. In this respect, complementarism is therefore promoted as superior to pragmatic, isolationist and imperialist approaches (for an overview see Flood and Romm, 1996; Jackson, 1997). Complementarism seeks to respect the different strengths of the various trends in management science, encouraging their theoretical development and suggesting ways in which they can be appropriately fitted to the variety of management problems. Critical systems thinkers (Flood and Jackson, 1991; Flood and Romm 1996; Taket and White, 1996; Midgley, 1996; Jackson, 1997) stress that: methodological complementarism should be theoretically informed, with differences between methodologies respected and promoted; using and mixing different methodologies or parts of different methodologies in one intervention are encouraged; and decision making is facilitated, but the way in which decisions are made might vary.

Having reviewed how critical systems thinkers might have addressed diverse problems by employing diverse methodologies, we now want to move on to explore the possibility of applying systems perspectives to inform the management of change.

TOWARDS SYSTEMIC MANAGEMENT OF DIVERSITY IN ORGANISATIONAL CHANGE

In this section, based on contemporary systemic perspectives, a classification of organisational change and of approaches to the management of change (MOC) are developed to form a systemic framework to facilitate the best use of diverse MOC approaches to managing organisational change. Flood's (1995) four dimensions of organisations gives a classification of four ideal types of organisational change:

- Processual change - change in flows, and controls over flows.
- Structural change - change in functions, their organisation, co-ordination and control.
- Cultural change - change in mediation of behaviour in terms of people's relationship to social rules and practices.
- Political change - change in power and potency to influence the flow of events.

This ideal classification suggests that there are four types of organisational change, each of which refer to a particular dimension of an organisation. From the systems point of view, different types of organisational change are interrelated and interdependent: one type of organisational change might influence or be influenced by other types of organisational change within a given context. In order to deal with these four interrelated ideal types of organisational change, managing organisational change requires diversity in theories and methodologies. Unless all four types of organisational change are taken into account, and different MOC approaches are used in a way that the whole organisational change can be addressed systemically, then problem solving is very likely to be ineffective. The four types of organisational change further give rise to a categorisation of four approaches to management of change, since approaches are needed to manage processual change, structural change, cultural change, and political change.

Based on this categorisation, strengths and weaknesses of different approaches could be assessed in relation to the four ideal types of organisational change. The typical approaches to managing processual change, perhaps, include TQM and BPR. Referring to the four types of diversity, TQM and BPR mainly focus on organisational process, but little is said about the other types of organisational change. The transaction cost approach can be seen as one of the methods managing structural change, since it asks whether it is efficient that certain transactions occur inside or outside the organisation, thus searching for and adopting the design that optimises efficiency (Starkey, Wright and Thompson, 1991). However, it has focused on evaluating the efficiency of current available designs; it has little to say about new organisational designs (Roberts and Greenwood, 1997), and little about processual change, conflictual values, and the influence of power. For managing cultural change in organisations, some researchers emphasise the need for a unitary or strong culture (Beer, Eisenstat and Spector, 1990; DeLisi 1990), whilst others take the multi-culture view such as the interpretative approach (Aldrich, 1993) and the cultural diversity approach detailed earlier. None of them, however, says much about processual, structural and political change. One of the methods of managing organisational politics is the labour process approach (Reed, 1992), which explains the complex interaction between intra-organisational control practices and the wider structures of class power and domination in which they are utilised, developing an understanding of the antagonistic relationship between the interests of owners and managers on the one hand, and workers on the other. Since this approach focuses on power, domination, political bargaining and negotiation within organisations, little is said about organisational processes and cultures.

In summary, it can be seen that each MOC approach has its strengths and weaknesses, and is suitable for addressing certain kinds of change problems, but definitely not suitable for all of the change problems. Therefore, a systemic framework that can help employ different approaches to deal with organisational change is needed.

A SYSTEMIC FRAMEWORK FOR THE MANAGEMENT OF CHANGE

From these categorisations of organisational change and MOC approaches, it is possible to develop a systemic framework of two dimensions (Table 1). One dimension is made up of the four types of organisational change, and the other is made up of the various approaches to management of change. The key idea of this framework is to assess which of the types of change each approach is most suited to addressing, in order to suggest a diversity of methodologies with which to systemically address the diverse problem context(s).

Firstly, it is suggested that this might help practitioners in developing a holistic view of problem contexts in relation to the four types of organisational change. What types of organisational change are involved? What are the possible interactions between them? Is it possible that the surfaced organisational change results in other types of organisational change within the change process? Are there any other explanations? These questions will help to get a better understanding of, and therefore a better chance of addressing, the problem situation. Secondly, the strengths and weaknesses of relevant approaches could be examined in relation to the types of organisational change. What extant approaches are suitable to tackle the surfaced

Table 1. A systemic framework for managing organisational change

MOC Types MOC Approaches	Processual change	Structural change	Cultural change	Political change
TQM				.
BPR				
Cultural diversity approach				
... Other MOC approaches				

types of organisational change? Are there any other alternatives? Do we have to use different approaches in one intervention? If yes, what kind of adjustment do we have to make? If the extant approaches are not appropriate, do we have to create our own method? How? Thirdly, decision making will be facilitated. What are the criteria to choose between different approaches, and why? What are the consequences of this decision?

CONCLUSIONS

This paper began by examining current approaches to managing organisational change. Typological and complementarist perspectives in contemporary systems thinking were then reviewed, and a classification of four ideal types of organisational change and four MOC approaches developed. From this, a systemic MOC framework is proposed, which, it is suggested, will help further research and practice in this area. Finally it is suggested that this framework needs further clarification and enrichment. For instance, relevant methods need to be examined critically in relation to the four types of organisational change, and the framework needs to be tested in practice, especially as how to operationalise it, so different parts of MOC approaches could be flexibly used in one intervention. These issues are currently being pursued as part of this ongoing research.

REFERENCES

Aldrich, H., 1993, Incommensurable paradigms? Vital signs from three perspectives, *in* "Rethinking Organisation: New Directions in Organisation Theory and Analysis," M. Reed and M. Hughes, eds, Sage, London.

Beer, M., Eisenstat, R. and Spector, B., 1990, Why change programs don't produce change, *Harvard Business Review,* 68(6):158-166.

Burgess, T., 1995, Systems and reengineering: relating the reengineering paradigm to systems methodologies, *Systems Practice,* 8(6):591-604.

Checkland, P., 1981, "Systems Thinking, Systems Practice," John Wiley & Sons, Chichester.

Chemers, M., Oskamp, S. and Costanzo, M., 1995, "Diversity in Organisations: New Perspectives for a Changing Workplace," Sage, California.

Dawson, P., 1994, "Organisational Change: A Processual Approach," Paul Chapman Publishing Ltd, London.

DeLisi, P., 1990, Lessons from the steel axe: culture, technology, and organisational change, *Sloan Management Review,* 32(1): 83-93.

Flood, R., 1995, "Solving Problem Solving", John Wiley & Sons, Chichester.

Flood, R. and Jackson, M. 1991, Total systems intervention: a practical face to critical systems thinking, *in* "Critical Systems Thinking, Directed Readings," R. Flood, and M.C. Jackson, eds, John Wiley & Sons, Chichester.

Flood, R. and Romm, N., 1996, "Diversity Management: Triple Loop Learning," Wiley, Chichester.

Gioia, D. and Pitre, E., 1990, Multiparadigm perspectives on theory building, *Academy of Management Review,* 5(4): 584-602.

Hammer, M. and Champy, J., 1993, "Reengineering the Corporation - A Manifesto for Business Revolution," Nicholas Brealey, London.

Hassard, J., 1993, "Sociology and Organisation Theory: Positivism, Paradigms and Postmodernity," Cambridge University Press, Cambridge.

Jackson, M., 1997, Towards coherent pluralism in management science, *Working Paper*, Lincoln School of Management, University of Lincolnshire & Humberside.

Jackson, M., 1995, Beyond the fads: systems thinking for managers, *Systems Research*, 12(1): 25-42.

Jackson, M. and Keys, P., 1984, Toward a system of systems methodologies, *Journal of the Operational Society*, 35(6): 235-248.

Midgley, G., 1996, The ideal of unity and the practice of pluralism in systems science, *in:* "Critical Systems Thinking - Current Research and Practice," R. Flood and N.R.A. Romm, eds, Plenum Press.

Pettigrew, A., 1985, "The Awakening Giant," Blackwell, Oxford.

Reed, M., 1992, "The Sociology of Organisations: Themes, Perspectives and Prospects," Harvester Wheatsheaf, London.

Roberts, P. and Greenwood, R., 1997, Integrating transaction cost and institutional theories: toward a constrained-efficiency framework for understanding organisational design adoption, *Academy of Management Review*, 22(2): 346-373.

Schultz, M. and Hatch, M., 1996, Living with multiple paradigms: the case of paradigm interplay in organisational culture studies, *Academy of Management Review*, 21(2): 529-557.

Starkey, K., Wright, M. and Thompson, S., 1991, Flexibility, hierarchy, markets, *British Journal of Management*, 2(3): 165-176.

Stickland, F. and Reavill, L. P., 1995, Understudying the nature of system change: an interdisciplinary approach, *Systems Research*, 12(2): 147-154.

Taket, A. and White, L., 1996, Pragmatic pluralism—an explication", *Systems Practice*, 9(6): 571-586.

Turner, B., 1993, The symbolic understanding of organisation, *in:* "Rethinking Organization: New Directions in Organization Theory and Analysis," M. Reed and M. Hughes, eds, Sage, London.

UNDERSTANDING INTERORGANISATIONAL HEALTH AND SOCIAL SERVICE GROUPS: WOULD ALTERNATIVE METAPHORS OF SYSTEMS HELP?

Pam Hearne

The Management School
Lancaster University
Lancaster
LA1 4YX

INTRODUCTION

Due to recent legal and environmental changes joint commissioning and inter-agency provision of services are seen to be the future for health and social service organisations. There is an emphasis on co-operation and collaboration to find a more holistic approach to social issues. This would seem to argue the value of a systems approach in thinking about this inter-organisational work.

However, in trying to apply systems ideas in doing some initial thinking about joint commissioning groups for the author's PhD work, problems were encountered. The core systems metaphor of an adaptive whole responding to its wider environment did not seem to the author to be particularly useful.

These groups, which may be thought of as virtual organisations, comprise representatives of Health Authorities and local authority Social Service departments. It is possible, and indeed likely, that any one Health Authority will, because of its geographical location, be working with all or part of two or more Social Service departments. The departments will, in turn, be working with multiple commissioning partners and provider agencies. Setting a boundary for the virtual organisation is relatively straightforward but locating these groups in *a wider context* is problematic.

Partners in joint commissioning groups can find themselves pulled in very different directions due to their other relationships. Despite this there is a practical need to address complex social problems locally. Following Atkinson and Checkland (1988) this paper will explore how various systems metaphors might be used for making sense of these groups. The vehicle for undertaking this exploration is the author's PhD work.

BEGINNING THE PhD WORK

The author of this paper began her PhD work by looking at current issues for health and social service organisations, particularly those in the public or voluntary

Synergy Matters: Working with Systems in the 21st Century,
Edited by Castell *et al.*, Kluwer Academic / Plenum Publishers, New York, 1999.

67

sectors. It became apparent that both the academic and practitioner literatures were concerned with joint commissioning, inter-professional and inter-agency working (Soothill et al., 1995; Huxham and Vangen, 1996; Osborne, 1996; Kooiman, 1996).

Given the espoused notion of taking a holistic approach to complex social issues it seemed that the systems ideas with which the author was familiar would be a good starting point for thinking about the issues. Further reading to focus the PhD on current issues led the author and her supervisor to consider joint commissioning groups via these ideas.

Initial Consideration Of Joint Commissioning Groups Using Systems Ideas

It was at this stage that the first problems with the idea of a system as an adaptive whole responding to its environment became apparent. In a situation where members of the joint commissioning group are permanent members of other organisations the question of how the environment or wider system could be defined arose.

Considering other aspects of the formal systems model (Checkland, 1981; Wilson, 1990) raised further issues concerning measures of performance and decision taking processes linked to the already identified difficulties defining the wider system. There were questions about setting relevant judgement criteria as well as about the limits of authority within which decisions could be taken.

On the subject of resources joint commissioning groups were seen to be reliant on the co-operation of participating organisations for identifying and enabling relevant people to become involved. Financially the situation was interesting as the groups themselves were unlikely to have their own resources so decisions on joint commissioning would have to be carried out using the budget of the organisation identified as being legally responsible for paying for the service concerned. It was not necessarily possible for organisations to contribute to a joint fund. Indeed, in the past, proactive attempts at joint commissioning in areas such as the London Borough of Lambeth had floundered when the previous UK government had declared that it was illegal for the public sector organisations concerned to attempt to pool resources.

Even the purpose or mission of the groups, joint commissioning, which initially seemed clear, proved to be rather more complex to define as different participating organisations undertook and viewed service commissioning in rather different ways. There was not necessarily a consensus about which services were to be commissioned. It seemed that joint commissioning groups were not just a matter of deciding how to commission services but thinking about what joint commissioning of services meant. This argued the case for a soft systems approach, which could deal with complex situations "in which known-to-be-desirable ends can not be taken as given." (Checkland, 1981)

Looking At Joint Commissioning Groups Using Soft Systems Methodology

The approach originally envisaged was to use SSM style human activity systems models to make sense of the situation (Checkland, 1981; Checkland and Scholes, 1990). The first model to be tried was "A system to jointly commission health and social services for a local population." In attempting to use SSM there were difficulties when considering CATWOE in defining the owners of the above system. Were they to be taken as one or more of the organisations with members involved? Or was some wider group such as the local population or the UK government, which has changed the legislation to require these organisations to work together, more appropriate? One possibility was to model the same system from different weltangschauungen linking these to specific organisations or groups to see what insights this generated.[1] In order to do this contact was made with various people working in the field.

Initial contact with people in the field revealed further issues such as a situation where, because of geography, one Health Authority would be commissioning in conjunction with two Local Authority Social Service departments. Each of these departments would in turn be working with at least one other Health Authority. The idea of a hierarchy of systems was not obviously applicable. A chain or even a net of systems seemed more appropriate. It was this that first led to consideration of Atkinson and Checkland's 1988 paper on alternative metaphors of systems.

Further investigation established that when the Health Authorities were working with a number of partners these organisations might have very different ways of working depending on the members' professional background and training. These in turn were likely to be different from those involved in health whether they were managers or health care professionals. One Social Services Director described the organisations as "speaking fundamentally different languages".

Despite the, in theory, common joint commissioning task the groups have, in practice, many barriers to overcome. Even where the organisations concerned are peopled by those trained in the same way the history of previous interactions between the organisations was an issue. Members of voluntary and public sector organisations who were trained in say social work might well have some common values professionally but very different values as regards management and decision-making. Some contacts talked of organisations being "at war" or "in conflict" due to these differences. This seemed again to point to the discussion in the Atkinson and Checkland paper.

ALTERNATIVE METAPHORS OF SYSTEMS

In their paper Atkinson and Checkland argue the case for considering various metaphors of systems. One way of doing this they argue is to "combine several purposeful wholes in more complex wholes which are not themselves purposeful in the sense of pursuing a single purpose in a unitary fashion." They suggest that "If purposeful wholes are combined a range of relationships between the parts is possible: parts may be combative, imperialistic, syndicalistic, in a parasite/host relationship, etc." They illustrate this with examples of the use of alternative systems images in SSM. One example was, the authors say, using these ideas, specifically that of a contradictive system, implicitly; the other, a study by Atkinson in a health care setting, does so consciously.

Discussion with one of the authors revealed that he had developed these ideas further for use in his own work on human/machine activity systems.[2] He now considers the value systems, management and learning processes of the organisations or groups that are involved. This has led him to identify variations on the Human/Machine System Concept where organisations: have multiple interests/management, are engaged in

[1] The author would like to acknowledge the comments of Peter Checkland (personal correspondence, July 1998) which helped move the thinking on this issue.
[2] The author would like to thank Chris Atkinson for the useful discussion she had with him at a conference (September 1998) and for the copy of his occasional paper (Atkinson, 1998) which he provided her with.

antagonistic/war like relationships and have incommensurate processes and value systems (Atkinson, 1998).

APPLYING THESE IDEAS TO DEVELOPING THE PhD WORK

At first glance Atkinson and Checkland's comment that the more complex whole would not have a single purpose would seem to suggest that the ideas are not immediately relevant to joint commissioning groups which do have an overt purpose of commissioning services for a specific population. However a closer reading of the paper revealed that a key aspect is that the groups concerned are not "pursuing a single purpose in a unitary fashion." It seemed that by considering the organisations whose members were involved in joint commissioning work as having previously established value systems and relationships it was possible to model a range of combinations of systems. These could then be used to structure further enquiries.

Images that proved particularly useful in considering joint commissioning included contradictive, combative and syndicalistic systems. In an ideal world these groups would exemplify "a syndicalistic system in which several autonomous wholes with different but compatible interests come together to form a whole which is of benefit to them all." (Atkinson and Checkland, 1988). Investigation of the perceived reality of those in the situation revealed that there was little agreement about what would benefit the organisations concerned or indeed the local populations they serve.

Thinking about a contradictive system opened up the thinking by leading to consideration of the values of the organisations involved, including their members' perceptions of the other actors involved. This allowed for a joint task (or transformation) being the subject of misunderstandings which had implications for the effectiveness of inter-agency working. Surfacing and making explicit the differences became an important role for joint commissioning groups. To go back to the earlier comment on the organisations involved "speaking different languages" the inter-organisational groups and their linking mechanisms to the individual organisations could be reconceptualised as value communication or translation systems!

Understanding others' values explicitly does not, of course, automatically lead to accommodations being possible. This led to further consideration of systems where the participating organisations have incommensurate processes and value systems. At the extreme this could have led to combative relationships – the organisations in conflict or at war that some people had referred to. The options in face of conflict or war range from withdrawal, through a negotiated truce or peace agreement to one party achieving domination. An important issue in joint commissioning appeared to be the relative power of the organisations involved and their freedom to decide whether to be part of the process.

In the case of health care in Britain at the time of writing there is, for instance, a legal requirement for Health Authorities to set up Primary Care Groups (NHS Executive, 1997) to commission (and eventually deliver) Primary Care services. There is a statutory requirement for the boards of these groups to contain GPs, nurses, a Social Services manager, a Non-Executive Director of the Health Authority and a lay person (who will probably represent a voluntary sector organisation). The Social Service departments and Health Authorities have no choice about engaging with these groups. Considering this led to focusing of the PhD on these groups. How were these groups going to work together to achieve their supposedly common task?

APPLYING THE IDEAS TO CURRENT AND FUTURE PhD WORK

The work on Primary Care Groups is in its early stages but already an alternative metaphor of system, that of an imperialistic system, is generating insights. The guidelines (NHS Executive, 1998 (a) and (b)) on establishing these groups have clear instructions on the numbers and types of participants who are to be members of the board. The instructions state that the board must contain four to seven GPs, one or two nurses, a Social Service manager, a Health Authority Non-Executive Director and a lay person. Though there is provision, indeed a requirement, for other health care professionals to be consulted as appropriate it is made explicit that GPs are to be in a majority on the board unless they waive that right. The chair is also to be a GP unless the GPs in the area agree otherwise. From this it can be seen that thinking of an imperialistic system in which one group dominates the governance of the group can help to make sense of the development of Primary Care Groups.

CONCLUSION

During the period of the development of the PhD work various alternative metaphors of systems have aided the sensemaking process. They have generated new insights into a complex situation helping the author to clarify her thinking. The work has been focused through their use. The author believes that continued conscious attention to "the largely unexamined assumptions" (Atkinson and Checkland, 1988) about systems in structuring and carrying out her work will facilitate understanding of an important current social phenomenon; interorganisational health and social service groups.

REFERENCES

Atkinson, C.J., 1998, Exemplars of human/machine activity systems based on varient systemic metaphors, Occasional Paper, University of Surrey.

Atkinson, C.J., and Checkland, P.B., 1988, Extending the metaphor "system", *Human Relations* Vol. 41 No. 10, pp. 709 - 725.

Checkland, P.B., 1981, *Systems Thinking, Systems Practice*, Wiley, Chichester.

Checkland, P., and Scholes, J., 1990, *Soft Systems Methodology in Action*, Wiley, Chichester.

Huxham, C., and Vangen, S., 1996, Working together: Key themes in the management of relationships between public and non-profit organisations, *International Journal of Public Sector Management* Vol. 9 No. 7, pp.5 - 17.

Kooiman, J., 1996, Research and theory about new public services management: Review and agenda for the future, *International Journal of Public Sector Management* Vol. 9 No. 5/6, pp. 4 - 6.

NHS Executive, 1997, *The New NHS - Modern: Dependable*, HMSO, London.

NHS Executive, 1998 (a), *Health Service Circular HSC 1998/065*, NHS Executive, Leeds.

NHS Executive, 1998 (b), *Health Service Circular HSC 1998/139*, NHS Executive, Leeds.

Osborne, S.P., 1996, The hitch-hiker's guide to innovation?: Managing innovation - and other organisational processes - in an inter-agency context, *International Journal of Public Sector Management* Vol. 9 No. 7, pp. 72 - 81.

Soothill, K., Mackay, L., and Webb, C. (Eds.), 1995, *Interprofessional Relations in Health Care*, Edward Arnold, London.

Wilson, B., 1990, *Systems: Concepts, Methodologies and Applications (2nd ed.)*, Wiley, Chichester.

BEYOND THE GOLDEN RULE: EMANCIPATORY PRACTICE AND CHANGE IN ORGANISATIONS

Marion Helme

Systems Department
Centre for Complexity and Change
The Open University
Walton Hall
Milton Keynes

INTRODUCTION

This paper explores a question that has arisen from my experience working and teaching in social work, and resonates with the experience of others: **Why is it that practitioners working with others in anti-oppressive and emancipatory ways so often talk about their own management as oppressive and disabling?** This question has surfaced now as I am preparing collaborative action research with practitioners and managers from a voluntary agency working with children's justice issues and participation. Recent conversations have included comments such as "how can practitioners work to empower and involve clients in decision-making processes if the decision-making in the organisation excludes them", statements that project teams feel undervalued by management, concerns that the number and length of meetings, written communications and administrative procedures impede the 'real work' with clients, and complaints about the opacity of central decision-making processes. I am not, however, framing the question as a problem. Problems call for solutions, and negative feelings expressed by one group in an organisation towards another may be considered part and parcel of organisational life. They may well arise from 'real' grievances[1]; they may function as necessary release valves for emotion (Fineman 1991), or as expressions of the tension between the control and co-ordination structures in organisations and the professional autonomy of practitioners. However this tension between talked-of experience as practitioner and experience as managed is important for 'projects of inclusion' - specifically organisations and agencies striving for the participation in organisational planning, decision-making and action of those whose interests the 'practice' is primarily designed to promote. As will be discussed, this is firstly because of the practitioner's role as 'gatekeeper' and intermediary, and secondly

---

[1] When procedures for the assessment and delivery of services for older people changed with the implementation of the NHS and Community Care Act 1991, social workers in one local authority complained that they had 26 forms to complete for meals on wheels to be delivered.

Synergy Matters: Working with Systems in the 21st Century,
Edited by Castell *et al.*, Kluwer Academic / Plenum Publishers, New York, 1999.

because prevalent ways of explaining this tension can reveal current understandings and assumptions about the relationship of client and agency.

This paper also inquires into how stories, in the form of 'explanations of practice', with their associated metaphors can reveal implicit relationships, and ways of talking about them, between practitioners, clients and managers. These stories recognise social and psychological theories but in "(becoming) understood by the very people of which these theories speak .. (the social theories)...can be said *to re-enter the very practices they claim to describe'* (Krippendorff 1996 p. 312) and to legitimise and make them unquestionable. The examples and stories used here will be drawn from my perspectives on social work, my personal and social context - 'the baggage' I bring to the research (Bell 1998), although the question I am exploring has resonated with the experience of people from related fields - community and voluntary work, counselling and teaching. Writing this paper is a reflexive process; the stories and metaphors I discuss are those that I have used in making sense of my experiences, and how I have understood the actions and talk of others, and taking account of the importance of a reflexive approach to an investigation of metaphors (Palmer and Dunsford 1996). Re-describing the stories in this paper is like being the observer entering the domain of observation, and unravelling the legitimising processes by which they become taken-for granted facts (for me); not just a recognition of a change of position but telling the stories for a differently constructed audience.

EMANCIPATORY PRACTICE, AND WORKING WITH 'OTHERS'

In this paper 'working with others in emancipatory and anti-oppressive ways' is used to describe practice that is explicitly, but not necessarily solely, designed to identify, and to counter, the disadvantage and discrimination experienced by the 'other'. Emancipatory practice assumes an ethic of fairness, equality and respect for difference[2], an understanding that society is structured in terms of unequal power relations that disadvantage some groups in society and privilege others, on grounds of inherent or constitutive characteristics (age, gender, race, class, dis /ability etc.), and a commitment to action. Working in emancipatory ways is by no means uncontested in practice. A Foucauldian analysis of power structures in society raises the dilemma that the practice itself is situated within taken-for granted structures and systems. 'Emancipate' and 'empower' raise questions about whether it makes sense to use the verb transitively - "in emancipatory dialogue people are neither alone nor can they be in charge" (Krippendorff 1995) p. 129). As a dialogical process, always open to being questioned, working in emancipatory ways does not 'fit' well with the demand for quantifiable output and performance indicators in organisations.

A puzzle here is how to refer to 'the other'. There is no one term that will stand for those who are not practitioners, but without whom there would be no practice[3]. The criterion I am using is what makes sense to me. Possible terms include 'client', 'carer', 'customer', 'user', 'service user', 'patient', 'consumer', as well as terms referring to the characteristic of the group with whom the practitioner is working - 'families', 'older people', foster carers', 'children and young people', 'the black community', 'the gay

---

[2] In Leonard's words " a recognition of the validity of the Other" (Leonard 1997 p. 164) as contrasted with the exclusion of the Other (through racism etc.). Talking of emancipation implies there are people who need emancipation 'from someone or something' ((Janks and Ivanic 1992)
[3] "However we conceive or speak of these Others, even when we omit explicit references to them, always directs our listening, our (re)searching, and our interacting with these unnamed and possibly unknown Others" (Krippendorff 1996 p. 312) - including our practising.

community' etc.[4]. This is problematic for working with others in emancipatory ways, firstly because terms applied in blanket fashion include some usages that can only be understood as metaphorical (or even paradoxical), for example 'customer'[5]. Secondly these terms are not necessarily those which would be chosen by those to whom they refer, for example because of the implied relations of dependency or exchange, or what they say about identity.

BEYOND THE GOLDEN RULE ('Do-as-you-would-be done-by')

By 'Golden Rule' I am not intending to invoke universal precepts such as Etzioni's 'virtual equilibrium' - the 'balancing universal individual rights and the common good' ([Etzioni 1997 p. xix), but as a way of inviting consideration of systems of mirroring relationships. In the organisations discussed here the usual model of 'supervision' of the practitioner is that in which the 'supervisor' is also the line manager, or senior in some way to the practitioner in the hierarchy. Several functions are combined in this relationship (e.g. 'management, education, support, mediation' (Hughes and Pengelly 1997), for which supervisor/managers are themselves accountable to their line manager, and by whom they might be supervised (a structure with Foucauldian echoes). The same term - 'supervision' that is applied to the relationship between practitioner and the person to whom they are directly accountable (their 'supervisor') is often also used to describe the relationship between the practitioner and client.

The participation (or at least, consultation) of service users in decisions about services which directly concern them, either as recipients or members of the 'community of interest' is established in many social care and voluntary organisations, and in some cases a legal right. The role of both agencies and practitioners to challenge societal power structures that disadvantage people is often explicit in policy documents and requirements for professional qualification. Of course, neither of these 'emancipatory practices 'are unproblematic or uncontested in implementation; the issues are complex, difficult and highly politicised. However there are ways of talking about this, publications about how to do it and practice examples that can be shared. The voice of service users in these organisations are usually mediated - by practitioners, by elected representatives, by 'interested parties' and advocates. Service users are, on one hand, spoken and written about as central (the organisation's 'core business') and on the other hand marginalised as having individual, special and local interests. How can we imagine a process in which service users are invited to participate in organisational learning and change? What will need to have changed for this to occur?[6] A question here that echoes one of the practitioner's complaints in the introduction is whether organisations need to first put into effect the participation of practitioners and teams throughout organisational decision making processes before including service users[7].

---

[4] As an illustration of the heterogeneity of usage: in 1996 I asked 25 final year DipSW students for their preferred term; there was a small majority for 'service user', followed by 'customer'. Almost all the cohort of students the following year preferred the term 'client'.

[5] For example, using the term 'customer' to refer to someone compulsorily admitted to psychiatric hospital; 'user' also has colloquial usage as exploitative or manipulative.

[6] Participation in organisations by employees is "an especially paradoxical form of change" (O'Connor 1995 p. 217) because it runs counter to organisational practices such as hierarchical decision-making, and selective information exchange, then the inclusion of service users calls for radical change !

[7] Heller argues for a holistic approach to organisational participation, referring to strong supporting data showing the ineffectualness of single isolated participative practices (Heller 1998)

METAPHORS AND STORIES

In writings about management and organisations, metaphors and stories are often used as tools of analysis or as instruments in achieving top-down change. What I am interested in here is how the special qualities (the 'meta-ness') of metaphors and stories can create shared languages of possibilities and how different stories can be used, not to 'test out reliability' (Boden 1994 p. 132), but to be emancipatory in revealing traps in ways of thinking and the possibility of different ways of understanding. Lakoff and Green showed how metaphors are embedded in everyday thought, action, language (Lakoff and Johnson 1980); they are "a primal means through which we forge our relationship with the world" (Morgan 1997 p. 276). Metaphors and stories are revealing -- "every metaphor is the tip of a submerged model (Black 1979). Narratives organise in a temporally meaningful way (Ricoeur 1986, Polkinghorne 1988 [8]). Metaphors incorporate complex or confused information into an organised whole (Ortony 1979). Both metaphors and stories can be generative (Schon 1979). They are innovative and synergetic - stories through the process of 'inventing plot' (Ricoeur 1986, Polkinghorne 1988) and metaphor "as the simultaneous experience of two exclusive properties in relation to the same identity" (Apter 1982 p. 56).

STORY ONE

Practice (social work, counselling etc.) involves working with clients who are experiencing, or who have experienced troubling situations. In order to be 'professional' in their relationships with clients, practitioners need to be objective and non-judgmental [9]. Troubling situations are those in which clients have powerful emotions (of fear, anger etc.). These are conveyed in some way [10] to practitioners, who may also have powerful feelings themselves about the client's experiences. Expression of feelings of oppression by their employing organisation is a way of practitioners dealing with feelings aroused from practice [11]. It is the role of the supervisor to allow for and 'monitor' these feelings [12].

A metaphor for this process is 'contagion' - the passing on of something potentially dangerous that must be cured, or disabled, or from which practitioners must be rescued [13]. In discussing a case study of consultation with staff of a drug dependency

---

[8] Boje tells it differently: "storytelling is a collective dynamic that scripts, sways and disciplines organisational learning" (Boje 1994 p. 435).

[9] "Many professional workers - counsellors, doctors, nurses, social workers and the like - are .. in effect paid for their skill in emotion management. They are to look serious, understanding, controlled, cool, empathetic and so forth with their clients or patients. The feeling rules are implicit in their professional 'discipline' (an apt term) - 'rational', 'scientific', 'caring', 'objective'. Benign detachment disguises, and defends against, any private feelings of pain, despair, fear, attraction, revulsion or love; feeling which would otherwise interfere with the professional relationship. There are costs if the mask slips.. ' (Fineman p 19)

[10] For example, through the psychoanalytic phenomenon of projective identification, or transference and counter-transference.

[11] In much the same way as going home and kicking the cat.

[12] Social workers must be helped to "recognise the effect achieved by the emotions being beamed out from the family...The method of supervision must identify how perceptions and feelings may be affecting work done, whether between field worker and family, or field worker and supervisor'. DoH 1991.

[13] A different telling of this story might evoke the metaphor 'mirroring'. " [Practitioners] who feel badly treated by their organisations may speak in a general way of this 'abuse' mirroring the abuse in the families with whom they work' ((Hughes and Pengelly 1997)p. 83).

clinic Moylan describes the outcomes as "being less caught up in projective identification with their clients, they were able to function in a more creative, satisfying and efficient way" (Moylan 1994. p. 58) and "by knowing about ways in which the institution can become "infected" by the difficulties and defences of their particular client group, staff are more likely to be aware when this is happening" (ibid. p. 59, my emphasis). Practitioners may have a predisposition to 'infection' in the form of "unresolved issues from our past", or ideals , which "have unconscious determinants, (which) can contribute to defensive institutional processes"(Zagier Roberts 1994, p. 110). What the metaphor of contagion entails is the possibility, in fact the necessity, of separating people, practice and feelings, an ambiguity towards clients, and an understanding of them as being outside the organisational system, and the supervisory relationship as being in part a 'disinfection' process.

STORY TWO

Practice takes place in a triadic system (or domain or context) between worker, agency and service user[14]. These three elements come together in three dyadic relationships. Workers, agencies and service users bring to these relationships expectations of each other and of the outcome of practice. Each dyadic relationship includes a number of aspects, including the relative power of each party, empathy, and 'the working relationship' which includes "the regulation of an appropriate emotional distance" and 'helping' (Evans and Kearney 1997 p. 68). The dissatisfaction experienced by practitioners, and their experience as oppressed, arises from difference or conflict between their aims and expectations[15] and the perceived aims of the agency, and the power of the agency to impose its aims in service user/agency and practitioner/agency dyads.

A metaphor for this process is (unrefereed) 'tag wrestling'. Everyone is in the same game, but only two 'contenders' are engaged at any one time, but may be 'observed' by the other[16]. Alliances can be formed off stage as well as on [17]. There are some rules of engagement, and also the possibility of different expectations of outcomes, and understandings of a ' fair fight', which can be jeopardised by power difference or unfair practices. Without extending this metaphor further, it does presuppose relationships being based on difference rather than mutuality, and that (the only) alternatives to power, or force are through political processes.

CONCLUSION

There are at least as many stories and metaphors about the question as there are people to tell them. Recognising the use of metaphors can lead to their undoing (Krippendorff 1995). Different descriptions and stories imply having a different way of relating "and (are) often what enables a person to say 'now I can go on' - to

---

[14] The 'central triangular context' (Evans and Kearney 1997), represented as points of a triangle

[15] For example the incongruence between the aim of the worker as meeting need, and the perceived aim of the agency as rationing resources

[16] Hughes and Pengelly describe staff supervision as a triadic process in which the service user is present as an unembodied 'participant'. ((Hughes and Pengelly 1997)

[17] Morrison applies a similar triad between workers, managers and trainers in child protection training, and describes the contextual nature of the shifting alliances. For example during training to develop practice during painful reorganisation, trainers and workers may form an alliance; where training is about implementing new agency policy, trainers and managers may be in alliance to change practitioners' ways of working .

understand in the Wittgensteinian sense" (Riikonen 1997 p. 103). Inquiring into stories and metaphors provides a way of considering what sorts of relationships they warrant. The stories told here, indicate that others are called for in order to make sense of how service users can be talked of in a way that includes them within the organisation. For example, a metaphor suggested by Riikonen and Maden Smith is that of a dance, which "includes the possibility of continuous responsiveness" (Riikonen p. 22) and cannot be evaluated "by studying the steps of (only) one of the dancers", two entailments which seem to me to fit with the project of inclusion. One of the research activities will be inviting practitioners and managers in the voluntary agency to explore their own metaphors and stories, and generate new ones, for the involvement of children and young people with the organisation.

REFERENCES

Apter, Michael 1982 'Metaphor as Synergy' in Miall, David A. (ed.) *Metaphor: Problems and Perspectives* Brighton: The Harvester Press

Bell, Simon 1998 'Self reflection and vulnerability in action research: bringing forth new worlds in our learning' *Systemic Practice and Action Research* 11:2

Black, M. 1979. 'More about metaphor' in Ortony, A. (ed.) *Metaphor and Thought*. Cambridge: Cambridge University Press.

Boden, Deirdre 1994 *The Business of Talk; Organizations in Action* Cambridge: Polity Press

Boje, David M. 1994 'Organizational storytelling: the struggles of pre-modern, modern and postmodern organizational learning discourses' *Management Learning* 25;3 433-461

Department of Health 1991 *Child Abuse; A Study of Inquiry Reports* 1980-89, London: HMSO

Etzioni, Amitai 1997 *The New Golden Rule: Community and Morality in a Democratic Society* London: Profile Books

Evans, D. and Kearney, J. 1996 *Working in Social Care: A Systemic Approach*

Fineman, Bob 1991 *Emotions in Organisations* London: Sage

Heller, Frank 1998 *Organizational Participation: Myth and Reality* Oxford: Oxford University Press

Hughes, L. and Pengelly, P. 1997. *Staff Supervision in a Turbulent Environment*. London: Jessica Kingsley.

Janks, H. and Ivanic, R. 1992. 'CLA and emancipatory discourse' in Fairclough, N. (ed.) *Critical Language Awareness*. Harlow: Longman Group.

Krippendorff, K. 1995. 'Undoing power'. *Critical Studies in Mass Communication* 12: 101-132.

Krippendorff, K. 1996 'A second-order cybernetics of Otherness' *Systems Research* 13: 3 311-328

Lakoff, G. and Johnson, M. 1980. *Metaphors We Live By*. Chicago: University of Chicago Press.

Leonard, P. 1997. *Postmodern Welfare: Reconstructing an Emancipatory Project*. London: Sage.

Morgan, G. 1997. *Imaginization*. San Francisco, CA: Berrett-Koehler Publishers Inc.

Moylan, Deirdre 1994 'The dangers of contagion: projective identification processes in institutions in Obholzer, A. and Zagier Roberts, V. 1994

Obholzer, A. and Zagier Roberts, V 1994 *The Unconscious at Work: Individual and Organisational Stress in the Human Services* London: Routledge

O'Connor, Ellen Swanberg 1995 'Paradoxes of participation: textual analysis and organizational change' *Organization Studies* 15:5 769-803

Ortony, A. 1979. 'Metaphor: A multidimensional problem' in Ortony, A. (ed.) *Metaphor and Thought*. Cambridge: Cambridge University Press.

Palmer, Ian and Dunford, Richard 1996 'Conflicting uses of metaphors: reconceptualising their use in the field of organizational change' *Academy of Management Review* 21:3 691-717

Polkinghorne, Donald E. 1988 *Narrative Knowing and the Human Sciences* Albany N.Y.: State University of New York Press

Ricoeur, Paul 1986 *The Rule of Metaphor* London: Routledge

Riikonen, Eero and Smith, Gregory Madan 1997 *Re-imagining Therapy* London: Sage

Schon, D. 1979. 'Generative metaphor: a perspective on problem-setting in social policy' in Ortony, A. (ed.) *Metaphor and Thought*. Cambridge: Cambridge University Press

Zagier Roberts, V. 1994. 'The self-assigned impossible task' in Obholzer, A. and Zagier Roberts, V.1994

SYNERGY BETWEEN HUMANS AND SOFTWARE AGENTS

Petri Jooste[1]

Lincoln School of Management
Faculty of Business and Management
University of Lincolnshire and Humberside
Lincoln, LN6 7TS

INTRODUCTION

It may seem odd to talk about synergy between humans and a computer tool created by humans, almost like synergy between a carpenter and his/her hammer. However, this paper recognises that advances made in the field of artificial intelligence in general and software agents in particular make this topic not so far-fetched. Software agents are designed with ascribed attributes such as 'autonomous', 'collaborative', 'mobile' and 'smart'. For an overview and classification of different types of agents see for example Nwana (1996).

This paper gives an account of some issues for the possibility of humans and software agents working together in a way that achieves more with less resources. Resources include time and effort. This synergy is possible at least in one way, namely by a division of labour: if humans do what they can do best while software agents do what they can do best. Moreover, the different parties should be able to take part in constructive collaboration. (In the author's opinion many software systems fail if they are designed to automate tasks which could be done manually and do not add value to the way it is being done.)

Determining what humans and software agents are respectively best at may not be so easy, at least for the purposes of generalisation. Let us use an intuitive distinction: humans are best at planning overall goals and giving direction, while software is best at doing repetitive tasks without getting bored, making calculations and comparisons fast and managing vast databases.

For investigating possible synergy we take as an example the task of searching for on-line information. Software agents are heralded to be the future tools for dealing with the proliferation of information available on-line. This paper argues that an effective strategy for using these automated information agents to deal with this problem, must take into account the possibilities for synergistic *collaboration* among agents on the one

---

[1] Also from: Department of Mathematics and Computer Science, Vaal Triangle Campus, Potchefstroom University for Christian Higher Education, Vanderbijlpark, South Africa.

Synergy Matters: Working with Systems in the 21st Century,
Edited by Castell *et al.*, Kluwer Academic / Plenum Publishers, New York, 1999.

hand and with human users on the other hand. Thus we are dealing with two types of synergy: agent-agent and agent-human. This collaboration requires agents to have suitable interfaces and linguistic structure for information transfer so that meaning and context are not lost. (See the discussion of agent abilities below.) Let us now briefly consider the possibilities for collaboration.

Agent-Agent Collaboration

There are at least two ways in which agent-agent collaboration can be envisaged. The first is where several agents are designed and built to work together in a system to solve problems. Research using this approach usually comes from the field of Distributed Problem Solving. The second is the possibility of a collection of possibly pre-existing heterogeneous autonomous agents that collaborate to solve a problem. (The type of problem can be of ad-hoc or recurring nature.) This approach corresponds to definitions of a multi-agent system for example: "a loosely coupled network of problem solvers that work together to solve problems that are beyond the individual capabilities or knowledge of each problem solver" (Jennings, Sycara & Wooldridge 1998: p.285).

Although collaboration may be accomplished leading to parallel processing, where the work load is effectively distributed over several agents, this does not automatically imply synergy. Synergy is achieved when the effect of collaboration achieves more (faster or better results) than the simple management or distribution of tasks and subtasks among several entities. For example when one agent can do something (e.g. collect information) in ten units of time then there is still no synergy if ten agents work together to do it in one time unit. Now, if these collaborating agents are distributed over the internet it may give them individual advantages to accomplish different subtasks (e.g. because of lower network latencies resulting from their locations) and the nett effect is an overall advantage. However, this advantage usually comes at a price, namely the overhead needed to coordinate efforts (load balancing among agents) and communicate results. If the effect of this disadvantage is less than the overall advantage, then we should have synergy.

Agent-Human Collaboration

Negroponte (1995) envisioned that agent-agent collaboration may be a powerful way of searching for information, but ultimately agent-human interaction needs attention: "... you will dispatch agents to collect information on your behalf. Agents will dispatch agents. The process multiplies. But [this process] started at the interface where you delegated your desires ." (Negroponte 1995: p.158) One metaphor sometimes used for this type of information agent is that of a personal assistant (Maes 1994). Other metaphors center around the idea of software robots (bots), for example Chatter Bots, Shopping Bots, Search Bots, Mail Bots and News Bots. These metaphors and applications can be helpful, but if the impression is created that all tasks delgated to software are now trivial then the metaphors do more harm than good. Artificial intelligence often works well in narrow, well-understood domains, but find it difficult to facilitate cross-domain integration of information. This integration is natuarlly done by humans as part of some common thinking activities(Dix, Finlay, Abowd & Beale 1993: p.35), but the trouble often is that humans need tools to cope with the increasing abundance of information. This suggests that if agents and humans can collaborate to supplement each other's weaknesses then synergy may be possible. It will however depend on the nature and level of collaboration.

One approach to overcome the difficulties with automated integration of information from different domains is the one taken by Cycorp Inc. in building a system called CYC. CYC is a universal schema of commonsense knowledge (general concepts spanning human reality). "One can think of CYC as an expert system with a domain that spans all everyday objects and actions ... it could help standardize – and make more efficient – information retrieval, integration, and consistency checking." (Lenat 1995: p.33). This approach assumes that the difficulties faced by automatic machine learning and natural language understanding can only be overcome by "manually crafting a million axioms" into large knowledge bases. Software systems using CYC should then be able to work together (Guha & Lenat 1994) in dealing with human knowledge and ultimately to collaborate with humans as well.

COMMUNICATION

One of the big issues in collaboration between agents (human or software) is to have an appropriate way of communication to support the collaboration. The three modes of communication considered here are: software to software; human to software and software to human.

Software to software communication is regulated by protocols and formal languages of interaction. At low levels of communication (such as those used in the Open Systems Interconnection model of computer networking) messages are passed between computer systems in a specified protocol according to predetermined functionality. However on the *knowledge level* where communications is desired by intelligent autonomous agents, the content of the messages does not belong to a pre-determined set of valid responses. Examples of such Agent Communication Languages (ACL) are the Knowledge and Query Manipulation Language (KQML 1993) (Labrou & Finin 1997) and the FIPA[2] ACL (FIPA 1997).

KQML was developed by the DARPA[3] Knowledge Sharing Effort in order to exchange knowledge between expert systems. The KQML standard did not specify the content of messages but regulated the intended meaning of the messages by prescribing a set of *performatives* [4] such as "ask-if(...)" and "tell(...)". A seperate standard called Knowledge Interchange Format (KIF) was developed to express the content of messages in first-order predicate logic.

KQML has been critisised (Cohen & Levesque 1995) because it lacks the *commissives* class of performatives without which agents cannot make commitments to each other. For example an agent would have no way of promising (guaranteeing) a product or service to another agent if it uses KQML. The FIPA-97 ACL is shown by Wooldridge (1998) to have unverifiable semantics for reasons of "complexity" and because the program semantics of the associated semantic language (SL) is "ungrounded".

This does not mean that these ACLs are useless. It just means that the functionality of systems using them cannot be guaranteed.

Human to software and **software to human communication** can be considered in the field of Human-Computer Interaction. One possibility which offers the

---

[2]FIPA = the Foundation for Intelligent Physical Agents

[3]DARPA = Defense Advanced Research Project Agency

[4]The notion of *performative* comes from Speech act theory (see Austin (1962) and Searle (1969) which indicates that using a language sometimes not only results in making statements but also in *performing actions*. The classes of speech acts used for computing purposes are listed by (Singh, Rao & Georgeff 1998: p.360): assertives (informing), directives (requesting or querying), commissives (promising), permissives, prohibitives, declaratives (causing events in themselves), expressives (for emotions and evaluations).

most expressiveness and flexibility is to make use of natural human languages. This approach has some difficulties which are being researched and improved by several efforts. A special issue of the International Journal for Human-Computer Studies (see the preface (McRoy 1998)) presents current research in detecting, repairing and preventing human-computer miscommunication where *natural language* is the primary modality of communication. Other human-computer interfaces can also be used for bidirectional knowledge transfer between humans and computers, However non-natural language interaction tends to be more artificial and usually requires much more user training and motivation (Dix et al. 1993: p.113). It is mostly up to the interacting human to anticipate, detect, repair and prevent miscommunication. So the possibility of communication failures exist and without proper communication synergistic collaboration is hard to imagine.

AGENT ABILITIES

It is now necessary to distinguish agents from other tools to show that it makes sense to consider them for taking part in synergistic collaboration. Tools in general are usually created to increase efficiency and effectiveness, but tools are generally not seen as having goals; a tool would also not normally work if it is not activated or controlled by someone. This is true even for software tools and some systems classified as agents[5].

The type of agents that in the author's view can be part of synergistic collaboration are intelligent autonomous agents. The author's working definition of an autonomous agent is *an agent whose behaviour in working towards a specific goal is non-deterministic*. This implies that an autonomous agent not only has control over his own internal state, but also over its actions. The term intelligent is used here with the meaning of *having the ability to reason explicitly*. Agents with this ability are sometimes referred to as cognitive, rational, deliberative or heavyweight (Singh et al. 1998: p.342). If *belief, desire, intention* (BDI), *know-how* (see Singh (1998)) and *commitment* is used as high level cognitive specifications then they can serve as scientific abstractions to define the current state of an agent, what the agent might do, and how the agent might behave in different situations. Jennings et al. (1998: p.277) argue that apart from autonomy and intelligence agents should also have *social* abilities to allow them to interact with other agents and humans "to complete their own problem solving and to help others with their activities".

Agent-Agent Synergy

As indicated in an earlier section of this paper, agent-agent collaboration can lead to synergy in situations where the advantages of being distributed outweighs the disadvantages. These agents do not need to be intelligent or autonomous in order to achieve this, in fact it can be done with conventional distributed software using low-level interaction protocols if the network conditions allow it. With problems which require reasoning the stakes are usually much higher, so if we can get a similar synergistic effect there is much more to be gained. We therefore need to look at how synergy can be achieved at the knowledge level.

In the field of cognitive psychology it has been pointed out that *goals* and *plans* are important aspects of human behaviour (Miller, Galanter & Pribram 1960). Goals and plans may thus also be required for adequate intelligent agent behaviour. (More recently

---

[5]The concept of "software agent" is very broad and there is little agreement on where its boundaries should be.

goal and plan analyses have also been used in the study of human-computer interaction (Black, Kay & Soloway 1989).) Planning is certainly necessary for *mobile agents* as indicated by the Plangent[6] approach (Ohsuga, Nagai, Irie, Hattori & Honiden 1997). In this system agents interleave the planning and plan execution steps while traversing a network searching for information. In BDI-type agents goals are set based upon the *desires* of an agent. Desires can be seen as the input to the deliberation process resulting in a choice which determines the set of consistent achievable desires called *goals* (Singh et al. 1998: p.343).

In a multi-agent system where agents depend on one another to solve problems there is a need for a mechanism to: (i) resolve conflicting objectives, (ii) correct inconsistencies and (iii) coordinate a joint approach towards a solution. In short, agents need to be able to negotiate. One way of achieving negotiation is through argumentation (the exchange of proposals, critiques, explanations and meta-information) (Parsons, Sierra & Jennings 1998).

Agent-Human Synergy

It is however not neccessary to let artificial agents resemble humans as closely as possible before they can be considered for synergistic collaboration with humans. After all, synergy is not only possible between equals but also for example in a master-slave relationship. Milewski and Lewis investigated "delegation" as a user-interface model for software agents. They share the generally accepted view that "software agents act on behalf of users by *autonomously* carrying out *delegated* activities made up of *multiple sub-tasks*" (Milewski & Lewis 1997: p.487). The benefits of delegation should not exceed the cost especially if synergy is desired. Typical costs of delegating to agents include: assesment of agent competence; monitoring of agents and progress; communication of desired outcome strategies; and anxiety associated with loss of control.

Delegation models usually place the human in the controlling position and leaves little room for agents to make suggestions to change the course of action. Another approach is to build agents with enough intelligence to infer appropriate high level goals from users actions and requests. This is the approach used in building *Apple Data Detectors* (Nardi, Miller & Wright:1998 1998). They can be classified as *interface agents* according to Nwana's typology(Nwana 1996). To build this type of agent requires a degree of user modelling as well as sophisticated recognition and parsing of documents and an analysis of user's interactions with different systems.

CONCLUSION

Builders of agents and agent systems have many tools and techniques to choose from. The functionality of some of these tools (alone or in combination) certainly looks promising enough to the extent that synergy can be expected to occur. There are however some practical challenges relating to performance issues. One of them occurs when agent mobility wants to be exploited: there is a trade-off between the functionality (which directly relates to the size of the agent) and the costs (e.g. speed of transfer and execution). Other issues include the level of collaboration that can be achieved with a particular user interface model and the possibility to use user modelling to streamline agent-human interaction.

---

[6]Plangent uses KIF for knowledge interchange between agents.

References

AUSTIN, J. L. 1962. How to Do Things with Words. Oxford: Clarendon Press.

BLACK, J. B., KAY, D. S. & SOLOWAY, E. M. 1989. Goal and Plan Knowledge Representation. Cambridge, Massachusets: MIT Press. p. 36–60.

COHEN, P. R. & LEVESQUE, H. J. 1995. Communicative actions for artificial agents. (Proceedings of the First International Conference on Multi-Agent Systems (ICMAS-95): . San Francisco, CA. p. 65 – 72).

DIX, A., FINLAY, J., ABOWD, G. & BEALE, R. 1993. Human-Computer Interaction. New York: Prentice Hall. 570p.

FIPA 1997. FIPA 97 Specification, Part 2, Agent Communication Language. Foundation for Intelligent Physical Agents. http://drogo.cselt.it/fipa/spec/fipa97/fipa97.htm (1/5/98).

GUHA, R. V. & LENAT, D. B. 1994. CYC: Enabling agents to work together. *Communications of the ACM, Special Issue on Intelligent Agents*, **37**(7):41–47.

JENNINGS, N. R., SYCARA, K. & WOOLDRIDGE, M. 1998. A roadmap of agent research and development. *Autonomous Agents and Multi-Agent Systems*, 1:275 – 306.

KQML 1993. Specification of the KQML Agent-Communication Language. draft edn. The DARPA Knowledge Sharing Initiative, External Interfaces Working Group.

LABROU, Y. & FININ, T. 1997. A proposal for a new KQML specification. *Technical Report TR CS-97-03*. Computer Science and Electrical Engineering Department, University of Maryland Baltimore County.

LENAT, D. B. 1995. CYC: A large scale investment in knowledge infrastructure. *Communications of the ACM*, **38**(11):33–38.

MAES, P. 1994. Agents that reduce work and information overload. *Communications of the ACM, Special Issue on Intelligent Agents*, **37**(7):31–40.

MCROY, S. 1998. Preface: Detecting, repairing and preventing human-machine miscommunication. *International Journal of Human-Computer Studies*, **48**:547 – 552.

MILEWSKI, A. E. & LEWIS, S. H. 1997. Delegating to software agents. *International Journal of Human-Computer Studies*, **46**:485–500.

MILLER, G. A., GALANTER, G. & PRIBRAM, K. H. 1960. Plans and the Structure of Behaviour. New York: Holt, Rinehart and Winston.

NARDI, B. A., MILLER, J. R. & WRIGHT:1998, D. J. 1998. Collaborative programmable intelligent agents. *Communications of the ACM*, **41**(3):96 – 104.

NEGROPONTE, N. 1995. Being Digital. London: Hodder and Stoughton.

NWANA, H. S. 1996. Software agents: an overview. *The Knowledge Engineering Review*, **11**(3):1–40.

OHSUGA, A., NAGAI, Y., IRIE, Y., HATTORI, M. & HONIDEN, S. 1997. Plangent: An approach to making mobile agents intelligent. *IEEE Internet Computing*, **1**(4):50 – 57.

PARSONS, S., SIERRA, C. & JENNINGS, N. R. 1998. Agents that reason and negotiate by arguing. *Journal of Logic and Computation*, **8**(3):261–292.
http://www2.elec.qmw.ac.uk/~sp/papers/journals/jlc.html (1/9/1998)

SEARLE, J. R. 1969. Speech Acts: An Essay in the Philosophy of Language. Cambridge, England: Cambridge University Press.

SINGH, M. P., RAO, A. S. & GEORGEFF, M. P. 1998. Formal methods in DAI: Logic-based representation and reasoning. (*In* WEISS, G., *ed.* Multiagent Systems: A Modern Introduction to Distributed Artificial Intelligence. Cambridge, Massachusets : MIT Press. p. 331–376).

SINGH, M. P. 1998. Know-how. (*In* RAO, A. S. & WOOLDRIDGE, M. J., *eds.* Foundations of Rational Agency. Kluwer).

WOOLDRIDGE, M. 1998. Verifiable semantics for agent communication languages. (Proceedings of the Fourth International Conference on Multi-Agent Systems, ICMAS-98).

THE APPLICATION OF INTERNET TECHNOLOGIES TO INFORMATION MANAGEMENT

Peter G. Lee & Dr. C.E.R Wainwright

The CIM Institute
Cranfield University
Cranfield
Bedfordshire, MK43 0AL

INTRODUCTION

Information management has changed immeasurably with the growth of the Internet. The global system of computers networked together not only provides the basis of its most popular application, the World Wide Web (WWW), but is increasingly being used as the facilitator in a new range of information applications.

Intranets have been well publicised for their power to keep corporate operations informed, but this tends to be on an internal basis. Over the past couple of years a new Internet technology has emerged- the Virtual Private Network (VPN). A Virtual Private Network uses the public and private network infrastructure to connect nodes using the Internet as the medium for transporting data. These systems use encryption and other security mechanisms to ensure that only authorised users can access the network and that the data cannot be intercepted. An added bonus is the use of a standard web browser that will allow access to even the most complicated of real-time applications.

A traditional Wide Area Network (WAN) requires leased communication lines, banks of modems, routers, network cards, software and many support personnel. The VPN utilising the public Internet obviates the need for much of the corporate investment in WAN since the hardware, communication lines, etc do not belong to the organisation. This means that the setting up of Global Information Systems can be achieved more quickly, more easily and most importantly – more cheaply.

Example applications of VPNs are growing daily and are typified by the remote control of telemetry equipment in the utility and process industries, the ready access of public records by the US Police Departments and Medical fraternity, and in the implementation of global manufacturing strategy.

This paper will therefore explain the technologies involved and will conclude with examples of the application of the VPN technology

Synergy Matters: Working with Systems in the 21st Century,
Edited by Castell *et al.*, Kluwer Academic / Plenum Publishers, New York, 1999.

ENABLING TECHNOLOGIES

This section will explain the technologies involved.

The Internet

Since much has already been written on the Internet, its history and its development as short description of what is meant by the Internet should be sufficient within this paper. The Internet is series of networked computers, banks of modems and routers, often non-homogenous systems using differing operating systems, across the world connected together by all current telephony technologies.

It is important to note that whilst the terms Internet and World-Wide Web (Web) are often used interchangeably, they are not the same thing. The Internet is the global network infrastructure that allows the Web to operate. The Web is a collection of documents linked together using a specific Internet Protocol known as Hypertext Transfer Protocol (HTTP). The Web is dependent on the Internet for its existence, but the Internet is totally independent of the Web. [1]

Virtual Private Network (VPN)

A Virtual Private Network (VPN) utilises the existing public and private network backbones and infrastructure of the Internet to connect nodes within a distributed organisation. Only authorised users have access to the information being transmitted by using a system of hardware and/or software encryption along with firewalls.

VPNs have been possible for a number of years with the usual rivalry between technologies. Microsoft developed a communication protocol known as Point To Point Tunnelling Protocol (PTTP) to enable VPNs and bundled it in Microsoft NT 4.0™. Cisco Systems produced a rival protocol known as Level Two Forwarding Protocol (LTFP). However, by the end of 1997 Microsoft and Cisco agreed to merge their respective protocols to producing Layer 2 Tunnelling Protocol (L2TP). [2]

Tunnelling protocols work by the establishing a secure tunnel between two computers allowing data to flow. The data cannot be intercepted and enters and leaves the computers through the tunnel. To ensure absolute security, encryption by either software or hardware may be used together with a firewall.

Tunnels are easily utilised by the use of tunnelling software on both the client and the tunnel server which is usually located at the main corporate site. It is possible for the tunnel server to exist at the Internet Service Provider (ISP)'s site which the main corporate site uses. First, the client software establishes the tunnel whilst the tunnel server ends the tunnel both allowing access on the basis of user identification and password protection. An ISP does not need to support tunnelling at all and data transfer will proceed without any requirement for monitoring or assistance from the ISP. By using the Internet in this way an organisation is utilising it as if it were their own private.

A VPN requires at its simplest implementation, browser software (normally supplied free, at least until the Anti-Trust action against Microsoft is resolved), an Internet Service Provider account, a telephone line and the tunnelling software. Remote workers, if the ISP account is with a multinational provider, can telephone local Points of Presence (POP) to access the VPN.

Benefits of VPN Technology Implementation

Banks of modems, network cards, routers, leased communication lines, and support personnel all represent considerable investments and costs to a distributed organisation operating a WAN. The VPN removes the need for much of the corporate investment in WAN since the hardware, communication lines, etc do not belong to the organisation. The benefits of VPNs are listed below:

- Major capital investment reductions required thus making VPNs ideal for greenfield, start-up or expansion
- Training cost reductions by using Web Browsers
- Local telephone calls for VPN access further reduce operating costs
- Maintenance costs are met by public and other private organisations
- Infrastructure upgrades to the Internet benefit the VPN user at no direct cost.
- The tunnel may improve Internet performance

Therefore benefits from VPNs can be summed up as both financial and operational which means effectively that an organisation is outsourcing its WAN when it uses a VPN.

Disadvantages of VPNs

- Loss of financial control of the network
- No guarantee of service quality when more Internet users load the system
- No control of down time
- Reliability of the service provider
- Open to exploitation by ISP

The above disadvantages may be enough to put off any organisation considering implementing a VPN. The loss of investments to improve the network, the chance that once the WAN equipment and accoutrements are dispensed with an ISP may put up its charges and the loss of control of the situation when an ISP goes down may pose serious problems. However, many of these can be ironed out through contractual agreements to compensate or provide a particular level of service.

VPN EXAMPLE IMPLEMENTATIONS

The next section will consider some of the many applications of VPNs currently in use.

Radiological Telemedicine

Since June 1998, the Tufts University School of Medicine has been using VPNs to allow specialists to view radiological images in the comfort of their homes or wherever they may be.[3] This means that the specialists' skills are used only when required preventing time wasting journeys for cases that more junior Doctors may be able to deal with. It also means that physicians are involved in decisions on images rather than just the radiographers.

The basis of the system is a collaborative effort between a VPN provider, Assured Digital Inc.(ADI), a service provider MediaOne and the medical imaging company JABR Technology Corp. MediaOne provide a Cable Modem rather than the traditional analogue modem. This allows transmission speeds anywhere between 1.544Mbps and 10Mbps, the fastest analogue modems currently transmit at 56Kbps. This speed of transmission is probably the reason for the success of the system, since typical file size may be well in excess of 1Gb.

Security is of great concern since patient records are confidential and this system operates using encryption and user authentication methods to ensure only authorised users have access. ADI supplies a VPN switching device that utilises embedded intelligent routing along with security routines that allow the system to adapt immediately to changes in the network topology. JABR Technology's contribution was a programme called "Synapse" which allows doctors to examine, diagnose, annotate, store and retrieve the images.

This system is a good example of the use of a VPN as an enabling technology advancing another area of science and information management.

Telemetry Using The Internet

"Telemetry is the technology which enables a user to collect data from several measurement points at inaccessible or inconvenient locations, transmit that data to a convenient location, and present the several individual measurements in usable form." (4)

Consider the means of transferring data between sensor and collection point in a telemetry system. A VPN could use the Internet as the medium of transference and the means of viewing and controlling the system could be facilitated by the Web.

However, an Internet Telemetry System in an existing situation may be better implemented by using the Internet as the means of disseminating telemetry data/information rather than the means of transmitting telemetry data. This is because most telemetry systems are already established with their infrastructures devoid of bugs. In the case of a large water utility company much investment has already been made and any new investment would reduce dividends to shareholders thus depressing the share price and therefore the company's value. Any additional money spent on an existing telemetry system would be better spent on making the telemetry data and the resultant information more readily available.

Telemetry data collected in the database could be displayed as a Web page using the various controls/programmes that allow software sockets in Web pages such as Java^{TM} and Microsoft ActiveX Controls^{TM}. Remote users could in effect use a VPN to connect to the Web page and therefore affect the control changes necessary. With the advent of Handheld Personal Computers (HPC) it is also possible to connect anywhere in the world to the control site Web page (providing your ISP has a local phone number).

A working example of this technology in process manufacturing can be seen at **http://www.csimonitoring.com** which is a demonstration site for Vendor Managed Inventory for the chemical industry run by Clover Systems Incorporated (CSI) of New Jersey, USA. [5]

A sample page from the site appears below:

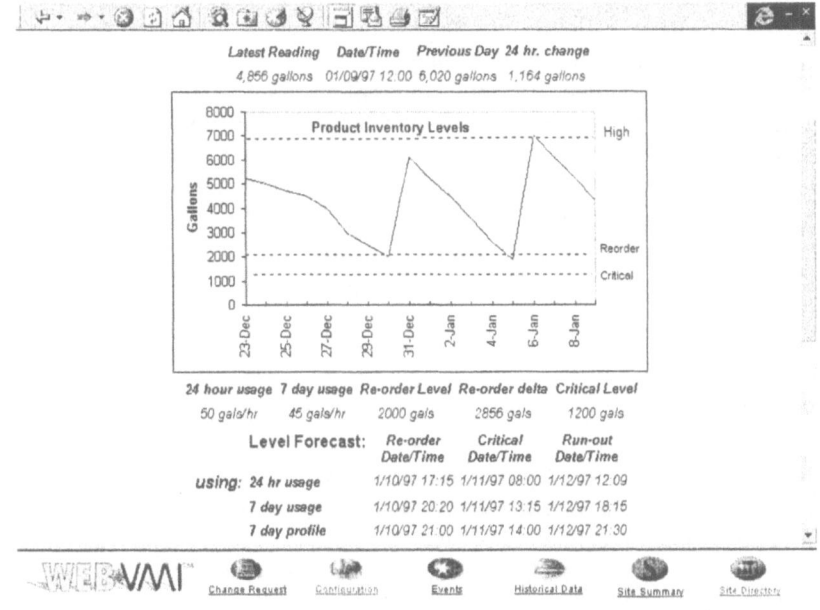

Figure 1. Web based telemetry system working demonstration.

Jetform Corp.- Illustrates The Savings

JetForm Corp., develops electronic forms automation and enterprise workflow systems, replaced its existing WAN with a VPN focussing all its sites on its headquarters in Ottawa, Canada.[6] Jetform has experienced massive cost savings and performance increases of double the speed of the old WAN. Previously their costs were US$4,000 per site per month, using AT&T as their service provider the cost has plummeted to US$1,500 per site per month.

Their system utilises hardware encryption and thus performance degradation is negligible. Performance has increased by routing data directly to the recipient whereas previously all traffic was routed to the headquarters and back out to the recipient. Accessing the network has also realised reduced costs. The average monthly telephone bill for a mobile worker to access the network was US$3,000. With local call access the average has been reduced by a factor of 10 to just US$300.

CONCLUSION

The above examples illustrate the immense benefits available in all industries by the use of Virtual Private Networks. As is usual with most types of technology and in particular IT, the United States has embraced VPNs quite readily. The reliability of service and the loss of control of network administration, may for many UK organisations, be two areas that prevent them from considering VPNs as a viable alternative to their current distributed network architectures.

The low capital expenditure required in implementing VPNs will appeal to those organisations following an agile manufacturing strategy within the extended enterprise. With the links between consumer and manufacturer becoming ever closer, VPNs offer the user shorter times to market and more efficient methods of operating.

The market for VPNs is expected to grow from $120m in 1997 to $6bn by 2001 in the US alone. This technology represents an opportunity for all organisations to reduce their operating costs and improve their productivity. The likelihood of UK organisations implementing VPNs is minimal if their track record of utilising new technologies is indicative of their commitment to VPNs. It is more likely that US organisations currently implementing VPNs will transport the technology to their UK subsidiaries and thrust the technology upon them.

REFERENCES

1. Stein, Lincoln D., How To Set Up And Maintain A World-Wide-Web Site – The Guide For Information Providers, Addison Wesley Publishing Company, 1995
2. Jennings, Roger, Using WindowsNT Server 4.0^{TM}
3. Internet Week, 15 June 1998, published at http://www.internetwk.com/case/study0615-1.htm
4. Strock, O.J., Introduction to Telemetry, Instrument Society of America, 1987
5. http://www.csimonitoring.com/   Clover Systems Incorporated, Berkeley Heights, New Jersey, USA
6. Internet Week 1 June 1998, published at www.internetwk.com/case/study0601-1.htm

AN INTRODUCTION TO CYBERNETIC VIABILITY STUDY (CVS)

Min Li and Wei-hua Jin

Lincoln School of Management
University of Lincolnshire and Humberside
Lincoln, LN6 7TS, UK

INTRODUCTION

This paper intends to introduce the Cybernetic Viability Study (CVS), a step beyond the traditional feasibility studies by utilising the systems approach. The CVS is underpinned by organisational cybernetics, which has long been advocated and practised by Beer (1979, 1981, 1985) and Espejo (1993, 1996, 1997). We start by explaining why this expansion of the systems approach becomes necessary. We will then introduce the basic notions of the CVS. Finally, we will propose where and how it can be used for practical research projects.

SYNERGY MATTERS

Traditional feasibility studies emphasise the financial returns of new projects. They "examine proposals in order to establish the degree to which these attributes, amongst others, are present", and provide reports "to highlight, evaluate and structure the advantages and disadvantages over time of alternative solutions to given problems" (Gruneberg & Weight, 1990, p.5). A Project becomes feasible if it can break-even and make a reasonable financial return within a prescribed period of time. However, financial matters are only one part of the project, while many other aspects should also be considered, for instance, people, physical equipment and their operations, etc. At present, an increasing number of analysts begin to include the additional use of organisational and technical studies. Social and environmental impact studies are also undertaken as part of the new project planning. However, it appears still difficult to study the feasibility of a new project without going through a number of specific studies. It becomes problematic to reach a consensual conclusion if different pictures emerge in different studies. There seems to be a need to synergise the various forms of studies which have been used for the purpose of determining the feasibility of new projects. In the systems field, this kind of

Synergy Matters: Working with Systems in the 21st Century,
Edited by Castell *et al.*, Kluwer Academic / Plenum Publishers, New York, 1999.

91

problem has already been well debated for more than two decades. It is perhaps possible to borrow the insights gained there to see the future direction for feasibility studies.

CRITICAL SYSTEMS THINKING

In the systems field, the Critical Systems Thinking branch has devoted itself to the study of synergy among different strands of systems approaches. This paper will follow the work by Jackson, since he has been the main advocator of Critical Systems Thinking. Jackson (1991; 1995; 1997) pointed out the importance to recognise that each individual systems approach possesses their unique strengths as well as weaknesses. This becomes a practical challenge to him in the notion of a critique of different systems approaches. At the same time, he insists that it is possible to utilise different systems approaches in a complementary way, so that individual weaknesses can be compensated and individual strengths can be exerted in an optimal way. For him, synergy among different systems approaches should be encouraged since variety is the key to preserving the vitality of the systems field as a whole.

The challenge for Jackson has been to identify a practical way to synergise various systems approaches. This has been an ongoing theme for research within the systems field for over two decades. Some proposals have been forwarded and tested in application. One suggestion is to match different systems approaches with different problem situations through a framework, which has come mainly in the form of Jackson's System of Systems Methodologies (SOSM) (Jackson and Keys, 1984; Jackson, 1991; 1995). The following Fig 1 is a depiction of Jackson's SOSM (adapted from Jackson, 1995, p.26).

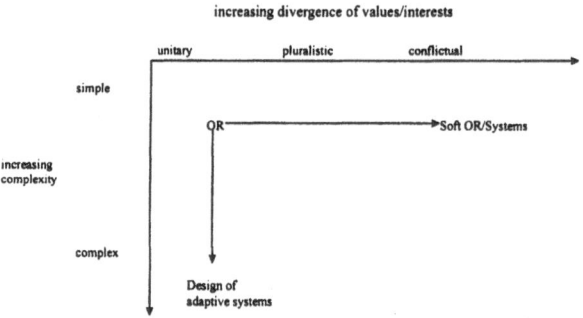

Figure 1 Jackson's SOSM

The two axes in Jackson's SOSM are used to reveal the possible 'ideal type' of problem contexts. The horizontal axis is concerned with increasing divergence of values among different stake-holders. There are three terms to depict the divergence, which are unitary, pluralistic and coercive (or conflictual). A unitary situation is where stake-holders share values and interests. A pluralistic situation is where stake-holders diverge in their values and interests, and yet they share enough in common to make it worthwhile their remaining members of the coalition in the form of the collective. A conflictual or coercive situation is where the stake-holders have irreconcilable diverged values and interests so that some groups get their way at the expense of others being coerced. The vertical axis depicts the increasing complexity of the problem context, from simple to complex. The complexity is measured in terms of the number of elements, rate and character of the interactions between the elements, attributes of the elements, nature of subsystems and the environment (Jackson, 1995, p.26). For Jackson, it becomes possible to choose different systems methodologies according to different problem contexts.

According to Jackson (1997), the greatest benefits from a synergy among different systems approaches can be realised if the systems field employs 'a meta-methodology to take maximum advantage of the benefits to be gained from using methodologies premised upon alternative paradigms together, and also encourages the combined use of diverse methods, models, tools and techniques, in a theoretically informed way, to ensure maximum flexibility in an intervention (Jackson, 1997, p.6). In this paper, we take on board the insights provided by Jackson's research for a synergistic use of different systems approaches while proposing the adoption of a different framework, as against Jackson's SOSM. This new meta-framework comes in the shape of Beer's VSM, which will be introduced in the following sections.

BEER'S VSM

Beer's VSM (Beer, 1979; 1985) is different from Jackson's SOSM. The latter proposes to match individual systems approaches with different problem contexts (e.g., simple-unitary, complex-unitary, simple-coercive, complex-coercive, etc., Jackson, 1995). This becomes almost impossible at the beginning of a new project, when perhaps the problem context it faces has not yet emerged. Beer's VSM, however, proposes to study all matters concerning organisational viability (viability can be seen as an ideal synergy both inside the organisation and in-between the organisation and its environment). For this reason, it seems to be an ideal candidate to become the meta-framework in the use of different studies concerning the feasibility of a new project.

Originally, Beer's VSM (Beer, 1985) is a model of an ideal organisation, as depicted in Fig 2 (adapted from Espejo, 1989),

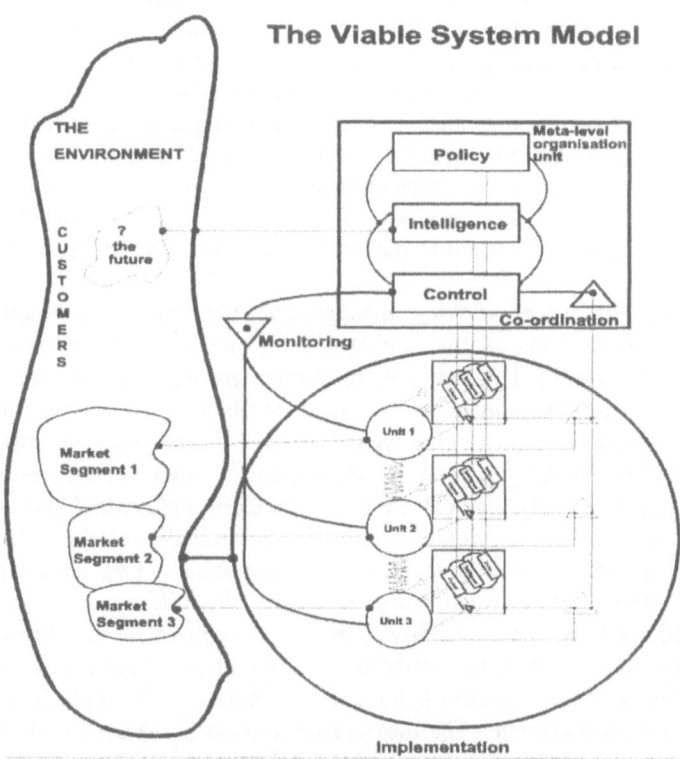

Fig 2 The Viable System Model

This organisation consists of five levels, which are implementation, control, co-ordination, intelligence and policy. Beer's main research concern is how organisational managers can go beyond their individual limited human competence to deal with the increasing complexity presented to them from the turbulent business environment and the ever-changing organisation itself. He identifies the importance of organisation structure, and indicates that only viable structures can enable the sustainable development of viable organisations. Structure, in Beer's sense, is mainly concerned with the communication within organisational members and the VSM is his ideal model for a viable structure. One of his most significant insights is that only information (not data) can serve as the negentropy which counter-acts with entropy, the force which is determined to blow the organisation apart like it does to all living systems. In this sense, Beer's VSM seems to have pointed out a way of sustainable development beyond the vitality of organisations as living systems. This particular way is the viable structure through the VSM with an emphasis on communication and control.

To utilise Beer's VSM as the meta-framework for the use of different methods for feasibility studies, the Cybernetic Viability Study (CVS) methodology is proposed. In the CVS, different methods can be used to study viability (synergy) factors of an organisation at three levels, i.e., the operational level, the strategic level, and the normative level, which have been proposed by Espejo and Schwaninger (1997). In the everyday organisational management, we seem to have developed our understanding and practice from operational management to strategic management. At the level of operational management, we stress the importance of an organisation's ability and performance to make profits with an overall aim to remain solvent at all times, by safeguarding the health of our balance and cash-flow and so on. According to Espejo and Schwaninger (ibid.), this attention on the operational management, leads to approaches to design operational systems to plan and organise the optimised use of resources (human, material and financial resources). This is similar to the traditional way of feasibility study which has orientation towards a search for inherent profitability and long-term liquidity (solvency) of potential projects through the study of various factors, such as discounted cash flow forecast, cost-volume-profit analysis and so on (Arnold and Hope, 1983). However, recent developments in management theories and practice means that more companies are paying attention now to strategic management. According to Espejo and Schwaninger (ibid.), strategic management indicates that a growing number of companies are keen to bring forth, maintain and develop the so-called 'Value potentials', which pre-controls profits. 'Value potentials' include a 'set of all applicable business-specific, latent resources, that must be realised when profits are to be achieved.' (ibid., p. 1)

Following this line of thinking, management attention is increasingly engaged in issues such as 'market share, relative market share, quality or customer benefit, cost, price, speed or flexibility' (ibid.). The shift from operational management towards the dual-level management on both operational and strategic levels is also reflected in the current approaches to feasibility study. One typical example is the research work done in the investment decision making process by stockbrokers, where their investment analysis seems to go beyond merely looking at the balance sheet and profit / loss account of the target companies. Today, they would consider more closely those 'value potentials' such as market share, patents, customer loyalty, and other competitive advantages (Natwest Stockbrokers, 1997; 1998). This may exemplify the difference between the usual feasibility study and cybernetic viability study, i.e., the cybernetic viability studies should go beyond what has already been covered by the usual financial feasibility studies. Espejo and Schwaninger (1993) have identified the need to develop a third phase of management called normative management. The measure for success is the synergy level among the different information gathered at different levels through channels of communication and control. Hence a project is only feasible (or viable, as the preferred term) when synergy is reached among different indicators. This is different from the existing pragmatic use of different studies, since it may be difficult to make sense of the complete picture without a metaframework for interpretation. It is important to enable different information to "talk" to each other, and to come alive through Beer's VSM.

THE CYBERNETIC VIABILITY STUDY

The VSM provides an insight into the so-far under-developed normative management. This model probes the underlying force, i.e. the organisational structure which may enable an organisation to sustain its development at operational and strategic levels and give it a platform to deal with the challenges of tomorrow in the ever-evolving complex environment. In carrying out normative management, managers need to be aware of both the 'inside and now' (its inward vitality) and the 'outside and then' (its outward potentials) so as to reach for a balanced and sustainable viability for the organisation as of today and of tomorrow, which is the focus of this cybernetic viability study. At present, this PhD project is being undertaken at the Lincoln School of Management, which aims to develop the CVS into practical application. The research is supervised by Professor Espejo, who has been developing the notion of organisational fitness in the form of a cybernetic methodology, following Beer's work.

Background reading of the project has covered the subject of organisational viability through the works of Beer (1979, 1981, 1985) and Espejo (1989; 1993; 1996; Espejo and Schwaninger, 1993; Espejo, et al., 1996; Espejo and Schwaninger, 1997; Espejo and Reyes, 1997.. Espejo and Stewart, 1998). A literature survey of different methods for feasibility studies will be taken to find out the strengths and weaknesses of existing approaches and the overall situation in the field of feasibility studies. Then, the researcher will explore how different strands of feasibility study methods can be used through the framework of Beer's VSM with an aim to formulate the Cybernetic Viability Study (CVS). The next stage will be to test the difference between a cybernetic viability study and studies used by British financial institutions in the choice of 'suitable' (viable?) investment targets. The intended application for the new methodology is to utilise viability studies on British firms who intend to gain their presence into China. Furthermore, unlike the traditional feasibility studies, the CVS can be used by firms who have already entered the Chinese market as an on-board diagnostic facility to check the well-being of their Chinese venture.

The CVS should be a systemic study which looks at the emergent property of a whole range of factors such as financial, technical, political, social aspects, with regard to the survival, development and sustainability of an organisation. The result may be applicable to the diagnosis and design of the viable collectives across all sectors in the society so that there is a practical way to reach for sustainable development at all levels. Since 1980 (cf. World Conservation Strategy, 1980), sustainable development has become the pursuit for many countries in the world as well as many business organisations. Sustainable development can be seen as the latest development of a series of global initiatives to improve our endeavours beyond the pursuit of profitability. This began with the quality movement, which advocated a move towards Total Quality Management (TQM) through the global initiative of ISO 9000 series international quality management standards. The quality movement led to the initiation of the ISO 14000 series of environmental management standards. Considering synergy matters, the researcher hopes to search for a more integrated approach to take account of various aspects of our common concerns with respect to feasibility/viability studies.

REFERENCES

Arnold, J., and Hope, T., (1983). "Accounting for Management Decisions", 2nd ed., Prentice Hall, New York.

Brundtland, G. H. (1987). "Our Common Future: World Commission on Environment and Development", Oxford University Press, Oxford.

Beer, S. (1974). "Designing Freedom", Wiley, Chichester.

Beer, S. (1979). "The Heart of Enterprise", Wiley, Chichester.

Beer, S. (1 98 1). "Brain of the Firm", 2 d ed.,Wiley, Chichester/New York

Beer, S. (1985). "Diagnosing the System for Organizations", Wiley, Chichester.

Espejo, R. (1989). The VSM revisited, In Espejo, R., and Hamden, R. (eds.), "The Viable System Model", Wiley, Chichester.

Espejo, R. (1993). Domains of interaction between a social system and its environment, *Systems Practice* 6, pp.517-526.

Espejo, R. and Schwaninger, M. (1993) (eds), "Organisational Fitness: Corporate Effectiveness through Management Cybernetics", Campus Verlag, Frankfurt/New York.

Espejo, R. (1996)   Requirements for effective participation in self-constructed organizations, *European Management Journal,* pp.414-422.

Espejo, R., and Reyes, A. (1997) Responsive accounting: a grounding of informational domain in the operational domain of an organisation, in Achterbergh, J., Espejo, R., Regtering, H., and Schwaninger, M. (eds.), *Oranizational Cybernetics, Research Memorandum,* Nijmegen Business School, Nijmegen.

Espejo, R., Schuhmann, W., Schwaninger, M., and Bilello, U., (1996). "Organizational Transformation and Learning: A Cybernetic Approach to Management and Organization", Wiley, Chichester.

Espejo, R., and Schwaninger, M. (1997). *Strategic Management in Global Corporations: A Cybernetic Perspective,* Working Paper, No. 18, University of Lincolnshire and Humberside, Lincoln.

Espejo, R., and Stewart, N.D., (1998).   Systemic reflections on environmental sustainability, *Systems Research and Behavioural Science (earthwhile Systems Research),* forthcoming.

Gruneberg, S.L. and Weight, D.H., (1990). "Feasibility Studies in Construction", Mitchell London.

Jackson, M.C., (1991).   "Systems Methodology for the Management Sciences", Plenum Publishing Company, New York.

Jackson, M.C., (1995). Beyond the fads: Systems thinking for managers, *Systems Research,*

Vol. 12, No. 1, pp.34-60.

Jackson, M.C., (1997). *Towards Coherent Pluralism in Management Science,* Working Paper, No. 16, Lincoln School of Management, University of Lincolnshire and Humberside, Lincoln, UK.

Jackson, M.C., and Keys, P., 1984, Towards a systems of systems methodologies, *Journal of Operational Research Society,* 35, pp.473-486.

Natwest Stockbrokers, (1997). *Private Investor Review: September 1997,* Natwest Stockbroker Ltd, London.

Natwest Stockbrokers, (1998). *The Stock Review: February 1998,* Natwest Stockbroker Ltd, London.

Reyes, A. (1996). "Theoretical Framework for the Design of a Social Accounting System", Doctoral Thesis, University of Humberside, Hull.

A MULTICULTURAL PERSPECTIVE ON INTEGRATING NON-NATIVE SPEAKING STUDENTS INTO THE COMMUNITY IN VIENNA/AUSTRIA AND TEXAS/U.S.A.: A PRELIMINARY REPORT

Eva Linton-Kubelka

Institut fuer Erziehungswissenschaften
Universitaet Wien
C/O Habichergasse 50/21
A-1160 Wien
Austria

INTRODUCTION

The purpose of this paper is to give a preliminary report on a research project comparing the perception of middle school teachers toward non-native speaking students in the city of Vienna, Austria and the state of Texas, U.S.A. The project hopes to determine the effect of integration programs in both school systems on the way the teachers perceive the non-native speaking students. In Vienna, the students chosen for this study come from Turkey and the former Yugoslavia (currently the countries of Croatia, Serbia and Bosnia and Herzegovina); in Texas, the students are Mexican immigrants.

In July and August 1998, I conducted research in Texas/U.S.A. I compared the perceptions of teachers in the two different educational systems using narrative interviews. First of all I want to give a short introduction to the reasons why I chose the narrative interview as the research method in this field. Then, I would like to give a short description of the two systems' methods of integrating non-native speaking students, and finally, discuss the importance of a multicultural education for all students and improved multicultural training for the teachers.

THE NARRATIVE INTERVIEW: AN ALTERNATIVE APPROACH TO INTERVIEWING

Narrative Interview gives a possibility to expand the critical character of empirical social sciences. It does this because it enables researchers, as Spradley says, to get "the view of the event out of the event itself" (1).

Synergy Matters: Working with Systems in the 21st *Century,*
Edited by Castell *et al.*, Kluwer Academic / Plenum Publishers, New York, 1999.

97

A lot of qualitative research projects are confronted with a large amount of data gathered interviews, and researchers are often not able to gage this amount until it is time for the analysis. Then, due to time pressure, the material is often evaluated rather quickly and unsystematically. Another major problem is that often too much emphasis is placed on the researcher's own interpretation, rather than using interpretative methods (2).

In my study I wanted to get insight in the perception of teachers of immigrant students, so the Narrative Interview seemed the best way to capture the teacher's main thoughts and opinions.

Narrative Interviews are able to reconstruct subjective perspective of the occurrence. They underline the perspective of the interviewed rather than the interviewer, and therefore can be seen as "speech events" (3). This means that the interviews can be seen as a discourse between the speakers instead of a stimulus-response model.

Narrative Interviews require the interviewer to pay close attention to the informant's linguistic features. They have to watch out for certain expressions or overlaps and responses cannot be seen as simple answers to a few questions, but as a reflection of the mind of the informant. There must also be an assessment as to whether a respondent has said enough. So the main interest lies in certain linguistic symbols that Spradley says "form the case of the meaning system of every culture and with which we can communicate about all other symbols in a culture" (4).

The Narrative Interview shares some features with a friendly conversation, but with some major differences: The interviewer and the informant do not take turns in asking questions, instead the interviewer asks and the informant talks about the experience. There also will be certain questions that will be repeated throughout the interview, because the informant has to be brought to the point where he goes more into detail.

These are just some of the issues that had to be raised before beginning my research. Now I would like to give a short overview of the Austrian and the U.S. school systems and their approaches to educating non-native speaking students.

EDUCATIONAL MODELS FOR NON-NATIVE SPEAKING CHILDREN IN VIENNA, AUSTRIA

Since 1961 there has been a permanent increase of people from former Yugoslavia and since 1971 from Turkey in Vienna. They came here to work as laborers or in blue-collar jobs. They had the status of "foreign workers." The idea was to allow young men to work in Vienna for some years and then to return to their own country. This plan soon fell apart when many of the workers began moving their families to Austria or marrying Austrian women, creating an increase in the non-native speaking population. This caused an increase of demands on the government and the communities primarily to provide health insurance, housing and education for the new immigrants and their children.

The Viennese school system offers nine years of compulsory education, which starts when the children are six years old. After four years of elementary school they can enter two different types of school, either a lower general secondary school (*Hauptschule*) or a higher one (*Gymnasium*). After eighth grade there is a further choice between a vocational, polytechnical, or the upper level academic secondary school (*Berufschule, polytechnische Schule* and *Realgymnasium*).

One group frequently found in lower general secondary school are immigrant children. These children make up from 32% of the students in elementary schools and 40% in the lower general secondary schools. Because these schools were the first to be confronted with non-native speaking students, they were also the first to develop a program for handling this situation. Many "multicultural" school models are found in these types of compulsory schools. If the number of non-native speakers in the classroom is less than 30%, then the teacher can request an educational assistant specially trained working with immigrant students. If the number of students is more than 30%, then a more supportive model will be used depending on the grade level. For instance, in elementary schools, the aim is to integrate all students into the regular time schedule as soon as possible. The pupils stay in their classes and either specialist teachers are pulled into the classroom or students are pulled out for a couple of hours (5). These conditions are used throughout Vienna as well as in the other urban areas of Austria (6).

EDUCATIONAL MODELS FOR NON-NATIVE SPEAKING CHILDREN IN TEXAS, U.S.A.

Between 1965 and 1994, the percentage of the foreign-born population in the U.S. rose to 8.5% from 5%. The large metropolitan areas of Texas also harbor a large concentration of new immigrants. Today's immigrants are predominately young, coming to Texas to work in blue collar jobs. Since immigrants constitute more than one fourth of the growth of the national labor force, their preparation for work is very important.

Up to the level of middle school, immigrants are nearly as likely to be enrolled in school as natives in the same age group (7). Immigrant children, most particularly those of Hispanic origin, who enter the country after the age of 15 or so, are less likely to enter the U.S. school system.

There are two prevalent methods used in the U.S. for the education of non-native speaking children. The first is an ESL program, or English as a second language. Children in this program are taught all subjects in English with other non-native speakers, before they are finally placed into mainstream classes. The second prevalent model is the bilingual program, where immigrant children are taught a large proportion of their school subjects -- such as math and science -- in their native language, which is in this case, Spanish.

ROLE OF EDUCATION IN A MULTICULTURAL SOCIETY

In my study I compared the teachers' perception of immigrant students in middle schools. This age group was chosen because it is the time when the pupils are mostly integrated in mainstream classes.

Socialization is one of the primary purposes of school and the educational system of any country. The education system cannot exist in a vacuum and must to some extent reflect the benevolence and malevolence of the society at large (8).

Despite the need for people from diverse backgrounds to get along, intercultural communication is made difficult by widespread discrimination, stereotyping and distrust. For example, most immigrant students from Mexico were embarrassed about where they are from and in Vienna a lot of immigrant students only feel accepted once they receive an Austrian passport.

Scholars have more than a hundred definitions for culture (in general they see it as the way one group perceives or justifies their own behavior). "Since everyone has culture, problems arise when one group refuses to attribute equivalent worth to another group culture" (9).

The concept of culturalism gained strength after the Second World War when each group became relatively autonomous. Assimilation is one of the two positions discussed with reference to the role of the school. In both the U.S. and in Austria, there is still the main goal that the pupils learn the country's primary language as quickly as possible. At the same time, these immigrants are also measured by the same standards as the native speakers. This view of education is the melting pot ideal. The educational systems are constructed so that the pupils are socialized to the "middle class ideal" to participate in society.

In a multicultural education, the emphasis should be put on similarities and differences existing from one culture to the next. Prejudices about certain ethnic groups should be dispelled and pupils should be taught not to judge people based on learned stereotypes. In Vienna, for example, many of the female Muslim students are made fun of for continuing to wear the traditional veil. Teachers in Texas told me that many of the immigrant students from Mexico are called "wet backs" by other pupils, a pejorative term for a recent immigrant who has crossed over the Rio Grande from Mexico.

When students feel respected and start feeling better about themselves, they will also feel better about others. Therefore a great emphasis should be placed on educating teachers about the diversity of their students, and on developing a curriculum that is more relevant for the pupil's cultural background. Teachers should also develop a cognitive flexibility by gradually helping the pupils to find their preferred learning styles. By doing so the gap between theory and practice in multicultural education will be reduced.

Attention needs to be drawn to the necessity of preparing students of all different ethnic groups to live in a culturally pluralistic society. Experts have shown that in a diverse society it is important to accept different life styles and socialization. However, it is also very important that each individual keeps its own ethnic and cultural distinctiveness. Teachers in both countries told me that if they had a multicultural dance festival or song contest at their school, most of the pupils loved to participate, particularly the non-native students because it provides them the opportunity to show part of their own culture. Multicultural education can introduce knowledge about the different ethnic cultures. That can utilize diversity to assist changing stereotypes and prejudices. An understanding of the different historical experiences of each cultural group can be of tremendous value. Systems of education between people of different ethnic backgrounds can also be used to facilitate teachers-students interactions.

More teacher training is needed to give teachers the tools and the skills to discuss issues related to human diversity and multiculturality. Once teachers feel comfortable talking about the different ethnic issues, they can start discussions about the attitude of the dominant, which can widen the students' and teachers' perspective of different cultural, ethnical and gender issues. When teachers in Vienna were asked about the veil they said they were bothered by it, but during the conversation it got clear that a lot of them have not really put a lot of thought into the religious background of the immigrant students. My idea is that sensitive training can help teachers to include multicultural issues in their main topic areas and can help form a critical view of biased material.

Multicultural consciousness should be an important part of the curriculum. Teachers need more training to see how multicultural perspectives can fit into existing systems. In my opinion the simple reflection on different cultural images is not enough, and we must also carefully examine existing stereotypes.

REFERENCES

1. J.P. Spradley. "The Ethnographic Interview," Holt, Rinehart and Winston, New York (1979).
2. S. Aufenanger, Qualitative analysis of semi-structured interviews: a report, in: "Qualitative-Empirical Research," D. Garz & K. Kraimer, eds., Westdeutscher Verlag, Opladen (1991).
3. J.P. Spradley. *Ibid.*
4. J.P. Spradley. *Ibid.*
5. D. Baker and G. Lenhardt, Integration of foreigners: school and state, in: "Koelner Zeitschrift f. Soz. und Sozialpsych," 40:50 (1988).
6. G. Khan-Svik, Research on children and at-risk youth in Austria, in: "Children and youth at risk and urban education," Day, C., Veen, D.V. & Walraven, G., eds., Leuven, Apeldoorn (1997).
7. G. Vernez and A. Abrahamse. "How Immigrants Fare In U.S. Education," RAND, Santa Monica (1996).
8. T. Thomas. "Relationships Between Teacher Philosophy Of &Teacher Behavior In Education That is Multicultural: Study Of An Urban Middle School," University of Texas at Austin (1992).
9. L. Olsen. "Made In America. Immigrant Students In Our Public Schools," The New Press, New York (1997).

CHAOS AND COMPLEXITY - THE WAY TO TRANSFORM ORGANISATIONS READY FOR THE NEXT CENTURY: INSIGHTS FROM A CASE STUDY OF THE OPEN UNIVERSITY

Elizabeth McMillan-Parsons

Centre for Complexity and Change
Open University
Milton Keynes, MK7 6AA, UK.

INTRODUCTION

The new sciences of chaos and complexity have yet to make any real impact on the way we think about organisations which are still dominated by thinking derived from the classical, scientific paradigm. If we are to transform organisations ready for the 21st century then we need to move away from this way of thinking and develop ideas and theories drawn from chaos and complexity. This paper attempts to show some of the ways in which these may be used by referring to a number of writers and researchers who are using them to develop new management theories and transform organisations. Further, I provide evidence from research which I have undertaken on an organisational change programme which took place at the Open University from 1993 to 1996. This provides a case study of the use and adaption of ideas from chaos and complexity, and, in particular, provides insights into the use of self organising principles in a complex, bureaucratic organisation.

A POWERFUL PAST THAT STILL INFLUENCES THE PRESENT

In the 17th and 18th centuries a powerful new view of the world developed from the work of a number of outstanding scientists and thinkers. Galileo and Descartes devised a conceptual framework in which the world was viewed as a perfect machine controlled by mathematical laws (Capra 1996). Newton built upon Galileo's work and 'unleased on the Western mind a clockwork universe' (Kauffman 1996) characterised by 'materialism and reductionism' (Wheatley 1994). Descartes considered human beings as essentially machine like and argued that the mind and body should be treated as separate entities (Capra 1996, Morgan 1986). For some 300 years classical science and the 'tradition of rationalism and logical empiricism... has been the mainspring of Western science and technology' and 'is also regarded, perhaps because of the prestige and success that modern science enjoys, as the very paradigm of what it means to think and be intelligent' (Winograd and Flores 1991). Early writers on organisations like Weber and Taylor drew on classical scientific notions for their ideas and began an approach which is still prevalent in today's organisations (Handy 1990, Morgan 1986). Planning and forecasting activities, for example, still reflect a Newtonian view of the world (Handy 1990) with many data collection and analysis techniques reflecting in Wheatley's (1994) view a reductionist, cause and effect approach. Morgan (1986) reminds us of the large paper processing, factory style offices set up in the 1980s and 1990s to deal with insurance, or banking where the staff are in many ways expected to behave as if they were parts of some machine. How much has really changed?

Synergy Matters: Working with Systems in the 21st Century,
Edited by Castell *et al.*, Kluwer Academic / Plenum Publishers, New York, 1999.

Many organisations claim to be creating new forms and new ways of managing themselves yet the structures they create and their thinking and behaviours are too often rooted in the Newtonian-Cartesian paradigm and many of those attempting to devise new systems for organisations are only able to think mechanistically (Morgan 1986).

WHAT CAN CHAOS AND COMPLEXITY OFFER ORGANISATIONS?

A number of writers are now looking to chaos and complexity for ways of designing and creating effective organisations that will be fit for the next century. These writers, like Stacey, Wheatley, Nonaka, and Merry did not develop their ideas from organisation theories but drew on principles emerging from the new sciences. Nonaka (1988), draws on the work of Ilya Prigogine and advocates the concept of self-renewal derived from self organisation which arises from chaos and dissolution. He suggests that an understanding of chaos science widens the spectrum of options and forces an organisation to seek new points of view.

Stacey (1996) too refers to the work of Prigogine and other scientists when he explores systems dynamics and the nature of self-organisation in organisations. An understanding of complexity provides 'a fundamentally different model through which to interpret business behavior and design innovative management actions' (Stacey 1992). He writes:

'nonlinear dynamics and chaotic behavior apply literally to human business systems; they are not simply an analogy or a metaphor. If managers wish to adopt a scientific approach to understanding their organizations, they must now take account of the far-from-equilibrium behavior of nonlinear feedback systems because organizations are just such systems.'

In Stacey's (1992) view the current 'received wisdom on strategic thinking is primarily an intellectual exercise'. It involves information collection, complex analytical processes and forecasting, all based on the notion that the future behaviours of an organisation's systems are predictable. This, he suggests, is 'a pointless exercise'. In his view complexity offers a uniquely different approach to strategic management 'in which people do not know where they are going over the long term, but through interaction develop, discover, and create a new direction through their self-organising interaction' (Stacey 1996).

Wheatley (1994) believes that nature's principles apply to human organisations because they are made up of webs of complex relationships and in this way mirror the real world. Although associated with colourful images of butterflies and exotic fractal patterning chaos theory reflects real life and the complexity of today's organisations (Berreby 1996). Merry (1995) asserts that all human and social systems are open, far-from-equilibrium systems in a never ending process of change and that a new order self organises and emerges out of internal turbulence.

How do we use ideas or insights derived from the new sciences? Nonaka (1988) bases his ideas on observation of transformation in Japanese firms and proposes that for an organisation to transform itself then it should deliberately generate internal chaos linked to the external environment. Key to this is the flow and creation of information and a self organising movement using self organising teams. Stacey (1996) considers the use of self organisation and self organising principles derived from complexity are key for organisations that wish to continuously transform themselves.

THE NEW DIRECTIONS PROGRAMME AT THE OPEN UNIVERSITY, 1993 - 1996

The Open University is a large and complex organisation of some 3,500 full time staff and some 7,500 associate (part time) lecturers. It has 10 academic units, some 8 other departments or centres and a huge administration sub structure consisting of 10 divisions. There are also 13 regional centres in the UK each headed by a director. The Vice Chancellor is the chief executive and his senior management team consists of five Pro Vice Chancellors. The structure of the University fits very neatly with Handy's (1993) model of the role culture.

In 1993 a major change in the University's funding arrangements and a number of other important external factors meant that the University needed to change both strategically and in the way it managed itself. It produced a strategic action plan, *Plans for Change*, which detailed the actions needed to drive through all the necessary changes. But the role culture organisation finds it difficult to see the need for change and to know how to change itself (Handy 1993). How would the University transform itself?

Plans for Change recognised that if the University was to change then it should work to enable all staff to take a role in the achievement of the University's strategic objectives. It would achieve this through a process of awareness-raising, staff development and management action in which all employees would have an opportunity to participate. Here lies the original raison d'etre for the New Directions programme.

The programme began as a series of consultative workshops on *Plans for Change*. A diagonal slice of staff from different grades and locations were invited to attend by the Pro Vice Chancellor, Strategy. Senior managers were asked to brief their staff on the University's strategic plans before they attended and to follow up with support for ideas afterwards. After the first few workshops a pattern of process emerged which was to become the model for later workshops. An important feature of the workshops was that they were a mechanism for the Pro Vice Chancellor, Strategy to hear the views of a very wide cross section of staff. Thus they set up a listening and communications mechanism outside the existing formal systems.

By early summer the workshops had generated so much interest and enthusiasm that people were volunteering to attend. By the end of 1993 some 144 staff had participated. The ideas from the workshops were fed into senior management meetings and into the formal planning processes. The programme continued in 1994 and focused on the development of key themes which had emerged in year one. The programme was not operating according to some predetermined plan but responded to feedback from the participants and their ideas about what the University should do next.

The 1994 Conference

Early in 1994 it was suggested that a conference was held to pull together many of the emerging themes. New Directions workshop participants were asked to volunteer to form a team which would devise a one day event in May. The team which met for the first time in February 1994 included a clerk, two secretaries, an administrator, an editor, two regional academics and a warehouse manager. All decisions about the conference were left to the team. I acted as facilitator and provided information on themes and outcomes from the workshops. The new team got together and by a process of brainstorming, sharing of tasks and rapid exchange of information using electronic mail, planned the conference. It focused on three main themes: OU achievements to date; experiencing change; and the future. Each theme had five parallel workshops on offer led by OU staff or external speakers. An exhibition showed changes in technology over the University's 25 year history and demonstrated multimedia facilities. The Conference was oversubscribed with 100 delegates from all categories and grades of staff attending. It culminated in the presentation of action plans for the future to the Vice Chancellor, and other senior managers.

During 1994 there were a total of 10 workshops and 2 Lunchtime Briefings as well as the Conference. An estimated 533 staff attendances were recorded. Workshops continued through 1995 and 1996. Between 1993 and 1996 some 27 % of the staff were involved.

The Staff Survey

An immediate outcome of the Conference in 1994 was that delegates were asked to volunteer to work with the Director of Public Relations on the design and delivery of a University Staff Survey. This had been planned two years earlier. Eight volunteers (two secretaries, two administrators, an O&M officer, a senior academic, an editor, a technical manager) and myself as facilitator, formed the team which produced a survey that was distributed to all staff in October. In January 1995 the Survey Team met with the Vice Chancellor and senior managers to hear the preliminary report of the survey and to finalise details on a series of presentations to the staff on the findings. The survey confirmed that the views that had emerged from the workshops and the Conference were those of most of the staff (Russell and Parsons 1996).

USING CHAOS AND SELF ORGANISATION

Data for the case study was collected from 28 staff (including all members of the Conference Team and the Survey Team) via structured interviews, workshops and questionnaires in order to create a rich, detailed picture of the programme and the impression it made on the University. The respondents described New Directions as a 'movement' for change that had evolved spontaneously because it responded to the issues and ideas raised by the staff as they arose and did not follow some predetermined plan.

Stacey (1993) describes eight steps with which to create order out of chaos and promote the conditions where spontaneous self organisation can arise. The New Directions Programme had used five and may have influenced two *others of them as the table below shows.

Stacey's (1993) 8 Steps	New Directions' Activities
Develop New Perspectives on the Meaning of Control. Group learning encourages a self policing form of control. Managers have to let go.	PVC, Strategy and the workshops encouraged managers to take part in group learning & to empower staff & free them from unnecessary controls
Design the Use of Power The application of power over groups inhibits complex learning development	The workshops & the Conference facilitated the development of complex learning by encouraging open questioning and the public testing of issues
Encourage Self Organising Groups	Three self organising groups were set up
Provoke Multiple Cultures - move people around the organisation & bring in new blood	* The Conference identified the need for more IT skilled staff & in 1995 / 96 a new cohort of academic staff with high level IT skills was recruited.
Present Ambiguous Challenges instead of Clear Long Term Objectives or Visions	The workshops & the Conference presented strategic challenges and issues to provoke discussion and debate and the search for new ways of doing things
Expose the Business to Challenging Situations	*Workshops discussed innovative & risky strategies. Did they influence future plans eg. massive NT investment & global expansion plans?
Devote Explicit Attention to Improving Group Learning Skills	The workshops & the Conference created opportunities for managers working in groups to question deeply held beliefs & to develop new mental models
Create Resource Slack	

Self Organising Teams

Further evidence of self organisation at work is provided by the three teams that arose from the programme. They all fit Stacey's (1996) description of self organising groups as ones that arise spontaneously around specific issues, communicate and cooperate about the issues, reach a consensus and give a committed response to the issues. The Conference Team in just 9 weeks (including the Easter break) had organised a very successful conference with 100 delegates that offered 20 workshops with a mix of internal and external speakers and an exhibition. The Staff Survey Team was convened at the beginning of June. It agreed the contents of the survey areas, identified key issues such as confidentiality, helped chose a professional survey organisation after setting up a series of presentations, oversaw a pilot study, then agreed the final survey document, which was printed and distributed to all staff (excluding Associate Lecturers) in October. There was an excellent rate of response, 65%. The Survey Team had worked in very much the same way as the Conference Team - as enthusiastic volunteers and not people who had this as part of a permanent responsibility.

Both teams had been very successful in carrying out a task in a limited time period, working with volunteers drawn from a mixture of categories and grades who would not normally work together. They had voluntarily responded to an issue and set their own ways of working. To further explore the self organising aspects of these two teams I issued all the members a questionnaire based on Stacey's (1996) work on self organising and self managed teams and my own research. My research suggested that self organising teams have a strong sense of shared purpose, strong personal commitment, display creative and spontaneous behaviours, have high levels of energy and enthusiasm, and that an inherent order emerges from their activities. The questionnaire was designed so that the respondents could pick out any features they considered described their particular team. Analysis of the results would show whether the teams fitted detailed definitions of self organising teams or self managed / empowered teams.

The tables below show the actual score (or actual identification with an attribute or feature) as a percentage of the possible total score.

Self Organising	Stacey (1996) Definition	Own Definition	Total Match
1994 Conference Team Match	70%	80%	75.5%
Staff Survey Team Match	46%	63%	54.5%

Self Managed	Stacey (1996) Definition	Own Definition	Total Match
1994 Conference Team Match	5%	34%	16%
Staff Survey Team Match	25%	39%	30%

As the tables show both teams could be described a self organising although the Survey Team also has significant attributes of a self managed or empowered team. This would appear to reflect the nature of the Team's project, its links with the formal organisation of the University (ie. it was a project on behalf of the Staff Policy Committee).

The third team, New Directions Action Group was a ginger group which formed spontaneously in early 1995 in response to the issues that had been thrown up by the Conference and the Staff Survey. It defined its role as a group dedicated to motivate and enable change in the University by encouraging local action and creativity. It would listen to all the staff, keep them informed and empower them to take action. It fits well with Stacey's (1996) description of self organisation as 'a fluid network process in which informal, temporary teams form spontaneously around issues... and decide on who takes part..... and what the boundaries around their activities are.' A key feature of self organisation is that people empower themselves (Stacey 1996) and the new group did just that, for example, it devised and ran a series of workshops designed to challenge the 'blame culture'.

SIGNIFICANT CHANGES

How do you change a large, complex, highly structured, multi-layered organisation like the Open University? The respondents agreed that New Directions had been effective as a programme for organisational change and that it had used a spontaneous, open, self organising approach that directly involved the staff of the OU in the strategic change process. Staff felt enabled to explore ideas for the future and empowered to change their behaviours. The programme did not use the existing formal processes but created the space for self organising groups to form and for complex learning to take place. Further, the workshops stimulated and encouraged the free flow and creation of information which Nonaka (1988) considers necessary for the deliberate generation of chaos. They also matched Wheatley's (1994) description of a mixture of people working together to self organise and weave potent visions of the future.

Additionally, the bringing together of staff from all categories and grades and locations and giving them the opportunity to exchange ideas, to challenge each other and the status quo and to engage in interactive learning experiences created a rich fractal dynamic. This was created by two things: the energy and excitement that many of the respondents associated with the programme and the information which flowed in a variety of feedback loops between individuals in the workshops, between colleagues, and around different levels of management. The New Directions programme had a constant flow of energy and matter (information) and non linear interconnectedness, all characteristics of a self organising system (Capra 1996).

More than 3 years after the event the 1994 Conference was recalled as a successful event that had encouraged staff to explore change issues and to help prepare the OU for change. As a one off event itself it had set a number of important changes in motion. The issues it raised achieved a higher profile and the respondents agreed that there has been substantial progress on many of them. For example, the University set up a Knowledge

Media Institute to research new technologies and now has a Pro Vice Chancellor, Technology Development. There has been real progress towards equalisation of staff terms and conditions and a major restructuring of the University's marketing function. Communications have improved with all staff now having access to email and regular lunchtime briefings on key issues.

New Directions was described as an 'impetus' and 'a catalyst' for change and 'a new broom'. Many of the respondents had been inspired and enthused by the programme, several sufficiently so that they had volunteered to join the two teams and take on new roles and responsibilities. Many were junior staff who felt empowered by the programme. Most considered that they had learnt or discovered something as a result of their participation in New Directions. Many reported that they had changed their thinking and their working behaviours, others spoke of colleagues whose 'mindsets' had been 'shifted'.

A self organising process akin to Nonaka's (1988) model was begun by involving individuals in self organising groups which served as triggers for a continuous change process to begin. Thus what began as an exercise in consultation evolved into an innovative, self sustaining programme for cultural and organisational change which encouraged staff to think creatively and to feed their ideas back into the organisation (Russell and Parsons 1996).

An independent IES study (Tamkin and Barber 1997) reports that one of the outcomes of the New Directions programme has been that the University is changing from an organisation that was resistant to change to one 'where there is a growing awareness of the need to change.' The study also compared New Directions with other change programmes in a range of organisations and noted:

'The OU were the only respondents who spontaneously mentioned that they had learned about their organisation through their involvement in the New Directions programme...the more junior people that we interviewed [commented] that they had learned how to work with different people in pursuing an end... The OU interviewees also showed considerable commonality of response describing an organisation that had become more flexible and less hierarchical, more sharing and open with information.'

There is clear evidence from the different aspects of the New Directions programme of the effectiveness of ideas derived from chaos and complexity, and the use of self organising principles in particular, in effecting change in a Weberian organisation with attitudinal and behavioural roots in the rational, classical scientific paradigm. There have been major changes at the OU, but no-one would deny that there is still much to do.

REFERENCES

Berreby, D., 1996, Between chaos and order: what complexity theory can teach business, *Strategy and Business*, Spring.

Capra, F., 1996, "The Web of Life", HarperCollins, London.

Handy, C., 1990, "The Age of Unreason", Arrow Books, London.

Handy, C., 1993, "Understanding Organizations", Penguin, London.

Kauffman, S. ,1996, "At Home in the Universe: The Search for Laws of Self Organization and Complexity", Penguin, London.

Merry, U. ,1995, "Coping with Uncertainty", Praeger, Westport, CT.

Morgan, G. ,1986, "Images of Organization", Sage, Newbury Park.

Nonaka, I., 1988, Creating organizational order out of chaos: self renewal in Japanese firms, *California Management Review*, Spring.

Russell, C. and Parsons, E. ,1996, Putting theory to the test at the OU, *People Management*, Institute of Personnel and Development, 11th January.

Stacey, R., 1992, "Managing the Unknowable", Jossey-Bass, San Francisco.

Stacey, R., 1993, Strategy as order emerging from chaos, *Long Range Planning*. 26:1.

Stacey, R., 1996, "Strategic Management & Organisational Dynamics", Pitman, London.

Tamkin, P. and Barber, L.,1997, comments on First Draft of Report, Learning to manage, Institute of Employment Studies, University of Sussex. Not yet published.

Wheatley, M., 1994, "Leadership and the New Science", Berret-Koehler, San Francisco.

Winograd, T. and Flores, F., 1991, "Understanding Computers and Cognition", Addison-Wesley, Reading.

A DISCUSSION ON COMPLEMENTARISM AND ITS STANCE TOWARDS PARADIGMS

Andrés Mejía D., Petri Jooste

Lincoln School of Management
Faculty of Business and Management
University of Lincolnshire and Humberside
Lincoln, LN6 7TS

INTRODUCTION

In the last few decades many texts and authors have used and elaborated on the notion of *paradigm*, which was first brought to broad attention in 1962 by Thomas Kuhn in his celebrated book "The Structure of Scientific Revolutions" (Kuhn, 1970). However, the extensive use made of this term seems to have dramatically changed the original meaning Kuhn gave to it, and the variety of adoptions has even led to contradictory conclusions (for a discussion on this as applied to the social sciences, see Bernstein, 1976, part II).

One strand of thinking in management science that has adopted this notion is the one known as *complementarist pluralism* or, in short, *complementarism* (see Jackson, 1997, and Flood and Jackson, 1991). Complementarism is the position that "would respect the different strengths of the various trends in management science, encouraging their theoretical development and suggesting ways in which they can be appropriately fitted to the variety of management problems that arise"[1] (Jackson, 1997, page 2).

Some comments and criticisms on complementarism, as developed by Flood and Jackson, have appeared since it was first introduced. These have opened some issues that still need to be examined carefully, and this paper is an attempt to contribute to this discussion.

In the first section of the paper, we examine the notion of *paradigm* through the consideration of some linguistic elements that are very closely related to it. Making use of these notions, in the following sections we examine two issues on complementarism.

The first of these issues concerns the sense in which it can be said that people can manage different paradigms and be *multiparadigmatic*? Tsoukas has suggested that complementarism is

---

[1] Let us notice that Jackson actually uses the term *pluralism* to refer to that position. Indeed, for him complementarism is only one of the three main aims of critical systems thinking (see Flood and Jackson, 1991). However, acknowledging that there might be other possibilities for pluralism that fit in the term as used widely in other disciplines, we prefer to use the word *complementarism*. For a discussion on this, see Gregory (1996).

not feasible if these various trends in management science that it deals with are regarded as reality-shaping paradigms, because "there is very little that is common between them to allow them to be included in a contingency framework" (Tsoukas, 1993, page 61). As the notion of paradigm incommensurability lies at the heart of the matter, our analysis is based mainly on its examination.

The second issue is related to the status of complementarism and its relation to the paradigms it seeks to deal with. On the one hand Midgley (1995, pages 62-63) suggested that it is a paradigm on its own, based on the idea that the methodology developed by Flood and Jackson was based on a meta-theory with its own assumptions. On the other hand, Jackson has suggested that the idea of a meta-theory that can resolve the incommensurability of paradigms, should be dropped (1997, page 5); and has also argued that complementarism cannot be one paradigm on its own, because "one paradigm pluralism is simply not pluralism".

I. PARADIGMS, ASSUMPTIONS AND VOCABULARIES

The notion of paradigm seems to be an elusive one, even for Kuhn himself, as it is revealed by Masterman's account (1970) of more than twenty different meanings that Kuhn gives to the term, and Bernstein's claims of Kuhn's confusion between theories and paradigms (1991, page 75). The general idea of a framework of interpretation (and action) appears in the literature with several different names and variations in its meaning and context of use. Examples of these names are *framework*, *language*, *conceptual scheme* (see Davidson, 1973-1974), *discourse* (see Burr, 1995), *idiom* (see Lyotard, 1988[2]) and *vocabulary* (see Rorty, 1989).

In the postscript to "The Structure of Scientific Revolution" (written in 1969), Kuhn accepts that he has used the term *paradigm* with at least two different meanings: "On the one hand, it stands for the entire constellation of beliefs, values, techniques, and so on shared by members of a given community. On the other, it denotes one sort of element in that constellation, the concrete puzzle-solutions which, employed as models or examples, can replace existing rules as a basis for the solution of the remaining puzzles of normal science" (1970, page 175).

While it is clear that Kuhn refers only to science, some other authors have adopted the notion broadening its scope to talk about any other domains of action. For example, Guba explicitly says this in the working definition he proposes for a paradigm: "a basic set of beliefs that guides action, whether of the everyday garden variety or action taken in connection with a disciplined inquiry" (1990, page 17). Kuhn's purposes went much further than simply to suggest the existence of paradigms, because he wanted to talk explicitly about the history of science. However, for our discussion we are going to focus on their linguistic characteristics rather than on their belonging to scientific communities or their role in the history of any scientific discipline, and this will allow us to extend its use to other domains.

One first element is the beliefs, or assumptions, that militants in a particular paradigm share, and this is what a critical inquiry might help to reveal. These are propositions that may refer to (i) how things are or behave, or (ii) about how things should be, in which case they would be *normatives*[3]. Concerning the first option, it might be argued that Flood's and Jackson's efforts in the development of their Total Systems Intervention methodology, or TSI, (1991) is devoted to reveal assumptions about the situations that are made in different paradigms, around two variables which are to some extent observable variables: the level of complexity of the system to

---

[2] In his book "The Postmodern Condition: A Report on Knowledge" (1984), Lyotard works mainly with the notion of *language games*, that he takes from Wittgenstein, attaching to them some characteristics Kuhn attributed to paradigms, like incommensurability. In "The Differend" (1988), he has also used the terms *idiom* and *language* with a meaning more similar to what we talk about in this paper.

[3] However, we are not ruling out some other possibilities.

be intervened (which they call the "systems dimension"), and the nature of the way actors' purposes and intentions are accommodated in the system (which they call the "participants dimension"). Regarding the second option, the whole philosophy behind Werner Ulrich's Critical Systems Heuristics (1994) might, at least in some cases, help to reveal and question normative contents of paradigms[4].

However, there is something else besides these assumed propositions that makes paradigms different, and that is their *vocabularies*. Let us now say that a proposition is a linguistic phenomenon that does not happen in *words*, but in *distinctions*[5]. When making a distinction, we divide the world in two: something, and the rest. That *something* then becomes an object in our interpretation and action and, as part of them, it carries some meaning. That distinction becomes, then, constitutive of our action. A set of distinctions that are related to each other forming a coherent whole and in connection to a certain domain of action would then constitute what we might call, adopting Rorty's word, a *vocabulary* (1989). By saying that different paradigms use different vocabularies, we re-interpret what Kuhn claimed when he said, when talking about the comparison between pre and post-revolution normal scientific paradigms, that "old terms, concepts and experiments fall into new relationships one with the other", and that scientists "see different things" (Kuhn, 1970, pages 149-150).

Since propositions are formulated making use of distinction based on a particular paradigm, they can be only true or false within that particular paradigm; in other vocabularies, those propositions simply may not even exist[6]. This does not rule out the possibility that different paradigms may share some distinctions and therefore that the same propositions may exist in different paradigms. Consequently, a proposition cannot be neutral with respect to all possible vocabularies, because it is based on one. Let us pay attention again to the fact that, even if words are the same, different distinctions would make propositions different.

II. INCOMMENSURABILITY

One aspect that seems to be very appealing about Kuhn's ideas is not so much the assertion that there are such things as paradigms, but his claim that paradigms are incommensurable. Following Bernstein (1991), we will argue that it means neither incomparability nor incompatibility, and we will also claim that it does not mean untranslatability either.

In Kuhn's own words, "in applying the term 'incommensurability' to theories, I had intended only to insist that there was no common language within which both could be fully expressed and which could there be used in a point-by-point comparison between them" (1976, page 191). This idea of the common language has been interpreted by Davidson in a particular way when he takes incommensurability as untranslatability. He claims that "we may identify conceptual schemes with languages, then, or better, allowing for the possibility that more than one language may express the same scheme, sets of intertranslatable languages" (1973-1974, page 7). However, we would like to point out that translatability between languages is not exactly what we are talking about in this case. If someone does not make a distinction, we can help him/her to make it, and

---

[4] Ulrich's methodology is intended to reveal the normative content of social designs. However, it can be argued that the core notions of the methodology might be adapted to do it at the level of paradigms. Moreover, when social designs are made based on particular paradigms, by revealing the normative content of designs we would be revealing something of the normative content of the underlying paradigm. In some other cases, resulting social designs might be better represented as the product of an accommodation of interests.

[5] A distinction does not necessarily correspond to a word: for example, one word in Spanish might refer to the same distinction as one in French. Let us notice that our claim argues neither in favour nor against the assertion that not any sentence in any language may be translated into a sentence in any other language. Our use of the term *distinction* is taken from Spencer-Brown (1972).

[6] Note that these propositions are not the same ones that are assumed as true by the *militants* of a paradigm.

perhaps we can do it by using a certain common vocabulary we both share. And what has been done here, although in Davidson's argument might sound like a translation, is a kind of teaching-learning process so that now the learner makes the same distinction the teacher made from the beginning. This means that we can go from one paradigm to another through, in Kuhn's own words, a kind of "Gestalt switch", rather than a translation.

It can also be said that incommensurability does not mean incomparability. It is possible to compare some aspects of two or more paradigms at least in two ways: by means of an external set of distinctions (for example a vocabulary for critique), or by means of some common distinctions between them. An example of the first kind is when we look at two communities of scientists working in different paradigms, and we compare the nature of the experiments they make; doing this from a particular vocabulary that allows us to make distinctions about experiments. An example of the second kind is when two theories based on different vocabularies attempt to predict certain phenomena, and we take accuracy for prediction as our criterion.

Lastly, incommensurability is not the same as total incompatibility either. In our case, incompatibility between paradigms might exist if the assumptions made by two different paradigms are mutually exclusive *over the whole range of space-time*, to put it some way. However, if it is possible to have a kind of Gestalt switch, this would mean that they are not. One possibility for this is a separation of domains of action such that a complementarist would interpret and act from one vocabulary when s/he recognises some domain of action, and from other ones when s/he recognises other domains of action. The very possibility of the same person making use of different vocabularies is what we mean by the term *multiparadigmatic*.

To say that it is possible to be multiparadigmatic does not contradict that humans interpret and act[7] in language, and that therefore we are inseparable from it. However, different authors have used these notions in such a way that whenever they talk about a paradigm (or one of the alternative concepts like the ones shown at the beginning of this paper), we are at the same time talking about some person(s) whose paradigm is being talked about. In a sense, this could be understood as a way of observing or talking about people through the observation of language. Some examples of this way of construing language are Lyotard's work on *differend* (1988), Bruner's exposition of the notion of *culture negotiation* (1986), and Bohm's works on *dialogue* (1996). Complementarism would require that this be not the only possibility.

As part of this, it requires us to be able to learn new paradigms without necessarily having to *unlearn* the old ones[8]. However, a question that arises is the following one: if we are interpreting and acting at every moment in our lives, then how can we be doing this at the same time in different ways? If we create a particular space in which we inquire into a situation from one vocabulary first and then from another one, and so on *one at a time* then it seems to be possible. This is what may happen in an intervention in an organisation. However, if we want that this complementarist practice be adopted as part of everyday interpretation and action in the organisation, then something else is needed, and one possibility is the separation of domains of action mentioned above.

In any of these cases, a vocabulary is needed when it comes to separate domains of action (because distinctions are needed at least to separate those domains) or to choose which paradigm (or method or methodology associated with it) will guide each part of the intervention. This can be done with help from a particular vocabulary for comparison of paradigms.

Even though we think that research into this possibility is very much needed, it seems that in everyday life we seem to change our vocabularies for talking about the same things, and that would reveal that we are already multiparadigmatic to a certain extent. One example of this is

---

[7] We are taking a notion of action that includes a meaning in it; that is, *meaningful action*.

[8] Let us note that if we do not accept that assumption, then we might have a different kind of pluralism that would not be complementarist.

the change in the explanation given by many members of the Colombian society, according to Alonso Salazar's account, to the social phenomenon of the Medellín *sicarios* (youngsters who are hired for murdering people) (see the prologue of Salazar, 1990). The vocabulary used in the mid-eighties in people's explanations of this phenomenon was one that was constituted mainly by mental terms like intentions, will, ambitions, etc. However, after some years this vocabulary was replaced by one that made more use of demographic and economic variables as causally related. This would mean that people learned to interpret and act differently in relation to the same thing[9]. A hint of evidence that suggests that a process of unlearning did not occur is found on the fact that that the first vocabulary of those two is still nowadays central to a large part of the explanations of other similar social phenomena at different levels.

III. THE STATUS OF COMPLEMENTARISM. IS IT A PARADIGM?

We would like to discuss now the status of the complementarist position in relation to the paradigms that it manages when dealing with a situation. In particular, we now turn our attention to the claim that complementarism is neither a meta-paradigm nor another paradigm on its own, standing side to side with the other paradigms (see Jackson, 1997).

According to Jackson a meta-methodology (such as TSI) should be developed "to manage the paradigms not by aspiring to meta-paradigmatic status and allocating them to their respective tasks, but by mediating between the paradigms" (1997, page 6). However, if such a methodology intends to be theoretically well founded, as Jackson has claimed it should be, then it should help its users to make distinctions in methodologies and their corresponding paradigms to assess and compare them, and this requires a vocabulary. One example of such a vocabulary is precisely Flood's and Jackson's TSI, which makes distinctions in terms of complexity and the so called knowledge constitutive human interests. The fact that the critical practice of examination of paradigms is done using a certain vocabulary cannot be escaped because, as Davidson says, "speaking a language is not a trait a man can lose while retaining the power of thought. So there is no chance that anyone can take up a vantage point for comparing conceptual schemes by temporarily shedding his own" (1973-1974, page 7).

Some metaphors used for referring to those elements that we look for when observing paradigms might have led us to think that they might be uniquely given by the paradigms themselves. Jackson (1997) talks about their *constitutive rules* and in Fuenmayor's metaphor it is what is hidden because it is "under our feet" (1990, page 531)[10]. But any report on constitutive rules or what is under our feet is necessarily a report made from some vocabulary.

Now, does this mean that complementarism is a paradigm by itself? Accepting the suggestion that a paradigm has a vocabulary of its own, when put into practice complementarism would have to use one of a number of possible vocabularies, and would take the form of a paradigm.

The vocabulary used would be a set of distinctions about vocabularies or paradigms, and this characteristic might be understood as suggesting that a complementarist paradigm stands at a higher level in relation to the other paradigms, holding a meta-paradigmatic status. However, let us notice that this simply means that the domain of action on which interpretation/action takes place is *paradigms* as such. This conclusion would suggest that, in spite of the intention of not having the need for something that resembles in any way what Lyotard has called a *meta-narrative* (grand récit) (1984), complementarism in practice requires a vocabulary to interpret

---

[9] When we use the expression *same thing* we do not mean something which is that particular thing outside of our language, but something that is understood in a broader vocabulary (in terms of an agreement by a larger number of people) as the same. In our example, this would be the existence of people who are hired for murdering others.

[10] Again, let us notice that Fuenmayor does not talk about paradigms. However, some of the *objects* he does talk about share some of the characteristics that we have found in paradigms (see section 1).

and act on paradigms. However, what we suggest here is precisely that one such paradigm does not have an absolute character to "allocate [paradigms] to their respective tasks", but a relative one to talk about them. Besides, it is also possible to look at this type of paradigm not as standing on higher ground, but as generating a circular relation in language; one that allows us to talk, using a particular vocabulary, about vocabularies including itself.

IV. CONCLUSION

We have argued that complementarism seems to be a viable option inside the more general position of pluralism, although more research is needed into the possibility for people to be multiparadigmatic. We have done this through an examination of the notion of paradigm incommensurability, and of he assumption that we can talk about vocabularies not directly tied to people. Besides, we have argued that complementarism is a position that needs to make use of vocabularies and therefore be a paradigm on its own when put into practice, and that these paradigms, having as domain of action paradigms as such, may be in a relation that is circular.

Complementarism should recognise its paradigmatic nature and give up its pretensions of being outside all languages and paradigms, and continue in the path it already started when adopted an interpretivist position on the different paradigms in management science.

REFERENCES

Bernstein, R.J. (1976). "The Restructuring of Social and Political Theory". Basil Blackwell, Oxford.
Bernstein, R.J. (1991). "The New Constellation". Polity Press, Cambridge.
Bohm, D., Factor, D. (editor). (1996). "Unfolding Meaning: A Weekend of Dialogue with David Bohm". Routledge, London.
Bruner, J. (1986). "Actual Minds, Possible Worlds". Harvard University Press, Cambridge.
Burr, V. (1995). "An Introduction to Social Constructionism". Routledge, London.
Davidson, D. (1973-1974). On the very idea of a conceptual scheme. *In:* "Proceedings and Addresses of the American Philosophical Association 47" 1973-1974.
Flood, R.L., and Jackson, M.C. (1991). "Creative Problem Solving. Total Systems Intervention". John Wiley & Sons, Chichester.
Fuenmayor, R. (1990). Systems thinking and critique. I. What is critique? *In:* "Systems Practice", Vol.3 No.6.
Gregory, W. (1996). Discordant pluralism: a new strategy for critical systems thinking. *In:* "Systems Practice", Vol.9, No.6.
Guba, E. (1990). The alternative paradigm dialog. *In:* Guba, E. (editor), "The Paradigm Dialog". Sage Publications, Newbury Park.
Jackson, M.C. (1997). Towards coherent pluralism in management science. Working Paper No.16, Working Paper Series, Lincoln School of Management, Lincoln.
Kuhn, T. (1970). "The Structure of Scientific Revolutions". University of Chicago Press, Chicago.
Kuhn, T. (1976). Theory-change as structure-change: comments on the Sneed formalism. *In:* "Erkenntnis", vol.10.
Lyotard, J.F. (1984). "The Postmodern Condition: A Report on Knowledge". Manchester University Press, Manchester.
Lyotard, J.F. (1988). "The Differend. Phrases in Dispute". Manchester University Press, Manchester.
Masterman, M. (1970). The nature of a paradigm. *In:* Lakatos, I. and Musgrave, A., "Criticism and the Growth of Knowledge". Cambridge University Press, Cambridge.
Midgley, G. (1995). What is this thing called Critical Systems Thinking? *In:* Ellis, K. et al., "Critical Issues in Systems Theory and Practice". Plenum, New York.
Rorty, R. (1989). "Contingency, Irony and Solidarity". Cambridge University Press, Cambridge.
Salazar, A. (1990). "No Nacimos Pa´ Semilla". CINEP, Medellín.
Spencer-Brown, G. (1972). "Laws of Form". Dutton, New York.
Tsoukas, H. (1993). The road to emancipation is through organizational development: a critical evaluation of Total Systems Intervention. *In:* "Systems Practice", Vol.6, No.1.
Ulrich, W. (1994). "Critical Heuristics of Social Planning. A New Approach to Practical Philosophy". John Wiley & Sons, Chichester.

A SYSTEMS VIEW OF INTELLECTUAL CAPITAL

Gary S. Metcalf

Interconnections, LLC
2979 Terrace Lane
Ashland, KY 41102
U.S.A.

INTRODUCTION

The concept of intellectual capital seems to have arisen out of an awareness that the very nature of work in advanced economies has changed. Where production was once a matter of human labor, augmented by animals and later by machines, it is moving into a world of thoughts and ideas. This shift has been described as a move from an Industrial Age at the end of the last century, to an Age of Technology, and now to an Age of Information. Where wealth was once tied to ownership of physical property or other tangible assets, today information, knowledge and ideas, and even the capacity to potentially produce these, seem to be the type of intangible assets which distinguish the differences in valuation between organizations (Handy, 1990).

Drucker (1966/1993) used the term "knowledge worker" over three decades ago to describe this change in the nature of work. As he pointed out, "Modern society is a society of large organized institutions. In every one of them, including the armed services, the center of gravity has shifted to the knowledge worker, the man who puts to work what he has between his ears rather than the brawn of his muscles or the skill of his hands" (p. 3).

ACCOUNTING FOR INTELLECTUAL CAPITAL

Every organization which seeks to produce anything, e.g., information, services, or products, utilizes the intelligence(s) of its members in some way. The concept of an information-based economy is predicated on the idea of information, itself, as a commodity. Knowledge-work may be sold in the form of a computer software program, or a book or a speech, or it may be a solution to a specific problem, be it a new organizational design or the right way to handle a difficult employee problem. From a strictly financial perspective, the novelty of the idea of intellectual capital is the recognition and identification of knowledge as something of tangible value to an organization – something that can be captured on the balance sheets and used to leverage financial investments.

Much of the attention more recently focused on the topic of intellectual capital has been on attempts to "account" for the value of knowledge. This is most frequently attributed to efforts coming out of Sweden, a country with a long tradition of social responsibility and collaboration. Sveiby (1997) traces this back to efforts that emerged in the mid 1980's, which developed into what he refers to as the Swedish Community of Practice.

Apparently, much of the initial concern of the companies involved in the Swedish Community was over the high rates of return being required by investors, due to the lack of "tangible assets" owned by financial services companies. This led to their efforts to more clearly value "intangible assets", so that they could compete more evenly for investment capital with the traditional, industrial companies.

Variations in the definitions and concepts regarding intellectual capital often have to do with the particular aspects of the issue that an author is trying to capture. The range of concepts within intellectual capital is extremely broad. Some authors or organizations are primarily interested in identifying and harnessing the intellectual capacity of employees (often in terms of creativity, product innovation, "tacit knowledge" of organizational issues or customer needs, etc.) Others, such as Skandia AFS, have focused on finding ways to value such assets, from an accounting perspective. Within the realm of organizational learning, the emphasis has been on understanding the ways in which intelligence can become an institutional, rather than an individual, process, and thereby be leveraged for competitive advantage.

Intellectual capital obviously comes from people, in some way. But not all persons in an organization are necessarily recognized as a resource. The majority of writers on the topic of intellectual capital seem to take an elitist viewpoint. Stewart (1997) describes employees according to something of a "scarcity of resources" model.

DIFFICULT TO REPLACE, LOW VALUE ADDED	DIFFICULT TO REPLACE, HIGH VALUE ADDED
EASY TO REPLACE, LOW VALUE ADDED	EASY TO REPLACE, HIGH VALUE ADDED

Figure 1. Value of employees according to Stewart's model. (From Stewart, 1997, p. 90.)

In the lower-left quadrant are those employees he describes as unskilled or semi-skilled labor; those who require little training and are easy to replace. The upper-left quadrant includes both skilled labor, such as factory workers, experienced secretaries, and corporate staff groups like accountants, auditors, etc. These Stewart describes as being "hard to replace, and doing important work, but it's not work customers care about" (p. 90). Those captured in the lower-right quadrant do work which is valued by customers, but which is not dependent upon them as individuals. According to Stewart (1997), "A company's human capital is in the upper-right quadrant, embodied in the people whose talent and experience create the products and services that are the reason customers come to it and not to a competitor. That's an asset. The rest – the other three quadrants – is merely labor cost" (p. 91).

An essential issue within the arena of human capital is the question of intellectual property rights. As noted in Skandia's definition, individual intellect or creativity is not something which can be owned by a company. Yet it has long been the legal standard (at least in the U.S.) that any work produced by an employee within the scope of his or her workplace responsibilities, or on paid company time, was the property of the company, unless the company chose to share it otherwise. This was true whether the product was a routine part of the employee's responsibilities, or resulted in a major scientific or economic breakthrough (e.g., a patent for a new vaccine.)

In a very simplified fashion, then, intellectual capital, from a "production" standpoint, might be described as shown in Figure 2. Basic knowledge or information is used to produce both "human capital" (increased knowledge or capabilities of employees) and "structural knowledge" (such as new-product ideas or patents, i.e., knowledge owned by the company), which are used to produce products and services to be sold. Customer capital is produced in the form of information about customer preferences and desires, as well as "intangible assets" such as customer loyalty, which become more information that the company can use.

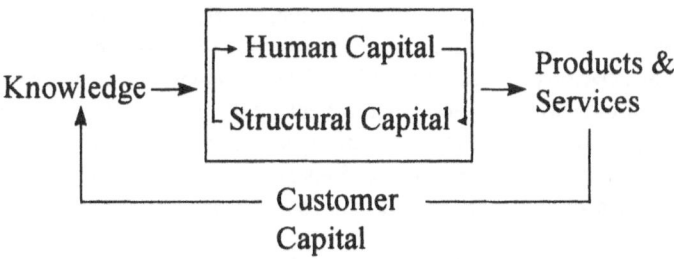

Figure 2. A cybernetic systems model of an intellectual capital process.

KNOWLEDGE AS SOCIAL AND SYSTEMIC

Viewed as noted above, knowledge is something which can be found, and "mined" or "harvested", then used as other natural resources by industries. But there are a number of fundamental challenges to this notion. In an over-simplified fashion, some of these can be summarized as follows.

In a post-modernist world, knowledge itself is a social construction, not a set of objective facts. According to Maturana & Varela (1987/1992), the nervous system is a structurally closed system which picks up sensations, turns them into "signals" and translates them into information. It is through language that we create the social world of humans, and neither language nor knowledge can be separated from action. "It is by languaging that the act of knowing, in the behavioral coordination which is language, brings forth a world" (p. 234). As noted by Gergen (1997) "Meaningful language is the product of social interdependence" (p. viii).

According to Maturana and Varela, a key characteristic of living organisms is that they are self-reproducing (*atuopoietic*) systems, at least on a biological level. Luhmann takes this notion a step further, proposing that social systems are in fact autopoietic systems, in which communication is the essential element (that of which the system is made.) As he points out, "Psychic and social systems have evolved together...This coevolution has led to a common achievement [which we call] 'meaning'" (1995, p. 59). Similarly, Lave & Wenger (1991), note that "learning, thinking, and knowing are relations among people in activity in, with, and arising from the socially and culturally constructed world" (p. 50).

In summary, from a variety of emerging perspectives (each of which deserve significantly more explanation than space here will allow), learning and knowing cannot be separated from the social contexts in which they exist. "Reality" is essentially the current, collective understanding which we help to perpetuate through our on-going participation in societies.

According to Perrow (1972), "A rational-legal bureaucracy is based on rational principles (rational in terms of management's interests, not necessarily the worker's), is backed by legal sanctions, and exists in a legal framework" (p. 3). It is this legal framework which establishes the current concepts of ownership and control within capitalistic systems, and which supports their enforcement through judicial systems. For Luhmann, the legal system is operationally closed, and capable of responding to issues only within its own frame of reference (e.g., based on fundamental notions of legal/illegal.) So, also, is the economic system, which as described by Luhmann (1995), "has its unity in money....The elemental autopoietic process, the ultimate communication that composes the system, the one that cannot be broken down any further, is payment" (p. 461).

Despite this operational closure social systems do influence each other, through a process of "interpenetration". In this way, different social systems become environments for each other, creating "perturbations" to which the system in question responds. All interconnected social systems, then, mutually evolve, not through a process of causation but through interpenetration or interdependency, in which each system continues to reproduce itself, adapting as necessary to what it perceives from its environment, and creating changes in the environment as it does so. Economic organizations, such as corporations, might then be defined as entities whose functions are based upon payments (or exchanges or transactions), but whose boundaries and relations are fundamentally of a legal nature. (The "elements" are those of the economic system, but the environment in which it exists is bounded and defined in the legal system.)

According to Perrow (1972), the bureaucratic structure is the essential form of all modern corporate organizations, and its primary goal is efficiency. One of the early hallmarks of bureaucracy was the scientific management movement, initiated by Frederick Taylor. As described by Perrow, within this process "the owners 'expropriated' the craft skills and craft system and designated as company property the ingenuity, experience, and creativity of the workers" (p. 19)[1]. One view of the current intellectual capital movement would be that this same process is being perpetuated, simply around a different skill-set than in prior times.

Based on the notion of autopoietic social systems, such a process would not only be "natural", but inevitable. An operationally closed, self-perpetuating system would not seek to change itself. It would continue to recreate its current structure and processes (for example, minimizing its own costs and maximizing its own profits), unless perturbed by something in the environment which triggered an internal reaction resulting in change. Therefore, "exploiting" the resources available for use in economic transactions, whatever they may be, is a natural process of such systems.

It is the emergence of the information-technology field as a new system which has brought the most dramatic changes to all social systems in recent times. Like the legal and economic systems, it is operationally closed, but its presence and activities have dramatically affected the environments in which all social systems exist. The "nature" of this system seems even less clear than others, due to its recent emergence. Somewhat like the military-industrial "complex" which developed before, its roots are in the scientific community, but its impact may be even more dramatic. Like the military uses of scientific research, its implications have direct social relevance. (The social world is in many ways a different place, by virtue of almost-instantaneous global communication capabilities.) Unlike the military developments, though, it power comes from increased participation, not limited control.

It is in this broad and changing environment that the issue of intellectual capital has arisen. In a world dramatically altered by the emergence of the information-technology system, how is the economic system to survive? How will it continue to perpetuate its

---

[1] It should be noted that not all management theorists view Taylor's intention or outcomes as exploitative, based on the current environment at the time of his work. See, for example, Weisbord (1987, p. 60).

activities of monetary exchange, based (in capitalism) within structures of limited ownership and control?

The answer is likely that it will have to do so differently than in the past. Just as changes from agrarian to an industrial economic systems brought changes in the ways that resources were managed and exchanged, so too will changes to a knowledge-based economy bring different needs for resources. Agriculture required fertile land, water, and hard work. Industry required large capital investments for equipment and facilities, and coordinated human labor. Knowledge work seems to require education and creativity (as examples.) If Peter Drucker is correct, though, it will only be knowledge put to use which is of value.

The key question would seem to be the way in which the economic system as we know it will change, especially as bounded within the legal structure. If we are able to observe the changes as they happen, we may at the same time learn much about the nature and evolution of social systems, as a whole.

Traditional economic organizations are likely to continue seeing all resources as they always have – for as long as they can; something to be bought at a wholesale price and resold at a higher one, into a market based on increased consumption. Even the most innovative must still function within the current market environments in order to survive. Some obvious changes have begun taking place, though, such as broadening the sharing of ownership (typically through stock) of the company and its profits. As this new world of work continues to evolve, it is likely that it will require a new form of economic relationship which we have yet to envision.

A CRITICAL VIEW OF INTELLECTUAL CAPITAL

According to Jackson (1991), critical systems thinking embraces five major commitments: critical awareness, social awareness, dedication to human emancipation, the complimentary and informed use of systems methodologies in practice, and the complimentary and informed development of all varieties of systems approaches. Jackson's own work in critical systems, as well as that of the "soft systems" approaches, support the basic notion of the social construction of reality. As stated by Jackson, "Social systems are the consequence, intended or otherwise, of the action and interaction of human beings" (p. 120).

Where critical systems thinking would differ from other systems approaches is in its espoused "value neutrality". Based on the philosophy of Habermas, and on Marx's critique of the political economy, critical systems theory seems to demand a stated position regarding issues of power and equality in social systems. By doing so, it also seems to take the position that the structure and functioning of social systems are things which can be consciously and purposefully changed. In its own way, it focuses on the essential role of communication in social systems, but primarily as an aspect of power and control.

Interestingly, both critical systems and Gergen's work in social construction reference the work of Michel Foucault in the use of language as a modern source of social power. Perrow's view of bureaucracy seems to support the need for such critique. As he states: "Bureaucracy is a tool, a social tool that legitimizes control of the many by the few, despite the formal apparatus of democracy, and this control has generated unregulated and unperceived social power" (1972, p. 5).

Current bureaucratic structures would not seem to be effective systems for dealing with the burgeoning Age of Information, from either an ethical or a practical standpoint. What is most unclear, though, is the way in which the new social systems which will take their place will evolve. For some systems thinkers, this can only be through an "evolutionary" process. Others envision a world in which they can be consciously and purposefully designed, and whose design should include all those involved with and affected by them (see Banathy, 1996).

SUMMARY

The intellectual capital movement, as it currently exists, generally seeks to harness human ideas the way that industry of the past harnessed physical energy. Given the nature of the economic system, this should not be unexpected. This is based, though, on a very mechanistic view of reality. What seems to lie outside the awareness of this movement is any conscious understanding of the fundamental nature of the economy as part of broader social systems, and the way in which knowledge and ideas, and their communication, are likely to reshape the economic system and its organizations, rather than become a tool of them. What is also in question at this point is the degree to which, in the presence of such awareness, new social systems can be consciously and purposefully created, in a way that minimizes exploitation and maximizes the dignity and value of each individual.

REFERENCES

Banathy, B., 1996, Designing Social Systems in a Changing World. Plenum Press, New York.

Drucker, P., 1966/1993, The Effective Executive. HarperBusiness, New York.

Gergen, K. J., 1997, Realities and Relationships: Soundings in Social Construction. Harvard University Press, Cambridge, MA.

Handy, C., 1990, The Age of Unreason. Harvard Business School Press, Boston.

Jackson, M. C., 1991, Social systems theory and practice: The need for a critical approach, in: R. L. Flood & M. C. Jackson, eds., Critical Systems Thinking: Directed Readings. John Wiley & Sons, Sussex, England.

Lave, J., & Wenger, E., 1991, Situated Learning: Legitimate Peripheral Participation. Cambridge University Press, New York.

Luhmann, N, 1995, Social Systems. Stanford University Press , Stanford, CA.

Maturana, H. R., & Varela, F. J., 1987/1992, The Tree of Knowledge: The Biological Roots of Human Understanding. Shambhala Publications, Boston.

Perrow, C., 1972, Complex Organizations: A Critical Essay. (3rd ed.). Random House, New York.

Stewart, T. A., 1997, Intellectual Capital: The New Wealth of Nations. Doubleday, New York.

Sveiby, K., 1997, The Swedish community of practice. Available: http://www.sveiby.com.au/CompaniestoLearnFrom.html [1997, June 29].

APPLYING MULLER-MERBACH'S FRAMEWORK TO SYSTEM ACCEPTANCE

Danita Morrison

Dept. of Management Systems and Information
City University Business School
Northampton Square
London EC1

INTRODUCTION

A significant number of IT systems fail to meet expectations after hand-over to the customer. The reason for such a high proportion of failure is unclear. Some studies (e.g. Lyytinen 1988) indicate that the majority are due to 'expectation failure'; the IT system delivered does not match the expectations of the customer. Historically, user input has been passive and infrequent whilst much effort was applied to the analysis of current processes and building of business cases. This discrepancy may therefore be due in part to the concentration on data and processing to the detriment of equally important social and organisational issues.

Beynon-Davies (1989) has identified other problems with conventional approaches, such as low productivity gains from automation, system maintenance issues and large amounts of documentation that have to be produced and maintained. It has also been difficult to scale traditional approaches to smaller projects; customisation becomes informal and quality is subsequently lost.

Martin (1984) argues that a shift is occurring from traditional, relatively large IT system development using a waterfall life-cycle to rapid application procurement techniques on shorter, 'fatter' projects that deliver immediate business benefit. Crinnion (1991) adds that once it is accepted that the conventional project life-cycle is no longer essential to the development process, then true evolutionary approaches can be considered.

WHAT IS RAPID APPLICATION DEVELOPMENT (RAD) ?

Although RAD is a generic term, its main theme is an approach based on the acceptance of changing business requirements and on customer ownership of the product. In essence, RAD is an approach which attempts to remove 'expectation failure', and shifts emphasis away from just data and processing to the consideration of organisational complexity.

There are proprietary methods of RAD such as James Martin's Information Engineering/Rapid Applications Development (IE/RAD) and EASE (Evolutionary Approach to Systems Engineering) which is a marketed version of the 'Systemscraft' methodology developed at City University, London. A non-proprietary example of RAD is Dynamic Systems Development Method (DSDM). All the methods are characterised by firstly having significant user involvement where the customer contributes a large part of the development effort and ensures the delivered system addresses the business need at the time of delivery. Secondly, the development is incremental, so that early delivery of the core functionality occurs, followed by evolution to the final product. The third trait is the use of prototyping to elicit needs and to design or procure the final operational system that will be accepted and will improve organisational performance.

WHY HAVE A SYSTEM ACCEPTANCE FRAMEWORK?

A systems acceptance framework is needed for customers to guide the assessment, approval and receipt of systems. This must interleave with the conventional project development methodologies which are primarily used to find technical solutions to the customer requirements and are driven by technology (Stowell and West, 1994). More importantly, the systems acceptance framework must map onto dynamic systems development methods, where the functional model as well as the design and build stages are iterative and driven by the business needs.

The customer has responsibility for the management of an effective analysis and evaluation of the business change process that has been commissioned and of which the IT system is integral. The evaluation carried out should identify any flaws or issues for rectification, but more importantly confirm a fit to requirements and readiness for implementation. This is dealt with by improving the way the assessment is managed by deliberately shifting from an ad-hoc to a planned approach. The customer is tasked with adopting a proactive approach that matches the rapid application development. For the evaluation task to have a structure, they must be both incremental and iterative so that new 'expectation failures' do not arise.

The evaluation exercise is a 'soft', unstructured problem situation with significant uncertainty. The lack of structure is due to the number of dimensions that the assessor has to consider; these include quality, functionality, integration, cost, the user interface, and business implications. The more rapid and iterative the product development, the more difficult the setting of the evaluation criteria.

WHY APPLY SYSTEMS CONCEPTS?

The broad systems approach has been found to enable contemplation and study of a wide variety of problems and situations (Checkland, 1993). The problem solving activity includes exploring and defining the problem situation, the evaluation of alternative solutions, and then taking action to alleviate the problem situation.

It is the fact that evaluation and acceptance is a 'soft' problem situation that systems concepts are applicable. Wilson (1996) suggests that 'Hard', structured problems, prompt the question, 'How'; we know what is required, the problem is how to do it. With 'soft', unstructured problems, there is more difficulty articulating the precise nature of the situation. This leads to a mix of 'What' and 'How' questions. A framework developed must help put structure to the problem situation, removing the 'What' and leaving just the 'How'. Consequently, a framework developed which uses systems concepts to provide structure to evaluation problems is described.

MULLER-MERBACH

Muller-Merbach (1994) suggests a typology of four approaches to systems thinking: Contemplation, which he refers to as systemic (implying holism and emergence) and three systematic (ordered and organised) approaches; Introspection, Construction and Extraspection. The approaches are characterised by different philosophies and methodologies and the discussion of each is dramatised and amplified through the dialogue between master and a scholar.

Building on the metaphor of the dialogue between the master and the scholar, a suggestion is made as to how each approach might be visualised in terms of a 'hard' problem situation and also how each might be applied to the 'soft' problem situation of evaluation in the RAD context. It refers to the four ways and hence takes a pluralist view as to how a customer might treat the acceptance requirement. A pluralist view is important because in the organisation, there may be conflicting interests. In the absence of a common set of goals, a coalition has to be negotiated and accepted (Flood and Jackson,1993).

The intention is to get the user interested in formalising the acceptance task through the use of systems concepts. Examples are provided as a base on which an acceptance framework can be developed with a common foundation. Most importantly, the approach should put evaluation issues into context.

CONTEMPLATION

To paraphrase the master's instruction, look at the system as an indivisible whole. The emergent whole is greater than the sum of the parts. Blend in, become one and meditate on this. All things are intrinsically connected and if separated would be destroyed. Look at the change process from the point of view of those greatly impacted by it. Empathise with their views, attitudes, cognition, sensitivity and physical aspects of the job. Model the situation as if in it and consider what factors represent success for the organisation. In this way, the customer gains knowledge about the system to be assessed by the process of holistic meditation.

To illustrate this outcome, consider the 'Hard' example of a fan watching a football match. Imagine how the individual is caught up in the atmosphere of the game. The fan shares the triumphs and defeats of the team. There are inseparable aspects to this system such the grounds, costumes, chanting, commentary and the players. They all come together to define the whole.

Building on this approach, in a RAD environment, use holistic meditation to study the feasibility of the business system change. The customer takes on responsibility for the 'big picture' and needs to have a high-level terms of reference in order to create a product definition for the organisation's benefit. As in the football game consider the 'grounds', here perhaps the technology opportunities , how the product might be arranged, that is, the user interface and what responses and reactions are given by the key stake-holders including any external suppliers.

INTROSPECTION

The master's instruction follows that the scholar should break down the things that are being investigated into parts, then study them and their inter-connectivity. See the system as a set of integrated sub-systems. Each sub-system is a major component that needs particular investigation and understanding in order to achieve the 'big picture'. Therefore the customer gains knowledge, assesses and evaluates the system by the process of introspective analytical reduction.

To illustrate this outcome, consider the 'Hard' example of a bicycle. Look at how the rider is able to give it a value and how momentum is achieved. Examine its attributes and how they make the bicycle fit for purpose. Separate the rider , making the rider distant and objective. The rider can dissect this system into its major parts such as: crank, pulley and binder bolts, tubes, gear, wheels, pedals, chain ring, frame, chain, derailleur, saddle, handlebar, sprockets and brakes. Some parts may be further dissected, for example, the wheel may have spokes, wheel trim, tyre, and hub. See the purpose each part serves, how they could be optimised and what the difficulties are. Consider what trimmings could be added for ease of use, for example, a padded saddle.

Building on this approach in the RAD environment, look at the business systems impact on the organisation. Split the organisation down to departments and sections. Study detailed interactions of processes, activities and resources. Model how information should enter and leave, how it is passed around, what changes it and where the information is stored. Take a product-based approach and consider the impact on the data mart. Use techniques such as Context Diagrams to partition and Data Flow Diagrams to further analyse and record the data inputs and outputs expected

Use analytical reduction during the functional model iteration. Examine usability in more detail by investigating aspects of the user interface such as colour and sound which do not affect functionality but will make the product easier to use.

CONSTRUCTION

The scholar is instructed by the master to dissect the system being examined into parts, then reassemble the parts into their meaningful context. Consider the background and context of each part in terms of internal and external influences. Examine in greater depth where discrepancy may occur, the feedback mechanism and coping strategies. Each system has components or   sub-systems whilst being part of more complex systems. Therefore the customer gains knowledge and verifies the system as a whole by creative design.

To illustrate the outcome, let us consider the 'Hard' example of the construction of a sculpture. The sculpture can be made from a block of solid material such as wood, marble or stone. The sculptor will need to select tools according to the building material. Steel points, gouges, claws and chisels are used with a mallet for shaping. Modelling can be used to experiment with the desired shapes using malleable materials such as wax, plaster and clay. For intricate models, the sculptor may need to embed an armature for internal support. The model may harden naturally or need firing in a kiln. If the model is cast in material such as bronze, it may be used to create replicas. The sculptor must also consider the 'big picture', the requirements and constraints of this activity.

Building on this approach in RAD environment, use creative design for the design and build iteration of the project. See how the IT prototype will fit into the business change process. Employ causal loops and re-visit Context Diagrams and ponder on the internal and external influences on the model. Whilst the prototype is designed using appropriate tools consider how these components may be verified and validated to confirm fitness for purpose.

EXTRASPECTION

The master's teaching suggests that the scholar should continue to put the parts into their meaningful context, building up parts into a distinct whole. All systems have components or sub-systems whilst being part of more complex systems. The integration is a mix of different insights and disciplines that particularly expose the external influences and the fit for purpose. Therefore the customer gains knowledge and validates the system as a whole through the process of synthetic integration.

To illustrate this outcome, let us consider the 'Hard' example of germination. The growth of the seed into a seedling, starts when the seed becomes active below ground and ends when the foliage leaves appear above ground. Germination only occurs if the seed has enough water, warmth, oxygen and in some instances, light. In the first stages, the seed takes in water and the embryo starts to use its food store. The radicle swells and breaks through the testa to grow downwards to form the root. The pumule is pushed or pulled upwards depending on the type of seed to create the first foliage leaves. Germination can be greatly impacted by other external but related systems such as flooding, world trade and agricultural policies.

To build on this approach in the RAD environment, use synthetic integration for the implementation of the final prototype. Assess aspects such as performance and capacity, ensuring the system can handle full loads. Define the acceptance or de-commit criteria so that the operational product can be sustainable given external influences.

CONCLUSION

The current approach to evaluation is not satisfactory. Traditional development methodologies partition and stage user acceptance as a tightly coupled activity at the end of the system development. In reality, social and organisational issues which impact project success are identified by evaluating 'soft' issues throughout the product development.

Waring (1996) suggests that the 'Hard' examples concentrates on how to solve apparently well-defined problems, while the 'Soft examples are about learning how to cope better with ill-defined problems concerning human relations. Morgan (1986) states that critical evaluation requires the exploration of competing explanations and arriving at judgements as to how they might fit together.

A systems acceptance framework can be derived through the application of systems concepts to the 'soft' problem situation of evaluation. The Muller-Merbach classification seeks to simplify this into four avenues, that encourages reflection, wider considerations and wholeness, as well as facilitating analytical partitioning. This approach elicits details at all levels for assessment, from evaluation activity about the organisation through to validating data inputs, processes and outputs.

Bronte-Stewart (1997) rightly suggests that the approaches can be thought of as reflecting stages in the education of those involved in the conventional IT systems development methodologies, particularly the move from larger centralised data processing problem solving to the more flexible user-centred approaches of application development. This can now also apply to setting the framework for assessment and system acceptance.

The 'What' must be evaluated has been derived using the Muller-Merbach typology so that only the question of 'How' remains.

REFERENCES

Avison, D., and Wood-Harper, A., 1990, "Multiview: an Exploration in Information Systems Development", Blackwell Scientific Publications.

Beynon-Davies, P., 1989, " Information Systems Development", MacMillan.

Booher, H., 1990, " MANPRINT: an Approach to Systems Integration", Van Nostrand Reinhold.

Bronte-Stewart, M., 1997, "Using diagrams and muller-merbach's framework to teach systems theory", *Systemist*, 19:3.

Checkland, P., 1993, " Systems Thinking, Systems Practice", Wiley.

Crinnion, J., 1991, "Evolutionary Systems Development", Pitman.

Evans, J. (Editor), 1994, " Ultimate Visual Dictionary", Dorling Kindersley.

Flood, R. and Carson, E., 1993, "Dealing with Complexity: An Introduction to the Theory and Application of Systems Science - 2nd edition", Plenum Press.

Flood, R. and Jackson, M., 1993 "Creative Problem Solving", John Wiley.

Humphrey, W. S., 1989, "Managing the Software Process", Addison-Wesley.

Lewis, R. O., 1992, "Independent Verification and Validation", John Wiley.

Lyytinen, K.,1988, "Stakeholders, information systems failures and soft systems methodology", *Journal of Applied Systems Analysis*,15, 61-81.

Martin, J., 1984, "An Information Systems Manifesto", Prentice-Hall.

Morgan, G., 1986, " Images of the Organisation", Sage Publications, 321-337.

Muller-Merbach, H., 1994, "A system of systems approaches", *Interfaces*, 24:4.

Patching, D., 1990, "Practical Soft Systems Analysis", Pitman Publishing.

Senge, P., 1990, "The Fifth Discipline: the Art & Practice of the Learning Organisation", Century Business.

Stowell, F. A. & West, D., 1994, "Client-Led Design: A Systematic Approach to Information Systems Definition", McGraw-Hill.

Waring, A., 1996, "Practical Systems Thinking", International Thomson Business Press, 220-256.

Warr, P. B., 1971, "Psychology at Work ", Penguin Books.

Wilson, B., 1996, "Systems: Concepts, Methodologies and Applications", Wiley.

CO-CONSTRUCTING THE METHODOLOGY:
LEARNING FROM PRACTICE WITH 'WIDE BAND' GDSS

Paulo Nunes de Abreu

Management School
Lancaster University
Lancaster, LA1 4YX

INTRODUCTION

The word 'groupware' brings together the concepts of a group of people and 'ware'. The term 'ware' is often described in dictionaries as 'manufactured articles, products of art or craft', 'an article of merchandise' or 'an intangible item (as a service) that is a marketable commodity'. Hence 'groupware' can mean tools, equipment or services to be used by groups. If we acknowledge this broad definition, then we probably could think of almost anything that is used to help groups to collaborate to be group ware, including writing boards, bulletin boards, slide or overview projectors as well as group process techniques, such as brainstorming or Synectics (Rickards, 1974). All these artefacts physical or abstract can be used to enhance or support group activity. However the designation 'groupware' has gained a new meaning within the context of Information Systems literature and more specifically within its specialised sub fields of Office Information Systems and more recently Computer Supported Co-operative Work (CSCW).

According to Johansen (1988), 'groupware' covers a set of computer assisted aids designed for the use of collaborative work within groups; these involve "hardware, software, services and group process support". Grudin (1991) claims that all kinds of Information Technology have always supported groups in a certain way. However, groupware's support differs from other kinds of IT support to groups in its being purposefully designed and conceived to support groups. The author argues here that this field of research is brought about under the influences of different stakeholders in the IT business. This makes it very difficult to reach an agreement on terms such as groupware or CSCW. Clearly the notion of 'co-operative work' needs to be properly addressed if one wants to make sense of the terms groupware or CSCW.

A different and promising new angle to the conceptualisation of CSCW was not so recently proposed by Lyytinen and Ngwenyama (1992). These authors have also identified the fragmented nature of CSCW as field (which can include contributions ranging from studies of organisational culture to the development of network protocols) and argue that it

Synergy Matters: Working with Systems in the 21st Century,
Edited by Castell *et al.*, Kluwer Academic / Plenum Publishers, New York, 1999.

127

lacks clear theoretical foundations. The fragmentation of the field can be seen as the result of applying the rationality of technological infrastructures (the computer support) to the experiential nature of human communication (the co-operative work), which is hard to conceptualise under a merely functionalist view (Shulman and Penman, 1990). To overcome such gap, Lyytinen and Ngwenyama (1992) propose the use of Giddens structuration theory as a conceptual framework upon which the distinctive nature of CSCW applications can be observed and analysed.

Structuration theory views human agency as continually reproducing a social structure, which is itself the product of social actions. Thus the notion of structure becomes that of a set of rules and resources or capacities that generate the observed regularities that become an instantiation of it. Social structure does not exist as an observable pattern of behaviour; it is at one time the producer and product of human agency (Giddens, 1984). This notion of social structure resembles the notion of autopoiesis (Mingers, 1996) and should be seen not as tangible but instead as a process of structuration, i.e. as the structuring properties of social systems.

Within this framework, co-operative work is defined as social practices (within a specific social system) that draw upon specific rules and resources that enable / constrain shared purposeful activity. This latter is performed by knowledgeable agents who constantly monitor their domain of action at two levels of rationality: the knowing about things in the world (discursive consciousness) and the knowing of the world (practical consciousness). Lyytinen and Ngwenyama (1992) propose a normative definition for CSCW based on Giddens' theory of structuration, one that is ontologically focused: "Computer Supported Co-operative Work applications are open evolutionary structures embedding organisational and linguistic rules and serving as resources that mediate and transform co-operative interactions via recurrent use-process (procedures and practices) within specific organisational contexts."

Let us briefly discuss the consequences of this definition. Probably the structuration theory framework will provide ways to organise inquiries enabling the reflexive anticipation of socially-constituted practices, and to assemble debates of possible courses of action (Lyytinen and Ngwenyama, 1992). As Mingers (1996) suggests, the examination of the process of structuration can be made by selecting the focus of attention on *strategic conduit* of the agents, suspending the reproduction of structure, or else to observe the reproduction of structural rules through social interaction via an *institutional analysis* that suspends the conditioning of social action.

Computer Supported Co-operative Work taps into very sensitive issues in the life of organisations. Issues concerning information sharing and power relations are extremely interconnected, so that an approach centred in the processes of organising may better than a functionalist IS development approach. "Groupware is not inherently democratic or authoritarian; it allows increased options in both directions... To varied degrees, groupware systems will embody beliefs about how people within a group ought to behave. The basic issue is this: Whose beliefs will be embodied?" (Johasen, 1988 p.9). This points to the need to develop a research approach or at least a theoretical framework that accepts different sets of beliefs as given in human situations and acknowledges its strategic importance, in the sense of Giddens' 'strategic conduct' or Suchman's 'situated action'. This directs attention not only to the understanding of groups or organisations but to the understanding of collaboration or co-operation (Blackler, 1994).

According to Lyytinen and Ngwenyama (1992) the study of co-operation as a work practice must deal with the cognitive dimension of signification-the shared understanding: "In most cases social and systems integration takes place on the level of agent's practical consciousness and does not necessitate written codes or guidelines. This [co-operative work] definition entails detailed and well-developed systems of significance and stocks of knowledge (interpretative schemes) on which the co-operative effort can rest. However it does not set any specific requirements on power relations (such as self-determination or democratic management), nor on legitimation (p.25)".

Clearly, these so called systems of significance and interpretative schemes are crucial for the understanding of co-operative work, from the level of practical consciousness, where they are tacitly known, to the level of discursive consciousness where they are explicitly articulated. A research approach that explicitly acknowledges different sets of beliefs in a given situation and deliberately addresses the nature of the explicit articulation of systems of significance can be relevant to groupware research.

A RESEARCH APPROACH TO GROUPWARE

The present work took place within a research consortium led by an Industry R&D organisation in Portugal and involved other researchers from Lisbon Technical University and from Lancaster University. The research concerned particularly the process of groupware adoption in business organisations. A selection of Portuguese host organisations that have implemented groupware were contacted and a major utilities corporation and a mid size IT company were successfully enrolled. The official aim of the research was to extract learning from actual cases of technology adoption that could be made explicit and usable by other organisations seeking to adopt groupware. However, research objectives are often a temporary construction that need to be clarified and constantly reconstructed (Keys, 1998).

Given this view, it is not important to locate the present research as social science or management research, or even as a Soft OR study. In fact, it can be perceived to be any of those since the inclusion of research practice under specific labels or categories is somewhat arbitrary. Each so called field of science is a socially constructed culture, whose members form distinctive learning communities (Checkland and Holwell, 1998). What is worthwhile then, is to shed light on the social process by which the so called 'scientific' knowledge is constructed within a given research community. Easterby-Smith, Thorpe and Lowe, (1991) conceive management research as a three instances model comprising philosophical, political and technical aspects. I propose to use this model to describe how knowledge about groupware adoption was constructed in the present research.

On the philosophical side, the present research follows a non-positivist approach (Morgan, 1983). It tries to understand and explain how people experience groupware rather than search for external causes and fundamental laws to explain it. Toward this end, a participant research approach based on Soft Systems Methodology (SSM) was perceived to be a reasonable strategy. However, with a non-positivist position it is not so useful to produce a formal account of the methodology employed (a crucial step to objectify knowledge under the positivist paradigm) but instead it is important to understand the processes by which the logos of the present research practice was socially constructed.

In the present case the researchers were not given the brief for intervention, at least not in the usual extent of SSM's use for driving change (Checkland and Scholes, 1990) or organisational design (Wilson, 1990). Instead, the agreed scope for this research was to assess several cases of groupware implementation undertaken by different companies within the same holding corporation and extract useful knowledge for further groupware adoption. SSM was used to assist the design of qualitative research. A conceptual model for the research was produced and used as a mean to support decisions concerning the research work, some of them made in collaboration with the host organisation. In this sense, the logos of research was partly co-constructed with the host organisation.

On the technical side, we privileged computer assisted group interviews as a medium for participant researchers to collect data deemed relevant for groupware adoption. In order to capture the full sense of groupware use, the research team decided to adopt an electronic meeting system (EMS). This was used for data collection purposes with participants from the collaborating organisations. Conceptual models of human activity systems were produced to assist discussion leading to the design of group interviews with the EMS. What later emerged was that the use of this EMS has produced an additional meta-level of research, where the researchers were themselves in the role of subjects which were using a novel form of data collection, which was groupware assisted. Contrary to the descriptive level of research on groupware adoption, at this meta-level, SSM was used as a wide-band Group Decision Support System (GDSS) for the co-operative design of research supported by electronic meetings (Eden, 1995). The conduction of electronic meetings for data gathering became in itself a rich locus of intervention, where reflections on the use of groupware by both researchers and participants produced useful knowledge. This latter is quite close to experiential research, i.e., research where the empirical basis is the experiential knowledge of persons in relation in their situation, in their world (Reinharz, 1981).

Finally, the present research can also be seen as a political process. If one acknowledges the ambiguous nature of groupware which is hosted by the organisation in a process highly vulnerable to political factors (Ciborra, 1996) then research on groupware can become itself an extension of the political struggle behind the adoption of groupware. In trying to understand the co-operative process that is subsumed by CSCW, we not only have to recognise the instability of what is being researched but also that we are using the very same process as that being researched during the research process. This brings to the fore the importance of the research relationship, since the nature and quality of the research findings will be undoubtedly influenced by the relationship established between researchers and researched.

DISCUSSION

Easterby-Smith et. al. (1991) argue that the way in which the practice of social science research is applied to management research leads to a rethinking of social research assumptions, which then will further help the practice of social research. The same argument can be made about research on groupware and in the present case, to advance the notion that SSM can also be seen as an embodiment of social science research.

SSM can be included in the set of alternatives to the objectivist approach to social science (Morgan ad Smircich, 1983). It can be seen as the logos of human inquiry based in systems thinking (Checkland, 1981), which encompasses a stable set of ontological principles and assumptions about human nature, conforms to a basic epistemological stance and uses some favoured metaphors. SSM embodies a unique inquiry approach that is characterised by 'interpretative flexibility' (Brown, 1992). Only the process of SSM is transferable never its interpretative content. This means that SSM validity needs to be redefined outside positivism, where validity means research that is well grounded in the procedures adopted to free its forms of knowing from distortion, and in the special skills involved in the knowing process (Heron, 1996).

To discuss the particular use of SSM made in this case it is necessary to focus upon the work of scientists and its context, recognizing that the practice of science is part of social practice (Keys,1998). This perspective emphasizes the subjective and sociopolitical aspects of scientific and technological activity alike, blurring the use of SSM with other process of inquiry, each embodied by a particular form of social practice or conversely of scientific practice.

So, in the present case, the understanding that is made of groupware is a temporary state, based on reaching some closure in a network of significance shared by concrete actors in a given situation. This actor-network comprises researchers and researched and enables the creation of representations and statements about situations that enable the aims of the research project to be realized. Once closure is achieved it is possible to treat parts of groupware reality as abstract representations and use those to inform decisions about subsequent actions to be taken. The knowledge produced about groupware then is dependent on the elements in the actor-network, which comprise actors associated with the organizations, actors (or resources) associated with the representation of the situation (in the present case SSM models and EMS software) and the means of communication that bind them together into a network of significance. This clearly shows that the research relationships are the vehicles through which the researcher (and the researched) come to understand a situation.

Heron (1996) asserts that special inquiry skills need to be developed if one is to produce valid knowledge from a research relationship in which the researcher is engaged with the experience of it. For example, "bracketing" conceptual labels and models embedded in the process of perceiving people and a world, "reframing" our assumptions of any conceptual context or perspective, "non-attachement" or the ability to bear without fixation the purpose which has been chosen as the form of action and "emotional competence" to identify and manage emotional states in various ways. All these seem to be tacitly accepted as crucial skills for successful use of SSM, but little explicit attention has so far been paid to their importance in practice.

Perhaps an ignored perspective on SSM inquiry is that of the clinical demands it makes on its practitioners. These demands result from any kind of research in which the researcher is directly involved with those researched in a social situation, where there is commitment to a process of self-scrutiny by the researcher and willingness to change theory or method in response to the research experience during the research itself (Berg and Smith, 1983) According to these authors the key work of the researcher in research is to establish a way to assist those researched in developing self-reflecting capabilities. Heron (1996) suggests that this subsumes empowering research participants within three distinct dimensions:
- cognitive or methodologically: giving away the methodology to the participants;
- politically: allowing participative decision-making in the use of it;
- emotionally and interpersonally: creating a climate in which emotional states can be identified and openly accepted and processed in the course of its use.

Clearly, all these three factors need to be addressed if SSM practitioners do not want to become hostage of their contexts of practice, alienating the researched from the co-operative inquiry process subsumed by SSM which is undoubtedly its major strength.

REFERENCES

Blackler, F. (1994) Post-Modern organizations: Understanding how CSCW affects organizations. *Journal of Information Technology*, 9, 129-136

Berg, D and Smith, K. (1983) The Clinical Demands of Research Methods. in "Exploring Clinical Methods for Social Research", David Berg and Kenwyn Smith (eds). Sage, Beverly Hills

Brown, A.D. (1992) Grounding soft systems research. *European Journal of Information Systems*, Vol.1, No 6, pp 387-395.

Ciborra,C. (1996) (Ed.) "Groupware and Teamwork: Invisible Aid or Technological Hindrance?" Wiley, Chichester.

Checkland, P.B. (1981). "Systems Thinking, Systems Practice," Wiley, Chichester.

Checkland, P and Scholes, J. (1990). "Soft Systems Methodology in Action," Wiley, Chichester

Checkland, P.; Holwell, S. (1998) "Information, Systems and Information Systems," Wiley, Chichester.

Easterby-Smith, M., Thorpe, R. and Lowe, A. (1991) "Management Research: An introduction," Sage, London.

Eden, C. (1995) On evaluating the performance of 'wide-band' GDSS's. *European Journal of Operational Research*, 81, pp. 302-311.

Giddens, A. (1984) "The Constitution of Society," Polity Press, Cambridge.

Grudin, J (1991) Groupware and CSCW: Why now?. in "Groupware'91: The potential of team and organisational computing." P.H.R. Hendriks (editor). Software Engineering Research Centre - SERC, Utrecht, The Netherlands

Heron, J. (1996) "Co-operative Inquiry: Research into the Human Condition." Sage, London.

Johansen, R. (1981) "Groupware: Computer Support for Business Teams." The Free Press, Series in Communication Technology and Society, Everett M. Rogers and Frederick Williams, Editors. New York.

Keys, P. (1998) OR as a technology revisited. *Journal of the Operational Research Society*, Vol. 49, Nr. 2 pp. 99-108

Lyytinen, K.J.; Ngwenyama, O.K. (1992) What does Computer Support for Cooperative Work mean? A structurational analysis of computer supported work. *Accounting, Management and Information Technology*, Vol.2, No1, 19-37.

Mingers, J. (1996) A Comparison of Maturana's autopoietic social theory and Giddens's theory of structuration. *Systems Research*, Vol. 13, No. 4, pp. 469-482.

Morgan, G. (1983) Ed. , "Beyond Method," Sage, Beverly Hills.

Morgan G. and Smircich L. (1980) The case for qualitative research. *Academy of Management Review*, Vol.5, No4, 491-500.

Rickards, T. (1974) "Problem-solving through Creative Analysis," Gower Press, Epping, Essex.

Reinharz, S. (1981) Implementing new paradigm research: A model for training and practice in "Human Enquiry: A Sourcebook of New Paradigm Research", Peter Reason and John Rowan (eds) Wiley, Chichester.

Shulman, A.D. and Penman, R. (1990) Putting Information Technology in its place: Organizational communication and the human infrastructure in John S. Carrol (ed.) "Applied Social Psychology and Organizational Settings," Lawrence Erlbaum, Hillsdale.

Wilson, B. (1984). "Systems: Concepts, methodologies, and applications." Wiley, Chichester.

AN EXPLORATION OF THE POTENTIAL OF SYSTEMS METHODOLOGIES FOR APPROACHING ORGANISATIONS OPERATING AT *THE EDGE OF CHAOS*

María Carolina Ortegón Monroy

Lincoln School of Management
University of Lincolnshire and Humberside
Brayford Pool
Lincoln LN6 7TS
United Kingdom

INTRODUCTION

Management literature seems to emphasise that management in the new global environment is considered to be the key to corporate success in the business world of today. According to Hardy (1994), new sets of challenges are faced by organisations previously developing activities through a local market approach: innovation, sustainable development, mergers and acquisitions, managing decline, among many others.

However, as it has been argued lately mainly by advocates of chaos (Complexity) theory (Kauffman 1993, Stacey, 1991, Wheatley, 1992), this messy global competition game is being explained using models which are based on order, stability, cohesion, and equilibrium. It is thought that not enough attention is being paid to the irregular, disorderly, chance nature of this game, probably because it is easier and safer than trying to explain the disorder, irregularity and unpredictability that characterise what some claim are the new realities for management today: 'managing chaos' (Stacey, 1992). Further elaboration on this theory by Stacey (1992, 1996) and Peters (1988), has tried to demonstrate that innovation and creativity, the keys to success in the global competitive markets of today, are deployed by organisations operating in a state of bounded instability, denominated as *"the edge of chaos"*. These, among other considerations, have the implication of evidencing a demand for suitable methodologies for approaching these so-called new dynamics of organisations.

This paper presents an attempt to study the perspectives of systems methodologies in approaching problem situations where the "mess" is said to be generated by situations of instability, unpredictability, uncertainty and ambiguity. The purpose is to explore about the possibilities of systems methodologies in dealing with the complexities expressed in relation to the *"edge of chaos"*, as presented by Stacey (1996) in his book: *Complexity and Creativity in Organisations*, where it is said organisations are capable of potential

Synergy Matters: Working with Systems in the 21st Century,
Edited by Castell *et al.*, Kluwer Academic / Plenum Publishers, New York, 1999.

133

innovation: a major challenge for managers today. The approach for conducting the exploration is systems thinking. Total Systems Intervention (TSI) (See Flood and Jackson, 1991) is used as philosophy of enquiry.

CHARACTERISATION OF THE *EDGE OF CHAOS*

The contention of Stacey is that every human system, every organisation is a non-linear feedback system (complex adaptive system), that operates in three zones. He considers an organisation to be in the stable zone when its shadow system favours the legitimate system and sustains it. It operates in the unstable zone when the legitimate system is not strong enough to contain *the basic assumption behaviour and psychotic fantasy[1]* of the shadow system. And it occupies the space denominated as the *edge of chaos* when its shadow system operates in tension with the legitimate system, seeking to replace at least parts of it. Furthermore, he considers that an organisation is only changeable in the phase transition at the edge of chaos, because it is only here that it can achieve double loop learning. (See Stacey, 1996).

GENERAL METHODOLOGICAL APPROACH

As indicated in Figure 1, TSI was used partially through the application of the creativity and choice phases, to support and enrich understanding of the so-called *edge of chaos* state for organisations and guide reflection upon the appropriateness of systems metaphors and methodologies to approach it.

Figure 1. General Methodological Approach

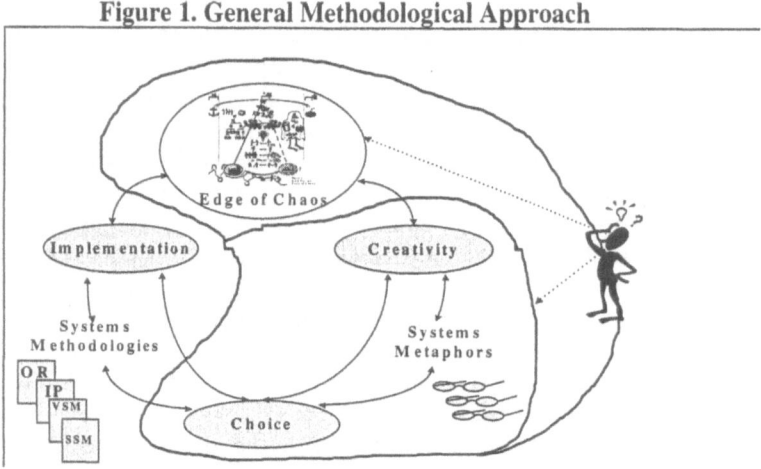

CREATIVITY PHASE: IMAGES OF THE *EDGE OF CHAOS*

The realities and complexities determined by *edge of chaos* state in organisations were 'examined' from different perspectives to throw different lights on the situation, enriching the whole picture. (See Carrizosa and Ortegon, 1998) This objective was

---

[1] Unconscious group processes, where members become very dependent, engage in fight-and-flight dynamics, seek salvation in pairing and try to fuse with each other in the "good" group. In this state behaviour is driven for lengthy periods by fantasies of a psychotic nature and rational group behaviour disintegrates. (Stacey, 1996).

accomplished mainly through the use of a series of systems metaphors, or "lenses", that induce such creative thinking.

As a result of this 'exploration' it was found that the psychic prison metaphor arises as dominant, pointing that the *edge of chaos* is essentially a psychic phenomena, while the brain and coalition metaphors arise as dependent, focusing on the intertwined learning and political activities that characterise the mechanisms at operation at this state to bring about innovation and change.

CHOICE PHASE:
SYSTEMS METHODOLOGIES FOR OPERATING *AT THE EDGE OF CHAOS*

In this part a critical reflection was developed towards choosing possible appropriate systems methodologies for approaching the "mess" present at the *edge of chaos*. It was achieved by determining about the problem situation, considering the two dimensions of the System of Systems Methodologies.

The Problem Context

The following characteristics provide evidence about the complexity of the problem context:

• There are many agents, (power groups, coalitions, cultures and subcultures, etc.) representing forces of stability and instability acting through many instances (covert and overt politics, defence mechanisms, co-operation, competition, etc.,) to either sustain, or replace the legitimate system or parts of it.

• In general, the behaviour of the agents cannot be predetermined, in the end it is a product of the individual's own process of construction and interpretation of the reality affecting him/her.

• Agents and instances appear and dissolve following different patterns, as part of a process of dialectical co-evolution, extended form the individuals to groups, to the environment and to the society.

The underlying dynamics present at this state are very likely to lead us to conceive the relationship between participants as coercive. Certainly, the following aspects confirm this view:

• there are two clear opposite schemas (among a series of many other) representing opposing interests that stand for conflictive value and belief systems, or systems of construction of reality in which opposing groups are "trapped" ;

• genuine compromise is not possible since people's minds are continuously bound to irreconcilable mental traps.

• the change in schema is achieved when the learning communities or power groups acquire enough power to overcome those who protect the status quo, indicating that the application of imposition or coercion is likely to define the situation.

However in terms of the "materialisation" of this situation, it was observed that political and learning processes are simultaneously taking place, whereby the relation of the participants could be referred to as pluralist, considering that:

• learning groups and power groups, as mechanisms to bring about changes in the recessive and dominant schemas, have the implication of groups of people with certain compatibility of interest, at a given time;

• the operation of these groups have the implication of contexts where participation is possible at least to a certain degree;

- the spontaneous self-organisation in groups of interest or coalitions, and the continuous "accommodation" of dominant and recessive schemas, have the implication of temporary agreements to achieve change or sustain the legitimate system.

From these considerations it was revealed that in the "psychic" domain of the organisation seems to lie the ultimate outcome of the situation, "the ends" (the author's denomination), achieved in a predominant coercive context. However, the conditions for the *edge of chaos* to take place are clearly said to lie in the "practical" domain, where the "means" (the author's denomination), are realised through learning and political activities acting as anxiety holders and activators of power differentials, in a pluralist context. Therefore, the problem situation was plotted onto the System of Systems Methodologies, by using the analogy of a pendulum continuously swinging between two zones: the complex-coercive, considering the domain where the "ends" are achieved, and the complex-pluralist, considering the domain where the "means" are achieved. This is shown in Figure 2:

Figure 2. The *Edge of Chaos* Problem Context Plotted on the System of Systems Methodologies

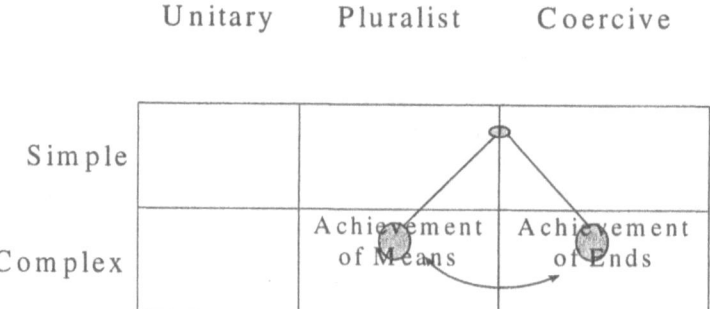

Methodological Choice

Considering the problem contexts within which the *edge of chaos* can "swing" it is evident that a soft systems approach is required. The System of Systems Methodologies suggests that presently there is no systems methodology suitable for dealing with complex-coercive problem contexts. However there is a selection of two possible methodologies making complex-pluralist assumptions about problem contexts. They are Interactive Planning (IP) and Soft Systems Methodology (SSM).

IP may seem as an attractive alternative to contribute to the dynamics operating at the *edge of chaos* by inducing a participative approach and addressing a wide range of organisational problems, with a 'dissolving' criteria (Ackoff, 1991), implying creativity. However, this methodology has been criticised for failing to come to terms with conflict and power dimensions of the organisation (Flood and Jackson, 1991). Moreover, it emphasises the need for all participants to plan and design an "idealised future", a process that in this context has serious limitations, considering the fact that the long-term future is seen as unpredictable, and that there are fundamental conflicts of interest among "stakeholders", situation that results incompatible with the generation of consensus.

SSM poses interesting possibilities for enabling intervention at this state. It is flexible, allowing different issues to be dealt with at any given time, with results favourable with the dialectical and self-organising nature of the dynamics at this state. A very important aspect that strongly favours this approach in this context is that it implies a learning process, where different points of view are considered, and where temporary 'accommodations' to *unstructured* problem situations are generated in a cyclic learning process.

Probably the most significant intervention at the *edge of chaos* is enabling exploratory dialogue to take place in other to contain anxiety, and develop creativity. This is facilitated by SSM in a process, where "myths and meanings" (Checkland, 1989, pp. 71-100), forming part of the debate are tested and examined, to generate change. The fact that: "If successful the debate will lead, if not to the creation of shared perceptions, at least to an accommodation between conflictive view points and interests so that desirable change can be implemented." (Jackson,1991, pp. 156), implies that a change in schema could be a possible outcome, thus double loop learning might be stimulated. Therefore in this very aspect, SSM presents an invaluable contribution.

It seems as though interpretive thinking and idealism underlying SSM, as stated by Flood and Jackson (1991), in a way explain one of the positions upon which the conflict at *edge of chaos* develops: ideological difference, contrasting world views and perceptions of reality. But at the same time this defines its limited character in operating under circumstances of a coercive nature, simultaneously present at this state.

Moreover, Flood and Jackson (1991) have pointed out that SSM can be considered as managerialist, reformist and unreflective, indicating that under these circumstances it would not necessarily guide to an emancipatory end result, but most probably, reinforce the existing situation by benefiting and serving those with power. As argued by these authors it neither propitiates reflection on the changes thus proposed, nor concern about prevailing structures of power, wealth and influence. [2]

It is then concluded that in the dimension of the situation where the means are realised, SSM can provide a suitable basis for intervention, generating temporary shared "accommodations" and...."providing creative solutions that enable organisational actors to escape the "traps" into which their current thinking has led them." (Flood and Jackson, 1991, pp. 190) . However it is not possible to address the coercive dimension of the situation, which is undeniably predominant, when it comes to defining outcomes.

CONCLUSIONS

Presently, the potential of systems methodologies for approaching organisations operating at the *edge of chaos* seems to be very limited.[3] The dynamics at this state seem to demand other forms of intervention that respect the self-organising nature of change and its unpredictability in terms of creative outcomes. The following quote corroborates this idea by providing a synthesis of Stacey's view of the role the human factor plays in this process of creative destruction: "It seems to me that we have no realistic alternative but to accept that we cannot be in control of the kind of complex co-evolutionary process that drives all non-linear feedback networks but can only participate in producing emergent patterns. No amount of human intelligence, self-awareness, and determination can alter the fundamental dynamics of non-linear feedback. In the end, creativity is inevitably destructive. This paradox will always make us anxious; it is a force that lies at the bottom of the dynamics of human systems, their irremovable nonlinearity, and their radical unpredictability." (Stacey, 1996, pp. 218)

---

[2] It is to be pointed out that in Checkland and Scholes (1990), SSM is developed to address power and politics, which were not contemplated in Checkland (1981). However this is done in an unsophisticated manner.

[3] However, it is to be noticed that work is being developed in this direction. See Mingers (1997), Veil (1997) and Jackson (1997).

Consequently, one is very inclined to think that the approach towards intervention must also differ. Morgan seems to be well aware of this, as expressed through the following reflection: "The postmodern world view, which, of interest, is paralleled in aspects of the new science emphasising the chaotic, paradoxical and transient nature of order and disorder, requires an approach that allows the theory and practice of organisation and management to acquire a more fluid form". (Morgan 1993, pp. 282-283). Stacey (1996, pp.281) seems to agree on this point when he suggests that: "... these new efforts, making up the science of complexity, provide an overall framework for pulling together many existing building blocks in the literature on management and organisation into a new way of approaching organisational life."

Hence in relation to approaching the *edge of chaos*, a dilemma is faced: systems thinking does give an account of the situation but its potential seems very limited in relation to influencing outcomes. Is it perhaps because a change in approach towards intervention is required? The *edge of chaos*, evidently, denotes a new paradigm towards organisational change. (Reaville and Scott, 1998). And, certainly, this new paradigm poses serious implications for intervention in organisations.

REFERENCES

Ackoff, R., 1991, "Creating the Corporate Future", Wiley, New York.

Carrizosa, A., and Ortegon, M.C., 1998, Using systems metaphors to interpret the edge of chaos, in *Systems Sciences and Systems Engineering- 3rd International Conference*, Beijing.

Checkland, P., 1981, "Systems Thinking, Systems Practice", Wiley, Chichester.

Checkland, P., 1989,Soft systems methodology, in: "Rational Analysis for a Problematic World," J. Rosenhead, ed., Wiley, Chichester.

Checkland, P., and Scholes, P, 1990, "Soft Systems Methodology in Action," Wiley, Chichester.

Flood, R and Jackson, 1991 "Creative Problem Solving," Wiley, Chichester.

Hardy, C., 1994, "Managing strategic action : mobilizing change: concepts, readings, and cases," Sage, London.

Jackson, M.C., 1991, "Systems Methodology for the Management Sciences," Plenum, New York.

Jackson, M.C. 1997, Towards coherent pluralism in management science, Working Paper No. 16, Lincoln School of Management, University of Lincolnshire and Humberside, Lincoln.

Kauffman, S., 1993, "The Origins of Order: Self-Organisation, Selection and Evolution," Oxford University Press, Oxford.

Mingers, J., 1997, Critical pluralism and multimethodology, post modernism, pp. 345-352, in "Systems for Sustainability, People, Organisations and Environments," F.A, Stowell, et al. Ed., Plenum Press, London

Morgan, G., 1993, "Imaginization," Sage, New York.

Peters, T., 1988, "Thriving on Chaos: Handbook for a Managerial Revolution," Pan, London

Reaville L., and Scott, E, 1998. The application of chaos theory to the management of change in organisations: a theory of humility: metaphor or reality?. "*42nd Annual ISSS Conference*", pp. 3084-3094, Atlanta.

Stacey, R., 1991, "The chaos frontier: Creative Strategic Control for business," Butterworth-Heinemann, Oxford.

Stacey, R., 1992, "Managing Chaos," Page, London.

Stacey, R., 1996, "Complexity and Creativity in Organisations," Berret-Koehler, San Francisco.

Veil, S., 1997, Social and organisational learning and unlearning in a different key: an introduction to the principles of critical learning, theatre and dialectical enquiry, pp. 373-380, in "Systems for Sustainability, People, Organisations and Environments," F.A, Stowell, et al. Editors, Plenum Press, London.

Wheatley, M.J., 1992, "Leadership and the new science: Learning about organisation from an orderly universe," Berett-Khoelher, San Francisco.

GROUP SUPPORT THROUGH RESTRICTED CONVERSATIONS

Hector Ponce

Lincoln School of Management
University of Lincolnshire and Humberside
Brayford Pool, Lincoln, LN6 7TS, United Kingdom
e-mail: hponce@lincoln.ac.uk

INTRODUCTION

The aim of this paper is to examine some research problems that emerge in the area of group support. The method proposed takes some aspects of language use and crystallises them in the notion of conversational structure. According to this method, one the elements that is considered in more detail is how to deal with groups in research, either as phenomena which are external to the process or, on the other hand, internal to the research process.

This paper attempts to elaborate a proposal in which group support can be examined from different angles. The paper is structured as follow: First, the method of study is presented which is based on aspects of language use, and the notion of conversation is put forward. Secondly, some relevant elements of science are briefly presented in terms of what is called restricted conversations. Third, the notion of restricted conversation is used to discuss some approaches that might be used to support groups through the research process.

Groups play an important role in people's life, principally because people form groups to fulfil a variety of purposes: to make decisions, to solve problems, to play music, to satisfy social needs, among others. In management, the importance of groups and the formation of teams have been recurrently stressed in several publications. It is reported that organisations which have changed to team-based work have improved their performance, although this is not always the case (Hayes, 1997). Therefore the question of how to support groups has become relevant in several disciplines: management, systems, information technology, social psychology. Thus, the consequences have been that the topic of group support has had distinct treatments in each of these disciplines, although there have been attempts to relate them. In the next sections we concentrate on the elaboration and application of the method proposed to examine group support.

Synergy Matters: Working with Systems in the 21st Century,
Edited by Castell *et al.*, Kluwer Academic / Plenum Publishers, New York, 1999.

THE METHOD

The framework developed in this paper emphasises observations about language, principally its pragmatic aspects. This method takes its source in the view of language use as twofold: language is not only a medium to describe the world but also an instrument to create it. Sentences such as "water boils at 100°C", "I pronounce you man and wife", "I promise I'll be there" have, each of them, different consequences in terms of communication among people: some of them play a descriptive role while others create new contexts for action (Austin 1962). We usually define the unit of analysis in which these communicative acts take place as conversations.

When a sentence is spoken in the course of a conversation, one can observe that a certain structure seems to emerge that allow us to make some hypotheses about conversational situations. This structure is in part described by Lyotard (1984), and also by Clark (1997). In a conversation, when a sentence is uttered, the next elements become visible: *the speaker* —the person who utters a sentence; *the addressee* —the person who receives the sentence; *the referent* —what the sentence deals with; and finally, *the conversational language* —the language from which the sentence is made. In this paper, it is intended to take this structure as an 'instrument' to discuss group support through research. Before going into the research discussion, let us explore two applications of this conversational structure, stressing aspects related to the referent; since it is what will be considered in more detail in the rest of the paper.

Describing the referent

Sentences such as "Mary is an engineer", "water boils at 100°C", "It's snowing out'" can be said to be *descriptive sentences or statements* in the sense that they represent or describe the 'world' or state some facts (Austin 1962). In the context of a conversation, these utterances situate the *speaker* in a position of knower, he is supposed to know that Mary is an engineer, or gives evidence that water does boil at that temperature. The *addressee* is put in a position of having to accept the statements by the speaker as true statements or, on the contrary, to challenge them —showing evidences that the statements are not true, such as "Mary is not an engineer but a lawyer, we went to the same School". The *referent* must be correctly identified and described by the statements that denote it (Lyotard 1984).

Creating the referent

On the other hand, other sentences exist such as "I declare the University open", "First we have to address the marketing problem" or "I promise I'll be there". These sentences when spoken in the course of a conversation create a new situation or a new course of actions for speaker and addressee (and others not involved directly in the conversation may be affected; e.g. a declaration of war). The referent in the sentence might change its status (e.g., the University) or, on the other hand, a new referent is constructed which did not exist before the conversation had taken place (e.g., the promise). This type of utterances is what Austin (1962) calls *performatives*. In the context of a conversation, the *speaker*, when uttering a sentence such as "I declare the University open" has to be invested with the authority (e.g., dean or rector) to make the declaration; and of course under the conditions that the utterance was understood and recognised by the *addressee* (otherwise there is no communication), the sentence in itself creates a new context of actions for the addressees (e.g., the university staff). They will now be able to begin with, for instance, the activities of a new academic year (Lyotard 1984).

RESTRICTED CONVERSATION

The notion of *restricted conversation* will be used to refer to a conversation in which the participants (speaker and addressee) agree to observe the rules of a conversational language, which allows the speaker and addressee to achieve some effects that otherwise would be difficult to undertake.

Why do we need to introduce the concept of restricted conversation? The reason is simple, most of the examples previously given are made in ordinary speech and taken from daily life experience; however, our interest is to explore some effects that can be achieved by restricting or structuring somehow what people exchange when they communicate. Some of these effects are for instance: producing reliable knowledge, creating common culture, speeding up problem solving processes, empowering individuals and groups. The hypothesis here is that one method by which these consequences might be achieved is structuring somehow the conversations in which the participants are involved. In the next section, I will start examining very briefly science as a restricted conversation, emphasising aspects related to *the referent*.

Science as a restricted conversation

Science can be seen and understood as a process of collective learning which reacts in a characteristic way to events outside it. That is to say, notification of such events takes place as *observations* (these include people's reports of their observations). The basic difficulty is to order the observations that are being made, whether in daily life or in laboratory settings. The aim is to get a high quality ordering of observations which is normally represented by a *scientific construct* (e.g., electron, magnetic field, personality). New observations might demand to re-accommodate the previous observations, again and again, until eventually a high quality order is achieved; making the construct more accurate and precise. Usually scientific constructs allow for predictions—although this is not a necessary criterion. Its study allows for high quality observations which can be used to optimise or improve various *actions*. These include interventions (e.g. heating a piece of metal to some desired degree), but also other activities such as throwing a piece of metal (De Zeeuw 1996).

Traditionally, scientific constructs have been thought to have a direct relation to observable facts, they are supposed to describe in very accurate and precise statements the phenomena (the referent) they represent, and the precision of these linguistic expressions make it possible to have critical debates (Graziano & Raulin 1993; Nagel 1979). Similarly, Cole (1996) argues that it is the 'structure of the world' that allows scientists to produce these high quality constructs. Therefore, the fundamental point is that the referent is assumed to be *external* to the scientific process[1]. This allows various observers to observe repeatedly the same phenomena independently of space and time. In other words, one might say that 'nature' acts as a guarantor of this observational stability. It is important at this point not to accept that this is the only possible account of science, since all what I have presented so far is the focus of an intense debate in which not only the status of the referent is being questioned but also other situations such as the relationship among scientists, the validity of statements generated and the institutions in which science is practised (Kuhn 1970;

---

[1] Although, Callon (1986) argues that the referent is internal to science in the sense that it is a socially-constructed system, which makes the results of science socially-constructed as well —including those of physics and chemistry.

Popper 1979; Bhaskar 1975; Callon 1986, Nagel 1979, De Zeeuw 1995; Lyotard 1984). However, some of the characteristics of science presented so far will allow us to explore some problematic features of group support.

GROUP SUPPORT

The previous discussion, although brief because of space constraints, may enlighten us on how to support groups. As a consequence of that discussion, one possible solution is to treat groups themselves and their activities, processes, structures, social relations, communicative interactions as 'external referents' to the research process; approaching them as standard research. That is, searching for scientific constructs that accurately represent group phenomena. Shaw (1981) follows this path and treats group phenomena as if they were structured as phenomena external to the research process, and as a result, one should expect to generate stable observations about the them. For instance, Shaw presents a series of hypotheses about groups following the linguistic structure of scientific statements, although they lack the accuracy and precision of statements made in physics or chemistry. If we examine the evidence that has been given to support this research, it is not difficult to realise that social groups have been carefully examined following this convention: "The data supporting this hypothesis derive primarily from observations of laboratory and natural groups..." (p. 116). Other researchers (Davis 1969, Argyle 1991, Turner 1987) have also adhered to the same convention. Despite the efforts made to maintain this convention, the main problem with this approach seems to be on how to deal with situations in which people's intentions and values play an important role, and consequently, the phenomena become difficult to characterise in terms of stable observations (Argyris *et. al.* 1985). Therefore, this dilemma makes it difficult to consider the referent as external to the research process whether we want to deal with these difficulties.

Inviting the referent into the conversation

Different solutions to the way groups are supported through research are proposed in the literature. Their fundamental characteristic is the attempt to make the referent *internal* to the research process, that is, doing research *with* the referent and not about it; allowing it to speak by itself and not through the scientists. Therefore, the referent becomes a speaker and an addressee in the research process, gaining an active role inside science. Let us for clarity use, rather loosely, the word *actor* to show that the referent is playing a new and active role inside research. The immediate consequences is that the conversations are no longer only among scientists, but, in the research process, both scientists and this new actor become a sort of partners in the process of enquiry. Following this and the conversation structure we have put forward as a framework, we are at this point faced with the question: What to do with this new actor in the research process?

Heron (1981) proposes to make the new actor a co-researcher in an instance of cooperative enquiry. Although the aim by Heron is still of explicative nature, it is interesting to observe that the solution is to have some type of restricted conversations with this new actor: 'The general form of this argument is that human beings are symbolising beings. They find meaning in and give meaning to their world, through symbolising their experiences in a variety of constructs and actions... To explain human behaviour you have, among other things, to understand this activity, and fully to understand it involves participating in it through overt dialogue and communication with those who are engaging in it' (p. 23). The proposal by Heron goes in the direction of extending research from 'propositional

142

knowledge' to include other types of knowledge such as 'practical knowledge' (knowing how to do something) and 'experiential knowledge' (knowing an entity such as a person, place, etc). The objective of the conversational structure of this research process is to create a value system between the scientists and the subject ('the new actor'); although the quality of the research support is not very much discussed.

Another plausible answer on what to do with the new actor is found in the area of system thinking through the important contribution by Checkland (1981) known as Soft System Methodology (SSM). It seems that this methodology also aims to generate a kind of structured or restricted conversation in which this new actor can cope with 'messy' or 'ill-structured problematical situations.' Consequently, the 'problem-situation' becomes the referent in a SSM conversation, and as a result the objective of the conversations will be to deal with this referent. A system-based language is also provided such that the referent and hence the conversations are structured in a systematic way. This new referent is of course not managed as external but it is locally relevant for those involved in the conversation, and it becomes the nucleus of the whole SSM process. The set of notions provided by SSM act as a language that is used to bring about the next process: First, it allows the participants to make observations about 'problem situations' (the referent of the conversation) as perceived by the individuals, using descriptive statements or what they call sentences that 'predicate the subject'. Following this, it allows them to exchange observations (rich picture) about the problem situation. Next, it provides the medium to discuss which observations belong together (root definitions). Subsequently, the process enable the participants to replace the original observations for new ones through conceptual models. Finally, based on the new observations (assumed to be of better quality), actions can be taken in the 'real world' to improve the problem situation.

What is interesting about this process is that, despite all the efforts to do a different type of research, an important part of the whole process looks very much like traditional science, however with a significant distinction: the referent is not taken as external to the process itself. Although this is significantly relevant it lacks other important characteristics of standard research. SSM was not originally developed as a research methodology, though it is claimed that its application is an instance of doing research (Checkland and Scholes 1990). If this is the case, SSM does not include explicit considerations of post-implementation problems, it does not consider the cost of future observations, it does not take into account the relation between solving the problem and the actions that should benefit from the solution (Higo 1996).

There are also other considerations to take into account if we conceive SSM as a conversation generator. First it is not clear when perceptions by participants in a debate on a specific problem situation take the form of descriptive statements or performatives, since if a person with authority is participating in the discussions, the other participants might 'perceive' some of those statements as performatives, creating 'confusion' in the process. Second, there seems to be not much consideration of the *local* languages already available to solve problems which might be faster and more effective to tackle problem situations. It may happen that those languages just need some additional *vocabulary* and/or just more precise *linguistic structures*, or additionally, the creation of contexts in which those languages become visible. All these aspects are clearly part of the research process and should also be considered.

CONCLUSIONS

The framework proposed so far seems to order the discussion about group support in an interesting way, since it allows us to concentrate on several issues which are difficult to visualise otherwise. It also provides a form of structuring a debate relative to research problems that are found in this area; similarly it gives us some tools to explore in more detail different types of group support when standard research seems to be not possible at all.

Finally, obvious space restrictions have not permitted to extend the discussion to other approaches that also aim to create some type of restricted conversations. Moreover, detailed explanations of the concepts introduced in this paper could not be provided. There is also a series of other elements that require special attention in terms of the framework proposed. First, the structure of conversational languages. Second, the relation between group structure and the structure of conversational languages. Third, the relations between speaker and addressee that can affect the course of a conversation, and hence the expected outcomes. Fourth, the organisational context that can influence the structure of a conversation, and finally, the effects for individual participants in a restricted conversation.

REFERENCES

Argyle, M. (1991) *Cooperation: The Basis of Sociability*. Routledge; London.

Argyris, C.; Putman, R. and Smith, D (1985) *Action Science: Concepts, Methods, and Skills for Research and Intervention*. Jossey-Bass Publisher; San Francisco.

Austin, J. (1962) *How to Do Things with Words*. Harvard University Press, Cambridge, MA.

Bhaskar, R. (1975) *A Realist Theory of Science*. 2nd Edition. The Harvester Press; Sussex.

Callon, M. (1986) *Some Elements of a Sociology of Translation: Domestication of the Scallops and the Fishermen of St Brieuc Bay*. In J. Law (Eds). *Power, Action and Belief: A New Sociology of Knowledge*. Routledge; London.

Checkland, P., (1981). *System Thinking, System Practice*. John Wiley & Sons; Chichester.

Checkland, P. and Scholes, J. (1990) *Soft Systems Methodology in Action*. Wiley & Son; Chichester.

Clark, H (1997) *Using Language*. Cambridge University Press; Cambridge.

Cole, S (1996) *Voodoo Sociology: Recent Development in the Sociology of Science*. In R. Gross, N. Levitt, M Lewis (Eds.) *The Flight From Science and Reason*. Annals of The New York Academy of Science, Vol. 775; New York.

Davis, J (1969) *Group Performance*. Addison-Wesley; Reading, MA.

Graziano, A. and Raulin, M. (1993) *Research Methods: a Process of Inquiry*. Second Edition; Harpen Collins; New York

Hayes, N (1997) *Successful Team Management*. International Thomson Business Press; London.

Heron, J. (1981) *Philosophical Basis for a new Paradigm*. In Reason, P. and Rowan, J. (eds.) *Human Enquiry: A Source of New Paradigm Research*. John Wiley & Sons; Chichester.

Higo, H. (1996) *Research-based Action in the Process of Socio-economic Development in Sudan*. PhD Dissertation; University of Lincolnshire and Humberside; Lincoln.

Kuhn, T. (1970) *The Structure of Scientific Revolutions*. Second Edition; University of Chicago Press; Chicago.

Lyotard, J (1984) *The Postmodern Condition: A Report on Knowledge*. University of Minnesota Press; Minneapolis (Translated by G. Bennington and B. Massumi from its French version *La Condition postmoderne: rapport sur le savoir*, 1979, by Les Editions de Minuit)

Nagel, E. (1979) *The Structure of Science: Problems in the logic of Scientific Explanation*. Hackett; Indianapolis.

Popper K., (1979) *Objective Knowledge, An Evolutionary Approach*. Revised Edition; Clarendon Press; Oxford.

Shaw, M. (1981) *Group Dynamics: The Psychology of Small Group Behavior*. Third Edition; McGraw-Hill; New York.

Turner, J. (1987) *Rediscovering The Social Group: A Self-Categorization Theory*. Basil Blackwell; Oxford.

Zeeuw, G. de, (1995). Value, Science and the Quest for Demarcation. *System Research*, Vol. 12, No. 1, pp. 14-24.

Zeeuw, G. de (1996) Second Order Organisational Research. *Working Paper No. 7*, School of Management, University of Lincolnshire and Humberside, UK.

MODELLING COMPLEX DECISION-MAKING

Susan A Smith

University of Paisley
High Street
Paisley

INTRODUCTION

This paper continues on from previous work carried out by Smith (1997). In that paper there was a description of the research which was being carried out into the modelling of the discharge decision-making process in the domain of Mental Health Care using the Appreciative Inquiry Method (AIM). Some of the lessons learnt about the domain and the method of inquiry AIM were described and illustrated

However the discharge decision-making process was used as the chosen domain in this research as it has been described as a complex decision-making process. In this paper there will be a discussion of the terms complexity and complex decision-making and the criteria for complex decision-making used in this research will be identified. These criteria will then be used to evaluate the discharge decision process to illustrate that it is an example of complex decision-making.

There have been many attempts by various authors to describe and classify 'models' and 'modelling' and in this paper there will be a brief outline of the arguments put forward and a definition and classification will be proposed. This classification and the criteria for complex decision-making will explain why AIM was chosen as the modelling approach in this research.

In the previous paper some of the lessons learnt from the research about the domain, AIM and the theoretical underpinnings to AIM were identified and discussed. However in this paper general lessons drawn from the research about the problems of modelling complex decision-making, decision-making and complexity will be outlined.

COMPLEXITY

In the literature authors, from different academic disciplines have discussed their ideas and attempted to define the nature of complexity (Klir, 1991; Checkland, 1981; Warfield, 1995; Ackoff, 1978; Kampis, 1990; Bremmerman, 1977). The definitions range from authors claiming that complexity can be broken down into constituent parts and measured to ideas that complexity involves people and thereby is very difficult to quantify. The author would argue that it would appear to be appropriate to categorise the literature on complexity under two headings.

Those that claim that complexity can be measured or quantified often by breaking it down into elements and the interrelationships between these elements. Among the writers of these books and papers is the idea that complexity is objective and reductionist methods can be used to analyse these complex problem situations (Klir, 1985; Kampis, 1991).

The authors that feel that an important element of complexity is that people have to be able to understand the situation (Ackoff, 1979; Warfield, 1995; Checkland, 1981). Therefore the involvement of people is a key element in complexity. If there is a group of people involved in a complex problem situation then this adds to the complexity as each person has their own perception and there is the interaction within the group. In the descriptions categorised under this heading two main themes have been identified by the author, namely that:

- Complexity involves people and this makes it uncertain and difficult to predict as people are initiators of action in their own right; and

- It is not possible to measure or quantify this type of complexity using a reductionist or scientific approach, as it is extremely difficult to quantify the context of any complex situation, and that a more holistic or systems approach might be a more appropriate approach.

COMPLEX DECISION-MAKING

Schoderbek, Schoderbek and Kefalas (1990) argue that the term complexity can be approached from many different viewpoints. They state that from a mathematical viewpoint, complexity can be explained in terms of the probability of a system's being in a specific state at a given time. From a non-quantitative viewpoint, complexity, they argue, can be defined as the quality or property of a system which is the outcome of the combined interaction of four main determinants, namely:

(i) The number of elements comprising the system;

(ii) The attributes of the specified elements of the system;

(iii) The number of interactions among the specified elements of the system; and

(iv) The degree of organisation inherent in the system. The existence or lack of predetermined rules and regulations which guide the interactions of the elements and/or which specify the attributes of the system's elements.

However, after reviewing the literature, the author would argue that a fifth determinant should be added to the list, namely:

(v) The involvement of and the interaction between people or stakeholders. The context of the situation may be difficult to understand by those involved.

However in complex decision-making the criteria for complexity, as previously defined, could be applied to the decision-making process and the decision makers have to review all the different elements and the relationships between them along with taking on board all of the ideas of the different stakeholders in the decision situation. It could be argued that all of the interactions makes each complex situation unique as people are initiators of action in their own right and are going to act differently in each situation.

EXPLORING AND MODELLING COMPLEX DECISION-MAKING IN THE DOMAIN OF MENTAL HEALTH CARE

The criteria for complex decision-making, identified in the previous section, can be applied to the domain of mental health care, and in particular the discharge decision-making process, to indicate why it can be regarded as complex in the following manner:

The number of elements comprising the system

There are a wide range of elements in the chosen system ranging from the issues to be considered in the decision itself to the fact that the process is a multidisciplinary team (MDT) process where a number of different stakeholders from different disciplines have input into the decision.

The attributes of the specified elements of the system

The attributes of the elements of the system vary.   The issues in the discharge process are partly rule-based and partly intuition gained from previous experience. The different stakeholders have their own ideas and perceptions that they bring to the decision as each person has their own set of values and beliefs.

The number of interactions among the specified elements of the system

After consultation with the other members of the MDT if the RMO decides that the patient could be discharged then a discharge plan or Care Programme is formulated to assess the services and care the patient will need in the community.   However the decision may be to move the patient to a ward with less security, such as an Open Ward, prior to discharge or to move him to a ward with more security, such as the State Hospital, if the patient is particularly violent. The patient may move through the different levels of security and care until ultimately he/she is discharged into Community Care.

The degree of organisation inherent in the system.

It is recognised that the practice of the act has to be balanced against the individual's civil liberty rights (Carson, 1990). The decision making is complex as it is a result of MDT decision making and it results from an assessment of a mixture of rules (derived from the Mental Health Acts) and 'personal' knowledge gained from professional judgement resulting from previous experience.

The involvement of and the interaction between people or stakeholders.

The Mental Health Act (1983) in England and Wales and the Mental Health (Scotland) Act (1984) in Scotland makes provision for the hospitalisation and treatment of individuals judged to be suffering from a mental disorder. When deciding if a patient should be discharged the Responsible Medical Officer (RMO) consults the various professionals who comprise the multidisciplinary team (MDT) which is responsible for the decision making process such as nursing staff, including the Community Psychiatric Nurse (CPN), Social Workers and sometimes Occupational Therapists and Psychologists.

PROBLEMS OF DEFINING MODELS AND MODELLING

A model can be described as a representation but it cannot be a perfectly accurate reflection of the real world. The main role of a model is to simplify the representation of reality so that certain aspects of the reality can either be examined or understood better. Therefore models can be used to help us to understand complexity (Flood and Carson, 1993). In all the ways that we think about describe and represent reality, we are always making simplifications for some purpose and therefore to some degree all our thinking and descriptive activities can be regarded as examples of modelling. So simplification of information is essential and any time that a situation is simplified then modelling is taking place. This indicates that models are used to change the participants' perception of the problem situation and this leads to the idea that models are subjective tools used to alter the inner mental models of the participants. Whatever the use of a model it is very important to remember that a model is an abstraction of the modeller's perception of 'reality' and it is neither 'reality' itself nor even an exact replicate of 'reality.' This is clearly identified in the definition as put forward by Wilson (1990) who defines a model as:

A model is the explicit interpretation of <u>one's</u> understanding of a situation, or merely of <u>one's</u> ideas about the situation. It can be expressed in mathematics, symbols or words, but is essentially a description of entities, processes or attributes and the relationships between them. It may be prescriptive or illustrative, but above all, it must be useful.

The important part of the definition is the one's (as underlined by the author to increase clarity) as it shows that the modelling is of the modeller's interpretation of a situation and that models will be different for each participant in a problem situation as each modeller has his own understanding which he will bring to the modelling process. In Wilson's definition he also says that a model has to be *useful* as it has to have a purpose. After reviewing the literature the author decided that the definition of model and modelling which is the most appropriate for this study is the Open University (1998) definition of modelling. This definition appears to encapsulate the ideas put forward by many of the authors in one definition. In this definition modelling is defined as "any process of abstracting and representing certain aspects of reality in a simplified form with some pre-defined purpose in mind" and implicit within this definition is the idea that a model is "a simplified representation of certain aspects of reality, constructed for some defined purpose." However this definition gives the impression that there is an external reality that is independent of the modeller and it does not include the ideas as put forward by Wilson in his definition. It would appear to be appropriate to add the ideas that modelling is of the modeller's interpretation of a situation and that the model is dependent on the modeller's perception and understanding, as put forward by Wilson, to the Open University definition.

Attempts have been made by various authors to develop a categorisation shema for modelling approaches (Ackoff, 1962, 1978 ; Wilson, 1990). An attractive approach might be to categorise modelling approaches into functionalist (objective) or interpretivist (subjective) approaches. Pidd (1996) used the terms 'hard' and 'soft' to categorise modelling approaches but as these terms are used frequently in the field of Information Systems it would appear to be appropriate to use more specific terms such as functionalist and interpretivist respectively. In a functionalist approach it is typically assumed that a model is a proper representation of part of the real world. It is accepted that the model will be a simplification and an abstraction of the real world that can be easily defined. In interpretivist approaches the models are developed to allow people to think through their own positions and to engage in debate with others about possible action (ibid.). If modelling approaches are categorised using this categorisation schema into functionalist and interpretive modelling approaches and then the criteria for complex decision-making applied to the two groupings functionalist approaches do not appear to comply with criteria (iv) or (v). However interpretivist modelling approaches generally appear to comply with the criteria for complex decision-making and it would appear that the modelling approach used should come from this grouping.

The Appreciative Inquiry Method (AIM) is an interpretivist modelling approach which has been developed to model the subjective intuitive elements as well as the more rule-based factors in expertise. AIM is an acronym for a set of elements arising from research by West (1991, 1995). AIM was chosen as it complies with the definition of modelling and the criteria identified for complex decision-making and builds on available expertise.

LESSONS LEARNT

This paper discusses the lessons learnt from the study about the modelling of complex decision-making, complex decision-making and complexity. The discussion of the lessons includes an outline of the difficulties in developing a single model for the complex decision-making process and the problems identified when attempting to model the different perceptions and opinions of all of the actors involved in the complex decision-making process. These lessons may help these issues to be addressed in a more appropriate manner in future studies.

REFERENCES

Ackoff, R.L., 1962, *Scientific Method: Optimizing Applied Research Decisions*, Wiley: New York
Ackoff, R., 1978,*The Art of Problem Solving*, New York: Wiley.
Bremmermann, H., 1977, Complexity and transcomputability. In: *The Encyclopedia of Ignorance: Everything you ever wanted to know about the unknown*. Pergamon Press.
Carson, D., 1990, *Risk-taking in Mental Disorder; Analyses, Policies and Practical Strategies*, SLE Publications Ltd.
Checkland, P.B. , 1981, *Systems Thinking, Systems Practice*. New York: Wiley.
Flood, R. and Carson, E., 1993, *Dealing with Complexity: An Introduction to the Theory and Application of Systems Science*.2nd ed., New York: Plenum.
Kampis, G., 1990, *Self Modifying Systems in biology and Cognitive Science: a new framework for dynamics, information and complexity*. Pergamon Press. IFSR Series on Systems Science and Engineering vol. 6.
Klir, G., 1991, *Facets of Systems Science*. Pergamon Press. IFSR Series on Systems Science and Engineering vol. 7.
Open University, 1998, *Working with Systems: Managing Change*, T247, Block 5, Part 2.
Open University, 1998, *Working with Systems: Representing Systems*, T247, Block 1, Part 3.
Pidd, M., 1996, *Tools for Thinking: Modelling in Management Science*, Wiley: Chichester.

Shoderbek, P., Schoderbek, C. and Kefalas, A., 1990, *Management Systems: Conceptual Considerations*, 4th ed., Irwin publishers.

Smith, S, 1997, Modelling the discharge decision-making process in the domain of Mental Health Care, In: *Systems for Sustainability: People, Organizations, and Environments*, F.A. Stowell, R.L. Ison, R. Armson, J. Holloway, S. Jackson and S. McRobb Eds., New York: Plenum

Warfield, J., 1995, Demands imposed on systems science by complexity, In: *Critical Issues in Systems Theory and Practice*, K. Ellis, A. Gregory, B. Mears-Young and G. Ragsdell Eds., New York: Plenum.

West, D., 1991, *Towards a Subjective Approach to Knowledge Elicitation for the Development of Expert Systems*, PhD Dissertation, Portsmouth Polytechnic (unpublished).

West, D., 1995, The appreciative inquiry method: a systemic approach to information systems requirements analysis, In: *Information Systems Provision: The Contribution of Soft Systems Methodology*, Ed.: F.A. Stowell, London: McGraw-Hill, pp183-204.

Wilson, B., 1990, *Systems: Concepts, Methodologies and Applications*, 2nd ed., New York: Wiley.

DISCOURSE, CALIBRATION, EXPLANATION AND EVOLUTION:

IMPLICATIONS FOR COMPUTER-BASED SYSTEMS DEVELOPMENT

David C Sutton

Lincoln School of Management
University of Lincolnshire and Humberside
Brayford Pool
Lincoln
LN6 7TS
dcsutton@systemsix.co.uk

INTRODUCTION

Requirements for computer based information systems (CBIS) are often influenced by rapidly evolving operating conditions so that either users do not know clearly what they want, or else constantly change their minds. In such situations, it may be impossible to ever arrive at a 'statement of user requirements' stable enough to be actionable by conventional information systems development methodologies. In addition, users and developers of CBIS have different agendas, working environments, professional backgrounds and even different languages. Misunderstandings may not be realised until the delivery phase of a system, by which time it is too late and too expensive to correct misunderstandings or to adjust in the light of redefinition of requirements.

The field of information systems has not yet established firm theoretical foundations for its 'best practice' (Dutton et al, 1995) or found its 'big idea' (Scarrott 1979; Checkland and Holwell 1995; Jones 1995; Checkland and Holwell 1998, xiii). Consequently it frequently fails to deliver on its promises (Kinsgton & Hassall, 1992; Dance 1994, p82). This paper explores possible implications for the conduct of CBIS development that arises from a synthesis of views on the manner in which people exchange and agree meanings. In particular, it combines a notion of 'levels of explanation' from Searle (1983, 1992) and 'somatic change' from Bateson (1972) to suggest how use of principles of natural selection may add a useful perspective on the problem of system development in contexts of rapidly evolving and/or unclear requirements.

THE STRUCTURE OF EXPLANATION

For people to arrive at a mutual understanding of anything, such as a statement of requirements for a CBIS, they need to establish a 'consensual domain' which is essentially linguistic (Winograd & Flores 1986, p50; Mingers 1993). Establishing consensus involves discourse to exchange and agree explanations and definitions (Lewis 1992). Asking anyone to explain anything without ambiguity can generate a never ending series of 'explanations' of successively greater detail or successively broader scope. This is termed by Searle the 'accordion effect' (Searle 1992, p88-89) and the same phenomenon is remarked upon by many other authors (Ayer 1956, p110; Heritage 1984, p158; Winograd & Flores 1986, p5; Maturana & Varela 1987, p135; Sutton 1987; Sutton 1988).

Basically, all explanations are a function of the perspective and interests of the speaker. Whatever is the focus of consciousness and attention is always awareness '*of* something *as* such-and-such', it is always seen in relation to a Background of knowledge and assumptions (Searle 1992 p132). This is equivalent to the Gestalt notions of 'figure' in relation to 'ground' (Perls 1973) and that "all knowing is knowing by an observer" (Maturana & Varela 1987, p34). In effect any 'appreciated world' is 'selected by our interests' (Vickers 1970, p97) and is a function of observer 'aspectuality' (Searle 1992 p131). The key is to note that all observation is 'aspectual' and so all explanations, and statements of requirements for CBIS, are observer relative.

If all explanations, and hence all definitions of meaning, are inescapably from points of view, they are all fundamentally properties of the unique actor concerned, and so subjective. This is where confusion can arise when, as in the case of CBIS development, users and developers are viewing requirements from very different aspectivities. Therefore there has to be some means to achieve alignment and calibration if there is to be any basis for 'objective' or at least consensual 'inter-subjective' (Mingers 1993) descriptions and definitions.

One approach to resolving this issue is discussed by Searle who uses the notion of 'emergence from the physics' to generate 'levels of explanation'. He proposes to use scientifically observable phenomena as the ground level for all explanations (Searle 1983 p268; Searle 1992 ch. 10). Searle's proposal sees both language and action as describable from three different perspectives. The gross observable phenomenon of 'what' someone did, 'how' it was done and 'why' they did it. Searle calls these three 'levels': Hardware, Functional and Intentional. This is a very powerful schema and fits in with other observations of the need for multiple levels of explanation (e.g. Vickers 1970, p113). These distinctions are illustrated in figure 1.

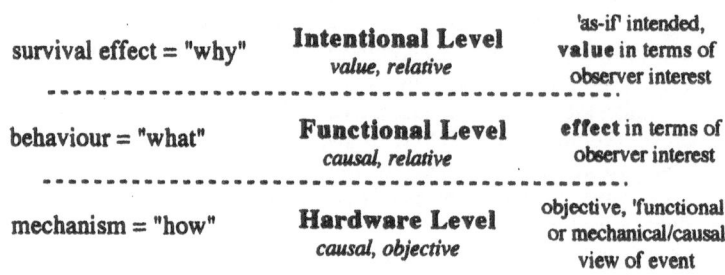

Figure 1. Distinctive features of 'Levels of Explanation' as developed in Searle (1992 ch 10)

This approach simplifies discussion of natural phenomena that are often discussed 'as-if' they were purpose driven. If the 'logical accounting' (Maturana & Varela 1987, p135) is kept clear, explanations can be derived with minimum influence from observer aspectuality. An example Searle uses of trying to separate explanations from observer aspectivity is the way plants turn their leaves towards the Sun. They do not do this 'in-order' to get more sunlight, their genotypically determined biological processes cause their leaves to turn towards the Sun. This is a phenotypic behaviour produced by the interaction of a genotype with its environment or 'medium'. Over long series' of plant lives, those whose secretion mechanisms happened to keep their leaves better oriented for maximal photosynthesis had better growth and reproduction dynamics that those which did not.   Their genotype survived, others died out.

It is important to realise that 'survival' is a value placed upon events by an **external observer with a particular interest,** and 'selection' is an 'as-if' functional description used by observers **with a particular interest** of a totally causally determined hardware phenomenon. These points are illustrated in figure 2. Basically, any 'pattern' observed in behaviour is already a description at a functional level and totally dependent upon the 'aspectivity' of the observer who 'sees' the pattern (Vickers 1970, p113).

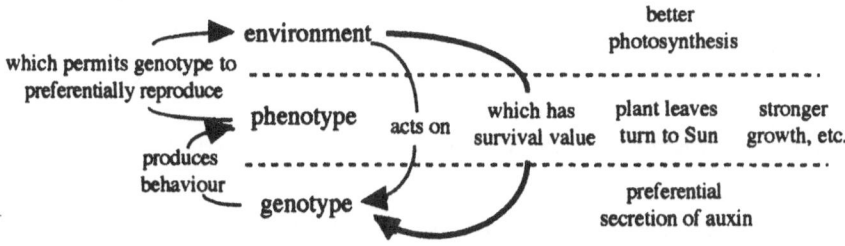

Figure 2. Illustrating the distinction between an explanation based upon a 'natural selection' perspective and one based upon 'emergence from the physics', as developed in Searle (1992 ch 10)

Before turning to how this material may be relevant to CBIS, it is necessary to elaborate upon how this scheme distinguishes between types of change and their effects. The *capacity* of an individual to behave is a function of its genetic makeup.   The *actual* behaviour is a function of the interaction of this capacity with the environment manifested as the phenotype.   Changes in actual behaviour are termed **somatic** changes whilst changes in capacity to change are termed **genotypic** change (Bateson 1972 p 346).

ECOLOGICAL ISD

The lesson from Searle and Bateson on 'somatic' versus 'genotypic' change for CBIS may run as follows: 'Genotypic change' in IS terms may relate to the rewriting of code and modification of computers and other infrastructure, probably by experts using rigorous methods. 'Somatic' change may relate to the utilisation of a system's inherent reconfigurability (hardware and software), probably by users in real-time.

If users, or other appropriate staff, are to be able to exploit 'somatic' systems reconfiguration as opposed to 'genetic' systems redesign, then CBIS must be 'genetically engineered' to provide radically increased levels of *capacity* to be reconfigured by the users. Naturally, also required will be other capacities required to maintain the cohesion and co-ordination of systems that can be tailored by individual users to avoid anarchy and maintain certain levels of standards.

As well as developing systems which support user driven 'somatic' reconfiguration, it may also be possible to apply the notion of 'natural selection' to phenotypical configurations derived from both user driven 'experimentation' (somatic variation) and longer time scale 'engineered change' (genetic change or mutation).   A top down view of systems development might consider the cycle to start at the Intentional level with a statement of requirements or the ends to be pursued. Design is then concerned with creating systems to bring about what should happen (the Functional level), the implementation and operation then become the actual mechanisms and events that actualise what does happen (the Hardware level).   These three levels can also be seen to relate well to the distinctions between 'computer' (hardware), 'program' (function) (Vickers 1970, p110) and the requirement or 'purpose' of the system (intention). A way of representing the classical 'systems development cycle' in this way is illustrated in figure 3.

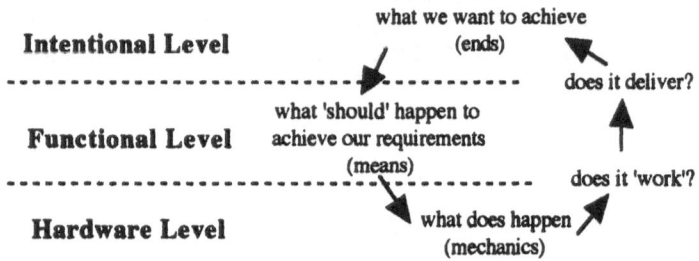

Figure 3. Classical systems development related to Searle's 'levels of explanation'.

On the other hand, an ecological approach to systems development would consist of trying a wide range of alternatives and 'seeing' what survives. This would entail encouraging experimentation and diversity (enabled by systems designed to be 'genetically' flexible, somatically reconfigurable yet organisationally controllable). These types of approach may be approximated by 'rapid prototyping', 'end-user computing' and 'user-led design'. This 'ecological systems development cycle' is illustrated in figure 4 which also suggests more 'managerial' terms for Searle's three explanation levels.

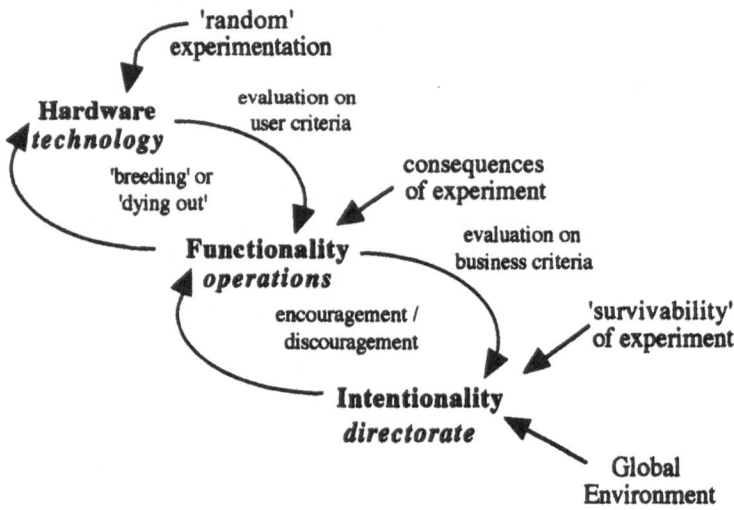

Figure 4. 'Ecological' systems development, employing Searle's 'levels of explanation' to relate facets of systems development to analogous features of natural selection.

A 'natural selection' approach may have a natural place in areas where 'end-user' computing is appropriate, but it would be unwise to fall into the usual IS trap of claiming it as the solution to all problems. An ecological approach would be unlikely to be appropriate for the design of, say, mission critical control systems. Nature is abundant but 'wasteful', in organisations a balance would need to be managed so that value adding 'mutations' predominate over unsuccessful ones. The required mix of control and support might be represented as shown in figure 5:

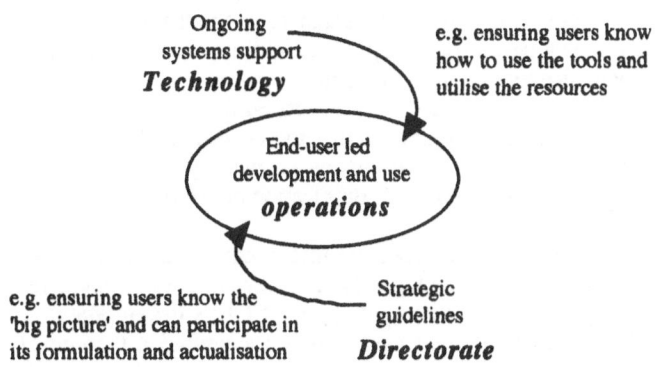

Figure 5. How 'Ecological' systems development may be supported by appropriate organisational processes and policies.

DISCUSSION

Clearly, the insight and intuition of experienced practitioners does move custom and practice into exploitable 'ecological niches' independently of the progress of formal theory. However, generally attempts are made to embed innovations into conventional formalised frameworks with resulting conflict and loss of much of their potential benefit. This is one danger of failing to recognise the ecological dimension to these 'improvements', another is that they become seen as a 'one size fits all' quick fix. The inevitable failures then destroy confidence, even in those situations where the advance is appropriate.

Apart from offsetting these disadvantages, recognising the ecological dimension can offer key benefits:

- **Ecological systems development may be faster and more inventive.** Controlled small scale experimentation can be conducted by far more people, potentially all users. This achieves deployment of far more developmental effort, far more rapid evolution and far more added value than if CBIS development is limited to a few assigned experts.
- **Ecological systems development may be more economical.** To exploit ecological principles requires consideration of how, in organisational terms, experiments (mutations) can be preferentially directed to advantageous directions rather than pursuing all avenues indiscriminately. If well informed about their overall and technical context, human judgement can achieve this in the same way as chess players 'instinctively' know which moves are worth exploring deeply and which to ignore.

That such additions to an enterprise's 'portfolio' (Rockart & Hofman 1992) of systems development approaches would add value is indicated by the 'evolutionary pressure' that has already given rise to innovations along the lines described above. These include: rapid prototyping, rapid application development, 'natural language' interfaces, 'evolutionary systems development' (Crinnion 1991), 'client-led design' (Stowell & West 1994), etc.

Naturally, there must be many managerial and cultural changes to exploit ecological systems development. This approach can be seen as a parallel in CBIS terms to the recommendation that, to operate in a chaotic environment, strategy must be allowed to emerge and must be formulated through contingent networks rather from hierarchical organised and closed 'project teams' (Stacey 1993).

CONCLUSIONS

All analogies are dangerous if pushed too far. However, consideration of the terminology of ecology and evolution suggests certain policy refinements with respect to CBIS development that could yield benefits:

- **Maximise the scope for user reconfigurability of systems.** The 'application backlog' is clearly an indicator that limiting the capacity to modify systems to a small group of specialists imposes a severe handicap on the ability of an enterprise's CBIS resources to coevolve with its operating requirements.
- **Maximise the scope for user sharing of successful innovations.** One advantage of centralised development of systems is that the chances of 'reinventing the wheel' are reduced. Steps to disperse development capability must be balanced by steps to promote sharing of learning to reduce duplication of effort.

Incorporating the ecological perspective into management and professional thinking must be more beneficial than allowing related innovations to struggle for survival in an environment that is geared to only one paradigm of CBIS development. Further, such innovations would present a stronger case if they recognised a theoretical basis for their ability to add value.

REFERENCES

Ayer, A.J., 1956, The Problem of Knowledge, Penguin (1990 reprint), London.

Bateson, G., 1972, Steps to an Ecology of Mind, Ballantine, New York.

Checkland, P., and Holwell, S., 1995, Information systems: what's the big idea?, Systemist 17(1):7.

Checkland, P., and Holwell, S., 1998, Information, Systems and Information Systems: Making Sense of the Field, Wiley, Chichester.

Crinnion, J., 1991, Evolutionary Systems Development, Pitman, London.

Dance, S.G., 1994, InfoPreneurs: the Hidden People who Drive Strategic Information Systems, Macmillan, Basingstoke.

Dutton ,W.H., MacKenzie, D., Shapiro, S., and Peltu. M , 1995, Computer Power and Human Limits: Learning from IT disasters, Policy research Paper No 33, PICT, SRC.

Heritage, J., 1984, Garfinkel and Ethnomethodology, Polity, Cambridge.

Jones, M.R., 1995, What is the distinctive nature and value of IS as a discipline?, Systemist, 17(1):31.

Kingston, R., and Hassall, J., 1992, Manufacturing estimating with graphical interface, J of Industrial Affairs, 1:2.

Lewis, P.J., 1992, The feasibility and desirability of a closer linking of SSM with data focused information systems development, Systemist 14(3):168.

Lewis, P.J., 1995, New challenges and directions for data analysis and modelling, in: Stowell F A (ed), Information Systems Provision: The Constribution of Soft Systems Methodology, McGraw-Hill, Maidenhead.

Maturana, H.R., and Varela, F.J., 1987, The Tree of Knowledge, New Science Library, London.

Mingers, J., 1993, An examination of information and meaning, Systemist, 15(1):17

O'Connor, J., and Seymour, J., 1990, Introducing Neuro-Linguistic Programming: The New Psychology of Personal Excellence, Mandala, London.

Perls, F., 1973, The Gestalt Approach, Science and Behaviour Books, New York.

Rockart, J.F., and Hofman, J.D., 1992, Systems delivery: evolving new strategies, Sloan Management Review, Summer 1992:21.

Scarrott, G.G., 1979, From computing slave to knowledgeable servant: the evolution of computers, Proc. R. Soc. ,London, A 369:1.

Searle, J.R., 1983, Intentionality: an Essay in the Philosophy of Mind, Cambridge Univ. Press, New York.

Searle, J.R., 1992, The Rediscovery of the Mind, MIT, Cambridge, Mass.

Stacey, R.D., 1993, Strategic Management and Organisational Dynamics, Pitman, London.

Stowell, F., and West, D., 1994, Client Led Design: a Systemic Approach to IS Design, McGraw-Hill, Maidenhead.

Sutton, D.C., 1987, Cognitive transactional analysis - towards a calculus of belief?, Systemist, 9(1):2.

Sutton, D.C., 1988, Cognitive Transactional Analysis - A Model of Purposeful Communication, in: Trappl R. ed., Cybernetics and Systems '88, Kluwer, London.

Vickers, G ., 1970, Freedom in a Rocking Boat, Pelican, London.

Winograd, T., and Flores, F ., 1986, Understanding Computers and Cognition, Addison Wesley, Reading, Mass.

WEAK INCOMMENSURABILITY
AND THE DEVELOPMENT OF SYSTEMS THINKING

David Watson and Paul Ledington

School of Information Systems and Management Science
Griffith University
Nathan Qld 4111

INTRODUCTION

Is "Systems" relevant any more? Is it a body of knowledge that is turned to when problems are encountered? Whilst the erudite arguments of learned academics in the area fill many shelves, where are the practical benefits of the ideas being realised? And, if the answers to these questions are not readily apparent, then how can the systems movement create a new dynamic in its development. How can we innovate Systems? It is this problem that we seek to clarify.

Innovation in this context is a rather ill-defined concept which can best be seen as a deliberate attempt to change the fabric of the discourse within the systems community itself. The emergence of Soft Systems Thinking (Checkland, 1981; Checkland & Scholes, 1990; Davies & Ledington, 1991; Wilson, 1984) and latterly Critical Systems Thinking (Flood & Jackson, 1991; Mingers & Gill, 1997) are clearly major innovations in the area. The focus of this paper however is not to present such a major innovation, rather it is to simply seek new areas in which discourse may emerge. We suggest that within the Systems domain the problem of "Paradigms" has constrained development because it has assumed a 'strong' definition of paradigm incommensurability. In this paper we propose using a 'weak' definition of incommensurability and the "Paradigm Interplay Model" (PIM) as a means to transcend the current impasse in the area. Finally the paper suggests that the PIM can open up fresh directions for development of the systems movement.

BACKGROUND AND MOTIVATION

Mingers and Gill (1997) provide a genealogy of the systems movement development from its, simple, operational research conception through to its current, embryonic, pluralistic paradigmatic and methodological perspective. It is our perception that the diversity of the systems movement has become a barrier to the progress of the movement; and whilst on the

Synergy Matters: Working with Systems in the 21st Century,
Edited by Castell *et al.*, Kluwer Academic / Plenum Publishers, New York, 1999.

surface it would appear that the systems movement has developed over the past two decades, arguably little progress has been made.

Hirschheim, Klein and Lyytinen (1995) mapped the elements of the Information Systems movement onto the Burrell and Morgan (1979) framework of thinking about social reality (Figure 1.); placing the various methodologies and approaches into a new and quite difficult discourse. Within the systems movement, the root problem stems from the different senses in which the word "paradigm" is used. In the original 'hard' - 'soft' debate (Rosenhead, 1989) the term was used as the dominant model of practice; but in the Burrell and Morgan terminology "Paradigm" is the set of (implicit) assumptions about reality, knowledge, and the social world. In this framework, each quadrant in the typology represents a different set of assumptions and hence each quadrant represents a paradigm. The paradigms are incommensurable, and, we suggest, the result of reframing the discourse of the systems movement in relation to one structure of social reality has been to create an impasse to the rational development of the area.

Sociology of Radical Change

	Radical Humanism (Critical Theory, SSM)	Radical Structuralism	
Subjective	Interpretive Sociology	Functionalist Sociology ("Hard" Approaches)	Objective

Sociology of Regulation

Figure 1. The Burrell & Morgan Typology

TOWARDS A NEED TO RETHINK THE 'SYSTEMS MOVEMENT'

We contend it is the notion of 'practice' that distinguishes the area of *'Problem-solving' applications of systems thinking to real-world problems* from the area of *Theoretical development of systems thinking,* within Checkland's (1981) conceptualisation of the Systems movement. Therefore, any representation of the 'Problem-solving' area should contain the notion of practice. Further it should also contain the notion of theory. Our concern is that the current state of the systems movement embodies a disabling tension between theory and practice. Practice remains complex and challenging, requiring both more sophisticated tools and more sophisticated practitioners, yet theory can offer neither a coherent set of tools nor a rational approach to the development of better practice. If theory cannot inform practice then practitioners will base their learning on individual intuition and experience. Practice will follow its own developmental path, perhaps producing an incoherent overall pattern. Therefore, there is an incipient crisis. If theory can no longer support practice they will diverge and theoretical work will be seen as increasingly isolated and irrelevant.

If 'strong' incommensurability is taken seriously then the development of the systems movement would have to be viewed in terms of changes in isolated methodologies. Development would depend upon independent research teams and the exigencies of research situations. On the other hand, if no importance is attached to underlying assumptions about social reality then the development of the systems movement would be seen as the elaboration of a single pattern which can result in a single body of knowledge expressing 'best practice'.

The actual process seems to lie somewhere between these two extremes. There is an interest in combining approaches yet also a recognition that achieving this aim is not necessarily straightforward, and that there are philosophical issues to be addressed.

In Table 1., below, the different extant methodological positions within the systems movement are affiliated with a dimension indicating the influence of philosophical assumptions on the nature of practice (cf., Jackson, 1997; Mingers, 1997). Thus Isolationism occurs when the influence of the philosophical assumptions is strong and they include 'strong' incommensurability. Total Systems Intervention (TSI) for example assumes choosing a single appropriate methodology for the problem being addressed. It therefore assumes strong incommensurability but that such assumptions only have a weak influence on practice. It treats methodologies as tools rather than as praxis. Whilst Multimethodology is a broad concept it does assume weak incommensurability and depending on the strength given to the importance of philosophical differences will produce either a Plural form of multiparadigm Multimethodology or an Imperialistic form.

Table 1. The Influence of Philosophical Assumptions on the Nature of Practice

		PHILOSOPHICAL POSITION		
		Strong Incommensurability	Weak Incommensurability	Commensurability
INFLUENCE ON PRACTICE	Strong	Isolationism	Pluralism, Multimethodology	"Best" Practice
	Weak	"Toolbox", TSI	Imperialism, Multimethodology	Pragmatism

The issue that this analysis raises is "what is the nature of weak incommensurability?" It is this issue that is addressed in the next section.

THE PARADIGM INTERPLAY MODEL (PIM)

In developing multiparadigm research in the domain of organisational culture Schultz and Hatch (1996) present a 'weak' incommensurability model as one of three possible metatheoretical positions. These three positions can be summarised as; 1. *Paradigm Incommensurability*, which is the position adopted by Burrell and Morgan (1979). It is the current shared frame of meaning and baseline for much of mainstream social science discourse; although it is coming under increasing pressure from the proponents of the multiparadigm approach to research (Gioia & Pitre, 1990). 2. *Paradigm Integration*, where a single metatheoretical position is sought within which to undertake research and discourse. This is essentially the position of natural science and leads to the notion of 'best' practice within management science. 3. *Paradigm Crossing* which is a position where: *"the researcher recognizes and confronts multiple paradigms rather than ignoring them (Paradigm Integration) ... or refusing to confront them (Paradigm Incommensurability)"* (Schultz & Hatch, 1996; p.533). This last position is similar to what has been termed 'weak' incommensurability.

Four variants of the "paradigm crossing" metatheoretical position are developed, these are; sequential, parallel, bridging, and interplay. In the "sequential" approach, mutually complementary paradigms are employed serially such that the output of one is used to inform the input of another. This is similar to the notion of "grafting" developed to understand the use of Soft Systems Methodology in the development of Information Systems (Miles, 1985). It is also the approach underlying Total Systems Intervention (TSI). Schultz and Hatch (1996) point out that this strategy is unidirectional and considers the boundaries of paradigms to be impermeable. The "parallel" approach, by contrast, is multidirectional and considers all paradigms to be of equal standing but similar to the sequential approach takes the paradigm boundaries to be impermeable. It allows the researcher to compare the differences and conflicts between paradigms but does not focus on the similarities. "Bridging" approach differs from both the sequential and parallel approaches by seeing the paradigm dimensions as blurred permeable continua rather than as precisely defined domains. Finally, the "paradigm interplay" approach is a strategy that simultaneously holds in tension "*both the contrasts and connections between paradigms*" (Schultz & Hatch, 1996; p.234).

If these ideas on 'weak' incommensurability are put into the context of the 'problem-solving' element of the systems movement then the issue can be seen to relate to a 'Learn from Use' experiential process. Paradigm isolation suggests that only one paradigm can be used to make sense of the experience arising in practice. Alternatively "paradigm crossing" suggests that it is possible to use more than one paradigm to generate knowledge from the experience. In the sequential and parallel strategies the problem of integrating the knowledge generated is left to be resolved at a metatheoretic level. It is only with the strategies of "bridging" and "interplay" that knowledge integration is addressed. In the "bridging" strategy a dominant paradigm is presumed and some complementary knowledge can cross the bridge between paradigms. There is a parallel here between the bridging strategy for paradigm interaction and the Imperialist developmental approach discussed by Jackson (1997). Finally, the "paradigm interplay" strategy suggests that the interpretation of experience occurs within a framework of similarities and contrasts generated by the interaction of differing ideal paradigm positions.

PIM and 'Zones of Indifference'

Gioia and Pitre (1990) reconceptualised the boundaries between paradigms. Rather than treat them as hard, impermeable walls they developed the concept of "transition zones". Transition zones are areas in which paradigmatic dimensions become indistinguishable to the researcher and hence in these zones bridging can occur. Similarly we suggest the concept of 'zones of indifference'. These are created by the juxtaposition of two or more opposing paradigmatic constructs whose influence varies throughout its length on a continuum from strong to weak. Thus, the strength of connection and the direction of influence of a particular construct will follow the degree of relative strength/weakness between the interacting paradigmatic constructs. An example is given in Figure 4., where the notion of 'Ontology' is conceptualised as the interaction of two dimensions - Nominalism and Realism. Each dimension varies in paradigmatic influence on a continuum from strong to weak. Paradigm interplay 'connections' are more likely to occur where neither construct is able to exert a significant distinguishing influence over the other. This area is called the 'zone of indifference' to distinguish it from the concept of "transition zone". At the extremes of the continua lie the areas where one paradigmatic construct exerts overwhelmingly strong influence over the weak opposing construct. These areas we conceptualise as 'zones of conflict' (which represent the areas of strong paradigmatic 'contrast' in the PIM strategy).

We suggest that this new concept of 'zone of indifference' differentiates the concept of "paradigm interplay" from "paradigm bridging", and by decentring the focus of discussion it creates space for a discourse on connections between paradigms - unencumbered by preexisting metatheoretical barriers.

ONTOLOGY

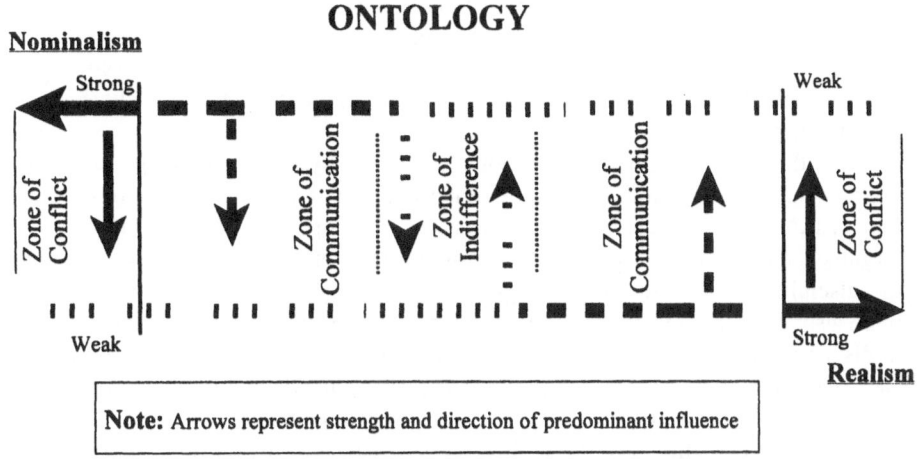

Figure 2. An example of the concept of "Zone of Indifference"

DISCUSSION

The idea of "paradigm interplay" conceptualises paradigms as ideal positions, or appreciative settings, within a broader discourse, or appreciative field. If for example development takes place close to an ideal setting then that setting will dominate the area and development will be similar to the notion of Isolationism presented in a previous section. Similarly "paradigm interplay" provides a mechanism whereby one dominant setting strongly influences development work, but knowledge from other areas can be incorporated, thus creating something parallel to the idea of Imperialism. It also provides however a development approach in which multiple paradigms interact with each other thereby creating a pluralistic mode of development. The PIM does not suggest that these are separate development strategies but rather that the strategies are interconnected. Isolationism develops a sharper definition of an 'ideal'approach, Imperialism sharpens the understanding of the connections and differences between one paradigm and work in others, together both Isolationism and Imperialism sharpen the definition of the zones of indifference within which work informed by multiple paradigms can occur. The PIM connects the disparate research approaches in the systems movement and suggests that they are complementary rather than exclusive.

CONCLUSION

In this paper we have suggested that there are tensions within the development of the applied systems discipline both within the theoretical aspects of the discipline and between theory and practice. The problem was traced back to the influence of the issue of 'strong' incommensurability between sociological paradigms on the metatheoretical development of applied systems thinking. We suggested that the idea of 'weak' incommensurability links to contemporary work on the development of multimethodology and of pluralistic conceptions of the development of the discipline. The work of Schultz and Hatch (1996) on multiparadigm research was introduced and was shown to provide concepts relevant to the notion of 'weak' incommensurability. Finally, the most advanced concept (that of "paradigm interplay") was used to suggest ways forward for the discipline.

The concept of "paradigm interplay" provides three potential sources of hope for the systems movement. First, it provides a potential mechanism for uniting and integrating work from across the discipline. Second, it legitimates Isolated, Imperialist, and Pluralist developmental approaches. Third, it suggests a possible pluralistic development strategy; that is, to develop an approach which utilises both the similarities and differences between systems approaches and paradigms.

Over the past 15 years, or more, Applied Systems Thinking has become compartmentalised into 'Hard', 'Soft' and 'Critical' domains. The practice focus of the area has withered as competing explanations, or approaches, developed but were not integrated. It is our argument that much of the current slow progress, and tension between theory and practice in the discipline can be traced to the problem of 'strong' incommensurability but 'weak' incommensurability is already implicitly accepted by the discipline and that opens up the arena for a discussion of models of 'weak' incommensurability. The Paradigm interplay concept creates space for new areas of discourse and consequentially offers the potential for exploration and therefore progress. There is a long way to go before it is sensible to talk of a 'Paradigm Interplay Methodology', but this paper has conceptually introduced the possibility of such methodology, and has thereby established the potential for progress.

REFERENCES

Burrell, G., & Morgan, G. (1979). *Sociological Paradigms and Organisational Analysis: Elements of the Sociology of Corporate Life*, Heinemann, London.

Checkland, P. (1981). *Systems Thinking, Systems Practice*, John Wiley & Sons, Chichester.

Checkland, P., and Scholes, J. (1990). *Soft Systems Methodology In Action*, John Wiley & Sons, Chichester.

Davies, L., and Ledington, P. (1991). *Information In Action Soft Systems Methodology*, Macmillan Education, Houndmills.

Flood, R.L., and Jackson, M.C. (eds.) (1991). *Critical Systems Thinking Directed Readings*, John Wiley & Sons, Chichester.

Gioia, D.A, and Pitre, E. (1990). Multiparadigm Perspectives on Theory Building. *Academy of Management Review*, 15, 584-602.

Hirschheim, R., Klein, H.K., and Lyytinen, K. (1995). *Information Systems Development and Data Modelling Conceptual and Philosophical Foundations*, Cambridge University Press, Cambridge.

Jackson, M.C. (1997). Pluralism in Systems Thinking and Practice. In Mingers, J., and Gill, A. (eds.), *Multimethodology The Theory and Practice of Combining Management Science Methodologies*, John Wiley & Sons, Chichester, pp.347-378.

Miles, R.K. (1985). Computer Systems Analysis: the constraints of the hard systems paradigm, *Journal of Applied Systems Analysis*, 11, 55-65.

Mingers, J. (1997). Multi-Paradigm Multimethodology. In Mingers, J., and Gill, A. (eds), *Multimethodology The Theory and Practice of Combining Management Science Methodologies*, John Wiley & Sons, Chichester, pp.1-20.

Mingers, J., and Gill, A. (eds.) (1997). *Multimethodology The Theory and Practice of Combining Management Science Methodologies*, John Wiley & Sons, Chichester.

Rosenhead, J. (1989). *Rational Analysis For A Problematic World Problem Structuring Methods for Complexity, Uncertainty and Conflict*, John Wiley & Sons, Chichester.

Schultz, M., and Hatch, M.J. (1996). Living With Multiple Paradigms: The Case Of Paradigm Interplay in Organizational Culture Studies, *Academy of Management Review*. 21, 529-557.

Wilson, B. (1984). *Systems: Concepts Methodologies and Applications*, John Wiley & Sons, Chichester.

HOW PREVALENT IS SYSTEMS THINKING IN THE METHODS, TOOLS AND TECHNIQUES USED IN PROJECT MANAGEMENT?

Diana White

Centre for Complexity and Change
Open University
Milton Keynes, MK7 6AA

INTRODUCTION

There is ample evidence that complex projects would be better managed by the application of systems thinking (see for example: Lyytinen et al., 1987; Kirby, 1996; Neal, 1995; Willcocks and Griffiths, 1994) but of the many methods, tools and techniques that are available for managing projects and assessing risk, few use such thinking. The majority of methods, tools and techniques encourage a reductionist approach, with the best known placing particular emphasis on quantification.

There is also a lot of data to show that many projects, especially IT projects, are not well managed. Widely known examples include the introduction of TAURUS (Currie, 1994) and the London Ambulance Service project (Hougham, 1996). However, it is impossible to start to judge the extent to which this is due to the inadequacy of any methods, tools and techniques used because the extent to which project managers make use of methods, tools and techniques is itself unknown.

This paper reports some of the findings of a recently conducted survey that was designed to capture the 'real world' experience of project managers and to establish which methods, tools and techniques they actually use. First, the methods, tools and techniques used by respondents are listed. The most frequently cited are then examined to try to determine how well they deal with the following factors:

- Complexity
- Environmental influences
- Change
- Uncertainty

Finally the paper considers the extent to which the survey finding support the argument that if they are to facilitate more effective project management, the methods, tools and techniques used must be particularly good at modelling complexity and at encouraging project managers to take account of the environments of projects.

Synergy Matters: Working with Systems in the 21st Century,
Edited by Castell *et al.*, Kluwer Academic / Plenum Publishers, New York, 1999.

163

THE SURVEY – A BRIEF DESCRIPTION

Following successful completion of a pilot survey, 995 questionnaires were sent to named project managers representing 620 organisations, in both the public and private sector across the UK. The main areas (relevant to this paper) explored by the survey were:

- Methods tools and techniques used for managing a recently completed project
- Limitations or drawbacks of the methods, tools and techniques used
- Any unexpected side effects or outputs arising from the project

A total of 236 questionnaires were returned (23.72% response rate) representing over 12 different industry sectors. The mode number of people involved in the projects was 10, and the mode duration of the projects was six months. 9% (21) of the projects did not run through to completion. Over 45% (108) of the projects were said to have given rise to unexpected side effects or outputs. Between them respondents used 1210 methods, tools and techniques and described 122 problems associated with their use.

METHODS TOOLS AND TECHNIQUES USED BY RESPONDENTS

The mean, mode, range and sum for the methods, tools and techniques used are tabulated in Table 1. 2% (5) of respondents stated that they did not use any methods, tools or techniques for managing their project or for assessing risk. 28% (66) of respondents did not use any project management method or methodology but over 95% (225) of respondents used at least one project management tool. 65% (154) of respondents did not use any risk assessment tools and 52% (123) of respondents did not use any decision making techniques.

Table 1. Methods Tools and Techniques Used by Respondents.

PROJECT MANAGEMENT METHODS, METHODOLOGIES, TOOLS AND TECHNIQUES	N		Mean	Mode	Range	Sum
	Valid	Missing				
Project Management Methods/Methodologies	236	0	0.87	1	3	206
Project Management Tools	236	0	2.61	1	7	617
Decision Making Techniques	236	0	0.73	0	4	172
Risk Assessment Tools	236	0	0.62	0	10	147
Computer Models/Databases/Indexes	236	0	0.17	0	3	40
Computer Simulations	236	0	0.05	0	2	11
Other Techniques	236	0	0.07	0	1	17
All Methods, Tools and Techniques	236	0	5.13	3	23	1210

Table 2 shows the individual methods, methodologies, tools and techniques used by the respondents and gives a count of the frequency of mention. 54% of respondents used their own 'in house' project management method. The most commonly used project management tool (77%) was 'off the shelf' software. Only one respondent stated that the Soft Systems Methodology had been used. This was included in "other project management methods/methodologies" category.

EXAMINATION OF THE METHODS, TOOLS AND TECHNIQUES

Methods and Methodologies

In house Project Management Methods. It difficult to comment on how well these

Table 2. Number of times the methods, tools and techniques were used.

PROJECT MANAGEMENT METHODS/METHODOLOGIES	
PRojects IN Controlled Environments (PRINCE)	23
PRojects IN Controlled Environments 2 (PRINCE2)	14
Structured Systems Analysis and Design Methodology (SSADM)	17
The European Risk Management Methodology (RISKMAN)	1
The RIBA Plan of Work	2
Other Project Management Methods/Methodologies*	16
In house Project Management Methods	128
In house similar to PRINCE	5
PROJECT MANAGEMENT TOOLS	
Critical Path Method (CPM)	70
Work Breakdown Structure (WBS)	75
Cash Flow Analysis (CFA)	43
Gantt Bar Charts	152
Graphical Evaluation and Review Technique (GERT)	4
Programme Evaluation and Review Technique (PERT)	24
Strengths Weaknesses, Opportunities and Threats (SWOT)	41
Other Project Management Tools**	21
Project Management Software	182
In house Project Management Tools	5
DECISION MAKING TECHNIQUES	
Cost Benefit Analysis (CBA)	88
Decision Analysis (DA)	9
Sensitivity Analysis (SA)	19
Expressed Preferences	23
Implied Preferences	11
Revealed Preferences	11
Other Decision Making Techniques	9
In house Decision Making Techniques	2
RISK ASSESSMENT TOOLS	
Life-cycle Cost Analysis (LCCA)	25
Event Tree Analysis (ETA)	8
Fault Tree Analysis (FTA)	6
Probability Analysis (PA)	34
Reliability Analysis	13
Uncertainty Analysis	3
Failure Mode and Effect Analysis (FMEA)	10
Hazard Analysis (HAZAN)	9
Hazard and Operability Studies (HAZOP)	9
Operation and Maintenance Risk Analysis (OMRA)	4
Preliminary Hazard Analysis (PHA)	5
Other Risk Assessment Tools	7
In house Risk Assessment Tools	14
COMPUTER MODELS/DATABASES/INDEXES	
CRUNCH	1
Lessons Learnt Files (LLF)	23
Expert Systems	4
In house Computer Models/Databases/Indexes	12
COMPUTER SIMULATIONS	
Hertz	1
Monte Carlo	10
OTHER TECHNIQUES	
Other Techniques	17

*Includes other methods used in Information Systems Development Projects
**Includes tools used in Information Systems Development Projects

deal with the factors listed earlier. Six respondents sent further information about their methods and although some of these did take account of complexity all but one could be described as 'hard', goal-directed approaches. A further four respondents stated that their in house methods were adaptations of the SSADM.

PRINCE and PRINCE2. These are complete, structured methodologies for the organisation, planning and control of projects. Projects are broken down into products and stages. Deliverables must be defined at the start of the project sufficiently precisely as to be measurable against pre-defined metrics. Emphasis is put on risk management rather than identification (CCTA, 1997). It is therefore suggested that environmental influences could be missed. Any project changes are handled as project issues, necessitating any proposed changes to be reported and logged. This paper submits that this could lead to changes being viewed in isolation and not linked to other activities or areas of change, furthermore, complex interactions between stages could be missed. Partington (1996) argues that a structured project management method that offers a standard approach is unlikely to be of value to a project that is characterised by uniqueness and uncertainty.

Structured Systems Analysis and Design Methodology (SSADM). This methodology was designed for the development of information systems. It is based on the premise that most information systems share a common life cycle. Projects are broken down into modules, stages, steps and tasks. It is a product driven approach that requires user requirements to be fully described and understood at the outset of a project (Weaver, 1994). It is suggested that this method would only be of benefit to projects where user requirements are certain and are unlikely to change as a result of environmental or other influences. Baccarini (1996) defines a technologically complex project as one requiring a great number of interdependent tasks and argues that technological complexity should be managed by integration. It is therefore suggested that technologically complex projects would not be well managed using a method that is based on separation.

Tools and Techniques

Project Management Software. The capabilities and features of project management software vary a great deal among the many products available, however most provide a means of planning, tracking and monitoring a project. The data format for describing the project to the computer is usually based on standard activity network typologies (Kerzner, 1998). Only a few software packages include the facility to track shared resources between projects (Cotterell, 1998), it is therefore suggested that the lack of this facility could constrain accurate modelling of complex multi-resource projects so that important aspects of the project's environment could be missed. Dawson and Dawson (1998) argue that the deterministic nature of the standard activity networks is not at all suited to projects where there is uncertainty. Rosenau (1992) suggests that the convenience of project management software in allowing any planning changes to be quickly and easily modelled may encourage frequent minor plan revisions which in turn could exacerbate the potential for different people working on the project to have different versions of the plan.

Gantt Bar Charts. These are a means of displaying simple activities plotted against time. An activity represents the amount of work required to proceed from one point in time to another (Kerzner, op. cit.). Complex projects with many interacting activities and dependencies would be difficult to plot on a Gantt chart (Maylor, 1996). They become difficult to update where there are many changes and are best suited to where the environment is static and there are low levels of uncertainty (op. cit., 1996).

Work Breakdown Structures (WBS). These are product-oriented family tree subdivisions of the hardware, services, and data required to produce an end product. A total project can be described as a summation of the sub-divided elements. For complex projects the interdependencies between activities can become so complex that meaningful networks cannot be constructed (Kerzner, op. cit.). Once the WBS is established it can become a very costly procedure to add or delete activities (op. cit., 1998). It is therefore suggested that WBS are not suited for use with complex projects characterised by uncertainty.

Critical Path Method (CPM). This method identifies the shortest line (in terms of elapsed time) through a network diagram of events and activities. The CPM uses a single time estimate for each activity and is deterministic in nature, resources are not easily rescheduled, and its use is best suited to projects where there are relatively small uncertainties (Kerzner, op. cit.). Hulett, (1995) argues that the use of the CPM in complex projects would result in many merge points, where high-risk activities could be missed. Hulett also suggests that the CPM does not explicitly deal with uncertainty arising from external factors.

Cost Benefit Analysis (CBA). In a CBA the expected benefits from a proposed activity are simply weighted against their expected costs. This assumes all adverse conditions for any given activity are known in advance (Cutter, 1993). Cutter argues that cost benefit techniques are not suitable for making risk management decisions. It is therefore suggested that high-risk projects characterised by change, environmental influences, uncertainly and complexity would be better suited to other decision making techniques

Expressed Preferences. This is a straightforward method for determining what people require or find acceptable by asking them directly to express their preferences. However people may not really know what they want, their values may rapidly change, and when dealing with complex projects it would be impossible to know and assess all alternatives. (Rowe, 1980). Once again it is suggested that this decision making technique is not suited to high-risk projects.

Life-Cycle Cost Analysis (LCCA). This takes place early in a project and is the systematic analysis of the total cost to an organisation of the R&D, production, operation and maintenance of a product. The disadvantage of using a LCCA is that it is highly sensitive to changing requirements (Kerzner, op. cit.). It also requires early estimates of project complexity and available resources (op. cit., 1998). It is therefore suggested that future changes or environmental influences could affect a LCCA, furthermore, in complex projects it may be impossible consider all costs.

Probability Analysis (PA). This is used to determine the probability of events and the consequences associated with their occurrence. The main limitations with a PA include ensuring completeness of the analysis and the adequacy of the data (Linnerooth-Baye et. al. 1991). The use of PA is therefore less appropriate for complex unstructured projects (op. cit., 1991).

Lessons Learnt Files (LLF). These are databases of lessons learnt from past projects. They rely on past failures having been recognised and recorded (Kerzner, op. cit.). Although LLF are valuable tools not all projects are the same. Environment influences could be different, complex interactions could cause different effects, and sources and types of risk evolve over time.

CONCLUSIONS DRAWN FROM SURVEY FINDINGS

41% (99) of respondents reported at least one limitation to the methods, tools and techniques they used. In total 122 limitations were reported. These were broken down as follows:

Inadequate for complex projects	32
Difficult to model the 'real world'	18
Too heavy in documentation, too time consuming	12
Failed to predict problems	9
Constrained activities, did not allow a holistic view	9
Too unwieldy, not cost effective	9
Others (6 different limitations mentioned)	33

The above taken together with the fact that nearby half of the respondents to the survey (46%) found that their projects gave rise to unexpected side effects and that most of the side effects (over 70%) could be attributed either directly or indirectly to lack of awareness of the environment, suggest that most tools and techniques currently used by project managers are poor at modelling 'real world' problems. Furthermore, over half of the respondents to the survey used 'in house' project management methods, suggesting dissatisfaction with standard methodologies. Comments made by respondents to the survey support the view that to be effective project management methods, tools and techniques must be good at modelling complexity and considering a project's environment. As one of the respondents put it: 'The requirement is for management tools capable of matching the realities of the accelerating range of complexity and change experienced in business.'

REFERENCES

Baccarini, D., 1996, The concept of project complexity – a review, *Int. J. of Project Mgmt.*, 14.1, pp. 201-204.

CCTA, 1997, "PRINCE 2 An Overview". HMSO, London.

Cotterell, S., 1998, Annual software review 1998, *Project Manager Today*, August. pp. 22-23.

Currie, W., 1994, The strategic management of a large scale IT project in the financial services sector. *New Technology, Work and Employment, 9:1* pp. 19-29.

Cutter, S., 1993, "Living with Risk: the Geography of Technological Hazards", Edward Arnold, Kent.

Dawson, R. J. and Dawson, C. W., 1998, Practical proposals for managing uncertainty and risk in project planning, *Int. J. of Project Mgmt., 16:5, pp. 299-301.*

Hougham, M., 1996, London Ambulance Service computer-aided despatch system, *Int. J. of Project Mgmt.*, 14:2, pp. 103-110.

Hulett, D. T., 1995, Project schedule risk assessment, *Project Mgmt. J.*, 26:1, pp. 21-31.

Kerzner, H, 1998, "Project Management: a Systems Approach to Planning, Scheduling, and Controlling", Van Nostrand Reinhold, New York.

Kirby, E. G., 1996, The importance of recognising alternative perspectives: an analysis of a failed project, *Int. J. of Project Mgmt.,* 14:4, pp. 209-211

Linnerooth-Baye, J. and Wahlstrom, B., 1991, Applications of probabilistic risk assessments: the selection of appropriate tools, *Risk Analysis*, 11:2, pp 239-248.

Lyytinen, K. and Hirschheim, R., 1987, Information systems failures - a survey and classification of the empirical literature, *Oxford Surveys in Information Technology*, 4, pp. 257-309.

Maylor, H., 1996, "Project Management", Pitman Publishing, London.

Neal, R. A., 1995, Project definition: the soft systems approach, *Int. J. of Project Mgmt.*, 13:1, pp. 5-9.

Partington, D., 1996, The project management of organisational change, *Int. J. of Project Mgmt.*, 14:1, pp. 13-21.

Rosenau, M. D., 1992, "Successful Project Management", Van Nostrand Reinhold, New York.

Rowe, W. D., 1980, Risk Assessment Approaches and Methods *in* "Society, Technology and Risk Assessment" J. Conrad, ed., Academic Press Inc. (London) Ltd.

Weaver, P. L., 1993, "Practical SSADM 4", Pitman Publishing, London.

Willcocks, L. and Griffiths, C., 1994, Predicting risk of failure in large-scale Information Technology Projects, *Technological Forecasting and Social Change*, 47, pp. 205-228.

SOFT SYSTEMS METHODOLOGY:
ITS POTENTIAL FOR EMANCIPATORY DEVELOPMENT

V.N. Callo[1] and R.G. Packham[2]

[1]Rizal State College, Philippines
(09783) 878 692
[2]UWS-Hawkesbury, Richmond, NSW 2753, Australia.
045-701324(Ph), 045-885538 (Fax), R.Packham@uws.edu.au

ABSTRACT

This paper discusses the use of Checkland's Soft Systems Methodology within the broader framework of Participative Systemic Action Research. It builds on an earlier critique of SSM, discussing particularly its use as a tool for participation and emancipation: The paper discusses how SSM was used as a method to help successfully achieve emancipatory development in a rural area in the Philippines. Important emergent features that ensured such success for SSM were (a) The role of the facilitator and the process of facilitation, (b). The commitment to learning with and from all participants (c). The contextualisation of the SSM, and (d.) A commitment to authentic participation of all stakeholders. Mention is also made of the different issues concerning the implementation of agreed change in situations where there are no clear power hierarchies.

INTRODUCTION

This paper will provide a critique of Soft Systems Methodology (SSM) as developed by Checkland (1981), from the perspective of its use in emancipatory development in a poor rural area of the Philippines. The focus will not be on the content of the research project from which this critique derives, but on the broad research methodology that was used in the project of which SSM was a part, and the potential for SSM in such contexts. The two aspects to be emphasised in successfully using SSM for emancipatory development will be *facilitation* and *learning*.

THE RESEARCH CONTEXT

The research upon which this paper is based was conducted in the rural community of San Fernando, an area in the south of the island of Luzon in the Philippines. It had a population of approximately 21,000 people and a land area of 8,700 hectares. Its economic activity was largely agricultural, with rice farming being the main source of income for 41% of all households. Most land is gently sloping from 0.5 to 20 meters

Synergy Matters: Working with Systems in the 21st Century,
Edited by Castell *et al.*, Kluwer Academic / Plenum Publishers, New York, 1999.

169

above sea level and much of this is subject to flooding. There is an area of hills that vary from 150 to 180 meters above sea level, and here coconut growing was the main agricultural activity. Rainfall is more or less evenly distributed throughout the year, but the Bicol Region (of which San Fernando is a part) is in a typhoon belt. At least once a year a typhoon or tropical storm passes through the region, generally in August.

Within the population, there was widespread unemployment (45%) and underemployment; most farmers and fishermen only had part-time jobs. This resulted in household income being low: 85% of all households having an average monthly income of not more than 3000 pesos (approx. $150 AUD). This situation had led to under-nourishment of children and in many children being unable to afford even secondary education which is free of tuition fees in government schools (Callo, 1990).

THE RESEARCH METHODOLOGY

In our choice for a methodology we were guided by the following questions:
- Will the methods used be useful in understanding the situation in its complexity?
- Will they enable us to explore possibilities for the poor?
- Will they allow various perspectives to be expressed and true interests of the participants to emerge?
- Will they allow a common vision to develop?
- Will they generate a social energy capable of creating action that moves towards the common vision of the desired change?

The *Participatory Systemic Action Research* that evolved in this research process was influenced by different intellectual traditions. It is Action Research that is sytemic and critical (Bawden and Macadam, 1990), emancipatory (Carr and Kemmis, 1986) and participatory (Fals-Borda, 1987; 19920.

As a conceptual framework, the Participatory Systemic Action Research has come out of the "marriage" of Participatory Research (Brown and Tandon, 1983) and Action Research (Rapaport, 1990) in its ideological perspectives. With the influence of Participatory Research, we recognise the conflicting interests of societal groups and that the plight of the disadvantaged as a critical problem. However, the consensus social theory underpinning Action Research also had a place in our methodology whenever there was a possibility that people could agree to some aspirations, or a common vision.

The approach was constructed using SSM as a *method* to guide the research. The research activities corresponded to the different stages of SSM (Checkland, 1981).

The inquiry about the situation and exploring possibilities for improvement engaged us in both the subjective and objective process or researching. The entire research was a continuous learning process for the participants. The account of the process can be found in Callo (1997), and will not be dealt with in this paper, as the focus here will be on the SSM; its effectiveness both in the systemic understanding of the situation, and in fostering debate on feasible and desirable changes and their implementation.

A key feature of the research process was the formation of a core action research group. The efforts of the core group enabled the research to be enriched by the broader community members sharing their personal experiences through dialogue and story telling, which also gave the community a sense of a collective experience. This created social energies for collective action.

A CRITIQUE OF SOFT SYSTEMS METHODOLOGY

Here we adopt the notion of Grundy (1990), where critique is used not for a sheer negative censure, but rather as *an expression of positive intention towards a rigorous discrimination.*

Our critique is based on that used by Flood and Jackson (1991, and follows their format:

On Theory

The first criticism raised by Flood and Jackson (1991) was the "restrictive nature of interpretive theory upon which SSM is explicitly based". It is this similar critique on the inadequacy of the interpretive theory in bringing out real transformation that has led to the development of a critical theory which underpins most participatory or emancipatory action research (Carr and Kemmis, 1986; Grundy, 1990). The influence of critical theory was evident in the intent and process of our research.

Participatory Systemic Action Research has a commitment to favour the less privileged as was the case in the research context of this paper. This was evidenced by our consistent search for the leverage for change (Senge, 1992) and the systemic response to the various needs of the people.

Another criticism of SSM was its "conclusion that the only way to change social systems is by changing people's world views or Weltanschaungen" (Flood and Jackson, 1991). Our research led us to conclude that:

- While a change in worldviews is considered important in SSM logic, there is no imposed restriction to limit transformation *only* at the level of the mind.
- The interaction between ideals (expressed in the Conceptual Models) and experiences of the real world leads to a debate on desirable and feasible changes. SSM is open to the never-ending flux between ideas and experience. The crux of the matter is "who" defines the change; The "who" and the level of participation in the debate are crucial in defining and implementing change;
- As borne out in this research, it is possible to use SSM underpinned by "multiple paradigms" and thus the implication of recognising reality as that which is constructed by the interaction of minds with concrete reality. For real transformation to occur, there must be a change in *both* the consciousness (the intent of a conscientization process of participatory research) *and* the concrete world (of politics and economy); both the "abstract world" of ideas and the "real world" of experience.

On Methodology

The supposed character of the changes arising from SSM are systemic desirability and cultural feasibility. But in practice, as observed by Flood and Jackson (1991), it is the cultural feasibility that guides the definition of change and this tends to favour the "dominant culture'. The debate on feasible and desirable changes, which are supposed to be "genuinely" participative, is often left in hands of the power holders. In regard to the issue of participation it was stated that "Checkland should insist that the debate stages are conducted as far as possible according to the rules for establishing "communicative competence' laid down by Habermas" (Flood and Jackson, 1991). Jackson (1990) finds that the "only possible justification for implementing the results of a soft systems study must therefore be that results and implementation have been agreed upon after a process of full and genuine participatory debate among all stakeholders involved or affected". Unfortunately, in situations where there are unequal power resources, Jackson (1990) contends that it is impossible to ensure a "genuine debate".

While this critique could lead to a possible strategy of developing communicative competence among stakeholders, there is another aspect that needs attention, that is the role of facilitation in the conduct of the "debate" or, much, better, of the whole research process. Based on our experience we would argue that the role of facilitation and/or facilitator has an equal, if not more important part in ensuring genuine participation. The personality of the researcher-facilitator will certainly have a strong bearing on how well participation can be achieved. As Brown and Tandon (1983) contend "it is doubtful how an authoritarian personality of the researcher can encourage (participation)..."

It must be noted that the difficulty, if not the impossibility, of a genuine debate has been observed mostly in corporate organisations where authority and power are strongly hierarchic. Also, other factors such as time constraints on the inquiry process, competence of facilitation and allegiance to the management who solicited the service of consultants, can tilt the balance of participation in favour of the privileged few.

On Ideology

Flood and Jackson (1991) argue that "the failure to establish the grounds for genuine participation means that SSM will always serve those with power." Their critique against SSM rested heavily on the issue of participation. But genuine participation is the very essence and ideological commitment of Participatory Systemic Action Research. This contradiction in the critique of SSM with our experience of its usefulness in exploring possibilities to improve the situation of a rural community leads to a different view of SSM effectiveness.

We believe that the ineffectiveness of SSM in dealing with power-related issues can only be highlighted when SSM is isolated from the whole interacting fabric of the research process. In this research, *it was the systemic integration of all other interacting factors that compensated for its inadequacy in dealing with such issues.* These interacting factors include the role of facilitation, the uniqueness of the relationship of the action research core group with the whole community, and the contextualization of a learning approach (Bawden, 1995; Packman and Roberts, 1996).

The organisation of Co-operative was an emergent outcome of our learning about how to participatively improve a situation considering prevailing constraints and possibilities. It was a response to the unfolding of events during the research inquiry process, and understanding these in the light of what people desired and what possibilities existed for their fulfillment. *Learning throughout the whole process was the key ingredient for our responsive and responsible course of action.*

Using SSM within the broad framework of and in the context of Participatory Systemic Action Research which *ensured participation* at all stages, achieved the improvement of the situation with this community. When the research formally phased-out, the community members still continued to participate as active members of the Co-operative.

THE ROLE OF FACILITATION

The action research core group, with its diversity of experience, interests and personality dynamics, developed an "emergent" facilitation competence not possible with a single facilitator. The competence of this facilitation was also seen in the way the "generative theme" was allowed to emerge and be identified; how various interests were encouraged to emerge before agreement was reached for a desired change. And most significantly, it was expressed in the way in which the action research core group facilitated the exploration of ways to implement changes by evoking the capabilities of the community.

In facilitation, a "round-table" type of communication was very important. This gave everyone a chance to speak and also created an atmosphere where they could speak openly. A feeling of equality is required for this open type communication. It was an advantage to our facilitation that the core group could identify with the various sectors of this community. The community acknowledged that they felt at home with and not "*alangan*" or inferior to the action research group.

The personality of the facilitator(s) is important in facilitation process. An important quality is the sensitivity, not only to the spoken language, but also to the unspoken thoughts, ideas and feelings that must be recognised and acknowledged to generate power. The rapport and good relationship of the core group with the community was present even before the start of this research. All of the action research core group, at one time or another, held some important leadership roles in social, religious and political activities in the community, and all on a voluntary basis.

The action research core group's relationship with the community allowed for a deep identification with the community that fostered open communication. One important point of identification was a shared history of being poor. While the action research core group members were all professionals, all had experienced poverty. It served as a unifying factor, particularly during the organization of the Cooperative, which was composed, of people from all walks of life. Their similar struggles and hopes in life and the solidarity with the people made their participation in the research a truly passionate engagement.

Contextualizing The Learning Approach

As this research was an exploration of possibilities, it was a continuous learning process. It was a type of learning that allowed a collective understanding and collective action, a process that developed collective praxis.

At the crux of our learning was the sharing of personal experiences which created an awareness of common experiences within the community, and that this was a collective experience of the poor. The primacy of personal experience and values as the basis of the learning process contains within it a dynamic concept of learning. As Reed (1984) puts it:

This approach (of learning based on values and experiences) makes full recognition of the fact that learners and society are constantly interacting, bound together in a dialectic of change.

The organization of the Cooperative had a political potential, which could not be overemphasized. There is a great potential that as the Cooperative continues to respond to the needs of the community, and as it widens the sphere of the problem context in its interaction with existing social structures, there will be a growing awareness that neither the individual, nor the organization operates in a social vacuum. Accordingly, the learning will continue through reflecting on the dialectic interaction of individuals and society. Looking at the learning process in a Cooperative emphasized the importance of collective learning for emancipation. Emancipation is a social process; it can only be achieved by collective efforts.

Considering this context, the "critical kernel" of the Methodology was manifest in the contextualization of the learning approach. This can have the quality of empowerment as long as the interests of the participants determines the direction of the learning process, not that of any agency or instrumentality. In addition, personal experiences that become collective experience through the process of sharing, is the basic content of the learning process. Making experience the object of the learning process had an empowering quality which can be directed to help overcome constraints in furthering social interests.

CONCLUSIONS

This research held the assumption that rural development situations are complex, and that their improvement should start from the exploration of the heterogeneous needs of the people. This led to a quest for a methodology that would allow various perspectives to be expressed and interests of the participants to emerge. The emergent methodology – Participatory systemic Action Research – proved effective in a collective understanding of the situation which led to strategic collective action. This research, while fulfilling its *raison d'être* of fostering change through action, also served as a verification of the value, usefulness and effectiveness of SSM in assisting systemic understanding of the situation and in fostering debate on feasible and desirable changes. However, a rigorous discrimination of contextual specificity must be noted. The achievement of the research outcomes can only be attributed to a number of interacting factors in a mutual causality, and not to any single factor.

Ensuring participation at all levels of the inquiry was vital in generating social energies capable of collective action. While participation may not be guaranteed by a certain methodology, it can be encouraged by competent facilitation. The relationship of the facilitator(s) and the research participants, and the relationships among participants, are key factors in ensuring participation. Authentic participation is possible and can be seen as an emancipatory process.

At the heart of this research process was learning by being open to surprises in the never-ending flux between ideas and experiences. At the crux of our learning process was its contextualization in the dynamic interaction of personal experiences (and collective experiences) allowing the community to mutually influence change.

The entire research process served as an affirmation that indeed, life can be bettered, and that we can facilitate others to better their lives through the process of learning and researching together. The research also illumined the potential of SSM in emancipatory process of development.

REFERENCES

Bawden, R.J. (1995) *Systemic Development: A Learning Approach to Change*. Occasional Paper No.1, Centre for Systemic Development, University of Western Sydney-Hawkesbury, Australia.

Bawden, R.J. and Macadam R.D. (1990) Towards a University for People-Centred Development. *Australian Journal of Adult and Community Education* 30 (3): 138-144

Brown, D.L. and Tandon, R. (1983). Ideology and Political Economy in Inquiry. *The Journal of Applied Behavioural Science*, 19 (3): 277-294.

Callo, V. (1997) *Towards Community Development: Exploring Possibilities with the Rural Poor in the Philippines through Participatory Systemic Action Research*. PhD Thesis, University of Western Sydney-Hawkesbury, Australia.

Callo, V. (1990) *The Municipal Socio-economic and Physical Profile of San Fernando, Camarines Sur, Philippines*. Commissioned Report.

Carr, W. And Kemmis, S. (1986). *Becoming Critical: Knowing Through Action Research*. Deakin University.

Checkland, p. (1981). *Systems Thinking., Systems Practice*. John Wiley Chichester

Fals-Borda, O. (1987). The Application of Participatory Action Research in Latin America. *Int. Sociol* 2(4): 329-347

Flood, R.L. and Jackson, M.C. (1991). *Creative Problem Solving: Total Systems Intervention*, John Wiley, Chichester.

Grundy, S. (1990). Three Modes of Action Research. In: *The Action Research Reader*. Third edition 1990. Deakin University, Victoria, Australia..

Jackson, M.C. (1990). The Critical Kernel in Modern Systems Thinking. Systems Practice 3(4): 357-364

Packham, R.G. and Roberts, R.J. (1996) Systemic Learning and Praxis – The Quest for a Meta-Methodology In M.Hall (Ed.) *Sustainable Peace in the World System, and the Next Evolution of Human Consciousness*, International Society for the Systems Sciences, Louisville, USA

Rapaport, R.N. (1990) Three Dilemmas in Action Research. In: *The Action Research Reader*. Third edition 1990. Deakin University, Victoria, Australia.

Reed, D. (1984). *Human Society. Evaluating Community Organizing in the Philippines*. La Ignaciana Apostolic Centre Printing Press, Manila.

Senge, P.M. (1992) *The Fifth Discipline: The Art and Practice of the Learning Organisation*. Random House, Sydney.

ISSUES IN THE CONDUCT OF RESEARCH THROUGH INTERVENTION IN REAL-WORLD AFFAIRS

Dr P.J. Dunning-Lewis

The Management School
Lancaster University
Bailrigg
Lancaster LA1 4YX

INTRODUCTION

Collaboration with external organisations is a long established strategy for academic researchers in many areas of systems thinking. Such collaborations may take many forms, but this variety is not evident when all are described by the overused label of 'action research'. The result is that too little attention been given to the special research issues and dilemmas that may arise when the researcher works at the most extreme levels of engagement in real-world affairs.

This paper suggests that an important dimension for differentiating between the various ways in which collaborative research may be conducted is the degree of responsibility of the researcher for bringing about changes to the real-world situation. This we term as the degree of engagement in the problem situation. When working at the extremes of engagement then the researcher will be faced with issues of politics and ethics. Recognition of this is important in discussing research, in planning research and in mentoring the work of students or new researchers.

We use the case of a 14-month study within a large financial institution to illustratehow the researcher may be required to interact directly with organisational politics, and so confront ethical considerations.

A GENERAL MODEL OF ENGAGED, COLLABORATIVE RESEARCH

There is much dispute about the nature of action research and the statement that: "What is and is not action research is a definitional problem of considerable magnitude" (Bryman 1983, p.181) is surely something of an understatement. If one examines the numerous papers that claim to report or discuss action research then one must conclude that there is little agreement about action research other than that it involves a collaborative effort between researcher and collaborating organisation: just about everything else is variable.

The researcher may be part of the organisation or not. The research may be quantitative or qualitative based. Organisational staff may participate or not. The outcome of the work may be advice, organisational learning or practical change. This leads to an unclear basis for discussion of

Synergy Matters: Working with Systems in the 21st *Century*,
Edited by Castell *et al.*, Kluwer Academic / Plenum Publishers, New York, 1999.

collaborative work, such that even prototyping activity may be claimed to fall within the definition of being action research (Baskerville & Wood-Harper). Rather than attempt any personal re-definition of action research we find it useful to examine collaborative research from the starting point of the simple, general model shown in Figure 1. This is based upon the 'FMA' model of Checkland (1985) but differs in certain aspects.

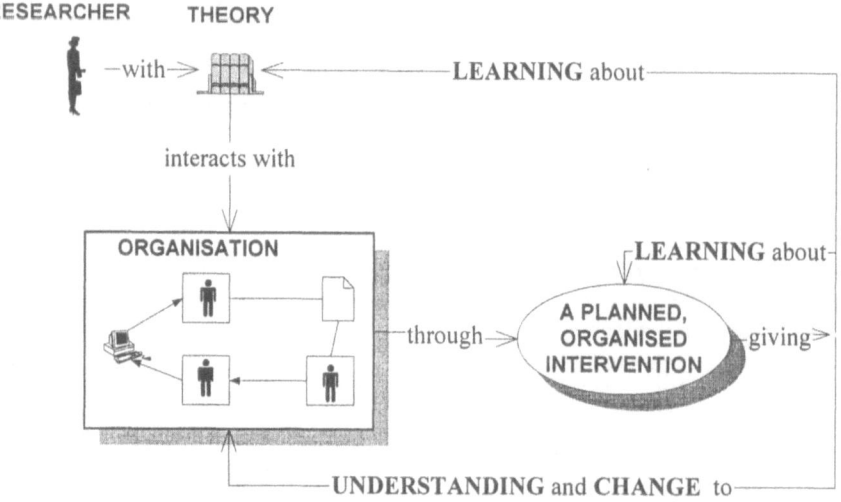

Figure 1.

The single most important feature of this model is that the researcher must enter the research situation with some declared-in-advance set of ideas that will be used in the research process. The membership of that set of ideas may change over the lifetime of the research; but to be 'research' rather than just 'messing about in the world' those ideas must be declared in advance. Jenkins (1983) makes the same point in relation to action research but conflates together the set of ideas and the means of their application into "Methodology,' a bunch of concepts which enable organisational problems to be tackled in a systematic manner"(p.37), one of his three components of action research Those ideas are then applied to a particular application area (in our model an organisational problem situation) through an intervention process that may or may not involve the use of a particular methodology such as SSM. By focussing upon an intervention process rather than a methodology the model allows that differing techniques, methods and paradigms may be used together if necessary and multimethodology (see Mingers & Gill, 1998) may be a possibility. The experience of doing this may lead to learning and insights in three areas, namely about the situation, about the ideas that have been used and about how to apply the ideas to such situations in practice. In addition, since this is engaged research we might expect some form of practical change to occur in the situation also.

This is a general model, which may be overlaid upon much management research in organisations. One would not though expect the same outputs in every case. For example, in a traditional style of operational research collaboration the researcher may apply quantitative techniques on behalf of the organisation. In doing so they expand understanding concerning a theory or technique, its application, and contribute to some change in the situation. It is much less likely though that the research will concern itself or contribute to any greater understanding of the organisation or its essential nature. An example of this form of collaborative research would be the use of an interactive algorithm for vehicle routing being used to minimise the costs of road gritting in winter for two counties in the North West of England. (Li & Eglese, 1996) In contrast, within an interpretative style of study the researcher may attempt to understand, through observation, interviews and other means, what is happening in a situation but, crucially, not usually themselves be actively involved in the shaping of events. In such a case there will be learning about the theory, how it can be used to give meaning to the history of actions and events, and learning about the situation. The latter may or may not be shared with those in the situation. An example of this style of research might be that concerning the role of management consultants where the authors used documentary sources, discussions with client organisations and non-participant observation of the consultants' work (Bloomfield & Danieli 1995). Here the emphasis was on reporting an appreciation of the role of consultants, not to affect their work practices or have any effects upon their work in the NHS.

IMPLICIT ADDITIONS TO THE GENERAL MODEL

In much of the Systems, Management Science and IS literatures one might detect two important, implicit, additions to the general model of Figure 1. The first of these concerns the view taken of the organisation, which is commonly presented as if it were a goal-seeking machine composed of functional components amongst which are the individuals that occupy particular defined job roles. If were to rely solely upon the written accounts of research one would believe this to remain the predominant position for most work in the IS field especially. A consequence of this is an almost total disregard of the organisational politics, with consideration of politics rarely mentioned in the accounts of the research, other than as being a 'noise' factor that must be eliminated through the research design. During informal discussions with the researchers the word politics is usually conjoined with adjectives in such phrases as 'bloody politics' or, more dismissively, '...that was just politics'.

In balance it must be said that there is a large and growing literature which does not take . this view. In this, organisations are taken to be social constructs formed by and affecting the interpretations of reality of the human actors. Politics are then seen as an integral part of the problem situation and of the intervention, such that the researcher is encouraged, even required, to understand the politics just as much as they should understand the such things as the flows of information and materials.

Even here though we find the second implicit addition to the general model of Figure 1. This concerns the role of the researcher themself, who is taken to be a rational, politically neutral party whose allegiance is to the pursuit of knowledge and disinterested in the course of events. Although we might find their role within the intervention classified as being change agent, emancipator, mediator , facilitator or the like, the question of why they are there, being a change agent etc is not discussed. It would seem that academics and researchers are shy of discussing the implications of the fact that they are not 'just there' doing the research but are human being driven by their needs and ambitions. A genuine desire for knowledge is probably the greatest of these, but researchers may also be motivated and directed by such considerations as:

- The need to publish in order to improve their professional status or keep their job.

- External pressure upon the research process. In the UK, the need to publish within the 4-year window of the Research Assessment Exercises has become a powerful deterrent to longer term studies and an influence towards shorter term research designs.

- That there may be financial payments involved. It has been recognised that to gain entry to the organisation and maintain credibility while working within the organisation the charging of a financial fee, commensurate with the organisation's normal expectations, should be charged. The researcher may not be indifferent to whether that source of income, whether to their institution or themselves, continues.

- Feeling of ownership for the research results. The researcher may believe that some change, identified by the research, would be genuinely advantageous to the organisation or groups within the organisation and therefore be unwilling to accept that whether it happens or not should be left to chance.

- A wish to emancipate or empower the weaker stakeholders, perhaps abhorring the suffering that may be caused by some change proposals

- A wish to satisfy expectations. Whether explicitly, as in performance against a declared Business Case, or implicitly, in the form of being seen as worthwhile to the sponsor, there may be felt a need for the researcher to satisfy expectations, meet promises and for their activity within the organisation to demonstrably deliver results.

The researcher themselves can therefore never be politically neutral. They will need to understand the organisational politics in order to cause the intervention to happen, to shape the course of the intervention and may also need to indulge in politics in order to ensure that it continues. So that, in engaged collaborative research, the researcher may be required to be attuned to and competent in handling several types of politics (Dunning-Lewis 1998). These are:

1. The politics of the situation.
   Understanding who has power, how it is used, what group and personal agendas underlie actions, and what battles are being fought under the guise of rationality.

2. The politics regarding the intervention.
   The ways in which the research intervention and the researcher may be used as part of (1).

3. The political consequences of the intervention.
   The shifts in power that may arise from the intervention having taken place and how the results of the intervention may alter both the formal and informal organisation.

And, least discussed in the management literature,

4. The politics of the intervention itself.
   The political actions necessary to engineer the intervention, to make it happen and to keep it alive in response to a turbulent organisational setting.

AN EXAMPLE OF ENACTING THESE POLITICS

To illustrate the type of actions that the researcher might be required to take we present the case of work recently conducted within a large UK corporate bank. This involved the use of a number of different techniques and approaches (Dunning-Lewis & Townson 1998). The principal concern of the 'ABC' project was the introduction of a new formal framework for business change projects and the integration of this with existing procedures for project management and benefits management. Responsibility for the ABC project lay directly with the researchers and the work had to be conducted against the timescales and deliverables defined by a Prince2 Business Case and Project Initiation document.

The purposes of the project were, from the perspective of the bank to improve delivery time and reduce costs of business change projects, and from the perspective of the researchers to provide material concerning the use of methods and their effects upon organisational work as part of a longer term study of this within the bank. The latter required a continuing relationship with the client so that careful attention had to be given to the planning of the ABC work, how the results would be used and the need to protect the existence of the project. A need for active, perhaps manipulative, involvement in organisational politics followed from this.

Much of our efforts were directed towards monitoring the position of the project vis-à-vis the rapidly changing circumstances within the bank. On reviewing our team meetings we discovered, somewhat to our surprise, that the majority of the time in these was spent discussing what had been happening and how we should respond. The root of the turbulence in the environment was that the bank was facing an exceptionally high level of mandatory projects. Of particular note were the need for millennium compliance of all computer based systems (the Y2K problem) and the need to redevelop large systems to cope with the two scenarios of the UK entering or not entering the European Monetary Union (the EMU problem). Both were costing very large sums of money and causing a great shortage of staff, with pressures to transfer staff from other projects. Already one person had been removed from the ABC project to EMU related work, we had fought off the attempt to remove another to COBOL programming, a task he had not done for over 10 years. In common with elsewhere, business improvement projects and infrastructure projects with longer-term paybacks were beginning to be cancelled or 'frozen'.

Recognising that careful attention should be paid to the context, culture and power relationships, we carried out a large number of interviews to build up a history of previous events and why initiatives such as this had failed in the past. These revealed, for example, that the business users (those who did corporate banking) acted in various ways to maintain full power over the change agents such as IT developers or business analysts. From this analysis of the politics of the situation we were then able to think sensibly about the politics of the intervention. Some of the results of this were as follows.

We planned the work to best fit the culture. There was extreme pressures for the delivery 'on time, on spec, in budget' of new IT products to the business units. There was also a system of performance related pay with each member of staff needing to satisfy their own Balanced Business Scorecard targets within the current year. This encouraged short term thinking and discouraged support for anything that might disturb the equilibrium and threaten the achievement of immediate goals. We therefore designed the ABC project as a series of sub-projects so that deliverables would arise at the end of each relatively short sub-project, giving rise to, in the language of the situation, 'quick wins'. This also influenced our recommendations to introduce a 'cut down' ABC to just a few areas first on a trial basis, rather than the comprehensive roll-out that had been originally asked for.

We made the project a solution to various 'problems' so as to strengthen the likelihood of the work continuing. There was for example a strong public commitment to quality in products and processes. We therefore ensured that the powerful and separate Quality & Standards Group saw that the ABC project was furthering their agenda and were closely involved in the project. In the later stages this proved important as they took over sponsorship of the work.

We built and nurtured networks of interest. Some parts of the larger banking group of which the corporate bank was a part had an interest in the work. We therefore set up a series of meetings with those parties to review and report what was being done. There was no direct requirement for doing this but it built up a network of individuals who would support the work (or at least not see it as any form of threat or problem) and whose influence might prove valuable in the future.

Given that the staff were well aware that we were academics we took any opportunity to demonstrate professional knowledge and capabilities. For example, when a chance opportunity arose to give comment upon a strategic planning document this was done. Credibility was enhanced when we pointed out that within this major financial institution not even the simplest form of discounted cost flow had been used in the justifications for work that would cost several million pounds.

We even attempted to assign public ownership of the project to senior management, through a very carefully worded article in the company magazine. This proved successful, for the project was subsequently referenced by the senior manager at the annual strategy review as an example of the excellent work being done.

CONCLUSIONS

We suggest that an awareness of the politics that are relevant to engaged collaborative research is essential for researchers in information systems and management, and that on occasions an active role in organisational politics may be called for, to enable the research to happen and to ensure that it continues. This brings with it certain problems of ethical practice.

We find these issues though little discussed, even though we observe that many renowned researchers are themselves extremely adept at interpreting and adapting to the politics of the situation. Politics cannot be avoided, since even the researcher who decides to avoid organisational politics is taking a stance that must be defended by political actions. So, with collaborative work within organisations becoming the predominant form of research in IS this is a topic to which more attention must be given.

REFERENCES

Baskerville, R., and Wood-Harper, A.T., (1998) Diversity in information systems action research methods. *European Journal of Information Systems*, **7** (2), pp.90-107.

Bloomfield, B. P. & Danieli, A. (1995) The role of management consultants in the development of information technology: the indissoluble nature of socio-political and technical skills. *Journal of Management Studies* **32** (1) pp. 24-46.

Bryman, A. (1989) "Research Methods and Organization Studies", Unwin Hyman, London.

Checkland, P. (1985) From Optimizing to Learning: A Development of Systems Thinking for the 1990s. *Journal of the Operational Research Society*, **36** (9), pp. 757-767.

Dunning-Lewis, P. & Townson, C. (1998) A pragmatic use of methodologies and techniques during engaged research: facilitating change in a corporate bank. *8th Annual BIT Conference*, Manchester, November 1998.

Dunning-Lewis, P.J., 1998, "Staying alive – the practice of politics in engaged management research"" *Conference of the Operational Research Society*, Lancaster September 1998.

Jenkins G (1983) Reflections on Management Science , *Journal of Applied Systems Analysis* **10**, April, pp. 15-40.

Li, L.Y.O., and Eglese, R. W. (1996) An Interactive Algorithm for Vehicle Routeing for Winter-Gritting. *Journal of the Operational Research Society*, **47** pp.217-228.

Mingers, J. & Gill, A. (eds.) (1997) "Multimethodology: Towards the Theory and Practice of Combining Methods", Wiley, Chichester.

ACRONYM HAPPY - APPLYING SSM TO THE CSA

Misha Hebel[1] and Adam T. Muggleton[2]

[1] Business School, University of Greenwich
Woolwich Campus, London, SE18 6PF

[2] JPS Environmental Services Ltd
26 Hemmells, Basildon, Essex, SS15 6ED

INTRODUCTION

The Commissioning Specialists Agency (CSA) was set up in 1988 to promote commissioning within the engineering and construction industry as a specialist profession. Commissioning in this context is the testing, balancing and setting to work commercial air-conditioning systems. It also entails mechanical installations for building services such as hot and cold water distribution systems. It is a young, specialised sub-industry of the wider building services industry. The first specialist commissioning company was formed only 30 years ago. In consequence there are very few "career" commissioning engineers. Typically they came from allied industries and adapted their skills accordingly. There are even fewer dedicated commissioning training courses and accredited qualifications. The CSA now provides a range of services and products from information to a full training and career development framework (CSA, 1997).

Unfortunately the CSA has experienced difficulty in attracting corporate members above its current level of around half the commissioning companies in the UK. It also faces obstacles attracting individual members for training and grading within the CSA scheme. Membership in all categories rose then fell to the current level where it has remained for four years. This recruiting problem inhibits the development of the association because without membership there is no power base to provide funding for projects and resources. Around 70% of CSA income originate from membership fees, the remainder from sales of goods and services. This paper summarises the analysis prompted by the problem situation.

ANALYSIS

As an organisation founded on traditional engineering principles it was felt that the situation might benefit from analysis using Soft Systems Methodology (SSM). The analysis combines features of both the original seven stage model (Open Systems Group 1972; Checkland, 1981; Open University, 1984) as well as later interpretations (Checkland & Scholes, 1990) as the authors felt this would provide the most insight. Analysis took two courses, firstly a direct mapping of SSM (Muggleton, 1997) onto the situation. This was followed by a comparison of the value systems (Hebel, 1998a) identified as inherent in the root definitions devised.

In general terms the problem owners were primarily perceived as members of the main committee (executive body) of the CSA. The main actors and customers were the CSA members and companies with potential membership. The underlying world-view of the analysis was one of a need for structure and discipline via an alliance of some sort if the commissioning industry was to be fully recognised. From the initial rich picture a number of primary tasks and significant issues became evident and the main ones are shown in Figure 1.

Primary Tasks
1. To promote the commissioning industry by raising its profile through a single authoritative body.
2. To raise standards of commissioning personnel through structured training, examinations and accredited grading.
3. To act as a communications hub between the commissioning industry and the building services industry.
4. To produce commissioning technical documents and services for its members and the wider industry as a fund-raising activity.
5. To recruit companies and individual field operatives as CSA members.
Significant Issues
1. Overall effectiveness – goals not being achieved i.e. commissioning still held in very low regard by the building services industry and to some degree by the non CSA commissioning industry, not attracting new members.
2. Tension and possible conflict between respective company members.
3. Tension and suspicion between the individual field technicians, engineers and the CSA.
4. There appears to be a crisis of identity. The CSA is not sure if it is a "trade" organisation or a "professional" body.
5. CSA membership and income not high enough for future development.
6. Building services industry is very competitive with a "short term" Weltanschauung. Companies tend to focus on the "right now" rather than have any long term strategy.
7. Training and qualifications not valued by field technicians and engineers.
8. Communication between the field operatives, CSA, commissioning companies and wider building services industry seems inadequate.
9. Opinions regarding the CSA and CSA secretary are very negative. There have been complaints and observations of poor administration, performance and attitude.

Figure 1. Primary tasks and significant issues

From the primary tasks and significant issues, a number of Relevant Systems (RS) were developed and these are shown in figure 2 alongside the item that suggested it from figure 1. To the authors RS1 appeared to indicate the highest order system, strongly supported by RS2. All others contributed to the well-being of RS2. These were further developed into Root Definitions and the SSM analysis continued but for the purpose of this paper the focus will be on the development of "A system to recruit and maintain CSA members". This was based on the premise that the organisation must continue to operate and if an organisation has no members it ceases to exist thus this system appears key to the operation of the CSA. Other systems explored are described in Muggleton (1997).

	Relevant Systems (RS)	Relating to
1	A system to promote the commissioning industry within the building services industry.	Primary task 1
2	A system to recruit and maintain CSA members.	Primary task 5
3	A networking and tension resolution system between the CSA, non-CSA companies, building services industry and field operatives.	Significant issue 2
4	A system to increase CSA income	Significant issue 5
5	A system to improve communication between field operatives, CSA, commissioning companies and wider industry.	Significant issue 8
6	A system to improve administration of the CSA and public relations between the CSA secretary, CSA members and the commissioning industry	Significant issue 9

Figure 2. Relevant Systems

RS2 was developed into the definition below. It is basically managerial and financially bottom line oriented. It is the heart of the matter as it generated income for the CSA. Interviews with the Chair and Company Director also revealed that both felt that income was a constraining factor by limiting the market.

A system run by the CSA and funded by its membership to recruit new members; to maintain their commitment through the products and services provided and the promotion of commissioning to the industry in order to secure the continuity of the CSA thus enabling it to grow and develop.

From this a conceptual model was devised which clarified the purpose and allowed shared understanding of the CSA among the committee members. The other conceptual models were seen to support specific aspects of this general one. Figure 3 draws together the key items to be debated, forming a distillation of the overall analysis aimed at a more general, client / real world level.

Agenda For Debate	Comments
1. Identify: CSA members opinion, expectations and requirements.	This does not appear to be done despite an individual member sitting on the CSA main committee. Does CSA membership have value?
2. Know / Understand: current situation and consider a SWOT analysis.	In view of the recent complaints about the CSA and its secretary a process of reflection could be vital.
3. Running: Of the CSA.	In view of the recent complaints and the lack of recruitment a review of the running of the CSA seems appropriate. This could be a control issue.
4. Plan: to improve the administration and internal/external public relations also for a recruitment drive.	There seems to be no adequate planning in the real world situation.
5. Develop: Plans to generate CSA income.	A logical progression from the planning stage.
6. Promote / Communicate: CSA aims, objectives, membership benefits, products and services both internally and externally.	This appears to be a weakness of the CSA with no apparent promotion schemes in place and inadequate levels of communication.
7. Recruit new members.	The CSA seems to have no active recruitment drive or policy.

8. Increase CSA income.	Is payment for CSA membership, products and services customer friendly? Increasing CSA income would facilitate more investment in development of products, services and promotional schemes.
9. Maintain CSA members.	What is done to ensure the new and existing members are retained?
10. Monitor and Control systems.	What monitor and control systems are in place and what are missing?
11. What are the CSA's Goals and are they attained?	Should the CSA be goal oriented?

Figure 3. Agenda for debate (Muggleton, 1997)

The analysis indicates that the situation in question is displaying inadequate levels of control i.e. in its administration, and communication i.e. internally and externally. These are symptoms of a failed system and using the Failures approach (Bignall and Fortune, 1983) was considered a viable option to take the project further. This may identify what incidents or inadequacies led to the CSA failing to recruit members and develop.

The key areas to come out of this analysis are both predictable and novel items for consideration and debate. In summary it was found that the CSA was effective at providing training via lectures and distance learning. It also encouraged long term development and provided valid accreditation. Unfortunately it also seemed inadequately managed, lacked goals, fails to promote itself and is in need of some self-examination and re-invention.

The analysis indicates that the situation in question is displaying inadequate levels of control in its administration and communication. Inevitably conducting such an analysis has had an effect on the problem situation whereby the owners and solvers began to address the problems prior to receiving the findings. The effect of openly discussing the problems has begun to solve them by describing where difficulties lie rather than just leaping in to solve them.

After the initial presentation of findings the secretary countered criticisms of his performance with a proposal to address the problems and develop his role. This was initially accepted primarily as there was no one else to take over. Despite the secretary's shock at being told what the main committee felt he was soon back to his old ways. At a recent meeting of the committee it was voted to fire the secretary and restructure the CSA with a full time administrator and more power and responsibility devolved to the sub-committees and their chairman. The CSA is also in the process of setting up facilities to accept credit card payments.

The analysis was beneficial in finding out what the problems were rather than what they are perceived to be in the beginning. At first the problem situation appeared to be simply one of recruitment, but the problem situation is more complex and the true issues ran deeper than just membership. From the authors' point of view more analysis was required in an attempt to further understand the issues and what seemed like almost inevitable conflict between owners and actors. The following section discusses the more theoretical issues.

Even in the early stages of development of root definitions there seemed to be some evidence of, or potential for, sub-optimising (Armson & Paton, 1994). What appeared most interesting (Muggleton, 1997) was the hierarchical order the RD's took given the different perspectives of member, company or CSA officer for instance. Even where the interest appeared philanthropic, such as caring for the operatives on site, self-interest or self-promotion seemed paramount. This narrow focus in turn appeared to cause confusion over the definition of the CSA's purpose.

For example discussions with the chairman, secretary and a company director following a CSA meeting indicated that a shift in the Weltanschauung of the secretary had occurred over time. Based on the development of RD6 it seemed that the CSA had a lower priority for him than it used to. Consequently he was not very open to discussion about the recent high number of complaints regarding levels of service. There appeared to be a need for some change in order to address the increase in complaints his change in Weltanschauung appeared to be an emergent property of the complaints. Conversely however the change in Weltanschauung could have been the cause of the complaints.

It was evident that individuals and groups were reacting in different ways to the same situation. Attitudes towards the CSA varied from downright distrust to active promotion of its services. Given that Checkland (1981) perceived attitudes as *"intangible characteristics which reside in the individual and collective consciousness of human beings in groups"* can an identification, let alone a change in attitudes really be feasible? Different definitions of value appeared to be placed on actions and sub-systems based on something imperceptible. We may not be able to name the characteristics themselves but we may ask where do they come from?

One possible explanation for this is that attitudes and behaviour emerge as a direct result of the particular combination of human value priorities present at any one time. Values, when articulated, are often linked to norms, culture and ethics. They are traditionally good qualities to be aspired to or possessed but have both positive and negative qualities (Hofstede, 1991). They are based on early experience, form the basis of individual personalities and consequently affect all human groupings. They are so subtle and core to our existence however that they very often alter in priority without us even knowing it. As each situation occurs an individual will compare that experience with values and the level of accord that exists will define the acceptability of the event. Hence attitudes and behaviour are not random but carefully constructed responses based on a narrow band of fortified values (Hebel 1998b: Hebel, 1998c).

It seems likely that if we learn by analogy it is almost inevitable we draw on our experience and therefore reinforce our values. If parallels are not obvious or favourable then we are likely to discard learning or avoid certain activities. This also applies to organisations that are influenced by the values present at its inception and early years. It is possible that value systems in organisations reflect those of founders and that values are reinforced over time both in individuals and in organisations.

In an attempt to gain a clearer view of what comprises a value system the following definition was composed (Hebel 1998b). This definition could apply to either people or organisations:

> *A subconscious system that regulates a sense of well-being and determines reactions to life events. Core values are established on early experiences and later events serve to adjust their priority or emphasis. The process is continually reinforcing unless a crisis occurs challenging high priority values and unearthing low priority ones.*

A comparison of this definition to the CSA scenario is summarised in figure 4. This comparison highlights conceptual as well as practical issues arising from the root definition and may be applied at all root definitions in an attempt to surface the authors perspective and influence.

Key areas of Value System definition	Comment on comparison
subconscious system	Greater income will generate practical well-being - appears so obvious it is not mentioned explicitly in RD but in comment afterwards
well-being	Supplied by recruitment and thus continuance of the CSA. There appeared no question that the CSA would not continue to exist
reactions	Grumbling expected - initial closing of ranks; Lack of support expected - kept sec. on for that reason ; Focus on CSA needs rather than that of membership
early experience	Engineering & profit making world-view steered CSA towards finance focus despite philanthropic stated aims; People in authority must dictate what is right and those below must co-operate
later events	Recruitment Issues; Recession & financial hardship; Lack of volunteers for committee
reinforcing	Low expectations = self-fulfilling prophecy; Continuance of commissioning services industry can only occur with a formal association such as the CSA
crisis	Decision to sack secretary

Figure 4. Looking for values systems in situation

CONCLUSION

SSM provided an original slant to a traditional engineering environment, prompting both novelty in intervention and world-views. Solutions were found as a result of the analysis and this in turn provided new ground to explore. The changes made initially were however relatively neat and tangible administrative changes and to tackle the more invidious influences of values and attitudes required intervention band understanding beyond the scope of SSM. By understanding the impact of the secretarie's values as a switchboard for information and progression a radical solution (his sacking) proved to be the only way of addressing the values incompatibility.

REFERENCES

Armson, R. & Paton, R., 1994, *Organisations, Cases, Issues, Concepts, (2nd edition)*, Paul Chapman Publishing

Bignell, V. & Fortune, J., 1983, *Understanding Systems Failures*, Manchester University Press

Checkland, P.B. & Scholes, J., 1990, *Soft Systems Methodology in Action*, John Wiley & Sons, Chichester

Checkland, P.B., 1981, *Systems Thinking, Systems Practice*, John Wiley & Sons, Chichester

CSA, 1997, *The What, Why and How of the CSA*, Commissioning Services Agency

Hebel, M., 1998a, Values Systems - A way to greater understanding, *Systems Practice and Action Research, Vol. 11(4), August 1998, pp.381-402*

Hebel, M., 1998b, Human Values in Technology, *presented at the UKAIS conference 15-17 April 1998 and published in Matching Technology With Organisational Needs (eds Avison & Edgar-Nevill), McGraw-Hill, UK*

Hebel, M., 1998c. *Exploring the Impact of Human Value Systems on Performance Measurement*, PhD thesis submitted to City University, London

Hofstede, G., 1991, *Cultures and Organisations, Intercultural Cooperation and its importance for survival*, Harper Collins, London

Muggleton, A.T., 1997, *Soft Systems Analysis of Membership Recruitment in Regard to Development of the Commissioning Specialist Association*, T301 Course Project, unpublished manuscript

Open Systems Group, 1972, *Systems Behaviour (third edition)*, Harper & Row, London

Open University, 1984, *T301 - Complexity, Management and Change: Applying a Systems Approach*, The Open University Press, Milton Keynes

THE POVERTY OF ACTION RESEARCH IN INFORMATION SYSTEMS

Judy McKay and Peter Marshall

School of Management Information Systems
Edith Cowan University
Perth, Western Australia

INTRODUCTION

The title of this paper is a blatant adaptation of one coined by Klein and Lyytinen, who in 1985, published a paper entitled *The Poverty of Scientism in Information Systems*. In that paper, they challenged the view prevalent at the time, that only through the rigorous application of the scientific method to research questions could the discipline of Information Systems (IS) be advanced. Klein and Lyytinen (1985: 131) presented arguments disputing this view, as follows:

> *"Information Systems will remain a dubious science as long as it tries to emulate the so-called scientific method as the only ideal of academic inquiry. The most visible symptoms of the poverty of scientism are paradigmatic anomalies - crucial research issues which cannot be resolved within the scientific tradition because they transcend its paradigmatic assumptions. The need for affirmative pluralism is offered as a fruitful avenue to improve the status of information systems in academia and practice."*

Since that time, there have been repeated calls for IS researchers to select from a range of suitable research methodologies depending on the subject and context of the inquiry (Shanks et al. 1993). Indeed, there has been an expansion of the research approaches adopted, with qualitative approaches such as ethnography (Davies & Myer 1993, Rusli & Marshall 1995), grounded theory (Atkinson 1996), and action research (McKay 1998) all gaining greater acceptance within the IS community. One could reasonably argue, therefore, that those who during the 80s were clamouring for a more pluralist approach to research methodology adoption, would now be feeling somewhat gratified that sections of the IS community had at last heeded their calls.

Synergy Matters: Working with Systems in the 21st Century,
Edited by Castell *et al.*, Kluwer Academic / Plenum Publishers, New York, 1999.

187

So why, more than a decade later, is there a need for a paper entitled *The Poverty of Action Research in Information Systems*? Could it be that the research pendulum has swung too far in the qualitative direction, prompting calls for it to swing back towards the orthodoxy criticised by Klein and Lyytinen?

'Poverty' can be defined in a number of ways: it can imply insufficient monetary or material resources; it can imply scantiness of something; it can mean a deficiency or lack of desirable ingredients or qualities (Macquarie Dictionary 1981). It is in the last two senses, especially the last, that poverty is used in this paper. The argument will be constructed that action research, as it is currently frequently practised in IS, is deficient in certain desirable ingredients or qualities, and at times displays scant attention to a number of key features of any rigorous and credible research. This paper is not a call for less action research, quite the contrary. It is not a call to "legitimise" action research in the eyes of positivist researchers by making it "look" more like positivist research. Rather, this paper does aim to rekindle a debate on the practice of action research, and attempts to encourage action researchers to enhance the maturity of their practice by becoming more introspective and reflective about their research.

In the sections which follows, characteristics, strengths and weaknesses of action research will be considered, together with some of the concerns that the authors have about the practice of action research.

ACTION RESEARCH AS A RESEARCH APPROACH

IS researchers have for some time now been exhorted to consider action research as a suitable candidate research approach amongst the repertoire of methodologies embraced by the discipline (West et al. 1995). Action research, after all, boasts many features which would tend to suggest it is ideally suited to study aspects of the planning, development and implementation of information systems within their human, organisational environments. In this section, it is proposed to consider some of the characteristic features of action research and to propose a suitable definition. From this can be drawn some of the strengths and limitations of action research as an approach to IS research.

Characteristics of Action Research

There is a sense in which the very essence of action research is encapsulated within its name: it represents a juxtaposition of action and research, or in other words, of practice and theory. Thus, as an approach to research, action research is committed to the production of new knowledge through the seeking of solutions or improvements to "real life" practical problem situations (Elden & Chisholm 1993, Shanks et al. 1993). However, it is more than just another approach to problem solving, for the action researcher is working from within a conceptual framework (Checkland 1991, Baskerville & Wood-Harper 1996) and actions taken to ameliorate a situation perceived as problematic should form part of and stem from strategies for developing, testing and refining theories about aspects of the particular problem context (Avison 1993, Susman & Evered 1978).

One distinguishing feature of action research is, therefore, the active and deliberate self-involvement of the researcher in the context of his/her investigation. Unlike the methods of objectivist science where the researcher is argued to be an impartial spectator on the research context (Chalmers 1982), the action researcher is viewed as a key participant in the research process, working collaboratively with other concerned and/or affected actors to bring about change in the problem context (Checkland 1991, Hult & Lennung 1980). Collaboration between researcher and what may be described as the "problem owner" is essential to the success of the action research process. A mutual dependence exists in that both researcher and problem owner are reliant on the other's skill, experiences, and competencies in order for the research process to achieve its dual aim of practical problem solving and the generation of new knowledge and understanding (Hult & Lennung 1980). In particular, the researcher brings an intellectual framework and knowledge of process to the research context: by contrast, the problem owner brings knowledge of context (Burns 1994). Thus action research evolves, in part at least, as a function of the needs and competencies of all involved (Susman & Evered 1978), with a key feature of this research approach a willingness to share and thus learn, a result of which are enhanced competencies of all concerned (Hult & Lennung 1980).

Underlying the action research process, therefore, is a rejection of many tenets of more traditional approaches to research which are embodied in the scientific method. The methods of natural science are viewed as both problematic and indeed, inappropriate, when applied in "human" disciplines such as IS, for intelligent human agents can (and tend to) take action which can effect both the phenomena under study and the outcomes of the research (Checkland 1991). "Facts" in a social context are viewed as being given existence and are interpreted within some socially constructed framework of understanding (Avison 1993). Hence, any scientific or systematic investigation of a social context cannot be regarded as value-free (Elden & Chisholm 1993), nor can it be divorced from the situational and historical context in which it is given meaning (Hult & Lennung 1980).

Given these features, a fulsome (albeit cumbersome) definition of action research is offered by Hult & Lennung (1980) who write that:

> "Action research simultaneously assists in practical problem-solving and expands scientific knowledge, as well as enhances the competencies of the respective actors, being performed collaboratively in an immediate situation using data feedback in a cyclical process aiming at an increased understanding of a given social situation, primarily applicable for the understanding of change processes in social systems and undertaken within a mutually acceptable ethical framework."

Within IS therefore, action research offers many features rendering it a powerful tool for researchers who are interested in finding out about the interplay between humans, technology, information and socio-cultural contexts. For example, unlike other research approaches, such as laboratory experiments, which struggle to maintain relevance to the real world, the "laboratory" of action research *is* the real world, thus avoiding the potential separation of research and practice (Baskerville & Wood-Harper 1992, Susman & Evered 1978, Avison & Wood-Harper 1991). Indeed, it could be argued that in applied disciplines such as IS, action research appropriately establishes action and practice as being the prime focus of research efforts (Shanks et al. 1993). It is ideally suited to gaining understanding of whether technology or methodology is perceived useful and helpful in practice, what problems and issues are perceived to arise, and to identify how practice can be improved within the value system of the problem owner (Avison 1993).

Its dual aim of being both a mechanism for practical problem solving and for generating and testing theory provides a win-win scenario for both researcher and participants in an action research study (Elder & Chisholm 1993). In addition, action research is viewed as a means for enhancing the skills and competencies of both the researcher and the participants (Hult & Lennung 1980). Its explicit requirement that an object of inquiry should not be divorced from the context in which meanings are ascribed supports a more holistic understanding of phenomena in changing contexts (Hult & Lennung 1980).

Nonetheless, action research is not without its weaknesses as a research approach, nor is it without its critics. Arguments are expressed, for example, which suggest that action research may be regarded as being little more than consultancy (Avison 1993). When interventions are deemed successful, some would argue that causal connections and explanations cannot be safely made (Baskerville & Wood-Harper 1992). Researchers are questioned over a perceived lack of impartiality and bias (Avison & Wood-Harper 1991). The supposed lack of scientific rigour and discipline in action research, the lack of validity of data (Baskerville & Wood-Harper 1996), and the difficulty of generalising results from action research studies have lead to it falling into disfavour in some academic circles, and in action researchers finding it difficult to attract research funds (Avison & Wood-Harper 1991).

THE APPLICATION OF ACTION RESEARCH

Arguably, appropriate research questions and research designs can do much to minimise the limitations and maximise the potential benefits of action research as discussed in the previous section. However there is scant attention within the IS literature on what constitutes an appropriate design for action research, and on how action research can be conducted (West et al. 1995). To support this claim, a review of fifty seven papers on action research was conducted. These papers were drawn not only from the discipline of IS, but from other relevant areas such as organisational studies, behavioural science, social science, nursing and education. The findings of the literature review can be summarised as follows:

Table 1: Outcomes of a review of the action research literature

	Yes	(%)	No	(%)	Total
Was a *definition* of action research provided?	34	(60)	23	(40)	57
Were *distinguishing features* of action research discussed?	33	(58)	24	(42)	57
Were *limitations* of action research discussed?	16	(28)	41	(72)	57

Were *areas suitable for study* via action research discussed?	**31**	(54)	**26**	(46)	57
Were *data collection techniques* appropriate for action research discussed?	**24**	(42)	**33**	(58)	57
Were *data analysis techniques* appropriate for action research discussed?	**7**	(12)	**50**	(88)	57
Did the paper contain a detailed description of *how to conduct an action research study*?	**0**	(0)	**57**	(100)	57

Table 1 suggests an interesting picture of writings on action research. It would appear that in the majority of papers reviewed, authors typically defined action research, discussed some of its distinguishing features, suggested areas within their discipline that were likely candidates for elucidation via action research, and in a number of cases, touched upon suitable approaches to data collection within an action research framework. However, it is interesting to note that in the papers surveyed, in only 12% of cases were approaches to data analysis mentioned, let alone discussed in detail. It should also be noted that of these seven papers that mentioned suitable approaches to data analysis, five suggested only statistical techniques, most likely implying a positivist orientation to action research. Interpretivist researchers, therefore, are unlikely to receive much assistance in the literature. In none of the papers reviewed were there detailed, "step-by-step" descriptions or guides offered to support the would-be action researcher.

CONCLUSION

The argument that we wish to put forward thus becomes apparent. It would seem that a disproportionate amount of effort has to date been put towards philosophical debate on the paradigmatic underpinnings of action research, on the exact origins of action research, on Lewin's intentions for action research (given his comparatively scant offerings on the subject), and on what exactly constitutes action research vis-à-vis consultancy. There can be little dispute that these constitute important issues for information systems researchers. However, there seems to have been much less importance placed on post-hoc reporting of the action research *process* itself, as opposed to the context and content of the research. In the same vein, it could be said that there has been insufficient academic scrutiny of the action research process and its underpinning data collection and analysis techniques rather than the outcomes of the research.

We would argue that this has a number of undesirable consequences. First of all, there seems sometimes a tendency for researchers to assert that they have conducted an action research study, and for results to be reported along the lines of "We did X and it worked"! The claims of an enthusiastic pseudo-researcher looking on at their own consultancy, and asserting that things were successful, or effective or improved, and so on, does not, in our minds, constitute sound research. Similarly, accounts of interventions and their contexts can not be regarded as a direct substitute for explanations of data collected and methods of analysing that data. The second consequence is that for new or less experienced researchers, there are comparatively few exemplars of the detailed workings of a rigorous and relevant action research study. If action research really is regarded as a suitable vehicle for the conduct of research in the field of information systems, then establishing benchmarks for practice seems a priority. It is questionable whether we have, to date, achieved such a level of maturity.

REFERENCES

Atkinson, D.J., 1996, A study of perceptions of individual participants of a client group undertaking a series of meetings supported by a group support system, Unpublished Doctoral thesis, Curtin University of Technology, Perth.

Avison, D.E., 1993, Research in information systems development and the discipline of information systems, in "Proceedings of the 4th Australian Conference on Information Systems", University of Queensland, Brisbane.

Avison, D.E. and Wood-Harper, A.T., 1991, Conclusions from action research: the Multiview experience, in "Systems Thinking in Europe," M.C. Jackson et al., eds., Plenum Press, New York.

Baskerville, R. and Wood-Harper, T., 1996, A critical perspective on action research as a method for information systems research, *Journal of Information Technology*, 11: 235-246.

Burns, R.B., 1994, "Introduction to Research Methods," Longman Cheshire, Melbourne.

Chalmers, A.F., 1982, "What is this Thing called Science?", University of Queensland Press, Brisbane.

Checkland, P., 1991, From framework through experience to learnig: the essential nature of action research, in "Information Systems Research: Contemporary approaches and emergent traditions, H.E. Nissen et al., eds., Elsevier, Amsterdam.

Davies, L. and Myer, M., 1993, Scholarship and practice: the contribution of ethnographic research methods to bridging the gap, in "Business Process Reengineering: Information System Opportunities and Challenges", B. Glasson, ed., North Holland, Amsterdam.

Elden, M. and Chisholm, R.F., 1993, Emerging varieties of action research: introduction to the special issue, *Human Relations*, 46(2):121-142.

Hult, M. and Lennung, S., 1980, Towards a definition of action research: a note and a bibliography, *Journal of Management Studies*, 17:241-250.

Klein, H.K. and Lyytinen, K., 1985, The poverty of scientism in information systems, in "Research Methods in Information Systems," E. Mumford et al., eds., Elsevier Science, Amsterdam.

McKay, J., 1998, Using cognitive mapping to achieve shared understanding in information requirements determination, in "Proceedings of the 9th Australasian Conference on Information Systems, University of New South Wales, Sydney.

Rusli, A. and Marshall. P. (1995) Using an interpretivist inquiry methodology in IS research: an ethnography experience in an Indonesian organisation, "Proceedings of the 6th Australasian Conference on Information Systems", School of Information Systems, Curtin University, Perth.

Shanks, G., Rouse, A., and Arnott, D., 1993, A review of approaches to research and scholarship in information systems, in "Proceedings of the 4th Australian Conference on Information Systems", University of Queensland, Brisbane.

Susman, G.I. and Evered, R.D., 1978, An assessment of the scientific merits of action research, Administrative Science Quarterly, 23(4): 582-603.

West, D., Stowell, F.A. and Stansfield, M.H., 1995, Action research and information systems research, in "Critical issues in systems theory and practice," K. Ellis et al., eds., Plenum Press, New York.

ATLANTIC CROSSINGS: A GENEALOGICAL ANALYSIS OF THE RELATIONSHIP BETWEEN SYSTEMS THINKING AND ACTION RESEARCH IN THE UNITED KINGDOM

Stephen. K. Probert[1]

[1]Computing and Information Systems Management Group
Cranfield University
RMCS Shriveneham
Swindon
SN6 8LA

INTRODUCTION

Few would argue that action research has had an important and endearing influence on the development of recent and contemporary U.K. systems thinking. Furthermore, many studies have been carried out in both organisational analysis and information systems (IS) research which employ (or at least claim to employ) an action research approach. Lau, (1997) and Stowell *et al.*, (1997) provide both useful descriptions of the action research method and give examples of its application in IS research. Various debates concerning both the practicality, utility and validity of this process can be found in both the IS and the organisational behaviour / management literature - e.g. the special editions of *human relations* (46/2) and *Management Decision* (21). However, this paper will not explore these issues – rather this paper attempts to take seriously Foucault's genealogical notion that we should seek the conditions (in terms of concrete and discursive practices) which create the circumstances under which action research / systems thinking became *readily acceptable* (Probert, 1993). This paper builds on the work - on the ready acceptability of Soft Systems Methodology (SSM) - developed in Probert (1996).

The genealogical analysis begins by outlining the problem of management in the U.S.A. of the 1930's; "situating" Lewin's early work in the Psychological Institute of Berlin University, and then moves on to consider the reception of Lewin's work in the U.S.A.; Lewin's collaboration with the (U.K.) Tavistock Institute will then be considered. For brevity, it is assumed that the reader is sufficiently well aware of the recent history of systems thinking in the U.K. (from socio-technical approaches, to SSM, to CST – notwithstanding some other important developments in cybernetics) for it to be omitted from this, necessarily brief, discussion. Conclusions will then follow.

Synergy Matters: Working with Systems in the 21st Century,
Edited by Castell *et al.*, Kluwer Academic / Plenum Publishers, New York, 1999.

EARLY ORGANISATION THEORY AND ITS RELATIONSHIPS WITH THEORIES OF THE NATION STATE

The first part of this work will be primarily informed by recent genealogical studies by Miller and O'Leary (1987; 1990). One of Miller and O'Leary's concerns is to develop a critical history of how it came to be that modern-day managerial authority became construable in a "democratic-participative" discourse (whether or not managerial authority in the real world utilises such principles is entirely another matter). It would, I would argue, be churlish to deny the appeal of such "democratic-participative" legitimatory discourse for (e.g.) many managers, and students of management (and systems), in the real world.

In this respect, Miller and O'Leary argue that the "democratic-participative" practices of "modern" management emerged from a variety of circumstances, an important number of which occurred in the American Depression of the 1930's. In this period, a fundamental change occurred in the conception of the organisational employees - from the disciplined individuals, seen in earlier conceptions, to the autonomous and rational subjects typical of modern day managerial discourses:

> The responsible, decision-making individual has not always existed and did not emerge automatically. Such a conception of the individual was constructed out of a particular understanding of the [American] Depression. In an influential body of social thought, the Depression was not viewed at root as an economic affair. Its economic face was considered to be but a symptom of a deeper malaise within the framework of America and of all the world's industrialised nations. (Miller & O'Leary, 1990, p. 490)

The problem was characterised in the U.S.A as "cultural lag", and this problem was considered to be endemic in American society of that era by many academics. Miller and O'Leary argue that Donham and Mayo (early Organisation Theorists) considered that, in pre-industrial societies, individuals had been bound by loyalties to family, church, community etc. The conclusion that these academics sought was an "allied" notion of leadership; the question became one of how to make leadership more effective. Behaviour in the firm was reconstructed as part of a general problematic of behaviour in any democratic political economy.

EMPLOYEE BEHAVIOUR AND EXPERIMENTAL INTERVENTIONS IN ORGANISATIONS

As the political space was, and always had been, in a sense "experimental" (governments conducting a kind of "action research"), so the organisation could become a "social laboratory". It is at this point that one condition for the emergence of the acceptability of action research: the point at which experimenting on the organisational actors becomes legitimated by the ethic of the impartial acquisition of knowledge (Miller & O'Leary, 1987, p. 259).

The changes in the approach to the legitimisation and practice of managerial authority (as have been outlined) lead to an increasing emphasis being placed on developing an understanding of the *subjective* (i.e. the psychological) motivations of employee behaviour. Such a view is convergent with the view of organisations most frequently found in (e.g.) SSM discourse:

...[T]here has been a social revolution in advanced western societies which is still underway. This may be summarised in the words 'democratisation' and 'participation'... this is manifest in many different aspects of life in our society, in work, in play, in politics, in education ... (Checkland and Scholes, 1990, p. 85)

It may be wondered as to how the ideas discussed above, which have their origin in the U.S.A of the 1930's, become extant in pariticipative managerialist discourse of the 1990's. The key "conductors" (in the electrical sense) of these ideas about management - and their relationship with research (especially regarding the concept of 'action research') – are Lewin (U.S.A.) and, later, Rice (U.K.):

The origin of action research is usually taken to be Kurt Lewin's view of "the limitations of studying complex real social events in a laboratory, the artificiality of splitting out single behavioural elements from an integrated system" ... This outlook obviously denotes a systems thinker, though Lewin did not overtly identify himself as such - he was in fact a psychologist of the Berlin Gestalt group who worked in America from 1933. (Checkland, 1981, p. 152)

Paul Lewis has recently identified the work of the Gestalt group as important precursors of "fully-fledged" systems thinking:

Some elementary systems thinking may be found in the work of the Gestalt psychologists such as Wertheimer, Koffka and Kohler, who emphasised the study of perception, learning and thinking in whole units, not by analysis into parts... (Lewis, 1994, p. 16)

Lewin's work is explicitly identified as a key influence on Soft Systems Thinking:

Lewin's field theory regarded any observed behaviour as the outcome of the operation of a large number of factors which it was the task of the social scientist to identify and analyse. The idea of action research follows from this, and it may be regarded as one particular response to the problem of the special nature of social phenomena ... (Checkland, 1981, p. 152)

However, Lewin's work can also be seen to be inextricably bound up with all recent U.K. systems thinking (Levin, 1994), with the exception of the kind of systems thinking that is still fundamentally functionalist at heart.

LEWIN'S WORK

During 1918-33 Lewin worked in the Psychological Institute of Berlin University. He was trained as a psychologist and was heavily influenced by two contemporary influences: Freudian psychoanalysis and Gestalt psychology (de Board, 1978). His reason for leaving for the U.S.A was – in common with several other influential German academics of his day – his Jewish origins. This had an impact on the focus of his work:

Perhaps, more than most great men, the life of Lewin is as important as his work, in the sense that his work reflects so greatly his own personal situation and the environment in which he found himself... As a German Jew, experiencing anti-

Semitism in his early academic career ... and finally forced to flee from the Nazis in 1933, Lewin's work shows a passionate concern for democracy in general and minorities in particular. (de Board, 1978, pp. 49-50)

If one passionate set of concerns for Lewin was *democracy and participation* (or inclusiveness), another passionate set of concerns were *for general social improvements to be obtained through the practical application of social science*:

In an article published in 1939, 'Experiments in Social Space' ... Lewin made explicit his belief that the social scientist could make a real contribution to society's improvement and welfare. (de Board, 1978, p. 63)

So, in a somewhat perverse manner, Lewin's concerns for the practical application of psychoanalytical theory (to management) - and for democracy - received a warm response in the U.S.A of the late 1930s. Indeed, "By 1940 Lewin was regarded as one of the leading experimental and theoretical psychologists in the United States." (de Board, 1978, p. 58). His work took him into the heart of the U.S.A managerial *problematique* characterised by Miller and O'Leary earlier:

He had already published a paper in 1920 concerning Taylor's system of 'scientific management' in which Lewin said that work has a 'life value' and that every job should sustain or enrich this. Prophetically he said that the task of making jobs richer and more satisfying is a job not only for the efficiency expert, but also for the research psychologist. In 1939 he had the opportunity to put his ideas into practice in the Harwood manufacturing company... in Marion, Virginia. (de Board, 1978, p. 60)

Lewin and his associates carried out action research in this organisation for several years; indeed the work continued after Lewin's death in 1947. The general conclusion from the work was that *participation* leads to *increased economic efficiency.* (Coch and French, 1948).

Before Lewin died, important collaborations with U.K. researchers had been established, but his influence on management thinking in the U.S.A cannot be underestimated:

At the Centre for Group Dynamics, Lewin rapidly built up a team of colleagues who had been working with him over the years ... He also worked closely with other eminent social scientists including McGregor and Allport. Lewin's influence was enormous, both on his colleagues and on the many doctoral students who came to study under him. (de Board, 1978, p. 62)

LEWIN AND THE TAVISTOCK INSTITUTE

Meanwhile, over in the U.K., members of the Tavistock Institute of Human Relations had taken a keen interest in Lewin's work:

...Trist and Wilson wrote to Lewin asking him if he would consider establishing the journal [*Human Relations*] jointly, that is between the Tavistock Institute and Lewin's Centre for Group Dynamics at Massachusetts Institute of Technology. Lewin agreed and the first issue in June 1947 (published after his death) commenced... (de Board, 1978, p. 64)

So, the influence of Lewin's ideas began to spread throughout U.K. universities after World War Two, particularly (at first) in Leicester University:

> In 1957 a conference was organised jointly by Leicester University and the Tavistock Institute of Human Relations ... The conference was strongly influenced by the National Training Laboratories at Bethel, United States [established by Lewin and colleagues in 1947]... The conference lasted two weeks... The task of the study group members was to examine their own behaviour in the 'here and now' and to observe their development over the period of the conference... A second conference was held in 1959 ... From 1962 to 1968 Rice [of the U.K. Tavistock Institute] directed all the Leicester Tavistock Conferences... (de Board, 1978, pp. 75-76)

It is clear that members of the Tavistock Institute such as Emery, Rice, Trist and Bamforth all shared a common concern for both *participation* and *economic efficiency*. The whole (U.K.) socio-technical school of thought that emerged from the 1950s has a clear genealogical legacy to Kurt Lewin's discourse - which provided an academic framework for work that had already been conducted with the aim of achieving greater economic efficiencies – particularly poignant in the U.K. of the 1950s. It was generally considered at the time that the U.K. was in a state of relative economic decline – partly as a result of long term global economic trends, and partly as a result of the actual costs of fighting World War Two. Lewin's work gave academic respectability to the search for ways of improving economic efficiency by encouraging greater workforce participation. As Stowell *et al.* argue:

> Parallel but independent work to that of Lewin's took place at the Tavistock Institute of Human Relations in London ... In some respects Lewin's work appears to be more academically orientated than that of the Tavistock group, which was more professionally orientated, with industrial organisations being a prominent area of application in which co-operative relationships between members were encouraged ... (Stowell, *et al.*, 1997)

The "reindustrialisation of England" (Levin, 1994) became an academically respectable pursuit – thus enabling successive U.K. governments to argue, cajole and (perhaps) coerce U.K. academics to conduct research with the explicit aim of improving the economic efficiency of "U.K. PLC".

Rice (at the Tavistock Institute) was to make two further important contributions to the action research / socio-technical systems movement. Firstly, the explicit importation of systems theory and secondly, the (re) introduction of certain psychoanalytical concepts - ostensibly to better explain human behaviour in organisations:

> An individual may be seen as an *open system*. He exists and can exist only through processes of exchange with his environment... The *mature ego* is one that can define the boundary between what is inside and what is outside and can control the transactions between the one and the other. (Rice, 1969, p. 574 [emphases added])

Rice also (re) introduced various other psychoanalytical concepts (such as the *primary task**), however work in progress indicates that these moves may not have had such a positive impact as is often assumed. At any rate, it was Trist and Bamforth (again from the

---

* This term is still used in SSM jargon, however its meaning has been hopelessly distorted from its original psychoanalytical origins.

Tavistock Institute) that first applied open system theory (in an action research framework) in the U.K. in 1951 (de Board, 1978).

CONCLUSION

It has been argued that the conditions for the ready acceptability of action research arise from (albeit different) macro-political (socio-economic) imperatives – both in the U.K. and in the U.S.A. In the U.S.A of the 1930s, perceived imminent social breakdown (and its economic corollary) was the "driver"; whilst in the U.K. of the 1950s, perceived relative economic decline was the "driver". Both of these drivers led to macro and micro initiatives to "install" action research *on the shop floor*. Academia was used as one of the vehicles by which such installation was to take place; academics' (widespread) "liberal agenda" being especially conducive to the notions of participation and inclusivity.

Further research could investigate how it came about that the metaphor of the *economically efficient capitalist organisation* became the dominant one for evaluating the effectiveness both of the U.K. and the U.S.A. *as nation states* – and for evaluating activities within those nation states. At any rate, as it could now be argued that democracy leads to increased economic efficiency (by reference to both Lewin's and the Tavistock Institute's work), so the corollary is now propounded with increasing vigour and certitude; i.e. that the "true" *function of democracy* should be "properly" seen as that of increasing *economic efficiency* under capitalism. Consequently, research agendas which overtly aim to produce results which improve the economic efficiency of the funding nation are now increasingly valorised, whilst those that do not are increasingly denigrated.

REFERENCES

de Board, R., 1978, "The Psychoanalysis of Organizations," Routledge, London.
Checkland, P., 1981, "Systems Thinking, Systems Practice," Wiley, Chichester.
Checkland, P., and Scholes, J., 1990, "Soft Systems Methodology in Action", Wiley, Chichester.
Coch, L., and French, J. R. P., 1948, Overcoming resistance to change, *Hum. Relat.*, 1:512-532.
Lau, F., 1997, A review on the use of action research in information systems studies, *in*: "Information Systems and Qualitative Research," A. S. Lee, J. Liebenau, and J. I. DeGross, eds., Chapman and Hall, London.
Levin, M., 1994, Action research and critical systems thinking: Two icons carved out of the same log?, *Sys. Prac.*, 7:25-41.
Lewis, P., 1994, "Information-Systems Development," Pitman, London.
Miller, P., and O'Leary, T., 1987, Accountancy and the construction of the governable person, *Accounting, Organizations and Society*, 12:235-265.
Miller, P., and O'Leary, T., 1990, Making accountancy practical, *Accounting, Organizations and Society*, 15:497-498.
Probert, S. K., 1993, Interpretive analytics and critical information systems: a framework for analysis, *in*: "Systems Science: Addressing Global Issues," F. A. Stowell, D. West, and J. G. Howell, eds., Plenum, New York.
Probert, S. K., 1996, A genealogical analysis of managerial authority (as portrayed in soft systems methodology) and its consequences for information systems analysis, *in*: "Information Systems Methodologies 1996: Fourth Conference of the British Computer Society Information Systems Methodologies Specialist Group," N. Jayaratna and B. Fitzgerald, eds., BCS Publications, Swindon.
Rice, A. K., 1969, Individual, group and intergroup processes, *Hum. Relat.*, 22:565-584.
Stowell, F., West, D., and Stansfield, M, 1997, Action research as a framework for IS research, *in*: "Information Systems: An Emerging Discipline," J. Mingers and F. Stowell, eds., McGraw-Hill, Maidenhead.

EXPLORING LINKS BETWEEN ACTION RESEARCH (AR) AND "ALTERNATIVES"

Norma R.A. Romm

University of Hull
School of Management
63 Salmon Grove
Hull HU6 7SZ
United Kingdom
email: N.R.Romm@mgt.hull.ac.uk

INTRODUCTION

Action Research (AR) is normally defined as research committed to doing research "with" rather than "on" or "about" people (Heron, 1996, p. 9). It is aimed at encouraging the development of democratic relationships in the knowledge-production process (McKay and Romm, 1992; Reason, 1994; Smaling, 1998). It is furthermore committed to generating improvements in "action" as part of the definition of its validity or credibility. It is suggested that it is through its ability to develop people's capacities for "insightful action" that it enhances its credibility/value (Weil, 1998, p. 51). Its quality can therefore be judged in terms of the processes that are instituted to generate cognitive participation of co-inquirers involved in the inquiry, as well as in terms of whether these processes are regarded as contributing to (valued) "social and organisational changes" (Smaling, 1998, p. 11). These two ways of considering its quality are not necessarily seen as separable, in that the value of the changes themselves can be seen as linked to developing dialogical partnership in the knowledge-construction process. As Smaling puts it: "Insofar as the aimed at social and organisational changes imply that better dialogical relationships should be implemented in a social structure or an organisational system, striving for dialogical partnership within the action research project will give the research project more *exemplary value*" (1998, p. 1).

Certain proponents of AR (for example, Fals-Borda, 1991; Reason, 1994; Heron, 1996) claim that "normal science" cannot be of the same worth to people in society as is AR. This claim is made on the grounds that research undertaken "on" people (as, for instance, in experimental designs or survey research) or "about" people (as, for instance, in ethnographic research) is unnecessarily exclusive. Those "being researched" are excluded from the process of designing the study, from participating in defining the meaning of "findings" generated through the study, and from the opportunity of developing themselves and their situations through the study (Heron, 1996, p. 17).

In this article I question the suggestion that all research of quality in, and for, society has to be viewed as involving co-researchers and co-participants at each moment of inquiry (as in AR). While there is some cause for concern about the way in which "normal" science might function to perpetuate monopolies of knowledge-production (through the way

Synergy Matters: Working with Systems in the 21st Century,
Edited by Castell *et al.*, Kluwer Academic / Plenum Publishers, New York, 1999.

199

scientific communities have tended to authorize their activities - see Fals-Borda and Rahman, 1991) these monopolies can be undercut by according a different status to the accounts generated in the process of so-called normal science. This can be done by seeing the accounts as resources that can be drawn upon, re-assessed, made relevant, interrogated, contested, etc., by (other) inquirers at any point (Romm, 1997, paragraphs 7.3-7.7, 1998, p. 72).

My argument is that as long as provision is made *at some point* for cognitive participation of participants/audiences, the research can be seen as having the potential to contribute to the democratization of knowledge relationships and to the possibility of "insightful action" that is informed by dialogical learning processes. Provision for cognitive participation implies that the constructions generated in any inquiry process (whether "normal" science or AR-oriented) should not be seen as offering "truths" independently of how these are played with and made relevant in specific contexts where people take responsibility for ways of seeing and acting (Romm, 1997). A case is made in the article for re-exploring the value of forms of inquiry such as survey research and ethnographic study in terms of the participative principle put forward by AR proponents.

SURVEY RESEARCH

The relationship between AR and survey research (as a form of inquiry in society where variables are isolated and their relationships explored - see Romm, 1991, p. 68-72) can be seen to be uneasy. The required isolation of conditions leading to supposedly regular (predictable) effects seems not to do justice to possibilities for (action) inquiry aimed at locating more concrete ways of connecting the design of actions with intentions as relevant for actors' decision-making in local contexts (Cartwright, 1978; Argyris and Schön, 1996; Eden and Huxham, 1996).

Survey research aimed at finding out how variables are regularly related, allows scientists (and others in society) to take the position that certain patterns "out there" in reality can be ascertained. One of the political implications of this is that the results discerned often come to be defined as unnegotiable statements that cannot be "worked past" (Romm, 1997, paragraph 7.5). However, actors in society may be faced with options for decision-making which require them to link their actions to a variety of possible effects in situations of ambiguity (Weick, 1995, p. 186; Weil, 1998, p. 42). It is never clear how "the situation" should be defined in the first place and what the effect of acting in "the situation" might be. Therefore, it is argued by AR proponents that it is preferable to work with people in exploring the dynamics of their situations, with the aim of developing insightful ways of considering the arena of possible action. The aim is also to develop ways of appreciating "what could be, rather than what is" (Elden and Chrisholm, as cited in Eden and Huxham, 1996, p. 83). Or, as Argyris and Schön put it: People "are in the situations they try to understand, and they help to form them by coming to see and act in them in new ways" (1996, p. 36).

But the question arises whether forms of research other than AR could become relevant to the kinds of judgments faced by actors in their lived experience (in which they may wish to create innovative responses). Can such forms of inquiry not be treated as invitations to explore "what could be"?

We can consider as an example the study reported by Winch, Millar and Clifton (1997) on the behavior of French and British managers working on the construction of the Channel Tunnel. Winch, Millar and Clifton note that it has often been considered (since Hofstede, 1980) that cultural values are a predictor of the organisational behavior of individuals; and hypotheses have been constructed about the way in which culture is deemed as affecting such behavior. Winch, Millar and Clifton used the example of the responses of managers involved in the construction of the Channel Tunnel in order to investigate two hypotheses. The first was that the British respondents would be likely (in terms of certain cultural characteristics) to indicate a preference for a more organic organisational structure; while the French would tend to be more bureaucratically oriented (1997, p. 239). The second was that the British would be more individualistically oriented in their work behavior; while the

French would be more likely to be motivated through a sense of belonging and security (1997, p. 239). Winch, Millar and Clifton drew on a "battery of measures" developed by Van de Ven (1980), while emphasizing the importance of interpreting the "quantitative data derived from questionnaires" in the light of what they call a "deep ... knowledge of the organisational context and dynamic" (1997, p. 238).

Winch, Millar and Clifton argue that on the basis of the evidence that they gathered, "although the cultural values of the two groups were largely as would have been predicted by Hofstede's work, the expected behavioral effects were not found" (1997, p. 246). Some opposite effects were identified. The French appeared to be more autonomous than the British and less reliant upon procedures; and also more individualistic (more competitive in their work). Winch, Millar and Clifton indicate that this could be a function of the research instruments, which rely on self-reporting of values held by individuals and also on self-reporting of perceptions of workplace activities (1997, p. 246). They also note that "more work needs to be done on the interpretation of the Hofstede indices [of cultural characteristics], their change over time and place and between different groups of people" (1997, p. 247).

Winch, Millar and Clifton make it clear that they are allowing the reader (and anybody concerned) to ask questions about the way the research has been set up to create "evidence" and about the interpretation of this "evidence". They also make it clear that it is possible that the indices of variables considered can change over time and place, which might render the measurements utilized less relevant and less applicable to the context of the research. They do not wish their account of their own "findings" (or that of other research in this field) to be left unquestioned. People can and should participate in deciding how to treat the information offered. This, in turn, it may be argued, allows them (those concerned) to consider ways of working around any presumably located "tendencies" discerned through the research. It is possible for located tendencies/patterns to be used as resources which contribute to our sense of how we can address issues of concern in the workplace, such as, in this case, ways of structuring our relationships and ways of defining our "individuality" or "collectivity". We (as participants/audiences) need *not* be captured by findings created and discerned within a community of scientists, but are able to work around what is presented.

ETHNOGRAPHIC RESEARCH

The way in which action researchers might regard ethnographic research (aimed at offering rich in-depth accounts of the texture of everyday life) is captured in a metaphor supplied by Heron. Heron suggests that in such research a "half-way house" is reached between controlling research *on* people (which he sees as linked to experimental or survey research) and research *with* people (1996, p. 27). Ethnographic research - as a halfway-house - is research *about* people.

Heron is concerned that the "unilateral research designs" that have been developed by ethnographic researchers, do not invite the research subjects to become co-researchers (1996, p. 29). They are still seen as subjects of some ethnographic researcher - who has decided to study "them" in some everyday setting. The advantage of their being studied in this way is that, at least, this is research *about* rather than *on* people. One of Heron's complaints, however, is that the subjects in ordinary qualitative research "may tell you a great tale which could be a packet of conscious or unconscious lies" (1996, p. 29). He believes, in contrast, that the relationship between co-researchers in AR allows all people involved to challenge and interrogate one another as part of the research process. They can thus check their assertions against the test of confrontational encounter. In this way, "packets of lies" can be interrogated in the course of the inquiries. (This is consistent with Hölscher and Romm's (1986) suggestion that a dialogical interventionist research strategy is defined by its encouraging all involved to confront one another's constructions through the research process.)

But the question can still be asked whether all ethnographic research is to be regarded as merely a halfway-house on the way to better quality (action) research. Let us reconsider this by drawing upon the example of Golding's (1996) ethnographic study of workplace practices. Golding (1996) offers an account of the way in which chains of command became perpetuated in the texture of relations of a company which he calls the "Wellblown manufacturing company" (1996, p. 79). He indicates how in the course of his involvement therein he discerned "violence to the spirit" through "intimidatory schemes", and he refers to other "rituals of oppression" (1996, p. 89). He thus offers an account which Heron would consider as being *about* the people concerned.

In his conclusions Golding acknowledges (to the reader) that he has imposed a particular "observer derived framework" for shaping his analysis. He also indicates that he is aware that "literal descriptions and analyses can ... produce ossification" (1996, p. 93). He thus acknowledges the temporal nature of his own (and any) analysis. Is such research to be regarded as merely of half-way house status in its quality? Or does the fact that Golding indicates that he expects that others will (now) participate in re-defining the import of his account mean that he does make provision for research *with* people? This could apply to the "subjects" involved and others in society who may wish to consider the relevance of Golding's study of ritual for some setting in which they are engaged. In each case a kind of participation in the design, analysis and interpretation could take place post facto by re-visiting each of these moments in the research and considering what changes, if any, might be imagined in order to reshape the "results" (or rather, the meaning of the results as envisaged by people). Must the (initiating) researcher insist on everybody concerned becoming co-researcher in the project at the point at which s/he is involved in designing projects, analyzing and interpreting - or can this cognitive participation be left to other times?

One of Heron's objections to ethnographic research is that perhaps some of the participants involved might trump up a pack of lies, consciously or unconsciously. According to him, it is important that researchers engage in (immediate) confrontation with these, rather than presuming simply to offer some record of the life in the setting under study. Can this objection of Heron's be met by Golding? Golding (as others) would say that the ethnographic researcher has the opportunity to offer new angles on any of the statements issued by participants - and that a qualitative research study can show how the researcher is engaging with, rather than simply trying to reproduce, their accounts. Hence, for instance, Golding's location of rituals of oppression is a way of offering a specific voice (way of coding) that extends, and at the same time confronts, the different subjects' experience of, and talk about, their way of life.

Heron's reply to this kind of argument is that in the researcher's efforts at such an extension, the cognitive rights of the subjects to contest the researcher's categories are not accounted for (1996, p. 29). The issue of contention thus seems to rest on the question of at what point in time a confrontation between "researcher" and "others" in society needs to be encouraged as a way of setting up a relationship of, indeed, "co-researchers". The argument here is that co-research does not have to exclude ethnographic-styled research, in that at *some* point provision is made for people (those interested) to "participate" with the original researcher in reviewing the design and examining the relevance of the results achieved through the design. Through this process, people can address themselves towards taking into account various understandings of the meaning of the (co-)exploration.

DEVELOPING AN OPEN SYSTEM FOR KNOWLEDGE ASSESSMENT

An open system for knowledge assessment can be proposed as a way of conceiving knowledge-construction processes and products in society. The development of such a system requires an epistemological awareness which makes provision for people to participate not only in reworking of any knowledge claims produced at points in time, but also in re-defining criteria for assessing knowledge-production (Romm, 1998, pp. 71-74). The openness of the system consists in its being marked by disjunction as different criteria for knowledge assessment are entertained within the system. This contrasts with a closed system (of knowledge assessment) which aims to maintain a given set of protocols for going about the process of social inquiry.

In an open system of knowledge assessment, relationships between different forms of inquiry can be explored, without any being used as a device to persuade the other of its own (sole) rationality (Romm, 1990). All forms of inquiry are regarded as processes which can be (re)appropriated through co-researcher/co-participant assessment of the process and attendant products. AR as a form of inquiry-in-action may be regarded as a special case exemplifying the democratic potential of all knowledge construction by facilitating a recognition of the possibility for cognitive participation on the part of those concerned with considering the value of the research.

Assessment in society of knowledge-construction processes and products is itself never completed (in the sense that incontrovertible answers can be attained). Recognition of this incompleteness is important so that practitioners/actors can act without using the legitimization of some authorized package of knowledge to remove from vision (their own and others') the choicefulness of their way of proceeding. They can account for their choices of vision (ways of seeing situations and connections/patterns) and of action (judgments made about how to address their situations), without recourse to the authority of "the facts" or "the realities" - as if these can be discerned in an unproblematic, unmediated and static way (Romm, 1996b, 1997). They can account for their involvement with "the world" by recognizing that "knowledge" is not innocent and that the way they choose to see situations might itself have certain impact therein (Romm, 1995). As Eden and Huxham indicate: "The language used to frame the theory will seriously influence the future thinking and actions of the consumer of the research" (1997, p. 80). Eden and Huxham hope to use the action research process to mitigate against the self-fulfillment of researcher-oriented discourse at the expense of practitioner discourses in directing courses of seeing-and action. An alternative, however, is to suggest that accounts generated through any form of inquiry should be subjected to an ongoing conversation in society concerning ways of assessing research processes and products (Romm, 1998, p. 74).

CONCLUSION

One way of lending credence to the democratizing potential of AR and to its aim of developing "insightful action" of practitioners is to attempt to create/vitalize an open system for knowledge assessment. An open knowledge assessment system implies that participants/audiences do not to take as "given" any research designs or results thereof developed in the process of inquiry. People (as co-inquirers) are enabled to develop their capacity to come to terms with, confront, engage, and re-assess, different processes of inquiry and the value of any insights generated therein. They can then become co-researchers with the original researchers, and with each other, at any point in time - while recognizing that there is disjunction in the system of knowledge assessment that cannot be neatly controlled.

An open system for knowledge assessment requires keeping alive an awareness in society that there are no clear-cut criteria to define when sound knowledge-production protocols have been adhered to. This lack of clear-cut criteria to define sound knowledge-production processes (and attendant products), supports those who wish to see the appropriation of knowledge as part of a continuing dialogue in society. Dialogue here is defined as a propensity to encounter opposition (to an initial way of seeing) and through the process to become more (discursively) accountable (Romm, 1996a,b). Dialogue can also be seen as linked to what Weil (1997) calls "working out of contradiction". Working out of contradiction in this context implies surfacing and rendering explicit some of the different ways of defining quality in knowledge-production. This way of addressing the knowledge problematic does not preclude people having to make judgments at points in time concerning ways of living with any understandings that they might have developed. This is part of the process of taking responsibility for ways of knowing and acting (see, for instance, Romm, 1996c). Responsibility here is seen as tied to processes of accountability, where people are able to account for (while never absolutely defending) their orientation in/with the world, and their engagement with others' arguments and concerns.

REFERENCES

Argyris, C., and Schön, D.A., 1996, "Organizational Learning II: Theory, Method, and Practice," Addison-Wesley Publ. Co., New York.

Cartwright, D., 1978, Theory and practice, *Jnl. Soc. Issues*, 34,4:168-180.

Eden, C., and Huxham, C., 1996, Action research for management research, *Brit. Jnl. Management*, 7:75-86.

Fals-Borda, O., 1991, Some basic ingredients, *in*: "Action and Knowledge: Breaking the Monopoly with Participatory AR," O. Fals-Borda, and M.A. Rahman, eds, The Apex Press, New York.

Fals-Borda, O., and Rahman, M.A., 1991, "Action and Knowledge: Breaking the Monopoly with Participatory AR," The Apex Press, New York.

Golding, D., 1996, Management rituals: maintaining simplicity in the chain of command, *in*: "Understanding Management," S. Linstead, R.G. Small, and P. Jeffcutt, eds, Sage, London.

Heron., J., 1996, "Co-operative Inquiry: Research into the Human Condition," Sage, London.

Hofstede, G., 1984, "Culture's Consequences: International Differences in Work-Related Values," Sage, London.

Hölscher, F., and Romm, N.R.A., 1986, Development as a process of consciousness, *in*: "Development is for People," J.K. Coetzee, ed., Macmillan, Johannesburg.

McKay, V.I., and Romm, N.R.A., 1992, "People's Education in Theoretical Perspective," Longman, Cape Town.

Reason, P., 1994, Human inquiry as discipline and practice, *in*: "Participation in Human Inquiry," P. Reason, ed., Sage, London.

Romm, N.R.A., 1990, Gouldner's reflexive methodological approach, *in*: "Sociology and Society," C.J. Alant, ed., Southern, Johannesburg.

Romm., N.R.A., 1991, "The Methodologies of Positivism and Marxism," Macmillan, London.

Romm, N.R.A., 1995, Knowing as intervention: reflections on the application of systems ideas, *Syst. Pract.*, 8:137-167.

Romm, N.R.A., 1996a, Inquiry-and-intervention in systems planning: probing methodological rationalities, *World Futures*, 47:25-36.

Romm, N.R.A., 1996b, Reflections on an action research project: women and the law in Southern Africa, *in*: "Critical Systems Thinking: Current Research and Practice," R.L. Flood, and N.R.A. Romm, eds, Plenum, New York.

Romm, N.R.A., 1996c, Systems methodologies and intervention: the issue of research responsibility, *in*: "Critical Systems Thinking: Current Research and Practice," R.L. Flood, and N.R.A. Romm, eds, Plenum, New York.

Romm, N.R.A., 1997, Becoming more accountable: a comment on Hammersley and Gomm, *Soc. Research Online*, 2,3. (<http://www.socresonline.org.uk/socresonline/2/3/2.html>)

Romm, N.R.A., 1998, Interdisciplinary practice as reflexivity, *Syst. Pract. and Action Research*, 11:63-77.

Smaling, A., 1998, Dialogical partnership: the relationship between the researcher and the researched in action research, *in*: "The Complexity of Relationships in Action Research," B. Boog, H. Coenen, L. Keune, and R. Lammerts, eds, Tilburg University Press, Tilburg.

Van de Ven, A.H., and Ferry, D.L., 1980, "Measuring and Assessing Organizations," John Wiley, New York.

Weick, K.E., 1995, "Sensemaking in Organizations," Sage, London.

Weil, S., 1997, Social and organizational learning and unlearning in a different key: an introduction to principles of critical learning theatre and dialectical inquiry, *in*: "Systems for Sustainability: People, Organizations and the Environment," F.A. Stowell, R. Ison, S. McRobb, J. Holloway, R. Armenson, and S. Jackson, eds, Plenum Press, New York.

Weil, S., 1998, Rhetorics and realities in public service organizations: systemic practice and organizational learning as critically reflexive action research (CRAR), *Syst. Pract. and Action Research*, 11: 37-62.

Winch, G., Millar, C., and Clifton, N., 1997, Culture and organization: the case of Transmanche-Link, *Brit. Jnl. Management*, 8:237-249.

CHANGING ORGANIZATIONAL CULTURE BY MEANS OF THE ACTIVITIES OF HIERARCHICAL POLY-AGENT ORGANIZATIONS

Hirokazu Tanaka

The Information Engineering Department
Kanagawa Institute of Technology
Kanagawa 243-0292, Japan

1.INTRODUCTION

New information technology is increasing its tremendous impact in various fields (organizational forms, working styles, communication patterns and so on) within the firm. Because it is possible, for instance, to share information as well as to manage information by using Internet (or Intranet) rapidly, traditional methodology or means of designing organizational form based on communicational flow related to responsibility – authority is not suitable. In firms where most members recognize the importance of organizational knowledge clearly, the specific organization's culture plays an important role in collaborative working style (team or project oriented organization) in order to achieve the organizational goals (or mission).

By focusing on such organizations mentioned above, the purpose of this paper is to develop a new methodology on how the traditions and culture of the main part of the organization can be not only changed but also spread throughout the organization. The main point is to demonstrate that the proposed methodology which is characterized by the activities of Hierarchical-Poly-Agent-Organizations is effective (to be described later), and to show the result of an action research applied to a Japanese trading company.

2.FRAMEWORK OF " Hierarchical Poly-Agent-Organizations"

The features of the Poly-Agent-Organizations proposed by Kijima[1] are as follows; First, individual agents, that is the employees in this case, have original mental models (rules reflecting the knowledge of the work to be performed) of their own. Second, it is possible that organizational properties (for example, organizational behavior, norms, work style, so on) are constructed by the activities, which can be done by *identification*, *investigation*, and *creation* related to the mental models on the basis of discussions with each other. As members are likely to make decision-makings more freely, rapidly, and timely compared

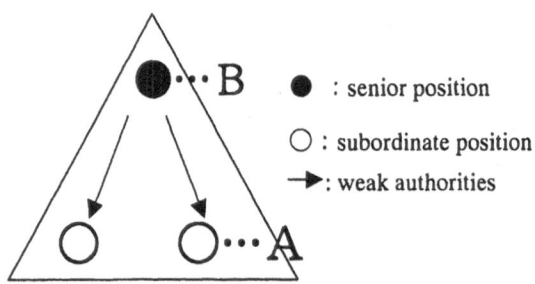

Fig.1 **Hierarchical Poly-Agent-Organizations**

with the traditional organization model (i.e. pyramid-type), it is possible to increase the opportunity to develop and lead to a new product/market strategy in the face of the environmental "chaos" pressures successfully. But there are some problems; first, expanded authorities are likely to be functioned freely independently of each other without relation to the organization's object or mission. So, the author proposes to introduce very "weak" authorities to Poly-Agent-Organizations, as shown in Fig 1. "Weak" authorities mean that any subordinate position member(A),in the Hierarchical Poly-Agent-Organizations（to be called HPAOs from now on）,is willing to accept the mental model that the senior position member(B) has possessed as his own , and also to replace his mental model with it ,if A is able to feel that B's mental model is reasonable as well as understandable. Such situations are often observed in many Japanese firms where the individual authority/responsibility is ambiguous and more priorities are given to the organizational level rather than the individual one.

3. CULTURE FOLLOWED BY BEHAVIOR

Generally, senior leaders are brought in and given positions of power in order to change unsuitable cultures in the traditional hierarchical (pyramid-type) organization, and are likely to use job-descriptions in many cases which says what kind of behavior individual member should perform and which behavior should not [2]. This means that the revision of the existing culture and the socialization of members in the culture rely on learning processes to ensure an institutionalized reality by means of the actual behaviors.

■should explore the new/product market
△ visit new customers more many times
△ analyze the products and customers of the
competitors
△ use the relationship with existing customers

■should improve the quality of existing products
△ analyze the bad news gathered from customers
△ use the cost analysis based on the reports

Fig .2   Example of the **Descriptive Model**

The paper proposes that the job-description referred to as principles, rules, or norms by members in making actions should be created by the activities of the HPAOs . Regarding this point, a descriptive model should be developed as the job-description that represents desirable principles, rules, or norms. This model is illustrated with an example from an actual firm in Fig. 2 . The characteristics of this model are the descriptive expression that is composed of a pair of main sentences (in Fig. 2 , the mark ■ is attached to the head of a sentence) and subsidiary example sentences (in Fig. 2 , the mark △ is attached to the head of individual examples). The former plays the role of explaining the desirable principles to be performed (that is the culture to be revised) roughly, and the latter to identify the principles by actual representations of the desirable behaviors. The proposed model makes it possible to be applied in flat-type organizations (collaborative work style) appropriately located intermediately between "ambiguity "and "strictness".

4.ACTION RESEARCH APPLIED TO THE FIRM

An action research was applied to a certain Japanese firm and the validity and possibility of the new proposals were examined. Table1 shows the properties of the firm(to be called firm A from now on). The problem that firm A faced several years ago was to create a new culture and to spread it throughout the firm, for sales and profits were decreasing continually and the employees' motivations towards the work were declining year after year.

So, the top manager made up his mind to organize a project team to revise the existing culture, and the author's group had an opportunity to advise the project team as a consultant as well as to make observations about the activities of the project team.

Table 1   Properties of the firm applied in the action research

Number of employees	About 400
Sales	About 350 million dollars
Type of industry	Trading of electrical & machinery Products
Inauguration date	1950's
Head office	In Tokyo
Number of branch offices	About 10 in Japan only

The project was performed in the following four phases:

(1) Selection of the project members

Desirable members should have sufficient skills and abundant experience as to their jobs, and also should have the skills to present the mental model of their own and to make discussions with each other in order to create a new culture. Based on the above conditions, the members were selected from among the organization to organize the HPAOs (shown in Fig.3). Some attention should be paid that HPAOs are organized in every division.

(2) Activity of individual HPAOs

Individual HPAOs were expected to create the new culture as a draft independently; at first, two or three subordinate position members made a draft to discuss with each other by using the descriptive model proposed by the author's group. And after that, a senior position member discussed with them about the draft. There were some gaps in the members' mental model between the subordinate and senior members. If there was a necessity, the senior exercised his authority to replace the mental models of the subordinates with his own.

(3) Activity of the senior position members in HPAO

After every draft made by the HPAO was submitted, a meeting was held, in which attendants were senior position members (each was a representative member of his division). And each member represented the mental model of not "his own" but "his division", because the mental models of an individual and division was turning to become the same in the second phase.

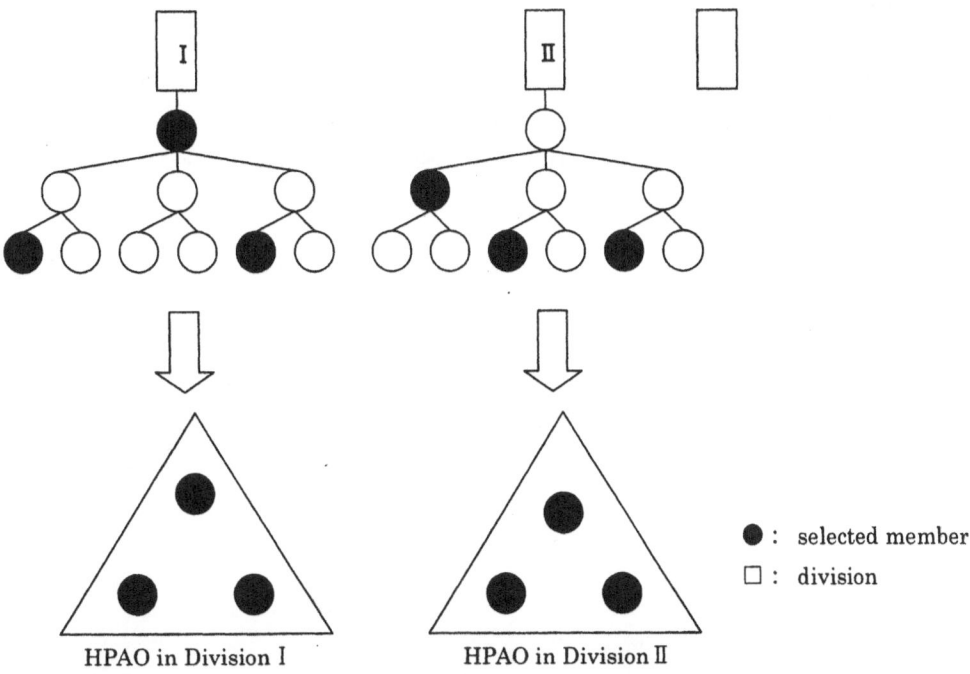

HPAO in Division I   HPAO in Division II

● : selected member
□ : division

Fig. 3 Organizational Form and Organized HPAOs

(4) Spread throughout the firm

All the members of the project were expected to play the role of "missionaries". And they talked with other members about the desirable behavior reflecting the new principles, rules, or norms in the meetings held in every workplace. This process made it effective to spread the revised new culture throughout the firm, because the descriptive documents were used as the standard in assessing the members as well as the members' being easy to understand the meaning in the actual behaviors.

5.CONCLUSION

The author has argued that the HPAO approach is suitable for organizations taking a middle point style between the strict pyramid-style and the loose network-style. It has also been argued that organizational culture is possible to be expressed partially using a descriptive model composed of a pair of main statements and example statements.

Furthermore, action research has been applied to a certain Japanese firm and the validity and possibility of the new proposals have been examined.

REFERENCES

1. Kyoichi Kijima, intelligent poly-agent learning model and its application, information and systems engineering, 2,47-61(1996)
2. Beer.M & others, MANAGING HUMAN ASSETS, The Free Press, A Division of Macmillan(1984)

ALL IN THE MIND - HEURISTICS, MODELS, SYSTEMS AND BUSINESS ADVICE

John Hassall, Kevin Mole

University of Wolverhampton
Management Research Centre
Shropshire Campus
Shifnal Road, Priorslee
TELFORD TF2 9NT
UK

INTRODUCTION

This paper looks at how professional business advisors, typically employed by business links and banks, determine the viability of small and medium sized enterprises (SMEs). Asked to make judgements with limited information, people tend to employ a small number of heuristic principles to reduce complexity (Tversky and Kahnemann, 1974). The study reported aimed to document some aspects of business advisor heuristics starting from a grounded perspective.

BUSINESS ADVISOR HEURISTICS - GROUNDED RESEARCH

The initial aim of the research was to find out how business advisors went about evaluating SMEs. Experience prior to the commencement of the project had suggested that reviewing formal methods of evaluation could prove un-helpful and it was decided to start from a grounded basis using a qualitative research design in the first instance. In adopting this approach it was hoped both to see the evaluation problem through the eyes of the business advisor and to better contextualise the outcomes of the research (Bryman, 1988), (Strauss and Corbin, 1990).

In an attempt to identify and express the business advisor perspective in a heuristic fashion some 30 semi-structured interviews with persons active in the advisor role were carried out between November 1996 and March 1997. This initial sample included advisors working for business links, other economic development organisations and banks. A common feature was the diagnostic element to the job. Open-ended question prompts were employed to allow the direct views of the advisors to be collected. Initial analysis of the data followed a

Synergy Matters: Working with Systems in the 21st *Century,*
Edited by Castell *et al.*, Kluwer Academic / Plenum Publishers, New York, 1999.

211

phenomenologically based process (Hycner, 1985) and was designed to derive the key constructs business advisors use to assess SMEs.

Principle outcomes, based upon codings derived from full transcriptions of all the interviews were, firstly, a concern by business advisors that SME managers should be "in control" of their businesses, secondly, that there should be some coherence between the aims of the business and the means of carrying them out as expressed within its organisation, procedures and operations. It was also clear that, whilst it was possible to see different advisors basing their judgements upon different data (bankers emphasising financial control, other advisors citing 'tidy workplaces' for example), the underlying approach was to look for evidence of a control system (in effect a homeostat) linking objectives, strategy, implementation and control within a beneficial feedback arrangement. This was expressed, using Capra (1996) as inspiration, in the form of the system diagram shown above.

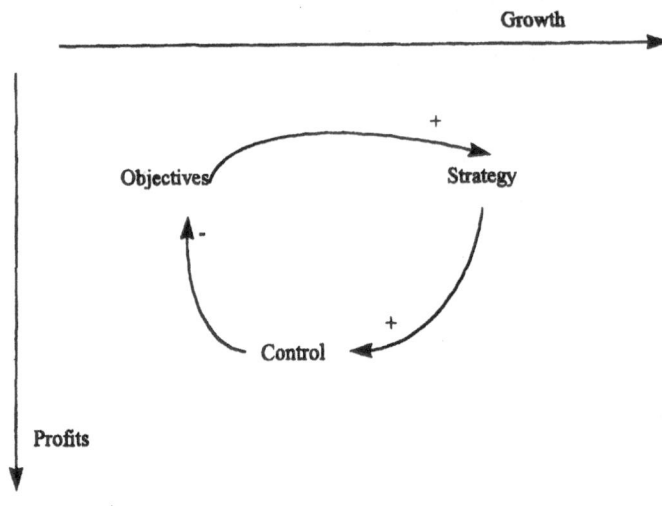

The closed system, Profits and Growth

HEURISTICS AS SYSTEMIC IDENTITIES

As a result of the initial stage of research business advisor heuristics have been interpreted as "sense making" systems. This is true from the perspective that the business advisors will seek to elicit evidence of systemic behaviour within SMEs; in effect to identify the presence of absence of a particular developed systemic identity. It is also true from the perspective of the systemic process in which the business advisor is an actor. We refer to these two heuristic mechanisms (systems) as "systemic effectiveness system" and "systemic sensing system".

The systemic effectiveness system incorporates the means by which the business in view links its objectives, strategy, operations and control. A primary outcome of the research is that business advisors seek signals which are indicative of the presence of this system in a developed or emergent form. The system is a construction, a holon (Checkland, 1981, 1990) employed by the business advisor to help determine whether the SME under consideration is "under control" and thus likely to prosper. The systemic sensing system, in contrast, is the means by which the advisor acquires and interprets signals which are indicative of an appropriate systemic identity within the SME.

Since the systemic sensing system would appear to offer an appropriate explanation of at least one heuristic employed by business advisors it was decided to seek some corroboration that it was widely employed. Accordingly a structured survey was carried out with 175 respondents currently working as business advisors across the national network of business links during the spring and summer of 1998. For the data presented below 2 sets of 4 questions were selected, one to test the null hypothesis that business advisors were not seeking to identify a systemic effectiveness system in SMEs and one to test the hypothesis that they were. Direct evidence of the action of the systemic sensing system in the work of business advisors was thus sought.

SURVEY RESULTS

Respondents were asked to score a number of questions/statements on an interval scale of 1 to 6 where 1 was "strongly agree" and 6 was "strongly disagree". The relevant questions are summarised below, the first 4 tending to support the null hypothesis (against the action of the systemic sensing system) and the second 4 supporting the hypothesis (for the action of the systemic sensing system).

NULL Hypothesis Support
"Poorly Managed" (even poorly managed companies succeed quite often)
"RightPlaceRightTime" (being in the right place at the right time is important)
"BusCtrlMgrs" (businesses control managers rather than vice-versa)
"Luck" (luck is important to success)

Hypothesis support
"FormalBusPlan" (successful businesses are likely to have a formal business plan)
"ClearFinControls" (successful businesses will have clear financial controls)
"Goals&Planning" (successful businesses will set goals and plan)
"ObjectivesMatchOps" (objectives will be matched by business operations)

The results from both sets of questions are presented below as frequency density scattergrams in 4 dimensions. In both cases the more general questions have been placed on the major axes and the more specific/focused questions on the minor axes. The axes have been adjusted to reflect the clustering of responses.

213

Null Hypothesis Support

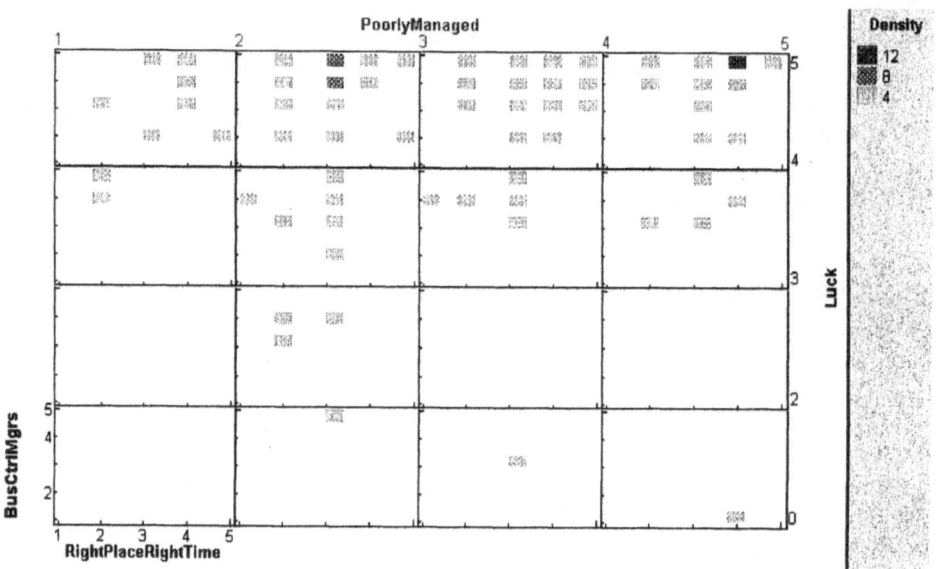

Hypothesis Support

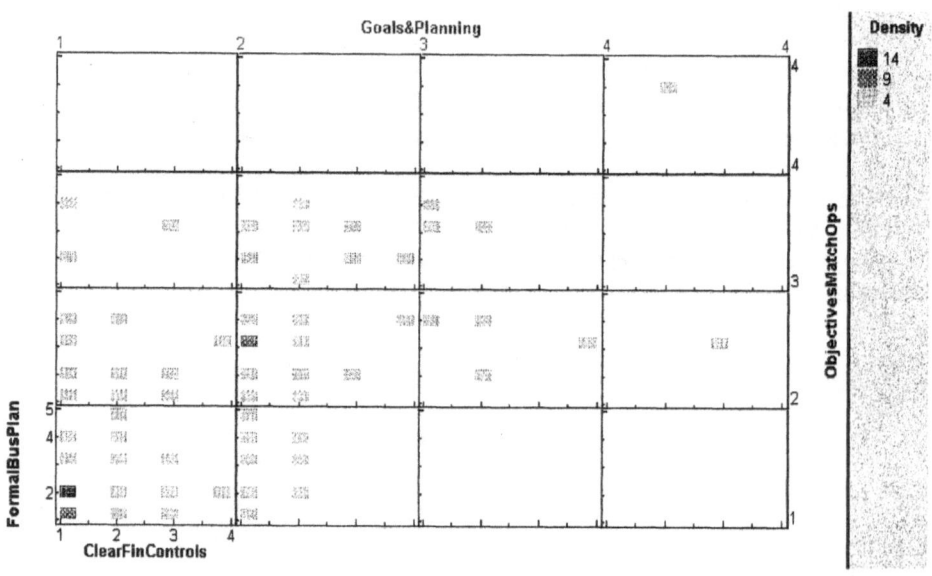

Inspection of the first (null hypothesis) scattergram reveals little support for the suggestion business advisors believe poorly managed companies can succeed but some support for the counter suggestion as evidenced by the clustering of responses in the top right hand side of grid elements. Other points of interest are the 'ambivalent' responses as evidenced by the clustering around the 2/3 square of the major grid and the generally strong disagreement with the idea that business success depends upon luck.

Inspection of the second (hypothesis) scattergram reveals clustering in the bottom left hand corner of both the major and minor grids. This is consistent with the idea of the systemic sensing system being in action when advisors are evaluating companies.

CONCLUSIONS

The paper has described both systemic effectiveness and systemic sensing system models based upon empirical evidence relating to business advisors. The approach has sought to explain evaluations which business advisors make in other than reductionist terms and to offer a vision of business practice which is richer and more systemic than a purely numbers based analysis.

REFERENCES

Bryman, A, (1988), *Quantity and Quality in Social Research*, Routledge, London, ISBN 0-415-07898-9.

Capra, F.,(1996) *The Web of Life* Harper/Collins

Checkland, P, (1981), *Systems Thinking Systems Practice*, John Wiley & Sons, Chichester, ISBN 0-471-27911-0.

Checkland, P, Scholes, J, (1990), *Soft Systems Methodology in Action*, John Wiley & Sons, Chichester, ISBN 0-471-92768-6.

Hycner, R.H., (1985) Some Guidelines for the Phenomenological Analysis of Interview Data. *Human Studies* 8: 279-303

Strauss, A, and Corbin, J. (1990) *Basics of Qualitative Research: Grounded Theory Procedures and Techniques* Sage, London.

Tversky and Kahnemann (1974) reprinted in Kahnemann, Slovic, and Tversky eds (1982) *"Judgement under Uncertainty: Heuristics and Biases"* Cambridge University Press.

A SYSTEMIC APPROACH TO CHANGE MANAGEMENT IN ANGLIAN WATER ENGINEERING

Brian Keegan[1], Rod Athey-Pollard[2]

[1]Change Agent
[2]Resource Manager
Anglian Water Engineering
Huntingdon
England

INTRODUCTION

The aim of this paper is to redress the balance between theoretical and practical papers by explaining how systems thinking has enabled the authors to develop a systemic approach to organisational development within an Engineering department. We start by describing how Anglian Water has moved from Public Sector environment via a behavioural cultural change programme (*The Transformation Journey*) to a customer focused organisation. Next how systems concepts, e.g. metaphors were utilised to enable new roles to be defined, from which a transformation matrix was designed to manage a cultural change programme which included a development matrix for addressing the people development 'gap'. We end by concluding that a systemic approach can be successfully used in practice to manage the people development needs of an organisation.

Anglian Water was formed in 1974 from a large number of disparate bodies which previously carried out the functions of river catchment management, water supply and sewage disposal. At it's formation it had 22 separate single function Divisions, each with their own Engineering Departments following their own approaches to delivering engineering solutions. Between 1974 and privatisation in 1988 the 22 Divisions had been reduced to 5 multifunctional Divisions with 5 Engineering Departments run very much along command and control lines. After privatisation a traditional approach continued until the Managing Director returned in 1992 from a spell at Harvard enthused by the culture of new managerialism encompassed in the empowered learning organisation most closely associated with the work of Peter Senge (1990). The MD re engineered the Company based on learning & empowerment principles, and introduced the concept of coaching as a role for managers. Although this paper concentrates on the *People* side, the Company was reengineered from a geographically based structure to a process based one.

This resulted in managers with a lifetime in a Public Sector environment being exposed to radical thinkers enabling the focus of the organisation to change from inward looking to outward looking. The next section gives more information about the influence of the *Transformation Journey* on the authors.

TRANSFORMATION JOURNEY A VEHICLE FOR CHANGE

The Engineering Management Team *Transformation Journey* (including Keegan) was led by Ronnie Lessem Director of the City Business School Management MBA Programme and this focused on what the role of engineers might / could be in the 21st Century. How could engineers, by examining their roots, understand how to put themselves at the heart of their business?

The Lincoln Engineering Management Team journey involved working collaboratively with Lincolnshire & Humberside University to develop an MBA in International Water, during which Keegan was exposed to systems thinking and cybernetics, both of which heavily influenced his thinking. In collaboration with two colleagues he wrote a people plan which utilises systems thinking to identify the training and development needs for the Engineering Department. In the next section we will describe how metaphors have been used to help address some of the soft issues faced when defining roles.

USE OF METAPHORS FOR ROLE DEFINITION

A crucial part of the People Plan was a redefinition of the roles of those in the Department to facilitate a move towards an empowered learning organisation. To explore these role changes metaphors were used (Morgan 1993). For example the Director of Engineering's role was defined as the Navigator, emphasising that his role was about the future, not dealing with the here and now. The next tier (Engineering Managers) were defined as Leaders, emphasising their role as that of ensuring that all in their departments were prepared for and could meet future challenges. By moving from the real world (Director) to the metaphor (Navigator) and back, a clearer definition was developed which has enabled the senior managers to change from a traditional command & control approach to an empowered model where the Engineering Managers deal with the day to day running of the Department hence freeing the Director to Navigate the future. In today's competitive markets the delivery of engineering services is about more than just technical skills. Great importance is now placed on **how** the service is delivered, hence the need for a cultural change in behaviour. To bring about cultural change in a department of 600 people is quite a challenge, so a systemic approach was adopted using a transformation matrix described below.

TRANSFORMATION MATRIX

This **MATRIX** identified the changes required in three ways as shown below:

LEVEL	TRANSFORMATION	OUTCOME
Individual	Cultural/Role	Team Player
Team	Behavioural/Attitudinal	Self Managed Teams
Management	Skills	Business/profit Generators

This enabled a more complete picture of the skills required for each Job Role to be developed. From this a gap analysis was carried out to identify training/development needs. An example of the new job role definition (Projects Manager) is shown below:

Job Role	e.g.	Enabler
Attributes	e.g.	Relationship Builder
Qualifications	e.g.	Degree, Chartered Engineer, DMS
Competencies	e.g.	Empowering, Motivating/Developing,
Management Standards	e.g.	Project Management, MCI Standards
Experience	e.g.	Seasoned Engineering Professional

Some of these skills were easy to identify, but a gap analysis showed that in two areas there were development needs. These were in the area of attributes and experience. To deal with attributes a change model was developed, and a change programme called engineering the future. To deal with experience a skills data base. The change model is described in the next section whilst the skills data base is described later.

CHANGE MODEL

It was realised that cultural and behavioural change is difficult to describe so metaphors which can imaginatively help people conceptualise changes were used. The metaphors used were as follows:

From	To
Managed Teams	Self Managed Teams
Engineer	Water Technologist
Performer of Technical Tasks	Creative Entrepreneur
Things	People
Change Resistor	Change Manager
Work Soaker	Business Creator / Getter
Adversarialist	Relationship Builder
Risk Avoider	Risk Manager
Engineering World	World Engineer
Heads Down	Heads Up

To enable all staff to adapt to the new culture described in the attributes a behavioural change programme was developed for the whole Department. This was called *Engineering the Future* and delivered in 3 phases. The rationale for a holistic programme was that previous Cultural Change programmes had failed because of lack of ownership by the Senior Management.

Phase 1 was aimed at the middle group of staff, the Senior Engineers, who were seen as a priority, because their managerial training needs had been previously neglected. This course comprised of 4 modules totalling 9 days.

Phase 2 was aimed at the Management of the Department including the Director and the Senior Management Team, and comprised of 3 modules over 7 days.

Phase 3 is an appreciation of the course content and is to be delivered to the rest of the Departments staff by a self managed team of Senior Engineers in 2 day modules.

The course material was drawn partly from Development Dimensions International (DDI) for the skills content, but was supplemented by a Creative Thinking Module designed and delivered by Gill Ragsdell from Lincolnshire & Humberside University.

The DDI modules included:	The Creative Thinking included:
Core skills	Brainstorming
Helping Others to Succeed	Metaphors
Teamwork	De Bono 6 Thinking Hats

Participating in Meetings
Handling Conflict
Trust
Setting & Reviewing Objectives
Partnerships

Rich Pictures
Idealised Design
Nominal group Technique

The courses were delivered by four Members of the Management Team (including a Director) who became accredited DDI trainers by attendance at a weeks DDI Trainers Course. This high level commitment was felt to be crucial in the perceived success of the Course. They were assisted by a self managed team of Senior Engineers who designed phase 3. Engineering the Future dealt with the behavioural aspects of the transformation, but a method of dealing with technical/skills development was also needed, this is dealt with in the next section.

ADDRESSING THE GAP

Define the 'Gap'

Selecting engineers with the appropriate skills to undertake technical projects was, until recently, based upon local knowledge and, if lucky, some out of date CVs. The level of competence of an individual in applying their skills was at best held on a paper matrix, more often selection was based on the position held within the company hierarchy.

Experience Record

In line with the objectives of the People Plan a decision was taken to create a live data base of skills together with the means of measuring of competency in applying those skills. The database comprises two parts.

Section 1 holds personal details, professional qualifications, key experience, generally the same information one would expect to see in a CV.

Personal details

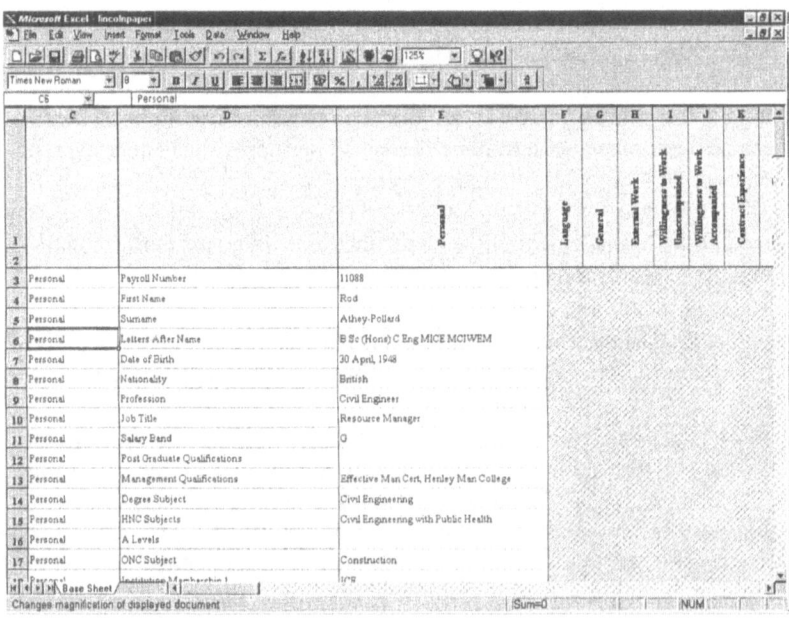

Section 2 is in the form of a development matrix. The column headings are assigned to general areas of technical expertise e.g. Process Design. The rows cover typical projects undertaken by engineers on behalf of the business. The matrix is scored in order to demonstrate the skill/competency level an individual holds in applying an area of technical expertise to a particular type of project.

Development matrix

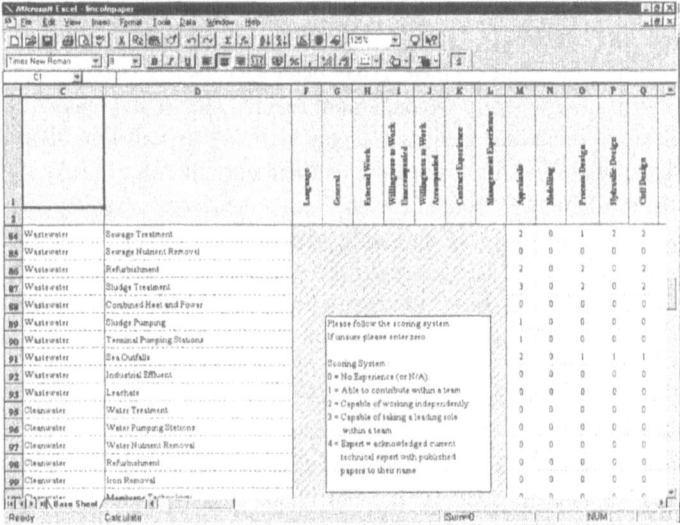

The development matrix can be used in a variety of ways. For example, it is useful when searching for the right skill mix for a particular project, measuring the development of an individual or seeking norms for various job roles. The use of 'performance webs' has also proved invaluable in setting targets, tracking performance and measuring development.

Performance web

Performance webs are a graphical means of setting development competencies as shown below.

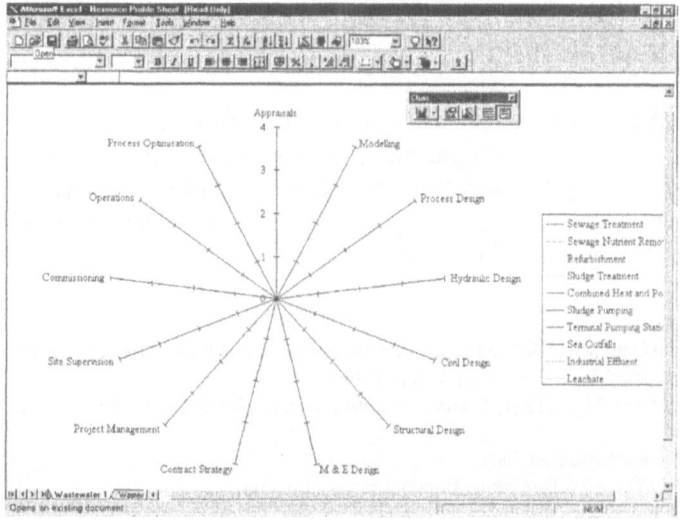

It is an extremely flexible system that can accommodate business change and variations in staff performance expectations with very little effort. As the Anglian Water Group moves further away from its local authority roots, towards its goal of being a 'Winner in the competitive markets of the twenty-first century', we see the Skills Data Base/Performance Matrix as being a major asset in helping achieve this goal.

EVALUATION

Engineering The Future

Engineering the Future Phase 1 has been seen as a successful cultural change programme by the Engineering Management team. This is evidenced by a range of positive indicators, including feedback sheets showing overall that 80% of attendees felt it was beneficial, and 75% relevant to their current roles, as well as anecdotal comments, the most frequent being *"I wish I had done this course 20 years ago"*. One Client Manager who was over 50 suddenly 'converted', and signed up to do an MBA at the Lincoln School of Management. Frequent use of the techniques helped to improve relationships with both internal clients and with contractors. The skill improvements have enabled the Department to restructure its strategy based on Partnering.

Skills Data Base

Successful searches have revealed previously 'unnoticed' people when looking for technical skills, and the database has been used as a basis for technical/experience development as part of the six monthly appraisal system, and is also being used to clarify job roles.

CONCLUSIONS & FUTURE WORK

The authors have concluded that the systemic approach has enabled a coherent people development programme to be successfully implemented within Anglian Water. Three areas have been identified for further development they are measurement of cultural change, revision of the People Plan as a vehicle for describing the desired new culture in the restructured partnership orientated Engineering Department, and revision of the Job Roles in respect of experience so they are based on the skills web rather than illuminating descriptions such as *Seasoned Professional*.

ACKNOWLEDGEMENTS

The authors would like to thank Ronnie Lessem (City University), Professor Mike Jackson, Dr Gill Ragsdell (University of Lincolnshire & Humberside) who have influenced our thinking. We also thank Phil Butler & Maureen O'Donnell (Anglian Water) who helped with the People Plan, and Phil Watson and his son who have made a valuable contribution to the skills data base.

REFERENCES

DDI, Development Dimensions International, Management Training Programme *Skills for An Empowered Workforce*, Pittsburgh, Pennsylvania.

Flood, R.L. and Jackson M.C. (1991) *Creative Problem Solving: Total Systems Intervention*, Wiley, Chichester.

Morgan, G. (1993) *Imaginization*, Sage.

Senge, P.M. (1990) *The Fifth Discipline*, Random House, London.

UNDERSTANDING HOW CONCERNS FOR THE NATURAL ENVIRONMENT ARE INTEGRATED INTO PRACTICAL STRATEGIC DECISION MAKING: AN APPLICATION OF ASHBY'S LAW

Gerard J. Lewis,
Faculty of Economics
Dresden University of Applied Sciences

Neil Stewart
Manchester Business School

INTRODUCTION

There is general agreement that the natural environment is of fundamental strategic importance to business. Consequently, one would expect to find clear evidence that the natural environment is being actively managed by business organisations at a strategic level. However, a systemic analysis of 7 leading[1] blue-chip corporations found an absence of strategic capability in the management of green issues (Lewis, 1997). Other academic research also suggests that corporations are not integrating green issues into their strategic thinking (Hart, 1997; Ketola, 1993), a finding which also receives support in the practitioner press (ENDS, 1996). Through the application of Ashby's Law (1956) of Requisite Variety this paper makes a contribution to explaining why business organisations are not integrating the natural environment into business strategy. A contingency approach is used to relate the perceived environmental uncertainty of executives and their degree of rational strategic decision making (RSDM) to their organisation's performance.

ASHBY'S LAW AND CONTINGENCY THEORY

To enable an assessment of organisation performance to be made, the open systems concept of Ashby's (1956) Law of Requisite Variety has been applied. Ashby's Law states that "only variety in the control mechanism can deal successfully with variety in the system controlled " (Beer, 1967). The application of Ashby's Law to organisations means that the organisation must generate at least as much variety within itself in order to 'control' the variety in its environment. It is possible to think of the control of organisational variety as

---

[1] Leading means that the corporations were all regarded as best practice organisations with respect to environmental management in their respective industries.

driven by two dominant decision techniques which are algorithms and heuristics (Waelchli, 1989). Algorithms correspond to what Simon (1987) terms rational decision making using analysis; heuristics correspond to Simon's non-rational decision making techniques such as judgement and intuition (Janis, 1989). Thus the variety in the organisation can be represented by the degree to which algorithms and heuristics are employed in strategic decision making.

Variety in the environment is a synonym for complexity in the environment, and complexity is one of two main parameters, along with stability, which are objective measures of uncertainty (Duncan, 1972). Thus it is possible to compare variety in the organisation and variety in its environment by comparing the modes of decision making and the perceived degree of situational uncertainty. The assumption which underlies this is that businesses will attempt to use RSDM where possible so that decisions are made using the best available intelligence at any time. Although these proxy measures have limitations, complexity and the management of complexity are central to strategic decision making; it is this sense that the use of Ashby's Law has validity (Waelchli, 1989).

We have conceptualised our approach to this in Figure 1. Here perceived situational variety is compared with organisational variety as measured by the degree of rationality in decision making. The diagonal line represents the matching of these two factors at different levels of complexity in a way similar to that required by Ashby's Law. This approach is also consistent with the contingency concept of fit between an organisation and its environment (Tushman and Nadler, 1978) and that organisation effectiveness (i.e. performance) is dependent on the degree of fit achieved (Duncan, 1973). Using this model we can explore how decisions about green environmental issues are made and how these compare with strategic decision making in the commercial arena.

Figure 1

Matching Organisational and Situational Variety

Organisational Variety (measured as extent of rationality in strategic decision making)

METHODS

In his comprehensive study of managerial work Mintzberg (1973) states that the most appropriate use of a questionnaire is in the study of executives' perception of their jobs. A self report questionnaire was designed which contained four research scales was devised to test the relations in Figure 1. Two scales measured perceived business and green uncertainty, and two scales measured the degree of rationality in business and green decision making. The business uncertainty and rationality scales were based on the work of Miller (1993) and Miller (1987) respectively. The green uncertainty and rationality scales were developed by one of the authors. A pilot study was undertaken with a company in the healthcare industry to test the overall research design, methods and research instruments.

In this way, a 'decision-based perspective' to strategic process research was taken. This approach overcomes the problem that not all organisations have formal planning systems (especially SMEs). Analysis was undertaken at the senior manager/ CEO level. An industry level of analysis was chosen since the nature of both environmental uncertainty and strategic decision making are industry context specific. The UK textile industry was chosen for a number of practical considerations (e.g. access and support) and its exposure to a number of different green issues. The questionnaire response statistics are shown below in Table 1.

Organisation performances were then mapped onto Figure 1. By comparing values of perceived uncertainty and RSDM from the questionnaire responses, an assessment of performance can be made. Measures of performance were made for a number of different contingency variables (Tosi and Slocum, 1984) within the textile industry which are shown in Table 2. Thus, aggregate measures of performance were measured for each contingency variable for each organisation in the sample

An important question which Ashby's Law raises when used this way is "what if the organisation's variety, measured by its capacity for RSDM, is either too much or too little when compared with the perceived situational variety?" If the organisation appears to possess too much variety (say position B in Figure 1), then it is wasting resources and financial performance is lower than it should be - it has redundant assets. Ashby's Law suggests that it is structured to respond to more situational variety than actually exists. If the organisation appears to possess too little variety (position A) then it is likely to make the wrong decisions because it has not taken formal account of all the issues (variables) in its environment; it is exposed to environmental risk. The gap, which allows its organisational variety to rise to that of the environment as required by the Law, is filled by intuitive (non-rational) decision-making processes and this brings in risk.

Thus by looking at how uncertainty relates to RSDM an assessment of the organisations' performance becomes possible. Furthermore, by looking at the values of uncertainty and RSDM for different contingency variables (Table 2) it is possible to make an assessment of where good and bad performance is occurring. Using this approach the problems associated with financial (lack of accurate data) and green (lack of appropriate metrics) measures of performance can be overcome.

RESULTS

In the scatter plot for the commercial environment, most points on the matrix occur below the diagonal on Figure 1 - such as position B. Therefore there is evidence of too much organisational variety in the textile industry when considering RSDM when only commercial business issues are involved. The further away a point is from the diagonal, the poorer the performance and we did obtain a view on which parts of the industry were better at matching their RSDM to the perceived business environment and which were the worst. Best business performance (i.e. well-matched requisite variety) was shown by machinery manufacture, making-up and retail sectors of medium size and with executives and CEOs with experience varying between 6-10 and 31 years. Conversely, the worst business performance (too much organisational variety) was in the raw materials sector, and organisations of very large size. Age of the executives and senior managers was not a factor in this latter instance.

Table 1 Questionnaire response rate statistics.

Parameter:	N	% of cases
Organisations contacted	332	100.0
Organisations agreeing to participate	293	88.3
Organisations responding to questionnaire	165	49.7
Questionnaires sent out	445	100.0
Questionnaires returned	209	47.0
Questionnaires filled out correctly	206	46.3
Questionnaires filled out correctly by a suitable respondent[a]	198	44.5

[a] A suitable respondent is someone who has the appropriate executive status.

Table 2 Contingency variables within the textile industry used for analysis of organisation performance.

Executive	Organisation	Industry Sector
Chairman/CEO	Small	Raw Materials
Director	Medium	Spinning
Senior Manager	Large	Weaving
Member Executive	Very Large	Dyeing
Operation	With EMS	Making-up/Retail
Marketing/Sales	Without EMS	Machine Manufacture
HSE/R&D		
HRM/Finance-Accounting		
Experience (years)		
Age (years)		

In the commercial environment, CEOs have the ability to match resources to needs either by obtaining more resources or utilising them more effectively ("variety-increasing strategies") or by rationalising the external scope of the firm's operations to meet organisational capability ("variety-reducing strategies"); indeed this is a key role of a CEO. Executives in the machinery and making-up/retail sectors appear to be more capable of matching variety. In general more senior managers have difficulty in matching resources to needs. They appear to be risk averse and possess too much variety. Similarly, executives in very large organisations also possess too much variety.

A scatter plot for decisions about the natural environment was produced. The striking observation is the shift of the points to above the diagonal line in Figure 1, such as position A, indicating that there is too little variety available for RSDM involving green issues. Organisations are therefore open to risk that environmental events will catch them unprepared at various levels. One factor included in our survey was the extent to which an Environmental Management System ("EMS") was in place and used. The source of best green performance were identified as the raw materials sector, possessing EMS, of very large size and with executives and senior managers with experience 21-30 years. Worst green performance (i.e. too little organisational variety) was found in the machinery, making-up and retail sectors, involving organisations without EMS, of both medium and large size. Age of the executives and senior managers was not a factor. Possession of an EMS was also significant factor in the equivalent analysis of performance in the commercial environment, but here organisations with such systems performed worse than those without - the implication being that an EMS adds to organisational variety without adding competence to manage the commercial environment.

Sustainable business practice was explored by inspecting the correlation between business and green performance from the scatter plots and the following observations emerged (a) that only executives in the senior manager group appear to have found the balance between green and business performance but this occurs at apparently a very low level of performance. This presumably reflects the degree of intuitive judgement used in both domains at these levels, (b) that most points on the matrix show executives in the textile industry to be concentrating on business performance rather than green performance, and (c) that the trends in the combined data suggests that there may exist some underlying structural problems in trying to meet both performance criteria simultaneously.

DISCUSSION

Organisations in the industry under study have too much variety to manage the complexity in their commercial environments; that is, they are inefficient, carrying too much cost. In itself, this is a surprising result given the amount of BPR and de-layering which has taken place in British textile industry, but this may reflect the long-term decline in the industry. The results also show that rationality decreases when executives engage in green strategic decision making; that is, the variety in the organisation measured by estimating RSDM decreases. The gap in performance can only be bridged by the use of intuition and judgement to meet the requirements of Ashby's Law.

Unfortunately for corporate performance intuitive decision making relies on training and experience (Simon, 1987). Green issues are relatively new and executives have very little training in, and experience of, strategic decision making involving green issues. The implication is that analytical decision making *per se* will not provide enough organisational variety to enable executives to make effective strategic decisions which include green issues. Our findings do show that organisations with EMSs possess enough requisite variety for effective green RSDM but EMSs are grounded in classical management procedures and there is concern that they are not suitable for the management of green issues (Moxen and Strachan,1995). In both the commercial and the green environments, the possession of an EMS moves the organisation in an almost vertical line in Figure 1 adding to organisational complexity but not contributing to strategic processes. Hence it is apparent that EMSs have limitations when it comes to managing the levels of complexity involved in green issues.

When the two sets of data are combined (business plus green) we find that organisations are not achieving balanced business and green performance (i.e. sustainable performance) except at very low levels of both. This finding places a question mark against previous research which has used combinations of business and green criteria in order to develop portfolio approaches to measures of sustainable performance (Lee and Green, 1994; Schaltegger and Sturm, 1996).

CONCLUSIONS

We have shown that Ashby's Law can be applied to measure organisation performance and inform our understanding of strategic decision making. This is a positive first step since although research on the development of environmental performance indicators currently occupies a major area of activity in environmental management, it will be some time before criteria and metrics will contribute to organisational comparisons in this area In the meantime, a more promising way forward would be to develop the application of Ashby's (1956) Law to cover non-rational (heuristic based) decision making. This would provide a more accurate picture of requisite variety in the organisation.

REFERENCES

Ashby, W.R. (1956). *An introduction to cybernetics*. Chapman and Hall Ltd., London.

Beer, S. (1967). *Cybernetics and Management*. 2nd ed., The English Universities Press, London.

Duncan, R.B. (1972a). Characteristics of organisational environments and perceived environmental uncertainty. *Administrative Science Quarterly*, 17, pp. 313-327.

ENDS (1996). Chemical Industry eases towards comparability in environmental reporting. Industry Report No. 65. *ENDS*, 259 (August 1996), pp 19-22.

Hart, S.L. (1997). Beyond greening: Strategies for a sustainable world. *Harvard Business Review*, Jan-Feb 1997, pp. 67-76.

Janis, I.L. (1989). *Crucial decisions: Leadership in policy making and crisis management*. Collier Macmillan Publishers, London.

Ketola, T. (1993). The seven sisters: Snow whites, dwarfs or evil queens? A comparison of the official environmental policies of the largest oil corporations in the world. *Business Strategy and the Environment*, 2, pp. 22-33.

Lee, B.W. and Green, K. (1994). Towards commercial and environmental excellence: a green portfolio matrix. *Business Strategy and the Environment*, 3, pp. 1-9.

Lewis, G.J. (1997). A cybernetic view of environmental management: The implications for business organisations. *Business Strategy and the Environment*, 6 (5), pp. 264-275.

Miller, D. (1987). Strategy making and structure: Analysis and implications for performance. *Academy of Management Journal*, 30 (1), pp. 7-32.

Miller, K.D. (1993). Industry and country effects on managers' perceptions of environmental uncertainties. *Journal of International Business Studies*, 24, pp. 693-714.

Mintzberg, H. (1973). Strategy-making in three modes. *California Management Review*, 16 (2), pp. 44-53.

Moxen, J. and Strachan, P. (1995). The formation of standards for environmental management systems: Structural and cultural issues. *GMI*, 12, pp. 32-48.

Schaltegger, S. and Sturm, A. (1996). Managerial eco-control in manufacturing and process industries. *Greener Management International*, 4 (1), pp. 79-91.

Simon, H.A. (1987). Making management decisions: the role of intuition and emotion. *Academy of Management EXECUTIVE*, February 1987, pp. 57-64.

Tosi, H.L. and Slocum, J.W. (1984). Contingency theory: Some suggested directions. *Journal of Management*, 10 (1), pp. 9-26.

Tushman, M.L. and Nadler, D.A. (1978). Information processing as an integrating concept in organisational design. *Academy of Management Review*, 3, pp. 613-624.

Waelchli, F. (1989). The VSM and Ashby's Law as illuminants of historical management thought. In Espejo, R. and Harnden, R., eds. *The viable system model: Interpretations and applications of Stafford Beer's VSM*, pp. 51-75. John Wiley & Sons, New York.

A PLURALIST APPROACH TOWARDS INTEGRATION OF QUALITY AND ENVIRONMENTAL MANAGEMENT, LEADING TO SUSTAINABLE DEVELOPMENT

T. McEwan and D. Petkov

University of Natal, Pietermaritzburg
Private bag X01, Scottsville 3209,
South Africa

INTRODUCTION

This paper describes the case for extending ISO9002 procedures to include an ISO14001 programme, as part of an on-going Total Quality Management programme, at a large paper pulp mill in South Africa from the standpoint of its practising managers. This synergistic development is seen as an organisational change "that involves all the staff, not only those in the production department, in a process of continuous improvement" (Frehr, 1997). In addition to these changes in the internal environment, the company also recognises the demands of various stakeholders in its external environment. This is strategically prudent because, according to Roberts (1994), "it is clear that the nature and scope of the business and environment relationship is now a matter of greater public concern and attention than in previous years. The chief reasons for this increased attention can be identified as:
- the growth of a common awareness that the growth of economic activities carries with it considerable direct or indirect implications for the environment,
- the realisation that the uncontrolled operation of business not only impacts upon the environment, but also creates very real costs for the business itself,
- the desire of policy makers to regulate environmental standards in order to improve the environment and also to ensure a reasonable place for business operations,
- the development of the theory and practice of corporate responsibility. This new dimension of business incorporates both socio-economic and environmental concerns."

Managers at the paper mill were aware of the "pursuit of excellence" strategic vision for the company, and the three workshops, conducted by the authors, were intended to explore how this goal could be realised through the systemic introduction of ISO9002 and ISO14001 standards. A second goal for the company can be formulated along the lines of the idea of "good citizenship". This is partially related to the implications of non-compliance with ISO14001 and partially due to the growing importance of the social responsibilities of companies, world wide, particularly in the new, democratic South Africa.

Synergy Matters: Working with Systems in the 21st Century,
Edited by Castell *et al.*, Kluwer Academic / Plenum Publishers, New York, 1999.

The justification for introducing both sets of standards is based on different yet related economic imperatives in the external environment. Underpinning this proposed strategic development is a series of 'trade-offs' between the pursuit of higher productivity, on the one hand, and accommodation of 'external' socio-economic and political factors, on the other. The situation is influenced by various stakeholders, including:

* Government, which imposes environmental standards and collects taxes;
* Consumers who exert both market demands and political pressures;
* Other organisations which may influence consumers in a particular direction.

There is increasing acceptance globally of the need for agreed standards to protect the wider interests of civil society. ISO9000 is a well-known example of an international standard for improving organisational performance by developing more consistent quality management procedures. Based on the latter's success, many companies have since attempted to introduce the ISO14000 set of standards on environmental management (Lamprecht, 1997), which is increasingly being enforced through market forces and government regulations.

More broadly, the introduction of both sets of standards is seen as a step towards a sustainable development framework developed by Welford et al. (1998). This is based on six dimensions, consisting of the environment, empowerment, economics, ethics, equity and education. The framework complements the goals of this research because it is consistent with increasing acceptance of the need to address both the social and environmental dimensions of sustainable development (Welford et al., 1998:48).

AN OVERVIEW OF THE METHODOLOGY APPLIED

This section attempts to clarify the problem under consideration and provide a justification for the selected set of approaches to be adopted by the management of the paper mill. The methodological discussion is followed by a report on results and related emerging issues in implementing the proposed organisational learning framework during the transition from compliance with the ISO9002 standard towards ISO14001 requirements.

The research is of practical importance for the mill as it aims to uncover intangible problem areas hindering the integration of ISO9002 and ISO14001 standards in the proposed TQM programme. This is also a novel research project in terms of the practical testing a specific combination of systems techniques, MCDM and statistics, following the recent Multimethodology meta approach (Mingers and Gill, 1997).

Moving towards compliance with ISO14000 is a complex "messy" problem, where the relationships between the stakeholders can be classified as either pluralist or sometimes even as coercive. This complexity justifies the different approaches. First a mixture of soft systems techniques like stakeholder identification, elements from Soft Systems Methodology (SSM) (see Checkland and Scholes, 1990) like rich pictures and CATWOE analysis on the one hand and a multi-criteria decision support technique, the Analytic Hierarchy Process (AHP) (see Saaty,1994), on the other, was used as a vehicle towards better problem solving and organisational learning, combining the strengths of elements from these methodologies, following (Petkov and Petkova,1998). The first part of the framework dealt with systemic issues and their prioritisation. The findings of the brainstorming sessions and stakeholder identification, as well as the applied elements of SSM advanced the problem formulation stage of AHP, which is assumed to be the least formalised stage of AHP.

The second part of the framework provided a basis for a more detailed analysis of underlying trends and factors which influenced the behaviour of the organisation with respect to the transition from ISO9002 to ISO14001.

Slack et al. (1997) identified seven cost factors at the operational and strategic levels in organisations which need to be evaluated before ISO9000 procedures can be successfully transformed into an effective Total Quality Management policy. To achieve this standard at the operational level, Prevention, Appraisal and Internal failure costs must be assessed, whereas at the strategic level, the costs of External failure, Unclear Management Strategy, the degree of Top Management Support and Socio-Political issues must also be known.

The achievement of acceptable environmental standards depends on similar cost/benefit considerations, according to Welford and Prescott (1992). They identify six cost areas, three in evaluating an environmental strategy at the company level; namely, the costs of Operational and External failures, and Future Threats to the organisation; and a further three costs in achieving sustainable development as a company response, namely the costs of environmental management at the local plant level, at the strategic level; and their auditing as a source for potential competitive advantage. These findings were used in this research as shown below.

In the second stage of the analysis were investigated the perceived costs in the transition from ISO9000 to ISO14000. A questionnaire on the major cost areas related to the introduction of these standards, developed by McEwan et al. (1998) was used together with another original survey instrument to measure progress in six policy areas derived from a systemic framework for achieving sustainable environmental development . The latter is based on a framework of policy areas and tools for sustainable development which includes six broad policy areas: environment, empowerment, economics, ethics, equity and education (Welford et al., 1998:48). In both cases a seven point Lickert type scale was used. The purpose of these instruments, which are available on request, is to reveal possible problem areas in the process of implementation of both sets of standards.

The above framework combines elements from two systems approaches with an MCDM method, plus a management model for identifying the main cost factors associated with the introduction of the two sets of standards within the company. According to Multimethodology, models and techniques as parts of different methodologies, from different paradigms can be brought together according to the requirements of a particular intervention (Mingers and Gill, 1997). Additional justification for the combination of approaches in this intervention can be found in Petkov et al. (1998). However, further theoretical work might be needed for the achievement of coherent pluralism in Management Science, as formulated by Jackson (1997).

IMPLEMENTATION OF THE FRAMEWORK FOR TRANSITION TO ISO9000 / ISO14000 COMPLIANCE

After identifying the need for such an intervention, a preliminary discussion with the top management of the mill was conducted and authorisation for a pilot study in the mill was obtained. Two months later a double session workshop was conducted with twenty representatives of senior and middle management. As a result of a brainstorming and prioritisation session which identified the critical factors in the transition from ISO9002 to ISO14001 standard, the least problematic issue was identified as management commitment, followed by an understanding of the need for compliance with ISO14000 to maintain market competitiveness. It is interesting to note that softer systems and human aspects of the problem occupy the lower part of the table. The fact that the company has embarked on such a development might account for these findings.

The task of the first workshop was to promote organizational learning through:
-Identification of stakeholders and their assumptions, as well as their subsequent rating.
-Provision of a graphical description in the form of rich pictures on the essence of the problem of transition from ISO9002 to ISO14001, which is supplemented by definitions of those elements affected by the process, stakeholders who have a role to play in its implementation, the nature of this transformation, the world view guiding it, the owners who can stop the process and the constraints that may disrupt the process.

231

Finally, consensus was also reached on the key environmental issues from the standpoint of various stakeholders, including senior management. These comprised the need to identify the potential environmental impact of pollutants, assessment of the risk for environmental disasters, the importance of community involvement in discussions about environmental pollution control and the need to improve training of all employees in ISO9002 and ISO14001 procedures.

The above steps are seen as necessary for the development of some common understanding before the preparation of action plans and prioritization of activities associated with the on-going process of linking of ISO9002 and ISO14001 standards at the paper mill.

However before that stage is reached, further work is needed to identify the attitudes of relevant stakeholders towards the evaluation of various operational and strategic issues related to the introduction of ISO9000 and ISO14000 from a cost perspective.

The second and third workshops comprised equal numbers of managers in two cohorts from departments which had either adopted ISO9002 (ISO), or had partly implemented these standards or had not yet managed to do so (semi/non -ISO). Participants at the second workshop completed an original survey instrument (McEwan et al, 1998) of 75 statements on the seven main Quality Assurance and six main Environmental Management cost factors involved in the transition from ISO9002 to ISO14001 standards, before providing pairwise comparisons of the factors from both sets as a separate activity.

Results from the second workshop indicated that paired t-test applied to the responses of the ISO and semi/non -ISO participants demonstrated that the differences in the mean values of their opinions were statistically significant for only four of the overall thirteen cost factors. That is, for two factors in the ISO9002 Quality Assurance instrument (viz. the costs of External Failures by, and Future Threats to, the company) and two in the ISO14001 Environmental Management instrument (viz. the costs of Future Threats to, and Environmental Auditing of, the company). This indicates that there were no substantial differences in the evaluation of the different cost factors by the two groups of managers. The latter is an indication that the process of transition towards ISO compliance in those units that are still perceived to be lagging behind is not hindered by negative attitudes of the respective managers. This is a source of optimism for the success of the overall programme throughout the paper mill.

In the prioritizing exercise, participants provided pairwise comparisons of the importance of the cost factors to obtain their rankings. The ISO compliant group of managers ranked Prevention costs of highest importance (0.291) followed by Top Management Support achievement (0.220) and Appraisal costs of lowest importance within the ISO9001 Quality Assurance cost factors. Those from the non-ISO or semi-ISO group ranked as most but equally important the same two factors as the ISO-group at 0.198 and socio-political costs as least important factors. Regarding the ISO14001 Environmental Management cost factors, both groups of managers ranked as most important Internal Prevention or Operational costs, followed by Environmental Management at Strategic Level costs while as least important Environmental Auditing as a Strategic Source of Competitive Advantage was considered. The only difference in the opinions of both groups was in the fact that ISO compliant managers considered Future Threats and Opportunities as more important than External Failure costs while the non-ISO and semi-ISO group was of the opposite opinion. However, again, the similarities between the views of both groups of managers show that there are similarities in the responses of both groups of managers which is a good indicator for the potential successful completion of the transition from ISO9002 to ISO14001 throughout the paper mill.

Participants at the third workshop completed a second original survey instrument of 45 statements based on the six Sustainable Development issues identified by Welford et al (1998). Here again, participants were asked to prioritise these issues separately using AHP and the group decision support environment Team Expert Choice.

Results from the third workshop (see Table 1) confirmed those of the second as t-test comparisons between ISO and SEMI/NON-ISO participants showed that the means for the responses of the two groups of managers were statistically significant for only one of the six

Table 1. Assessment of policy areas affecting sustainable development by two cohorts of managers and paired t-test of the mean values for both groups

Policy area	Cohort	Mean	Standard Deviation	t-value	Statistical significance
Environment	ISO	4.73	0.19	1.61	Non-significant
	Semi/non-ISO	4.33	0.19		
Empowerment	ISO	4.08	0.20	0.87	Non-significant
	Semi/non-ISO	4.31	0.19		
Economy	ISO	4.83	0.17	0.38	Non-significant
	Semi/non-ISO	4.90	0.16		
Ethics	ISO	3.73	0.25	0.16	Non-significant
	Semi/non-ISO	3.80	0.24		
Equity	ISO	4.45	0.21	0.83	Non-significant
	Semi/non-ISO	4.22	0.17		
Education	ISO	4.80	0.18	2.15	Significant
	Semi/non-ISO	5.40	0.16		

Sustainable Development factors; namely, Education requirements. The pairwise prioritization of the policy areas provided results shown in Figure 1.

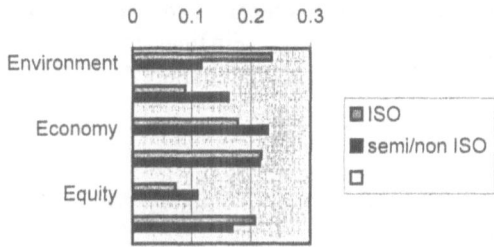

Figure 1. Ranking of policy areas by ISO and semi/non ISO managers

It is interesting to note in two cases there were substantial differences. While the ISO compliant group considered the environment to be the most important policy area, the semi /non ISO group ranked it in fifth place. The latter shows that educational work is needed within the semi/non ISO departments. The composition of the two groups is reflected also in the slight differences between their opinions on other factors.

CONCLUSION

This paper is a summary of initial steps in clarifying the complex problem associated with the transition to ISO14000 from ISO9000 at a paper pulp mill. This cannot be treated fully by a single methodology or a model containing only a small portion of associated factors and a combination of several techniques based on Multimethodology is proposed here.

Findings from the second and third workshops helped to alleviate reported concern about the possible negative impact on the proposed ISO9002/ ISO14001 programme between the two cohorts of managers. In reality, participants acknowledged that the statistically significant differences reported in the second workshop were to be expected since these refer to cost factors which cannot be evaluated accurately until after ISO14001 had been implemented. Similarly, the significant difference on future Educational requirements in achieving sustainable development was anticipated, since it is widely recognised as a major socio-political issue in the 'new' South Africa. It was also acknowledged that as proposed Government policy is still only at the White Paper discussion stage, limited time remains for extending the evaluation of ISO9002/ ISO14001 transition programme to supervisors and other historically disadvantaged employees in the paper mill so that their outstanding training needs can be identified and provided urgently in the 'The Pursuit of Excellence' TQM programme.

A suitable vehicle promoting ecological modernisation in the enforcement of international standards governing quality and environmental issues like ISO9000 and ISO14000 is recommended. At a local level these standards need to be operationalised into a set of working proposals and measures to ensure effective compliance. The initial results indicate that the proposed framework combining elements of SAST, SSM and AHP includes the diversity of perspectives needed, promotes participation in the process and gives a sense of direction in the multifaceted management intervention on this problem. However further work is necessary to apply these techniques with other stakeholders like the local community and operational level staff from the mill.

The suggested framework for a systemic evaluation of the transition process to ISO14000, while implementing ISO9000, and the evaluation from a cost point of view of the factors affecting that process is aimed at assisting the company in the transformation of its social relations. These are seen as a prerequisite for the process of compliance with these standards to be implemented in a holistic manner, leading to sustainable development. A similar approach is suitable for other chemical companies for which compliance with quality and environmental standards is of importance. Findings indicate that the paper mill under consideration here should be able to implement the entire TQM programme by the year 2000.

REFERENCES:

Checkland P. and Scholes J.,1990, *Soft Systems Methodology in Practice*, J. Wiley, Chichester.

De Waele, M.,1998, The management of policy making and implementation: conceptualizing development, *Human Systems Management*, 17:1-4.

Frehr, H.U., 1997, From ISO9000 to Total Quality Management, a rough road, *Human Systems Management*, 16:185-193.

Jackson, M C.,1997, *Towards a Coherent Pluralism in Management Science*, Technical report No 16, Lincoln School of Management.

Lamprecht, J.,1997, *ISO14000*, Amacom, American Management Association.

McEwan, T., Petkov D., O'Neill C., and von Solms S., 1998, Organisational and personal competence and the implementation of ISO9002 and ISO14001 programmes, *Proceedings of the 1998 Eco-Management and Auditing Conference*, University of Sheffield, ERP Environment, Shipley, 123-130.

Mingers J. and Gill A. ,1997, *Multimethodology: the theory and practice of integrating OR and Systems Methodologies*, J. Wiley, Chichester.

Petkov D. and Petkova, O.,1998, The Analytic Hierarchy Process and Systems Thinking, *Trends in Multiple Criteria Decision Making*, Proceedings of the Int MCDM97 Conference, Springer Verlag, 243-252.

Petkov, D. McEwan, T.,von Solms, S. and Vezjak, M.,1998, Moving towards compliance with standards for environmental protection - an example of mixing approaches to "messy " problems, *Cybernetics and Systems'98*, Vol 2: 252-257, Vienna.

Roberts P.,1994, Environmental sustainability and business: recognition the problem and positive action, in Williams C C and Haughton G. (Eds): *Perspectives towards sustainable environmental development*, Alverbury, Newcastle.

Saaty, T.,1994, *Fundamentals of Decision Making and Priority Theory with the Analytic Hierarchy Process*, RWS Publications, Pittsburgh.

Slack, N., et al., 1997, *Operations Management*, 2nd Edition, Pitman, London, pp. 770-780.

Welford , R. and Prescott, K., 1992, *European Business: An Issues-Based Approach*, Pitman, London, pp. 232-44

Welford R J.,Young W. and Ytterhus, B.,1998, Towards sustainable production and consumption: a literature survey and conceptual framework for the service sector, *Eco-Management and Auditing*, 5 (1):38-56.

PROCESS MANAGEMENT

A SYSTEMS PERSPECTIVE

Eugene H. Melan

School of Management
Marist College
Poughkeepsie, New York 12601, U.S.

1. INTRODUCTION

Process Management is an improvement methodology developed in the private sector in response to the difficulties encountered in the implementation of the empirical approaches to TQM. TQM, ungrounded in theory and research, evolved from practices in an industrial milieu. In philosophy it reflects a customer orientation in providing a product or service in the context of organization-wide continuous improvement. The three primary approaches to TQM, namely the prescriptive, project, and multiple point, described elsewhere (Melan, 1998), do not explicitly deal with planned change or with organizational realities that influence the success of the intervention. Of particular note is the fact that these approaches do not address an organization as a system (Ackoff, 1993). The concept of viewing an enterprise in terms of an ensemble of interrelated processes composing a productive system emerged in response to limitations of the traditional approaches. The approach is inherently systems based in contrast to the others thereby increasing its horizon management and became the fourth approach to improvement (Melan, 1986). It is the purpose of this paper to explicate the methodology, discuss its grounding in systems theory, describe an application and suggest its position as a supporting or complementary systems based methodology.

2. PROCESS MANAGEMENT

The concept of a process has been known for hundreds of years. Alchemists of the Middle Ages were preoccupied with finding a process for converting lead to gold. In 1800 the Soho Engineering Foundry in England was applying the process idea in describing sequentially performed tasks for fabricating iron (George, 1968). It is evident that notions underlying the process concept such as transformation, input and output predate that of General Systems Theory. We define a process as a repeatable, bounded group of purposeful and interrelated activities, characterized by one or more transformations, resulting in a flow of "work" (resulting from activity) proceeding from an input boundary, Figure 1. Not all activities are transformations in the sense of conversion—as for example a monitoring-control activity (A^1_3 in Figure 1). Since a process in itself is not viable in terms of its ability to sustain a separate existence it can be considered a component of a subsystem and possesses several systemic features discussed below.

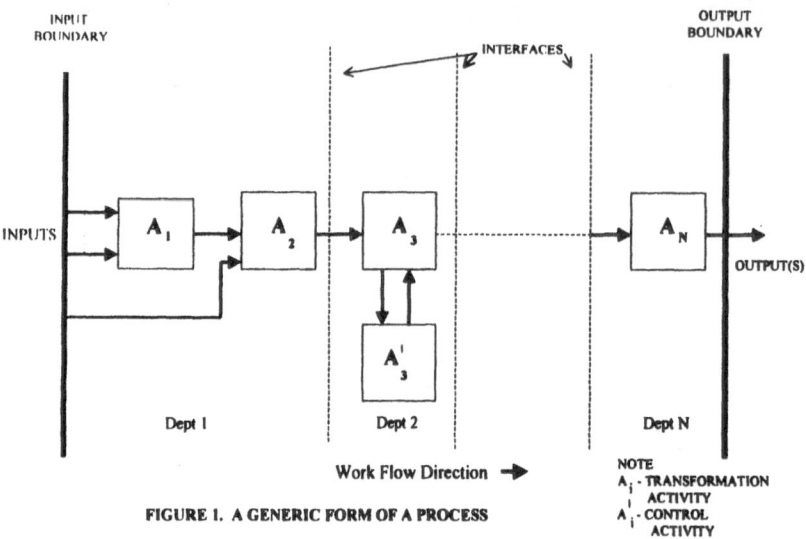

FIGURE 1. A GENERIC FORM OF A PROCESS

Process management is based on the notion that a purposeful system such as an enterprise may be viewed as a group of interrelated processes largely independent of organization structure. Complex processes frequently contain several functional demarcations and possess the inherent problems of ownership, boundary spanning and suboptimization due to the existence of functional 'silos.' A model for managing complex processes can be derived by examining a well-managed industrial process. Industrial processes, driven by economic considerations, are intended to be efficient as well as effective and possess the following features: unambiguous ownership, defined boundaries, documented operations, prescribed control points, and measures to serve as a basis for regulatory action. The realization that key processes of a firm often lack on or more of these features (thereby providing opportunity for improving efficiency and effectiveness) provided a rationale for developing a methodology for managing and improving business processes (Melan, 1986).

Process management as a methodology consists of four phases: initialization, definition, control and improvement. The initialization phase consists of articulating objectives, determining process mission and ownership and defining its input and output boundaries. The definition phase involves graphically mapping the various activities as currently performed to provide a baseline for improvement. In this phase, the actors who are the individuals actually performing the activity describe the process in terms of the real world, as currently practiced. In the third phase, control, points where actual control is exercised are determined and measurements of effectiveness and efficiency assessed with view towards improvement. The final phase, improvement, is where comparison of the actual process relative to conceptual processes is performed, and where obvious process deficiencies are debated and improvements deemed feasible agreed to. In practice, the methodology is intended to be team based and facilitator driven with the participation of the owner, actors and stakeholders.

3. APPLICATION EXAMPLE

Executive management responsible for the product development of a critical electronic subassembly for a computer manufacturing firm faced schedule uncertainties in view of the newness and complexity of its design process. The manager, being familiar with the capability of process management methodology, requested an analysis and assessment of the design process to determine which sectors would affect the schedule and what general improvements in effectiveness and efficiency could be made. The author was appointed to execute the analysis and assessment

236

and served as a facilitator as well as a coordinator and team leader. Team members, both managers and professionals, representing various functions supporting the design process were selected for their expertise in different aspects of the process such as electronic and mechanical design, fabrication and testing of the prototypes and reliability evaluation. Others provided stakeholder viewpoints.

During the initialization phase the objectives of the assessment and the purpose (mission) of the process were debated, viewpoints presented and consensus as to direction achieved. Because of the organizational complexity of the process the key issue requiring resolution was that of ownership. In this case, because of organizational structure and the perceived primacy of the electronic design function, a second level (project) manager was designated as the process owner. The remaining activities of this phase were devoted to establishing input and output boundaries.

The second phase involved developing a detailed graphical description of the process as practiced. As with many complex processes, the process itself was undefined causing confusion and misperceptions among the actors. As the process unfolded by means of the facilitator eliciting activity descriptions from the actors performing the activity, self-realization of process deficiencies occurred, a condition that commonly occurs during this phase. The total process comprised over fifty key activities contained within five subprocesses. Interfaces (interorganizational boundaries) demarcating the flow of work and information from one organizational entity to another were also identified. Experience derived from analyzing various processes shows that many of the problems governing the effectiveness of the workflow occur because of misunderstanding in the producer-receiver relationship (Melan, 1993 p. 32). Realization by the actors of these interface deficiencies served to complement the set of improvement items to be debated in phase four. Phase three involved determining the existing points of control, assessing their adequacy and defining areas for improvement in control of the design work flow.

The completed process description enabled a cycle time assessment to be made which showed that, without reiteration, a design would be released in 39 months – a fact which produced surprise and dismay to all involved. The improvement phase consisted of assessing the deficiencies discovered in phases two and three by comparing against views of effective and efficient subprocesses. By facilitating a free exchange of views among the team members, ten key improvement recommendations (both incremental and reengineering) were developed and proposed to the owner (who also participated in this phase). In all, the design cycle time was reduced by 51% to nineteen months and direct savings of $2 million accrued by changing test and design methods. Indirect benefits in earlier time-to-market because of reduced cycle time was not computed.

As process management became more widely applied among US firms other benefits of using the methodology, became evident. Practicing the methodology became a vehicle for understanding the activities among the actors involved, promoted boundary spanning and communicated the views of the stakeholders (Melan, 1993). It became one of the key components of a quality system as defined by the Malcolm Baldridge Award criteria[1]. The notion of processes operating in a horizontal fashion, bridging organization boundaries is now reflected in a 'new' view of the organization as a "portfolio of dynamic processes" (Ghoshal and Bartlett, 1995).

4. PROCESS MANAGEMENT AND SYSTEMS THEORY

Studies by Jaques (1951), Gouldner (1955) and others suggested the desirability of examining enterprises systemically and contributed to the approach of 'organizations as systems' (Jackson, 1991). Katz and Kahn (1966) proposed that an organization, viewed as an open system, consists of five primary subsystems: production (technical), support, maintenance, managerial and adaptive subsystems. Business processes are in effect subsystems of the organization and can exist in any one of the five types. Considering the similarity of process and subsystems, it is evident that the Katz and Kahn view of systems components is not inconsistent with the process management view of a system as composed of interrelated processes. It should be noted that the five primary subsystems are not unto themselves viable; viability exists at an aggregate level.

---

[1] 1998 Criteria for Performance Excellence, p 16. Published by the National Institute of Standards and Technology, Gaithersburg, MD.

Based on the work of General Systems theorists such as Boulding and von Bertalanffy, Katz and Kahn define the characteristics of open systems as possessing input and output, transformation, homeostasis, differentiation, cyclicality of events, negative entropy, feedback, and equifinality – properties which are also observed in processes. The primary features of a process as noted in the definition in section 2 are input, transformation, output and repeatability (cyclicality). As with systems, the attributes of emergent properties and hierarchy are also possessed by processes. The removal of the transformation activity of a process step, for example, changes the properties of the output. In terms of hierarchy, processes are, for purposes of analysis, divided into sub-processes or key groupings of activities. In turn, subprocesses are composed of activities (both conversion and control). Finally, activities are composed of tasks or elements of action. The time stamping of an invoice, for example, is a task. The level of recursion within the hierarchy is referred to as the 'process in focus' for the subprocess in question, analogous to the term 'system in focus' used in systems theory.

Checkland (1992) states that the concept of the holon "implies processes" which contribute to the behavior of the system and ultimately the perceived world, inferring that processes are viewed as an element of a system construct. From a cybernetic perspective, Espejo and Schwaninger (1997) view an organization in terms of primary processes that produce its products and services and define its identity. The proposition that processes exist at various levels of recursion suggest similarities to the basic precept of process management noted in Section 2.

In terms of applicability of the process concept in a systems context, it is appropriate to examine process management using the Jackson-Keys 'systems of systems methodology' (SOSM) six cell grid (Jackson and Keys 1984, Jackson, 1991) which employs as dimensions for assessing the positioning of a methodology system complexity and participant social environment. In practice, process methodology shows that a range of process complexity exists from simple (contained within a department) to highly complex (spanning business units and locations). Participant environment (problem context) ranges from unitary in the case of a cohesive group of actors and stakeholders to 'simple coercive' (Midgley, 1972) where there is domination by virtue of power structure or a coalition of actors possessing referential power and where debate is not closed. The realities of business organizations suggest that because of position power existing in management structures 'simple coercion' exists to varying degrees both overtly and covertly which influences unity of view and tempers pluralism that promotes creativity and innovation in problem solving. The management style of the process owner and the sensitivity of the facilitator to the social structure of the participants are contingency factors that influence problem contexts and the position of a methodology within the grid. Given the behavioral contingencies that may be operative, there is a relatively wide zone of applicability in the SOSM framework.

Regardless of the contingencies that influence the positioning of the methodology within the SOSM framework, the issue of paradigm incommensurability (Kuhn, 1970, Dando and Bennett, 1981, Burrell and Morgan, 1979) suggests viewing system methodologies in the light of complementarism (Jackson, 1991, Flood and Jackson, 1991) or, currently methodological pluralism (Jackson, 1997, Midgely, 1997). Process management can be viewed as a complementary methodology in supporting a methodology such as SSM in examining the 'real world' (Checkland, 1981, Checkland and Scholes, 1990, Wilson, 1984) or VSM in examining subsystem at a lower level of recursion. The process approach can be used as an alternative methodology for intervention in organizations in the context of methodological pluralism (Gill, 1997).

5. CONCLUSION

Process management, developed as an alternative to traditional TQM approaches, possesses systemic features found useful in the analysis of organizations. Many attributes of processes possess identical open systems attributes. As a systems based methodology, process management possesses proven capability of being used in organizational interventions and can therefore be used in the context of complementarism as a supporting methodology or, in terms of methodological pluralism, as an alternative methodology for organization improvement.

References

Ackoff, R., 1993, Beyond total quality management, *J. Quality and Participation*, March, 1993.

Burrell, G., and Morgan, G., 1979, "Sociological Paradigms and Organizational Analysis," Heinemann, London

Checkland, P., 1981, "Systems Thinking, Systems Practice", Wiley, Chichester.

Checkland, P., 1992, Systems and scholarship: The need to do better, *J. Oper.Res. Soc.*, 43: 1023.

Checkland, P., and Scholes, J., 1990, Soft Systems Methodology in Action, Wiley, Chichester.

Dando, and Bennett, P., 1981, A Kuhnian crisis in management science, *J. Oper. Res. Soc.* 32:91

Espejo, R., and Schwaninger, M., Strategic management in global corporations: A cybernetic perspective, working paper no. 18, Lincoln School of Management, Lincoln, U.K.

Flood, R., 1995, "Solving Problem Solving", Wiley, Chichester.

Flood, R. and Jackson, M., 1991, "Creative Problem Solving: Total Systems Intervention", Wiley, Chichester.

George,C., 1968, "A History of Management Thought", Prentice Hall, Englewood Cliffs, N.J.

Ghoshal, S. and Bartlett, C., 1995, Changing the role of top management: Beyond structure to processes, *Harvard Business Review*, Jan.-Feb., 86-96.

Gill, A., 1997, Mixing methods: Developing systemic intervention, in "Multimethodology," J. Mingers and A. Gill, eds., Wiley, Chichester.

Gouldner, A. 1995, "Patterns of Industrial Bureaucracy", Routledge and Kegan Paul, London.

Jackson, M., 1991, "Systems Methodology for the Management Sciences," Plenum, New York.

Jackson, M., 1997, Pluralism in systems thinking and practice, in "Multimethodology", J. Mingers and A. Gill, eds, Wiley, Chichester.

Jackson, M. and Keys, P., 1984, Towards a system of systems methodologies, *J. Oper. Res. Soc.* 35:473.

Jaques, E., 1951, "The Changing Culture of a Factory", Tavistock, London.

Katz, D., and Kahn, R., 1966, "The Social Psychology of Organizations," Wiley, New York.

Kuhn, T., 1970, "The Structure of Scientific Revolutions," University of Chicago Press, Chicago.

Melan, E., 1986, Process Management, *Transactions 40th Annual Quality Congress*, American Society for Quality Control, Milwaukee, Wis.

Melan, E., 1993, "Process Management: Methods for Improving Products and Service," McGraw-Hill, New York.

Melan, E., 1998, Implementing TQM: A contingency approach to intervention and change, *Int Jour of Qual Sci*, 3:126.

Midgley, G., 1997 Dealing with coercion: Critical system heuristics and beyond, *Systems Practice,*10, 1:37-57.

Midgley, G., 1997, Mixing methods: Developing systemic intervention, in "Multimethodology", J. Mingers and A. Gill, eds., Wiley, Chichester.

Wilson, B., 1984, "Systems: Concepts, Methodologies", Wiley, Chichester.

THE PROBLEMS OF THE LYRIC THEATRES IN LONDON
- A SYSTEMS PERSPECTIVE

Lawrence R. P. Reavill

Department of Management Systems and Information
City University Business School
Northampton Square
London EC1V 0HB UK

INTRODUCTION

"Lyric Theatres" are those which concentrate on performances of dramatic works with a musical basis, such as opera and ballet. The works are generally large scale, with substantial casts of singers or dancers, or both. They require orchestras of symphonic size, and usually have spectacular scenery and costumes. A major opera or ballet, such as Tosca or Swan Lake would have three or more sets, hundreds of costumes, and require singers, dancers, musicians and technicians probably numbering about two hundred to stage the performance. The activity is therefore massively expensive, though opera is generally twice as expensive as ballet. Performers of international standing, for example opera stars such as Luciano Pavarotti and Kiri Te Kanawa, and ballet stars such as Sylvie Guillem and the late Rudolph Nureyev attract large audiences and therefore command very high fees. However, the great international companies tend to be the major attraction in themselves.

Performances cannot be supported by Box Office receipts alone. A high level of subsidy is required wherever these art-forms are presented, be it Paris, Moscow, Vienna, or New York. The subsidy can be from national or local taxation, by sponsorship from individuals or organisations, or by a combination of both.

In London, there are three major Lyric Theatres, the Royal Opera House (ROH), home of the Royal Ballet and the Royal Opera; the Coliseum, home for most of the year to the English National Opera; and the Sadlers Wells Theatre, a "receiving house" for visiting companies, mainly dance companies. Sadlers Wells is just completing a two-year major reconstruction which has been made possible by a grant of £30 million from the National Lottery, with matching funding from commercial, charitable and private sources.

The ROH is the flagship of the London Lyric Theatres. It stages opera and ballet of international standard, and is half way through a major reconstruction of its stage and public areas, and a major refurbishment of its theatre, costing about £216 million. £78 million has come from the National Lottery and the rest by private donations. The Coliseum is London's largest theatre, and has very good sightlines. Unfortunately, it is an old theatre in a poor state of repair, and is in need of substantial modernisation and refurbishment.

THE ROYAL OPERA HOUSE SYSTEM

This paper will consider the activities of the ROH, and examines these as a system. General Systems Theory indicates that the ROH system of interest relates to a supra-system and has sub-systems. It will be argued that the objectives of the system of interest are clear but unattainable, as they are inconsistent with those of the sub-systems, the supra-system, and other systems in the greater system of interest with which they interact.

The ROH has been in trouble for most of the last decade. The major resources the system needs are: money; personnel with specific specialist expertise, such as singers, dancers, musicians, theatrical technicians, producers, and designers; and administrators. There are surprisingly few significant problems emanating from the "artistic" side, and availability of talented artists does not appear to present a serious difficulty. However, money and administration have been problematic for many years, and it will be argued that the level of input to the system of the necessary financial resources is insufficient for the stated objectives of the system. Furthermore, the control of the system has been inadequate, particularly in addressing this major incompatibility and its implications.

The objectives of the ROH are clearly stated in a *mission statement* "Aims and Aspirations" (Royal Opera House, 1996). Most significantly to this paper, these include:

* To present high-quality opera and ballet to the highest possible standard;
* To offer a wide range of repertory.....;
* To encourage and promote creativity in production, choreography and design
* To provide the widest possible access to opera and ballet
* To secure financial viability and to eliminate income and expenditure deficits;

The clarity of these objectives is admirable. Unfortunately it is also transparent that the ROH has succeeded in the first stated objective and failed dismally in the last two. The width of the repertory is well regarded for the Royal Opera, but that of the Royal Ballet has been criticised in recent years for limited choreographic innovation.

SYSTEM CONTROLS

The formal control system for the ROH consists of a Board of Directors which has a Chairman and which appoints a General Director or Chief Executive Officer. The Chairman may assume a more or less pro-active role in the operation of the theatre. The General Director/Chief Executive Officer is similar to the "Intendant" in charge of opera houses in continental Europe. There are also Boards of Directors for the Royal Ballet and the Royal Opera. The Royal Ballet has an Artistic Director and an Administrative Director. The Royal Opera has a Director and a Music Director. The major components of the Royal Opera House are the theatre itself, the ballet company, the opera company, the orchestra, the technical staff, the support staff and the administration.

The ROH is a major recipient of public funding, so the Arts Council is a major player from the greater system of interest. The Arts Council is the body appointed by the Government to oversee the distribution of tax derived money to subsidise the arts. It is part of the "government system of interest". In 1992 it set up a committee chaired by Baroness Warnock to review the management of the ROH. She found "inadequacy of management, both executive and non-executive", and "the tendency to decide what is right artistically first and to count the cost later" (Everitt, 1997). Although the ROH Board, recruited from the 'great and the good', included many able and successful people, according to ex Arts Minister David Mellor they tended to "leave their brains outside the boardroom door".

Later investigations, such as those of the House of Commons Select Committee under the chairmanship of Gerald Kaufman in 1997, and the working party set up by the Arts Minister, Chris Smith, under the chairmanship of Sir Richard Eyre in 1998, were even less convinced of the competence of the Royal Opera House management. Many senior members of the management team have resigned or been fired. In the aftermath of the scathing Kaufman report, (Kaufman, 1998), the Chairman and the Royal Opera House Board resigned. A new Chairman has been installed and new Board members appointed. Massive changes are in train at the Arts Council. However, though mismanagement is very much in evidence, this is mainly financial. The effectiveness of the ROH is generally agreed, it is its efficiency especially in the control of resources which is questioned.

THE FINANCE SUB-SYSTEM

A root cause of the problem, and probably the major cause, is inadequate finance, the Macawber problem of an excess of expenditure over income. Attainment of the system objectives is incompatible with the current level of funding. The current subsidy for the ROH is about 14.5 million pounds, about half the subsidy the Victoria and Albert Museum receives with a much less labour intensive product (Maddocks, 1998). This, together with Box Office receipts and sponsorship, supports the fabric of the theatre, the Royal Ballet and Royal Opera companies, the orchestra, and all the support staff. The subsidy is between a half and a third of that received by other major European opera houses.

The level of subsidy has fallen in real terms consistently since the beginning of the decade. Both the current Labour and the previous Conservative governments have been unsympathetic to any increase in public subsidy. Both governments try to avoid increases in public expenditure which might lead to tax increases and electoral unpopularity. In all government thinking, it is forgotten or deliberately disregarded that the Lyric Theatres pay VAT on ticket prices, are a major component of the London theatre scene, and a principal tourist attraction in London. It has been estimated that the London theatres provide more than 41,000 jobs. Londoners spent 250 million pounds on theatre tickets and a further 433 million pounds on restaurants, hotels, transport and merchandise (Bridge, 1998).

The recent Conservative government believed that the ROH should raise a larger proportion of its income from sponsorship, in the manner of the American model. The comparison is invalid, as unlike the UK, American individuals and organisations can offset charitable donations to the arts against tax. Another problem is that sponsorship is more difficult to find when the economy is in recession. The Tory government had no truck with accessibility, and was generally lukewarm to subsidising the arts. Sponsorship saved public money, so was acceptable. However, sponsors support the ROH because it is prestigious, and to some extent exclusive. Exclusivity enhances the value of membership of the club, and the value of giving financial support. Also, corporate entertainees will be more impressed and flattered to be part of a club which is exclusive.

Some members of the current government are opposed to what they regard as a subsidy for "toff's opera". This is quite illogical, as the government policy sub-system has as one of its tenants the principle of accessibility. This is incompatible with the objectives if the sponsorship sub-system. As the system of interest receives major funding derived from taxation, the performances should be accessible to all members of the public, regardless of their income level. As the current prices at the ROH are very high, about twice the price of the equivalent seats at the Paris Opera or the Vienna State Opera, access is largely available to the well-off and the rich. "Peoples' opera" can only be available if the people are prepared to pick up the bill, as ex-ROH General Director, Sir Jeremy Isaacs forcefully stated. Either accessibility must be abandoned, or a more realistic level of public subsidy must be made available, as Sir Richard Eyre has proved (Eyre 1998).

Faced with the dilemma of a decreasing subsidy income while trying to maintain standards in an increasingly expensive international market, the ROH was forced to increase seat prices and search for yet more sponsorship. With increased seat prices, the accessibility was much reduced, particularly for the opera where the seat prices are generally twice the cost of those for the ballet. Increased sponsorship tended to involve increased corporate entertaining, with greater pressure on seat availability for popular performances, and public areas of the theatre roped-off to allow corporate entertainees to sip champagne and munch salmon sandwiches without having to rub shoulders with the general public. Even commercial sponsorship has not been forthcoming in sufficient quantity to balance the budgets, particularly when paying customers developed resistance to high ticket prices, especially for the less attractive items in the repertory. This, together with lack of money for new productions was instrumental, particularly with the Royal Ballet, in recent seasons devoid of significant innovation. The tendency to trundle out the old war-horses caused audiences to become bored with over-exposure to previously popular productions. Inability to resolve these financial problems generated increasing debt.

THE LOTTERY AWARD - A CHANGE FROM THE ENVIRONMENT

With an increasingly technically antiquated theatre which had also been without significant structural refurbishment for many decades, the need for modernisation was becoming desperate. It was more doubtful each year whether the theatre would obtain safety and fire certification. An appeal was launched for over £200 million for redevelopment. Part of the funding was the application for a grant of £78.5 million from the National Lottery. When the award from lottery funds was made, there was an outcry in the tabloid press. The use of "public" money to subsidise the specialised activity of the "elite" was denounced, and numerous comparisons made, such as the number of kidney machines the money might otherwise have provided. The violence of the reaction caught the ROH management unprepared, and they made an inept defence of the award when they might better have mounted a vigorous defence or ignored the comments altogether.

This was not the only misjudgment of the ROH management. The major re-development of the site required that the theatre should close for two years. Some means was required to keep the system running. The ballet company, the opera company, and the orchestra needed to keep going during the rebuilding period. An alternative large theatre was required from the moment rebuilding commenced in 1996. The management wasted valuable time considering a totally unrealistic plan to build a new theatre. By the time the naivety of this option was as apparent to the ROH management as it was to everyone else, the possibility of hiring a commercial theatre of appropriate size was gone. Here the system of operation of the West End Theatres of London comes into play. There are very few theatres with stages large enough to accommodate the substantial productions of the ROH, or with auditoria of sufficient seating capacity to accommodate enough customers to make the events viable. Almost all these were contracted to long running musicals. Drury Lane would perhaps have been the best option, but "Miss Saigon" showed every sign of running beyond the millennium. The newly rebuilt Lyceum had an adequate stage, but a smaller capacity than was ideal. It was available when the ROH first considered a newly built theatre, but later acquired "Jesus Christ, Superstar". The opera and ballet companies had to perform in houses of various degrees of unsuitability such as the Apollo Hammersmith, the Shaftesbury, the Barbican, and the Albert and Royal Festival Hall.

The extra funds needed to subsidise this itinerant period, likened by Gerald Kaufman to "wandering around London like the Flying Dutchman", were grossly under-estimated, the potential receipts from performances in the new venues grossly over-estimated, and the resulting cash crisis brought the ROH within a whisker of bankruptcy. At the eleventh hour it was rescued by large donations from two of its principal charitable supporters. The conclusion of Gerald Kaufman's committee report was that it would be better that the ROH be "run by a philistine with the requisite financial acumen, than by the succession of opera and ballet lovers who have brought a great and valuable institution to its knees". A new Chairman, Sir Colin Southgate, was appointed, and attempted to stabilise the financial system, as it was a prerequisite of the government considering an increase in public funding. Whether he is the philistine Gerald Kaufman had in mind is not clear, but his approach has made the proverbial china-shop located bull appear a model of tact and subtlety. The outcome at this time is that Bernard Haitink, the Royal Opera's world-renowned musical director, has resigned, following the cancellation of the 1999 opera season without consultation. All employees have been threatened with redundancy unless they sign contracts at much reduced salaries. The Royal Ballet might consider decoupling from the ROH, as the Birmingham Royal Ballet wisely did a few years ago.

The only aspect of this pantomime with few laughs is that the refurbishment of the theatre appears to be more or less on budget and on schedule. Whether the matching funding from sponsorship will continue to arrive is questionable, with the overall outcome in serious disarray. The final result could well be the arrival of a superb state-of-the-art opera house, a jewel for London's theatrical crown, and a complete white elephant. This will stand resplendent in the wreckage of the opera and ballet companies that might have made it all worthwhile. This has been achieved by a ROH management of breathtaking incompetence, an Arts Council content to wring its hands but take no decisive action, and a government with no coherent policy for arts subsidy, besides the notion of "peoples' opera", provided the people do not have to pay for it. Huge amounts of public and private money have been invested, and the outcome is little opera or ballet of international standard in London for two years or more, and the probable destruction of a ballet company of international standing, an opera company which can attract the major international stars, and the dismemberment of a first class orchestra. All that has been built up over many past decades is likely to be lost. Would-be patrons for substitute culture will have to take the Eurostar to Paris, where they manage these matters much better.

DISCUSSION OF SYSTEMS INTERACTIONS

The objectives and resources of the ROH are incompatible. Either production of opera and ballet of international standard must be downgraded to a good national standard commensurate with the resources available, or the resources must be increased to correspond satisfactorily with the objective. For almost a decade the ROH management have tried to provide a Rolls-Royce product, for opera in particular, on a Ford Granada budget. When this inevitably failed, they arrogantly assumed that their status as a royal foundation and a premier arts institution would ensure that they were baled-out by the government. The failure of the Arts Council to resist or control this behaviour has led to justified government impatience, and plans for massive surgery within the Arts Council. The government policy sought a standard of product for which it was not willing to pay, and a mish-mash of incoherent objectives such as accessibility, and "peoples' opera", with the bill collected by commercial sponsors for no obvious advantage to their shareholders.

If we look at the supra-systems and sub-systems which relate to the system of interest, more incompatibilities are apparent. An international standard in ballet is achieved by particular companies achieving a high standard. It is rather like the football teams which qualify, by superior performance, for international competitions. At present, the "premier division" might include the Royal Ballet, Royal Danish Ballet, the Paris Opera Ballet, the Bolshoi, the Kirov, and New York City Ballet. International opera is more akin to the professional tennis circuit, where the stars perform at Paris, Wimbledon, New York and Sydney. To maintain its position, despite its immense prestige, the Wimbledon Tournament had to offer top league prize money. The international opera supra-system has equivalent top venues: the Paris Opera, Vienna State Opera, Bolshoi Opera, Bayreuth, and the

Metropolitan Opera, New York. International opera singers derive status from their appearances at these opera houses. The ROH has difficulty maintaining its membership of this premier operatic division, as it needs to call upon the top singers to grace its productions, and their market price is determined by global demand. The opera system of interest is part of a global opera supra-system with which it is not compatible. The supra-system is the determinant, so our system of interest will have to adapt to conform.

This brings into play two other systems. The first is the public perception of the relative value of the art forms of ballet and opera. Although public feeling may not be so extreme as the tabloid press tends to suggest, the elitism, arrogance and insensitivity of the ROH have left a serious public relations problem which might deter the government from giving help. The government has also inherited from the previous Conservative government a sponsorship sub-system that is incompatible with its accessibility policy sub-system, and does not appear to notice this incongruity.

Finally, there are sub-systems within the political system, for example the Kaufman Committee, with an overt objective of protecting the use of public money, but with a hidden agenda of deflecting political criticism away from the Government. In this respect, Kaufman has been brilliantly successful, but has done nothing but make a difficult situation much worse. Another sub-system is the Eyre Working Party, (Eyre, 1998), which tried genuinely to offer viable solutions. Its proposals deserved to be taken seriously by the Government. Finally, there is the media, particularly the tabloid press, who have had a field-day exploiting and reinforcing the philistine and anti-elitist attitudes of their readers. This system is clear in its objectives: good stories sell newspapers.

CONCLUSION

A simple use of general systems theory suggests that the ROH system of interest has a supra-system and sub-systems which are inconsistent with one-another, and incompatible with the known objective of the system of interest. There is no synergy between the interacting systems, indeed there appears to be 'anti-synergy', if such a concept is possible. Unless these incompatibilities are reconciled, the ROH system will fail, and it is proving this point by heading rapidly towards disintegration. The problem can be resolved if the resources from public funds are increased to a level compatible with that of other international opera houses. Another option would be to reduce the standard of the product to a national standard which would be consistent with the resources input. A third option would be to abandon the principle of accessibility, regardless of the use of tax revenue. Any of these changes will render the system of interest and its sub- and supra-systems compatible. There is no other viable option. Put in simpler terms, if you want international opera and ballet, you have to pay the international market price. Basic economics and simple systems science go hand in hand. A failure to take a holistic view of the overall funding of the ROH is leading to a massive waste of money, and serious damage to hard-won artistic standards.

REFERENCES

Bridge, R. (1988) *The Times*, July 15 1998.
Eyre, R. (1998) *Lyric Theatres Working Party Report*, July 1998.
Everitt, A. (1997) *Financial Times*, December 4 1997.
Kaufman, G. (1997) *House of Commons Select Committee Report,* December 1997
Maddocks, F. (1998) *The Observer*, September 13 1998.
Royal Opera House, Covent Garden; Theatre Programmes, May 1996.

SOME REFLECTIONS ON PRACTICE: WHAT *DID* I THINK I WAS DOING WHEN I JOINED SYSTEMS DESIGN CONVERSATION GROUPS?

Sylvia M. Brown

Management Learning Research Unit at The Open University Business School, Walton Hall, Milton Keynes, MK7 6AA, U.K.

1   Introduction

This paper is part of a personal journey towards greater understanding of the author's attempts at international collaboration, viewed from an Idealised System Design (ISD) perspective. It begins with a brief review of a trend towards concern with "global" issues that can be construed as altruistic or neurotic, depending on perspective. Outlined next are some key concerns of the International Systems Institute (ISI) social system design conversation groups with which the author is involved. What the author sees as paradoxes in the ISD conversation approach are noted, followed by a brief discussion of what these groups can and cannot be, in terms of "Processes to....". Some personal dilemmas are posed.

2   Background review

Synergistic collaborations, partnerships and stakeholder involvement all are seen as increasingly important in a wide range of fora where change is discussed, whether inside business enterprises or during formulation of government policy.

A related and concurrent trend is "globalisation"; in recent years the systems movement has encouraged thinkers to take a broad perspective. For example, successive UKSS conferences have invited participants to consider "The next ten years of systems research", then "Addressing Global issues", then "Systems for Sustainability". On the other side of the planet from the U.K., conferences of Anzsys (The Australia and New Zealand Systems Society) also have concerned themselves with societal, ecological and other high system-level issues. In similar vein, the Vice-President of IFSR (International Federation for Systems Research) recently stated that a major role of this body is "to facilitate a continuing project on global crisis". (Yong Pil Rhee June 1998). ISI (International Systems Institute) includes in its mission statement the purpose "3) to design models and methods for social and societal systems".

The commonest means adopted by those who express these well-meaning sentiments is the international conference; another model is the conversation; others include workshops and research collaborations.

What is a conference? A typical conference experience is a large scale event with several streams allowing such a plethora of papers that even dissemination of ideas to those attending is limited by forced choice. Questions, interaction, debate and discussion at the actual event tends to be crowded out by a very full schedule. Formation of new, synergistic relationships is inhibited by the "here today, gone tomorrow" format; as with holiday romances, addresses are exchanged and communications promised but not often forthcoming.

The author is not quite clear what a conference is, in terms of "A process to..." or "A process that..." but has noted some worrying trends at conferences where wider dissemination might be seen as "the whole point". For example, the ratio of "practitioners" to "academics" has dropped at successive British Academy of Management conferences to a very low level; this means that even as, "A process to disseminate findings that might assist business practice", this system is not succeeding.

What is a conversation? A conversation, according to Bela Banathy, (IFSR newsletter June 1998) is a collectively disciplined enquiry, an exploration of issues of societal and social significance, is engaged in by scholarly practitioners in self-organised teams and is initiated in a preparation phase that leads to an intensive learning phase. In the ISI version, the experience is of an international group meeting face to face on an infrequent basis but continuing to communicate at a distance between. At the face-to-face encounters, small groups (as with training groups, more than eight members reduces group effectiveness) concentrate, uninterrupted, on an enquiry of their choice for several days, usually five. The author is unsure what this system "is" either but is moved to explore that question.

What does the term "Idealised System Design" mean? "Ideal" describes a conceptual model for the future of the system of interest, i.e. this exists only as an idea; it is arrived at by a process of transcending the present. Banathy (1996) views this as a mental leap outside one's knowledge and views about the present, in order later to view the present from that ideal perspective. "Design" in this context does *not* mean a plan but describes the activity of design*ing*. The whole term is most usually found in discourse on social systems, hence a key concept is evolutionary co-designing by a community of the stakeholders in the social system of interest.

3   What are the key concerns of ISI conversation groups?

The author suspects that Idealised system design is likely to appeal to idealists. If true, this implies that the vision achieved by transcendence is likely to be ideal in a Utopian sense, as well as a conceptual modelling sense. Since they are international research groups, ISI conversations probably tend to represent those who have some experience of the ability to exercise civil rights, who believe strongly that others should enjoy such rights and in principles of autonomy, especially self governance, principally from W1. From this perspective, ISD can be seen as a statement of a designing ethic as much as a methodology.

The principles of ISD are principles in two senses. As well as a carefully steered sequence of processes that are intended to result in dynamic co-creation of sustainable, viable systems, they embody a deep commitment to human rights to freedom from oppression. Co-creation by a community of the systems that both control and support the life of that community is appropriate; imposition of a system plan is inappropriate.

Maintaining both senses of the principles simultaneously and in practice is not, however, an easy matter. The historical development of Banathy's model (ibid) can be interpreted as a struggle to balance real potential for social improvement and human rights to freedom of choice of one's destiny. The result has been increasing complexity of the model as "further back" and "links between" stages of the process are developed as responses to the perceived shortcomings both of this model and of the models of others. Thus a first stage of co-designing of an Ideal System as a practical principle to secure stakeholder commitment and thereby system viability, (see, e.g. Ackoff's Ideal System 1974) has evolved backwards several stages. The first stage in recent versions of the Banathian model now is creation of a design culture that will foster propensities to the conversations that will generate the mutual understanding that will enable co-design. In Banathian terms, the concept of "Design" itself also has evolved backwards to a stage where it precedes both System Planning and System Building; Idealised System Design as currently conceived is not a product; one may not speak of *an* Idealised System design, as one speaks of *an* Ideal System or *a* Conceptual Model, rather is it the set of processes that occurs within a design culture. On this view, it is only ever permissible (or possible) to engage in design for oneself, in community with, but not for, others. At the next system levels up, a group, then a design community may co-design only for itself. The human rights of the individual are paramount; it may have been overlooked in this framework that these logically include the right to be wrong. This designing "for oneself" is not selfish, rather is it an acknowledgement that only I know what I need and that I should not presume to know what is best for you. You should also design for yourself. This World View contrasts sharply with earlier approaches to change imbued with missionary zeal - or does it? That is the conundrum that vexes the author.

4   Avoiding paradoxes: what ISD conversation groups can be

Implicit in the ISD literature is the belief that exercise of civil rights during the designing process is a goal that should be given higher priority than rapid and efficient achievement of system "improvement". (For a discussion of the tension between natural rights and civil rights when using IDS, see Brown 1998b.)

In practical terms, Banathy himself agrees that any co-design process based in the ISD approach, based as it is on human rights, could take a very long time indeed. This timescale presents a paradox, as well as logistical problems. Any group engaged in a design conversation must not, realistically, expect to see immediate co-design effects in the system, probably not until the distant future, by which time changes will have taken place in the system environment and stakeholder set membership also will have changed greatly. Any actions taken by members of the group when designing for themselves will have changed the range of options for others. Under these circumstances, how can the group's members not design for others?

The biggest paradox of all in relation to timescale is, perhaps, that some states of the human condition that need improvement are too pressing to wait for ISD, whereas it may be the case that those same states cannot be addressed in a way that can produce sustainable systems by any other principles than those of ISD.

In addition, an ethical, co-designing community is, potentially, vulnerable to the more predatory system "next door".

In the case of ISI conversation groups, therefore, the insistent principles of ISD necessarily push back their international, collaborative effort, vital to synergy, as well as greater understanding and shared vision, into consideration of theoretical frameworks, at a very high level of generalisation, that may be accepted or rejected by particular design communities, as "guidance systems" in their co-design process, assuming that they can come to know about them by some means.

Herein lies another paradox; if the chosen metaphor is system levels, whilst conversations are possible at group level and groups can and do communicate within communities that share purposes and values, what must happen at the level of the wider society in order to realise co-design in ISD mode is very mysterious indeed, both in terms of boundary crossing (agency) and contextual receptiveness (design culture). The importance of a design culture is relative to ethical value sets and its permeation is problematic in relation to boundaries even within a system of ethics, let alone between them. Historical examples with principles close to ISD, such as religious communities, either have closed themselves off from the rest of society in order to maintain their ethical system's integrity or have become missionary. Some questions of agency and ethics are discussed in another paper (Brown 1998b).

It would seem that the only practical and ethical option for an ISI conversation group attempting to operate as a co-design community designing only for itself is to do theoretical work and disseminate it somehow, in the hope that "real" co-design of "better" social systems by their stakeholders will be aided. There are several W1 ways to interpret that option. One is as a retreat into the academic "ivory tower", another is as avoidance of my responsibilities to my fellow citizens of the world, a third is as treatment of means as ends. Furthermore, the most ready means for dissemination of a conversation group's outputs, of course, are the conference and the journal - it seems impossible to escape these at some stage.

A fourth interpretation is as a piece of W1 crassness or callousness, depending on perspective, that might be described as "Fiddling while Rome burns." The author is reminded of a story of black children "bussed" during forced racial integration in the USA. One little boy was asked by his new white teacher the sort of question she usually put to her class, which was "How many legs does a caterpillar have?" The reply she got, after a telling pause, was "Miss, you should have *my* problems."

Issues such as these have caused the author to question her own motives and frustrations in participating in ISI conversation groups; missionary zeal is suspected.

Other methodologies, such as Action Research, even Participative Action Research, do not present such paradoxes for those with an urge to help the world to improve itself, since they are cheerfully interventionist, albeit in an appropriately circumspect, facilitational manner.

5   Individual motives and ethnocentric perspectives; knowing, if, what and when you are compromising

Ultimately, what I do is my decision. If I use ISD as an ethical system for system design I must recognise that it can tell me what I ought to do in principle but not what I can and must do in any given situation. Practical, effective action may involve some compromising of principles.

My situation is that I find myself *wanting* to seize opportunities to make a difference in systems I perceive to be self-defeating, unethical (by my values) or misery-creating, and to help others to do so. At such times I am right to remind myself:
that I should question my own motives as much as my capacity for effectiveness, before, during and after my, yes, *my* intervention.

If I join, entice, exhort or encourage others in a co-design process, I must recognise my agenda for doing so and examine its ethics.

Do I then allow the severity of co-design principles to subvert that agenda? Banathy (and probably Ulrich) would, unhesitatingly, say yes. I am not so confident that I always should or even could; based in the knowledge that I have prevented one individual self-destruction and tried to prevent another, I suspect that I, driven by my value system, would be just as interfering in the case of a whole group. The best I could do here might be to own up to my agenda, my W1 ethnocentricity and system of ethics, at least to myself and the other stakeholders.

Self examination and rueful recognition is not enough, however, to make a morally defensible decision about whether, when and what to compromise or not. To think too long and hard about what to do is paralysing to action, natural (spontaneous) or unnatural. If I am not to be paralysed by thought into inaction, a W1 view might be that perhaps I should give Utility Function Theory, which I have decried in the past, another chance. Zen being entails that trying to be natural or trying not to be natural are both affectations; ultimately I merely experience, I merely do. That, however, doesn't let me off the responsibility to become an ethical being, that "merely does" and trusts emergence.

6   Conclusions and options for action

These are presented as a simple list.

(i) Researchers collaborating mostly at a distance and meeting occasionally, even for intensive conversation, cannot use the ISD framework to "design models and methods for social and societal systems", because to do so is to design for others.

(ii) They can design models, methods and guidelines for designing de-contextualised models and methods for social and societal systems. How useful these are to co-designing communities in actual social contexts is debatable.

(iii) Even within ISD versions of system co-design, agency (i.e.intervention) cannot be escaped. Even the attempt to "do good" by dissemination of "take it or leave it" models is a form of intervention. The belief that this might be otherwise is W1, naive and ethnocentric; from a Zen viewpoint, we cannot act at all if frozen by ethical examination of every action before we make it and in any case, all our actions ultimately are "natural" (spontaneous) and all have effects - "we cannot live for a day without destroying the life of some other creature." (tr. Watts 1957)

(iv) Whilst attractive, the ISD ideal presents me with paradoxes that I cannot resolve from within the ISD research paradigm.

(v) Given the means of dissemination available at present, the world probably will remain unaffected by even the best efforts of ISI researchers.

This presents me as a researcher with a number of options:-

a) I could try to let go of my desire to interfere, to "sit quietly and say nothing". (Watts, Ibid)

b) I could try to develop myself as an ethical being, such that my natural responses to particular cases seen as in need of improvement are ethical. This resonates with Banathy's current position that creation of a "design culture" is a critical factor in societal design (improvement?).

c) If I want to improve myself, ISD can help, in that its conversations can immerse me in a design culture.

d) If I can become a person that acts spontaneously in an ethical manner, I may occasionally improve the world somewhat.

e) I can try to improve my reflection on practice and that reflection might prevent me from being too specious in explanations of my actions.

f) If I decide it is acceptable to want deliberately to improve the world, I must choose another research paradigm and other means of dissemination.

REFERENCES

Ackoff, R.L., (1974), *Redesigning the Future*, N.Y., Wiley.
Banathy, B. H., (1996), *Designing Social Systems in a Changing world*, N.Y & London, Plenum.
Banathy, B. H., (June 1998), IFSR Newsletter 17:2.
Brown, S.M., (1998b), "Some Issues of Agency and Utility in Idealised System Design", in *Creative Systems Practice*, Proceedings of the 4th Anzsys Conference, Hawkesbury, University of Western Sydney, Australia.
Rhee, Yong Pil, (June 1998), IFSR Newsletter 17:2.
Watts, A.W., (1957) *The Way of Zen* Penguin.

SYNERGY: A SPIRITUAL ISSUE IN THE CONTEXT OF CHANGE MANAGEMENT SYSTEMS

Cathal M. Brugha

Michael Smurfit Graduate School of Business,
University College Dublin, Blackrock, Dublin, Ireland

Nomology, a decision science system for explaining qualitative structures is shown to have three dimensions: *adjusting, convincing* and *committing*. Examples of the system include Maslow's hierarchy of needs, Jung's thinking types, the Systems Development Life Cycle, and the Oriental systems approach WSR. The *convincing* and *committing* systems combine into one system of *developing*, which is shown to be associated with autopoiesis and to have the energy of negative-entropy. Two cases of *adjusting* are shown to be linked: (1) body, mind, soul and spirit, and (2) direction, mission, vision and synergy. The *adjusting* and *developing* systems combine into a system of *change management*, examples of which include the Twelve Step programme of Alcoholics Anonymous and from Information Systems development.

NOMOLOGY: A CONTEXT FOR EXAMINING SYSTEMS

Nomology, the science of the laws of the mind (Hamilton, 1877, pp. 122-8) is a meta-model whereby issues such as change management, synergy and spirituality can be considered. The basis of Nomology is that decision-makers tend to analyse problems which involve qualitative distinctions by breaking them into activities, or categories of behaviour, which are each important in themselves and follow natural sequences. This is a natural approach that the mind uses when addressing a problem where there is no clear external frame of reference. The first categorisation is about the degree of uncertainty involved. What sort of problem is it? High uncertainty will require some sort of planning activity, low uncertainty some form of putting plans into effect. The second dichotomy relates to where is the main focus of the problem? Is it more to do with people, or more to do with structures, organisations, i.e. the "place" where some system is based? These categorisations and the language associated with them are very general, and are applicable to many different situations. The fundamental generic set of adjustment activities is shown in Figure 1. There are

Synergy Matters: Working with Systems in the 21st *Century,*
Edited by Castell *et al.*, Kluwer Academic / Plenum Publishers, New York, 1999.

253

numerous examples of adjustment in management based on these general activities (Brugha, 1998a) and on eight particular activities (Brugha, 1998b).

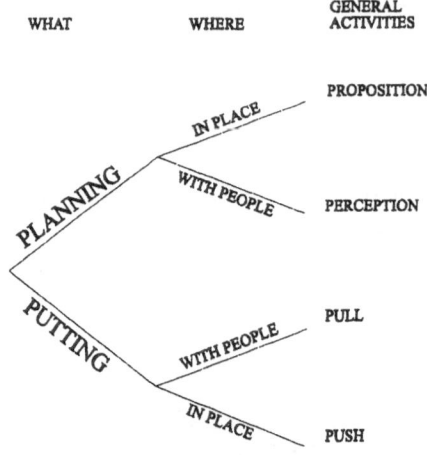

Figure 1. The four general kinds of activity

Adjustment is distinguished from development decision making in the case where the decision maker owns the process and consequently cannot "pull" himself or herself to make the decision. Hence the pull activity in Figure 1 disappears. The emphasis then is more on building on levels then on finding balance between different activities. Development decision making can be introverted or extroverted (Brugha, 1998c). The first introverted level is the somatic, and refers to tangible things such as needs. Then there are psychic (psychological) aspects such as preferences. Finally the pneumatic level refers to values or higher goals. Hamilton (1877) introduced the terms cognition, affect and conation as a triad of mental activities corresponding to the somatic, psychic and pneumatic levels. Thus Nomology takes a broader view than cognition. It incorporates feelings, but also includes the more neglected area of will corresponding to the highest level of introverted **commitment**. Soma, psyche and pneuma come from the Greek words for body, soul and spirit (literally wind).

The extroverted dimension corresponds to stages of **convincing** and starts with technical or self-orientated issues. Then it relates to other people, and finally it takes account of situations. The introverted and the extroverted combine as two dimensions and lead to the construction of nine levels, stages of activity and types of thinking and the reconstruction of Maslow's (1987) hierarchy of needs and Jung's (1971) orienting functions (Figure 2).

The Systems Development Life Cycle (e.g. Whitten, Bentley and Barlow, 1989) fits this nine phase structure of convincing within committing (Table 1). A central claim of Nomology is that adjusting, convincing and committing comprehensively describe the three dimensions of how the mind structures decisions. One proof is that so many independent and unconnected qualitative structures can be explained in terms of those dimensions. These come from different fields such as management, philosophy, marketing, multi criteria decision making cases, and information systems. A powerful validation comes from an East-West parallel. Brugha (1998d) has explored the commonalities between Zhu's (1998) Oriental systems approach, WSR *(wuli, shili* and *renli)* and corresponding western concepts. He suggests that *wuli* corresponds to a*djusting, shili* to *convincing,* and *renli* to *committing* and that Zhu's (1998) description of a hydro-engineering case in northern China showed that the project followed a Systems Development Life Cycle but with an adjustment process occurring within each of the nine stages.

		Convincing Stages		
		Technical Self	**Others** End-users	**Situational** Business
Committing **Phases**	**Somatic** Have / Need	Physical / Intuiting	Political / Recognising	Economic / Believing
	Psychic Do / Prefer	Social / Sensing	Cultural / Learning	Emotional / Trusting
	Pneumatic Are / Value	Artistic / Experiencing	Religious / Understanding	Mystical / Realising

Figure 2: Levels of developmental activities and types of thinking

Table 1: Systems Development Life Cycle activities

Introverted Orientation	Extroverted Orientation		
	Technical	**Others**	**Situational**
Somatic	Survey project scope and feasibility	Study current system	Define the end-user's requirements
Psychic	Select a feasible solution from candidate solutions	Design the new system	Acquire computer hardware and software
Pneumatic	Construct the new system	Deliver the new system	Maintain and improve the system

The distinction between adjusting as externally owned and developing (committing and convincing) as owned by the decision maker is seen to lead to development being constructivist and based on levels. This is easily understood in terms of the individual. If a country's politics is insecure an individual will be nervous about setting up a business; if one does not have a job one might be worried about taking on the responsibilities of a family. The breakdown of stability at lower levels and the collapse of one's situation is akin to entropy or the escape of energy. The building up of such layers of energy systems could be described as negative entropy. When the ownership belongs to a group the sense of building on levels and the need that the lower levels be secure becomes an issue for the group. The sense of integrity that a group needs to develop or to be self-creating has been named autopoiesis from the Greek words auto (self) and poiesis (creation, production). It has been applied both to biological systems and to social systems who are bound by the emotion of love (Maturana and Varela, 1992; Maturana, 1988). It is clear that the more mutually supporting people are in an organisation, i.e. the more committed and convinced they are about what they do, the less entropy or energy loss there will be. Central to the development concept of ownership is the idea that this cannot be forced from outside. This has ethical consequences for the management of systems (Brugha, 1998e).

THE LINK BETWEEN SYNERGY AND SPIRIT

Because Nomology is a meta-model of how the mind works it can be used to explore the meanings of terms such as synergy and spirit. In the context of Nomology language is seen as labels to be put on categories of activity so as to differentiate them from other categories. Labels for similar things can change between contexts from being more generic for a field to being more grounded for a particular case. Categories contain sub-categories; hence the level of abstraction is important. To explore a term such as **spirit** one would normally start by asking "spirit as distinct from what?" The answers might be **body** and **soul**. Doing it again with the word body might suggest the word **mind**. (See Maturana and Varela, 1992 for a consideration of the mind-body problem.) The question then arises: how do "body, mind, soul and spirit" differ from "soma, psyche and pneuma"? Is the former an adjustment concept and the latter a commitment concept? If they are the same why have the two languages co-existed for so long? The nomological exploration of "body, mind, soul and spirit" is done by relating them to the generic adjustment terms: proposition, perception, pull and push (Figure 1). If they relate properly there should be a consistent qualitative difference between each corresponding pair. This should reflect itself in meaningful phrases linking each pair such as "we propose in the body", "we develop perceptions in the mind", "we are pulled in our soul", and "we push with our spirit". Some of these phrases seem to clarify the links, others seem to challenge our understanding of terms such as soul and spirit which we may have tended to compartmentalise as "terms which have something to do with religion".

Where there is difficulty making a link between terms that seem to have a parallel it can be useful to find a bridging set. This is done by exploring what we mean by a phrase such as "we propose in the body". In this case it means that we consider what we are able to do, at a very basic level: our **capacity**. If this is a meaningful bridge there should be corresponding terms for the other three categories. Going beyond the capacity of the body into the area of the mind corresponds to working on our **capabilities**: the aspect of training and educating our perceptions. Our capacity and capability focus on the aspect of what we should or will do. To find a corresponding bridging term to elucidate the phrase "pulled in our soul" requires a deeper understanding of the dynamic that mediates this feeling of duty. In this case it is our sense of **community**. Extending it to the fourth quadrant in the same way produces a term such as **contribution**. Together capacity, capability, community and contribution provide a sort of road map for what one would do with a useful life.

Further bridges can help to fill out the meanings inherent in each sector. Using a shorter focus than the "useful life" one could see the bridge between capacity and body as one's **energy**. One's capacity is limited by the energy one has. Correspondingly, how much we are capable of doing at any time depends on the amount of **equanimity** that we have in our minds. A corresponding bridge between community and soul would be the effort that one can make. The more support we have from our community the more **effort** we are able make in response. Likewise, ultimately the contribution that we, at a deep level, i.e. in our spirits, make to society depends on our **effectiveness**. Another test of such sets is in the context of the balances that should exist within them. The effort that we can make cannot be out of balance with the energy we have. If we do not have equanimity we cannot be effective. A useful model for assessing the balance between the different categories of activity is given in Figure 3.

Another set derived from Johnsen (1993) is: **direction, mission, vision and synergy**. Before an organisation can have a mission it must agree on its direction. Clearly mission is a perception concept, and vision exists in the context of a community. If one thinks in terms of direction, mission and vision as a progressive focusing of one's energy, equanimity and effort, then it is clear that full effectiveness is reached when the many strands are brought together, as in **synergy**. This particularly makes sense if we see one's spirit as the ground in which this is done, where all the internal conflicts and contradictions that destroy synergy are resolved.

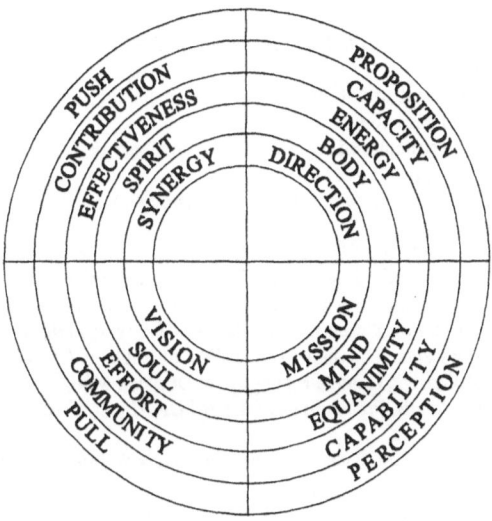

Figure 3: The cycle of general activities

CHANGE MANAGEMENT

Brugha (1998f) has describe categorisations of activities based on adjusting convincing and committing as **nomological maps**. An example is the Twelve Step Programme of Alcoholics Anonymous (Anonymous Authors, 1955). Here the dominating issue is adjustment. Each adjustment phase has three development stages. The same twelve steps can also be used to change people to a higher level of spiritual activity in the world, e.g. the Spiritual Exercises of St. Ignatius of Loyola which has been presented (Tyrrell, 1982; Fessard, 1956) as to reform the deformed, conform the reformed, confirm the conformed, and transform the confirmed. Such a representation could be applied to any adjustment process, depending on how broadly one interpreted the idea of being "deformed".

A development-led self-managed twelve step programme (Peace Pilgrim, 1981) involves four (somatic) preparations, four (psychic) purifications, and four (pneumatic) relinquishments. At each level there are steps corresponding to body, mind, soul and spirit. The spirit-based steps in this programme are (somatic) to simplify life to bring inner and outer well-being into harmony, (psychic) purification of motives, and (pneumatic) relinquishment of all negative feelings. Clearly synergy is facilitated by the elimination at the spirit level of anything that could conflict with bringing the vision to fruition.

Brugha (1998f) has shown how Galliers and Sutherland's revised 'stages of growth' model for information systems strategy (Galliers & Sutherland, 1991; Galliers, 1991) could be converted into a full nomological map containing seventy two individual steps. In that context one would recommend to an information systems organisation that is moving up the stages of growth, that it wait until it has completely adjusted at each stage before embarking on the next stage. In the context of what has been explored in this paper this would mean waiting until the synergies start appearing, as these would be a sign that the adjustment processes are completing.

REFERENCES:

Anonymous Authors, 1955, *Alcoholics Anonymous,* 2nd Edition, New York: Alcoholics Anonymous World Services, Inc.

Brugha, C. 1998a, The structure of qualitative decision making, *European Journal of Operational Research*, **104** (1), pp 46-62.

Brugha, C. 1998b, The structure of adjustment decision making, *European Journal of Operational Research*, **104** (1), pp 63-76.

Brugha, C. 1998c, The structure of development decision making, *European Journal of Operational Research*, **104** (1), pp 77-92.

Brugha, C. 1998d, "Considering WSR In The Context Of Nomology, A Generic Meta Model For Systems Studies", Proceedings of the ICSSSE conference, August, Beijing, China.

Brugha, C. 1998e, "Information Systems Development: Who Owns the Decision?", *Philosophical Aspects of Information Systems: Methodology, Theory, Practice and Critique*, Proceedings of the PAIS2 Conference, University of the West of England, Chapter 5.

Brugha, C. 1998f, "A Generic Change Management Decision Model", *Matching Technology With Organisational Needs*, Proceedings of the 3rd UKAIS Conference, Lincoln University, Chapter 39, pp 420-430.

Fessard, G. 1956, *La Dialectique des Excercises spirituels de saint Ignace de Loyola*, **1**, pp 40-41, Paris: Aubier.

Galliers, R.D. 1991, Strategic information systems planning: myths, reality and guidelines for successful implementation, *European Journal of Information Systems*, **1**, pp 55-64.

Galliers, R.D. and Sutherland, A.R. 1991, Information systems management and strategy formulation: the 'stages of growth' model revisited, *J. of Information Systems*, **1**, pp 89-114.

Hamilton, W. (1877, *Lectures on Metaphysics*, Vols. 1 and 2, 6th Ed., in *Lectures on Metaphysics and Logic*, Edinburgh and London: William Blackwood and Sons.

Johnsen, E. 1993, *Strategic Analysis and Synthesis - An Axiomatic Model*, Copenhagen Business School Management Research Institute.

Jung, C. 1971, *Psychological Types*, in The Collected Works of C. J. Jung Volume 6, London: Routledge & Kegan Paul.

Lavidge, R. and Steiner, G. 1961, "A Model For Predictive Measurements of Advertising Effectiveness", *Journal of Marketing*, (October) 59-62.

Maslow, A. 1987, *Motivation and Personality*, New York: Harper & Row.

Maturana, H. 1988, "Reality: The Search for Objectivity or the Quest for a Compelling Argument", *Irish Journal of Psychology*, **9**, pp 25-82.

Maturana, H. and Varela, F. 1992, *The Tree of Knowledge: The Biological Roots of Human Understanding*, Boston: Shambala Press.

Peace Pilgrim 1981, *Steps Toward Inner Peace*, Friends of Peace Pilgrim, 43480 Cedar Avenue, Hemet, California 92544.

Tyrrell, B.J. 1982, *Christotherapy II*, New York: Paulist Press.

Whitten, J., Bentley, L., and Barlow, V. 1989, *Systems Analysis and Design Methods*, 2nd Ed., Homewood, IL, IRWIN.

Zhu, Z. 1998 "WSR: A systems approach for information systems development", *Systems Research and Behavioural Sciences*, in press.

ADMINISTRATIVE SCIENCE QUARTERLY: CANARY OF WORLDVIEW SHIFT?

Eric B. Dent and Edward H. Powley

The George Washington University
Administrative Sciences Program
2136 Pennsylvania Avenue, NW Suite 300
Washington, DC 20052

INTRODUCTION AND SUMMARY OF STUDY

Theoretical Framework

In the past twenty years, a number of scholars and researchers have called attention to the importance of an individual's worldview, the set of deeply held beliefs and fundamental assumptions which serve as a mental map for providing coherence to what the world and life are and how they work (Slife and Williams, 1995). An individual's worldview is more deeply ingrained than attitudes, opinions, traits, and values. It is also more difficult to access and much harder to change than any of these attributes.

The contents of worldview are often philosophical in nature and may include the following abbreviated list: determinism, equality, belief in the transcendent, and hierarchical ordering. Although worldview includes a variety of such beliefs, a few assumptions make the greatest difference in ascertaining an individual's worldview. These assumptions pertain to observation, causation, and level of explanation (Dent, 1997). A fourth construct, interrelatedness will also be tested here. The literature suggests two primary worldviews which are labeled here the *traditional* (TWV) and the *emerging* (EWV) (Ackoff, 1981; Dooley, 1997). The term *emerging* connotes novelty, but the worldview is not new in many contexts. It is relatively new in organizational life in the Western world. Rather than being opposite worldviews, the EWV can be seen as the recognition that there is a polarity (Johnson, 1992) between the assumptions of the TWV and EWV. The TWV emphasizes only one side of the polarity. A number of writers have suggested that the acceptance of EWV assumptions is critical for a continuation in the increase of business performance and the quality of life (Begun, 1994). The contrasting assumptions of the TWV and EWV are shown in Table 1.

Synergy Matters: Working with Systems in the 21st Century,
Edited by Castell *et al.*, Kluwer Academic / Plenum Publishers, New York, 1999.

259

Table 1. Assumptions of the TWV and EWV.

Construct	TWV Assumptions	EWV Assumptions
Level of Explanation	Reductionism	Holism
Causation	Linear Causation	Mutual Causation
Observation	Objective Observation	Perspectival Observation
Interrelatedness	Competition	Cooperation

Because the terminology is not common in the literature, the specific assumptions used here are defined in Dent and Powley forthcoming. The word *construct* is used here to denote a phenomenon such as "causality." Different *assumptions*, then, can be made about a construct. For the construct *Causality*, the two assumptions labeled are *mutual* and *linear*. Henceforth, assumptions will be italicized and constructs will be capitalized and italicized.

When decision makers hold assumptions which are not appropriate to the choices at hand, they are less likely, for example, to select an effective decision. Mismatches in assumptions have been implicated in the failure of a major business (Jacobs and Jaques, 1987, p. 34), the implementation of government social policy which attempted to increase the stock of low-cost housing but actually reduced it (Dent, 1997), and inadequate descriptions of actual leadership behavior. Worldview assumptions have important implications for how work is performed. Person A making TWV assumptions and Person B making EWV assumptions will have completely different approaches to learning (Vaill, 1996), leadership (Wheatley, 1992), strategic planning (Begun, 1994, p. 330), performance appraisal (Dent, 1997), organization structure (Wilber, 1995), problem solving (Ackoff, 1981), and communication (White, 1990).

THE SIGNIFICANCE OF THE TOPIC

A potential contribution of this study is to begin documenting the extent of this alleged paradigm shift *in thinking*. Writers such as Begun (1994), Capra (1993) and Wheatley (1992) wax eloquently about the change that is occurring and point to variables such as the increasing use of holistic medical approaches and alternative organizational structures, but for the most part they still beg the question of whether or not people are *thinking* differently. People could be going to different types of doctors simply as a matter of a pragmatic search to find something that works. Such a search does not mean that a change in thinking has occurred. Consequently, this research will serve as one test of the question, what evidence is there to support the claims of writers such as Begun, Capra, and Wheatley, who believe that an emerging worldview is taking hold?

NARRATIVE ANALYSIS AND METHODOLOGY

The research method for this study is narrative analysis, which is similar to content analysis in that both go beyond the manifest meaning to a deeper level of interpretation. Narrative analysis differs from content analysis in that it is an attempt to capture ideas rather than words. Content analysis as often described (Weber, 1985) is inadequate for this purpose because it is overly reductionist, for example. Narrative analysis endeavors to balance *holism* and *reductionism* and combines aspects of qualitative and quantitative research. In general, though, the research design used here primarily makes assumptions of *reductionism* and *linear causality*.

Narrative analysis requires that the articles be read by people, rather than a computer. Singleton, Straits, and Straits (1993) suggest that unless the recording unit is the word, word sense, or phrases such as idioms and proper nouns, the computer software available at this time is not useful (p. 383). Computer software is of limited value because assumptions are implicit rather than explicit. The coders had to make judgments, for example, about terms such as "influence," "effect," "play a role," "determine," "driven by," and others which were sometimes coded as linear causal and sometimes not. Coders occasionally had to read further into an article to properly code an earlier paragraph(s). Consequently, the recording unit of analysis for this study was the embedded paragraph, the paragraph coded within the context of the entire article. This approach provides evidence about the assumptions that *the authors* of these articles made.

The eight assumptions of the four constructs listed in Table 1 above were coded when present in a paragraph. Assumptions were reduced to constructs for some of the analysis later in this paper. In Western society, the TWV is conventional wisdom, as described by Galbraith (1976). Nearly all of the Fortune 500 companies, for example, can be shown to be structured and acting in accordance with the TWV (Gergen, 1995). Consequently, nearly every paragraph could be coded for TWV assumptions depending on how "deep" an interpretation is sought. Our approach was to make more surface interpretations. With such an approach, it is fair to say that uncoded paragraphs primarily reflect deeply implicit TWV assumptions.

Sample

The sample consisted of full-length articles from the journal *Administrative Sciences Quarterly* (ASQ). ASQ is considered to be the leading organization journal, publishing path-breaking research in macro-level organizational analysis (Usdiken and Pasadeos, 1995). All full-length articles of the years 1957 and 1997 were selected for coding. The two time frames were chosen to capture points near the beginning and the end of the alleged paradigm shift (Banathy, 1998).

Our hypothesis is: The 1957 articles make primarily TWV assumptions and the 1997 articles still make predominantly TWV assumptions, but contain significantly more EWV assumptions than the 1957 articles.

Coding Notes

A coding manual, several pages in length, was devised for this research. A couple of important guidelines are discussed here. For example, an attempt was made to show the possible change in assumptions in the clearest light. Consequently, for the assumption of *competition*, we did not code instances in which competition between two organizations in the same industry was mentioned. Such instances were so frequent in both the 1957 and 1997 articles that it would have obscured the *Interrelatedness* assumption in other settings. Similarly, only pure instances of *linear causality* and *mutual causality* were coded. If an author assumed circular causality, for example, no score was tallied.

Reliability and Validity

The coding process was refined in many ways. Several practice codings were conducted and a set of coding guidelines were developed during this practice. Any problems with any aspect of the coding were surfaced and resolved. Intra-coder reliability and intercoder reliability data are reported in Dent and Powley (forthcoming).

The primary question of internal validity is, "does the classification relate to the causes or consequences hypothesized?" This study assured construct validity by selecting constructs from previous research and relevant theories, and by using assumption definitions from previous research and relevant theories Many of the classical threats are not applicable when there is not a knowing subject. The primary question of external validity is, "do the findings depend upon the specific data, methods, or measurements of this particular study?" If there is high agreement about definitions, "the problem of validity is no problem at all" (Berelson, 1952, p. 169). Clear definitions were borrowed directly from earlier work in which these suitable definitions were offered.

RESEARCH FINDINGS

The research findings are divided into three parts. First, we analyze the mean scores for the assumptions individually. This allows comparison of EWV and TWV assumptions separately, and thereby avoids the assumption that, for example, *holism* and *reductionism* are separate poles on a linear continuum. We then compare the incidence of each assumption, and compare the difference for each assumption and construct longitudinally. Finally, we collapse the eight assumptions into their respective constructs. In so doing, we seek to understand the direction these constructs are headed. Given that the universe of 1957 and 1997 ASQ articles were used, it is not appropriate to perform an analysis using inferential statistics. The findings, supported by descriptive statistics, sustain a portion of our hypothesis. The 1957 articles make more explicit TWV assumptions than EWV assumptions with the exception of the *Observation* construct. The 1997 articles do show increases in the EWV assumptions of *holism, perspectival observation,* and *cooperation.* As expected, the TWV assumptions are still strongly present, particularly *reductionism* and *linear causality.* The 1997 articles still make predominantly TWV assumptions, but contain significantly more EWV assumptions than the 1957 articles.

The mean scores for the eight assumptions are displayed in Table 2. One striking point is the huge increase in *linear causality* from 1957 to 1997. We believe there may be one primary alternative explanation for this effect. In 1957, research in the fields of social, organizational, and business inquiry, was emerging. Articles in ASQ's early years, for the most part, do not reference a base of existing knowledge. In 1997, each of the articles introduces the topic of the paper with an extensive literature review which generally takes the form of specifying linear cause and effect relationships found in prior research. For a more complete comparative analysis, including adjustments for stylistic differences and number of paragraphs per article, see Dent and Powley (forthcoming).

To eliminate the effect of the literature review and the greater number of paragraphs in the 1997 articles, we shifted the unit of analysis to the article rather than the paragraph and scored each article for the presence or absence of the assumptions, yielding the absolute frequencies. These data are presented in the right hand columns of Table 2.

Table 2. Scores for assumptions and differences between 1997 and 1957.

	Mean score (by paragraph)			Total count (by article)		
	1997[1]	1957[2]	Diff.	1997[3]	1957	Diff.
Reductionism	3.2	2.4	0.8	15.0	14.0	1.0
Holism	1.9	1.0	0.9	10.8	8.0	2.8
Linear Causation	24.0	4.0	20.0	18.3	18.0	0.3
Mutual Causation	0.2	0.3	-0.1	1.7	3.0	-1.3
Objective Observation	0.5	0.6	-0.1	5.8	9.0	-3.2
Perspectival Observation	2.0	0.9	1.1	11.7	11.0	0.7
Competition	0.2	1.0	-0.8	2.5	6.0	-3.5
Cooperation	1.8	0.2	1.6	4.2	2.0	2.2

[1] 1997 N=24.
[2] 1957 N=20.
[3] Adjusted for larger N.

These data, perhaps, more easily lend themselves to helpful interpretation. Again, the TWV assumptions predominate in the 1957 articles with the exception of *Observation*. In 1997, *reductionism* and *linear causation* predominate and so do the EWV assumptions of *perspectival observation* and *cooperation*.

The third level of analysis consisted of collapsing the assumptions into their respective constructs. We did this by subtracting the EWV construct from the TWV to arrive at a construct score for each article. Table 3 shows the mean construct scores for both 1957 and 1997. Positive scores indicate a greater use of EWV assumptions, and negative scores a greater use of TWV.

Table 3. Assumptions Collapsed into Respective Constructs (EWV-TWV) by year and unit of analysis.

	1997	1957	1997-1957 by paragraph	1997-1957 by article
Level of Explanation	-1.3	-1.5	0.2	1.8
Causation	-23.8	-3.8	-20.1	-1.6
Observation	1.5	0.3	1.2	3.9
Interrelatedness	1.6	-0.8	2.4	5.7

Table 3 allows for analysis of the relative ascendance of two assumptions in a construct, assuming they are linearly related. The two right hand columns suggest that three EWV assumptions are on the rise relative to their TWV counterparts: *holism, perspectival observation,* and *cooperation.* With the article as the unit of analysis (far right column), the EWV gains are fairly modest, with *cooperation* showing the largest increase. *Causation* has moved more toward the TWV both because of a slight increase in *linear causality* and because of a very slight decrease in *mutual causality.* The actual instances of *mutual causality* are infrequent enough that this finding should not be considered conclusive.

CONCLUSION

Several authors have suggested that a paradigm shift is occurring in the worldview of people in organizations. A comparison of ASQ articles from 1957 to 1997 provides limited evidence of such a shift. Our results imply that a modest shift is occurring in the assumptions of *holism, perspectival observation*, and *cooperation*. The TWV assumptions of *reductionism* and *linear causality* show very slight increases over the same time frame. Proponents of *mutual causality* will be especially disappointed that our research identified only five paragraphs with this assumption in the entire set of 1997 issues. Consequently, the TWV remains a powerful force in the set of assumptions used by authors of ASQ journal articles.

BIBLIOGRAPHY

Ackoff, Russell L. (1981). *Creating the corporate future: Plan or be planned for.* New York: John Wiley & Sons.

Banathy, Bela H. (May-June 1998). "Evolution guided by design: A systems perspective," Systems research and behavioral science. Vol. 15, 161-172.

Begun, James W. (December 1994) "Chaos and complexity: frontiers of organization science" *Journal of management inquiry.* Vol. 3, No. 4, 329-335.

Berelson, Bernard (1952). *Content analysis in communication research.* Glencoe, IL: The Free Press Publishers.

Capra, F. (1993) "A systems approach to the emerging paradigm." *The new paradigm in business: Emerging strategies for leadership and organizational change.* Ray, Michael and Rinzler, Alan, eds. New York: G. P. Putnam's Sons.

Dent, E. B. (1997). *The design, development and evaluation of measures of individual worldview.* Dissertation, The George Washington University, The School of Business and Public Management.

Dent, E. B. and Powley, E. H. (forthcoming). "Paradigm shift in progress?: Evidence from *Administrative Science Quarterly* and *Harvard Business Review.*"

Dooley, Kevin J. (1997). "A complex adaptive systems model of organization change," *Nonlinear Dynamics, Psychology, and Life Sciences.* Vol. 1, No. 1, 69-97.

Galbraith, John Kenneth (1976). *Affluent society* (3rd ed.). Boston: Houghton Mifflin Company.

Gergen, Kenneth J. (Aug/Nov 1995) "Global organization: From imperialism to ethical vision." *Organization.* Vol. 2, no. 3/4, 519-532

Jacobs, Thomas Owen and Jaques, Elliott (1987). "Leadership in complex systems." Human productivity and enhancement: Organizations, personnel, and decision making, vol. 2. Zeidner, Joseph, ed. New York: Praeger.

Johnson, Barry (1992). Polarity Management: Identifying and Managing Unsolvable Problems. Amherst: HRD Press, Inc.

Singleton, Royce and Straits, Bruce and Straits, Margaret Miller (1993). *Approaches to social research* (2nd ed.). New York: Oxford University Press.

Slife, Brent D. and Williams, Richard N. (1995). *What's behind the research: Discovering hidden assumptions in the behavioral sciences.* Thousand Oaks, CA: Sage Publications, Inc.

Usdiken, B. and Pasadeos, Y. (1995). "Organizational analysis in North America and Europe: A comparison of co-citation networks." *Organization Studies.* Vol. 16, no. 3, 503-526.

Vaill, P. B. (1996). *Learning as a way of being: Strategies for survival in a world of permanent white water.* San Francisco: Jossey-Bass, Inc.

Weber, R. P. (1985). *Basic content analysis.* Beverly Hills: Sage Publications.

Wheatley, M. J. (1992). *Leadership and the new science: Learning about organization from an orderly universe.* San Francisco: Berrett-Koehler Publishers.

White, O. F. (1990). "Reframing the authority/participation debate" Chapter 6 in Refounding Public Administration by G. L. Wamsley et al. Newbury Park, CA: Sage Publications. 182-245.

Wilber, Ken (1995). Sex, ecology, spirituality: The spirit of evolution. Boston: Shambala.

DOES THE QUEEN HAVE MAGICAL POWERS:
LEADERSHIP & SYSTEMS IN THE 21ST CENTURY

Ernest L. Hughes, Ed.D.

The Systems Thinking Company
40 NW Alder Place, Suite 203
Issaquah, WA 98027 USA
Phone: 425.557.9946
FAX: 425.557.9947
ehughes@systemsthinking.com

INTRODUCTION

In October of 1995, my family and I attended the opening ceremonies of the University of Lincolnshire and Humberside. When I was explaining to my daughter, Melina, then four years old, that Her Majesty Queen Elizabeth II would be coming as well, Melina remarked, "Does the Queen have magical powers?" When I shared this story with Mike Jackson, Dean of the Lincoln School of Management, he remarked soberly, "I wish she did." Perhaps Her Majesty does, in at least one way significant to organizations in the 21st century.

This paper will explore leadership from a systems perspective, and systems from a leadership perspective. Both leadership and systems thinking will be examined as transformational processes, and related methods for improving organizational effectiveness and efficacy. The construct of transformational leadership will be related to the systems thinking methodology, Total Systems Intervention. Organizational contexts viewed from a leadership perspective–structural, human resource, political, and symbolic frames, will be compared with popular systems metaphors. The leadership and systems thinking skills of framing and reframing a problem context for creative problem solving will be contrasted. Lastly, a conception of systemic leadership will be discussed.

SYNERGY MATTERS IN THE 21st CENTURY

In 175 days from the close of the 6th International Conference of the United Kingdom Systems Society, this century will end. "A century's end," according to Schwartz (1996, p. xiv), "is taken as the end of a generation, the end of an epoch, and the end of a hundred years of progress or misery. The beginning of a century had better be the start of the NEXT GENERATION, the commencement of a REVOLUTIONARY NEW epoch, and the wiping clean of the slate for another, MORE BRILLIANT hundred years."

Synergy Matters: Working with Systems in the 21st *Century,*
Edited by Castell *et al.*, Kluwer Academic / Plenum Publishers, New York, 1999.

As Schwartz suggests, many challenges–known and unknown, lie ahead in the 21st century. This section will outline these challenges, and suggest what leadership and systems theory and practice have to offer to meet them.

Challenges of the 21st Century

Allen, Bordas, Hickman, Matusak, Sorenson, and Whitmire (1998) believe four prominent trends are shaping thought and action for the future: (1) globalization; (2) increasing stress on the environment; (3) increasing speed and dissemination of information technology; and (4) scientific and social change. The challenges implied by these trends include: increasing diversity and complexity, increasing tensions around value differences, an increasing gap between the rich and poor, and an increasing requirement for continuous learning (pp. 41-46). Allen et al. see these trends and implications as being highly interdependent, demanding a total systems approach for resolution. "The challenge and implication for leadership," they argue, "will be to initiate and practice a systems perspective" (p. 45).

Leadership Theory and Practice for the 21st Century

According to Allen et al. the purpose of leadership in the 21st century is to:
- Create a supportive environment where people can thrive, grow and live in peace with one another;
- Promote harmony with nature and thereby provide sustainability for future generations;
- Create communities of reciprocal care and shared responsibility, where every person matters and each person's welfare and dignity is respected and supported (1998, p. 41).

Leadership theory and practice have much to offer in achieving this purpose. "Leadership," according to Carlson (1996), "produces change, often to a dramatic degree, and has the potential of producing extremely useful change" (p. 136). Bass (1990) asserts, "Leadership is often regarded as the single most critical factor in the success or failure of institutions" (p. 8).

More specifically, Burns (1978) provides a theory and process of transforming leadership in which "leaders and followers raise one another to higher levels of motivation and morality" (p. 20). "Leadership over human beings," contends Burns, "is exercised when persons with certain motives and purposes mobilize, in competition or conflict with others, institutional, political, psychological, and other resources so as to arouse, engage, and satisfy the motives of followers. This is done to realize goals mutually held by both leaders and followers" (p. 20). The transforming leader achieves "significant change" (Burns, 1978, p. 425).

Extending Burn's work, Bass and others (Bass, 1985; Bass and Avolio, 1994; Bass, 1998) developed the construct of transformational leadership. "Transformational leadership is seen when leaders stimulate interest among colleagues and followers to view their work from new perspectives, generate awareness of the mission or vision of the team and organization, develop colleagues and followers to higher levels of ability and potential, and motivate colleagues and followers to look beyond their own interests toward those that will benefit the group" (Bass and Avolio, 1994, p 2).

Transformational leaders behave towards followers in one or more of four ways, categorized as the "Four Is" by Bass and Avolio: (1) idealized influence; (2) inspirational motivation; (3) intellectual stimulation; and (4) individualized stimulation (1994, p 3).

Transformational leaders gain idealized influence with followers by being consistent, demonstrating ethical and moral conduct, sharing risks with followers, and considering the needs of followers over their own. They provide followers inspirational motivation by being enthusiastic and optimistic, communicating expectations to followers, providing meaning and challenge to their followers' work, involving followers in creating a desired, future shared vision, and demonstrating their own commitment to goals and the shared vision. Transformational leaders provide intellectual stimulation to followers by questioning assumptions, reframing problems, encouraging creativity and new approaches, and supporting differing ideas, especially from their own. Lastly, they provide followers individualized consideration by listening, enabling two-way communication, creating a supportive climate, accepting individual differences, meeting individual needs, developing personal relationships, coaching, and mentoring (Bass and Avolio, 1994, pp. 3-4).

Couto (1997), in an analysis of Burn's transforming leadership and Bass's transformational leadership, argues that transforming leadership, with its emphasis on social change, may be beyond the grasp of most leaders. He suggests that transformational leadership, which focuses on the capacity to change conditions and to change culture at an institutional level, is within reach (p. 152-3).

Systems Thinking and Practice for the 21st Century

The philosophy and principles of current and emerging systems thinking and practice, particularly critical systems thinking, are similar in nature to the purpose of leadership for the 21st century. These principles include critical awareness, social awareness, and human emancipation (Jackson, 1991). Jackson (1995) asserts that systems thinking "carries the hopes of all system participants for bringing about improvements in organizations and societies which can benefit all stakeholders" (p. 39).

Flood and Jackson (Flood and Jackson, 1991; Jackson, 1991; Flood, 1995) have developed a methodology, Total Systems Intervention (TSI) that can be used by those who wish to transform organizations, and follow the philosophy and principles of critical systems thinking. TSI uses range of systems metaphors to encourage creative thinking about organizations and their problems. These metaphors are linked by a framework that is used to select one or more systems interventions to guide problem solving and ensure that organizational issues are addressed (Jackson, 1991).

The TSI methodology is divided into three phases: creativity, choice, and implementation. The outcome of the creativity phase of TSI is the selection of one or more metaphors that highlight the primary interests and concerns of the organization. The organizational metaphors that may be used include machine, organism, brain, culture, and coercive-system (Jackson, 1991). The outcome of the choice phase of TSI is the selection of a specific, dominant systems methodology to be used, and perhaps one or more secondary ones. Flood and Jackson have developed a decision framework using assumptions about the problem context, and the underlying organizational metaphors to determine which systems methodologies are appropriate as interventions. The methodologies included in this framework are System Dynamics, Viable System Diagnosis, Strategic Assumption Surfacing and Testing, Interactive Planning, Soft Systems Methodology, and Critical Systems Heuristics (Flood and Jackson, 1991). The outcome of the implementation stage of TSI, Jackson asserts, "is coordinated change brought about in those aspects of the organization currently most vital for its effective and efficient functioning" (Jackson, 1991, p. 275).

LEADERSHIP AND SYSTEMS IN THE 21st CENTURY

A synthesis between leadership and systems theory and practice is suggested by the previous discourse. Leadership researchers and practitioners are augmenting leadership praxis with a systems perspective, and to a lesser degree, vice versa. Leadership from a systems perspective, and systems from a leadership perspective are explored in this section, and an initial conception of systemic leadership will be discussed.

Leadership from a Systems Perspective

Few systems researchers and practitioners have much to say about leadership. Senge (1990), elaborating on the disciplines he believes are essential for a learning organization, asserts, "Systems thinking, personal mastery, mental models, building shared vision, and team learning–these might just as well be called the *leadership* disciplines as the learning disciplines" (p. 359). "These disciplines span the range of conceptual, interpersonal, and creative capacities vital to leadership," he concludes (p. 360).

Flood (1995) argues that leadership style influences the way a problem solving system such as TSI operates. Leadership style, he states, will affect the likelihood of successfully matching up to the principles of TSI, and following the TSI process (p. 59-60). He briefly examines four styles of leadership: (1) authoritarian, (2) supervisory, (3) laissez faire, and (4) participative, and suggests that a participative style, in which leaders encourage members to participate and make decisions, is best suited for TSI and bringing forth the principles upon which it is based.

Systems from a Leadership Perspective

Some leadership researchers and practitioners are incorporating a systems perspective into leadership thinking and practice. Bass (1990) credits Katz and Kahn (1966) with pointing the way on leadership and systems. He states, "A systems approach looks at the leader as someone embedded in a system with multiple inputs from the environment, the organization, the immediate work group supervised, the task, the leader's behavior, and his or her relationship with subordinates and outputs in terms of effective performance and satisfactions" (p. 908).

Luke (1998), in defining what he calls catalytic leadership for complex, public issues, sees systems thinking (primarily System Dynamics) as one of four distinct sets of analytic skills essential for leaders to see interconnections and linkages, and identify strategic leverage points (p. 175).

Carlson (1996) recognizes the need for a "total and comprehensive systems approach" for organizational transformation. He claims that Flood and Jackson's Total Systems Intervention methodology "offers potential as a strategy for leadership as a transformational process" (p. 143).

Bolman and Deal (1984, 1991, 1997) articulate a leadership theory and process that is analogous to the choice phase of TSI. They contend that they have consolidated the major schools of organizational thought into four frames: structural, human resource, political, and symbolic. These frames are schemata, maps, images, or metaphors that can provide leaders with different perspectives. Bolman and Deal see these frames as forming a holistic framework that encourages inquiry into a range of significant issues that managers and leaders must think about—people, power, structure, and symbols (1991, p. 18).

The structural frame, according to Bolman and Deal (1991), emphasizes the importance of formal roles and relationships. Structures are created to fit an organization's environment and technology, and problems arise when the structure does not fit the situation. In the human resource frame, the "key to effectiveness is to tailor organizations to people—to find an organizational form that enables people to get the job done while feeling good about what they are doing" (1991, p. 15). The political frame views organizations as arenas in which different interests compete for power and scarce resources. Conflict is everywhere, and bargaining, negotiation, coercion, and compromise are all part of organizational life. In the symbolic frame, organizations are cultures powered by rituals, ceremonies, stories, heroes, and myths. Organizational improvements come through the use of symbol, myth, and magic (1991, p.15-16). The Queen's visit mentioned in the introduction to this paper, although not magical, would be an example of leadership using the symbolic frame.

Successful managers and leaders, according to Bolman and Deal, "frame and reframe until they understand the situation at hand" (1991, p. 16). Using focusing questions developed by Bolman and Deal, a leader can identify–frame, an organizational issue, then reframe that issue to suggest "new questions to ask and new options for action" (1984, p. 245). They define reframing as the "sequential application of each frame to the same event or issue" (1984, p. 255). The first step is to ask which frame or frames are being used, and which are being ignored. The second step is to "expand the horizon by applying each of the other frames" (1984, p. 243).

Like Bolman and Deal, Terry (1993) proposes that "leadership depends on an ability to frame issues correctly" (p. xvii). Terry identifies seven generic features, or frames of leadership action: fulfillment, meaning, mission, power, structure, resources, and existence (1993, p. 58). He contends, "Every human act reveals these seven generic features, all the time, every time. These features are explicitly or implicitly present whether the action has happened, is happening, or will happen" (p. 60). Terry believes Bolman and Deal's frames parallel his (p. 162). He outlines several framing questions, and identifies the particular language that leaders might hear to position an issue within a diagnostic and action framework he developed.

Systemic Leadership

The previous discussion outlines the potential and possibility of systemic leadership. As suggested by Carlson (1996), leaders can use Total Systems Intervention as a transformational process for organizational change. Additionally, the leadership frames, framing questions, and framing and reframing processes identified by Bolman and Deal (1984, 1991, 1997), and Terry (1993) can augment the choice phase of TSI. These frames can be used as additional metaphors themselves, or as a "frame behind the frame" (Terry, 1993, p. 88). Since 1995, the author has applied these techniques with fruitful results as part of a leadership and systems practice.

CONCLUSION

Meeting the challenges ahead will require significant organizational and social change, and calls for synergy between leadership and systems theory and practice. Bass (1990) contends that the application of a systems approach to leadership is a substantive issue for the 21st century (pp. 896-909). This paper has taken some first steps in doing so by suggesting a possible linkage between transformational leadership and the systems methodology known as Total Systems Intervention. Other linkages are undoubtedly possible and likely to emerge in the next century.

ACKNOWLEDGMENTS

The author would like to thank Mike Jackson, Dean of the Lincoln School of Management, for extending an invitation to visit the university, Doreen Gibbs for recommending that my visit coincide with that of Her Majesty Queen Elizabeth II, and Amanda Gregory and Giles Hindle for their stimulating conversation and hospitality.

REFERENCES

Allen, K.E., Bordas, J., Hickman, G.R., Matusak, L.R., Sorenson, G.J., and Whitmire, K.J., 1998, *Leadership in the Twenty-First Century, in Rethinking Leadership: Kellogg Leadership Studies Project 1994-1997*, S.W. Webster, ed., The Burns Academy of Leadership, University of Maryland, College Park, MD.

Bass, B.M., 1998, *Transformational Leadership: Industry, Military, and Educational Impact*, Lawrence Erlbaum Associates, Mahwah, NJ.

Bass, B.M., and Avolio, B.J., 1994, *Improving Organizational Effectiveness Through Transformational Leadership*, Sage Publications, Thousand Oaks, CA.

Bass, B.M., 1990, *Bass & Stogdill's Handbook of Leadership: Theory, Research & Managerial Applications*, 3rd edition, Free Press, New York.

Bass, B.M., 1985, *Leadership and Performance Beyond Expectations*, Free Press, New York.

Bolman, L.G., and Deal, T.E., 1997, *Reframing Organizations: Artistry, Choice, and Leadership*, 2nd edition, Jossey-Bass, San Francisco, CA.

Bolman, L.G., and Deal, T.E., 1991, *Reframing Organizations: Artistry, Choice, and Leadership*, Jossey-Bass, San Francisco, CA.

Bolman, L.G., and Deal, T.E., 1984, *Modern Approaches to Understanding and Managing Organizations*, Jossey-Bass, San Francisco, CA.

Burns, J.M., 1978, *Leadership*, Harper & Row, New York.

Carlson, R.V., 1996, *Reframing & Reform: Perspectives on Organization, Leadership, and School Change*, Longman Publishers, New York.

Couto, R.A., 1997, Social Capital and Leadership, *in Transformational Leadership Working Papers: Kellogg Leadership Studies Project*, S.W. Webster, ed., The Burns Academy of Leadership, University of Maryland, College Park, MD.

Flood, R.L., 1995, *Solving Problem Solving: A Potent Force for Effective Management*, Wiley, New York.

Flood, R.L., and Jackson, M.C., 1991, *Creative Problem Solving: Total Systems Intervention*, Wiley, New York.

Jackson, M.C., 1995, Beyond the fads: systems thinking for managers, *Systems Research*, 12:25-42.

Jackson, M.C., 1991, *Systems Methodology for the Management Sciences*, Plenum, New York.

Katz, D., and Kahn, R.L., 1966, *The Social Psychology of Organizations*, Wiley, New York.

Luke, J.S., 1998, *Catalytic Leadership: Strategies for an Interconnected World*, Jossey-Bass, San Francisco, CA.

Schwartz, H., 1996, *Century's End: An Orientation Manual Toward the Year 2000*, Doubleday Currency, New York.

Senge, P.M., 1990, *The Fifth Discipline: The Art & Practice of the Learning Organization*, Doubleday Currency, New York.

Terry, R.W., 1993, *Authentic Leadership: Courage in Action*, Jossey-Bass Publishers, San Francisco, CA.

TAMING THE PROBLEMS OF HYPERMOBILITY THROUGH SYNERGY

C. Jotin Khisty[1], and P. S. Sriraj[1]

[1]Dept. of Civil & Architectural Engineering
Illinois Institute of Technology
3201 South Dearborn Street
Chicago, Illinois 60616-3793, USA

INTRODUCTION

Our interest in space and time has existential roots. It stems from a need to bring meaning and order into a world of events and action. Modern concepts of time and space are expressed through the desire to achieve higher mobility and increased accessibility, and this in turn is associated with social progress. To meet this modern-day demand, an extensive transport system was created, developed, expanded, and persistently "improved" in this century, in the hope of integrating society in time and space, all in the name of enhancing economic and social "progress".

In the last two or three decades, however, people have been confronted with the darker side of the expansion of transport system and the hypermobility (excessive and imbalanced mobility for the most part) associated with it. For example, this extensive system, with the objective of providing higher mobility coupled with increased accessibility, has endangered the quality of life and the ecological sustainability of modern society. More importantly (and ironically), this very expansion, designed for providing high speeds, has resulted in traffic congestion that has drastically reduced mobility and decreased accessibility, thereby lowering business productivity, increasing fuel consumption, increasing pollution, and adversely affecting safety. In the USA alone, Americans are losing 2 billion hours a year to traffic gridlock, equivalent to losing $ 20 billion a year plus business losses equivalent to $40 billion a year (Khisty and Lall, 1998). There is no question that a specter is haunting the world of the transport policy maker - the specter of the "future". Indeed the challenge is so bewildering that for some it's the only game in town.

In view of this predicament, the search for transport systems that can provide efficient movement of people and goods compatible with the long-run quality of the environment has intensified in the last decade. Fortunately, breathtaking advances in communication technologies and information systems provide a useful construct for exploring their convergence with Intelligent Transport Systems (ITS). The main objective of this paper is to examine the problems

Synergy Matters: Working with Systems in the 21st *Century,*
Edited by Castell *et al.*, Kluwer Academic / Plenum Publishers, New York, 1999.

271

created by hypermobility and to demonstrate ways of taming it by integrating telecommunication and information systems to support ITS through synergy.

THE TRANSPORT SYSTEM

The transport of people and goods is a highly complex system that has not yet been completely understood. It was only in the mid-1950s that some of the early pioneers of traffic science began seriously studying the relationship between the speed and the flow of a moving stream of vehicles. In more recent years, attempts have been made to come to grips with the entities comprising transport systems, and a fairly simple model showing the basic relationship between three of its essential components is shown in Figure 1. The activity system comprises the movements of persons and goods between two or more points or positions in space relative to the infrastructure. This activity system can be conceived as a market for movement. The transport system consists of persons or goods needing some kind of vehicle (besides their own personal power) to move from one position to another. Each movement is a transport service, and this service is matched by an equivalent supply of services by vehicles and their operators using roads or tracks. Lastly, we have the traffic system, consisting of the physical movement of vehicles in time and space, along physical networks. Each vehicle (or a set of vehicles connected to one another) is considered as a traffic unit, and the resulting flow is measured as the number of vehicles per unit of time on a specific link of the infrastructure (Khisty and Lall, 1998).

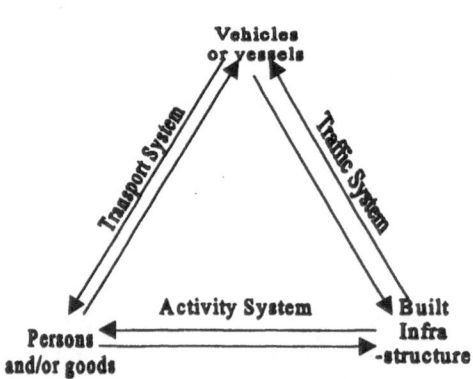

Figure 1. A Basic Model Connecting
Vehicles/Vessels, Persons/Goods
and the Built Infrastructure

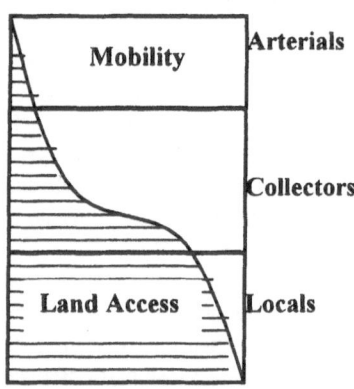

Figure 2. Relationship of Functionally
Classified Systems in Service
Traffic Mobility and Land Access

In the last two decades there has been a coordination of these three subsystems and a concerted effort has been made by policy makers to attempt to move people and goods by the right mode, in the right quantity, to the right place, and at the right time. If anything, this is a utopian goal, as we will see in the next section.

WHAT IS HYPERMOBILITY?

The development of modernity, with its stress on rationality, social and economic "progress", and the search for order was closely wedded to the Cartesian notions of space and the universal absolute ideas of time. But the shift away from modernism led to new notions of

272

space and time associated with a wider set of paradigm shifts within the social sciences and urban studies. The two most dominant parameters that have been most affected in this context are the concepts of accessibility and mobility.

In its broadest sense, accessibility refers to the ease of movement between places. Accessibility increases, either in terms of time or money, when movement becomes less costly. In fact, it is a measure of how easy or how difficult it is to reach different land use activities via the transport network binding the city together. Naturally, the greater the accessibility the greater the power over the resources in the area. The urban area can thus be seen as a "surface of opportunity" and we can measure this accessibility by determining the access opportunity scores of different zones in the area (Khisty and Lall, 1998).

Mobility, on the other hand, is a measure of the capacity and speed of a system to handle movement of traffic, both people and goods. For example, a 6-lane motorway has high mobility as compared to a 2-lane local road. Figure 2 illustrates the connection between accessibility, mobility, and the functional classification of our highway system. Notice that while freeways have high mobility they have poor accessibility to adjoining land use activities, while local roads offer high accessibility but have poor mobility. Policy makers in the past have tried to maintain a healthy balance between mobility and accessibility in a transport system by providing a hierarchy of highway and street systems possessing a variety of accessibility and mobility characteristics, with the hope of achieving a proper balance (Khisty and Lall, 1998).

Coupled with accessibility and mobility is the entity of efficiency which connects the total cost (direct and indirect) of transport and the productivity of the system as a whole. Direct costs are composed of capital and operating costs while indirect (or external) costs represent environmental impacts and other unquantifiable costs, such as safety.

With the enormous increase in commuter traffic in recent years one observes many stretches of the transport system operating far beyond their designed capacity, resulting in severe traffic congestion and an adverse impact on sustainability, accessibility, and the quality of life. This phenomenon created by excessive and imbalanced mobility and correspondingly reduced accessibility is termed as hypermobility.

HOW CAN HYPERMOBILITY BE TAMED?

Although traffic engineers, urban planners, and economists have in the past 30 years suggested a myriad of solutions to tackle the hypermobility problem, some highly radical and others more mundane, the problems associated with hypermobility are not likely to be mitigated, leave alone solved, by say, road pricing strategies or by restricting the use of the automobile (by forcing people to use non-motorized modes of transport). Because of space and budget constraints coupled with environmental concerns, building new or better highways and streets is out of the question. Economists have for the past three decades repeatedly said that adding a lane to an existing congested freeway would increase congestion still further.

To increase the efficiency of the transport system in urban areas, it is necessary to use systems thinking by modifying old and stereotyped ideas and assumptions about the development, planning, and management of the modern, industrial city. Also, the current notions regarding the nature of space, time, and distance need to be revised in light of developing technology. Transport systems planners are now focussing on promoting more efficient use of existing capacity. In fact, the central theme of these efforts is more efficient integration of existing transport components through the use of telecommunication and information technology.

Remarkable leaps in the capability of telecommunication systems have resulted in altering and adjusting space and time barriers, in ways that approach 'real time'. 'Telematics', which refers to services and infrastructures that link computer and digital media equipment over telecommunication links, are providing the technological foundation for rapid innovation in computer networking and voice, data, image, and video communication (Graham and Marvin, 1996).

In addition to telematics, the deployment of an information system is essential, consisting of the following components: (1) layout and utilization of the telematics network; (2) capture and organization of information; and (3) development of application software. A judicious combination of telematics and information systems through Intelligent Transport Systems is the key to taming the problems of transport, as will be described in the next section (Brascomb and Keller, 1996). This integration alone has spurred interest and implementation of the substitution of physical transport movement by tele-based activities such as tele-commuting, tele-education, tele-shopping, and tele-banking (see Figure 3). Indeed, from an environmental standpoint, there is considerable interest in telematics and information networking being considered as the "alternative fuel" of the future. Several studies done in recent years have indicated that such substitution is likely to result in about 20% of transport by conventional means in the USA; eliminating the daily drive to work for 6 million commuters; about 3 billion shopping trips per year; nearly 13 million business trips per year; and more than 600 million truck and airline miles per year.

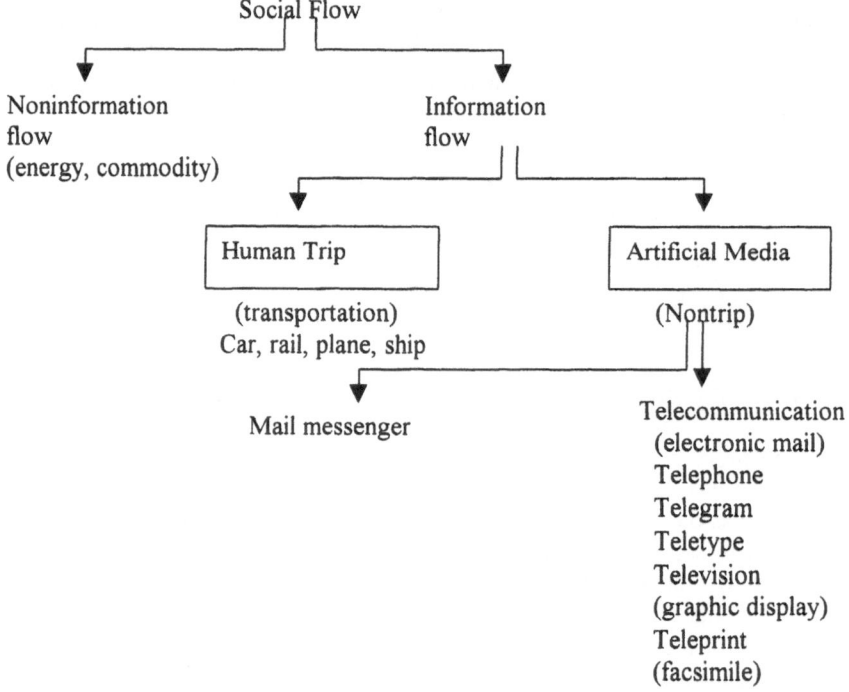

Figure 3. Relationship between Human Trip and Artificial Media with Social Flow of Information

ACHIEVING SYNERGY THROUGH INTELLIGENT TRANSPORT SYSTEMS (ITS)

Synergy implies the capacity of two or more forces, entities, goals, functions, or structures of information to optimize one another and achieve mutual enhancement (Hampden-Turner, 1981). It is a term used to describe the emergence of unexpected, surprising and interesting properties manifested in group work. It is argued that the synergy of a group leads to much greater creativity in problem solving (Flood and Carson, 1993).

Intelligent Transportation Systems (ITS) integrate advanced communications, computers, sensors, satellites, and information processing technologies into the existing transport systems. Major advances in computer, communication and electronics technology make it possible to implement ITS at a low cost. The mission of ITS is to improve the safety, efficiency, and capacity of surface transport systems through the use of information technology, consisting of inputs from such sources as highway sensors, vehicle Global Positioning Satellite (GPS) systems (Branscomb and Keller, 1996).

In addition, many ITS applications for the motorists are anticipated, such as traffic management, vehicle navigation and tracking, electronic toll collection, augmentation of driver perception, automated emergency intervention, real-time traffic and travel information, trip planning, and eventually automation of parts of the driving process.

The following elements provide a useful construct for thinking about ITS: (a) the transportation system as previously described in Figure 1; (b) the telecommunication system; and (c) the information system. All three of these systems can integrate to support ITS through a synergistic system. This system is in fact the glue binding heterogeneous networks, computers, data bases, and applications, critical for ITS's success. There are at least seven ITS applications, listed below, that can be implemented in the near future:

- Travel and Transport Management providing driver information, route guidance, traffic control, and incident management[*].
- Travel Demand Management supplying pre-trip travel information, and ride-sharing matching information[*].
- Public Transport Operations helping transit management companies with routing information, management guidance, and public security information.
- Electronic Toll payment[*].
- Commercial Vehicle Operation providing electronic vehicle clearance, automated roadside vehicle safety inspection, on-board safety and security monitoring, hazardous material incident monitoring[*].
- Emergency Management Control and Mitigation[*], and
- Advanced Vehicle Control and Safety System consisting of various vision enhancement and safety techniques.

*An asterisk denotes projects that have been experimented to some
degree in the form of pilot projects*

Currently, the United States' Department of Transportation has developed a national ITS program Plan which incorporates these seven applications. While very few evaluations of these projects have been conducted, reports indicate that they tell a compelling story about the value of such deployment. The beginnings of synergistic integration are thus already underway.

CONCLUDING REMARKS

Although the three subsystems comprising ITS are being used in a preliminary way in the last few years, the process of having a truly synergistic system in all areas of ITS has yet to happen. But this is not the problem, because in course of time when further pilot projects are implemented and systematically evaluated, policy makers will be able to augment the synergy to the three subsystems.

In addition to the integration of the three subsystems, there is yet another area that needs to be attended to. In a modern society, the provision of an efficient transport system is undoubtedly the responsibility of the government. However, this responsibility needs to be shared with the public at large. Unfortunately, at the present time, there is no formal mechanism connecting the public, the government agents, and the interests of ITS. Implementing ITS is a massive undertaking consisting of several forces and several inputs of knowledge, working together over time, to produce the behavior which we identify as "knowledge utilization". This calls for synergy, which goes beyond just coordination and integration. A variety of messages, technical literature, and practical demonstrations explaining what ITS is all about, almost to the point of "teleological redundancy" needs to be worked out for the information of the public, all focussed toward the same goal: adoption of innovation and diffusion. At the current time, interest in applying ITS to the existing transport system is at an all time high as far as the general public is concerned, and this interest rests on fertile ground for the advancement of ITS on technical, economic, sustainable, and social fronts.

REFERENCES

Brascomb, L. M., and Keller, J. H., 1996, Editors "Intelligent Transportation and the National Information Infrastructure", MIT Press, Cambridge, Mass.

Flood, R. L., and Carson, E. R., "Dealing with Complexity – An Introduction to the Theory and Application of Systems Science", 2nd Ed., Plenum Press, New York, NY.

Graham, S., and Marvin, S., 1996, "Telecommunication and the City", Routledge, London, UK.

Hampden-Turner, C., 1981, "Maps of the Mind", Macmillan, New York, NY.

Havelock, R. G., 1971, "Planning for Innovation", The University of Michigan, Ann Arbor, MI.

Khisty, C. J., and Lall, B. K., 1998, "Transportation Engineering", Prentice Hall, Upper Saddle River, NJ.

CHANGE MANAGEMENT AND INFORMATION SYSTEMS

Roger Stewart

School of information Systems
Kingston University
Penrhyn Road
Kingston upon Thames
Surrey, KT1 2EE

INTRODUCTION

The implementation of most new information systems involves more than just a replacement of the technology platforms with little change to operational procedures, but will involve new working practices and possible changes to the culture of individual organisational units. It is not just the selection of appropriate technologies and the design of systems to support the necessary business processes that will ensure the successful implementation of the new information system, but also the requirement to successfully manage the resulting 'soft' changes. For example, identifying and managing the users and stakeholders expectations of the new system, and understanding the emotional and cultural impacts linked with operational changes.

Systems thinking and practice is playing an increasing role in areas such as: designing appropriate systems; increasing user participation; building user ownership of change and the project managing of socio-technical systems. Some examples are: The Dynamic Systems Development Method (DSDM) is described as being "holistic" and "iterative" in its approach (Stapleton, 1998); the use of system dynamics in project management (Rodrigues and Bowers, 1996); the Systems Failure Method has been used in risk management (Fortune et al., 1997);

This paper describes how short duration soft systems workshops based on the Soft Systems Methodology (Checkland, 1981; Checkland and Scholes,1990) can assist in the process of change management within the areas of: information systems analysis and design; project management; cultural change.

Synergy Matters: Working with Systems in the 21st *Century*,
Edited by Castell *et al.*, Kluwer Academic / Plenum Publishers, New York, 1999.

277

WORKSHOP APPROACH

The Soft Systems Methodology (Checkland 1981, 1990) was developed in order that a complex "messy" situation, characterised by human values, emotions and different perceptions of the situation by the involved parties, could be analysed to identify organisational changes that could be adopted in order to improve the situation. An essential facet of the methodology is the close involvement of the analyst with the people in the situation. The appropriateness of this methodology in situations where information systems, and possibly complex technological platforms, are being implemented in areas rich in human activities is therefore obvious. There are, however, situations in which the dominant vehicle for change already exists, has been chosen for the future, or particular environmental constraints affect the chosen approach. This is often the case for information systems design where organisation standards, such as SSADM or DSDM, are company norms. Soft systems analysis in these cases can be a complementary activity feeding the design process, as in Multiview where conceptual models and derived information flows are used to create Data Flow Diagrams and Function Event Matrices (Avison and Wood-Harper, 1990).

Soft systems analysis, using concepts from both the 1981 (Checkland) and 1990 (Checkland and Scholes) versions of the Soft Systems Methodology, can be used to achieve a wider range of objectives other than feeding the structured functional design process. The following soft analysis approach using short duration workshops has been developed and successfully used in a variety of environments. For example, as part of a project evaluating the effectiveness of Naval on-board existing command and control technology a series of short workshops using groups of serving officers to use the SSM stages from rich pictures to conceptual model comparison. The results enabled different stakeholder perceptions of existing and desired command and control functions to be examined.

The following example shows two different types of workshops that were used in a Community Health Trust to identify and manage the cultural and operational changes involved in the installation of a new Trust-wide information system. For a more detailed explanation see Stewart (1999). The Community Health Trust provides primary and community care services in close association with general practitioners, social services, voluntary organisations and other local health and social care organisations. The services provided are diverse and include areas such as: district nursing; health visiting; chiropody; physical and learning disabilities; family planning and community child health services. Each of these areas operates as a team. The teams having different responsibilities and focus of activities had developed their own individual cultures and values whilst subscribing to the overall aim of the Trust to provide community care. The multi-team environment was therefore complex with some clashes of priorities and values. Each team had developed their own information system and therefore replacing them with a Trust-wide system was a challenging task. The new information system was a package provided by a third party supplier and significant prototyping in pilot areas was undertaken.

It was realised that in order for the project to be a success cultural change issues as well as operational changes would needed to be addressed. Involvement of the users in the process of functional piloting was not sufficient, the 'soft' human values would also need to be considered in order to build a sense of ownership of the new information system. The following two soft systems workshops were therefore designed , each of half a day duration, and consisting of representatives of a selection of user teams, for example, family planning, administration, IM &T. Figure 2 shows the Awareness workshop model, the focus of which was to gain an awareness of the problems currently faced in each of the user teams and the impact of implementing the information system in the pilot areas. This information could then be used to identify, for example, stakeholder expectations, problem drivers, cultural drivers and critical success factors.

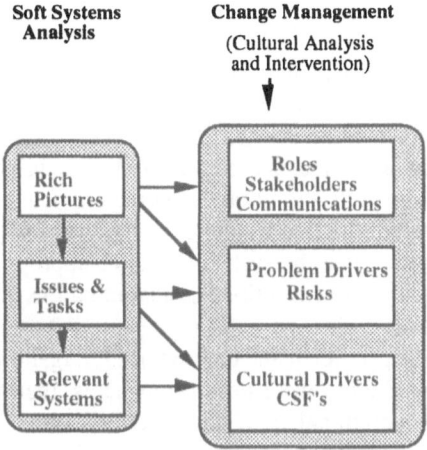

Figure 1. Awareness workshop model

Awareness workshop

An explanation of each stage was given to the workshop attendees followed by a practical session in which the user teams applied the technique. For example, the first stage was to draw a rich picture on a flip chart encompassing: how they saw their situation including their activities in the performance of their job; the problems they faced; conflicts with other jobs and people in the Trust; what they believed should happen after implementation of the new information system. Stage 2 consisted of the construction of a list of problem themes identified in the rich picture, and again to write these on a flip chart. Stage three was to name on a flip chart relevant systems that could address these problem themes. Each team then placed their flip charts on the walls and gave a verbal explanation to the group as a whole which was then discussed by the whole group. A real benefit that resulted from this approach was that team problems were made explicit and cross-team learning occurred. A subsidiary aim of the Awareness workshop was to develop a sense of ownership of the new IS system by not just identifying the requirements of change in operation but also an attitude of how can we get the best out of the system.

The findings of each stage were collected and formed the basis for subsequent analysis and used to feed into the management of change process shown in figure 2 and in the examples below.

Rich Pictures

The rich pictures gave an holistic picture of how the workshop attendees saw their situations and in the pilot sites. They provided the structural elements of peoples' jobs, their relationships to the environment, the tasks or processes they undertook and the problems that they encountered both in their existing job and in the pilot running. This technique, unlike more traditional structured analysis techniques, enabled soft information, such as human feelings, conflict between individuals and between teams to be depicted. The information gained in this way was subsequently used in role analyses, comments on the key stakeholders and their expectations being met or otherwise, the communication requirements, and problems that had been experienced.

Issues and Tasks

Each team examined their rich picture, and produced a list of tasks that they were trying to achieve and a list of issues or problems they experienced. These were then clustered into groups of related tasks and their associated issues. These clusters were subsequently examined for the information requirements and compared to the information provided by the information system. Where the tasks and information requirements, identified as a result of the soft systems analysis, were satisfied by the existing and new prototyped information system, it provided confirmation that the system would be appropriate for the users needs. Where a major task was missing it was identified as a required IS system change and fed into the project management process.

The issues reflected the operational and cultural needs. Again where the comparison to the prototyping phase showed there were no discrepancies, then there were no changes needed to the culture and operational procedures. Where issues were not addressed it indicated that cultural or operational changes were required. These clusters of issues formed the Problem Drivers (operational changes) and Cultural Drivers (attitude changes or cultural constraints such as user Norms and Values) that were needed to be addressed. These were then placed as risk areas into the project management process.

Relevant Systems

The Problem Drivers and Cultural Drivers indicated from the previous analysis were classified in two ways. Those that required simple changes were compared to the risks and associated contingency plans contained within the existing project plan. Where these were not covered they were given a name and plans constructed for the implementation of the changes. Critical Success Factors were identified to monitor the progress of the changes.

Action planning workshop

Although no major omissions were identified and only minor changes to the project plan were identified, major problems identified would have been the subject of the second type of workshop, the Action planning workshop as shown in figure 2.

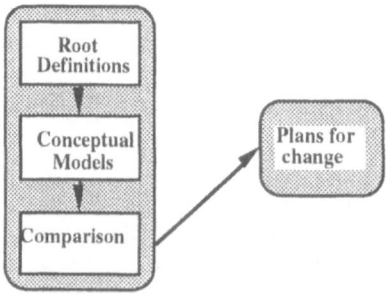

Figure 2. Action planning workshop.

The format of this type of workshop is similar to the Awareness Workshop in that an explanation of each stage is given to the workshop attendees followed by a practical session in which each attendee uses the technique. Verbal explanations are given to the group as a whole and discussed.

Major problems identified in the Awareness workshops would have been 'named' and provide the focus for Action workshops to develop an action plan. The process is as follows. For the named area of change a root definition, or statement of what the system has to do, is constructed, and conceptual models of the activities involved developed. A comparison to what happens at present reveals those changes required to put this culture or operational change into practice. The agreement of these changes and subsequent implementation can be the subject of an intervention meeting with involved parties and an action plan constructed. Similar to the Awareness Workshop these changes can be input to the risk analysis process and critical success factors identified.

In retrospect short duration soft systems workshops have worked extremely well in identifying the human, cultural and task issues associated with the implementation of technology and information systems. In addition to providing information to the planning and change management process, an unexpected by-product was the identification of more general organisational issues not directly connected to the information systems domain. In fact the workshops had worked holistically. The use of these workshops has been in specific circumstances and for specific objectives, however, more general observations about their use can be made.

ANALYSIS, DESIGN AND SYSTEMS DEVELOPMENT

Analysis and design approaches not of the traditional waterfall type, such as Rapid Application Development (RAD) and the framework approach of DSDM can also benefit from using the techniques involved in short duration soft systems workshops. Two examples are as follows.

DSDM

As discussed by Stapleton (1997), one of the fundamental principles of DSDM is that systems development is a team effort utilising the business skills of users and the technical skills of information systems people. One of the mechanisms for achieving this are Joint Application Development teams (JAD teams) who can be part of the design process from start to finish. In many cases this involves the extensive use of facilitated workshops. Through the use of timeboxing and the mix of requirements of must, should and could haves, the deliverables are achieved. Hopefully in a continuous sequence. Is quite clear that discussion, debate and conflict occur in the JAD sessions. Differences and feelings among the participants (clients, problem owners, problem solvers?) are a part of the sessions. The use of soft systems analysis workshops could help the facilitation of these workshops by providing a structure in which to examine differences and feelings.

System requirements engineering

"Requirements engineering deals with activities which attempt to understand the exact needs of the users of a software intensive system and to translate such needs into precise and unambiguous statements which will subsequently be used in the development of the system" (Loucopoulos and Karakostas 1995).

There are various definitions as what is contained in requirements engineering, a useful example of which is quoted above. What is common to many of them are the elements of understanding a problem, describing the problem and attaining agreement on the nature of the problem. All of these are problematic and will involve the perceptions of the individuals involved. We are back to the thorny question of how to deal with the human element which may or may not be rational in dealing with identifying the needs of a software intensive system. There is a body of work in requirements engineering that believes a socio-technical approach must be used in eliciting requirements and that users must be involved in the development process and development teams (Eason, 1987).

The soft systems workshops could be used as part of the development teams approach to handling the problems above.

SUMMARY

Replacing and updating information systems in an organisation involves not just a technological change but also a potential change to the culture of the organisation. This is particularly true in organisations in which teams operate as individual units providing different services but are to an extent dependant upon other teams. Analysis of the changes engendered by new ways of working is extremely important, involving the understanding of each team's culture in terms of the ways in which they operate and the values and norms that prevail.

Short duration soft systems workshops, combining in an interactive manner an analysis of the human aspects in the situation with the technological and functional analyses, will, reduce the 'soft risks', facilitate cultural change and develop a sense of shared ownership of the changes.

REFERENCES

Avison D.E., Wood-Harper, A.T., 1990, "Multiview: An Exploration in Information Systems Development", Blackwell, Oxford.

Checkland, P.B., 1981, "Systems Thinking, Systems Practice", Wiley, Chichester.

Checkland, P. B., Scholes, J., 1990, "Soft Systems Methodology in Action", Wiley, Chichester.

Eason, K., 1987, "Information technology and organisational change", Taylor and Francis, London.

Fortune, J., Peters, G., Stewart, R.W., Dodd, W., 1997, Using the systems failures method within the prince methodology to explore risk, *International Journal of Project and Business Risk*, 1,2:205-220

Loucopoulos, P., Karakostas, V., 1995, "Systems requirements engineering", McGraw-Hill, London.

Rodrigues, A., Bowers, J., 1996, The role of system dynamics in project management, *International Journal of Project Management*, 14,4:213-220.

Stapleton, J., 1998, Giving RAD a good name, *The Computer Bulletin*. 10,6:28-29.

Stapleton, J., 1997, "Dynamic Systems Development", Addison Wesley, Harlow.

Stewart, R.W, 1999, Information systems and cultural change in a multi-team environment, in: "The Gower Handbook of Teamworking", R.W. Stewart, ed., Gower, Aldershot.

SERVICE INTEGRATION, HUMAN SERVICES COLLABORATION, AND COMPUTER SUPPORTED CO-OPERATIVE WORK

Ken Udas

Fakulta Managementu
Univerzita Komenského
Odbojarov 10, P.O. Box 95
820 05 Bratislava 25
Slovak Republic

INTRODUCTION

In this paper questions are posed regarding the disparate adoption of computer supported co-operative work (CSCW) and groupware in commercial organizations and human service collaboratives (HSC). CSCW and groupware are introduced and potential business and human service applications of groupware are compared. An abstract of an interprofessional collaborative case study is presented to illustrate the need for groupware in HSC and potential benefits are identified. Factors associated with group structure and client privacy are discussed in terms of HSC's slow adoption of groupware.

There is a relatively complex web of relationships that tie together computer supported cooperative work, and the acceptance of CSCW in commercial organizations, traditional human service oriented organizations, and interprofessional human service collaboratives. The development of global markets and the desire to take advantage of newly opened business opportunities are often cited as critical factors associated with the demand for and adoption of CSCW tools (Coleman, 1997). This is particularly true among commercial enterprises that compete in saturated local markets, markets with high price elasticity of demand, are subject to rapid product and development life cycles, are facing pressure to reduce the cost of organizational overhead, and that require coordination among experts and other organizational stakeholders who are distributed organizationally and geographically. An additional application area for CSCW that has received a lot of attention is business process reengineering (Malhotra, 1993). CSCW embraces a set of activities and processes that help organizations succeed in nontraditional market conditions and also helps managers transform organizations to better take advantage of new and developing markets. While the motivation in commercial enterprise for financial investment in CSCW and the groupware that supports CSCW are relatively easy to identify, the lack of acceptance in HSC is less obviously identified, at least in terms that point toward solutions to relevant organizational problems.

Synergy Matters: Working with Systems in the 21st Century,
Edited by Castell *et al.*, Kluwer Academic / Plenum Publishers, New York, 1999.

283

CSCW AND GROUPWARE IN BUSINESS AND HSC

Goupware is the hardware and software technologies that facilitate the work of groups through providing certain functionality to the CSCW environment. CSCW can be broadly defined as a field of study or set of disciplines that investigates the development, design, adoption, and application of groupware. Because computer supported collaborative work environments are highly coupled and integrated socio-technical systems, CSCW relies on contributions from experts in numerous disciplines including computer science, organizational psychology, cognitive science, sociology, and a host of other social and applied sciences. Unlike many types of technology-based systems in which social factors can be marginalized, albeit with varying degrees of risk, groupware demands centrality of social considerations precisely because groupware applications are inherently social.

There are numerous benefits that have been associated with CSCW and groupware. The advantages that groupware promises over single user and less robust communication technologies can be categorized as facilitating higher value added functionality to existent types of supported communication processes and as enabling communication processes that would not otherwise exist. The benefits of groupware will likely fall into different categories for different organizations. Commonly cited benefits include support for shared resources, enabling telecommuting, cost savings associated with training and coordination of group work, support for business process reengineering and TQM, and the provision of new organizational services that distinguishes the organization in the marketplace. In general, the principal benefits of groupware use is enhanced communication leading to the creation, dissemination, and storage of more timely accurate, and relevant information. This presumably translates into enhanced organizational efficiency and effectiveness, better decision-making, increased customer satisfaction, and distinct advantage in a particular market.

Human service collaboratives are organizational structures that in many ways reflect the same general goals as business organizations that have adopted new communication enhancing configurations. HSC developed in response to environmental pressures, changing demands, and a perceived imperative to better meet the needs of consumers. Like their commercial counterparts, HSC are designed to enhance communication to improve information quality. HSC are diverse multi-organizational systems of social service and educational service providers that link with children, families, and communities to design and provide more effective human service products and service delivery processes. At the very heart of these complex collaborative work systems are the notions of information and resource integration and sharing to support collaborative work.

Over the past decade in the United States, momentum has been building on the federal, state, and local levels for social service integration and the formation of HSC. The movement coalesced in April 1993 when U.S. Department of Education, Secretary, Richard Riley and U.S. Department of Health and Human Services (USDHHS), Secretary, Donna Shalala wrote and co-signed the forward to *Together We Can*, published by the Office of Educational Research and Improvement (OERI) (1993). *Together We Can* prescribed service integration and interprofessional collaboration as the most promising means to social service reform. On the local level, hundreds of grassroots collaboratives have been started among service providers and clients. Because this movement is receiving attention on all levels of government, it is critical that communication not only be enhanced among service providers through horizontal integration, but also through vertical integration across levels of government. Similar reform in human service development and delivery is evident in the form of the *Partnership* movement in the United Kingdom (Hastings, 1996), and the economic transformation process in Russia and throughout Central and Eastern Europe (Mikhalev, 1996; Smith, 1996). Furthermore, the European Education Research Association has taken service integration and interprofessional collaboration as primary research themes for the past two years, providing additional evidence that enhancing communication in human service organizations is a phenomena building momentum in North America, the United Kingdom, and Continental Europe.

The term collaboration refers to an array of activities in which two or more institutions jointly engage to reach consentual common goals. Furthermore, collaboration extends beyond trivial or incidental activities to include training, policy making, resource blending, staffing, and other consequential decisions. Borrowing from Ladd (1969), members of a collaborative become each other's agents, so a level of trust, understanding, and common mission must be achieved that allows one member to speak for another member in all collaborative matters. In contrast, cooperation must reach a much lower standard of mutual trust, extending only to those joint activities that do not significantly impact major policies or practices of the cooperating institutions. This distinction between collaboration and cooperation separates human services collaboratives, as discussed in this paper, from more common cooperative arrangements in which organizations share information, make operational arrangements for the efficient exchange of resources, or share common sources of authority. Many published examples of computer supported cooperative work are best categorized as co-operative as opposed to collaborative (see Gronbaek, Kyng, & Morgenson, 1993).

In addition to a cadre of thoughtful and system aware human service professionals, the interprofessional collaboration effort requires innovative problem solvers who can bring their backgrounds in end-user computing, interface design, participatory system design, and CSCW to address the information needs of interprofessional collaboratives. The reasons why interprofessional collaboratives might be of special interest to practitioners and scholars of CSCW include the facts that human services collaboratives are receiving significant attention and funding; they strive philosophically and practically to be truly collaborative environments; and they could serve as excellent design laboratories for research in collaborative computer support systems.

For decades it has been recognized that certain factors tend to cause tension and conflict in social systems. Ineffective communication, due to information exchange, language, and information system structures, has been cited as a significant barrier to interprofessional group activities (Corrigan & Udas, 1996). These communications issues are among those that must be addressed by human service administrators who manage service integration projects and information systems professionals who design and model communication systems. In addition, they are among the issues that groupware supporting CSCW is often designed to address and are the same types of barriers that managers of commercial enterprises confront as they reengineer their organizations to meet new environmental challenges. While business entities have been investing in groupware to support CSCW, human services collaboratives have made relatively modest investments in communication technologies that support information sharing beyond e-mail. The following case study abstract illustrates a situation in which an apparently appropriate CSCW application founders without the adoption of groupware (Udas, 1996).

DIXIE CITY CHILDREN'S PARTNERSHIP BOARD CASE STUDY ABSTRACT

With a small grant from the Danforth Foundation, Jeannie Heller established the Children' Partnership Board (CPB). The CPB was the operational structure of the Children's Partnership Project whose mission was to better meet the human service needs of children, youth, and families in Dixie City. The CPB provided an appropriate forum to initiate and support enhanced collaboration among social service, health, and educational agencies in the community. The Board was composed of eight leaders from agencies judged to be most critical to the support of teen parents and their children.

During the Board's first administrator's meeting, leaders from the founding agencies renewed their commitment to participation of their staffs and scheduled five full-day meetings to take place over the next eight months. The meetings were primarily vehicles for members of the participating agencies to learn about each other's procedures, polices, services, needs, and barriers and for members to discuss potential means to establish collaborative systems among participating agencies. Among the many activities in which members engaged, they broke into small groups and listed pros and cons of collaboration. The primary cons included time investment, differing organizational priorities, competition for budgetary funding, and issues of client confidentiality. Privacy and confidentiality remained a major concern throughout the meetings. Pros included enhanced communication among agencies and reduced duplicate services to mutual clients.

By the final scheduled meeting the Board had developed and revised a mission statement, agreed on a set of goals, developed and signed letters of agreement to solidify collaboration among the members' agencies, developed a universal release form, an interagency referral form, and a universal intake form. These products represented a tremendous amount of collaborative work, as agency members had to give and take on language used in the forms, resolve issues of ownership, and come to terms with their obligations to client confidentiality and develop trust in their fellow collaborators. The use of the universal forms would require significant changes in policy and procedure affecting not only the boundary spanning roles of the organizations, but also internal administrative functions.

The universal forms and new procedures were tested extensively using case scenarios. Eighteen months after their first meeting, the system and forms had been "laboratory" tested many times and it was time to start using them in earnest. In preparation of implementing the unified delivery system and using the forms on real clients, all members of the CPB signed an agreement stating that they would uphold the confidentiality of families, and failure to do so might result in removal from the CPB. Although the practical application of the unified delivery system and use of the universal form was important for several reasons including perceived credibility of the CPB and likelihood of further funding, the universal forms were not adopted by any of the collaborating agencies. The anticipated improvement in service delivery, decreased paperwork, and decreased expenses was offset by the time and energy needed to physically move the universal forms from agency to agency. Because no shared electronic communication system and integrated database was adopted by the collaborative, the forms had to be physically transported as needed. Even though the explicitly stated goals of the CPB included the use of the universal forms and a computer networking system to best take advantage of the potential benefits of collaboration, no system was ever adopted.

WHY HAVE HSC BEEN SLOW TO ADOPT GROUPWARE?

If the Dixie City case is typical, and I would suggest that it is given the lack of reported groupware applications in HSC, why have social service agencies and HSC been slow to adopt collaborative technologies? The support and services commonly cited as qualities of CSCW and groupware are consistent with the communication and collaboration needs of HSC. As with the Dixie City CPB, participants in a HSC could potentially benefit from the use of groupware in at least three distinct ways. Most obviously, a HSC could benefit on an operational level from the enhanced communication groupware can provide. HSC often need to share documents, make group decisions, communicate synchronously and asynchronously, integrate the expertise of individuals who are geographically distributed, and support individuals who functionally telecommute because of their non-traditional work arrangements and time spent in the field. These needs have been reported in a number of case studies and articles (see Lawson and Lawson-Briar, 1997). A second area of potential benefit that groupware could offer to HSC is providing an environment for the actual design and development of the HSC. Because the design of a collaborative organization requires input from all participants, it is often difficult to arrange meetings when all members can meet. Large amounts of information about each other's organizations must be shared and collaborative products must be dynamically updated to reflect new documents and artifacts being created. Additionally, the types of information that must be communicated may be uncomfortable to express and may require anonymity to be brought into the open for consideration. Groupware can provide these types of functionality. Finally, groupware can be used as a tool to facilitate process reengineering in organizations. Internal process change in agencies participating in a HSC is common. Groupware could be used to connect internal process changes, design of the HSC, and operational collaboration of the HSC to help ensure consistency and leverage the likelihood of successful implementation.

There are differences between the ways HSC and commercial organizations would typically use groupware to support CSCW. Two factors that might have bearing on the disparate adoption of groupware are the configuration of technology users and the involvement and nature of the customer or client of the organization. These differences are projected on a background of inherent tension associated with the tradeoff between the efficiency of information use and the market value of information. Perfect information is free, complete, instantaneous, and universally available. Its exchange is encumbered by no barriers in a perfect market. On the other hand, it is also recognized that failure to restrict the flow of information across some boundaries can result in the loss of market value. An additional tension that arises relates to the ethical duties relating to privacy various professionals have to their clientele.

CSCW in commercial organizations tends to be limited to internal applications. Although commercial competitors may engage in some co-operative activities, especially in the form of joint marketing of complementary products, collaboration as defined above falls somewhat outside of the structure and belief system that defines the macroeconomic concept of a free, open, and competitive market. In fact, many collaborative activities may transgress the bounds of anti-trust law. At the heart of HSC are the transboundary relationships that fall outside of typical groupware applications among commercial participants. Although individual member agencies of HSC generally do not compete for customers, they often do compete for budgetary resources that create a competitive atmosphere if not a competitive market. Such budgetary competition has been recognized as a collaborative barrier in Udas (1996) and Corrigan and Udas (1996) and has been addressed in terms of blended funding schemes. While commercial enterprises have tended to limit their CSCW to internal groupware applications, HSC have made efforts to extend their boundaries to include several agencies. Such reconfiguration of agency boundaries poses unique challenges to collaborative work (Udas 1997).

Specific and personal detailed information about the customer or client tends to be the primary element around which HSC processes are organized, while products, projects, and markets tend to be more central to commercial groupware applications. This distinction strikes at another feature of HSC that might slow or limit adoption of CSCW. There are strong codes of professional conduct that spell out the responsibility of healthcare, social work, justice, and education professionals to protect the privacy rights of clients. This factor has been cited as a major concern of human service practitioners in general (Gardner, 1992; Lawson & Briar-Lawson, 1997; Greenberg & Levy, 1992) and was presented as a major and persistent barrier in the Dixie City CPB case. The special role of privacy is fundamental to the relationships that clients and human services professionals share and much effort is expended in making sure that private information does not enter the public arena and because subject to commodification. The legal and professional impetus to restrict the free flow of client information in HSC will influence the use of groupware the use of integrated and shared databases. The inherent reduction of control of information flow associated with CSCW will likely continue having a cooling effect on the adoption of groupware in HSC.

CONCLUDING THOUGHTS

Although HSC share numerous communication needs with business organizations, groupware technologies have been adopted at a lower rate and to lesser extent in HSC than their commercial counterparts. The interorganizational focus of HSC coupled with the extremely high priority many professionals place on client privacy demands more than technological solutions to address concerns of HSC participants. Adoption of groupware in HSC will require the collaborative effort of thoughtful human services professionals, mangers, and information systems professionals who understand the unique mission, structure, cultural context of HSC.

REFERENCES

Bruner, C. "Thinking collaboratively: Ten questions and answers to help policy makers improve children's services," Education and Human Services Consortium, Washington, DC. (1991).

Coleman, D. "Groupware: Collaborative Strategies for Corporate LANs and Intranets," Prentice Hall, New York (1997).

Corrigan, D. C. & Udas, K. Chapter 41, Creating collaborative, child and family centered, education, health, and human services systems, in: "Handbook of research on teacher education," J. Sikula, T. Buttery, & E. Guyton eds., Macmillian Publishing, NY (1996).

Gardner, S. L. Key issues in developing school-linked, integrated services. *The future of children*, 2:85 (1992).

Greenberg, C. W., & Levy, J. "Confidentiality and collaboration: Information sharing in interagency efforts" Education Commission of the States, Denver (1992).

Gronbaek, K., Kyng, M., & Morgenson, P. CSCW challenges: Cooperative design in engineering projects. *Communications of the ACM*, 36:67 (1993).

Hastings, A. Unraveling the Process of 'Partnership' in Urban Regeneration Policy. *Urban Studies* 33:253 (1996).

Ladd, E. "Sources of tension in school-university collaboration," Emory University, Urban Laboratory in Education, Atlanta (1969).

Lawson, H., & Briar-Lawson, K. "Connecting the Dots: Progress Toward the Integration of School Reform, School-Linked Services, Parent Involvement and Community Schools," Oxford, OH, The Danforth Foundation, and The Center for Educational Renewal at Miami University, (1997).

Malhotra, Y. (1993). Role of Information Technology in Managing Organizational Change and Organizational Interdependence [WWW document]. URL http://www.brint.com/papers/change/ [1998, October]

Mikhalev, V. Social Security in Russia Under Economic Transformation. *Europe-Asia Studies* 48:5 (1996).

OERI. (1993). "Together we can: A guide for crafting a profamily system of education and human services," Author, Washington, DC. (1993).

Quinn, J. B., Anderson, P., and Finkelstein, S. (1996). Leveraging intellect. *Academy of Management Executive*, [Online] Vol.10 No. 3. Available: http://ame.bschool.unc.edu/ [1998, August]

Udas, K. Chapter 37, Interprofessional, Participatory Design of Management Information Systems, In: "Expanding Partnerships" K. Hooper-Briar, and H. A. Lawson eds., CSWE Press, Alexandria, VA (1996).

Udas, K. Lewin's conflict in marriage revisited and expanded: Implications for interprofessional collaboratives. *Systems Practice*, 10:509 (1997).

A SYSTEMIC ANALYSIS OF PROCESS ORIENTED CHANGE

Jan T.Wilton and Lawrence R.P.Reavill

Department of Management Systems and Information
City University Business School,
London, EC1V 0HP

INTRODUCTION

Large Multinational Corporations (LMCs) are increasingly operating on a global scale to deal with the pressures of change. The growth of new markets, the development of new technologies, fierce competition, as well as the increasingly demanding customer have all fuelled these changes. To remain competitive, LMCs are having to rapidly improve their quality, speed and service whilst reducing their costs, at ever faster rates. Consequently, LMCs have had to become leaner and more flexible and have employed process oriented change programs such as quality management and process re-engineering to achieve this.

Earlier papers by the authors have shown how change programs ultimately move beyond the organisation system boundary to include the organisation's suppliers. The authors, and others, argue that this expanded network constitutes a supply system, rather than the traditional supply chain, and the greater complexity requires a system of alliances and networks that is constantly adapting and evolving. The level of complexity necessitates an approach that can deal with the performance improvements and issues involved.

This paper gives details of two case studies of LMCs and their suppliers, one in the automotive industry and the other in the telecommunications industry. Both are experiencing major changes in the relationships with their supplier organisations, many of which are smaller organisations, Small/Medium Enterprises (SMEs). The paper shows that the two cases have significant similarities, particularly in the nature of the supply system, its complexity and dynamics. However, there are major differences relating to the nature of the industrial sectors involved, and particularly to the maturity of the product, both in technological and market terms.

THE BUSINESS AND COMMERCIAL HURRICANE

Any brief foray into the business media and literature will suggest that organisations today are experiencing an unprecedented degree of turbulence in the form of change. The increasing globalisation of markets and fierce competition; the advances in information and

Synergy Matters: Working with Systems in the 21st Century,
Edited by Castell *et al.*, Kluwer Academic / Plenum Publishers, New York, 1999.

289

communications technology; and the changes in customer expectations are often cited as core catalysts to change, although other contributions clearly add to this list. Whilst recognising that the phenomenon of accelerating change is itself not new, in either society or business, the speed with which it is prevalent today has been argued to be unique (Ackoff 1981). What transpires from the research is that this ongoing turbulence is increasingly challenging the fundamental structures, models and philosophies of business today.

To succeed, LMCs face the equivalent of 'giants learning to dance' as Moss Kanter (1989) proffers. They need to offer continually improved and innovative products and services; exploit the advances in information and communications technology, and other technologies; offer lower costs with shorter lead and delivery times; to ever more demanding customers in an increasingly globalised market. To adapt to these pressures LMCs from almost every industry, from manufacturing to services, all have similar inter-linked and complementary objectives; to continuously improve quality; to achieve leaner logistics; to lower costs; and to provide better value and faster innovative solutions.

To achieve these levels of competitive advantage or mere parity, LMCs have been forced into reviewing their operations and often adopting process oriented improvement initiatives. As such, there has been considerable trade, media and academic literature written about process improvement practices such as quality management (QM) and business process re-engineering (BPR) which have sought to tackle these issues. Yet, it is no longer sufficient to merely focus upon these practices within the boundaries of LMCs alone, these issues and practices should be considered in terms of how they can effectively be carried out across the traditional organisational boundaries to involve such stakeholders as their suppliers. As the complexity of operating in this turbulent environment continues, the importance of interacting with all these stakeholders becomes strategically critical.

FROM SUPPLY CHAIN TO SUPPLY SYSTEM

The reaction of many leading LMCs to these competitive pressures has been to focus on their core competencies, to the point that they are increasingly in a process of vertical disintegration. They are adding less value to the final product or service and instead are relying on their suppliers to provide it for them, whilst they focus on tasks such as the final assembly, marketing, distribution and protection of their intellectual property.

As this form of interaction develops and continues, the need for client LMCs to work effectively with value adding, flexible suppliers becomes ever more critical. One of the most popular models that depicts the more traditional interrelation between organisations is that of Porter's (1985) Value Chain. Yet as the complexity of the age comes under greater scrutiny and the true patterns of operation emerge there is growing dissatisfaction with this model. Alternative approaches are being presented on an increasingly frequent basis (Clewer, 1995; Normann and Ramírez, 1994; Schön in Normann and Ramírez, 1994). In its place, it is argued, the value chain concept is more accurately replaced with what is termed a supply system, whereby there is interaction between all stakeholders through a network based system involving both feed-forward and feedback which includes other parties such as the education system and the science base.

A SYSTEMS VIEW OF PROCESS IMPROVEMENT

The paper argues that if LMCs and their suppliers are to operate within a supply system, rather than a supply chain, then a more systemic evaluation and perspective of their process improvement programs is also required. This is not to say that the concept of holistic, systemic thinking itself is not new or alien to the field of process oriented improvement within organisations. Within the QM literature, W. Edwards Deming was a firm advocate of taking a more systemic outlook (Deming, 1993), exemplified through his model, 'Production viewed as a System', referred to as "the spark that in 1950 and onward turned Japan around"(Deming, 1993). The need for a more holistic approach and its links to systems thinking are also evident in the BPR literature (Harvey, 1994). However, the paper argues, the degree to which this approach has been applied in the context of operating within a supply system is limited and consequently justifies further research and analysis.

Critical Systems Theory (CST) and Total Systems Intervention (TSI)

In adopting a systems perspective of process improvements, it is proposed that Critical Systems Theory be used as a basis for the analysis. It will be conducted through what has been termed, its practical face - Total Systems Intervention (TSI). It is proposed that by carrying out such an analysis a more systemic understanding of the methodologies might be ascertained whilst highlighting areas of possible development with regard to process improvements in the supply system.

Following its original publication (Flood & Jackson 1991), the TSI meta-methodology has been put into practice, evaluated, critiqued, and further developed (Flood, 1995). The outcome has resulted in TSI developing into a tripartite process encompassing three modes: the critical review mode; the problem solving mode; and the critical reflection mode. Each of these modes now in turn makes use of the three phases of creativity, choice and implementation in their operations in a recursive manner.

Still at its core, TSI is concerned with thinking about organisations as whole entities. In achieving such an enriched view of organisational dynamics, Flood (1995) proposes four key dimensions that need to be considered. The first is to have a clear understanding of the organisation's processes. This is augmented by understanding the organisation's design through which the processes run. In addition there is the notion of the corporate culture and the need for it to be understood. The final consideration concerns the political dynamics involved, in terms of power and influence over how events occur.

TSI in its revised form presents itself as a problem solving system that takes these four dimensions into account. Flood (1995) argues that TSI achieves this by developing a system of methods that is able to deal with the problems arising from each of the four dimensions. The appropriateness and evaluation of methodologies available for use in the problem solving mode is carried out by the critical review mode within the TSI process of operation, as presented by Flood (1995) and Flood and Jackson (1991).

A SYSTEMIC ANALYSIS OF PROCESS IMPROVEMENT PRACTICES

A recent paper, by one of the authors (Wilton, 1998), gives an outline of both QM practices and BPR along with a review and comparison of their principles, developments, theoretical origins and proposed methodologies. It is evident that one systemic analysis of quality management practices (Flood 1993) has already been carried out, although this is not to say that it is beyond any improvements itself. It is nevertheless proposed that, due to restrictions on space, the paper should consider the more recent developments in discontinuous improvement practices such as BPR with relation to the supply system. In undertaking such an analysis, it is evident that some systemic analysis of BPR has also been carried out by Flood (1995) using the Critical Review Mode. However, it is argued that it only considers a limited and internally focused evaluation of re-engineering and not the wider organisational view that would be more applicable to the supply system as a whole.

Based on this review, in addition to further analysis of the extant literature, a number of conclusions can be presented regarding the current position of BPR practices within the supply system. The first highlights the disparity of BPR efforts that have been initiated, which range from the improvement of a single low level inter-departmental process right up to the radical redesign of a whole organisation. This suggests that re-engineering can either be carried out at either an operational or strategic level. As Talwar (1994) points out, the risks, scope and rewards involved within this range vary considerably. Therefore if a LMC decides to initiate a BPR programme, it must consider not only the appropriate scope but also the potential impact it will have vis-à-vis their suppliers and the subsequent effects it will have upon the final product or service.

Secondly, it is argued that the theoretical basis on which such practices have developed has been pieced together in a retrospective manner. It is therefore of a pragmatic nature which has led to many different approaches being proposed, often resulting in a lack of understanding and subsequent failures. As a methodological approach and ideology it does tackle both the technical and practical elements and originally viewed certain emancipatory aspects as achievable in the sense that an individual would become more responsible for their actions. In practice this has turned out to be an area of particular weakness with regard to the approach since it was adopted in a very mechanical manner. In early attempts this could be attributed to both the over-reliance on I.T. as well as the cost cutting agendas that many organisations had in the early years of its promulgation.

A third conclusion from the study is that many of the concepts of re-engineering have focused almost exclusively on the development of the internal design and processes that achieve greater competitiveness. Yet as previously highlighted, the changing organisational structure has placed increasing emphasis on the interactive relationships required with suppliers. Additionally, if considerable improvements are to be achieved then all suppliers both large and small must start to adopt these practices.

A fourth conclusion relates to which dimension re-engineering falls into regarding TSI. Flood (1995) suggests that BPR is merely an operational methodology that falls neatly into the process dimension of the organisation. Yet an organisation undertaking a BPR initiative including the supply system would be more strategically positioned, encompassing all aspects of the organisation, and clearly provoking the changes in the other dimensions.

The fifth conclusion from the analysis is to consider how the re-engineering fits the four basic principles of TSI. As much as re-engineering claims to be holistic by taking an organisation perspective, it is still evident that many initiatives are still relatively narrow in their scope as depicted in Talwar (1994). In contrast, TQM has always actively sought to involve customers and suppliers, yet the proponents of re-engineering have rarely advocated such participation. However, it is argued that their involvement is of even greater importance in BPR initiatives, since the growth of any significant disparities between performance levels would be detrimental to the product/service being offered by the LMC.

With regard to the other principles of TSI, of being participative, reflective and human oriented, BPR has had a mixed review. It certainly espoused certain aspects of these principles, yet it would be difficult to argue that they were particularly strong aspects of the approach. The BPR approaches have certainly encouraged bottom-up initiatives, but invariably for them to be successful, such changes have required the sponsorship from senior management. Most of the early BPR models also failed to consider the reflective nature of change, often promoting a very linear approach. This is in contrast with quality management principles which proposed, from an early stage, the need to reflect on actions taken. Finally, in analysing BPR's human orientation it can be argued that this factor has been increasingly addressed in recent developments. From being a 'fad that forgot people', BPR and its core principles of process orientation and dramatic improvements has developed into an approach that increasingly concentrates on the inter-relationships between inter- and intra-organisational teams and the knowledge they create as a source of its success.

The final conclusion reviews how this analysis has contributed to a critique of the BPR approach itself as well as to the TSI methodology. It is evident that, even through this briefly presented analysis, the Critical Review Mode has highlighted a number of issues that need to be tackled if adopting a BPR approach within the supply system. Notably, it highlights the need to identify the scope of the BPR approach to be adopted with relation to the effects and efforts needed to ensure that the suppliers can adapt appropriately. Regarding the development of the TSI approach itself, the analysis highlights a number of areas of improvement. The first regards TSI's view of quality management and BPR. It is proposed that these approaches are merely methodologies to adopt when the organisation requires improvements in its operational and strategic processes. Yet it is proposed that this only takes an operational view of what has arguably been adopted by organisations at a strategic level. This implies that TSI fails to clearly differentiate between solving strategic or operational problems, despite its claims to be an all encompassing approach.

The second area of improvement questions the view that an organisation can be viewed holistically under the four dimensions as presented earlier. Whilst these dimensions do provide a good framework to begin with, there are clearly other dimensions that could and should be included. Notably in the case of supplier development with the supply system the dimension of resources must be considered, encompassing such factors as the financial, technical and personnel availability to them in order to facilitate the changes required.

The third area of possible development revolves around the issue of TSI's effectiveness as a systemic problem solving approach within an inter-organisational setting such as the supply system. To date TSI appears to focus upon internal organisational problem solving rather than between organisations and there appears to be little to suggest that work in this field is being carried out, and yet as the business environment increases in its complexity, such developments appear to be appropriate. This concludes the critique of both the current process improvement and the TSI methodology itself.

TWO CASE STUDY LMCs AND THEIR SUPPLY SYSTEMS

The proposals made in this paper, of LMCs increasingly operating within supply systems and having to adopt appropriate process improvement strategies, are further strengthened through some in-depth case study research that one of the authors has conducted within two LMCs. These are the Ford Motor Company and Ericsson Mobile Telecommunications, and a selection of their medium sized suppliers within their respective supply systems.

In terms of industry characteristics, the two sectors have some interesting comparisons. The automotive sector is characterised as being an established and relatively mature industry with medium length product life cycles. Its markets are highly competitive and, in the case of Europe, fairly saturated. Whilst remaining competitive, the industry has had to dramatically change its traditional structure and means of operation, through rationalising its supply base and restructuring. In contrast, the mobile telecommunications sector is characterised as being a relatively new market. Its established market regions are comparatively new themselves and are still experiencing dramatic growth, whilst on a global scale it has been exponential. The life cycle of a mobile phone has been expressed as a division of halves. Each year the product life cycle for mobile phones halves, as does the price, as does the size of the components needed to carry out the same tasks. To remain competitive, therefore, has required considerable research and development as well as volume production to satisfy the market's demands. The question of rising demand in mobile phones versus the saturation of the automotive market and the strategic question that faces the mobile phone manufacturers can be seen in Figure 1. It is evident from the research that this is a strategic question that has yet to be resolved. Despite the differences in the industries, there are a number of common themes. Both are increasingly operating on a truly global basis and are under pressure to produce quality goods, in volume, at increasingly faster rates, yet at consistently lower cost. They have also had to adapt their organisational structures to suit these market pressures as they intensify, one through consolidation and one through growth in their respective markets.

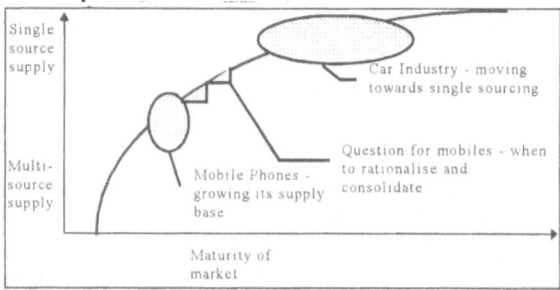

Figure 1: Supply Base vs. Market Maturity

And finally, both LMCs have increasingly relied on outsourcing contracts with suppliers to provide them with component and service systems. In order to achieve these goals, both LMCs have had to invest considerable resources in selecting, developing and working with their suppliers, of whatever size, in order to obtain the appropriate standards of quality, cost, speed and service. The evidence also suggests that the client and suppliers are concentrating on inter-organisational process improvements that focus on the interactive aspects between the organisations, rather than merely their own internal improvements.

Whilst the paper length restricts the degree of detail and findings presented, it concludes that as LMCs increasingly co-ordinate their supply systems, rather than chains, the process improvement methods being applied need to be more systemic and in line with the complex and dynamic inter-organisational interactions that take place.

REFERENCES

Ackoff, R.L., 1981, Creating the Corporate Future, New York: Wiley & Sons.

Clewer, G.R., 1995, Understanding supply- chains and forces or feedback and systems? Time for a rethink, *Proceedings From The European Conference On The Management Of Technology*: 93-100.

Deming, W.E., 1993, The New Economics For Industry, Government, Education, Cambridge MA: MIT

Flood, R.L., 1995, Solving Problem Solving , Chichester: Wiley & Sons.

Flood, R.L., 1993, Beyond TQM , Chichester: Wiley & Sons.

Flood, R.L., Jackson, M. C., 1991, Creative Problem Solving - Total Systems Intervention, Chichester: Wiley & Sons.

Harvey, D., 1994, Re-Engineering: The Critical Success Factors, London: Business Intelligence

Moss Kanter, R., 1989, When Giants Learn To Dance, London: Routledge.

Normann R., Ramírez, R., 1994, Designing Interactive Strategy, Chichester: Wiley & Sons

Porter, M.E., 1985, Competitive Advantage, New York: The Free Press.

Talwar, R., 1994, Re-engineering - A wonder drug for the 90s?, *in*: Business Process Re-Engineering: Myth And Reality, By C. Coulson-Thomas, London: Kogan Page, pp. 40-59.

Wilton, J.T., 1998, Process oriented change within the industrial supply system - a systemic evaluation of current practices, *Proceeding of the ISSS98 conference*.

CONVIVIAL INFORMATION SYSTEMS

Ian Beeson

School of Information Systems
Faculty of Computer Studies and Mathematics
University of the West of England
Coldharbour Lane
Bristol BS16 1QY

INTRODUCTION

One theme in the recent information systems literature is the emergence of 'emancipatory' approaches to information systems development. Hirschheim, Klein, and Lyytinen, for instance, name emancipatory approaches as the seventh and latest 'generation' of information systems development methodologies, and see them as deriving mainly from Habermas's work on communicative action (Hirschheim et al., 1995). The general aim of emancipatory approaches is to free users of information systems to use them creatively for their own purposes. Another author whose work can help us develop ideas on emancipatory information systems is Ivan Illich. In *Tools for Conviviality* (Illich, 1990), Illich sets out an analysis of the relationship between technology and society which shows technology being pushed beyond humanly acceptable limits by over-powerful institutions. He proposes a 'convivial reconstruction' to tackle an accelerating crisis:

> 'This crisis can be solved only if we learn to invert the present deep structure of tools; if we give people tools that guarantee their right to work with high, independent efficiency, thus simultaneously eliminating the need for either slaves or masters and enhancing each person's range of freedom.'

(Illich, 1990, p.10)

In contrast with Habermas, Illich is primarily concerned not with the nature of the users or their communication, but with the structure of the tools and technology available to them. Illich envisages a future, convivial, society in which modern (but no longer 'industrial') technologies will serve politically interrelated individuals rather than managers. He uses 'convivial' as a technical term to designate 'a modern society of responsibly limited tools'; and further defines the term to mean a disciplined and creative playfulness which will exclude enjoyments which are 'distracting from or destructive of personal relatedness' (Illich, 1990, p.xii).

Synergy Matters: Working with Systems in the 21st Century,
Edited by Castell *et al.*, Kluwer Academic / Plenum Publishers, New York, 1999.

295

This paper examines some aspects of Illich's analysis and relates his view of the dynamics of technological development to Winner's discussion of 'autonomous technology'. Following a presentation of Illich's ideas about how a convivial reconstruction might be achieved, attention is given to the existential situation of the citizen confronted by an incomprehensible technology, and a critique of Illich by Borgmann is used to supply a deficit in Illich's proposals. The combination of ideas from Illich, Winner, and Borgmann is finally applied to make some suggestions about what convivial information systems should be like.

THE SECOND WATERSHED AND REVERSE ADAPTATION

Illich's analysis includes a characterization of major industrial institutions and social agencies as passing through two watersheds, the first with positive and the second with negative consequences. At the first watershed, the institution or agency becomes firmly established in society by applying new knowledge to the solution of a well defined problem and proceeding to account for the consequent increase in efficiency with an acceptable scientific measure. Further development of the institution and its methods, however, lead it inexorably toward a second watershed, at which point the previously demonstrated progress itself becomes a rationale for the ongoing exploitation of society in the service of values determined and revised by a self-certifying professional élite (Illich, 1990, ch.I). Illich has famously analyzed institutionalized education and medicine, as well as other areas such as transportation, in these terms. It is his contention that once institutions pass the second watershed, they are driven by values of self-perpetuation and expansion to an ever increasing production of the goods and services that they need society to consume, regardless of whether there is independent evidence of continuing need. This trend in modern society destroys conviviality by a relentless process of commodification and homogenization. Institutions become locked into their products: the commodity 'education' needs the institution 'school' to produce it, and the institution in turn depends for its existence on demand for the commodity, which it must therefore stimulate.

Langdon Winner's analysis of 'reverse adaptation' has points of similarity with Illich's notion of the second watershed. In *Autonomous Technology* (Winner, 1977), he describes how large-scale technological systems become 'self-generating, self-perpetuating, self-programming' mechanisms which achieve an autonomous existence within society, out of political control. By reverse adaptation, he is referring to an inversion of the normal relationship between means and ends: instead of a system's or institution's means being defined to achieve its ends, ends are adapted to suit the means available (Winner, 1977, p.238). His basic hypothesis is that:

> '..beyond a certain level of technological development, the rule of freely articulated, strongly asserted purposes is a luxury that can no longer be permitted.'
> (Winner, 1977, p.238. Original italicized.)

Winner sketches out ways in which systems make their ends serve their means: by achieving control of markets, by controlling their own regulators, by defining a mission which matches their technological capabilities, by propagating the needs they serve through promotion and advertising, or by discovering or creating a crisis which justifies their further expansion.

CONVIVIAL RECONSTRUCTION

Convivial reconstruction depends for Illich on an 'inversion' of institutions, so that they become:

> '...institutions that would foster the use of individually accessible tools to support the meaningful and responsible deeds of fully awake people.'
> (Illich, 1990, p.16)

Illich does not expect reform of the institutions to come from their managers, but hopes it might come instead bottom-up, from a reconstruction of the tools. Convivial tools, which he thinks will at least provide new options, have the following character:

> 'Tools foster conviviality to the extent to which they can be easily used, by anybody, as often or as seldom as desired, for the accomplishment of a purpose chosen by the user. The use of such tools by one person does not restrain another from using them equally. They do not require previous certification of the user. Their existence does not impose any obligation to use them. They allow the user to express his meaning in action.'
>
> (Illich, 1990, p.22)

Illich gives as examples of convivial tools the telephone, most hand tools, the postal service, and Mexican markets. Conviviality is not tied to the level of technology of a tool, but to the conditions of its use.

Illich does not expect nor demand that all centralized production or large-scale tools would have to be excluded from a convivial society, but talks rather of achieving an acceptable balance between 'manipulative', planned tools - which can deliver efficiencies, but at the cost of abstraction - and a set of 'complementary, enabling tools which foster self-realization' (Illich, 1990, p.24).

Convivial reconstruction will draw, according to Illich, on two forms of 'counterfoil research' (normal research is caught up in the 'growth mania' of industrial development): one will provide guidelines for detecting 'the incipient stages of murderous logic in a tool'; while the other will discover 'systems of institutional structure which optimize convivial production' (Illich, 1990, ch.III). He provides an outline of five key areas where the efficiency of tools can become destructive, and where convivial reconstruction must therefore strive to keep a balance: environmental exploitation, radical monopolies of a class of product (eg, cars over other forms of transport), 'overprogramming' (where training and certification drive out learning), social polarization, and engineered obsolescence. At the least, Illich's analysis here gives a useful checklist for builders of tools.

Winner is less sanguine than Illich about finding a remedy for the problems of technological development. The victory of technics over politics that he describes is not, he asserts, an ideological one. He believes that putting technology and tools under workers' control in a Marxist solution would produce substantially the same results observed under capitalism. He asks whether it could be possible to achieve the high levels of productivity associated with modern technological systems and at the same time establish a communal life in which division of labour, social hierarchy and political domination were eradicated: his answer is a firm 'no' (Winner, 1977, p.275).

'BETWEEN NARROWNESS AND BEDAZZLEMENT'

Winner goes on to consider how well people in a technological society understand their artificial environment (Winner, 1977, Ch.7). He believes that 'the gap between the complex phenomena that are part of our everyday experience and our ability to make such phenomena intelligible and coherent' is growing; that despite the wealth of information available in the modern world, relative ignorance is increasing. The complexity of modern society is at the root of this problem: knowledge is highly fragmented and specialized, different specialists trade on one another's ignorance, and it is impossible for anyone to learn enough to make the world comprehensible. People are unable to give an adequate account of man-made phenomena even when they form part of their own direct experience. On top of and exacerbating a *manifest social complexity*, Winner maintains, there has in recent times been added a *concealed electronic complexity*:

'Relationships and connections once part of mundane experience (in the sense that some person had to attend to them at each step) are now transferred to the instrument. The unintelligible mass of sociotechnical interconnections is enshrouded in abstraction. ... The most important of the technical systems of the modern age are structurally complex and spatially extended and make the handling of daily business through face-to-face relations increasingly irrelevant.'

(Winner, 1997, p.285)

The situation confronting modern citizens in a technological society is for Winner worse than that faced by the newborn child trying to make sense of its surroundings: at least the child comes gradually to understand its environment, while the citizens, swamped by complex and extraordinary phenomena, 'never escape a fundamental bewilderment'. People are caught between 'the narrowness of their everyday concerns and a bedazzlement at the works of civilization' (Winner, 1977, p. 296).

To the objection that we can cope with complex technology without understanding it, Winner replies that this can only produce a docile, uncreative utilization. Nor does he see any hope that some overarching account will shortly be available to make sense of all the complexity at a higher level. He agrees with Valéry's observation that our means of investigation and action have outstripped our means of representation and understanding. No tools of intellectual synthesis are available: systems thinking, for example, has become just another specialism - or, worse, a technique in the service of the 'mega-machine' (Winner, 1977, p.289).

In the absence of an explanatory theory of the complexity of modern society, let alone any assurance that the incremental actions of large technical systems will somehow combine to produce harmonious social relationships, how can people find their way about the extended technological networks they now find themselves living in? Winner only has a tentative answer to this question: he suggests that intellectuals, looking for explanatory models, move from one partial theory or tool of inquiry to another, never finding a permanent answer; while ordinary citizens, reliant on the mass media for information, expect occasional items of real news to be signalled within the stream of entertainment. What Winner does point to as a consequence of life within technological networks is a loss of moral agency: on the one hand, the possibility diminishes of directing technological systems towards chosen and shared goals; while on the other, the very architecture of highly complex systems undermines any sense of moral responsibility and instead constitutes 'vast webs of extenuating circumstances' which can be used to avoid blame or deny responsibility (Winner, 1977, p.303).

FOCAL PRACTICES

Albert Borgmann finds Illich's definition of convivial tools too narrow. What is essentially lacking in Illich's formulation is an appreciation of the alienation Winner identifies:

'...we may lead the disengaged and distracted life that is typical of advanced technology in the midst of conviviality as defined by Illich. The electronic and video marvels that we are being promised meet his definition of conviviality. Illich tries to secure the good life by establishing boundaries that would keep the dehumanizing technology outside and allow the good life to flourish within. But such limits are always drawn too narrowly and too broadly at once.'

(Borgmann, 1984, pp.167-8)

This does not refute Illich's analysis, but makes a point that Winner might make, that the accessibility of tools does not prevent vapidity of use or content. Borgmann contends that technology becomes convivial not by putting a boundary round it, but by relating it to a 'centre'.

Borgmann goes on to develop a proposal for reform of technology through the establishment of 'focal practices' (and focal things). In the context of technology, Borgmann means by a focal practice an orienting activity which clarifies our relation to the technology. In order to comprehend and begin to master technology, rather than be consumed, distracted, or bewildered by it, we must engage with it. In order to engage with technology, one must be able to stand apart from it so as to be in a position to value and evaluate it in relation to some other personally significant action. Borgmann gives examples of runners and musicians, who find a centre or focus to their life in a particular activity or practice and who through that gain a distance from and a sensitivity to the technologies they may use in that practice. He suggests that a focal commitment leads to an intelligent limitation of technology (Borgmann, 1984, p.222). For instance, if we engage in running or music ourselves, we become more sensitive to televised performances:

> 'Given the counterweight of an engaging practice, televised performances need no brutality, carnal danger, promises of new records, or the spice of financial rewards for which the performers are made to fight.'
> (Borgmann, 1984, p.284)

More generally, he concludes:

> '...engagement provides resonance for those commodities that represent and support excellence, and finding no echo in the trivial and frivolous, it ignores banal commodities and helps to reduce them.'
> (Borgmann, 1984, p.284)

CONVIVIAL INFORMATION SYSTEMS

From the analysis above, what can we now say about how convivial information systems might be built? We start from an observation that as far as information technology goes, it has been moving in the right direction for convivial use, in that the accessibility and usability both of computers and of networks has been growing steadily in recent years. IT has spread beyond formal organizational contexts into society at large. Despite the best efforts of computer departments and professional bodies, end users have increasingly been able to build, acquire or adapt their own systems, and computing has remained a relatively open (uncertified) work arena. The Internet is encouragingly free of centralized control. From an Illichian perspective, if we are interested in creating convivial information systems, we should work to keep control of the underlying technology diffuse, decentralized, and unprofessionalized. We are helped in this by the pace of change in IT, which looks set to prevent it, as a separate technology or specialism, reaching Illich's second watershed.

At the level of information systems development, we should set out along the emancipatory path, but our efforts should be devoted not so much to the establishment of free channels of communication (through, say, a computer-supported cooperative work initiative), as to the fostering of diversity and the provision of a plurality of tools, in the expectation that 'A pluralism of limited tools and of convivial commonweals would of necessity encourage a diversity of lifestyles' (Illich, 1990, p.15). We will be practically limited by the nature of our engagement in the development process, but should strive where possible to use available technologies to reflect the richness of users' experiences and to permit the articulation of different voices. In an IS development project, it will not generally serve the cause of conviviality well if we produce a system which 'solves' an organization's 'problem'. What is rather needed is the provision of a technological resource adaptable to users' changing needs and circumstances.

We can use Illich's multiple balances to check whether the information systems we are building contain any destructive potential. If a system appears likely to lead to environmental degradation, radical monopoly, overprogramming, social polarization, or engineered obsolescence, we may have the opportunity and the responsibility to limit it.

Winner's notion of reverse adaptation alerts us to the likelihood that technological systems will tend to subvert means into ends. In the development of information systems, and especially in redevelopment from an earlier base, it will be important to try to establish the purposes of the system as authentically independent of the capabilities of the technology. Winner's analysis of social and electronic complexity, and the ensuing loss of comprehension and moral agency, has a number of implications for the construction of information systems. First, we should be careful not to produce systems which merely add to the burdens of information overload by providing more information, more models, or more functions than anyone can comfortably learn or use. Second, we should make sure that in building an information system, there is no absorption or dissipation of responsibility, but instead a clear retention or amplification of specific responsibilities for actions of the system by individual users or operators of it.

Finally, to take account of Borgmann's observations on focal practices, builders of information systems should try to ensure a critical engagement with the technology on the part of the users. Undifferentiated, packaged technology will generate uncommitted, homogenized users, who will find it difficult to understand the system they are using or assume responsibility for its actions. The implication is that the engaged user must have some independent commitment to and understanding of the process represented in the information system: the user of the accounts package must have some separate understanding of accounting processes, and the operator running a computerized industrial process must have an interest in that process which goes beyond the information on the screen. It would be a new departure for information systems designers, but the clear need here is to include in the information system design some specification of focal practices outside but complementary to the system which will anchor it to users' lives.

REFERENCES

Borgmann, A., 1984, "Technology and the Character of Contemporary Life," The University of Chicago Press, Chicago.

Hirschheim, R., Klein, H.K., and Lyytinen, K., 1995, "Information Systems Development and Data Modeling," Cambridge University Press, Cambridge UK.

Illich, I., 1990, "Tools for Conviviality," Marion Boyars, London.

Winner, L., 1977, "Autonomous Technology," The MIT Press, Cambridge MA.

ENABLING OLDER PEOPLE: DEVELOPING A SOCIAL SERVICES-OWNED SYSTEM

Martin Booy and Gail Boniface

Department of Occupational Therapy Education
University of Wales College of Medicine
Cardiff, UK

INTRODUCTION

The stated desire of Birmingham Social Services to "move away from managing independence to promoting and supporting independence" required a review of service provision to determine what was creating this 'managing dependence' environment. Their particular concern was for the large numbers of older people older people who are discharged from hospitals within the city, of which 90% return home.

Birmingham's concern for services for older people is in line with an emerging national trend. Pressure on the National Health Service for the early discharge of patients from hospital into the community, impacts on social service departments who are struggling to provide basic personal care. The 'rehabilitation' function has therefore been lost as it falls between the two agencies. Occupational Therapists are employed by both agencies but feel hindered in their professional 'enabling' role by regulations and procedures that demarcate responsibility between Health and Social Services.

The Department of Occupational Therapy Education in UWCM has based its philosophical view of rehabilitation on the need to create enabling environments so that those using the service can carry out those occupations (activities) they either want to or have to carry out. It is our view that the occupations, their motivators and organisers are vital to the health of the individual, regardless of physical or psychological disability and can be encouraged by altering the environment. Although these occupations are subdivided into activities of daily living, productivity and leisure sub-categories, organisational constraints and service user expectations mean that the 'activities of daily living' category is concentrated on to the detriment of others. By using a conceptual model such as the Occupational Therapy Model of Occupation through Adaptation (Reed and Sanderson 1992), occupational therapists can ensure that a holistic emphasis is placed on assessment, planning and evaluation of their intervention with clients.

The researchers, as occupational therapy educators, used this 'Model of Adaptation through Occupation' to ensure that the investigation focused on what older service users 'did' or

Synergy Matters: Working with Systems in the 21st *Century,*
Edited by Castell *et al.*, Kluwer Academic / Plenum Publishers, New York, 1999.

301

'needed to do' themselves to promote their independent living, rather than concentrating simply on 'care needs'.

METHODS OF ENQUIRY

Four research tools, semi-structured interview, focus group discussion, documentary audit and soft systems methodology (SSM), were used to gather data from four different sources of: *Staff, Users, Carers and Documentation.* It was important to gather both quantitative and qualitative information to ensure the richness of the study.   Also, as Birmingham City Council has the largest social services department in England with over 7000 staff, two of the twelve area offices were chosen as the sample for this study.   During the four-month project the views of fifty-one members of staff, forty-seven service users and six carers were obtained.

The membership of all four focus groups was multi-professional, representing Social Work, Occupational Therapy and Home Care teams working with the older service user, but each group represented a different organisational level: Managers, Assistant Managers, Practitioners and Assistants.   Thus these groups enabled the 'iterative consultation' of SSM to take place with a representative sample of 'actors' in the scenario.   Each focus group met twice and at the initial meeting a 'rich picture', (figure1), based on information obtained from other strands of the investigation, was tested and augmented by the group members. The focus question *'To what extent do YOU feel Birmingham Social Services department ENABLES older people?'* was also posed at this stage.

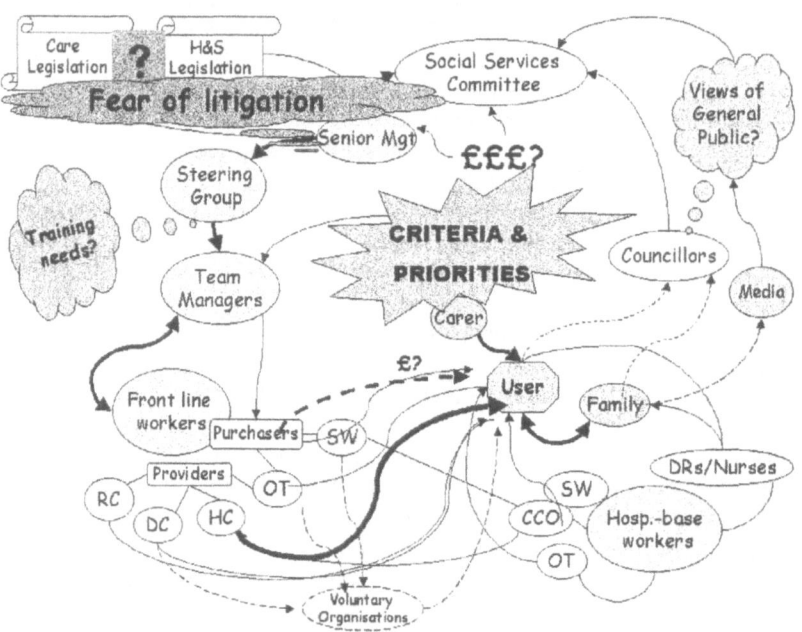

Figure 1: A Rich Picture of the influences effecting Birmingham Social Services' intervention with older people.

In presenting the researchers' initial understanding of the unstructured scenario to the groups, the rich picture was built up in stages, first identifying factual information such as the 'actors' involved and major organisational relationships such as hospital and social services. The support network surrounding carer and service user was then added. Finally issues of concern and their perceived emphasis were added in the form of 'dark clouds' or starbursts to reflect the importance given by staff.

DEVELOPING THEMES

Following the staff interviews and the first focus groups, the data was initially analysed by one person in order to identify apparently dominant themes listed below in what appeared at the time to be their order of importance:

1. Communication - barriers to and in hierarchical form.
2. Criteria and priorities of Birmingham Social Services department.
3. Time - there seemed to be an expectation of staff to do as much as possible in as short a time as possible.
4. Focus on personal care and doing it for them (older people).
5. Safety predominates, as does a fear of litigation.
6. Whither rehab? - Hospitals were perceived as having dispensed with rehabilitation and leaving a gap that seemed not to be being filled.
7. Training - is it appropriate for staff needs?
8. Client - expectations vary and they have a lack of power.
9. Geographical differences across Birmingham.
10. Staff - a picture of frustrated staff whose roles have changed.

In order to verify them, these themes were then presented at the second meeting of each focus group and group members were asked to comment on them. Two additional themes of Ethnicity and Assessment were subsequently added. Staff were asked to consider *'What are the needs of 'front-line workers' to be addressed in order to enable them to enable older people?'* (Front-line workers in this context referred to those staff directly in contact with service users in perceived contrast to those who were more involved in management).

THE DEVELOPMENT OF AN ALTERNATIVE 'WELTANSCHAUUNG'

A major result of the focus group discussion and the documentary audit was the discrepancies that emerged. These were in relation to the emerging reasons for the research project and the apparent difference between Birmingham City Council's stated service philosophy and its action in implementing that philosophy. Initially, the research project was set up to attempt to answer the question: "enabling older people, can we do better?" Later the commissioning team wanted staff training needs in relation to this question to be addressed. This seemed to assume that if the answer to the question was yes, then staff would need to be trained in order to improve. Interestingly, the response from the staff surveyed to this question was more in the realm of "if we are given the opportunity to", which was summed up in a staff member's comment that the Department needed to "enable the enablers (i.e. staff)". This implies a discrepancy between the views of the managers' commissioning the study, and the staff view.

A second discrepancy emerged in relation to the Department's stated philosophy and its actual action in relation to older service users. This philosophy was found to be very enabling as indicated by the following extracts from their policies and procedures documentation.

> "Introduction to the Management of Care: Section 1.H.1.2
> At every stage of the management of care process, workers should focus positively on the strengths and capacities of the user and should aim to maximise his/her independence." (Birmingham City Council 1995)

> "User and Carer Empowerment (Section 1.H.5.9)
> Questions must be phrased in such a way to ensure they are unbiased and obtain the real feeling and wishes of the user and carer'. 'Users and carers should be advised to think broadly by being asked open questions such as *"what would you like to do that you are not able to do now?"* (Birmingham City Council 1995)

Yet the action taken at the level of service delivery is predominantly in the context of 'doing to' and 'providing for' older people, linked to the extent to which they 'meet certain criteria' in order to trigger service provision. Consequently the nature of the service provided has evolved towards inflexibility, dominated by departmental criteria and priorities which 'manage' the older person's needs rather than 'enabling' the older person to capitalise on his/her strengths.

There was a need to encourage an alternative 'W' for this aspect of the work of Birmingham Social Services. This need was not so much connected with a reluctance to 'be enabling', but rather with the barriers to 'being enabling', created by the clouding of the Department's strategic intent within a tangle of criteria and priorities. In order to create a process through which this alternative, enabling 'W' could be both **identified** and **used** in a practical way, the research team generated an alternative, Birmingham Social Services-owned, conceptual model of practice.

BUILDING A SOCIAL SERVICES-OWNED MODEL OF PRACTICE

Though the 'Birmingham Model' (figure 2), emerged directly from the issues raised, a possible root definition for the proposed model could have been:

'A Birmingham Social Services-owned system to enable front-line workers to enable older service users through effective commissioning, service provision and communication within allocated resources'.

Seven major actions involved in the work of front-line workers in social services are identified at Level 1, then at a greater level of complexity in Level 2 this human activity model comprises fifty-three actions within these seven sub-systems. (Table 1 refers) At this level of detail front-line staff, managers and strategic planners can use the model to monitor current practices and procedures of Birmingham Social Services in respect of their 'enabling older people' function.

Table 1: Activities in the Birmingham Model

Function	Level 1	Level 2 sub-activities
Assessment and care-management	Commission Intervention	1. Receive referral 2. Allocate case the appropriate priority 3. Conduct assessment 4. Identify user's strengths and needs 5. Check criteria for social services' intervention 6. Devise Care Plan with user 7. Consider strengths and needs of carer 8. Request services for user 9. Monitor and review
	Allocate Resources	1. Know resource allocation 2. Be aware of BSS strategic policy 3. Consider requests for resources 4. Check funds available 5. Decide priority 6. Allocate resources 7. Monitor use of resources
Service provision	Provide Service	1. Receive request from referring agency 2. Clarify nature of request 3. Re-assess if required 4. Check whether user's contribution is appropriate 5. Check resources available 6. Provide service 7. Monitor service and advise assessor of changes in circumstance 8. Continue, reconfigure or withdraw service as necessary
	Ensure staff can do the job	1. Assess service(care) needs of users 2. Appraise skills of current staff team 3. Identify staff training needs 4. Identify training priorities 5. Consider training options for staff 6. Select staff who fit training criteria 7. Allocate training resources 8. Carry out training 9. Recognise new competencies.
Staff training and development	Encourage staff effectiveness	1. Be aware of the strategic purpose of supervision 2. Obtain skills or qualifications in supervision 3. Have supervision skills or competencies 4. Advise staff on the expectations of supervision 5. Carry out staff supervision sessions 6. Monitor effectiveness of supervision process
	Enable front-line workers	1. Understand and subscribe to enabling philosophy 2. Check that documentation can accommodate enabling values. 3. Communicate enabling philosophy to all concerned (incl. users/carers) 4. Ensure staff use enabling concepts when assessing 5. Ensure service providers receive an enabling care plan 6. Monitor evidence of enabling via supervision and documentation.
Communication	Communicate information	1. Clarify strategic vision of Birmingham Social Services 2. Have appropriate documentation available 3. Promote realistic expectations of social services by general public. 4. Train staff in enabling philosophy 5. Receive enabling information from other agencies 6. Ensure all relevant staff have access to enabling information 7. Monitor via supervision process

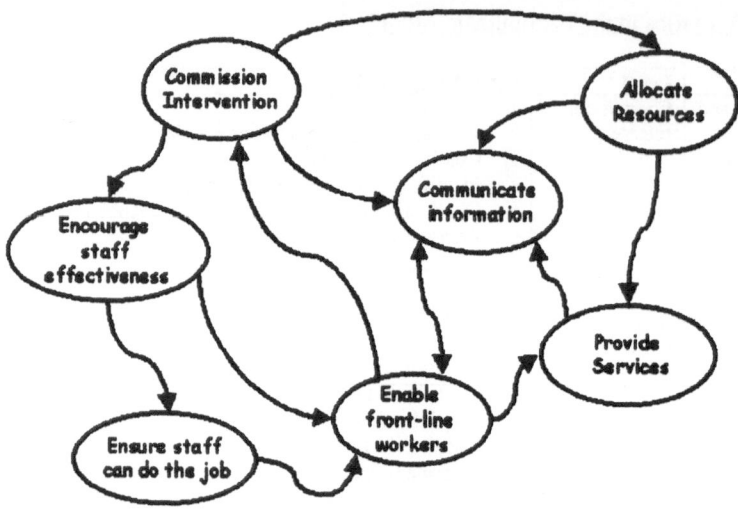

Figure 2: a social services-owned system to enable those front-line workers who enable older service user.

CONCLUSION

As the researchers have only recently reported their findings and recommendations to social services management in Birmingham (October 1998), it is anticipated that this proposed 'Birmingham Enabling Model' will require further iteration and modification in consultation with staff within the organisation before wider implementation.

REFERENCES

Reed K.L. and Sanderson S. R. 1992 "Conceptual Models of Occupational Therapy," Williams and Wilkins, Baltimore

Birmingham City Council 1995 "Adult Services Procedures: General, Home Care Service, Community Service," Birmingham Social Services Department, Birmingham, UK

Checkland P. and Scholes J. 1992 "Soft Systems Methodology in Action," John Wiley and Sons, Chichester

ACKNOWLEDGEMENTS

In trying to conduct 'action research' it was vitally important to identify the views and ideas of staff within Birmingham Social Services. Acknowledgement is given to the members of staff who participated formally in the study and to other members of the organisation who helped with supporting communication and organisation.

ACHIEVING POLICY SYNERGIES: THE CASE OF REFORMS TO THE COMMON AGRICULTURAL POLICY

Susan Carr

Systems Discipline
Faculty of Technology
The Open University
Milton Keynes, MK7 6AA

INTRODUCTION

As a concept used by the systems discipline, synergy is often explained by referring to the familiar saying: 'the whole is greater than the sum of the parts'. The word stems from the Greek *synergos*, meaning 'to work together'. Definitions of synergy given in the *International Encyclopaedia of Systems and Cybernetics* include: the effect 'when the combined action of two or more agents, substances or policies [are] greater than the sum of the separate individual actions'; and, 'the fusion between different aims and resources to create more between the interacting parties than they had prior to the interactions' (Francois, 1997).

On the basis of any of these definitions, synergy sounds like the answer to a policy maker's prayer. It holds out the promise of policies that fulfil all three 'E's of the policy audit mentioned by Richards (1998): economic, in that additional benefits may be achieved at little or no extra cost; efficient, in that the ratio of output to input is likely to be greater than if no synergy occurred; and effective, in that more policy objectives are likely to be met than would otherwise have been the case.

Synergy is associated with a host of positive and politically fashionable words, for example, partnership, co-ordination, co-operation, interaction, interdependence. But synergy is not an automatic or inevitable outcome, especially when it comes to achieving policy objectives. It rarely happens by chance. More often than not, the overall effect of different policies is less than the sum of the parts, as a result of departmental rivalries, uncoordinated efforts, fragmented responsibilities, and conflicting objectives and values.

This paper examines the response in the UK to the EC's Agenda 2000 proposals for reforming the Common Agricultural Policy (CAP) as an example of the potential for, and difficulties of, achieving synergies among policy objectives relating to agriculture and the rural environment. In particular, it examines the objective 'to increase the competitiveness of European agriculture', in relation to the other objectives of CAP reform. The analysis draws from policy documents, the media and the academic literature, as well as from

Synergy Matters: Working with Systems in the 21st *Century,*
Edited by Castell *et al.*, Kluwer Academic / Plenum Publishers, New York, 1999.

307

interviews with stakeholders in CAP reform in the UK. Ideas from a theoretical approach called the 'advocacy coalition framework', a tool for analysing policy change that has many features in common with soft systems approaches, are used to inform the analysis.

AGENDA 2000

The European Commission set out its proposals for further reforms of the CAP in a document called *Agenda 2000* (CEC, 1997). The objectives of the reforms are:
- to improve the competitiveness of European agriculture
- to guarantee food safety and support quality products
- to provide a fair standard of living in the rural community
- to develop the role of farmers in protecting the environment
- to encourage employment in rural areas, and
- to simplify the CAP (CEC, 1997, p. 29).

The principal measures proposed to achieve these objectives are: first, further cuts in production-linked subsidies (for example, a 20% cut in the intervention price for cereals in 2000 to bring their price closer to those of world markets); second, some compensation to the farmer for these price cuts through direct payments not linked, or 'coupled', to production; and third, reinforcement and extension of existing measures for environmental protection and rural development, re-organised under one simplified and integrated set of rural development measures (to incorporate, for example, agri-environment schemes, afforestation, early retirement, aid for younger farmers, less favoured areas scheme, and training to help farmers change their farming practices to incorporate conservation, animal welfare and hygiene objectives) (NFU, 1998a).

In addition, Agenda 2000 proposed three conditions under which the entitlement to direct payments might be restricted or withheld: very large payments to individual farms should be contained by reducing the rate at which direct payments are made above certain limits (a measure referred to as 'capping'); member states should be allowed to reduce direct payments to farmers who fail to employ the number of labour over a year considered appropriate for a particular context ('modulation'); and direct payments should be reduced or withheld from farmers who fail to meet certain environmental conditions, to be determined by individual governments ('cross-compliance'). Money saved from cross-compliance and modulation can be used by governments to fund additional agri-environmental measures, while money saved from capping is to be withdrawn from the CAP budget altogether.

Externally, the CAP reforms are driven by two main pressures: the need to prepare for expansion of the European Community to include new members from central and eastern Europe, and the need to prepare for the next round of negotiations on international trade, due to begin at the turn of the century. Both these needs put pressure on the Commission to reduce the production-related subsidies it pays to farmers, in the first case to avoid a potentially huge increase in CAP budgetary commitments, and in the second case to meet the conditions for trade liberalisation imposed by the 1994 Uruguay round of GATT (General Agreement on Trade and Tariffs). These conditions are likely to be further reinforced by the forthcoming international trade negotiations of the World Trade Organisation (WTO). By reducing or removing production-related subsidies, while offering farmers some support in the form of direct payments and payments for rural development measures, the European Commission hopes to avoid WTO restrictions on subsidised exports while providing a safety net to maintain the livelihoods of its farmers.

Internally, the reforms are a response to a variety of competing pressures from member state governments and many other stakeholders, including organisations representing farmers, environmentalists, consumers, and the food and farming industries.

Not surprisingly, in the light of all these competing pressures, the Agenda 2000 proposals have been described as 'ad-hoc and nothing more than political deals and compromises' (by members of the UK House of Commons Agricultural Select Committee, cited in Farmers Weekly, 27/2/98). The Commission has acknowledged that 'rural policy in the Union still appears as a juxtaposition of agricultural market policy, structural policy and environmental policy with rather complex instruments   and lacking overall coherence' (CEC, 1997, p. 27) but intended Agenda 2000 to address these policy deficiencies.

The competing demands at EU-level, of maintaining competitiveness, retaining rural livelihoods, providing employment and safeguarding the environment, are apparent to a greater or lesser extent at all decision-making levels, from the international, through the national and local, to the individual farmer. As an example of the opportunities for, and constraints on, policy synergies that the Agenda 2000 proposals have provided, later sections analyse the response of officials and advocacy groups in the UK at the national level. That analysis is informed by the theoretical framework discussed in the next section.

APPROACHES TO THE ANALYSIS OF POLICY SYNERGIES

A preliminary assessment of the potential synergistic and antagonistic effects among policy objectives may be based on intuitive judgement backed up by policy makers' professional experience. A more thorough assessment ought to involve a systematic and systemic analysis, for example, for each objective in the agriculture and rural development policy system: checking the policy makers' beliefs and assumptions about the problem and solution; comparing, and if possible, reconciling, them with those of other stakeholders; considering potential slow, rapid or sudden changes in the system's context that may constrain or favour the adoption and success of particular options; for all these aspects, exploring the links with other objectives at the same level, as well as the links between different levels (for example, the links between the objectives of policy makers and the objectives of the individual farmer, which will affect policy implementation).

Encompassing most of these variables, the advocacy coalition framework of Sabatier and Jenkins-Smith provides a useful source of ideas for the structured analysis of potential policy synergies, even though it is intended mainly as a tool for investigating policy change (Sabatier and Jenkins-Smith, 1993). The authors liken public policies to individual belief systems, in that policies incorporate implicit theories about how to achieve policy objectives. Like belief systems, policies incorporate value priorities (core values) and beliefs about important cause-effect relationships, about the seriousness of the problem, and about the state of critical contextual factors. The framework has close links with systems ideas, for example, in that its focus is a policy subsystem, studied in relation to its environment.

A policy subsystem is defined as 'the interaction of actors from different institutions who follow and seek to influence governmental decisions in a policy area (Sabatier and Jenkins-Smith, 1993, p. 16). To avoid the need to study many different actors and institutions, it is assumed that actors can be aggregated into a limited number of advocacy coalitions composed of people who share a set of basic values, causal assumptions and problem perceptions, who often act in concert. Policy change over time is viewed as a function of three sets of processes:

- the interaction of competing advocacy coalitions within a policy subsystem;
- changes external to the subsystem, in socio-economic conditions, system-wide governing coalitions, and opportunities and obstacles for competing coalitions that arise from other subsystems;
- effects of stable system parameters such as social structure and constitutional rules on the constraints and resources of subsystem actors.

The following analysis is informed by these ideas, particularly those linking policies with particular sets of values, beliefs and assumptions.

RESPONSE IN THE UK TO AGENDA 2000

In general terms, the UK Government shares the overall objectives of the European Commission for agriculture and rural development. It views competitiveness as a priority: 'our agriculture must become more competitive if it is to prosper and continue to be a major force in international markets' (UK Minister of Agriculture, 1997, cited in NFU, 1998b, p. 13). At the same time, it supports environmental objectives: 'In addition we need agricultural policies which are consistent with better care and protection for the environment. We must aim to switch support away from production to rural development to help country communities adapt and compete at home and in world markets'.

A key subject of debate, both at EU and UK levels, is whether the dual objectives of improving competitiveness and protecting the rural environment can be synergistic and can be combined in a set of integrated policy measures, or whether they are incompatible and have to be addressed separately. This question is being debated within the context of an even larger, strategic question: in response to WTO pressure concerning trade liberalisation, should the EU cut or eliminate production subsidies so as to increase its exports but leaving its farmers to sink or swim in international markets with limited protection (as proposed in Agenda 2000); or should it ignore or challenge WTO pressure, giving priority to protecting rural livelihoods with subsidies and to meeting domestic consumer and environmental concerns and markets instead of aiming for all-out production?

The lead ministry in the UK for the CAP negotiations is the Ministry of Agriculture, Food and Fisheries (MAFF). Indeed, because the negotiations commenced in 1998 during the UK's presidency of the European Union, MAFF officials initially chaired all the relevant EU negotiating committees and working groups.

The UK Government's negotiating position is based on its support for the first of the strategic options just mentioned. It is in favour of removing all production-linked farm subsidies, although it would like member states to be given more discretion about how best to compensate farmers (Farmers Weekly, 29/5/98). Explaining the Government's position, a MAFF official said:

> 'Subsidies are economically distorting. The Government and the public think they are not a good use of taxpayers' money. There is a view that UK farmers would survive and prosper in a more liberal system. Over time they would benefit. The reasons are independent of WTO pressures, but WTO pressures reinforce them. If we didn't reform we'd have to face more voluntary constraints on export markets. When we reach that GATT limit, we either have to export, having removed the subsidies, or we have to restrain production. With a relatively competitive agriculture, we see immense disadvantages in restraining production.' (MAFF interview, 14/7/98.)

The UK opposes all three proposals for restricting direct payments: 'From the UK policy point of view we don't support cross-compliance because it will obstruct the phasing out of coupled payments. ... We're dead against payment ceilings and modulation. (MAFF interview, 30/6/98.) MAFF may dislike cross-compliance partly because their previous experience (reducing headage payments for stock where there is over-grazing) is that it is complicated to administer, and can lead to bitter disputes with farmers (Lowe et al., 1998).

The UK would prefer to remove all payments that might be viewed by the WTO as coupled to production: 'Better to get rid of arable aid and focus on enhancing environmental features' (MAFF interview, 30/6/98.) Asked how the objective of enhancing environmental features might be reconciled with that of improving competitiveness, the MAFF official envisaged two separate groups of farmers: 'A farmer could either go flat out for production, or in marginal areas it could be worthwhile to go for payments to conserve or enhance countryside features'.

Forming an advocacy coalition with MAFF (although less united than in the past) are the farming unions: the NFU, NFU Scotland and NFU Wales, and to a lesser extent the Country Landowners Association, the Tenant Farmers Association and the Family Farms Association. According to MAFF, these are the principal organisations that responded to their consultation document and attended their open meetings about Agenda 2000: 'In many respects they speak with a common voice. The need for cuts is more or less agreed. They would like compensation. There is nervousness about cross-compliance. On payment ceilings they're a bit more divided, for example the Family Farms Association would welcome them.' (MAFF interview, 14/7/98.)

However, the NFU has cautioned against uncritical acceptance of the assumption that UK farmers will remain amongst the most competitive in Europe (NFU, 1998b). Other member states are restructuring their agriculture rapidly and in many cases their yields and productivity are increasing more quickly than in the UK. The NFU has expressed some doubts about the objective of competitiveness based only on food production: 'UK farmers may find that in terms of food production alone, they cannot compete when protection from foreign competition is reduced; as a provider of a wider range of services within the rural community, including food production, they might' (NFU, 1998b, p. 13).

Other organisations, who might be viewed as being in a different advocacy coalition, question more strongly the priority given to competitiveness defined in terms of productivity and world commodity markets. They would like to see a much more integrated rural development policy adopted. For example, English Nature, a public body that advises the UK Government on nature conservation matters, wants to encourage all farmers to 'operate sustainable, less intensive, environmentally sensitive regimes'. It supports Agenda 2000's proposals for an integrated set of rural development measures and, while direct payments remain, cross-compliance. It would to see CAP priorities move away from a narrow focus on production, and more adequate funding for agri-environment schemes, which currently receive only 3.6% of the CAP budget (English Nature, 1998). The Countryside Commission, another government advisory body, is also pressing for agri-environment schemes to be more central (Farmers Weekly, 6/3/98).

Environmental groups such as the Worldwide Fund for Nature (WWF) and the Royal Society for the Protection of Birds (RSPB) are more specific in their demands. WWF wants three quarters of CAP funds to be spent on sustainable rural development, with one third going on agri-environmental measures (Agra Europe, 5/6/98). The RSPB wants an increase to 25%, with all farm subsidies eventually coming from agri-environment schemes. It questions the view that cuts in production subsidies will bring environmental benefits through a reduction in inputs: 'It's difficult to see how a farmer could be environmentally friendly if competing on the world market. Lower prices might lead farmers to revert to older cheaper products, which may be less specific and more environmentally damaging (RSPB interview, 10/6/98).

A wide range of environmental and countryside groups, supported by English Nature and the Countryside Commission, have joined forces to propose a set of environmental conditions on direct payments to farmers (CPRE et al., 1998). They propose that farmers should be asked to meet three conditions in order to receive payment: adhere to basic environmental conditions (existing environmental regulations and codes of good agricultural practice); maintain a proportion of the farm (say 10%) as wildlife habitat and landscape features; and produce a whole farm environmental plan. They propose that 10% of the total budget for direct payments should be kept aside and used as 'additional financial incentives', for example, to subsidise the production of farm plans and to provide grants for adopting more environmentally sustainable farming practices. The proposals as outlined apply specifically to arable farmers, since intensive arable farming is considered to create many environmental problems, but the intention is that similar conditions should apply to all farmers.

In terms of potential synergies, the 'integrated approach' of the countryside coalition has a number of advantages over MAFF's 'dual approach' (which envisages that competitiveness and environmental objectives will be pursued by separate groups of farmers). In the integrated approach, the potential for synergy exists between objectives at different levels, as well as across objectives at the same level. For example, at international level, direct payments are less likely to be challenged by WTO as being production-related if they are conditional on environmental benefits. In addition, environmental conditions can be meshed with local, regional and national initiatives aimed at fulfilling commitments to international agreements on biodiversity and sustainable development, such as Biodiversity Action Plans.

One of the most important features of the integrated approach is the opportunity it provides for project officers to work alongside farmers to design whole farm environmental plans to suit the individual, local and regional context, within the framework of Agenda 2000's objectives. Implementation of the plans is intended to be optional, but at least the plans could increase awareness of the environmental possibilities for the farm, and of sources of further advice and financial assistance. Measures under the proposed rural development regulation, and related schemes available to farmers from non-government sources, could be co-ordinated by the project officers so that farmers need only contact a single organisation.

In terms of the potential synergies across the level of Agenda 2000's objectives, arable farmers could still choose to compete on the basis of productivity and price, provided they met the basic environmental conditions, or they could decide to compete in terms of quality based on social and environmental criteria, encouraged by the additional financial incentives and advice provided. Either decision would contribute to both competitiveness and environmental objectives. The second choice might also contribute to improved food quality and increased rural employment.

The synergies designed into the integrated approach would encourage UK farmers towards an agriculture more attuned to a changing social context. Needed now is a closer integration of the government departments who determine the policy context.

REFERENCES

Agra Europe, 5 June 1998, WWF urges bigger CAP spend on environment, p. EP8, *Agra-Europe*, London.

CEC, 1997, "Agenda 2000", Office for Official Publications of the European Communities, Luxembourg.

CPRE, The Game Conservancy, RSPB, Suffolk Preservation Society, The Wildlife Trusts, WWF, 1998, "CAP Reform and the Arable Environment - a New Way Forward", CPRE, London.

English Nature, 1998, CAP: time to change?, *English Nature Magazine*, 38:8-10.

Farmers Weekly, 27 February 1998, MPs critical of EU CAP proposals, p. 7, *Farmers Weekly*, London.

Farmers Weekly, 3 March 1998, Countryside body wants more EC aid, p. 8, *Farmers Weekly*, London.

Farmers Weekly, 29 May 1998, Farm ministers conclude radical CAP reform is 'vital' for future, p. 15, *Farmers Weekly*, London.

Francois, Charles, ed, 1997, "International Encyclopaedia of Systems and Cybernetics", K.G.Saur, Munchen.

Lowe, P., Hubbard, L., Moxey, A., Ward, N., Whitby, M. and Winter, M., 1998, United Kingdom, in: "CAP and the Rural Environment in Transition: A Panorama of National Perspectives", pp. 103-140, Floor Brouwer and Philip Lowe, eds, Wageningen Pers, Wageningen.

NFU, 1998a, Agenda 2000 Proposals: Implications for Agriculture, Revised Edition, Preliminary NFU Technical Analysis March 18th 1998, NFU, London (unpublished briefing document).

NFU, 1998b, "Is UK Agriculture Competitive? A European Perspective", NFU, London.

Richards, S., 1998, Wicked problems and clever solutions: sustainable development and the institutional framework for UK agricultural and rural policy.

Sabatier, P.A. and Jenkins-Smith, H.C., 1993, "Policy Change and Learning: an Advocacy Coalition Approach", Westview, Boulder, Colorado.

SUSTAINABILITY OF PRIMARY HEALTH CARE : A SYSTEMS APPROACH

Gautam Chakraborty, MBA

Indian Institute of Health Management Research
1 Prabhu Dayal Marg, Sanganer Airport
Jaipur (India) - 303906

The present paper is an attempt to analyze the dynamics of community, applying the systems perspective. The attempt is not to ponder on any theory, but to put empirical evidence in a systems perspective so as to identify the problem areas for future intervention. The paper discusses the case of primary health care delivery system in the rural Indian context. The health care system at the community level comprises of human and physical systems, and the present endeavor is an attempt to synthesize these two systems by linking the elements in a chain, instead of juxtaposing the two. It shows that the physical elements remain static until the human elements work on them, imbuing dynamism. It reveals that the interaction among the human elements is political in nature, determined by the control over the physical elements. The paper suggests that emphasis should be laid on this dynamic interaction in any future intervention for improving community systems.

INTRODUCTION

In the Alma Ata declaration, all the countries of the world agreed on the goal of "Health For All" that was based on equity, respecting people's perspective. Later the nature of health services delivery changed to Specialized Primary Health Care, which is a top-down approach and the service mix is decided by the providers. This approach hinders sustainability to a great extent as is evident from the example of Ghana, where 100% immunization was achieved when the programme was being run by UNICEF, but the moment it withdrew, the coverage fell to 40% (Godlee, 1995). This approach directs resources and technical expertise to specific diseases, at the cost of primary, secondary and tertiary care (Emmel, 1998). This shows that a top-down approach lays too much emphasis on expensive campaigns at the cost of building community level health care delivery systems.

When community level primary health care is put in a systems perspective, it is clearly seen that most of the flows are unidirectional. The flows are the resource flow and

Synergy Matters: Working with Systems in the 21st *Century,*
Edited by Castell *et al.*, Kluwer Academic / Plenum Publishers, New York, 1999.

the direction of control. Within the system the flows are directed from the providers to the community. There is a possibility of a reverse flow, emanating from the community and flowing to the providers. In the context of resources, it involves recovery in the form of user-fee or community contribution. In the context of control, it involves community control of the health care delivery system. Existence of this reverse flow constitutes a loop.

In the Indian context, in most of the cases, loops exists mainly for providing resources (monetary and physical). But community involvement is assured only when there exists a loop for community control (Kanjilal et al., 1997). The following discussion is built on the premise that **sustainability of a system is assured only if feedback loops, emanating from beneficiaries, exist for resources and control.**

THE PRIMARY HEALTH CARE SYSTEM

The primary health care system in India is predominantly owned by the state governments and provide free services. The flows are unidirectional as shown in Fig.1 below.

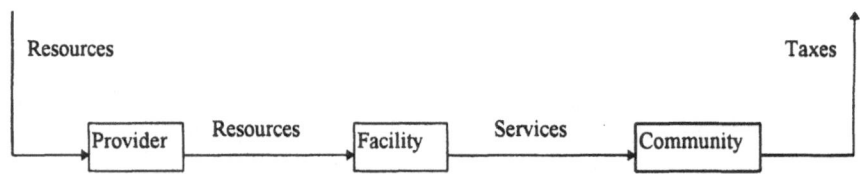

Figure 1. A Simple Model of Primary Health Care Delivery

It is always he who commands the resources who controls the flow. As the system is an open system and depends on external elements for all the resources, the stability and sustainability of the system are subject to outside influences. Moreover, as it is the provider who controls the physical elements (facility), community needs and perceptions are ignored. Although the community pays in terms of taxes, the funds change a number of hands before being put back into the system, causing a big time lag leading to depreciation in the value of the resources.

So, even in order to maintain the type of services being offered, more resources need to be put into the system, inflating the expenses in the process. This highlights, as shown in Fig.2, the need to generate some resources within the system so as to maintain the minimum services offered.

Resource Feedback Loop

In India the Resource Feedback Loop is chiefly characterized by the mechanism of user-charges going to the provider (Kanjilal et al., 1997). The resources may be monetary or physical. The physical resources take the form of voluntary labour, donating physical space, etc. This system is able to generate a small portion of the operating cost of service delivery and some capital cost like physical space, furniture, etc. But, control over the physical facilities and resources is still unidirectional.

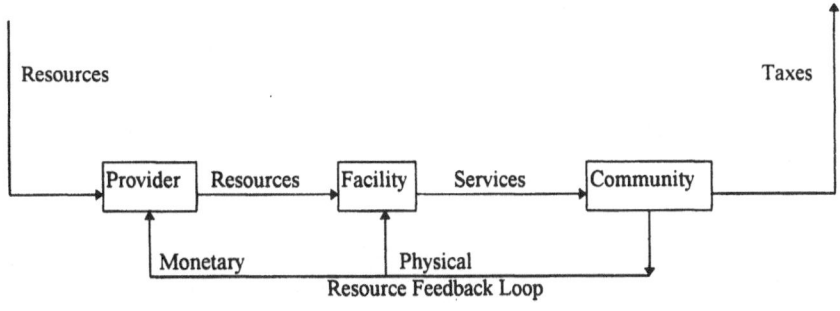

Figure 2. Resource Feedback Loop

This system of charging user-fee is characterized by engaging the community individually i.e. it is a particular individual who pays for the services. Obviously, his negotiating power is less than that of the provider because of inequality in command over the resources. Also, the community is involved after the delivery of services. This makes the community merely a recipient, thus neglecting the community perspective.

The community plays a dominant role in deciding the nature and type of services only when it is engaged collectively and there exists a control feedback loop, as in Fig.3.

Control Feedback Loop

A Control Feedback Loop is characterized by the existence of collective forums. The participation of the community members in the collective forums is either direct (village committees) or indirect through representatives (village *Panchayats*). The control is exercised by regular review meetings to decide future activities (Kanjilal et al., 1997). As shown in Fig.3, the community exercises indirect control over the utilization of resources through the control feedback loop.

The resource feedback loop helps in recovering a certain portion of the spent resources, and the control feedback loop helps addressing the community needs and perceptions. These ensure a degree of stability and sustainability to the system.

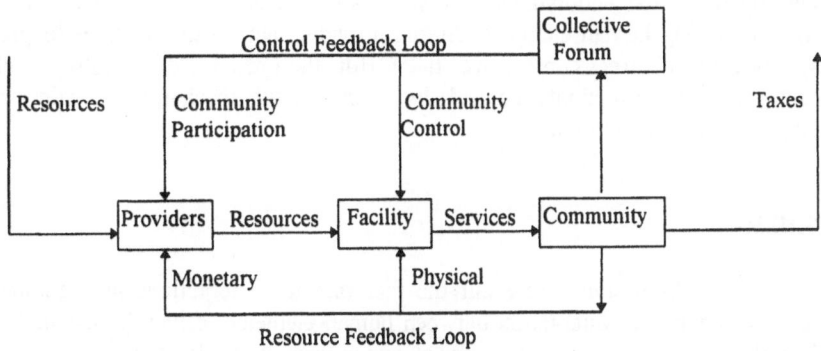

Figure 3. Control Feedback Loop

ISSUES

Analyzing the process of community level primary health care delivery in the systems perspective brings out the issues mentioned below:

Sustainability

As seen in the above discussion, the primary health care delivery system is an open system and there is dependence on external elements for resources. The resource feedback loop ensures a certain amount of recovery of spent resources, thus reducing the dependence on external agencies to that extent. The control feedback loop ensures indirect control of the community over the resources, thus reducing the monopoly of the providers and funding agencies. This provides stability to the system, as there is no mechanism to address the community's needs and perception without these loops. The quest for sustainability should not lead to the creation of a closed system, as the community is a micro component vis-à-vis the society and cannot provide for all the resources required.

Interaction of Human and Physical Elements

The physical elements are both real (physical facilities) and virtual (funds). In the system the providers act upon the facilities to provide services to the community, utilizing the funds provided by external funding agencies (also from the community). Through the linkage of providing funds, the funding agencies get an opportunity to direct the health delivery activities. Thus all the interactions can be reduced to the active interaction of the human elements acting through the passive physical elements. This deduction highlights the need to focus all interventions on the human elements instead of the physical.

Nature of Interactions

As it is the human elements which act among themselves through the physical elements, the interaction takes the form of negotiation with the motive of control over the physical elements. This renders the interaction political in nature (going by Laski's definition of politics being the act of struggle for power). This is because the interaction involves the element of decision making on how to utilize the resources to make the system run, which requires control over the resources. The outcome of the negotiations is determined by the initial endowment of resources (physical, financial and information) among the interacting human elements. As the providers act on the facilities to provide services, they exercise direct control over them. But, the presence of a control feedback loop ensures indirect community control, by acting as the mechanism to address the community needs and perception.

CONCLUSION

The above discussions bring out the fact that the interactions in a community system can be reduced to interaction between human elements, driven by the motive of controlling the resources (physical, financial and information). It (the outcome of the interactions) is determined by the initial endowment of resources among the participating human elements. The actor (provider or the beneficiary) possessing relatively more resources, can direct the course of interactions according to his preferences. It also shows that until there is a mechanism to address the community opinion through control feedback loop, and recovery of some part of spent resources through resource feedback loop, the system will remain unstable and cannot achieve sustainability. So any future intervention designed to improve community systems must address the issues of control, endowment of resources and the negotiation process, instead of the physical elements.

ACKNOWLEDGMENT

The author would like to take this opportunity to acknowledge the guidance and support provided by Dr. Barun Kanjilal in putting the original concept into this paper form and making available the required reports. Here the author would like to mention that the paper is based on the empirical evidences drawn from the project titled "Community Financing Experiences in India" undertaken by Dr. Kanjilal and his team, of which the author was a part. The author would also like to thank Dr. K. Srinivasan for reviewing the paper and providing moral support. The author also acknowledges the valuable contribution of Dr. G.R.Rao in editing the paper.

REFERENCES

Emmel, N.D., 1998, Health-for-all for 21st century - demise of primary health care, *Economic & Political Weekly*, 33(11):577.

Godlee, F., 1995, WHO's special programs : undermining from above, *British Medical Journal*, 310(6973):178.

Kanjilal, B., Sharma, S., Chakraborty, G.,1997, "Community Financing Experiences in India," IIHMR, Jaipur (India).

IS THE SYSTEMS APPROACH A PRECEDENT OF

THE DEVELOPMENT APPROACH ?

Gustavo González C.

Universidad de Navarra	Universidad de los Andes
Departamanto de Filosofía	Facultad de Administración
Edificio de Bibliotecas	Apartado Aereo No. 4976
31080 Pamplona (España)	Bogotá (Colombia)

THE ARRANGEMENT OF WORK IN THE NORTH

By the end of the Second World War different methods of approaching problems in organizations became a subject of scientific enquiry. The magnitude and complexity of operations such as the fast production of weapons and equipment, the coordination of a large volume of resources coming from very different sites and their distribution to the correct war front on time, together entailed a complete shift in warfare strategy from that used previouly. The profound change in world view that the construction and use of the atom bomb brought with it should be borne in mind. (Man's possibility of destroying all life, including his own, meant a radically different standpoint on war, science and technology too, on thought, social and personal life in general.)

Science and newly developed technology were put to the service of war and violence. It was an effort that left in its wake unintended results such as : intellectual disciplines such OR, highly elaborate craftmanship, new resource extraction and transformation techniques, global communication and the distribution of goods, etc. These were accompanied by innovative engineeering and management.

As Drucker (1993) has noted, how work had been performed until then changed substantially. Apprentices fled from the monopoly of unions into being taught by vocational schools or by on-site activities. It brought about, and still does, improved working capacity plus versatility on the part of the workers. Such change helps to account for the leaps of increased productivity, featured in the North, more than do updated and sophisticated machines, technology and highly elaborate engineering.

Synergy Matters: Working with Systems in the 21st Century,
Edited by Castell *et al.*, Kluwer Academic / Plenum Publishers, New York, 1999.

NEW THOUGHT FOR NEW WORK ARRANGEMENTS

How work was arrranged also underwent real change. Both intellectual and manual operations were coordinated such that high precision could develop. Accurate measurement in manufacture (as today in microscopic surgery), for instance, could allow for miscalculations only in the range of microns (as opposed to millimeters). Regulation of time demanded that activities take place in the precision of minutes and even seconds (air traffic control, trains that follow strict time tables, etc.)

Most medium-sized businesses, corporations, government agencies, associations and cooperatives embarked on research and the application of methods that would contribute knowledge and thus improve 'management operations' (rather than warfare operations). This course of affairs paralleled for intellectual activities the above-mentioned tendency in manual operations. Management was then amenable to scientific inquiry. The prestige of science and its methodology, took over the whole of organizational activities. New forms of energy (steam, oil and electricity) and new artifacts account for the coming of the industrial revolution. The systematic study of how work is performed accounted then, as now, for substantial increases in productivity. We can thus argue that science, applied to the cause of permanent improvements in management, accounts not only for the Allies' victory, but for the coming of a new age. As Ackoff (1991) has noted with insight, the world entered the systems era.

A group of researchers, belonging to those who contributed to the improvement of manufacture and management using experimental sciences, pioneered the examination of their own shortcomings. In doing so, they helped to develop a whole new approach, not only to problem solving, but to thinking and acquiring knowledge in problem understanding and course setting for action in organizations. *Systems* concepts and *systems thinking* came then to the forefront of management science. In this latter half of the century, several versions of the *systems approach* (theory and practice) in management have been developed. Many of these versions can be considered through a conceptual framework whereby their contributions complement each other.

CONVERSATION AMONGST COLLEAGUES

Jackson and Keys' "Systems of systems methodology" (1991) conceptual framework, can be thought of as offering issues for a hypothetical conversation amongst authors whose versions of the *systems approach* they study. They propose important epistemological differences, while offering means that allow the authors reviewed to come, if they so desired, to an agreeable exchange of ideas. They accomplish this by concentrating on two dimensions of problem contexts. First, is the extent of assent amongst decision makers. Second comes the degree of complexity of the relevant systems that ecompass problems. Following Jackson and Keys' idea, I suggest a number of issues (themes), different from the above mentioned, with which a hypothetical conversation could possibly take place amongst a handful of UK and Spanish colleagues. The issues I have selected are not usually discussed in the literature. Nevertheless, I have viewed them as aiding such a hypothetical conversation in order for understanding work arrangements in the South. It is hoped that sound critique will suggest other issues to be taken into consideration for the future.

Management Paradigms

A first theme is to think of management science as a paradigm that overturned what had been, until then, the secular *management* equals *shrewd steward* paradigm. It transformed mangement science into what may be referred to the *management* equals *measurement* paradigm. While the latter is generalized in the North, the former is predominant in the South. A word of caution: this proposal does not aim to assert that no shrewd stewards are found in the North, nor that measurement is non-existent in the South. The equations are meant as analytical tools only.

Following this idea I propose that Geoffrey Vicker's (1995) contribution be thought of his claim that *management* equals *appreciative system*, while Peter Checkland's (1990) contribution be thought of *management* equals *accomodation*. Finally, on the part of UK contributors, Flood and Jackson's (1991) efforts be thought of their claim that *management* equals *liberating people from power and other constraints*. For the first round of hypothetical conversations our Spanish guests would offer the equation: *mangement* equals *people's improving themselves*.

Ways of Knowing

An unavoidable issue is that of an observer's capability to grasp reality (epistemology). It establishes a defining characteristic which allows for the study of two trends. In the first trend (UK contributors), social reality is not only constructed by the observer, but thought of as being so complex and dynamic that 'description of what is out there' is close to impossible in the actual state of knowledge. For this trend, agreement amongst actors' perceptions is the most that can be aspired to, while for the latter trend (Spanish colleagues), true (not absolute) knowledge can be aquired by ever-better concepts of lived social reality. Such concepts can be developed and improved with the course of time.

When a concept (e.g., *system)* is used in different contexts it is prone to conveying several meanings. If the concept is meant only to name some reality without any ulterior motive, learned discussions might settle discrepancies, as the hard system - soft system dichotomy can be be thought of doing. With the *systems approach* something more than knowledge is striven for; setting courses of action and intervention are intended, too. My contention is that this being the case for such an approach, then two possible outcomes can follow. Firstly, different meanings might name different aspects of the *same* reality; each one adding understanding. Trend two claims this. We can refer to them as 'real systems'. Hard systems (HS) are included in them. However there exist 'real systems' (e.g., parents and their teenage children: a 'system out there' independent of our perceptions of it) that entail conflicting objectives (e.g., parents-teenagers relations). Secondly, there is confusion about *what* 'reality' we intend to name and, furthermore, on *how to act* in face of that 'reality' we are naming. All versions in trend one come to grips with this through 'ideal systems', i.e., perceptions not necessarily descriptive of what is out there- : appreciative systems (AS), soft systems (SS), critical systems (CS), etc.

Trend two would then ask: "How can you manage a business based only on people's perceptions? Isn't there danger for both naïve or deceitful, but eloquent and imaginative, actors to misguide 'accommodation' ?"; while trend one, in turn, would ask: "How dare you, Mr. Manager, impose your point of view claiming 'true' knowledge about reality? What makes your knowledge about things 'out there' better than my knowledge based on my daily experience in the shop floor ?".

How to Act and How to Intervene

A second round of a hypothetical conversation amongst colleagues could then be carried on by looking at suggested ways for action. Within the first trend, in spite of some diferences, 'acting by agreement' would encompass all of them. For trend two this is not excluded, but what is stressed is continuous self-improvement in personal qualities of the agents, with special emphasis laid on managers. From this standpoint personal example is a determining factor. "Inspirational leadership" is one of many aspects of trend two's proposal.

Table 1 below summarizes the above mentioned and following issues, all of them being candidates for further hypothetical conversations. A comparative analysis of each author's notion of complexity, history, and culture brings out differences between UK and Spanish colleagues, while insisting on the already mentioned parallel and complementary trends. Furthermore, such a comparison highlights expected differences amongst UK contributors.

Subjetive Dimension and Objective Dimension of Work

Another theme is that offered by John Paul II's apostolic letter on Work (1981). In this, he calls attention to the West's secular emphasis on the objective dimension of work (*what* is produced: goods and services) over and above the subjective dimension (*who* produces: a person). Most HS give priority to the objective dimension. SS and CS bring the subjective dimension to the forefront. However, it takes entrepeneurial systems (ES) to give priority to the subjective dimension over and above the objective dimension. This prioritization does not imply careless work or unreliable stewardship. Some applications of Total Quality Control emphasize the subjective dimension.

Trend one gives priority to the productive system and the interweaving of ideas. For instance, SS gives priority to human aspects in many ways, but only oriented towards production and not necessarily to the full development of man. There is definite orientation of the versions belonging to this trend towards studying how the productive dimension interrelates with organizations and societies. There is more concern with changing the context that surrounds man and work, while trend two gives priority to changing the person from within. In other words it gives priority to the subjective dimension of work.

Their Notion of History

With respect to the notion of history, an issue usually overlooked, trend one views its efforts as contributing to improved states of society starting with the development of rational thinking (experimental science for HS, action research for SS, CS) and the industrial revolution. This trend is more concerned with the context within which ideas arise and individuals act. For the second trend, rational thinking is a quality of the human person. One of the first written accounts of it in the Western World was that in Greek philosophy. While Christianity can be thought of playing a determining role for contemporary notions of history. This trend considers the person a complex interrelated system (CIS) to be the only innovator left in the present state of evolution.

Complexity

The notion of complexity not only differs between the two trends, but also within trend one. It is possible to describe all the notions within an inclusive order, the latter including the former. Subjectivity (AS, SS) includes complexity derived from an increase of elements and relations (HS). Power relations cause problems for the free interaction of subjectivity; also, they cause problems in the learning about elements and relations. Power relations (CS) is the new added aspect of complexity considered in trend one. Two notions of complexity in trend two can be considered, in spite of epistemological differences, to include those notions present in trend one: first, non-unitary complexity (CIS), by which is meant: disruption of time (non-linearity); anomie (moral uprooting); non-intended consequences (non-predictability); involution of institutions (work, family, State). Secondly, man, by himself or interacting, constitutes the highest source of complexity (ES).

Table 1. Notions that give rise to the trends of viewing of some English and Spanish contributors to the systems approach.

Versions of systems approach considered in relation to the following notions:	TREND 1	TREND 2
Source of *hope*	The artifact: HS The social context: AS,SS,CS	The person: ES, CIS
Ways of *knowing*	Real systems: HS Ideal systems: AS, SS, CS	Real systems: ES, CIS
Ways of *acting*	By command: HS, AS by accommodation: SS by inclusion and consensus: CS	Exercise of personal qualities (AS), and mutual correcting between instructing and complying: ES
Productive dimension and subjective dimension of human work:	Emphasizes productive dimension: HS distinguishes but does not give priority: SS, CS	Emphasizes subjective dimension: AS, ES
Notion of *history*	Development of rational thinking. Experimental science and industrial revolution as greatest historic events: HS. Interweaving strands of history: of ideas and of events: AS (= history of an individual or a society), SS, CS.	The person as the only source of innovation:CIS
Notion of *complexity*	Increase of elements and relations: HS. Human subjectivity and the unsuitability of analytical thought for understanding society: SS. Power relations: CS.	Non-unitary complexity: CIS Man: ES.
Notion of *culture*	In-built process for changing norms: AS. Possibility of accommodation of world views: SS. A useful metaphor: CS.	*Continuatio naturae* : CIS

HS: hard systems; AS: appreciative systems; SS: soft systems; CS: critical systems; ES: entrepreneural systems; CIS: complex interrelated system

Culture

Appreciative systems can be thought of as offering a theory on how cultural values influence and are, in turn, influenced by the exercise of each individual's appreciative system. Soft systems bring culture to the forefront. In contrast to hard systems they contribute significantly to an understanding of social situations avoided by the hard tradition. They accept different cultures without making value judgements about them or by assuming standpoints. Critical systems offer the use of metaphors as means of enhancing the creativity of actors in setting courses of action. For trend two, culture is more than is a metaphor. Polo (1993) stresses Aristotle's systemic observation that language is conventional in that it distinguishes man from other highly developed animals and allows for the construction of the world of symbols. Such a world is man's (CIS) dwelling place. Men's and women's creativity continually adds to the natural world; they transform already existing things and add ever-new symbols. This 'cultivation' is culture; through it nature is 'continued'. Culture like language based on sound and art based on nature, can be referred to as *continuatio naturae*. Action then, embedded within a logic of rules and practices, made meaningful by constitutive meaning, require more than a metaphor for its understanding.

The need of conversation around the above mentioned issues and others to be proposed in the future, hopefully would add intelligence to one of the most pressing problems of our time: the different work arrangements between North and South.

WORK ARRANGEMENTS IN THE SOUTH

The development approach also in vogue since the end of the Second World War resembles theoretical and practical aspects of the systems approach. While the latter can be interpreted as a kind of therapy for organizations, the former aspires to a very precise form of societal therapy for 'developing countries'. Its source of hope is the artifact. By it I mean technical assistance, technology, and their related *artifacts* (foreign capital, machines, computers, birth control devices, etc.) Only until the past two decades has the development approach considered other sources of hope: restructuring social intitutions and listening to the 'development' recepients. My contention is that until then this approach had considered none other but its 'hard systems' version, based on economy and demography and their related public policy, where objectives and goals unilaterally set in the North had to be compulsory complied with in the South. (e.g., food aid or short term loans tied to population control measures, etc.)

By examining the subjective dimension of work, Polo (1993) claims its meaning-giving attribute. It is then naïve to think that an *artifact* disembodied from the culture that gives it meaning, together with arrangements of work (space and time) that improve already existing products, a political order that relies on socialized inventions and holds labour in high esteem, can transform a culture where part or all of these conditions are missing. The development approach would do well to study the already well beaten systems path.

REFERENCES

Ackoff, R., The future of operational research is past, *in* : "Critical systems thinking-directed readings", Flood, R. L., Jackson, M. C. eds., Wiley, Chichester (1991)
Checkland, P., Scholes, J., Soft systems methodology in action, Wiley, Chichester (1990)
Drucker, P., "Post-capitalist society", Butterwort-Heinemann, Oxford (1993)
Jackson, M. C., Keys, P., Towards a system of systems methodologies, *in* : "Critical systems thinking-directed readings" , Flood, R. L. , Jackson , M. C., eds., Wiley, Chichester (1991)
John Paul II, Apostolic Letter: Laborem Excersens, Rome (1981)
Polo, L., Quien es el hombre, Rialp, Madrid (1993)
Vickers, G., The art of Judgment, Sage, Thousand Oaks (1995)

TOWARDS A METHODOLOGY FOR THE MANAGEMENT OF COMMUNITY ARTS PROJECTS

Richard Kamm

Community Information Systems Centre
Faculty of Computer Studies and Mathematics
University of the West of England
Coldharbour Lane
Frenchay
Bristol BS16 1QY

INTRODUCTION

Popular imagery places management and art at opposite poles. One is the province of organisation, using rational techniques to administer bodies of people within instrumental values, while the other rests on individual inspiration and creativity (Smith, 1988): art is driven by pressures that are not purely material and should not be subject to organizational constraints. But increasing attention is being paid to more consultative or participatory views of art which covers "work which is grounded in community practice" (Dickson, 1995), which implies that artists need not be seen as isolated individuals but as working within a social context.

It is in this spirit that efforts that are being made to develop a methodology which summarises some principles of project management for a specific area of artistic activity, that of community arts. The work is a collaborative effort between academics from the areas of fine arts and community information systems and professionals from industry and community arts administration. The focus of this paper is on the extent to which systems ideas can contribute to the development of such a methodology and on whether there are ideas which are familiar in conventional managerial contexts which can be directly translated while retaining the spirit in which community arts have been developed.

THE CONTEXT OF COMMUNITY ARTS PROJECTS

The community arts movement has a short but tortuous history. A distinctive community arts movement developed during the late 1960s with a radical critique established arts practices and methods of teaching. What followed was a series of debates about the precise

Synergy Matters: Working with Systems in the 21st *Century,*
Edited by Castell *et al.*, Kluwer Academic / Plenum Publishers, New York, 1999.

nature of "community art": did it simply mean an effort to involve in the creative process those who were excluded by the conventional arts establishment, or, to qualify, did it have to contain some critique of conventional social relationships? What was its relationship, if any, to "public art", subsidised or sponsored by official agencies? Should it keep its distance from the established areas of culture? (Braden, 1978; Morgan, 1995)

Since then, community art has become extremely diverse. The line between it and public art has blurred as funding agencies have acknowledged the value of the products and methods which are characteristic of the community arts movement. The educational benefits of participation in creative activity have led to greater communication between community artists and fine arts institutions such as galleries (Allen, 1995). Possibly most significantly, the argument that art can contribute to the regeneration of deprived localities has become influential. Recent well-publicised examples of urban renewal that include cultural projects have strengthened the efforts to develop art projects as an integral part of the development of city spaces (Bianchini and Landry, 1995; Worpole, 1992).

A result of this interest is that attention needs to be paid to the practicalities of managing a small-scale artistic project because of the responsibilities involved. If communities are to benefit from the development of art works then the artist needs to be aware of their interests and capabilities. If funding is to be obtained then the producers need to be accountable for that subsidy and to be able to draw on some source of management guidelines to enable them to justify the faith placed in them. Introducing ideas of project management is intended to be means of both stimulating contacts between artists and communities and also of guiding those contacts to be maintained through the work that leads to a finished artefact.

THE RATIONALE FOR PROJECT MANAGEMENT

The community arts organisation which is the prime sponsor of the methodology has been developing its work as an agency, working on behalf of both artists and community organisations to bring the two together. In the past it has overseen a number of projects and performed most of the work in developing contacts between artists and potential participants, which raises the question of why formal guidance for artists is necessary at all: why not leave the artists to concentrate on creative work and leave the administration of projects to administrators just as some freelances, notably authors, will usually have their finances sorted out by professional accountants?

One answer to this question is the size of the task that would have to be undertaken by what tend to be small voluntary sector groups. The local body in this case, for example, has had the number of artists on its database increase from 7 to over 90 since it began to act as an agency, an indication of the number of creative professionals who can benefit from guidance on the management of community-related projects. Similarly, the interest in community arts from funding bodies, probably related to the desire to involve the arts in urban regeneration, has extended the number and variety of projects that might be attempted. Recent work has included the integration of art into the architecture of urban renewal schemes funded by the city council. Other projects involve the provision of workshop-based training to individuals who might not otherwise have access to arts education.

All of this makes it difficult for a small, charitable organisation to provide full support for projects on top of the effort to facilitate contacts in the first place. Producing the methodology is intended to allow projects to be established and run without the organisation having to be involved in the detail of operations.

But even if the voluntary sector agency had access to larger funds, personnel and resources there would still be benefits arising from producing a defined methodology. The motivation for most people to enter art college and to attempt to make a living from their talents is their initial ability and imagination as artists. To continue to make a living as an artist, however, it is necessary to develop a variety of organisational skills. These can include the capacity to obtain funding and negotiate with official bodies which is needed all forms of art. Work with communities requires additional capacities: to educate, to promote participation and to act responsibly to repay the trust that the end product will enhance the community itself, as well as the artist's own career.

The purpose of producing the methodology, then, is to provide accessible and comprehensible guidance for artists in those skills: managing successful projects will help them remain in art rather than give up and move into other forms of employment. The objective of the current efforts to define a methodology for it is to identify the principles on which it is to be based: whether derived from conventional forms of project management, from less orthodox systems thinking or from the development of ideas which are specific to the running of community arts.

ISSUES IN PLANNING AND PARTICIPATION: THE RANGE OF STAKEHOLDERS

Even the most commercial project management approaches recommend exploration of stakeholder interests and perceptions, at least in principle (Mumford, 1995). Community arts were developed to encourage participation in cultural activity and similarly need to enlist the commitment of different partners who may have varying priorities. Some of the vocabulary of conventional systems methodology concerning the definition of aims and scope of projects therefore has some relevance for community arts, but with the distinction that the range of stakeholders to be considered is extremely wide and the interests that they hold are inherently difficult to quantify.

Taking the communities and community organizations that are intended to be the most direct beneficiaries of such schemes, a complex network interests is apparent. The community arts movement has always prioritised the stimulation of creativity among non-artists, which means that professionals must be capable of attracting people to the project and conveying the necessary skills as well as being talented artists. The outcome, for these participants, is less the product than the process, working on the assumption that individuals' lives are enhanced by involvement in artistic activity and by being something other than passive consumers of culture. The methodology therefore needs to offer guidance to artists in their dealings with potential participants by suggesting ways in which their interests and capabilities can be elicited.

But "participants" can also be defined in a wider sense a wider sense, as an audience or as the inhabitants of site which provides a home for the art work. There may be some overlap between these categories, participants and hosts, but it is unlikely that they will ever be identical. Where a work of art is seen or heard by people who were not directly involved in its creation then the content becomes extremely important: something that Mulgan and Worpole (1986) argue some earlier community arts practitioners lost sight of. This is particularly significant for structures that have a long-term existence: they will affect not only the environment of the current members of the host community but also those who are going to live there in future. Clearly future generations cannot be consulted, but an aim of project management in this area would be to encourage artists to be aware nature of the communities in which they are working so that the eventual outcome is appropriate.

Beyond that, stakeholders would also include the organizations or individuals who sponsor and fund the project. Again, the particular priorities of sponsoring bodies will vary but there are likely to be concerns both with the nature of the project, whether it is something with which they can be associated, and with the management of any resources that are granted to it. It should also not be forgotten that it is much easier to regard the artist as a stakeholder with a legitimate interest in the outcome than, for example, a systems designer on an IT project: artistic development is widely accepted as an important element of the choice of work that an artist takes on.

What connects this diverse and complex collection of stakeholders is that the extent to which their "stakes" can be realised is inherently difficult to measure: although awareness of the different nature of their interests is important for those organising arts projects. Although the evaluation of commercial projects against quantitative objectives is controversial, methods using some form of cost-benefit analysis have some pedigree (Remenyi et al, 1997). In community arts, however, the nature of stakeholder concerns is such that they need to be appreciated rather than measured. For participants, the benefit is primarily that of increased capacity and confidence in their creativity, leading to an enhanced quality of life. Sponsoring bodies necessarily spend rather than save or make money on artistic activity, so the most realistic objectives for a project are that it should meet some wider aim such as status, for private funders, or the achievement of a remit, for government-backed arts councils.

Even the value of a lasting piece of art in the community is hard to assess. For most visual arts, price is derived from some market value that assumes that ownership, and usually location, can be altered. Community art, however, derives much of its importance from the connections between a work and the host locality. Moving it elsewhere would necessarily deprive it of meaning, even where the work is a permanent artefact. The benefits arising to all stakeholders will usually be of this kind: intangible and closely related to individual or collective states of mind. Guidance to the organisers of arts projects will therefore need to include suggestions for ways in which to understand and empathise to an even greater extent than is the case in commercial development.

THE PROCESS OF COMMUNITY ARTS PROJECTS: PHASES AND FEEDBACK.

Within the activities of producing a work of community art, the relevance of conventional project management becomes stronger. The framework, in simple terms, is one of plan-do-review and was originally developed through examination of commercial project organization. It retains some features of that approach, notably the division into phases or "stages". These were labelled *Definition*, *Planning*, *Development* and *Post-Project* but with links between each of them formally laid out: thus, Planning might be expected to refer back to the Definition of a project's aims, while obstacles encountered during Development would lead to adjustment of the Plan. An overall Post-Project review would be expected to assess the outcomes which, as noted above, might lie in the value of the process to the participants or of the end product of the host community.

The precise terminology is not particularly descriptive of art work since the main purpose was not to coin particular words but to make any users of the methodology aware of the importance of the whole project cycle. Both professional artists and participants will gain from the process of creation that takes place within *Development*, for example, but work of planning and review are equally important in ensuring that the creative activity can actually take place.

The history of community arts suggested that it was important to establish guidance about the *Definition* phase. The informed community artists' contribution to the discussions surrounding the methodology suggested that that this often consisted of a passage from research into tangibles, such as site and materials, to formal definition without consultation with such stakeholders as the inhabitants of the locality which contained the site. What needs to be stressed in the framework for projects with a community orientation is the necessarily consultative nature of the process, which in turn means that phases such as *Definition* are seen as iterative processes and as revisitable over the life of the project as a whole.

The planning phase is clearly essential in obtaining funds and approval, particularly for projects which require areas of land, and the kind of guidance which is most useful here is based on the experience of those who have previously worked with the different bodies and authorities with a direct influence. While the overall framework of the methodology is intended to be stable, there will need to be regular updating of the contents of this guidance on planning: even casual observers of the arts world will know that the organisation of funding has been extremely unstable in recent years, both for large and small projects.

The *Post-Project* review is similarly useful, particularly in view of the range of community arts stakeholders mentioned above. If the different benefits to the participants and to the wider host community are to be realised then formal appreciation of what all of those involved have gained from the work. Compared to commercial design activity, review not only concerned with an evaluation of the results of the endeavour (although this may be a useful educational exercise in itself). It also involves a working out of how those who have developed new horizons or aptitudes as a result of the project can take them from there. In theory, any project cycle, if it is to be cyclical, should look to future development. In community arts this becomes particularly important because the stimulation of creativity in non-professionals, its principal objective, should be an effect which continues after the project. Review of outcomes will include those that are educational and cultural.

This raises the important question of the communication of feedback. Both the qualitative nature of artistry and the necessity of stakeholder involvement suggest that a purely linear approach would be ineffective. At time of writing, therefore, the approach comprises a network in which most phases are linked to each other: deliverables proceed from one phase to the next but communication of other kinds takes place from each phase to most of the others. Interestingly, the participants with professional and academic interests in community arts were comfortable with this framework: having argued from the start that a linear process would not meet the needs of the area of activity. The enthusiasm for principles of industrial project management was matched by a healthy realisation of the need for different forms of feedback between different phases.

What has effectively emerged is a methodology that explicitly prompts both single and double loop learning. The single loop is represented by the formal definition of adjustments to the project plan that arise from obstacles or complications encountered during *Development*. Suggestions for the type of information that this implies would include financial resources. The idea of a regular *Post-Project* phase which informs the perceptions of *Definition* and *Planning* would be closer to Argyris' definition of double loop learning as an evaluation of aims and objectives (Argyris, 1993; see also Senge, 1996, on adaptive and generative learning). The conduct of the particular project is not directly affected but learning takes place through the responses of both artists and participants to the work that has been produced and in reflection on the conduct of the implementation of the original proposals.

Although feedback has always been a central part of systems thinking, not every methodology includes it in quite such a formal way. In Soft Systems Methodology, for example, feedback is implied by the recommendation that the approach should not be seen as a rigid set of stages but as a learning cycle with informal but frequent iteration (Checkland and Scholes, 1990): practitioners are assumed to be capable of realising that regular rethinking is inherent the methodology's application. Here, however, the specification of feedback was thought worthwhile because it is an area where community arts practice has tended to be weak.

As noted above artists are rarely as highly trained in the activities of monitoring costs and resource use as they are in producing the art work, so the suggestion of information that they might consider while the project is underway may be a valuable *aide memoire*. With respect to post-project reflection, community arts has, according to some who have had responsibility for co-ordinating support for it, often neglected evaluation of the end product (Mulgan and Worpole, 1986). Where emphasis has been placed on process, and the creativity of the participants, it has sometimes been possible for the wider stakeholders' view of the end product to be ignored. Guidance on the eliciting and analysis of views might usefully be provided to allow this evaluation to feed into future developments which also lead to lasting artistic contributions to communities.

CONCLUSION

At the time of writing, much of the detail of the guidance concerning particular phases has yet to be developed. What has been learnt so far in the process of developing the methodology is the ease with which concepts which are common to systems approaches can be incorporated into a general framework, notably participation and feedback. This indicates that both the patterns of work and the culture surrounding community arts have points of contact with some relevant approaches to management. The learning process moves in the other direction too, with community arts demonstrating some ways in which collective learning can be put into practice.

REFERENCES

Allen, F., 1995, Gallery Education, *in*: M. Dickson, ed., "Art with People", AN Publications, Sunderland.
Argyris, C., 1993, "On Organizational Learning", Blackwell, Oxford.
Bianchini, F. and Landry, C., 1995, "The Creative City", Demos, London.
Braden, S., 1978, "Artists and People", Routledge and Kegan Paul, London.
Checkland, Peter and Scholes, Jim, 1990, "Soft Systems Methodology in Action", Wiley, Chichester.
Dickson, M., 1995, Introduction, *in*: M. Dickson, ed., "Art with People", AN Publications, Sunderland.
Morgan, S., 1995, Looking back over 25 years, *in*: M. Dickson, ed., "Art with People", AN Publications, Sunderland.
Mumford, Enid, 1995, "Effective Systems Design and Requirements Analysis", Macmillan, London.
Mulgan, G. and Worpole, K., 1986, "Saturday Night or Sunday Morning", Comedia, London.
Remenyi, D., Sherwood-Smith, M., and White, T., 1997, "Achieving maximum value from information systems: a process approach", Wiley, Chichester.
Senge, P., 1996, The leader's new work: on Organizational Learning *in*: K. Starkey, (ed.) "How Organizations Learn", ITP, London.
Smith, Bernard, 1988, "The Death of the Artist as Hero", Oxford University Press Australia, Melbourne.
Worpole, Ken, 1992, "Towns for People", Open University Press, Buckingham.

SYNERGY IN 'LEARNING CITIZEN-GROUPS' USING TELEOGENIC SYSTEMS METHODOLOGY

C. Jotin Khisty[1], Lena L. Khisty[2], R. Keith Ellis[3], and P. S. Sriraj[1]

[1]Civil &Architectural Engineering Department
 Illinois Institute of Technology, Chicago, IL – 60616.
[2]College of Education., University of Illinois at Chicago, IL – 60607.
[3] 14 Pendennis Ct., Harpenden, AL51SG, UK.

INTRODUCTION

A 'Learning Citizen-Group' (LCG) is one that is continually making an effort to expand its capacity to create its future. At the heart of the LCG lies the importance of systems thinking, and embedded within this system perspective is the element of citizen involvement (Checkland & Scholes, 1990)

The emergence of effective citizen involvement through the vehicle of LCGs is relevant and necessary in every sector of public decision making, be it education, health services, forest conservation, water supply, or transport. When such involvement is practiced successfully, it transforms and enriches the decision making process to a higher level, and promotes sustainable development (Khisty, 1996, Khisty and Leleur, 1997). In fact, the citizen involvement process is a journey toward organizational wholeness. Without it, a citizen group would not be able to operate coherently toward shared goals and objectives. But the successful outcome of such a journey comes at a price - the price of revising our mental maps or models of reality, and by promoting synergy through the use of teleogenic systems methodology. Teleogenic systems promote synergy by eliminating or reducing "allergy" or "dysergy". In contrast to teleonomic and teleozetic systems, teleogenic systems are ones that are capable of generating and using their own goals, in combination, if necessary, with other external goals (Coulter & Johnson, 1987).

The objectives of this paper are: to briefly describe a teleogenic system that is particularly suited for tackling conflictual problem situations connected with LCGs and decision makers; to demonstrate how LCGs can be synergized through consciousness-raising; to lay out the connection between Arnstein's ladder of citizen participation and synergy; to demonstrate the interplay of virtuous and vicious cycles in the learning process that may reinforce or retard the collaboration needed for collective decision making; and finally, to show how an assessment of teleogenic systems methodology could be done.

Synergy Matters: Working with Systems in the 21st *Century,*
Edited by Castell *et al.*, Kluwer Academic / Plenum Publishers, New York, 1999.

SYNERGY & TELEOGENIC SYSTEMS

For the purposes of this paper, synergy implies the capacity of two or more forces, goals, functions, or structures of information to optimize one another and achieve mutual enhancement (Hamden-Turner, 1981). Synergy also represents redundancy, particularly teleological redundancy. It is a term used to describe the emergence of unexpected, surprising and interesting properties manifested in group work. It is argued that the synergy of a group leads to much greater creativity in problem solving (Flood and Carson, 1993). The global aim of LCGs (with the help of a change-agent if necessary) is to promote synergy through a variety of inputs, generated in a number of different channels, and in a number of different formats, all coordinated to achieve the set of goals that the citizen group has formulated, possibly blended with those formulated by the decision maker.

Most of us are familiar with engineering control systems whose goals are externally set by operators controlling the system and fed from top power hierarchies. On the other hand, living systems are teleological (goal-seeking), in the sense that they possess the capacity to measure their own performance and have a decision-making function (Hutchins, 1996). It is useful to distinguish three types of teleological systems, each more advanced than the one before. These are: (a) Teleonomic systems capable of goal-seeking behavior, whose goals are fed externally into the system, (b) Teleozetic systems which are not only teleonomic, but which can choose their own goals from a repertoire with which they were initially endowed, and (c) Teleogenic systems which are not only teleozetic, but which can create their own new goals, and are capable of participating in human network systems through communication and interaction, a quality absent in teleonomic and teleozetic systems (Coulter and Johnson, 1987).

A teleogenic system methodology involves teleogenic components such as citizen groups and organizations they deal with. It can be used by any party (individually or collectively) whose purpose it is to change the system in some way. The methodology is applicable to such problems as conflict resolution where transformation of adversarial relations into synergic mode situations is desired. Indeed, one of the principal functions of this methodology is to promote the initiation, growth, and development of evolutionary processes of a system that tends to enhance synergies among its various functional components and the system's global synergy, through communication and interaction.

CITIZEN GROUP-BUILDING THROUGH SYNERGY

Citizen group-building is essential for optimal performance and learning capabilities. Without it, citizen groups cannot work coherently toward shared goals and objectives. Citizen group-building can be best thought of as a journey toward organizational wholeness, where decision-makers, planners, and citizens can synergistically work together (Gozdz, 1995). Consciousness-raising is an essential part of citizen group-building. It refers to the process of learning to perceive social, economic, and political contradictions systemically, and to take action against the oppressive elements in society. Freire's words are most appropriate in this context. In the "Pedagogy of the Oppressed" he wrote that education of citizens is the answer, and on that basis he advocated that:

"every human being, no matter how ignorant or submerged he may be in the culture of silence', is capable of looking critically at his world in a dialogical encounter with others. Provided with the proper tools for such an encounter, he can gradually perceive his personal and social reality as well as the contradictions in it and become conscious of his own perceptions of that reality, and deal critically with it" (Freire, 1972).

The twin benefits of consciousness-raising are: to gradually transfer the responsibility for monitoring the learning process of team-members from planners, managers and engineers to citizen groups themselves, and to promote positive self-perception, affect, and motivation. In this respect, consciousness-raising provides personal insights into one's own thinking and fosters independent learning (Paris and Winograd, 1990).

MAPPING THE TERRAIN OF CITIZEN GROUP DOMAIN AND CONTROL

Almost thirty years ago Sherry Arnstein (1969) defined citizen involvement in terms of the degree of actual control citizens could have over policy decisions. Her 'Ladder of Citizen Participation' depicted three categories of participation ranging from no control to complete citizen control as shown in Figure 1. Arnstein believed that without actual redistribution of power, citizen participation was an "empty ritual" and that the only way of achieving any significant social reform was to get citizen groups involved in activities high up on the ladder. Notice that as one climbs up the ladder the involvement demands a higher level of synergy. Also, notice that the three highest rungs of the ladder which include partnership, delegated power, and citizen control, require progressively enhanced application of teleogenic systems methodology. The bottom two rungs of 'nonparticipation' and the middle three rungs of 'tokenism' correspond to teleonomic and teleozetic systems respectively.

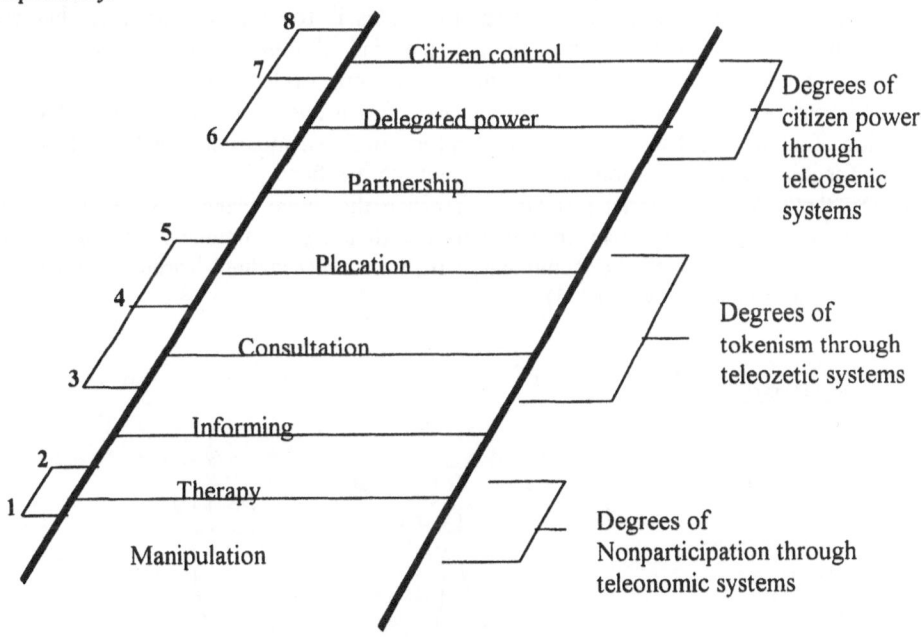

Figure 1. The Ladder of Citizen Participation

TELEOGENIC SYSTEMS, SYNERGY & DISTRIBUTED COGNITION

The idea of knowledge being distributed in a group is intuitively compelling - who can doubt that the knowledge that a team possesses in some sense exceeds that possessed by any single member of the group? Clearly then, the collective knowledge of a group must be greater (or at least not less) than the knowledge of any individual member.

Recent research in cognition has shown quite conclusively that knowledge is socially constructed through collaborative efforts to achieve shared objectives. It is strongly believed that such cognition helps participants to exhibit intellectual curiosity and persistence, to be inventive in their pursuits of knowledge, and to be strategic in their problem-solving ability. Under these circumstances, it is quite appropriate for the citizen group to distribute the tasks and responsibilities among themselves. This distribution of shared work and shared knowledge leads to a high degree of cohesiveness and raises the synergy among participating members.

HARVESTING THE POTENTIAL OF LEARNING CITIZEN GROUPS

In analyzing Arnstein's ladder and connecting it to the three teleological systems, it is evident that the inherent challenge of LCGs is in sustaining cooperative and collaborative interaction with planners, engineers, managers, and their decision-makers. A structure for such collaboration is to initiate Learning Citizen Groups (LCGs). LCGs are more likely to arise when certain critical conditions are present within a group, such as curiosity, the desire to learn, commitment, and a desire to act collaboratively with a spirit of experimentation. In a team situation, learning occurs through the effort of leaders who lead because they want to serve the group, generally without any reward.

One way to enrich the environment for LCGs is to create a structure that can support the points cited above through new patterns of thinking, communicating and interacting. Figure 2 shows two reinforcing or virtuous cycles that can promote such learning. Briefly stated, shared vision, through shared goals and objectives, can stimulate the formation of an initial collaborative design/plan for a problem situation, which can form the nucleus for joint experimentation, discussion, and more refined planning. This can lead to reflection that generates shared insights to improve the initial plans. Of course LCGs involve an ongoing commitment to collaborative thinking, communicating, and acting together in the interest of the community. It is a never-ending learning process of continuous improvement (Ryan, 1995).

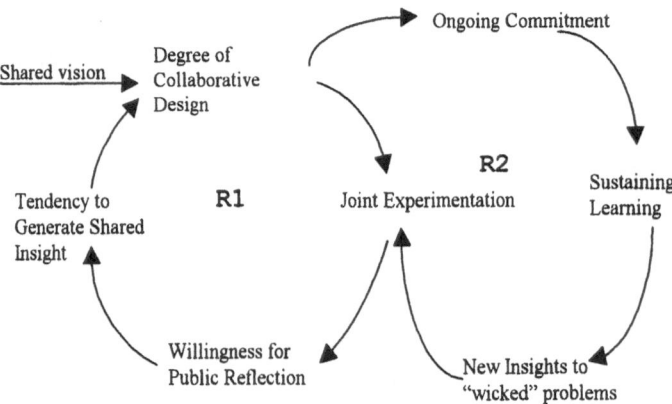

Figure 2. Cycle of Collaborative Learning

Naturally, this brings up the question: "What are the obstacles to LCGs?" They come in several forms: fear of failure, inability to deal with complexity, inability to see the big picture, in-fights with members of the team, hindrances posed by planners and their decision makers, level of participation in a planning process, late entry to the process and so the list goes on. Figure 3 shows some of these braking cycles.

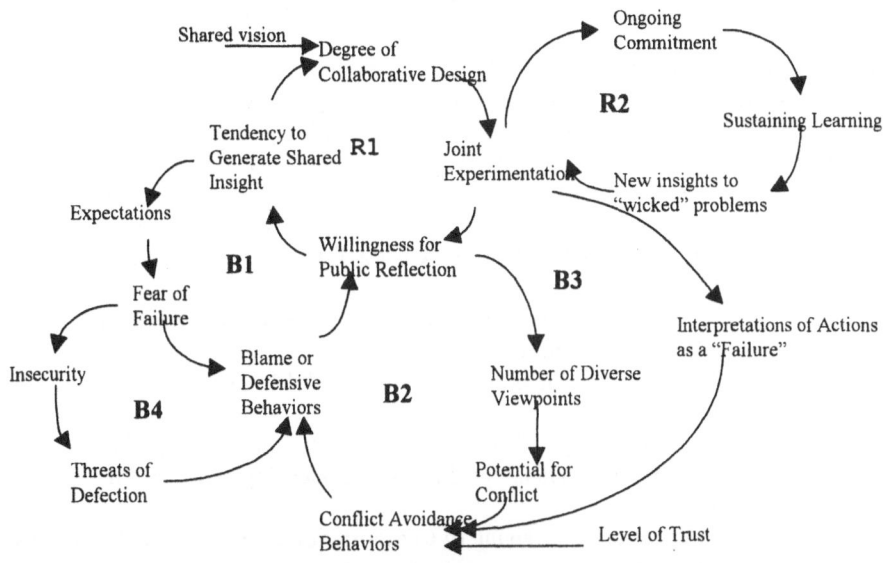

Figure 3. Reinforcing (**R**) and Braking (**B**) Cycles

ASSESSMENT & CONCLUDING REMARKS

From time to time it would be beneficial for all parties concerned to evaluate the application of a teleogenic systems methodology. Such an evaluation may help to correct for: a democratic planning process; undeserved resentment and mistrust of managers, engineers, and planners; and unintentionally counterproductive planning and managing practice. It may also be effective in practice to apply Habermas' four validity claims of truth, rightness, sincerity, and comprehensibility (Habermas, 1979; 1987). An evaluation using these validity claims guarantees that a rationally grounded consensus can emerge from practical discourse. Some evaluators have also used specific questions to supplement their judgements, such as, (1) Are the values, interests, goals, and perspectives of all involved adequately considered and respected? (2) Is there an effective focus on promoting synergy among the parties concerned, through communication and interaction? (3) Do decision-makers and their representatives actively modify their actions to promote the goals and interests of all concerned? (4) Do parties have empathy for each other? Are there positive signs of consciousness-raising to be observed in team members? These are just a set of sample questions that one could ask and many more could possibly be added.

If we look back at Arnstein's ladder, it will be obvious that there are no definite 'yes' or 'no' responses in answer to the questions raised above. It is only the collective honest approximations that will provide a clue as to where an LCG stands with respect to the rungs of the ladder, and what actions if any are necessary to make it more synergistic. The bottom line is whether an LCG has been able to motivate itself to promote synergy among its members and more importantly whether it has acquired the capability to formulate its own goals in the broader interests of the community.

REFERENCES

Arnstein, S., 1969, A ladder of citizen participation, Jour. Of Amer. Inst. of Plan. 35:216-44.

Checkland, P., and Scholes, J., 1990, "Soft Systems Methodology in Action", John Wiley, Chichester, U.K.

Coulter, N. A., and Johnson, A., 1987, Teleogenic Systems Theory in "Decision Making about Decision Making", Abacus Press, Cambridge, MA.

Freire, P., 1972, "Pedagogy of the Oppressed", Penguin, London, U.K.

Flood, R. L., and Carson, E. R., 1993, Dealing with Complexity – An Introduction to the Theory and Application of Systems Science, 2nd Ed., Plenum Press, New York, NY.

Gozdz, K., 1995, Creating learning organizations through core competence in community building, in "Community Building", K. Gozdz, Ed., New Leaders Press, San Francisco, CA.

Habermas, J., 1979, "Communication and the Evolution of Society", Heinemann Educational Books, London, U.K.

Habermas, J., 1987, "Theory of Communicative Action. Vol 1.", Beacon Press, Boston, MA.

Hampden-Turner, C., 1981, "Maps of the Mind", Macmillan, New York, N.Y.

Hutchins, C. L., 1996, "Systems Thinking", Professonal Development Systems, Aurora, CO.

Khisty, C. J., 1996, Education and training of transportation engineers and planners vis-a-vis public involvement", Transport Res. Rec. 1552, National Academy Press, 1996:171-176.

Khisty, C. J., and Leleur, S., 1997, Citizen participation through communicative action: Toward a new framework and synthesis, Journal of Adv. Transport, 31:2, 119-137.

Paris, S. G. and Winograd, P., 1990, How meta-cognition can promote academic learning and instruction, in "Dimensions of Thinking and Cognitive Instructions, B. F. Jones and L. Idol, Eds. Lawrence Erlbaum Assoc., Hillside, NJ.

Ryan, S., 1995, Emergence of learning communities, in "Community Building" K. Gozdz, New Leaders Press, San Francisco, CA.

AFTER JUSTICE: ON THE CONDITIONS OF POSSIBILITY OF JUSTICE IN THE PRESENT

Hernán López-Garay,[1] Ricardo Sotaquirá [1,2]

[1]Departamento de Sistemología Interpretativa
Universidad de los Andes
Mérida 5101, Venezuela
[2]Facultad de Ingeniería de Sistemas
Universidad Autónoma de Bucaramanga
Bucaramanga, Colombia

ABSTRACT

In so called "developing countries" justice, or rather the lack of it, has become a dominant theme in the lives of millions of citizens. Both, in national and international arenas, the citizens of these countries have to face situations of brutal violation of human rights, extremely unjust distribution of their nations' wealth, generalized corruption of their leaders and institutions (e.g. the administration of justice, public health care, etc.), unfair trade with the economic powers of the world, etc.

These conditions make one think if perhaps *the torch of justice has been extinguished in these nations*. But then, *how this has come about, i.e., what are the conditions of possibility of such unjust social order? Can we recover the torch of justice? How can we keep its flame alive? Do we need a new social order for this to happen?*

In order to tackle these and other related questions, we started a project two years ago, based on the principles and concepts of interpretive systemology (López-Garay, 1986; Fuenmayor, 1991a,b). Our research has been focused on the phenomenon of injustice in Venezuela. Our starting point has been the current social debate to reform justice (both social and penal justice) in Venezuela. In particular, we are investigating what conception of justice is embodied in the reform proposed by the state. We want to find out what is the "order of things" (i.e., the cultural context of meaning) to which such a notion of justice might be paying tribute. The preliminary outcomes of this research have opened the way to the design of a counter reform, based on an entirely different conception of the world. A debate between the two reforms will be conducted and its outcomes will help us not only to gain a holistic understanding of justice, but also to bring forth some of the features of what it looks as the emergence of a new epoch. The paper intends to be an illustration of a new systemic way of "managing" complex social issues, a way that is in contrast with some current practices in management and systems sciences.

Synergy Matters: Working with Systems in the 21st Century,
Edited by Castell *et al.*, Kluwer Academic / Plenum Publishers, New York, 1999.

337

1. INJUSTICE: ANOTHER SYMPTOM OF THE MALAISE OF MODERNITY?

According to Charles Taylor (1991) there is an increasing feeling in Western Civilization that modernity has gone astray. The project of the Enlightenment, whereby reason would liberate man from the oppression of nature and cultural forces, seems to have failed. What are the symptoms of this failure? One of them is concerned with *lack of holistic meaning*; another has to do with local and global orders perceived as *unjust*. The latter is acutely manifest mainly in so called underdeveloped societies. The former is present mainly in developed ones. Perhaps the two are interrelated and have to do with the sense of a lost of justice. The case is that for underdeveloped societies the malaise of modernity seems to manifest itself more in a strong concern with *a lack of justice* than with a lack of sense in life: people in those countries claim everyday for a just social order (i.e., a claim to the right to education, health care, work, respect to their human dignity, equal sharing of common resources, etc.). Thus, living in these countries and being aware of these issues create the proper conditions to question the social order and wonder where has justice gone.

Following we are going to present some preliminary results of our project, but first we will summarize the main systems ideas which are guiding the inquiry.

2. SYSTEMS THINKING REVISITED

How does a systems thinker approach the study of any complex phenomena? A systems thinker is someone that has made the commitment to see the world from a *holistic perspective*. So, what does it mean to take a *holistic view* of, say, justice? For Interpretive Systemology it means *to unfold the holistic meaning* of that which has been "distinguished" as the research object. As we will explain in a moment, this view assumes an antireductionistic posture: "things" are never standing by themselves, rather they are like "holograms" i.e., things make their presence always within a rich web of cultural and historical practices, or "forms of life", that constitute their very essence (like the rays of a hologram constitute the object which, an innocent bystander might take as standing by itself). In other words, the "essence" of the object is neither within nor without. The two form a unity.

In showing up, in "appearing" a thing, it actually co-discloses the "web" of "rays" or the "background" which constitutes it and sustain it, thus keeping what shows up from "falling", "sinking" or dis-appearing. The co-disclosure of the background can be illustrated by the act of drawing a circumference. In such an act not only a circle is "created" (distinguished) but also its background (i.e., what is outside the circle). Thus, we can say that any phenomenon is a *figure-background distinction* (distinction both as a verb and a noun). In these terms, a *system is a figure-background unity*. A holistic approach, then, focus its efforts on the problem of understanding the *figure-background* nature of any phenomenon.

Three comments must be made at this point. First, notice that systems are neither a set of interrelated elements, nor their unity can be thought as the emergent property of its parts. Second, the system's environment is neither a set of elements, nor is something which influences the state of the system. The reader more familiar with classical notions of systems might think the notion of "background" is equivalent to that of "system's environment". We must warn him/her, that for reasons that follow[1], "background" is not concept translatable into the mechanistic or the organismic paradigm of systems. In fact, and this is our third comment, the background has a peculiar nature. In López-Garay & Suárez (1998), is explained that the background is an *indistinct* and *homogeneous* "outside" (e.g., the circle's outside in our example of the act of drawing a circle), not something made of elements. It is also *elusive*, i.e., each time we want to explore what it is, we are forced to draw distinctions, thus creating new figure-background units and therefore, dissolving the original background. Another characteristic is its flux-like nature. This is because every act of

---

[1] Cooper (1990) has also extensively argued about the nature of the background (boundary drawing) and its importance for the study of human organizations. Churchman (1968), Checkland (1981), Flood &Ulrich (1990), have identified similar notions (but not the same) to that of "background". Their notions of *Weltanschauung* and *constitutive meanings*, have played an important role in the development of systems thinking.

distinction is an *occurrence*, and occurrence implies change (otherwise it would not be possible to identify it as something happening). This means that systems are not static but continuously in the process of becoming, which in turn means that their occurrence is a continuously coming from a past which is actualized in the present. Consequently, systems are *historical*[2].

Another important characteristic of the background is this. In every specific situation where a distinction is made (e.g., a system's boundary is drawn), the background brought together to light with the distinction seems to be a background of familiarity and concern. Hall (1993, p.132), making reference to Heidegger's work express it thus: "...things show up for us or are encountered as what they are only against a *background* of familiarity, competence, and concern that carves out a system of related roles into which things fit."

What is the nature of this implicit *background* by means of which things are intelligible or meaningful, and which we seem to be unaware of? Taylor (1997, p.69) points out that we are not simply unaware of it, in the same sense that one can be unaware of what is happening in the moon right now, simply because the background makes intelligible what one is aware of in a given moment. Hence, it is inextricably related to what is the focal object at a given moment. Furthermore, this *background*, can be articulated i.e., I can bring it out "...of the condition of implicit, unsaid contextual facilitator... In this activity of articulating, I trade on my familiarity with this background."

The picture one gets from these descriptions is that things always show up associated with an implicit web of practices "...Take a humble entity like a jug. As it shows up in the world of the peasant, still unmobilized by modern technology, it is redolent of the human activities in which it plays a part, of the pouring of wine at the common table, for instance. The jug is a point at which this *rich web of practices can be sensed, made visible in the very shape of the jug and its handle*, which offers itself for this use." (Taylor, 1997, p. 122, italics added). Hence, what is essential to the jug, its use and form, is shaped by the rich set of practices of the peasants. Certainly, these peasants are not fully aware of this "socio-historical" background each time they pour a glass of wine from the jug, yet they are neither completely unaware of this.

We can now summarize several characteristics of the background with the following quote from Taylor (1997): the jug as such "...stands on and emerges out of a vast domain of still *unformed* and *unidentified* reality. This is a field of potential future forming, but it is *limitless, inexhaustible*. All forming is surrounded by and draws on this *unformed*. If we are not closed to it, the jug will also speak of its *history* as a formed entity, of its emergence from *unformed* matter, of its continuing dependency on the *unformed*, since it can only exist as an entity *as long as it is supported by the whole surrounding reality*." (p.122, italics added).

The above implies that the primary task of a systems approach is to display, or reveal how the figure is constituted by its "background", and vice versa. As we will explain in a moment, this task consists in the "opening" or disclosing of the background, i.e., unfolding its *historical interpretive* nature. It is interpretive because, every attempt at opening it, demands building a *context of meaning*, in order to see how the figure "fits" such a context. But given that people have different "backgrounds" (in the everyday use of this word), different contexts of meaning can be designed for a given figure. In other words, we can build different *interpretations* (i.e., different ways of displaying how the figure "fits" its background). Its *sense* as a whole (its *holistic sense*) is the unity underlying these different interpretations.

3. AN *INTERPRETIVE SYSTEMIC* STUDY OF JUSTICE: SOME PRELIMINARY RESULTS

Based on the previous framework, we are conducting a systemic research on justice in Venezuela. Accordingly, we have initially "distinguished" such a complex socio-historical phenomena named justice as "an unjust social order". Following, we are going to explain how to disclose the "background" of such a distinction. We will keep in mind that such disclosings are aimed at helping to explore three basic questions guiding our research, namely: has the torch of justice been extinguished? Can we recover it? How can we keep its flame alive?

---

[2] About the historical constitution of phenomena see for instance, Fuenmayor (1991a,b); López-Garay (1986, ch.7).

3.1. A "narrow" opening of the "Background": *Logic-based* contexts

Our first attempt to disclose the background of the phenomenon "unjust social order" demanded questions such as: what is considered a "just social order"? What is the notion of "justice" underlying this accusation? To pursue these questions, we began to build what we call a *logic-based* interpretive context. This context has the form of a "theory", with a central concept, say justice, being developed in rational conceptual system bounded together exclusively by logical relationships[3]. In this connection we have sought inspiration in Rawl's (1971) theory of justice and in MacIntyre's (1988) theory of liberal justice. We have chosen the former because it is a good example of what we call here a logic-based context. The latter puts in historical perspective the former and show the vital importance of doing so in order to understand the modern notion of justice. Our initial attempts to interpret our current social order in this liberal context of meaning have resulted in two important outcomes. The first outcome is that the logic-based context has been questioned by the "background" itself for not taking into account the fact that its notion of justice is historically and culturally bonded. For one thing, liberalism did not fall from Mars. From the present cultural context of Western societies, liberalism is the outcome of the project of modernity. The second outcome is, as we might expect (based on the conceptual framework presented in previous sections), a demand from the background for something more specific to the particular phenomenon we are studying, namely, Venezuelan unjust social order. This does not mean that logic-based contexts are not useful as a first disclosing of the background. The point to keep in mind is that the process of disclosing *demands* what is needed, and we are not supposed to interfere with any methodological precepts such a process. This last comment may be considered a third outcome of the research, namely, the object of research is constituted by the research and vice versa, something absurd from the point of view of classical scientific research (López-Garay, 1998).

3.2. A "wide" opening of the "Background": *Historical-ontological* contexts

These two outcomes coincide with outcomes from other interpretive systemic studies on public institutions in Latin America (see special issue of *Systemic Practice and Action Research*, December 1998). They reveal that in order to understand our complex social phenomena, first we need to understand the history of those cultural *contexts of meaning* that have led us to want to become "moderns" (e.g., wanting to implant Western institutions, such as (Western) justice, universities, democracy, etc.). Certainly, this history cannot be independent from the history of the cultural contexts of meaning that led the Western civilization to "invent" modernity as such.

In other words, to answer the question *has the torch of justice been extinguished*? we need to perform two fundamental tasks. One is to build the history (a narrative) of the transformations of the cultural contexts, what is called an *historical-ontological* interpretive context (Fuenmayor & Fuenmayor, 1998). The notion of history embodied in the historical-ontological concept is explained thus: "The historical question is not then simply: 'Which events have led us to the present?', but 'What has been the *series of cultural contexts....* that have led us to experience reality (including history) in the way we do at present?'" (ibid., italics added). One very important aspect of this history is how we have come to appreciate the world as "moderns" of a particular kind, namely, as people tutored by the real modern ones (i.e., the Western cultures)! The second task is to ask persistently to this historical interpretive context the question about the sense of justice in the present (in Venezuela). Notice this second task is quite different from one pursuing the following questions: why have the institutions of modernity failed in Venezuela (among then the institution of justice) and what can we do to make them work properly? A research led by these questions would probably aimed at finding a causal theory of the failure of liberalism in Venezuela. Once such a theory is built, actions to correct the situation would be taken. As the reader might have noticed it, our research goes in a radically different direction. For instance, we want to show, not only that we have not managed to copy the Western system of justice in Venezuela, but also, and more important, to provide a context that explains why such a "failure" is an actual "success"!

---

[3] Certainly, the idea of logic is based on a cultural background. Within the systems movement, the assumed purity of logical relationships in the construction of theories and conceptual models is questioned by Checkland & Tsouvalis (1996).

Furthermore, we want to go further and argue, from the historical-ontological context, that justice itself has died!

Although we have not completed these tasks, already we are handling a hypothesis related to our second research question, namely: *can we recover the torch of justice?* Again, notice this is a question we are posing to the historical-ontological context, not to a logic-based context. Our hypothesis is that at a time of fragmentation and the lost for a *will to holistic sense* (López-Garay 1994; Fuenmayor, 1997), the light of justice has to died out. Therefore, there seems to be the need for a preliminary task, namely, to recuperate the will for holistic sense. How can our project help in this direction is, at this point of the research, not clear yet[4]. What it is clear is the following. Assuming the conditions for the recovery of the torch of justice are given (these conditions are implicit in the historical-ontological context), an important task is to design a reform that *can keep the flame of justice alive*. The design of a justice reform will be advanced in the light of the historical-ontological context. Inasmuch as such a context is neither fully developed nor understood, the design will help to unfolded and enrich it. The starting point of this recursive process will be the current debate on justice animated by the desired of the Venezuelan state to carry out the project of building a more just society. Two poles around this debate are: neoliberals and globalizers, on one side, and those who attack the neoliberal project on the other side. We will enter this debate with a counter reform. Notice that our design is not merely a "blue print" for a new social order, where the torch of justice will light again. It is, fundamentally, a device to help us obtain a more profound understanding of the present. Let us explain this point in the final section of this paper.

4. MANAGEMENT AS THE SCIENCE OF *MEANING-FULL* DECISIONS

We would like to conclude drawing some quick comparisons between the systemic approach to "manage" complex social phenomena presented in this paper and more conventional approaches of management and systems sciences. The comparison suggests that management science, at least in relation to its application to manage complex social systems, may have to follow a different road, one related to *meaning-full* decisions rather than merely *effective* and *efficient* decisions.

Some streams of management and systems sciences (Flood, 1998), embody a view of the manager as a hard-headed practical pragmatic realist. Accordingly, the realm of managerial expertise is defined as one purport to be objectively grounded, a realm whose aim is to provide solid knowledge to predict and control human organizations. Success in this realm is based on scientific knowledge (i.e., a sound stock of law-like generalizations). A scientific manager is then a man that uses the knowledge of management science to organize and control some course of events (MacIntyre, 1985, p.101). Planning, and design are some of the tools provided by management science to achieve these purposes.

But the interpretive systemic perspective provides a different possibility. As we have seen, according to this perspective human action is always embedded in a cultural web which constitute action, man, and phenomena in general. Therefore, we can say that, from this perspective, humans are always *engaged* beings (Taylor, 1997, ch.4). Moreover, we are engaged not only to our cultural contexts but also to the sediment of their gradual transformations. Inasmuch as these transformations are not of our own making (i.e., we do not decide we are going to see the world in "modern" or "postmodern" terms), we can hardly talk about controlling our lives at will. In fact, since we are historical and culturally engaged beings this means that we can never find a sort of neutral, a-historical ground from which we can see the world and "control" it. On the contrary, since we are always constituted by a cultural background, then we might be more truthful to our human condition if we say that it is the latter that actually controls us! In another context, Flood (1998) quoting Peter Reason has made a similar point: "phenomena as wholes can never be fully known for the very reason that we are part of them, leading us to acknowledge and respect the great mystery that envelops our knowledge" (p.95). But if this notion of man and reality were seriously taken then what would be the sense of a science of management, which by definition is the science of control (Beer, 1959)?

---

[4] In our Department, the project: *Education for the Twentieth First Century*, led by professor Fuenmayor is precisely aiming at helping our culture to recover this will.

Churchman, a management scientist that has made important contributions to the systems movement, could perhaps be interpreted in the direction offered by our interpretive systemic perspective. In his well known book *The Systems Approach*, he says that this approach is in the business of *deceptions*, by which he means that every time we think we have caught reality in our perspectives, we learn, more to our dismay, that they are too limited. (Churchman, 1968, pp.228-230). According to this view, management science is then in the business of liberation, that is, it is in the business of helping us to get rid of the illusion that science (or any other perspective) has access to special knowledge that can be used to predict and control social phenomena.

The interpretive systemic view opens the door, then, to a completely different way of dealing with social phenomena, a way based on *appreciation* of the mystery of being, its multiple manifestations and our finitude. In its light the task of management science could become, then, one of helping to make *meaning-full* decisions, i.e., decisions based on some degree of holistic appreciation of that background which constitute us and our actions. Clearly, many of its concepts would have to change. Take for instance one of them, the concept of design. In the new perspective design would not be merely a tool to find the most appropriate means to achieve some given objective. Rather, it could be the means to reveal our constitutive engagement and, consequently, that of social phenomena in general. This is, precisely, the use we intend for design in our project.

REFERENCES

Beer, S., 1959, "Cybernetics and Management", The English Univ. Press, London.

Cooper, R., 1990, Organization/disorganization, *in:* "Theory and Philosophy of organizations", J.Hassard y D.Pynn, eds., Routledge, UK.

Checkland, P. B., 1981, "Systems Thinking, Systems Practice", Wiley, Chichester.

Checkland, P. B. and Tsouvalis, C., 1996, Reflecting on SSM: The link between root definitions and conceptual models, *in:* The Lincoln School of Management Working Paper Series, 5, University of Lincolnshire & Humberside, Lincoln, UK.

Churchman, C. W., 1968, "The Systems Approach", Dell, New York.

Flood, R., 1998, Action research and the management and systems sciences, *Systemic Practice and Action Research*, 18:1.

Flood, R., Ulrich, W., 1990, Testament to conversations on critical systems thinking between two systems practitioners, *Systems Practice*, 3: 1.

Fuenmayor, R., 1991a, The self-referential structure of an everyday-living situation: A phenomenological ontology for interpretive systemology, *Systems Practice*, 4: 449-472.

Fuenmayor, R., 1991b, Truth and openness: An epistemology for interpretive systemology, *Systems Practice*, 4: 473-490.

Fuenmayor, R., 1997, The historical meaning of present systems thinking, *Syst. Res. Behav. Sci.*, 14:4.

Fuenmayor, A. and Fuenmayor, R., 1998, Researching-acting-reflecting on public health services in Venezuela: II. Community action and critique, *Systemic Practice and Action Research*, 18:6.

Hall, H., 1993, Intentionality and world: Division I of *Being and Time*, *in:* "The Cambridge Companion to Heidegger", C. Guignon, ed., Cambridge University Press, Cambridge.

López-Garay, H., 1986, "A Holistic Interpretive Concept of Systems Design", Ph.D. dissertation, University of Pennsylvania, The Wharton School, Philadelphia.

López-Garay, H., 1994, Proyecto de una plataforma de base para pensar sistémicamente el problema del desarrollo de América Latina, *Sistémica 94*, Instituto Andino de Sistemas, Lima, Perú.

López-Garay, H., 1998, Interpretive systemology and systemic practice, guest editorial, *Systemic Practice and Action Research*, 18:6.

López-Garay, H., and Suárez, T., 1998, The holistic sense of prison phenomena in Venezuela: III. The unity of the research, *Systemic Practice and Action Research*, 18:6.

MacIntyre, A., 1985, "After Virtue: A Study in Moral Theory", Duckworth, London.

MacIntyre, A., 1988, "Whose Justice? Which Rationality?", University of Notre Dame Press, Notredame, Indiana, USA.

Rawls, J., 1971, "A Theory of Justice", Oxford.

Taylor, C., 1991, "The Malaise of Modernity", Canadian Broadcasting Corporation, Canada.

Taylor, C., 1997, Heidegger, language and ecology, *in:* " Philosophical Arguments", Harvard University Press, Cambridge.

IDENTITY AND BOUNDARY MANAGEMENT IN SYNERGISTIC RELATIONSHIPS

Yasmin Merali

Warwick Business School
The University of Warwick
Coventry
CV4 7AL
United Kingdom.

INTRODUCTION

Notions of dynamism, regeneration, change and innovation are often cited as important in competing in the future (Hamel and Prahalad, 1994). This paper is concerned with the development and management of synergistic relationships between social and organisational systems operating in dynamic contexts. The first part of the paper develops a definitional framework for examining the nature of synergistic relationships in terms of

- the role of identity in defining the synergistic scope
- the nature of boundary relationships
- the development and management of boundary and identity constructs in dynamic contexts over time.

Autopoietic concepts (see Maturana and Varela, 1973) are used to describe the boundary phenomenology (see also Merali, 1998b). Issues of survival of individual identity over time and the dynamic of the boundary are discussed in the context of Heidegger's (see Heidegger, Being and Time) notion of *Dasein* (see also Merali, 1997). This treatment considers issues about

- the persistence of identity,
- the authenticity of *being*, and
- the alignment of inter-organisational boundary components

in synergistic relationships.

The discussion that follows attends to these questions, and suggests that the sustainability of such relationships is a function of the congruence between individual and societal (organisational) self-concepts.

Synergy Matters: Working with Systems in the 21st *Century*,
Edited by Castell *et al.*, Kluwer Academic / Plenum Publishers, New York, 1999.

the reasons that move us, the values we try to realise, the plans we admire, and the people we seek out. Avoidance of such decisions slows down the development of attractions, reasons, values, plans and associates'.

For an organisation that is operating in a dynamic environment in which knowledge acquisition and evaluation are significant activities, establishing and understanding its own identity is particularly important. In the sense of this paper, *self-concept* arises from being and gives rise to doing. For the observer, identity is manifest in the observable actions of the unit members.

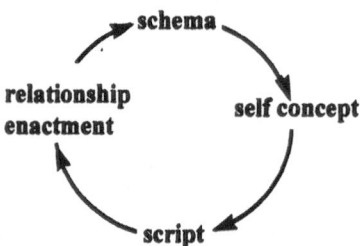

Figure 1: Cognitive Processes Underpinning the Enactment of Boundary Relationships (Merali, 1998a)

Relationships. Relationships between individuals determine the content and structure of intra- and inter-organisational knowledge networks. Extended to the societal level, the individual relationship network is an important mechanism for sharing and diffusing ideas and the intersubjective realisation of the synergistic relationship.

The Script. In considering the enactment of *relationships* with regard to mutual perceptions and shared meanings, it is useful to characterise the roles that are played by the participants in the *relationship*. Each distinctive *relationship* is mediated through, and evidenced by, its transactions. Transactions are influenced by the perception of each participant about him or herself in relation to the other participants. These perceptions are analogous (see Merali and McGee, 1998) to the so called *scripts* referred to by adherents of *Transaction Analysis* (Berne, 1961):

An individual's script is an internal conceptual structure containing a set of rules and norms and a highly cross-linked set of data about self, the world and interactions between the two. The script acts as a filtering mechanism for fresh data : nothing is accepted into a script unless it can be made to fit with what already exists. The script evolves over time as new information is incorporated, and socialisation modifications occur as a result of its involvement in meaningful relationships. The nature of this evolution (e.g. in terms of what can or cannot be incorporated into the script) is itself determined by, and is characteristic of, the existing script.

The collective enactment of relationship scripts is the mechanism by which the *self-concept* is realised, capabilities are leveraged, experiences are formed and learning takes place (i.e. the collective enactment of *scripts* is the manifestation of the organisational *being*).

The Schema. The *schema* is the total cognitive construct which contains the organised collection of interconnected beliefs and perceptions about self and the universe together with their spatial, temporal and semantic *relationships*. The schemata embrace *self-concept, scripts* and environmental sensemaking together with mechanisms for their linguistic and semiotic articulation.

SYNERGISTIC RELATIONSHIPS

Recent strategy literature on the resource based view of the firm and core competence management highlights the dynamic nature of competitive contexts and suggests that

- for individual firms competitive rents derive from leveraging co-specialised firm-specific capabilities (Peteraf, 1993),
- the competitive context will be defined in the future by firms competing over competences (for example, see Rumelt, 1994; Teece, Pisano and Sheun, 1997) and that
- the successful organisations of the future will be the ones that have the ability to continually "reinvent" themselves to be competitive within this context (Prahalad and Hamel, 1994).

This implies that firms will need to have access to a diversity of capabilities from which they can dynamically generate "valuable resource bundles" that can be deployed to compete in the changing context. To obtain access to the requisite variety of capabilities necessary for viability in this scenario an individual firm would need to

- develop the requisite variety of capabilities internally or
- acquire a company that has a complementary set of capabilities.

The organic development of capabilities carries a consequent time implication that may render it ineffective in a fast-moving competitive context. Similarly, the acquisition option may be attractive in the short-term, but the need for reinvention over time and the problems associated with post-acquisition integration and divestiture rates (Rumelt 1994) make it less attractive as a long-term proposition.

The synergistic relationship model suggests an alternative. Within this model a firm can develop a synergistic relationship with another firm such that it can have access to part of that firm's capability pool. This alternative potentially offers a greater degree of flexibility and choice in terms of the diversity of options available. By participating in the relationship the firms can leverage their selectively combined capabilities and resources to accrue far greater rents than would be possible on the purely intra-firm resource-based competitive model. The emergent literature on inter-organisational networks and alliances illustrates this concept.

IDENTITY AND COGNITIVE PROCESSES IN BOUNDARY MANAGEMENT

The Concept of Identity and the Enactment of Boundary Relationships

The identity of the social unit in terms of distinguishing the "self" from the "other" is embedded in cognitive structures and shared models recipes and routines within the organisation (Nelson and Winter, 1982).

The model (derived from Merali, 1998a) described below outlines the role of identity and cognitive processes underpinning the enactment of boundary relationships:

Self-Concept. Organisational *self-concept* is of fundamental import in developing, realising and leveraging distinctive capabilities. *Self-concept* is a perception of that which we call our identity (see Weick, 1995 for a treatment of identity). Weick highlights the importance of identity as a focus for sensemaking and as a basis for congruent action:

'Each....organisation chooses who it will be by first choosing what actions, if any it needs to explain, and second by choosing which explanations for these actions it will defend. An inability or unwillingness to choose, act and justify these leaves people with too many alternatives and too few certainties. Binding decisions affect the tasks we are attracted to,

Boundary Definition. Boundary definition is controversial (Weick, 1977): in this paper we use Maturana and Varela's (1973) notion of the "autopoietic unity". In these terms, the "autopoietic unity" is observed to be a unity by an external observer because there is a visible cleavage from the environment of the components that constitute the unity. Maturana and Varela use the notion of *structural coupling* and *neighbourhood relations* to speak of the boundary relationship and the internal/external partitioning.

Components of the autopoietic unity have strong neighbourhood relationships with other components belonging to the same unity. Boundary components at the interface of the unity with the external environment have recurrent interactions with environmental elements, but these interactions do not display the strength of sustained linkages that characterise internal neighbourhood relations. Individuals engaged in the enactment of the boundary relationship are potentially key actors in the process of modifying the organisational schema: their significance depends on the extent to which they are meaningfully networked with other individuals (inter- and intra- organisationally).

ENACTMENT OF THE SYNERGISTIC RELATIONSHIP MODEL

Sustainable rents are earned by imperfectly mobile, imperfectly imitable resource bundles. Causal ambiguity is an important mechanism for meeting the imperfect imitability criterion. In the synergistic relationship model, inimitability is related to
- the degree of specificity
- the strength of the structural coupling and
- the relative complexity

of the relationship between the participating firms.

Enactment of a Loosely Coupled Synergistic Model

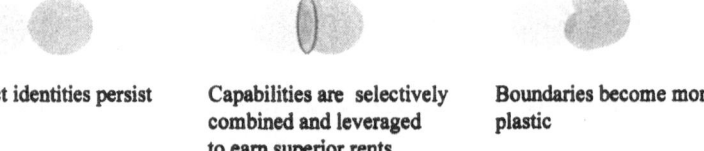

Distinct identities persist | Capabilities are selectively combined and leveraged to earn superior rents | Boundaries become more plastic

Figure 2: Characteristics of a Simple Enactment of a Synergistic Relationship

Figure 2 illustrates the characteristics of the least complex enactment of the synergistic model. Firms engaged in such a relationship would retain their individual identities, and synergistic leveraging of the capabilities would be of the type readily described by Williamson's (1975) transactional analysis. The organisational boundaries may be plastic, with processes and information being shared across boundaries, but the observable actions of individuals interacting at the boundary can be clearly identified as belonging to the distinctive autopoietic phenomenology of the individual firms.

In the competitive context, this type of relationship is fairly transparent to the observer, and therefore readily imitable by others. In terms of the resource-based theory, rents earned by this type of collaboration can be easily eroded by fast followers.

Tight Coupling and the Persistence of Individual Identity

One aspect of the synergistic model is that the relationships may be quite amorphous, ideally only lasting for as long as the collaboration proves to be mutually beneficial. An interesting situation arises when individuals interacting at the boundary level become engaged in activities that are tightly coupled (e.g. inter-organisational teams engaged in joint projects). Team members are boundary components (i.e. points of interaction between the two firms). The success of the team depends on the successful alignment of these components and is observable in the internal congruence of team behaviours. While the project is active, the strength of the structural coupling between the team members may exceed that of the individual team members to their non-team "compatriots". The issue of what the organisational identity is during this particular "time slice" is an interesting philosophical question. This blurred interface boundary will be considered in greater detail in a future paper. Here it is addressed in terms of the definition of boundaries in autopoietic systems.

Maturana and Varela distinguish between *"organization"* and *"structure"* in defining autopoietic systems. Mingers (1997) provides a definition:

the *organization* of the autopoietic unity is "…..the subset of the relations between components that determine the properties of the unity as a whole, and thereby its identity, type or class. All unities of a particular type have the same *organization*. A change of *organization* implies a change of identity and vice versa."

the *structure* of the autopoietic unity is "…the total set of actual components and relations belonging to a particular concrete example or instance….. the *structure* of a particular entity may change without its organization changing. *Organization* is abstract, *structure* is concrete".

Asymmetric Synergistic Relationships

In the closely coupled enactment of the synergistic model the distinction between *organization* and *structure* (in the specific sense defined here) is blurred and identification of the "autopoietic unity" is uncertain. This is not problematic if such arrangements are symmetric relationships. If the relationship is asymmetric (i.e. one firm develops a resource dependence on the other) the dependent firm risks exploitation by the other. This occurs commonly in consultancy arrangements where high calibre client personnel work alongside consultants in project teams, often for high profile projects billed as learning partnerships but where the consultancy firm is perceived as having the more marketable expertise. Often, the client firm becomes vulnerable to poaching of capable staff at the end of the consultancy assignment (particularly in sectors where there is a high demand for sector-specific consultant expertise).

The enactment of the relationship is a societal process where individuals share common experiences, engage in sensemaking and develop new meanings. In this process, people *are* boundary components. Tight coupling between client and consultant personnel in the "here and now" creates a transient "project identity", and the project space is the shared *clearing* (see Heidegger, Being and Time) within which *Dasein* (*being there*) is situated, pressing into new possibilities, embodying time. At the termination of the project the individual *script* of the apparently "defecting" client personnel has become discontinuous with and decoupled from the original client firm self-concept, and the individual realises a new script derived from the enactment of the project identity. In Maturana and Varela's terminology, defection of individuals is a change in the *structure* of the firms. It may also represent a threat to the *organization* (and hence the identity) of the firms as the individuals may take "unique" cognitive capabilities and structures with them.

Symbiotic relationships

Long term synergistic relationships may result in the atrophy of redundant capabilities in each of the participating firms. In the extreme case, the two firms may develop mutual dependencies in a manner that is analogous to the symbiotic relationships found in the ecosystem. The symbiotic nature of the relationship constrains the movement of both firms in the competitive space. This may render the firms vulnerable by limiting their transformational capability in the face of a dynamic competitive environment.

CONCLUSIONS

The sustainability of the synergistic relationship is a function of the congruence between the individual and collective perceptions of the organisational self-concept. If the enactment of the synergistic relationship destabilises individual scripts to such an extent that the authenticity of either of the parent organisational identities is threatened, the coalition is unlikely to survive.

REFERENCES

Berne, E. (1961). *Transactional analysis in psychotherapy*. Grove Press, New York.

Hamel, G. and Prahalad, C. K., 1994, "Competing for the Future", Harvard Business School Press, Boston.

Heidegger, M., "Being and Time", (translated by Macquarrie, J. and Robinson, E.), SCM Press Ltd.

Maturana, H. and Varela, F., 1973, Autopoiesis and Cognition : The Realisation of the Living : Organisation of Living, Reidel, Holland.

Merali, Y., 1997, Information, Systems and *Dasein* in Systems for Sustainablity: People, Organisations and Environment, Stowell, F. *et al*, ed. Plenum, New York.

Merali, Y., 1998a, The role of information in the leveraging of capabilities, *Fourth International Conference on Competence Based Management*, June 1998, Oslo, Norway.

Merali, Y., 1998b, Application of Autopoietic Concepts to a Social System: an Exploration and Reflections, *Colloquium on Organisations as Complex Evolving Systems*, Warwick, December 1998.

Merali, Y. and McGee, J., 1998, Information competences and knowledge creation at the corporate centre, *in* "Strategic Flexibility: Managing in a Turbulent Environment", Hamel, G., Prahalad, C.K., Thomas, H. and O'Neal, D, ed., Wiley, Chichester.

Mingers, J. (1997) "Systems Typologies in the Light of Autopoiesis: A Reconceptualization of Boulding's Hierarchy, and a Typology of Self-Referential Systems" *Syst. Res. Behav. Sci.* **Vol 14**, pp 303-313.

Nelson, R. and Winter, S., 1982, "An Evolutionary Theory of Economic Change", Harvard University Press, Cambridge, MA.

Peteraf, M. A., 1993, The cornerstones of competitive advantage: a resource-based view, *Strategic Management Journal*, **14**, 179-191.

Rumelt, R., 1994, Forward *in* "Competence-Based Competition", Hamel, G. and Heene, A., ed., Wiley, New York.

Teece, D.J. Pisano, G. and Sheun, A., 1997, Dynamic capabilities and strategic management, *Strategic Management Journal*, **18**, 509-533.

Weick, K. E., 1995, "Sensemaking in Organisations", Sage Publications, Thousand Oaks, California.

Weick, K. E.., 1977, Enactment Processes in Organisations in "New Directions in Organisational Behaviour", Staw, B.M. and Salancik, G.R.. ed. , St Clair Press, Chicago.

Williamson, O. E., 1975, Markets and Hierarchies, Free Press, New York.

SHAPING A LOCAL COMMUNITY INFORMATION SYSTEMS SUPPORT INFRASTRUCTURE USING A COLLABORATIVE APPROACH

Nick Plant, Morris Williams and Anne Moggridge

Community Information Systems Centre
Faculty of Computer Studies and Mathematics
University of the West of England
Frenchay Campus
Bristol BS16 1QY, United Kingdom

INTRODUCTION

The authors share an interest in information systems (IS) practice in community organisations (COs), which research (Williams, 1998; Ticher & Firth, 1997) suggests have, in recent years, become technologically sophisticated when compared with -say- the early 1990s (Plant, 1992; Rowan, 1994). Use of small computer systems is prevalent, albeit focused primarily on basic office automation applications and lacking adequate investment, user expertise and strategic alignment. Interest in new media technologies is growing, although as yet a small number of COs are taking advantage of the potential of Internet and multimedia technologies. National level policy debates and initiatives are anxious that the community sector avoids exclusion from the information society in the 21st century by joining the communications revolution (IBM, 1997; Day & Harris, 1997).

In this paper, we aim to show how our local collaborative research and practice has, whilst these policy debates have developed, taken us beyond the individual and particular concerns of individual COs, to focus on the collective needs of community organisations. We discuss the IS needs of COs, explore the gaps found to exist between needs and provision in our own locality (Bristol), and establish the requirements for a support infrastructure that could improve the situation overall. We give examples of how various collaborative projects with which are engaged could be considered to be helping to shape such an infrastructure, and the conclusions to the paper raise some pertinent issues relating to our role as academics within the emergent local community network.

THE LOCAL COMMUNITY NETWORK MODEL

It has already been shown (Plant, 1998a) that IS situations in COs can be "messy", and work in progress is concerned with *sustainable* approaches to community information systems (CIS). The term sustainable is used here to incorporate not only the quality and life

Synergy Matters: Working with Systems in the 21st Century,
Edited by Castell *et al.*, Kluwer Academic / Plenum Publishers, New York, 1999.

349

expectancy of the technical components of CIS, but also the degree of autonomy in the use of IS exhibited within COs, and the extent to which IS is successful in strategic and management terms (Plant, 1997a). One strand of our examination of sustainable CIS has led us to explore the extent to which a holistic set of IS needs within COs is being addressed, and to hypothesise that sustainability might be improved by (amongst other things) redressing any gaps between internal needs and external expertise.

A model of IS needs has therefore been derived from previous work (Plant, 1997b) as well as academic sources (Galliers, 1995; Wilson, 1991) and voluntary sector research (Plant, 1994; Ticher & Firth, 1997; Gilchrist, 1995). A key characteristic of this needs model is the inclusion of management and inter-organisational or collaborative needs, as well as the technical needs which we have found to be the dominant concern previously in COs.

A recent project described this model in lay terms and mapped out the availability of external expertise in comparison with the requirements suggested by the needs model (Plant, 1998b). This study found that needs in technical areas are relatively well catered for by external service providers, although it is not always "sector-friendly" in its character. More significantly, however, supply of external expertise in *all* of the non-technical areas (which include IS planning, management, and information analysis) is very limited indeed. Furthermore, there is hardly any of the inter-organisational support and collaboration that, in areas of work other than IS, is at the heart of the modus operandi of the voluntary sector.

These findings, and other work in conjunction with local partner organisations within Bristol's voluntary sector, have led us towards a vision of a local infrastructure to improve upon the situation outlined above. Such an infrastructure would require to be sector-led, and to be aligned with and sensitive to the working practices and culture of COs. It would need to be highly collaborative, and have the ability to coordinate disparate projects and initiatives. It would also promote cooperation between partners, in order to reduce fragmentation of services and ad hoc or short-term approaches, and build understanding and coherency systemically within a network of supply and demand. It would have to accommodate a range of different interests in a complementary, multi-agency fashion, because (for example) commercial as well as philanthropic service providers as well as "users" would need to be recognised as an essential part of the overall picture, and in order to manage the likely political tensions. The principle of reciprocity would be essential, and a commitment to empowerment and capacity-building within COs (Skinner, 1997) would be needed, especially on the part of the more powerful partners. A strong commitment to evaluation, the ability to learn from and build on past experiences, and an ethic of inter-organisational learning, would be needed. It would finally itself need to exhibit sustainability.

Such a local community network model, as we term this vision, might more formally be defined as *a system to promote the effective development, delivery and use of appropriate local CIS services and deliver mutual benefits to a complementary and diverse range of partners*. Its activities have been preliminarily identified as those shown in figure 1, and participants include COs, CIS service providers, CO umbrella and (non-IS) support organisations. Its ownership would be in the collective hands of its participants, but dominated by voluntary sector partners, unlike in previous models such as described in Plant (1994). In the light of the action research orientation of our own work, and our commitment to academic endeavour which leads to practical benefits in partner organisations (Lynch, 1998), this model also includes ourselves as University academics as participants.

TOWARDS THE OPERATIONALISATION OF THE MODEL

Our local community network model has not yet, to our knowledge, been operationalised in the real world, and there has been a somewhat chequered history of

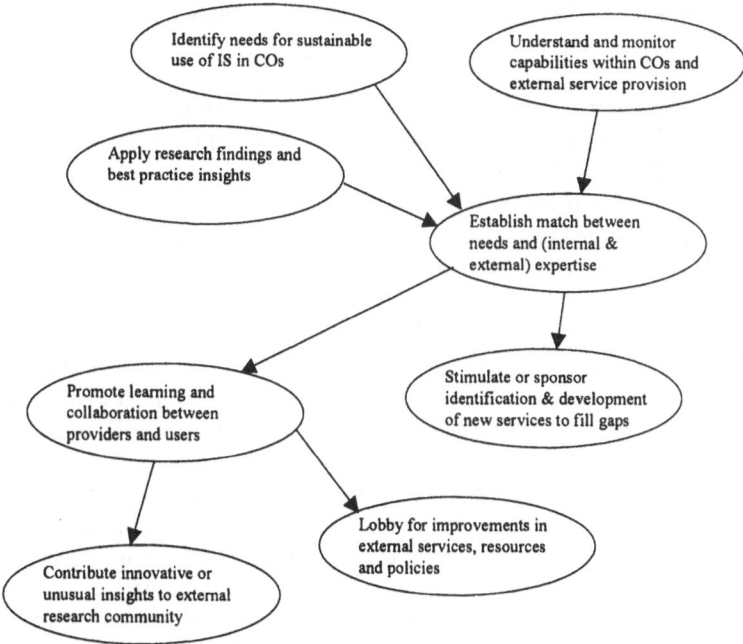

Figure 1. Local Network Model

attempts to establish similar entities in the UK community computing movement, some of which have been the subject of reflection and evaluation (Plant, 1994; Ticher & Firth, 1997). However, the problematic nature of the gaps between CIS expertise requirements and available support, highlighted above, is now being increasingly recognised, with growing concern, during various action research and consultancy engagements with which we are involved in the Bristol area.

We will therefore give three examples of projects we are involved in that together offer some evidence that a movement towards the operationalisation of the model is already taking place, and provide insight into how this process could be continued or completed. The examples chosen are also intended to draw out some interesting issues about the relationship between a Centre within a University and the local voluntary sector.

Resources Guide. Following on from student placement work in the voluntary sector some years ago (Smith & Plant, 1994), this project as already mentioned (Plant, 1998b) involved the research and documentation of a profile of providers "of computer or computer-related services" to community organisations in Bristol. The resulting guide also attempts to convey to COs the practical implications of the sustainability principles, and makes observations about the mapping of supply and demand thus created. Details of the outcomes of this project have already been summarised, but for the present purpose some observations about project processes are more significant. Firstly, whilst initiated by ourselves, the project was constituted on a partnership basis with a local voluntary sector umbrella organisation, and our work was informed strongly by a sector-led Steering Group. Secondly, whilst this engagement could not yet be said to be sustainable, plans are being laid for the research, production and publication of the next edition of the guide to be carried out within the sector. We therefore aim, with the active agreement of the partner organisation, to transfer ownership from the academic to the community environment at the earliest opportunity, though this process is likely to require intermediate capacity-building steps before being completed. Thirdly, a considerable degree of mutual learning has accrued to the partners as this project was carried out and evaluated.

"Connections". This project is a rare example of a University-owned project funded by the government's Single Regeneration Budget, and was set up to investigate the potential for new media technologies in Bristol's communities. The University's role in the pilot phase of the project was as lead partner, though from the outset a strong and increasingly empowered Advisory Group dominated by active voluntary sector partners was established and has been maintained. The findings of a detailed evaluation of the pilot phase (Williams and Plant, 1998) include convincing evidence of the need for a community development approach (Taylor, 1992), not technical expertsie, to be dominant in this sort of work. It also reinforces the need for a holistic approach to service provision and CIS needs, in showing that the potential of emerging technologies in COs (Williams, 1997) can be realised only with infrastructural support which is fully integrated with other CIS and organisational development & management provision. Such integration is, worryingly, *not* evident in the provider expertise mapping exercise already mentioned (Plant, 1998b), which found that new media technology support was in most cases isolated from general CIS technical support and management provision. The greatest challenge that this leaves the project in its current phase is to attempt to establish, or stimulate the establishment of, service provision which makes good such deficiencies. Meanwhile, the locus of control of this project has already (deliberately) shifted significantly from the University to the major community-based partner organisation, and our aim is to continue and complete this transfer before the end of the funding period, so that our engagement ultimately results in sustainable change having occurred within the voluntary sector.

Student Consultancy Projects. A significant source of experience and insight for us, as academics committed to both teaching and research, is the management of student IS consultancy projects undertaken with local COs (Moggridge, 1994; Williams et al, 1998). In our role as managers of these projects, we are active in promoting learning and collaboration between students, as providers of IS expertise, and COs. A critical success factor in such collaborations is the development of a set of shared understandings and realistic expectations about what needs to be, and what can be, achieved through a particular engagement. We see our role here as applying our experience to preparing the ground with both students and COs so that they can work effectively together to generate the kind of meaningful, practical outcomes that we expect to meet sustainablility criteria. Debating our own understandings of best practice with the broader academic community helps shape our approach but we also need to check that our own assumptions are valid on the ground. To this end, we employed an external consultant with considerable CIS expertise to evaluate the experience and outcomes of student projects from the points of view of the COs who have participated. The findings from this evaluation (Clarke, 1998) confirm the importance of each of the sustainability criteria identified earlier, but also suggest new ways in which we as facilitators, as well as occasional providers, might contribute to the development of the local network. The research confirms the gaps in CIS expertise already discussed, and highlights the point that the sustainability of the tangible outcomes of student work rests to a great extent on the extent to which client COs have internal capabilities to continue student work or procure appropriate follow-up expertise from outside. More encouraging however is the evidence that the less tangible outcomes of student interventions, in the form of individual and organisational learnings from the experience, play a valued role in informing future IS initiatives within participant COs.

CONCLUSIONS AND DISCUSSION

The examples given above help us begin to see how as members of a University we might contribute to many of the ongoing activities which will be necessary in order for the model to be realised as a network which is itself sustainable. Our regular connections with a range of COs at any point in time enable us to continue to monitor the broader picture in terms of capabilities and needs, and through the Resources Guide we offer concrete assistance to COs attempting to establish a match between their current needs and local sources of expertise. As researchers and occasional practitioners, and through our students, we are able to provide some of the expertise required, and perhaps more significantly we hope that our own *IS* expertise may contribute by encouraging a broader stance than the *IT*-focused interpretation of CIS problems that has so far been more typical. Research already described elsewhere shows promise in filling what we see as a key gap in the current pattern of CIS service provision, via an innovative workshop offering practical guidelines on IS management in COs (Plant, 1997b). The management of student consultancy projects to ensure that the service delivered is responsible and appropriate has required and inspired ongoing innovation in education, and also enables us to contribute to research into IS practice generally and CIS practice specifically. The Connections project illustrates the potential role for a University in facilitating and stimulating innovation in community contexts, and more specifically will we hope lead to the establishment of new provision for community new media technology support that is integrated with general CIS support.

Returning to the model showing the requisite activities of the local community network in figure 1, it can therefore be seen that most of the theoretical activities therein have a real-world counterpart. Furthermore, we have attempted to demonstrate that the general principles set out in our vision of a local CIS support infrastructure are being addressed through the manner and style of collaborative working that is developing within our actual local network. This we would argue promises to contribute the necessary coherency and systemic "glue" to the shaping of the model in the real world.

At a more practical level, during dialogue with voluntary sector partners, increasingly explicit attention has been given to "the big picture" (as the local network model tends to be labelled in those circles), and a number of specific and practical action steps towards operationalisation are being taken. A key feature of this dialogue and the collaborative projects we have outlined has been that we as academics have been accepted into the emergent inter-organisational network because of the practical benefits associated with our various contributions. Genuine commitment to, and practical skills in, partnership working (Wilcox, 1994) have also been essential.

The realisation of the local community network vision is therefore not only systemically feasible but has become an active and practical proposition that is viably shared between us academics and our community partners.

In demonstrating the potential of the model and some of our current contributions to the necessary activities, we have raised many issues that require further research. Whilst we are convinced that we cannot and should not play a dominant role in the future operation of the local network envisaged, it seems clear that inter-organisational learning will be key to its sustainability. Our professional commitment to education, and to reflective practice in IS, qualifies us as contributors in this regard, we hope. However, a fuller understanding of the principles and practices of community development, and how these can be integrated methodologically with our best endeavours towards empowerment-based IS practice, is yet to be developed in more detail. Meanwhile, the continued development of our knowledge and experience of CIS, and any further successes in rendering that knowledge accessible, practical and meaningful to COs at the grass roots will, we hope, continue to be amongst the rewards.

REFERENCES

Clarke, M., 1998, "Student Computer Consultancy in the Voluntary Sector; an evaluation from a client perspective, Community Information Systems Centre, University of the West of England, Bristol

Day, P. and Harris, K., 1997, "Down-to-Earth Vision; community-based IT initiatives and social inclusion", IBM United Kingdom Limited, London

Galliers, R., 1995, Re-orienting information systems strategy: integrating information systems into the business, *in* "Information Systems Provision - the contribution of soft systems methodology", F. Stowell, Mc-Graw-Hill, Maidenhead

Gilchrist, A., 1995, "Community Development and Networking", Community Development Foundation, London

IBM, 1997, "The Net Result; social inclusion in the information society; report of the national working party on social inclusion (INSINC)", IBM United Kingdom Limited, London

Lynch, M., ed, 1998, "An Introduction to the Community Information Systems Centre at UWE", Community Information Systems Centre, University of the West of England, Bristol

Moggridge, M., 1994, Linking Research and Practice in Community Information Systems Projects, *in* "Proceedings of the sixth conference on information systems teaching: improving the practice", University of Hertfordshire, Hatfield

Plant, N., 1992, "Community Computing in Avon - Taking Stock; a survey and report", (then) Bristol Polytechnic in association with Avon Community Computing Network, Bristol

Plant, N., 1994, Local People Networks and the Community Computing Challenge - an evaluation of the ACCN experience, *in* "Local Support for Community Computing, Proceedings of the Community Computing Network Conference", R. Smith and R. Ennals, eds, Community Computing Network, Manchester

Plant, N., 1997a, Sustainable Information Systems in Community Organisations, *in* "Philosophical Aspects of Information Systems", R.L. Winder, S.K. Probert and I.A. Beeson, eds, Taylor & Francis, London

Plant, N., 1997b, Towards Information Systems Sustainability in Community Organisations Using an Organisational Learning Approach, *in* "Systems for Sustainability: People, Organisations and Environments, UKSS Conference July 1997", F.A. Stowell, R.L. Ison, R. Armson, J. Holloway, S. Jackson and S. McRobb, eds., Plenum, New York

Plant, N., 1998a, 'Messes' in Community IS: The Story of 'Briscare', *in* "Matching Technology to Organisational Needs, UKAIS Conference April 1998", D. Avison, and D. Edgar-Nevill, eds., Mc-Graw-Hill, Maidenhead

Plant, N., 1998b, "Community Computing Resources in Bristol; a guide for local voluntary organisations", 2nd edition, Community Information Systems Centre, University of the West of England, Bristol

Rowan, P., 1994, "What is happening out there? IT support needs of community sector organisations", Community Development Foundation, London

Skinner, S., 1997, "Building Community Strengths; a resource book on capacity building", Community Development Foundation, London

Smith, S., and Plant, N., 1994, "Guide to Local Community Computing Resources; a guide for voluntary organisations on behalf of Avon Community Computing Network, UWE Department of Computing, Research & Consultancy Papers 94(1), Bristol

Taylor, M., 1992, "Signposts to Community Development", Community Development Foundation and National Coalition of Neighbourhoods, London

Ticher, P. and Firth, L., 1997, "Information Technology Support Needs of Advice Agencies", unpublished report to London Boroughs Grants Committee

Wilcox, D., 1994, "The Guide to Effective Participation", Partnership Books, Brighton

Williams, M., Local economic and community development: the growing potential of information & communicatiion technologies, *Local Economy* 12:1 (1997).

Williams, M., 1998, Computer Use in Bristol's Voluntary Sector; a survey and report, Community Information Systems Centre in conjunction with VOSCUR, University of the West of England; Bristol

Williams, M. and Plant, N., 1998, "New Media Technologies in Bristol's Communities; an evaluation of the 'Connections' project pilot phase", Community Information Systems Centre, University of the West of England, Bristol

Williams M., Moggridge A. , Plant, N. & Betts J., 1998 (forthcoming), Student Community Information Systems Projects in Bristol: Collaboration between undergraduates, staff and local community organisations, *in* "Student-Community Service Learning in Higher Education", S. Buckingham-Hatfield, ed., Department for Education and Employment, London

Wilson, B., 1991, Information Management, *in* "Systems Thinking in Europe", M.C. Jackson, G.J. Mansell, R.L. Flood, R.B. Blackham and S.V.E. Probert, eds., Plenum, New York

INFORMATION WITHIN THE (HUMAN) SYSTEM: CONSIDERATIONS OF FREEDOM, POWER AND ORDER

Antonín Rosický

Department of Systems Analysis, University of Economics, Prague
e-mail: Rosicky@VSE.CZ, fax: ++ 420 2 24 095 499

INTRODUCTION

For most technicians the concept of freedom is commonly understood as an affair of romantic poets or political orators. Politicians themselves connect it with democracy and human rights, many economists regarding it trust in "the invisible hand" of a free market and perhaps all journalists vigorously guard freedom of speech. Only in the last case is freedom loosely and confusedly identified with information. Thanks to the traditional understanding of information, attention is directed rather to liberty and accessibility of information sources than to dealing with information.

Many managers and technicians but also many politicians, economists, journalists and other orators value the increasing importance of information within society to such a degree that they commonly name its recent evolutionary stage as the information society. In the academic field we should appreciate rather perceiving or emerging significance of information in society or better in an evolutionary process of the universe including civilisation. The acceptance of this – only seemingly marginal – shift creates a space for better understanding not only information but some problems of this days.

INFORMATION AND EVOLUTION

From many different points of view information as well as matter and energy constitutes the nature of this world, its order and whole being. Probably nobody gave particular and evident attention to this fact in the case of genetic information. A general concept arising and permeating through many disciplines seems to be difficult, too "theoretical" and only of little use for daily practice to most. Moreover such a concept over-reaches the traditional positivistic (mechanistic) paradigm. However from a systems view many authors (Stonier[19], De Vree[21], Kampis[10] ...) point to the natural association between information and evolution in recent times. One of them, Bela Banathy[3] argues clearly:

> ... in living systems in general, and social systems in particular, evolution and information are so tightly interrelated that it is not appropriate to discuss one without the other.

It is just the coherence of information and evolution which brings new views of such issues as freedom, order and power and their mutual associations and consequences. All these abstract notions are important aspects in the synergy of social systems and its actually emerging properties.

Most people comprehending rather designed development of society than its evolution confuse it with the vague term "progress". Face to face with the definition of it we hesitate (perhaps we remember such noble goals as welfare for everybody, peaceful world... and freedom). Living daily life we

Synergy Matters: Working with Systems in the 21st Century,
Edited by Castell *et al.*, Kluwer Academic / Plenum Publishers, New York, 1999.

355

unconsciously consider, rather, values developing and developed during the last stage of the evolutionary stage of society from the enlightenment times to these days: welfare, certainty & positiveness ... and liberty.

Actually the nature of the evolution of living systems increases system's complexity having for it only one purpose – *"survival of the species alone and no more"* (Konrad Lorenz[12] p. 20). Regarding this judgement the evolutionary concept of the universe appears useful for better understanding information and its meaning within the system. Also a meaningful trend of systems thinking accepts idea of the evolution including evolution of the universe. Let us notice the concept of Salk[16], useful for a broader understanding of information and its meaning within (social) systems. He distinguishes three essential stage of this evolution briefly characterised in table No. 1.

Table 1. Three Salk's evolutionary stages of the universe (rearranged and completed)

Aspect / LEVEL	PHYSICAL SPHERE	BIOLOGICAL SPHERE	SOCIAL SPHERE
Matter	Physical matter	Living matter	Individual & social matter
Emergence	Matter and energy	Life	Consciousness & culture
Elements	Atoms	Cells	Minds & communication
Attributes	Interactions	Rationality	Creativity & discourse
Determinants	Probability	Replication, selection	Intention and choice

Salk entitles the last column as "human matter"; the title used accepts both human consciousness on the individual level and the social character of human beings.

CHARACTER OF COMPLEXITY

However the notion of complexity is a difficult topic to comprehend and moreover depending on *"the eyes of the observer"* (Klír[11], p. 114). Klír also characterises complexity in the epistemological and methodological sense:

> Complexity is thus associated with systems, that is, some abstractions distinguished in objects that reflect the way in which the objects are interacted with.

Just the way of interactions - that is, the distinctive quality of interactions and not only their number – is an essential feature of complexity. In spite of it many researchers and engineers strive to measure the degree of complexity by the quantity of information. Although Ashby[2] concurs with such opinion he points out that *"system's complexity is purely relative to a given observer"* at the same time and as a result of this paradox he casts *"the attempt to measure an absolute, or intrinsic, complexity...".* Regarding this distinction, and similarly to Klír, suggesting two general principles of complexity, we could comprehend two types of complexity:

- Descriptive complexity depending on the beholder and his skills,
- Intrinsic complexity proportional to any uncertainty of the system.

Two kinds of complexity divide the observer and reality in the same sense as the well-known Heisenberg principle of uncertainty. The identification (or better non-distinction) of both types of complexity is the principal aspect of the worldview, that anchors it resolutely to a mechanistic position with large consequences. Actually accepting the recent state of scientific thought is the human possibility of discovering absolute complexity of system (world) not only "limited" or "imperfect" but natural.

ORDER: COMPLEXITY & INFORMATION

Merely an uncertainty is an essential condition of natural selection, the system's adaptation (towards survival of the species) is an important condition of the whole evolution in the end. The complexity of human and especially social systems brings new quality and meaning to information (mentioned later). Information relationships acquire novel significance thanks to both aspects – vastly higher quantity and principally new quality. We must accept not only their temporal variety and fluctuating "intensity" but primarily – resulting from symbolically presented information – also ambiguity, vagueness, equivocalness as well as the influence of human intentionality. Consequently the system's complexity as well as the variety of system cannot be finite.

Ulrich[20], distinguishing purposeful systems from concepts of living (organic) systems argues (page 330):

> *The internal variety generation of social systems is inextricably rooted in the semantically and pragmatically meaningful experience of subject.*

Whenever we reflect on social systems we must accept the unique ("higher" type of) complexity in which the novel and natural uncertainty arises from the subjectivity of human individuals. The evolutionary process in these systems introduces original phenomena such as a humanity, moral, culture, competence, power and others resulting in the effort to create suitable order to reduce infinite complexity. Understanding these systems evokes new and relevant issues as character of the mentioned order, tension between individual freedom and responsibility and others...

Information always a constitutive role in the affairs sketched above. In spite of its accepted importance the actual meaning of information remains confused and a lack of understanding of this attends many including professionals from "information science" and information systems.

INFORMATION OR INFORMATIONS ?

Face to face the known significance of information and the increasing attention assigned to it reveal more and more apparent "information confusion". Many academics – mainly from the information systems area – point to its improper concept based upon the Shannon-Weaver theory. Explicit critics point to the lack of meaning for man and associate information with processes and products of the human mind as "capta" (Checkland[4]) or "wisdom" (Ackoff[1]). Some others (for example Stamper[18]) suggest to use semiotic concept of information (presented as symbol) and trace the way towards an interpretative approach and hermeneutic.

Often, outside of these trends rather than in general system's level, these remind the traditional cybernetic ideas mentioned above. But another group of researchers regarding rather living systems (Kampis[10]) connect information with the adaptations of living systems (organisms). They emphasise the holistic connection of relationship - exchange of information – between receivers of information and their environment. The idea of the autopoises (Mingers[13]) brings novel and important stimulation for this way and ultimately opens the question of the character of social systems (Ulrich[20], Banathy[3]).

Each of the above briefly sketched ways of consideration contributes to better understanding information and improves dealing with it in a more successful way. However some of them present a certain coherence. It seems that they contemplate the same generic idea but actually deal with different affairs. A probable stumbling block lies in the philosophy of the absolute (invariable) character of information. Regarding Salk's three-faced sphere resulting from universal evolution and creating new types of complexity, we can distinguish three emerging aspects of information (see table 2.)

Table . Three aspects of information depending on system's complexity

	Physical sphere **Physical system**	Biological sphere **Living systems**	Human (social) sphere **(human) social systems**
Emergence	Order	Meaning	Value and culture
Novel presence	Information in system	Information from system	Information about system
Sense	**NEGENTROPY**	**INFORMATION**	**KNOWLEDGE**
Kampis / Banathy terms	State referential information (as organization or order)	Referential information (as action)	Non-referential information (as knowledge)
Characteristic	It characterises system's state and originate its organisation	Received by organisms from their environment, bears meaning and initiates actions	Product of human mind, shared between people in symbolic form (as data)

The table covers similar distinctions of information given in an other context by Kampis[10] which demarcates two types of information – referential (as action) and non-referential (as knowledge). Regarding these Banathy[3] supplies a third type naming it as "State referential").

THREE FACES OF INFORMATION

In the physical sphere information is an intrinsic property of systems; it is named as **information in** system. It is the internal property of a system (Stonier[19]) and as objective affairs it mirrors the state of system and originates its organisation or order.

Accepting the principle of a system's hierarchy we must in the biological sphere distinguish two systems, commonly labeled as "receiver" and "environment". Both systems are determined by their internal information, however the presence of new information emerges: Complexity of organisms includes special sensors as elements that receive **information from** the higher **system** (environment). This information is presented by physical signal mirroring the variety of environment, but depends upon the properties of the receiver and therefore hasn't an objective character. Received information initiates actions (behaviour) of the organisms appropriate to its purposes (that are conducive to the survival of the species). The variety of such actions arises from the complexity of the organism and the increasing complexity (of nervous system) and increases with the expanding abilities of an organism to learn.

The highest complexity in the human (social) sphere produces consciousness and the vast learning capability resulting in the ability to originate very complex images of the higher systems embracing one's own accounting. Many other human attributes such as intention, creativity and anticipation of the future emanate from consciousness. Among them the genius of abstract thinking takes a fundamental place and originates a capability to use symbols for a representation of demarcated system's states and language as a symbolic system for dealing with the infinite complexity of the world. In other words: man is endowed with the competency to present information that is the result of individual perceiving and cognition – **information about** higher **system.**

INFORMATION TANGLE

Thanks to the shifting meaning of traditional terms understanding the phenomenon of information isn't easy, notably when three information appearances (aspects) interrelate, change roles and modify a system's complexity. Regarding the human being in the world, the information tangle is schematically demonstrated in figure 1. The simplification of such a sketch stands out if we take account of the infinite variety of human beings, range as well as ways of communication and the number of recursive relationships. Depicted relations create some close information circle, or better continuous circulation of information as a "source" of synergy and evolution.

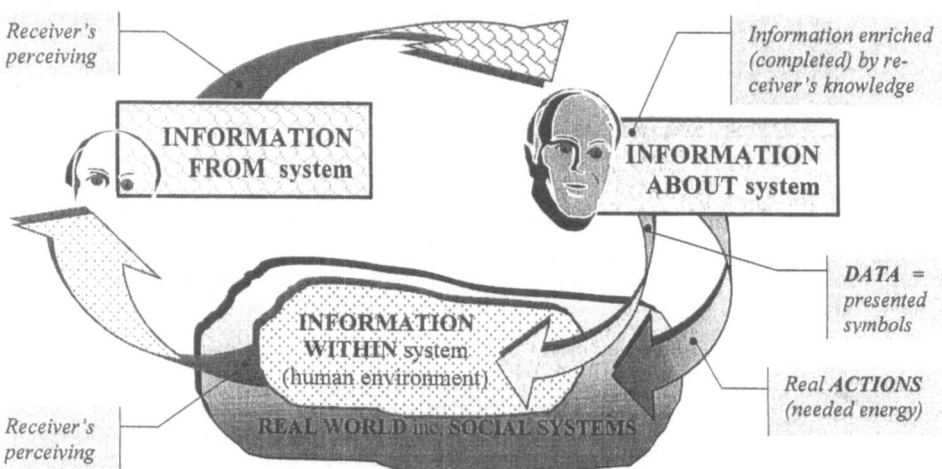

Figure 1 Information circle: interrelations of different types (appearances) of information

Man changes the states of his environment by real actions that result from meaningful information about system and also modifies internal information of the "higher system" in this way. Using language man implements information about systems into these systems. This information is represented by symbols and it is enriched by his knowledge. Consequently internal information in the human environment also covers human information (knowledge). Man also receives from systems both information types: information presented as signal (non-symbolic) on the one side and information presented as data (symbolic) on the other.

The intention of this paper is to emphasise the actual (or used) meaning of two mentioned terms: Data are truly symbols expressed in formalised form with respect for particular syntactic rules – from printed letters and spoken sentences to coded signals used in computers. Deep-rooted illusion of represented facts is false and dangerous. Just the opposite – the representative role of symbols admits a presentation of opinions, beliefs, intentions… but also errors, mistakes and falsehood (dis-information).

However the issue of knowledge is more covert and more confused too. Commonly we understand it as an affair (information net) kept in human memory. But an individual competence to create knowledge seems to be the better imagination. This competence is based on a cognitive structure (Hayes[8]) or rather dynamic cognitive systems constituting complex and meaningful information about the environment mentioned above. Information – as well symbolic as signals – is a stimulus for these cognitive structures in the learning process. Just the learning from personal experiences enacts a primary role: creates and develops conceptual ability to deal with the language (firstly the mother tongue) and enriches its semantic aspects with value association (pragmatic aspects). Only on this basis the other type of learning ("teaching"?) from "knowledge" presented by symbols arises (Rosický[15]).

Just the human knowledge (or competence to emanate knowledge) constitutes an important part of man's phenomenon and its individuality; together with genotype gives a framework for individual adaptation (competition). Similarly Von Glasersfeld[5] points *"The knowledge is actively built up by the cognising subject. The function of cognition is adaptive…".*

Together with Winograd & Flores[22] let us stress the fact that human knowledge isn't purely subjective however its individual nature plays an essential role: Also the meaning of information interpreted by the receiver has an individual character as well as initiated actions. John Mingers[13] (p. 189) supports as follows:

…information is very much dependent on the meaning structure or cognitive domain of the originator and receiver.

Thus information isn't either an objective entity and the result of a representative process or a matter which can be simply transported and stored.

INFORMATION AND ORDER

Reality and the communication of human information necessarily generate uncertainty within social systems. Often there is no decrease in uncertainty for a receiver system and also its higher knowledge increases the variety of possible actions and makes optimal decision less clear. On the opposite side such uncertainty affords broader space for evolution and offers higher freedom. To deal with this dilemma is a - for ever open - issue of a system's character often named "order of system".

Perhaps we try to create an order of social system by information put in systems and (create the order in this way) as standards, defined information flows or defined (rational) ways of transformation… However these ways raise an essential question of power and consequently reduce the system's adaptability in the long term. Moreover the infinite complexity of human systems refers to the limited possibility of such solutions in advance.

The opposite way opens a space for individual knowledge and supports a culture (as a sharing of meaning) and actual communication (not only data transmission. Hayek[8] (page 37) labels it as "spontaneous order" and its emergence is characterised as *"the patterns of interactions of many men can show an order that is of nobody's deliberate making…".* Hayek's concept of spontaneous order is far from absolute liberalism and connects freedom with organisations (designed order) and responsibility too.

Hardly anybody doubts that power is associated with information but only a few consider the order arising from dealing with it. Also prevalent concept information as a source turns the attention in this way – many, suggesting appropriate laws, try to freely open more information (and protect some others). However useful and clear way it is, , the danger of *anonymous power* (Havel[6]) emerges and arises from a wrong understanding of information and inappropriate dealing with it.

Commonly information is reduced to its symbolic form and data manipulation (transformation and transmission) as a substitute for actual dealing with information (interpretation, communication). Thanks to vast amounts of used data sources (databases, public media) we are overloaded by data, but we lack appropriate competence (knowledge) to interpret it as meaningful information. Similarly by the use of advanced information technology and designed information systems we use also covered knowledge (Salomon[17]) overstepping our intrinsic experience. Rational solutions also give as many opportunities as dangers to their users and radically change freedom and the possibility of adaptation.

The contemporary situation presents a new stage in the evolutionary process and the way back does not exist. Understanding the nature of information we can design such systems (order) so that the free dealing with information will be based upon individual knowledge. Hayek[7] reasons about systems in that *"individual can choose his pursuit and consequently freely use his own knowledge and skill"*.

However – will he actually do this?

References:

1. Ackoff, R., "From data to wisdom", *in: Journal of Applied Systems Analysis, 19, pp. 3-10 (1989).*
2. Ashby, W-R., Some peculiarities of complex systems", in: Cybernetics Medicine, 9:2 (1973).
3. Banathy, B. "Information-based Design of Social Systems", *in. Behavioural Science, Vol. 41, pp. 104 - 123; (1996).*
4. Checkland, P., "Information Systems and Systems Thinking: Time to Unite?" *in: International Journal of Information Management*, Vol. 8, p. 239-248 (1988).
5. Von Glasersfeld, E., *"An Exposition of Constructivism: Why Some Like it Radical"*, in: Klír, G. Facets of Systems Science, Pergamon Press, Oxford, 1991;
6. Havel, V. "Politics and Conscious", *in: Living in Truth* (Vladislav, J. eds.), Faber and Faber Limitid, London, (1989).
7. von Hayek, F.A, "The use of knowledge in society", *in: The American Economic Review, 35:4 (1945).*
8. von Hayek, F.A, "Law, Legislation and Liberty," Routletge, London, (1979).
9. Hayes, R., "The measurement of information", in: Vakkari and Cronin (eds.) Conceptions of library and information, Taylor Graham, London (1992).
10. Kampis, G., "Self-modifying Systems in Biology and Cognitive Science," Pergamon Press, Oxford, (1991).
11. Klír, G. "Facets of Systems Science," IFSR International Series on Science and Engineering, vol.7, Pergamon Press, Oxford, (1991).
12. Lorenz, K. "Der Abbau des menschlichen", Piper, Munich, (1983).
13. Mingers, J. [1995b]: Self-Producing Systems - Implication and Applications of Autopoises, Plenum, New York;
14. Rosický, A. "Information within the (Human) System", in: *Systems for Sustainability: People, Organizations and Environments, Plenum, New York (1997).*
15. Rosický, A. "The Danger of an Anonymous Power", *in: Proceedings from IDIMT '98 conference, Universitatsverlag Rudolf Trauner, Linz (1998).*
16. J.E. Salk, "Anatomy of Reality", Greenwood Publishing, Westport, CT (1983).
17. Salomon, J.J., "Le Destin technologique", Balland, Paris (1992)
18. Stamper, R. "Organisational Semiotics", in. *Information Systems: An Emerging Discipline?* (Mingers, J. & Stowell F. eds.), McGraw Hill, London (1997).
19. Stonier, T, "Information and the Internal Structure of the Universe", Springer-Verlag, London (1990).
20. W. Ulrich, "Critical Heuristics of Social Planning, John Wiley, Chichester (1994).
21. De Vree, J.K., " Information in nature, human behaviour and social life, in: Behavioural Science, in. Behavioural Science, 39:117 (1994).
22. Winograd,T., Flores, F, "Understanding Computers and Cognition", Ablex Pub. Norwood (1986).

COMMUNITY MATTERS: COMMUNITY OPERATIONAL RESEARCH WORK ON HEALTH NEEDS ON THE STONEBRIDGE ESTATE.

Leroy White

Faculty of Health
South Bank University
London SE1 0AA, UK

INTRODUCTION

Current UK health policy (NHS Executive 1997, 1998) has heightened interest in communities promoting their own health and actively participating in the decision-making process around needs assessment, priority setting and resource allocation. The mechanisms being suggested to achieve the policy aims, that is, Health Improvement Programmes and Health Action Zones, will be or should be seeking ways of involving the community and will most likely adopt community development as one component in its strategy.

Community development is not a new concept. However, in health, it is only since the World Health Organisation proclaimed the need for community involvement in health systems policy development, has the concept been taken more seriously. The aims of community development approach are:
- redressing the inequalities through action on the underlying social and economic determinants of inequalities in health and health related behaviour
- seeing community participation in policy and planning and development
- promoting inter-sector collaboration between health service and other statutory and voluntary agencies

Often programmes that have been set-up based on community development have failed. The reason for this is that they have been conceived in a paradigm which views participation as the 'magic bullet' to solve health problems. There is need to use a different paradigm that views participation as an iterative learning process allowing for a more eclectic approach to be taken and more realistic expectations to be made.

This paper will describe a process that can be used in a community development context. It was developed to enable communities to be involved in assessing their own needs and to act on the meeting them. It is a framework that aims at participation and learning. The framework will be described in the next section. This will be followed by a description of a case-study based on work carried to identify health needs of the black and minority ethnic communities on an estate in London, UK. The final section will outline some conclusions on this experience in using the framework in a health context.

Synergy Matters: Working with Systems in the 21st Century,
Edited by Castell *et al.*, Kluwer Academic / Plenum Publishers, New York, 1999.

361

METHODOLOGY: Participatory Appraisal of Needs and the Development of Action.

Practitioners working in the developing countries have developed a series of methods for needs assessment and for working on community issues. The most popular is Participatory Rapid Appraisal (PRA). PRA was developed in response to problems associated with outsider's intervention and it takes a different or alternative view of an outsider's role (Chambers 1994). It sees the outsider as a facilitator who works together with the people and he/she ensures that the resulting knowledge is owned and shared by local people.

White and Taket (1997) summarise the principles of PRA as the following:
- learning rapidly and progressively.
- offsetting biases.
- optimising trade-offs.
- triangulation.
- seeking diversity.
- facilitating - they do it.
- Critical self-awareness.

PRA has been mainly criticised for not being able to move beyond appraisal (White 1994, White and Taket 1997). That is, it is an extractive process and is unable to ensure shared analysis, planning, prioritisation of possible solutions, and lead to a commitment to act. It may be a good process for data collection but it does not explore the issues nor assess different options and choices. It simplifies and overlooks the inherent problems in developing a plan of action. For example, when there is a conflict of opinions, or different parties pursuing their own interests. It has been shown that these tensions can be addressed by incorporating tools from management sciences and operational research.

There is a set of tools developed in the systems and management science field, called here 'issue structuring methods' (ISM), which incorporates ideas on participatory decision-making (White, 1994, White and Taket 1997). These methods have been seen to be useful in conjunction with PRA methods, where the participatory and transparent nature of the techniques and the process facilitated learning. More importantly, however, it was found that the combined approaches can move appraisal into action. The approach is called 'Participatory Appraisal of Needs and the Development of Action (PANDA). It can be seen as another contribution to the evolution of participatory approaches to learning and action.

Another key reason for the development of our participatory approach is that we understand that multiple 'socially constructed' realities may vary in any given situation. It has been argued elsewhere (Popay and Williams 1994) that methods must be developed to ensure that a genuine synthesis of different intent and perspectives be created. PANDA is a framework that aims to link the multitude of voices. This is addressed through what has been called 'Pragmatic Pluralism' (Taket and White 1996, White and Taket 1997a). Here, in the spirit of pluralism, each person, professional, tools and method are able to find an individual pathway for meeting the challenges of participation, learning and action. They all work together to help reveal local realities and to empower communities. Through working with communities, PANDA has evolved through developing new approaches and different combination of methods. It is continuously developing and shifting shape.

Pluralism in PANDA works in three connected ways:
- Pluralism in the roles of the facilitators: That is, people who facilitate do not dominate, and can take on guises and perspectives at different times throughout the process
- Pluralism in the use of methods: Methods are varied and can be combined in different ways at different times. The methods can allow different modes of representation of the problems
- Pluralism in the nature and perspectives of the community, i.e. allow the use of different rationalities. The aim is to build up shared values and partnerships with the different players.

Our use of Panda (White 1994, White and Taket 1997a, Taket and White 1998) has led to a number of interesting insights (Figure 1).

- On facilitation: We have found an increase in a commitment to working together, in collaboration across different boundaries. Bringing the best out of the participants is achieved through being open, critically self-aware, sensitive to the needs of others, having a rapport and having fun.
- On methods: The plurality of methods allows diagrams to build up a picture of local realities, and through this we get visual sharing which can verify understanding, be cross-checked, be amended to and owned by the community. Also, it enables complexity to be expressed. Methods and processes can be combined in many ways and applied in many forms. Other uses of PANDA have seen new combinations and applications that are continuously being improvised and invented (Gibbon, 1998). Different modes of representation are necessary because realities differ. We now talk of multiple realities, and so we need to privilege the multiple realities of the participants in order not to pronounce on them singular or mis-representative views. This also needs careful and reflective facilitation.
- On Partnership: We found that by linking relationships, a commitment to participation increases. All the participants learn through the process, that communities, individuals and institutions can be empowered. The process is important, because empowerment comes through the process, i.e., it can enable people to express and analyse their own reality.

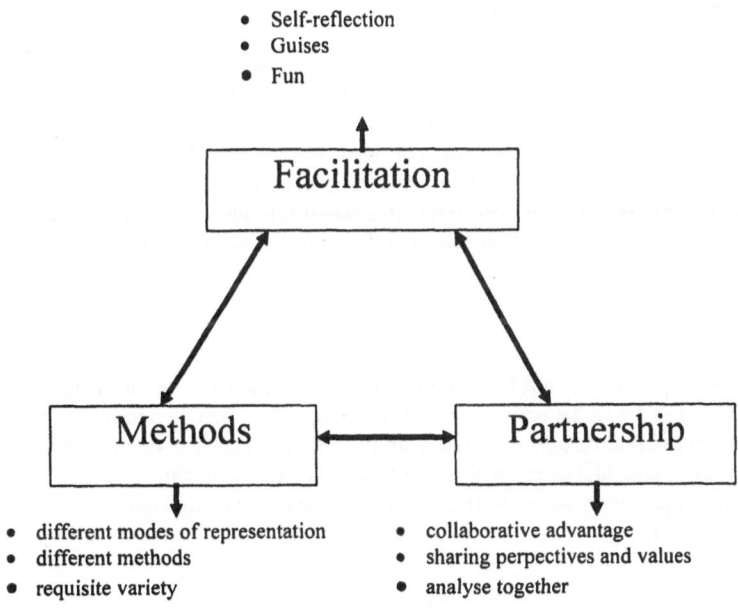

- Self-reflection
- Guises
- Fun

Facilitation

Methods

Partnership

- different modes of representation
- different methods
- requisite variety

- collaborative advantage
- sharing perpectives and values
- analyse together

Figure 1

Systemic nature of PANDA:

It seemed sensible to seek solutions through methodological pluralism, flexible and continuous learning and adaption in conditions of fast change in order to prompt learning and rapid adaptive response. Our use of PANDA is in the mode of the complex adaptive system model of fast feedback in conditions of rapid change, i.e., solving problems under pressure as adaptive change. Ashby's Law of Requisite Variety (Ashby 1965) also informs the use of PANDA, which implies that only variety can deal with variety. The problems of needs assessment can be characterised as exhibiting complexity, diversity and dynamism, thus the methods would need to match these if understanding and empowerment are to be achieved

We have also learnt that complexity, diversity, creativity and adaption will be greatest at the local level. Our approach focuses on the particular and the local, in that localness often has better information, responds faster and is relevant. Thus, our way of working is contingent on localness and emergence. This will be demonstrated in the case-study below.

CASE-STUDY

The case-study is based on some recent work to identify the health needs of the black and minority ethnic communities on the Stonebridge Estate, London, UK. The work was commissioned by the Stonebridge Housing Action Trust (HAT). The HAT is responsible for regeneration of one of London's worse housing estate. In fact the estate was reported recently as having the highest incidence of crime and the highest level of deprivation in the UK.

The HAT's aim was to develop a Health Strategy in partnership with the local health authority. It wanted some evidence of needs of the Black and Minority Ethnic communities, particularly the Afro-Caribbean community. The HAT was aware that it had a poor relation with this community and that the community felt it was being neglected in the HAT's development plans. PANDA was proposed as a possible way of getting amongst the issues and encouraging both the HAT and the community to participate in the needs assessment and planning in order to try and ensure that the most important needs are met.

The need to use participatory methods for needs assessment arose because the HAT felt it needed to understand more about how to work with the community and be more involved in community development. In the past it tried to use processes such as focus groups and community conferences with very little success. This was put down to the fact that these processes did not really involve the community. It was also found that the processes were used rather mechanically and were not adapted to the situation. The HAT also stated that the communities were weary of being consulted to then find that no action followed the work. PANDA was described as a way to alleviate these problems.

The process used involved individual interviews, group workshops and validation workshops. The first stage of the work was to identify key the broad key issues and identify the key informants. Meetings with the HAT led to these being identified and the individual interviews began with the key members of the community. This is a necessary first step to focus on the set of issues to address and to build rapport and trust with the key informants in order to gain access to the community. Key issues identified at this stage were related to health needs of the elderly, single parents, and key problems were disaffection of the youths and mental health. Full details can be found in White (1998).

Workshops were then arranged and organised with the community. Essentially with the elders, young parents, youths and mental health group users. The workshops used processes such as Nominal Group Technique, repertory grid analysis, concept mapping and Strategic Choice Approach.

There were many issues identified and discussed. I will only concentrate on issues relating to the elderly here. The workshop with this group used a combination of nominal group technique (NGT) and repertory grids. The emphasis of the workshop was to use visual forms of representation to facilitate debate on identifying their needs. NGT allowed the group to share concerns and prioritise which were important. However, the repertory grid analysis brought out more interesting issues.

The group was presented with a list of elements representing the estate, e.g. resource centre, shops and so on. Three were chosen at random and they were asked to associate two, then discuss why they did so. This process was repeated several times. At the end a list of key constructs were produced which began to shape into a story about their needs. The major one being that they were reminiscing about the 'old-days' on the estate when there were a number of amenities, that brought the community together. They believed that a lot of the problems were based on a lack of facilities to bring people together and the general health of the community was deteriorating due to the isolation and fear that has come about due the break lack of community facilities. This issue was explored further in the validation workshop.

The final stage of the process was the validation workshop. Several validation workshops were set up to discuss and allot priority to the problems identified and to explore potential interventions. Representatives from the groups and individuals seen in the previous stages of the work were invited. A Concept map was used, as *aide-memoire* for the group, drawn from the key problems that emerged from the previous stages of the work. The group used it as a focus to discuss the issues further and to explore prioritising the needs as they saw them. The process used elements of Strategic Choice, thus the groups were guided to thinking about what action could be taken which groups would be useful to work with and contact and how and in what way the next stage of the work could be conducted. Thus a commitment package of next steps was produced. People and groupings were identified to help in the development of the processes for the next step.

In the validation workshop with the elderly, the HAT representatives and representatives from the local health authority and community health council were present and the key issues were presented for discussion. The process involved working on the key issues identifying links between them and exploring action points leading to a commitment package. In this validation workshop, one of things the group decided to do was to set up a 'reminiscence group' for the elderly as a way of bringing them together. It would provide a way for the HAT to find out about the facilities that had existed on the estate in order to see whether it would be possible to rekindle some of them again.

DISCUSSION AND CONCLUSIONS

The focus of this section is to discuss what was learnt from the projects. This will mainly be on the process of working and on how PANDA is shaping up and developing.

Is PANDA a methodology? We have claimed that PANDA is more of a framework and a guide to mixing different methods and processes with the aim to maximise participation, not only to find out or consult with people but to get them actively involved in determining their future activities. What we have found that any attempt to codify PANDA as a methodology or to describe the application of concepts or offer accounts of methods at certain points can easily be jettisoned on its next application or taken up in an entirely different fashion. Thus any codification is best regarded as a *summary* that revisits process and analysis after the event, rather than a rationalistic plan decided upon before the analysis and put into practice.

This, we think, allows for 'emergence' at a number of levels. First is at the level of the process, where we could try to use lots and lots of different things in new ways, until a pattern of behaviour is created which will tempt the participants to adopt it, thereby causing them to look for appropriate new forms of behaviour. I think this works systemically and pragmatically. We see this process as providing a space for emergence, in that we could be groping towards something the meaning of which is unsure. Since we are in a social system some meaning will emerge, although that meaning will remain localised and contingent. For example, early on in the study it was realised that a health forum would be a useful body to the HAT. Over the course of the project this was established and it has a wide representation of residents as well as local community agencies.

The second level is content. PANDA enhances non-linear thinking about the problem, through a process of 'backing' and 'forthing', i.e. the framework encourages continuous thinking about new issues and reviewing old ones, both sequentially and all at once. This was seen in case-study, particularly, in the work-shops with the elders. Four dynamic behaviours emerged in this process, these were: reminiscing, desiring, chancing and deciding.

Another issue to reflect on is the process of validating the findings from the work. In participatory research a test of validity is the use of triangulation. Triangulation is perceived in different ways, it can be the use of several different research methods in combination or it can be results from different groups of people being asked similar questions to see if there is agreement. A view that is emerging in our work is to move away from triangulation to the concept of crystallisation. Where, "validity" is not fixed in a triangle. Rather, the image is a crystal, which combines symmetry and substance with a variety of shapes, multi-dimensionalities, and angles of approach.

Finally, in many accounts on the use of participatory methods from the management sciences in the context of community work and development, there is little to no discussion on evaluating the impact on the groups or communities of the decisions, strategies or the action resulting from the intervention. Recently, there has been a growing interest in exploring the impact of participatory approaches in increasing community capacity. I have been particularly interested in exploring a community's capacity through social networks and social capital (Smith and White 1996). Accumulating social capital is increasingly being seen as a good criteria for community development (Gilles, 1997), and there is a burgeoning literature suggesting that increasing a community's social capital will lead to an increase in the prosperity in Social, health and economic terms) of that community. I think that an increase in social capital is a good criterion to judge the impact of the kinds of interventions we are interested. In the case-study, the result of the interventions was the formation of new organisations that brought groups of people together which had the potential to increasing their social networks. In the long-run this may have an effect on their well-being and their situation.

REFERENCES

Ashby,W., 1965, "Introduction to Cybernetics," Chapman & Hall, London.

Chambers, R., 1994, The origins and practice of PRA. *World Development.*22:953-965.

Gibbon, M., 1998, Cycles within cycles. Proceedings of the FICOSSER Conference, Cuernavaça, Mexico, August 1998.

Gilles, P., 1997, Social Capital and Health. *Health lines*

NHS Executive, 1997, "The New NHS: Modern and Dependable," HMSO, London.

NHS Executive, 1998, "Our Healthier Nation," HMSO, London.

Popay, J., and Williams, G., 1994, "Researching the peoples health," Routledge, London.

Popay, J., et. al., 1996, Public Health Research and lay Knowledge. *Soc.Sci.Med* 42:759-768

Senge, P., 1990, "The Fifth Discipline," Doubleday, New York.

Smith, G., and White, L., 1996, "Making the connection," Proceedings of NCVO conference, Birmingham, 1996

Taket, A., and White, L., 1996, Pragmatic pluralism: an explication. *Systems Practice.* 9:571-585

Taket, A., and White, L., 1998, Experience in the Practice of one Tradition of Multimethodology. *Systems Practice and Action Research.*11: 153-168

White, L., 1994, Development Options for a Rural Community in Belize - Alternative Development and Operational Research. *Int. Trans. Opl Res.* 1: 453-462

White, L., 1998, "Same Difference: Report on health needs of the BME community on the Stonebridge Estate," Stonebridge HAT, London.

White, L., and Taket, A., 1997, Beyond Appraisal: Participatory Appraisal of Needs and the Development of Action (PANDA), *Omega.* 25: 523-535

White, L., and Taket, A., 1997a, Multimethodology as metamethodology, in: "*Multimethodology*," Gill, A., and Mingers, J., ed, Wiley, London.

THE LIFE FLUX, SYNERGY AND MULTIPLE PARADIGM METHODOLOGY

Mr. Jack A. Castle

Senior Lecturer, School of Operations Management,
The University of the West of England

INTRODUCTION

This paper reflects on the practical and theoretical progress with an action research programme dedicated to the development of a new multi-paradigm methodology, CVAM. The methodology has been shown to be capable of integrating and enabling strategic and operations strategy processes through a multi-paradigm debate. Over the past five years it has been improved and enhanced through co-operative projects in engineering product development: leisure management: telephone service provision: building society communications strategy: and engineering quality systems. Current applications are progressing in aerospace manufacturing strategy: car component manufacturing strategy: the prison service: government agencies, and devising new strategies for a large charitable organisation in Bristol.

CVAM A LIFE FLUX MODEL

CVAM (see-vam) is a mnemonic for circumstances, values, activities and means. The methodology was inspired by the idea of an appreciation system graphically explored by Checkland and Casar (1986) from the writing of Geoffrey Vickers. Vickers represented life as a flux of events and ideas and the reforming of value systems as interaction with the life flux progressed. Within CVAM the flux was initially conceived as a four threaded rope which was constituted by the threads of circumstantial forces (C's), values and viewpoints (V's), activities (A's) and the means of their prosecution (M's). CVAM was initially developed as an interpretive methodology to enable complex strategic and operations problems to be addressed (Castle 1995, Castle and Spurrell 1997). As our work progressed it was clear that we were inadvertently crossing paradigms. To achieve greater clarity, the whole methodology was reconstructed with purposeful engagement of multiple paradigms. To this end the fundamental model has been reconstructed around objective, subjective and radical/strategic paradigms. It now models the life flux as three intertwining threads of a hard objective, subjective, and strategic/radical nature. This has facilitated a rapid advance both in revisiting functionalist operations and strategic theory, and in addressing radical concerns of power and communication constraints that surfaced in several of our studies. In recent years the United Kingdom systems movement has displayed a progressive movement towards pluralist approaches. The virtues of embracing methodological pluralism have been strongly advocated by Jackson (1997) and others.

Castle (1998b) has argued that a methodology will be more accessible to users if it exhibits ontological complicity. That is to say it engenders in the user some recognition as a reasonable and useful representation of his or her lived experience. So for example analysis with linear

Synergy Matters: Working with Systems in the 21st *Century*,
Edited by Castell *et al.*, Kluwer Academic / Plenum Publishers, New York, 1999.

367

programming implies that we can express a challenge as a set of linear equations. Use of total systems intervention calls up metaphors, perhaps seeing life as a prison or a machine (Flood and Jackson 1991). There is greater ontological validity in seeing life as a time based flux. Life is not a metaphor. Here we are concerned to model the life flux as experienced by agents and not to map the paradigmic outcome of theoretical attenuation. Constructivist theory informed by the work of Giddens (1984) and Bourdieu (1990) suggests that there are three intertwining and spiralling life fluxes. These fluxes are seen to interact and shape each other. Bourdieu's ideas of habitus and field seeks to try to overcome the traditional divide between object and subject in the life flux, each is shaped and constituted by the other. There is an objective hard flux with which we interact on a day by day basis. Investigation would proceed in this domain using positivistic approaches. There is a subjective flux that is to be discovered in the viewpoints and thinking of subjects as individuals, groups, and organisations. In this flux knowledge is gained through phenomenological approaches, involving communication and engagement.

The final flux is an ongoing strategic/radical concern to resolve positive and negative power issues. Agents engage in strategic adjustment of the relationship of the capital within the objective and subjective fluxes. In this model power is seen as flux property and hence strategic adjustment is a power re-ordering activity. The model supports the Foucaulvian view that power cannot be structurally stripped out of a challenge situation as it is systemic. It favours the contingent relational view of power:

> "In the contingent, relational view, .. structural forces of power are grounded in and constituted by the self understandings of all agents involved .. the approach is contingent, but in a critical sense. It conceives of power as having a positive as well as a negative potential". (Oliga,1996:83)

One cannot stand aside and view power as a divorced structuralist constraint. These ideas have informed the CVAM'ic view of power. Power may be envisaged as generated by the processes within each of the life flux threads. In other words the force that the flux can generate may attract or repel. It can be positive or negative for the subject in his or her habitus. It can be tapped like an electrical power line, or behave like an electric fence imprisoning agents within their fields and habitus. Thus power is a total flux property, an emergent property depending on the dynamics and structure of the flux at any moment of time. Sometimes the hard flux dominates with 'C' elements within that flux spiralling to the top. At other times when resources are scarce 'M' issues may surface. We may experience subjective flux domination as we discover that others conceptualise situations differently and are able to exert a defining influence. These two fundamental power flows generate a third control activity. The human agent is concerned at all times to understand the flux flows and to attempt actions to redefine and reorder power relationships to their advantage. Sometimes to tap the flow and use the capital, sometimes to reconfigure the flow to gain access to the power or to liberate themselves from its demands. The agents knowledge and power interests are interrelated. They may only feel the power in a flux as recognised through their habitus. In other words the power of the flux depends on the agents ability to recognise and collect information, and depends on their knowledge of the flux in time. The model seems to show empathy with Midgleys (1997) view that we are concerned here with power knowledge formations. Bouchikhi (1998) has also pointed out the that agents in organisations do not react functionally and unquestioningly to events. They decide to give support or obstruct depending on the nature of the issues they encounter.

The structure or capital intrinsic to each flux was derived from a study of existing systems approaches. The view point activity debate (V's and A's) intrinsic to the rich pictures and holonic modelling developed for Checkland's (1981) SSM was seen as paying insufficient attention to the resources or means (M's) that are so central to much of operations theory (Castle and Spurrell 1997). If we are to resolve trade offs and resource allocation challenges this needs additional attention. Beer (1979) has made it clear that the variety or complexity of the environment is potentially infinite. The attention given to circumstantial issues within SSM seems once more to be limited. Their identification as 'E' elements within the CATWOE check often produces a limited response. To counter this, and provide increased attention to the variety associated with the environment a circumstantial (C's) element was added. The constituents or capital intrinsic to each flux is seen as constituted of uncontrolled elements described as circumstances (C's): viewpoints and values (V's): activities (A's): and the means (M's) of their prosecution. The three fluxes and their intrinsic capital spiral across time so that at any moment one flux and its particular capital may dominate the agents concerns. The idea is shown in figure 1 below.

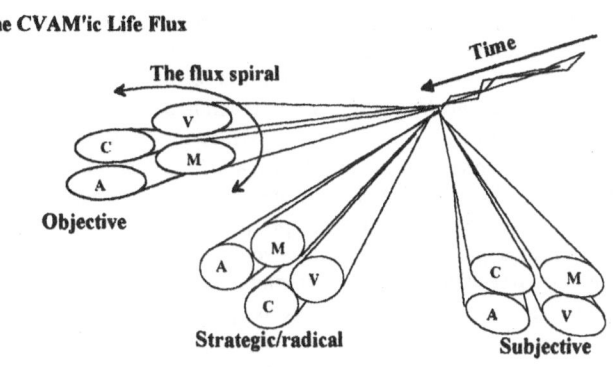

Figure 1: The idea of the life flux as a CVAM'ic spiral

THE AVOIDANCE OF DECEPTIONS IN PERSONAL AND OPERATIONS STRATEGY

This section provides a highly selective overview of the main theoretical turns intrinsic to the CVAM methodology. Here we define a methodology as '... an attenuated and attenuating guide, to enable users to avoid deceptions in the achievement of improvement action through the application of practical reason'. These deceptions include those associated with variety imbalance and complexity: deceptions of reductionist thinking: time based deception: communicative deceptions: deceptions of boundary judgement: isolationist deception: and methodological deception. The CVAM process is summarised by a sequence of five formal steps which may require systemic loops or returns to earlier stages depending on findings. These are labelled B-F, and the three fluxes 1-3 in the overview diagram presented below. Specially designed sub-methodology is located within this framework, but it may be supplemented or replaced with imported alternatives as explained later. As the process moves from start to end the variety engaged by complex challenges is progressively attenuated. The mapping approach was inspired by Mingers (Mingers and Gill 1997). The stages differ somewhat from those he proposed:

	Disturbance (unplanned ?)	Systemic Discovery	Attenuation	Validation	Analysis	Improvement action
Objective	A1	B1	C1	D1	E1	F1
Subjective	A2	B2	C2	D2	E2	F2
Strategic	A3	B3	C3	D3	E3	F3

Figure 2: The CVAM process framework

Following some disturbance in the flow of capital in the objective or subjective fluxes agents are stirred into strategic reflection at A3. Variety deceptions occur when complex challenges are engaged using methodology with inadequate variety. In these situations the laws of requisite variety means that success cannot be achieved (Ashby 1964). CVAM addresses this deception by the use of systemic data capture with potentially infinite variety at stages B1, B2 and B3. Here evidenced statements (objective flux), viewpoints (subjective flux), power and communication issues (radical flux) are surfaced. This process is carried forward to saturation. It may involve hundreds of data cards, verbations and other contributions. Collection is systemic, and not systematic as so vividly discussed by Beer (1979). Our studies have indicated the importance of systemic viewpoint collection. For example a debate amongst experienced engineers centred on future aerostructures manufacturing strategy produced hundreds of individual views and ten distinctly different strands of strategy (O'Brien 1998). In this case it was clear that the capture of the variety of the challenge was essential to effective progress.

Reductionist deception causes slicing of the challenge territory and the neglect of key elements and variables. This is a common complaint about functionalist operational theory (Bessant 1991). In CVAM this is avoided by the requirement that the assembled team address and collect information in all three fluxes holistically identifying C's, V's, A's, and M's.

Short term thinking has been raised by many writers in operations and quality as a deception see Deming (1982). We have found time based deception appears to be a common phenomenon in the action research applications of CVAM. Several of our studies revealed that managers had never been asked to reflect on past history and hence pre-judices, or to envision future scenarios. Time flux reflection is an intrinsic part of systemic discovery and validation stages.

The emancipatory potential of effective communication is at the heart of Habermas's (1984) theory of communicative action. Communicative deception can occur when team members, aware of power relations, play language games. This may suppress learning. In two of the practical applications of CVAM we have noted situations where participants were unable to communicate their concerns for fear of organisational coercive control. We assert that this form of O-I behaviour is not uncommon (Argyris and Schon 1978). Steps are taken to ensure an effective speech situation at B3 using external observers, and in a social validation procedure at D3.

Ulrich's (1983) seminal work has pointed out that the identification of a challenge and the formation of a team are processes which are steeped with 'a priori' boundary judgements. Initial stages of CVAM encourage systemic debate about the affected and their witnesses as the first stage of critical boundary expansion at position B3. As the process proceeds the team may be re-structured to include those discovered from boundary expansion, and to surface their viewpoints at D3. There can be little doubt that operations strategy can have deep implications for the environment and local communities. These social aspects of planning demand deeper consideration.

Isolationist deception occurs when the analyst privileges knowledge from only one (usually the hard) paradigm. CVAM avoids this by encouraging a multiple paradigm enquiry right from the start. This position is also taken by Mingers (Mingers and Gill 1997). Within CVAM all three are engaged, but the user is free at any time to revert to revert to single paradigm (isolationist) or dominant (imperialist) paradigm approaches see Reed (1985). Users may choose to become isolationist; this is enabled but discouraged by the CVAM process.

In designing any methodology, theoretical selection must occur. Hence every method is an attenuated and potentially misleading view of the constellations of theory available (Bernstein 1991). It is a feature of functionalist management theory that it arrives pre-attenuated. For example Porter's five forces method encourages the user to look for five variables from the infinite variety of the competitive environment. It therefore discourages systemic analysis and attenuation. This method is clearly a hard 'C' analysis and could be imported at stage E1. On the other hand rich picture building in SSM is clearly a systemic 'V' capture approach which could be imported at stage B2. The process framework can guide the user in the use of other tools and technology by acting as a meta-methodological framework for the import of other approaches. CVAM can overcome its own attenuation by adding to its variety infinitely if required.

ATTENUATION, VALIDATION, ANALYSIS AND SYNERGY

Stages C, D and E are concerned with sensible attenuation and analysis. We may regard the whole CVAM process as a battle with complexity. This is achieved in the subjective flux in keeping with its life flux ontology by employing Heideggerian hermeneutical processes to complete the identification of themes employing procedures informed by the work of Diekelmann, Allen, and Tanner (1989). The epoche of Husserlian approaches intrinsic to SSM is rejected. Hermeneutic themes are identified in a reflection of past experience and prejudice using independent observers to challenge interpretations. External observers are also used at C3 to help attenuate identified power and strategic themes and these will be exposed to the team. They may ask for team reconstitution or the reconsideration of the challenge statement depending upon findings. In the usual CVAM procedure, the main strategy strands to be pursued are identified as the hermeneutic themes from the subjective attenuation. In objective mode strategies may have been developed at C1 using hard strategy models. Our research has indicated that strategy strands require careful testing and validation by the teams. This stage seems to pass without question in other methodology. It appears as part of a historical reflection on what has been discovered. Does the strategy meet the challenge of hard, strategic and subjective discoveries? Hard validation simply requires the agents to examine the strategy developed by exposing it to the hard facts collected in the systemic hard flux. We constitute it as a field by dipping it into the hard data and seeing which CVAM'ic hard data 'sticks to it' (Bourdieu and Wacquant 1992:94). The hard data selected depends on a classification of the strand statements in a new approach to the validation of value trees and themes see Castle and Spurrell (1997). Subjective validation proceeds at D2 using the sub-methodology MRAM (mapped resolution of activities and means). Users model the conceptualised activities and means needed to action the strategy strand. This raises the strand validity as an issue of sufficiency. MRAM once more employs Heideggerian approaches reflecting on captured data in order to identify possible holons, in place of the epoche practised in SSM. The time dimension is further explored to test the strategy against future scenarios as conceptualised by the team.

The identification of hard or subjective strategies enables, radical validation to be achieved at D3. Once the broad thrust of our strategy has been identified we are in a better position to identify those affected. Their views may then be polemically engaged in the emancipatory challenge of our strategy (Ulrich 1983). Boundary expansion is facilitated by the attenuated power themes identified at the stage C3, and by chain questions engaging the chains of capital captured in the other fluxes. Sometimes starting with the means of our own production and chaining back to the suppliers. Consideration of products and services as the means used by customers for their own activities has enabled expansion to the customers of our customers. These discoveries have enabled us to map new boundaries, and new·agents. The problem is that this expansion brings with it possibly massive variety expansion. It may be necessary, in this case, to return to stage B and employ the systemic stage to re-engage this variety.

Validation of the strategy justifies deeper analysis. Hard models may examine issues like capability, capacity or flexibility at E1. Hard strategic and marketing analysis may be employed around the strand imperatives. Subjective analysis continues within CVAM to the conceptual identification of means and the modelling of key processes as chains of activities at E2 (Castle 1998a).

The identification of the main strategy strands and their hard validation and soft modelling informs a further questioning of the habitus as strategic validation at E3. Individual strategy strands may be identified and furnished with both objective and subjective capital as part of a first identification of the power issues attaching to a strand. The team engage in holistic examination of the strand as a guide to a new field, and hence to the power relations intrinsic to this new field. This is called domination analysis within CVAM. This corresponds to Bourdieu's insights that fields only emerge when noticed by the agents that inhabit the habitus. In essence this is the start of an ordered search for new fields, and new power relations. Strategic validation pulls together the hard and soft data to explore for synergy and power relationships. The resolution of the strategic validation enables many conclusions to be drawn in terms of the re-ordering of relationships within the strategic flux. This together with the hard goals and subjective validation enables formulation of improvement action at stage F.

CONCLUDING REMARKS

There can be no correct answer to the ontological form of the life flux. The model here attempts to consider modern social theory and provide a guide which will help users avoid many of the deceptions intrinsic to existing approaches. If strategic and operations theory is to become more relevant to the user, it is time to abandon the security of our isolationist tendencies. It is suggested that we should engage the challenge of constructivist/pluralist approaches to provide solutions that recognise the world as a complex place. Differences in viewpoints, and the influences of power are the essence of organisational life. Simplistic approaches offered by existing functionalist models are themselves deceptions waiting to trap the unwary.

The CVAM'ic life flux model implies that multiple paradigm thinking is an intrinsic part of everyday life. That isolationist tendencies are forced upon agents by the relational power of the organisations that employ them. If this is the case we need to help agents reveal and challenge these power formations to enable a return to critical thinking within organisational life.

REFERENCES

Argyris, C., and Schon, D., 1978,"Organisational Learning, a Theory of Action Perspective," Addison Wesley, Reading, Massachusetts.

Ashby, W., 1964, "An Introduction to Cybernetics", Methuen, London.

Beer, S., 1979, "The Heart of Enterprise," John Wiley and Sons, Chichester.

Bernstein, R.J., 1991, "The New Constellation", Polity Press, Cambridge.

Bessant, J., 1991, "Managing Advanced Manufacturing Strategy," NCC Blackwell, Manchester.

Bouchikhi, H.,1998, Living with and building on complexity: A constructivist perspective on organisation, *Organization*. Vol. 5(2): 217-232. Sage, London.

Bourdieu, P., 1990, " In Other Words: Essays Towards a Reflexive Sociology," Trans. Adamson, M., Polity Press, Cambridge.

Bourdieu, P., and Wacquant, J.D., 1992, " An Invitation to Reflexive Sociology", Polity Press, Cambridge.

Castle, J.A.,1995,The Development of an integrating methodology for strategic and operations management,. *Proceedings of the British Academy of Management conference*, Sheffield University.

Castle, J.A.,and Spurrell S., 1997, CVAM a new systems methodology, and systemic sustainability, in an Engineering Company, *in:*"Systems for Sustainability," F.A. Stowell, R.L. Ison, R. Armson, J. Holloway , S. Jackson, and S. McRobb, ed., Plenum, New York.

Castle, J.A., 1998a, New methodologies for integrated quality management. *The TQM Magazine, Vol.10*, 2:83-88. MCB University Press.

Castle, J.A.,1998b, CVAM as a meta-methodology, theoretical and practical progress, *A paper presented at the Operational Research Annual conference*, OR 40. Lancaster University 8-10th September.

Checkland, P.B., 1981, " Systems Thinking Systems Practice," Wiley, Chichester.

Checkland P.B., and Casar A., 1986, Vickers' concept of an appreciative system: A systemic account. *Journal of Applied Systems Analysis*, vol. 13: 4-17.

Deming W.E., 1982, "Out of the Crisis", Cambridge, Mass., MIT.

Diekelmann, N., Allen, D., Tanner, C., 1989, A critical Hermeneutic Analysis, *The NLN Criteria for Appraisal of Baccalaureate Programs*, National League for Nursing, New York.

Flood, R.L., and Jackson, M.C., 1991, "Creative Problem Solving," Wiley, Chichester:

Giddens, A., 1984, "The Constitution of Society: Outline of the Theory of Structuration," Polity, Cambridge.

Habermas, J., 1984, "The Theory of Communicative Action: Reason and the Rationalization of Society," Beacon Press, Boston, Mass.

Jackson, M.C., 1997, Towards coherent pluralism in management science, *Lincoln School of Management. Working Paper* No 16.

Midgley, G., 1997, Mixing methods: developing systemic intervention, *in:* "Multi Methodology", Mingers, J., and Gill, A., ed., 249-290." Wiley, Chichester.

Mingers, J., and Gill, A., 1997, "Multi Methodology," Wiley, Chichester.

O'Brien, A. M., 1998, The application of advanced system methodology to investigate operational strategy, *MBA dissertation*, The University of the West of England.

Oliga, J.C., 1996, " Power ideology and Control", Plenum, New York.

Porter, M.E., 1985, "Competitive Advantage," Free Press, New York.

Reed, M., 1985, " Redirections in Organizational Analysis," Tavistock, London.

Ulrich, W., 1983, "Critical Heuristics of Social Planning," Wiley, Chichester.

CRITICAL THEORY AS A FOUNDATION FOR STRATEGIC MANAGEMENT

Steve Clarke, Brian Lehaney and Yongmei Nie

Department of Finance, Systems and Operations
Luton Business School
University of Luton
Park Square
Luton
LU1 3JU

INTRODUCTION

Since Ansoff's work in the 1960s, the idea that strategy can be rationally planned and implemented has remained in much of the practice and theory of strategic management. Through the 1980s and 1990s, however, such a view has been widely challenged, with growing support for a more holistic, participative perspective. An example of this is the thinking of Quinn and Mintzberg, whereby strategy is seen more as an incremental or emergent process, giving rise to unpredictable patterns of activity. However, it could be argued that the complexity of addressing pluralistic problem contexts has not been fully addressed within the theory and practice of strategic management; that an over simplistic approach to eliciting participant views has been taken, leaving both the pluralistic and coercive nature of such contexts underdeveloped. During the same period, significant effort has been expended within the systems community in addressing pluralism and coercion in business organisations, evidencing a common ground between the two domains, which has given rise to the notion that strategic management might be improved if informed from systems thinking.

This paper summarises work undertaken so far on a research programme which seeks to enhance strategic management by developing an approach based on contemporary systems theory. The basis of the work is a critique of functionalist, interpretivist, and radical humanist positions, through which the strong reliance on methodology is questioned. A future direction is identified through the critical social theory of Foucault, using current work in the systems community which focuses on action research to address the diversity of pluralistic and coercive problem contexts characteristic of many strategic management problem situations. The outcome is a framework for strategic management linking action research to relevant issues from critical systems theory.

CORPORATE STRATEGY: PLANS OR PATTERNS

Whilst there is a wide variety of categorisations of corporate strategy, two extremes frequently emerge as polarised strategic views: is strategy something which can be planned, or

Synergy Matters: Working with Systems in the 21st Century,
Edited by Castell *et al.*, Kluwer Academic / Plenum Publishers, New York, 1999.

does it just surface as the result of organisational activity for which no discernible plan is evident? Mintzberg (1987) characterises this debate as the distinction between a plan and a pattern, whilst other authors (e.g. Quinn, 1980; Johnson and Scholes, 1993) refer to planned, or emergent/incremental strategies.

The planning approaches to strategy may be seen as developed from the so called design school (Ansoff, 1964), which in turn can be traced to scientific reductionism. Ansoff refers to such an approach as: "a succession of different reduction steps: a set of objectives is identified for the firm, the current with respect to the objectives is diagnosed, and the difference between these (or what we call the 'gap') is determined." Strategy is then concerned with finding those 'operators' which are best able to close the gap. What is evident here is a process that is seen as objective, and as a result may be criticised for its limited attention to human activity.

The argument in favour of viewing strategy as patterns, is that the patterns of action which we see in organisations as strategic may not have derived from any discernible plan. Citing the example of Volkswagen, Mintzberg points to the problem that even if the organisation's plans are expressly written down, there is reason to suppose that these may not represent the true collective strategy of the organisation. Organisational theorists, agues Mintzberg, overcome this problem by the principle of attribution: "given realisation, there must have been intention, and that is automatically attributed to the chief" (Mintzberg, 1987). Empirical evidence further shows strategies emerging from the organisation without there having been deliberate plan. The pattern is a stream of realised actions, which may or may not have been intended. That there may be no formal plan behind the pattern gives rise to the categorisation of strategies as deliberate or emergent, and it is this emergent view that is seen in much of Quinn's work. Quinn (1980) has observed that, for many organisations, whilst strategic planning forms part of the bureaucratic control process, most important strategic decisions seem to be made outside this formal planning structure. This leads Quinn to challenge the standard 'rational-analytical' approach, which he sees as normative. The goals and objectives of strategic planning are seen by Quinn (in Mintzberg, Quinn *et al.*, 1998) as determining *what* is to be achieved and *when*, but not *how* the results are to be achieved. Quinn's logical incrementalism (Quinn, 1980; Mintzberg, Quinn *et al.*, 1998) addresses these problems through a collaborative view of the strategy process, within which strategy is seen as a "fragmented, evolutionary and intuitive" process. What emerges is not a strategic plan, but a new consensus from which the organisation's way forward emerges. Logical incrementalism combines the planning and behavioural approaches to strategy, and is particularly strong in its ability to enable managers to respond to unforeseen change. Viewed in this way, the domain of corporate strategy is seen to echo the functionalist / interpretivist debate that has been so fundamental within the development of systems science. The next section critiques this perspective on corporate strategy, and lays the foundation for an approach grounded on critical theory.

SOCIAL SYSTEMS THEORY AND CORPORATE STRATEGY

The limitations of functionalism identified in corporate strategy are equally relevant to the study of social systems, where predictive models may be seen to have only limited value. Social action does not readily lend itself to study by reductionist methods, but is determined by the meaning that individuals attribute to their actions. Hard (functionalist) and soft (interpretivist) methods are both cast in the sociology of regulation (see Figure 1), and radical approaches developed from critical theory have been demonstrated as offering a way forward from this regulative, uncritical position. This is the direction which has been pursued by part of the systems movement, from its origins in the so called Singer/Churchman/Ackoff school (Jackson, 1982; Britton and McCallion, 1994), through to present day systems thinkers. Jackson has argued that the soft methods of Ackoff, Checkland and Churchman all adhere to some degree to the assumptions of the interpretative paradigm, and identifies a third position which distinguishes hard, soft and emancipatory systems thinking (Jackson and Keys, 1984; Jackson, 1985). The argument is for a complementarist approach, which sees the strengths and weaknesses in each of the three areas and argues that each one must be respected for those strengths and weaknesses.

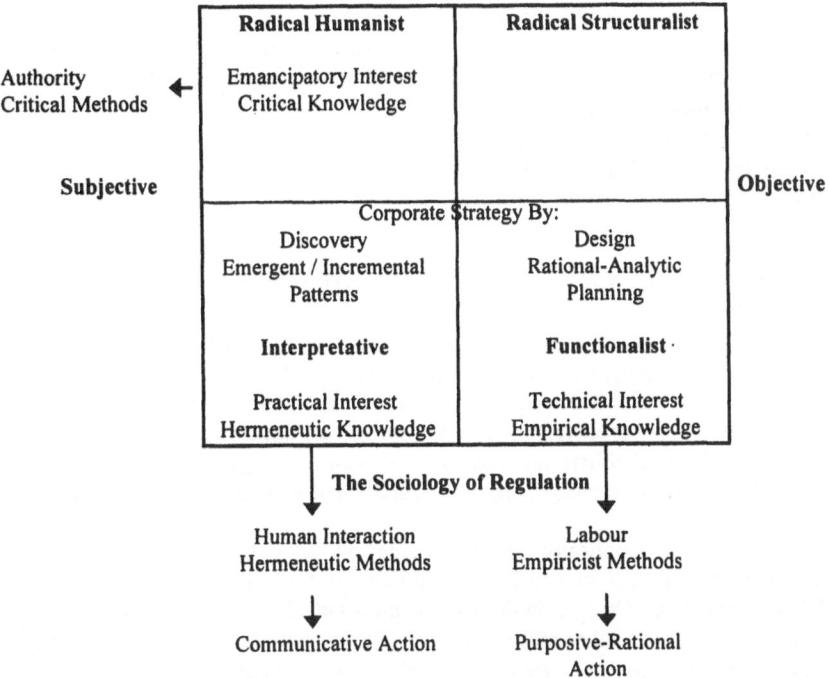

Figure 1. A Categorisation of Approaches to IS

All of this is mirrored in corporate strategy, where the argument is wrongly cast within the sociology of regulation. The effect of this is illustrated, in Figure 1, by positioning the approaches to corporate strategy on the Burrell and Morgan grid (Burrell and Morgan, 1979). From the perspective of social theory, current approaches to corporate strategy may be classified as functionalist or interpretivist, or premised on a view of strategy as a process of design or discovery, the design school being functionalist in orientation; discovery methods interpretivist. Critical social theory therefore offers the potential for combining functionalist and interpretivist approaches with a radical intent. Much work in this area has already been undertaken in the management science domain, and it is this work that is used in the following section to formulate an alternative framework for IS strategy.

CRITICAL SOCIAL THEORY

Space limitations dictate that this can be only a brief summary. For a more detailed discussion in the first instance, see [Brocklesby, 1996 #842].

Critical social theory (CSoT) can be traced from the work of Kant, through Marx and the Frankfurt School, the two most widely accepted modern theorists being Foucault and Habermas. CSoT applied to corporate strategy is appealing for its denial of the natural scientific principles on which study has largely hitherto been based. CSoT refutes this, seeing our understanding of the world as determined by *a priori* conditions which are uncritically accepted. Critical theory seeks to expose these, and thereby release human beings from their 'false consciousness' to a position from which true potentiality can be attained.

From CSoT, two streams of thought have emerged: Habermasian, and Foulcauldian. Significant development work based on the theories of Habermas (see, for example, Midgley, 1995) has given rise to the view that a methodological solution is possible, whereby a critical methodology (e.g. critical systems heuristics, Ulrich, 1983) can be used to progress beyond functionalist and interpretivist approaches. Problems have emerged with this position particularly in relation to its reliance on debate. Significant among these are the difficulty of what to do where coercion is so severe that debate is closed, and the contention that debating methods simply lead to another form of coercion, where power is held by those most able in debate. Furthermore, adherents to Foucault's theories deny the possibility of a methodological approach, since they see power and coercion as individual concepts, embedded in the social fabric: individuals can be helped to attain their own potentiality, but the idea of 'emancipating' a group of people is meaningless.

A CRITICAL FRAMEWORK FOR CORPORATE STRATEGY

Recent research points to a number of possible approaches which offer improvement based on Foucauldian ideas. Firstly, forms of action research, in particular co-operative inquiry (Reason and Heron, 1995) offer an action learning cycle which has open participation as a primary aim. The initial objective is to creatively investigate the system through critical analysis. Methods available to specifically facilitate this task include brainstorming, lateral-thinking, the use of metaphor, Ackoff's idealised design, and Checkland's soft systems methodology stages one to five.

A theoretically and practically informed framework for corporate strategy must therefore incorporate elements of planning and design within an overarching critical framework premised on participative analysis. 'Critical' in these terms means challenging both the assumptions made within a study, and the material conditions according to which those assumptions have been made. A diagrammatic representation of such a framework is given in Figure 2.

The first consideration in using this critical framework is the need to set a boundary for any investigation, which in itself should be set critically (see Midgley, 1992). Since the strategic system is to be seen in social terms, this boundary should consider primarily those involved in and affected by the system. The core of the strategic study is then seen in terms of the 'critical cycle of learning and action', whereby a mix of interpretative and structured analysis may take place within the determined boundary, having regard to the given organisational context.

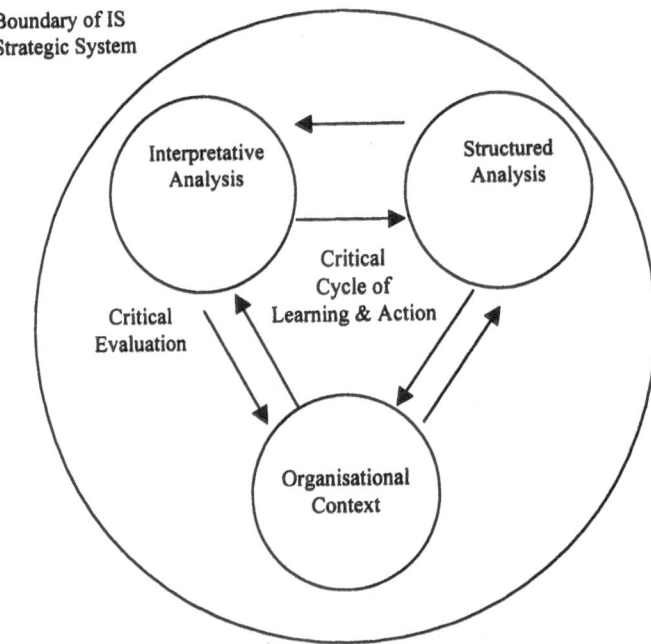

Figure 2. A Critical Framework for Information Systems Strategy

Critical evaluation is represented as a counter-clockwise activity to indicate the need to re-evaluate at every stage. So, for example, whilst interpretative analysis within a given organisational climate may be used to determine what structured approaches to apply, it will be necessary to reflect critically on whether the interpretative analysis accurately represents the organisational climate, and whether the structured analysis can be seen to flow from the interpretative study.

At any stage of an intervention, it will be necessary to select methods for that or subsequent stages. The work of Jackson and Keys (1984) proved a major turning point in the development of an approach to this within a critical framework. By looking at the range of problem contexts and at the systems methodologies available for addressing these contexts, Jackson and Keys provided a unified approach which draws on the strengths of the relevant methodologies, rather than debating which method is best, and argues for a reconciliation focusing on which method to use in which context, controlled by a "system of systems methodologies" (SOSM). A number of developments have followed this initial work, from which Midgley (1995) summarises the key approaches which may be seen as having adequate theoretical underpinning and practical potential as: total systems intervention (TSI) combined with SOSM; the creative design of methods; critical appreciation, and TSI reconstituted.

CONCLUSIONS

The planning or design schools of corporate strategy may be determined as functionalist in orientation, and as offering a limited perspective of the domain. The discovery approaches, resting on interpretative methods, offer improvement through a perception of strategic thinking as socially constructed reality, but are unable to overcome *a priori* conditions and false consciousness, and may therefore be classified as regulative.

Critical methodologies, hitherto based on Habermasian thinking, offer significant progress from this position, but are limited to situations where debate is not closed, and privilege those most skilled in debate.

Building on co-operative inquiry, it has been possible to construct a strategic framework which is equipped to address differential debating skills, and, by offering the possibility of embedding the approach within the social fabric of an organisation, is less exposed to the barriers to debate which mitigate against the use and success of methodologically-based activities.

REFERENCES

Ansoff, H. I., 1964, A quasi-analytical approach of the business strategy problem, *Management Technology* (IV): 67-77.

Britton, G. A. and McCallion, H., 1994, An overview of the Singer/Churchman/Ackoff school of thought, *Systems Practice* 7(5): 487-522.

Burrell, G. and Morgan, G., 1979, "Sociological Paradigms and Organisational Analysis", Heinemann, London.

Jackson, M. C., 1982, The nature of soft systems thinking: the work of Churchman, Ackoff and Checkland, *Applied Systems Analysis* 9: 17-28.

Jackson, M. C., 1985, Social systems theory and practice: the need for a critical approach, *International Journal of General Systems* 10: 135-151.

Jackson, M. C. and Keys, P., 1984, Towards a system of systems methodologies, *Journal of the Operational Research Society* 35(6): 473-486.

Johnson, G. and Scholes, K., 1993, "Exploring Corporate Strategy", Prentice Hall, Hemel Hempstead.

Midgley, G., 1992, The sacred and profane in critical systems thinking, *Systems Practice* 5(1): 5-16.

Midgley, G., 1995, Mixing methods: developing systemic intervention, *Hull University Research Memorandum* No. 9.

Midgley, G., 1995, What is this thing called critical systems thinking, *in:* "Critical Issues in Systems Theory and Practice", K. Ellis, A. Gregory, B. R. Mears-Young and G. Ragsdell, eds, Plenum, New York, 61-71.

Mintzberg, H., 1987, Crafting strategy, *Harvard Business Review* 65(4): 66-75.

Mintzberg, H., Quinn J. B., Ghoshal, S., 1998, "The Strategy Process: Revised European Edition", Prentice Hall, Hemel Hempstead.

Quinn, J. B., 1980, Formulating strategy one step at a time, *Journal of Business Strategy*: 42-63.

Quinn, J. B., 1980, "Strategies for Change - Logical Incrementalism", Irwin, Homewood, Il.

Reason, P. and Heron, J., 1995, Co-operative inquiry, *in:* "Rethinking Methods in Psychology", J. A. Smith, R. Harre and L. V. Langenhove, eds, Sage, London.

Ulrich, W., 1983, "Critical Heuristics of Social Planning: A New Approach to Practical Philosophy", Haupt, Berne.

DEALING WITH POWER IN ORGANISATIONAL INTERVENTION: THE PLACE OF METHODOLOGY

Brian Lehaney and Steve Clarke

Department of Finance, Systems and Operations
Luton Business School
University of Luton
Park Square
Luton
LU1 3JU

INTRODUCTION

This paper focuses on the problem contexts encountered within interventions. A classification of problem contexts is undertaken, and an assessment is made of the extent to which four key methodologies contribute to addressing coercive influences. Issues regarding current approaches are surfaced, and the discussion concludes with possible future directions.

A number of approaches to problem context classification have been undertaken, key among which has been the work of Jackson and Flood (see Jackson and Keys, 1984; Flood and Jackson, 1991a; Flood, 1995; Jackson, 1995). Jackson's (1995) approach is shown in Figure 1.

On Jackson's grid, problem contexts may be seen in terms of two dimensions: the vertical dimension of increasing systems complexity, and the horizontal dimension of increasing people complexity. The basis of Jackson's arguments is that early approaches to organisational problem solving focused on the upper left quadrant, seeing problems as simple, and the people dimension as unitary, or subject to no disagreement about aims, objectives or means. Traditional operational research techniques such as systems analysis and design have been placed in this quadrant. During the last twenty years or so, however, there has been an increasing recognition that 'problem solving' needs to deal with more than just simple/unitary problem contexts, leading to consideration of methods to address situations characterised by greater complexity and increasing divergence of views. In addressing coercion and power, this paper is essentially concerned with the horizontal axis of the Jackson grid, and is therefore focused on problem contexts within which there are varying levels of disagreement between participants. In addition to unitary contexts, Jackson developed two further people-complexity classifications: pluralism, in which it is

Synergy Matters: Working with Systems in the 21st *Century,*
Edited by Castell *et al.*, Kluwer Academic / Plenum Publishers, New York, 1999.

379

assumed that, although currently no agreement exists, there is the potential for agreement to be reached; and coercion, in which the lack of agreement cannot be resolved because of coercive influences or the exercising of power, distorting the problem context.

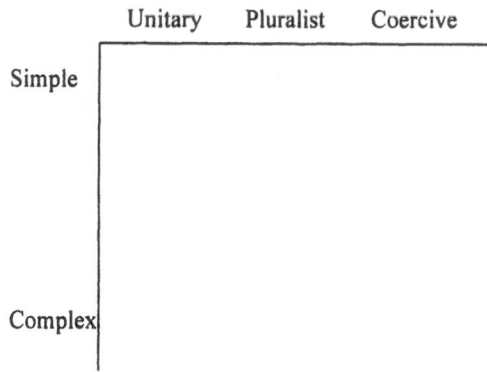

Figure 1. A Classification of Problem Contexts

There are (now) many methodologies which purport to deal with pluralism and coercion, and four have been chosen to highlight the approaches taken: soft systems methodology, interactive planning, strategic options development and analysis and critical systems heuristics. The key to understanding how soft methodologies work, lies in the recognition that all, to a greater or lesser extent, rely on a process of debate. Soft Systems Methodology (SSM: Checkland 1981; Checkland, 1986; Checkland, 1989; Checkland and Haynes, 1994), for example, develops a participant-centred view of a problem situation by developing agreed views of systems within a debating forum. Interactive Planning (IP: Ackoff, 1981) develops participant-informed scenarios by engaging participants in 'idealised design', a structured debating approach to determining the ideal future of the system of concern, and, by comparing this to the likely future implied by current actions, arriving at necessary changes to be made. Strategic Options Development and Analysis (SODA: Eden, 1988; Eden and Simpson, 1989) differs slightly, in that it relies in the initial stages on cognitive mapping, whereby perceptions are drawn from one-to-one interviews with participants. Once the cognitive maps have been determined, however, SODA again uses a debating approach to test the outcomes and determine an agreed way forward. Critical Systems Heuristics (CSH: Ulrich, 1983; Ulrich, 1991) relies explicitly on debate to surface within the system of concern what *is* and what *ought to be*, the latter being used to challenge existing perceptions of the system.

A BRIEF CRITIQUE

Where 'soft' methodologies have been applied to problem contexts in which power is a primary influence, the focus has been on the potential for changing the power structure through a process of debate. Critical systems thinking (see Jackson, 1987; Flood and Jackson, 1991b; Jackson, 1991a) has pursued these ideas, using as a foundation the work of Habermas (1971a;1971c;1976;1987). The main source of critique of 'soft' methodologies rests on Habermas' 'Theory of Knowledge Constitutive Interests' (KCI), in which it is argued that all human endeavour is undertaken in satisfaction of three interests: technical (prediction and control), practical (human communicative interaction) and emancipatory (social relations of power and domination).

Jackson (1991b p.170), challenges the view that "soft" methodologies in general, and Soft Systems Methodology (SSM) in particular, have the potential to deal with coercive influences, arguing that "... soft methodologies lack any social theory that might allow them to understand, let alone challenge, the social arrangements that produce distorted communication." Similarly Mingers (1984), suggests that Checkland's pure, strong subjectivism does not allow for the possibility of some extra-individual reality. This subjectivist stance raises criticisms of the ability of SSM to be utilised as a tool of radical change, as social norms might work to preserve the status quo, regardless of any evidence or arguments to the contrary.

Interactive Planning (IP) is ideologically premised on participation, continuity and holism, and, from a Habermasian perspective, is therefore dependent on all participants being able to contribute in open debate. Although focused on pluralistic and coercive issues, IP relies on the pre-existence of conditions for overcoming existing power structures. In its approach to process it deals with issues of culture, but only within existing political constraints

Strategic Options Development and Analysis (SODA), it is claimed, recognises the role of the individual in decision making within an organisation, rather than working with the 'department' or 'section', as if it were an individual, and by so doing tends naturally to consider power structures and relationships concerning individuals, seeking consensus and commitment, rather than compromise and agreement. The philosophy of SODA sees the organisation as a negotiated enterprise, in which a view can be built from the personal constructs of individuals. In its aim to build consensus from an individual level, however, SODA may be viewed as operating within the status quo, and any consensus achieved seen as a false consensus, grounded on a 'false consciousness' on the part of participants.

Where a communicative, debating forum exists, CSH is able to deal with coercive interests within it. Where coercion is "characterised by closure of debate" (Midgley, 1997), CSH may be seen to be of limited value, since debate is fundamental to the operation of the methodology. In addition, there is the issue of why those in power should listen to the powerless (Jackson, 1985; Flood and Jackson, 1991a; Mingers, 1992). Ulrich sees Habermas' ideas of rational argumentation as too idealistic and lacking practicality, he seems not to acknowledge that CSH will only work where "all participants are able to handle involvement in rational argumentation." (Midgley, 1997). CSH therefore risks replacing one form of coercion by another – coercion by those privileged in debate.

This critique has so far drawn mostly on Habermas for a theoretical foundation, a choice which is not arbitrary, since much of the development of the critical stream in management science has been based on Habermasian thought. There is, however, a wealth of critical theory available to the interventionist, emanating from the foundations laid by Kant (1724-1803), whose exposure of "synthetic *a priori* statements" shows how an uncritical approach may lead to a false consciousness, which a critical approach is able to expose. More recently this has been explored (see, for example, Midgley, 1995; Brocklesby and Cummings, 1996; Probert, 1996). Brocklesby and Cummings (1996) refer to a historical development through Hegel to Marx and thereby to the Frankfurt School, the main contributors to which they identify as Horkheimer, Adorno and Marcuse. Probert

(1996) queries the exclusion of Benjamin and gives more weight to the work of Adorno. The most current thinking has largely pursued the ideas of Foucault as an alternative to the Habermasian strand. Habermas has concentrated on a view of emancipation in which methods can be developed to emancipate people as a whole, which in management studies has been applied to the emancipation of groups, or emancipation within organisational interventions. Foucault, by contrast, sees emancipation as an essentially individual concept. In Foucauldian terms, emancipation of participants within an interventionist situation is simply not possible, since power, according to Foucault, should not be looked at in terms of domination: subjection is constituted in the subjects, and runs through the whole social body (Foucault, 1979).

Even if it were accepted as justifiable to cast Habermas as currently the most significant critical thinker, the choice of his knowledge constitution theory as a basis for the development of a critical approach to organisation studies can be questioned. Midgley (1995), for example, argues that the theory of knowledge constitutive interests supports a predict and control approach, thereby perpetuating a view of the human domination of nature which, it could be argued, will have detrimental consequences. Midgley proposes a solution based on Habermas' work on universal pragmatics, in which Habermas argues that communication aimed at reaching an understanding always involves the raising of four validity claims, which may be categorised as comprehensibility, truth, rightness and sincerity. Midgley (1995) has undertaken some initial work to develop these as an alternative basis for a pluralist theory. Truth is seen by Midgley as relating to the objective/external world, and thereby to hard, cybernetic methods; rightness to the normative, social world, and hence soft methods; and sincerity to the subjective, internal world, and cognitive methods such as cognitive mapping and personal construct theory (see Kelly, 1955; Eden, 1988; Eden, 1994). Similarly Oliga (1996) and Foong, Ojuka-Onedo *et al* (1997) have focused on Habermas' (1987) system-lifeworld concept, which conceptualises "society as a whole" as consisting of lifeworld: the inner needs of its members addressed via communicative action; and system: the outer needs addressed by material reproduction through labour. The outer needs are concerned with "system integration", and the inner needs with "social integration", and only if balanced, argues Habermas (1987 p.152), does society as a whole become ".. *systematically stabilised* complexes of action of *socially integrated* groups." In modernity, it is argued, system dominates, with the lifeworld undermined by "transfers of communicative infrastructures to the system" (Foong, Ojuka-Onedo *et al*, 1997).

There are two major areas of current research and practice which are particularly relevant. Firstly, the application of Foucauldian theories is focused on the consideration of power as part of the social fabric, and suggests that a methodological approach is fundamentally flawed. The fields of participatory action research and human inquiry may have something to offer in this respect. Secondly, ideas labelled "diversity management" (Flood and Romm, 1995; Flood and Romm, 1996a; Flood and Romm, 1996b; Gregory, 1996) offer a potential for future development. Rather than looking for methodological solutions, diversity management sees the problem as characterised by two forms of diversity: the diversity of problem contexts, and the diversity of choice ".. available for people to manage organisational and societal affairs." (Flood and Romm, 1996a p.xi).

CONCLUSIONS

The findings of this study point to issues of power and coercion being addressed through critical social theory. For intervention in business organisations, management science has used this as a foundation for the development of critical systems thinking, which, since its commencement in the early 1980s, has relied almost entirely on the theories of Habermas, and more particularly his theory of knowledge constitutive interests. Such approaches can be demonstrated as having a limited domain of application, with participatory action research, human inquiry, and diversity management offering a possible way forward.

REFERENCES

Ackoff, R. L., 1981, "Creating the Corporate Future". New York, Wiley.

Brocklesby, J. and Cummings S., 1996, Foucault plays habermas: an alternative philosophical underpinning for critical systems thinking, *Journal of the Operational Research Society* **47**(6): 741-754.

Checkland, P. B. (1981). Systems Thinking, Systems Practice. Chichester, Wiley.

Checkland, P. B., 1986, The application of systems thinking in real-world problem situations: the emergence of soft systems methodology, *IEE Colloquium on 'Controlling Complexity: Systems Theory and Practice'* **41**.

Checkland, P. B., 1989, Soft systems methodology, *Human Systems Management* **8**(4): 273-289.

Checkland, P. B. and Haynes M. G., 1994, Varieties of systems thinking: the case of soft systems methodology, *System Dynamics* **10**(2-3): 189-197.

Eden, C., 1988, Cognitive mapping, *European Journal of Operational Research* **36**: 1-13.

Eden, C., 1994, Cognitive mapping and problem structuring for system dynamics model building, *System Dynamics* **10**(2-3): 257-276.

Eden, C. and Simpson P., 1989, SODA and cognitive mapping in practice, *in:* "Rational Analysis for a Problematic World", J. Rosenhead, ed, Chichester, Wiley.

Flood, R. L., 1995, Total systems intervention (TSI): a reconstitution, *Journal of the Operational Research Society* **46**: 174-191.

Flood, R. L. and Jackson M. C., 1991a, "Creative Problem Solving: Total Systems Intervention". Chichester, Wiley.

Flood, R. L. and Jackson M. C., eds., 1991b, "Critical Systems Thinking: Directed Readings". Chichester, Wiley.

Flood, R. L. and Romm N.R.A., 1995, Diversity management: theory in action, *Systems Practice* **8**(4): 469-482.

Flood, R. L. and Romm N.R.A., 1996a, "Diversity Management: Triple Loop Learning". Chichester, Wiley.

Flood, R. L. and Romm N.R.A., 1996b, Plurality revised: diversity management and triple loop learning, *Systems Practice* **9**(6): 587-603.

Foong, A. L. F., Ojuka-Onedo A.E., Oliga, J.C., 1997, Lifeworld-system, juridification, and critical entrepreneurship, *in:* "Systems for Sustainability: People, Organizations, Environments", (conference proceedings), Stowell, F. *et al,* eds, U.K., Plenum.

Foucault, M., 1979, "The History of Sexuality, Volume I". London, Allen Lane.

Gregory, W. J., 1996, Dealing with diversity", *in:* "Critical Systems Thinking: Current Research and Practice", R. L. Flood and N. R. A. Romm, eds, New York, Plenum.

Habermas, J., 1971a, "Knowledge and Human Interests". Boston, Beacon Press.

Habermas, J., 1971b, "Theory and Practice". Boston, Mass, Beacon Press.

Habermas, J., 1976, "n systematically distorted communication, *Inquiry* **13**: 205-218.

Habermas, J., 1987, "Lifeworld and System: A Critique of Functionalist Reason". Boston, Mass, Beacon Press.

Jackson, M. C., 1985, The itinerary of a critical approach: review of ulrich's "critical heuristics of social planning", *Journal of the Operational Research Society* **36**: 878-881.

Jackson, M. C., 1987, "New Directions in Management Science". Aldershot, Gower.

Jackson, M. C., 1991a, Five commitments of critical systems thinking, *in:* "Systems Thinking in Europe" (Conference Proceedings), Jackson, M.C. *et al,* eds, Huddersfield, Plenum.

Jackson, M. C., 1991b, "Systems Methodology for the Management Sciences". New York, Plenum.

Jackson, M. C., 1995, Beyond the fads: systems thinking for managers, *Systems Research* **12**(1): 25-42.

Jackson, M. C. and Keys P., 1984, Towards a system of systems methodologies, *Journal of the Operational Research Society* **35**(6): 473-486.

Kelly, G. A., 1955, "The Psychology of Personal Constructs". London, Weidenfeld and Nicholson.

Midgley, G., 1995, Mixing methods: developing systemic intervention, *Hull University Research Memorandum* **No. 9**.

Midgley, G., 1997, Dealing with coercion: critical systems heuristics and beyond, *Systems Practice* **10**(1).

Mingers, J., 1984, Subjectivism and soft systems methodology: A critique, *Journal of Applied Systems Analysis* **11**: 85-104.

Mingers, J., 1992, Recent developments in critical management science, *Journal of the Operational Research Society* **43**(1): 1-10.

Oliga, J. C., 1996, "Power, Ideology, and Control". New York, Plenum.

Probert, S. K., 1996, Is total systems intervention compelling?, *in:* "Sustainable Peace in the World System, and the Next Evolution of Human Consciousness", (Conference Proceedings), Wilby, J., *et al*, eds, Budapest, Hungary, Omni Press, Madison, USA.

Ulrich, W., 1983, "Critical Heuristics of Social Planning: A New Approach to Practical Philosophy". Berne, Haupt.

Ulrich, W., 1991, Critical heuristics of social systems design, *in:* "Critical Systems Thinking: Directed Readings". R. L. Flood and M. C. Jackson, eds, Chichester, Wiley: 103-115.

INTERDISCIPLINARITY, SYSTEMS AND TOTALITY: SOME ANTI-SYSTEMIC MOTIFS IN THE THOUGHT OF BENJAMIN AND ADORNO.

M.W.J. Spaul

Design and Communications
Anglia Polytechnic University
Victoria Road, Chelmsford, CM1 1LL.

INTRODUCTION.

Kellner (1989:7) points out that 'interdisciplinarity' has at least two senses. There is a pragmatic sense, in which practical benefits may accrue when a mix of different perspectives and skills are brought to bear on a problem. There is also a more abstract sense, in which what is being expressed is a dissatisfaction with modes of thinking bound to specialised and restrictive forms. Systems thinking embraces interdisciplinarity in both senses; but this paper is principally concerned with the latter, the dimension of systems thinking which aspires to holism, or 'knowing the totality', as a normative or moral principle. This aspect of systems thinking appears strongly when it embraces German critical philosophy, whether it be the Kantian aspects of Churchman (1979), or the Habermasian thought utilised by various contributors to 'critical systems thinking' (see, e.g., Flood and Jackson 1991). This paper attempts to illuminate some aspects of the relationship between systems thinking and critical philosophy by showing points at which aspirations to holism and interdisciplinarity converge in the critical tradition, but also showing points at which they diverge. Historically, this divergence is exemplified by the approaches to interdisciplinary study embodied in Horkheimer's original programme for the Institute for Social Research, and the work of Benjamin and Adorno which came to have a greater influence with the Institute's migration to the United States - see Wiggershaus (1994) for a detailed history. The tension between these two approaches is given a contemporary relevance by revivals of Benjamin's and Adorno's work in the broad area of 'cultural studies'. The conceptual distance between this work and that common in systems studies suggests that 'interdisciplinarity', even at its most ambitious, cannot be equated with the transcendence of intellectual boundaries; but rather with their reappearance in other forms.

INTERDISCIPLINARY STUDY AND HOLISM.

Taylor (1989) argued that key contemporary dilemmas, particularly moral ones, are anchored in 'inescapable frameworks' - aspects of the modern identity which have been laid down in historical layers, and which necessarily resurface in contemporary discussion. In an earlier work (Taylor 1977:3-50) he traced one of these inescapable frameworks, unease over the disunity of modern life and its mutually-incomprehensible specialisms, to a crisis in European thought at the end of the 18th century. So, at least in part, when we respond to the rhetorical appeal of systems thinking with its images of holism and interdisciplinarity -

especially when couched in the humanistic terms of Churchman (1979) or Ulrich (1994) - we are responding with an aspect of the modern identity which may be archaeologically recovered from the romantic era. This linkage is more explicitly embedded in the thought of the Frankfurt School, since it responds directly to a complex of problems in German philosophy stemming from the late 18th century.

The complaint of the German romantics was that the enlightenment had destroyed a unity. Science had objectified nature and set it apart from humanity, draining it of its potential for inspiration and meaning; similarly human nature had been objectified and dissected into disparate faculties. For Schiller (Habermas 1987b:45-50) this dissection was paralleled by an increasing specialisation in social roles, and a narrowing of the consciousness of the individual. Ancient Greece took on an immense symbolic significance as a supposed exemplar of the unity which had been lost; but it was realised that such a social totality could not be recaptured by a simple regression. For those who were not able to reject the enlightenment conception of reason, foremost of whom was Hegel, the challenge was to reintegrate reason and an expressive unity with nature, so that all aspects of human life could find their proper place in a harmonious whole (Taylor 1977:28-35). Whilst the details of Hegel's metaphysics have limited relevance in contemporary debate, some aspects of his systematising philosophy bear comparison with the our ideals of interdisciplinarity. Hegel believed that all sciences, when they had reached a sufficient level of refinement, could be unified - and that philosophy could mediate that unification. This was a matter of rational necessity, since if all sciences reflect aspects of a single, intelligible entity (for Hegel, the world spirit), then they must share, at root, a common framework (Inwood 1992:265-267).

The lineage of Marxist thought brought this principle down to the Frankfurt School. In Marx's materialist transformation of Hegel's thought, the unification and transcendence of the sciences of the bourgeois era played a significant role. The division, specialisation and inequality which characterise the current stage of humanity's progress in the reworking of nature could be transcended when humanity had come to dominate nature and freely impose its designs upon it. This would somehow be achieved with a revolutionary jump in which current, divided forms of science and organisation would be sublated into new forms of knowledge at which Marx could only gesture (Taylor 1977:547-551; Marx 1988:99-114). Such a model is present - in a cautious form - in the programme of research set out by Horkheimer for the Frankfurt School (Horkheimer 1993:1-14). He envisaged a dialectical interplay between extant social sciences and 'social philosophy' (for which, read 'Lukacsian Marxism') which would lead to the transformation of those sciences and an enhanced knowledge of social conditions - in the limit, approaching the knowledge of the 'social totality'. As Bonss (1993) expresses it, Horkheimer's programme held out the prospect of a "cognitive and social learning process" which "leads to a more comprehensive form of constituting and appropriating reality". That it failed to do so (Honneth 1991:5-31) has not prevented it from achieving an iconic status alongside other holistic, utopian projects of social transformation from the era of the Weimar Republic, such as the Bauhaus (see, e.g., Forgacs 1995).

References to Horkheimer's programme reappear at strategic moments in Habermas contribution to the Frankfurt School tradition (Habermas 1978:301-350, 1987a:374-404); and the overall architecture of the programme is still imprinted on the theory of communicative action. A philosophical approach to the "universal and necessary presuppositions of communicative action" is to be developed in commerce with the "reconstructive sciences" (those which postulate generative mechanisms underlying the development and actions of thinking subjects). There are residual ambitions of synthesising a total view of society; in particular, Habermas (1990:18-19) holds out the (distantly Hegelian) hope that philosophy might "refurbish its link with the totality" by interpreting the discourse of the specialised spheres (which have drifted into mutual incomprehensibility), overcoming "the isolation of science, morals and art and their respective expert cultures" on behalf of the lifeworld. Such ambitions have made Habermas' work attractive to a strain of systems thinking in search of an articulated basis for an abstract holism which talks of "sweeping in" incommensurable factors in decision making processes.

READING FRAGMENTS AND RUINS.

The Frankfurt School harbours an approach to interdisciplinarity which contrasts with Horkheimer's early position and Habermas' revivals of it. Benjamin and Adorno derived from Kracauer an approach to critique which dispenses with the ambition of mirroring a social totality (Jay 1984:241-275) - an idea which they found incoherent - substituting a view of social reality as something discontinuous, to be studied in fragments. Part of this 'logic of disintegration' (Buck-Morss 1977:67) was the rejection of a key assumption behind systematising thought: that the whole must be something more than the sum of its parts (the systems principle of emergence). For this they substituted the principle that social reality exhibited a 'discontinuous finitude' (Benjamin 1977:37, 43) and was properly grasped as reflections in intensively-studied particulars - a derivation from Leibniz' monadology (1973:179-194). Importantly, this fragmentary approach still serves to integrate the work of different disciplines; the fragment is not to be equated with the restricted view of the single discipline, or reductive methods of study. It must also be noted that it is a supremely non-instrumental mode of thought; a medium for objective and detached critique which views the accumulation of 'useful' knowledge with scepticism and distaste (Benjamin 1977:29-30; Brodersen 1997:87-92).

Benjamin developed his monadological position in a literary context (Rosen 1988). There are artistic genres and works which defy simple characterisation: tragedy, comedy, the novel, etc., and specific instances of them, are not susceptible to 'conceptual definition' (in a defining sentence or paragraph) - so how is the critic to approach such elusive objects? A response developed by romantic critics was a species of Platonism: terms for artistic genres designate Ideas, which cannot be defined by any exhaustible process, but only by an infinite series of "incommensurable propositions" (Novalis, cited in Rosen 1988:155). Benjamin took up this suggestion and fused it with aspects of Kabbalist thought (Buck-Morss 1989:231-238) to produce a conception in which it is the task of the critic to search for elements of phenomena which can be arranged into images which restore the lost links between words and Ideas. In passages of notorious obscurity he introduced the term 'constellation' to describe the Idea - something which is "an objective, virtual arrangement" of phenomena, which contains things as constellations do stars (Benjamin 1977:27-41). As Rosen (1988:159) expresses it: "The total range of significance [of, e.g., a term such as 'tragedy'] represented objectively, and as a structure of its most distant relationships, is the Idea in Benjamin's sense". Ideas are to be represented as a configuration of concrete elements, which is also a constellation (Benjamin 1977:34). At work here is an image of the functioning of language which has re-emerged in poststructuralist thought; that words signify by means of an indefinitely extended process which, at the limit, involves an exhaustive history of those words and their contexts of use. The constellation constructed by the critic is a finite, revealing abridgement of the ideal structure. The end result - variously termed the constellation, dialectical image, or historical image - is an example of 'dialectics at a standstill', in which a contradictory and conceptually ungraspable reality becomes visible between the elements of the montage woven by the critic (Buck-Morss 1977:101-102; Buck-Morss 1989:218-219). The process is akin to literary or photographic montage - both avant garde activities contemporary with Benjamin's and Adorno's work (see, e.g., Ellman 1987; Pachnicke and Honnef 1992) - in which existing literary fragments or pictorial elements are given a fresh significance in a different setting, and the whole structure is intended to reveal something which a direct declaration could not hope to express.

What was developed by Benjamin as a literary device was used in music criticism by Adorno (Paddison 1996), and then developed by both of them for the critical analysis of a wide ranage of social phenomena. Although it is reductive to see the outcome as a developed 'method' in any mechanistic sense, the studies which make up Benjamin's 'Arcades Project' (Buck-Morss 1989) or the many essays which exemplify Adorno's 'negative dialectics' (Adorno 1973, 1974) have a discernible shape (Buck-Morss 1977:96-110). In this method a social object is chosen for micrological study; it might be one of the elements drawn from the Paris arcades - dust, gamblers, wrought-iron architecture, etc. - or a jazz song, or the idea of recreation. Archetypally the object is seemingly insignificant, far removed from the assumed 'central forces' at work in a society. Benjamin had a methodological preference for objects which had decayed and were incongruous in the contemporary world, on the grounds that a ruin (whether it be a forgotten drama or a

387

rusting arcade), as a lifeless object, highlights the illusions which once had surrounded it (Buck-Morss 1989:160-201) - in the manner of a photographic negative. The chosen object is studied in its particularity, and not as the representative of an abstract category, on the assumption that this single object mirrors the entire social world from its unique perspective. The image of the social world is not superficially available in the object and requires critical work to bring it to light; in particular it requires the contribution of existing disciplines to make sense of, and to break into comprehensible elements, the phenomena which surround the object. These elements are then available for configuration into a constellation. Adorno described this process as an "exact fantasy which abides strictly within the material which the sciences present to it, and reaches beyond them only in the smallest aspects of their arrangement" (Adorno, cited in Buck-Morss 1977:86).

There are structural parallels between Horkheimer's programme of interdisciplinary research and that practised by, especially, Adorno. For both the developed form of existing disciplines provide standards of rationality which must be respected by the critical thinker. For Horkheimer, social philosophy provided guidance on research topics and the interpretation of results; for Adorno - and, after some debate, Benjamin (see Buck-Morss 1977:136-163) - it provided a map on which the results of research could be pinned. Social philosophy provided conceptual grids, constructed out of dialectical oppositions. As an example, Buck-Morss (1989:47-204) argues that the overarching conceptual grid of the 'Arcades Project' is comprised of the Freudian opposition between dream and waking, and the Marxist opposition between nature frozen into reified categories and the dissolution of such ideological categories by historical decay. In addition, critical philosophy provides a range of heuristics to disrupt easy categorisations or habits of thought (Buck-Morss 1977:98-100): differentiations based on particularity which undermine abstract categories and force the idiosyncratic into view; juxtaposing extreme examples in a way which defies conceptual reconciliation, emphasing the contradictory nature of social reality; and by transforming ideologically-motivated representations into their opposite by subversive editing (see, e.g., Pachnicke and Honnef 1992:77, 290)

CONTEMPORARY MONADOLOGICAL STUDIES.

Benjamin and Adorno's techniques for the construction of dialectical images may be regarded as simply a piece of intellectual history: a characteristic of the inter-war period and of a piece with avant garde modernism in the arts. As a method it has well-known weaknesses; for example, Kolakowski (cited in Jay 1984:266) complained that negative dialectics was a "blank cheque", enabling the critic to make claims which were, in principle, immune from falsification by more systematic study. However, matters cannot be allowed to rest there, since there has been a contemporary resurgence of interest in the work of Benjamin and Adorno. This is partly the result of affinities with poststructuralism (Weigel 1996); but also, more broadly, because elements of the intellectual climate in which their method made sense have recurred (Paddison 1996:131-132).

One example of this resurgence of interest is in a problematically interdisciplinary area which floats behind a set of related signifiers: cultural geography, urban studies, heritage studies (see, e.g., Gregory 1994, Gilloch 1996, Lowenthal 1985). This area relates directly to Benjamin's central concern, one which grows from his urban upbringing in an era of rapid reconstruction: how individuals are to make sense of their surroundings and identity under the conditions of modernity (Brodersen 1997:4). Benjamin's response to the 'shock of the new' of early 20th century modernism is in process of being re-adapted to the shock of globalisation. In addition, the medium of his response - the construction of texts which themselves had the properties of cityscapes   (Gilloch 1994:94), a collision of different perspectives which have to be inhabited before they make sense - has a renewed significance in an era in which new media both complicate the urban experience and provide a means of representing that experience.

It is perhaps no accident that Benjamin's approach is more talked about than practised in academic circles (he was always a freelance critic and 'independent scholar'); and extended examples of monadological studies may be found most readily outside the academic sphere. For example, Wright (1991, 1995) took the Benjaminian expression "botanising on asphalt" quite literally, structuring a montage of small-scale critical essays around the experience of Dalston Lane and a short stretch of the B3070, respectively. In his

1991 study Wright uses the technique to approach a problem which defies systematic study: to make some sense of the contradictions of British public life in the 1980's, in which aggressive modernisation went hand-in-hand with nostalgic appeals to the supposed glories of an imperial past, and a heritage industry arose almost overnight to supply a national identity in the midst of cultural confusion. The elements of Wright's montage respect few of the disciplinary boundaries associated with academic rigour, as he draws on history, architecture, planning, politics - and even the fads of popular business theory. From an academic perspective such a work presents a conundrum: it is difficult to justify as a valid piece of research in any academic tradition, and it is difficult to say exactly what it has demonstrated since it exemplifies Benjamin's declaration that he had nothing to say, only to show (Buck-Morss 1989:222); but it is impossible to specify a form of disciplined study which might replace it.

When Benjamin and Adorno spoke of the concrete elements of phenomena appearing within their constellations then, practically, this meant elements of original texts (such as historical documents) or textual representations of direct experience arranged within the confines of a standard printed text. Some contemporary uses of their methods have dispensed with this restriction as they have been applied in the field of museums and exhibitions, and with the use of new media. As Urry (1996) has urged, Benjamin's way of exploring the relationship of a society to its past is a central theoretical resource for an understanding of the heritage industry. Those concerned with challenging, with Benjamin, linear and simplistic representations of the past have taken his principle of "blasting the object out of the historical continuum" (Buck-Morss 1989:218-219) and re-presenting it in novel ways as the basis for new forms of exhibition design. Porter (1996:118-119) describes this design practice as one which works "across the divides and disciplines of art and science, social history and natural history, work and home". Concrete artifacts are not used as instances of abstract, historical categories, but are arranged in representations which invite personal explorations from visitors. In a similar vein, the new media provide the means of giving a new sense to a potential which Benjamin located in photography, as an art form in which "multiple fragments are assembled under a new law" (Benjamin 1992:227).

CONCLUDING REMARKS.

Strictly speaking, this paper has no conclusions, since it has merely sought to show that by following the thread of interdisciplinarity within critical systems thinking to some of its more distant connections, one may end up in quite alien regions of thought - alien, at least, insofar as recognition in the critical systems literature is concerned. Perhaps this should not be so since, especially in the contemporary monadological studies indicated above, there are clear resources to underpin the kind of social learning processes which are central to models of citizen planning and decision making (see, e.g., Forester 1989; Ulrich 1994). Whether a sphere of academic debate can span such a wide conceptual range is quite another question.

REFERENCES.

Adorno, T., 1973, "Negative Dialectics," Routledge, London.
Adorno, T., 1974, "Minima Moralia," Verso, London.
Benjamin, W., 1977, "The Origin of German Tragic Drama," Verso, London.
Benjamin, W., 1992, "Illuminations," Fontana, London.
Bonss, W., 1993, The programme of interdisciplinary research and the beginnings of critical theory, in: "On Max Horkheimer," S. Benhabib, W. Bonss and J. McCole, eds., MIT Press, Cambridge, Mass.
Brodersen, M., 1997, "Walter Benjamin: A Biography," Verso, London.
Buck-Morss, S., 1977, "The Origin of Negative Dialectics," Harvest Press, Hassocks.
Buck-Morss, S., 1989, "The Dialectics of Seeing," MIT Press, Cambridge, Mass.
Churchman, C., 1979, "The Systems Approach and Its Enemies," Basic Books, New York.
Ellman, M., 1987, "The Poetics of Impersonality," Harvard University Press, Cambridge, Mass.
Flood, R.L. and Jackson, M.C., (eds.) 1991, "Critical Systems Thinking: Directed Readings," John Wiley, Chichester.
Forester, J., 1989, "Planning in the Face of Power," University of California Press, Berkeley.

Forgacs, E., 1995, "The Bauhaus Idea and Bauhaus Politics," Central European University Press, Budapest.
Gilloch, G., 1996, "Myth and Metropolis," Polity, Cambridge.
Gregory, D., 1994, "Geographical Imaginations," Blackwell, Oxford.
Habermas, J., 1978, "Knowledge and Human Interests," Heinemann, London.
Habermas, J., 1987a, "The Theory of Communicative Action Vol. 2," Polity, Cambridge.
Habermas, J., 1987b, "The Philosophical Discourse of Modernity," Polity, Cambridge.
Habermas, J., 1990, "Moral Consciousness and Communicative Action," Polity, Cambridge.
Honneth, A., 1991, "The Critique of Power," MIT Press, Cambridge, Mass.
Horkheimer, M., 1993, "Between Philosophy and Social Science: Selected Early Writings," MIT Press, Cambridge, Mass.
Inwood, M., 1992, "A Hegel Dictionary," Blackwell, Oxford.
Jay, M., 1984, "Marxism and Totality," University of California Press, Berkeley.
Kellner, D., 1989, "Critical Theory, Marxism and Modernity," Polity, Cambridge.
Leibniz, G., 1973, "Philosophical Writings," Dent, London.
Lowenthal, D., 1985, "The Past is a Foreign Country," Cambridge University Press.
Marx, K., 1988, "Economic and Philosophic Manuscripts of 1844," Prometheus, Buffalo.
Pachnicke, P. and Honnef, K., 1992, "John Heartfield," Abrams, New York.
Paddison, M., 1996, "Adorno, Modernism and Mass Culture," Kahn and Averill, London.
Porter, G., 1996, Seeing through solidity: a feminist perspective on museums, in: "Theorizing Museums," S. Macdonald and G. Fyfe, eds., Blackwell, Oxford.
Rosen, C., 1988, The ruins of Walter Benjamin, in: "On Walter Benjamin," G. Smith, ed., MIT Press, Cambridge, Mass.
Taylor, C., 1977, "Hegel," Cambridge University Press, Cambridge.
Taylor, C., 1989, "Sources of the Self," Cambridge University Press, Cambridge.
Ulrich, W., 1994, "Critical Heuristics and Social Planning," John Wiley, Chichester.
Urry, J., 1996, How societies remember the past, in: "Theorizing Museums," S. Macdonald and G. Fyfe, eds., Blackwell, Oxford.
Weigel, S., 1996, "Body- and Image-Space," Routledge, London.
Wiggershaus, R., 1994, "The Frankfurt School," Polity, Cambridge.
Wright, P., 1991, "A Journey Through Ruins," Radius, London.
Wright, P., 1995, "The Village That Died for England," Jonathan Cape, London.

LINKING METHODOLOGICAL COMPLEMENTARISM WITH THE THEORY OF JOINT ALLIANCES

Maurice Yolles

Liverpool John Moores University Business School,
Liverpool, UK.

1. INTRODUCTION

The complementary use of more than one methodology that can lead to intervention strategies for complex problem situations is not possible because they have paradigms that are incommensurable, and so cannot be used together. At least this is what we are told by the fundamentalists of paradigm incommensurability. There are moves to theoretically counter this argument. One approach lies in exploring a connection between the notion of methodological complementarism and strategic management theory of joint alliances.

In the field of management systems methods/methodologies can be seen as an organisation of inquiry (Yolles, 1998a, 1999) that can be applied to complex situations. It can enable inquirers to reduce that complexity and introduce desirable change. There is an argument that the joint or complementary use of a plurality of methods/methodologies can bring more to an inquiry process than the use of a unitary approach. However, there is also a *fundamental* argument against this that such a plurality involves paradigm incommensurability, that invalidates the joint use of methods/methodologies.

In the field of strategic management durable corporate organisation operate in complex situations too, and they operate in ways that will affect their effectiveness and durability, and will determine their futures. They too enter into a type of complementary activity called a joint alliance. If it could be argued that the two types of organisation are generically related, then we should be able to apply the notion of paradigm incommensurability. That this has not been done might suggest that any generic relationship that might exist has not been discovered. In this paper our interest will be to show how they can be argued to be generically related. If this can be accomplished, then it will follow that *fundamental* arguments of paradigm incommensurability should be applicable to joint ventures. However, this would appear to be contrary to much of the literature on strategic management and indeed in corporate practice. Our intention is to explore this.

2. COMPLEMENTARISM IN INQUIRY AND CORPORATE ACTIVITY

If an autonomous coherent group of people can be seen to form a purposeful adaptable activity system, then it will have associated with it a paradigm from which a behaviour will be manifested (Yolles, 1999). The paradigm will be different from another paradigm belonging to a different autonomous coherent group. The two paradigms will likely also be incommensurable because they will either have different conceptual extensions, or if not then similar conceptual extensions will be qualitatively distinct and thus take on different meanings. Paradigms are

Synergy Matters: Working with Systems in the 21st Century,
Edited by Castell *et al.*, Kluwer Academic / Plenum Publishers, New York, 1999.

391

describable as being commensurable when they are conceptually coextensive and qualitatively similar. Mostly we can think of different paradigms as being incommensurable to some degree.

This view derives from the *fundamentalists* of paradigm incommensurability, explained by Burrel and Morgan (1979) in their exploration of organisational analysis. This work has become part of the field of *management systems*. Within this field a group of would be inquirers will have defined a paradigm that establishes an approach to structured inquiry that may be called a method or methodology. In line with the ideas proposed by Checkland and Scholes (1990) and Yolles (1998), methodologies can be seen as purposeful activity systems that have paradigms and facilities that can control inquiry behaviour. Such methodologies are intended to be applied to complex problem situations that lead to feasible intervention strategies for the management of desirable change. There is an argument that different methods/methodologies exist that each have their own penchant, and these could with benefit be used jointly to improve the nature of the inquiry.

In the field of *strategic management* there is a parallel idea to that of joint inquiry ventures, where durable corporate organisations are seen to be able to benefit from a joint alliances (Kelly and Parker, 1997). It is not controversial to say that such organisations can be seen as purposeful adaptable activity systems, having been explored for well over a decade in terms of viable systems theory as defined by Beer (1985). Such organisations can be seen as autonomous and develop their own paradigms that are ultimately determined by organisational culture (Yolles, 1999). Thus, one would surmise that envisaging that two organisations become involved in a complementary way would invoke the spectre of paradigm incommensurability. Curiously, there do not appear to be any *fundamentalist* arguments that deny that joint alliances can legitimately occur.

Inquiry processes and corporate survival can both be embedded within viable systems theory, and enables us to claim that inquiry and corporate activity can both be seen as different species of the same genus. In particular since both type of organisation have associated with them paradigms, and paradigms are the subject of *fundamentalist* arguments, then these arguments should be equally applied to both species of organisation. Indeed, Kelly and Parker (1997) summarise the benefit and failing of joint alliances, and make no mention of the *fundamental* argument. That fundamental arguments make no appearance here while being decried in management systems would appear to be paradoxical.

3. WORLD VIEWS, WELTANSCHAUUNGEN, AND PARADIGMS

In order to explore the notions supported by our *fundamental* friends and thus to explore whether apparently valid arguments against methodological complementarism are also applicable to joint alliances, it will be appropriate to first consider the base ideas upon which rest their criticisms. This means that we shall explore the nature of paradigms. However, to do so we will shall broaden the exploration to that of world view.

According to Yolles (1996, 1998) there are two types of world view: weltanschauung and paradigm. The term weltanschauung was first introduced into systems by Churchman (1979), and is used as part of Soft Systems Methodology (Checkland and Scholes, 1990). It is seen by some as a view that is often personal and indescribable: that is it cannot be clearly described formally through language that enables a set of explicit statements about its beliefs and other attributes that enable everything that might be expressed about the world view to be expressed. In this sense we refer to *weltanschauung* as an informal world view. Different from weltanschauung is the paradigm, a term explored in some depth by Kuhn (1970). Weltanschauungen become *paradigms* when they are more or less formalised (Yolles, 1999). In the formation of a paradigm a formalised non-normative or semi-formalised shared weltanschauung is created (called a *virtual paradigm*) that may or may not become a paradigm (Yolles, 1996). While individuals and groups may behave in ways that are determined by their weltanschauung, paradigms emerge when the groups become coherent through a degree of formalisation.

Both weltanschauungen and paradigms are forms of world view that operate through culture (beliefs, values, attitudes and language), concepts established within "rational" organised structures called propositions, and norms (Ibid.). They have a relationship with each other, and with the behavioural world that is coupled to the physical or social forms that we see around us. The two types of world view that exist within a group may be identified as a cognitive domain (Yolles, 1999), and differentiated from the behavioural domain within which is defined by the "real" or perceived behavioural world. The two domains are distinguished from each other by a transformational domain. The three domains can be placed one inside the other to form a deep or cognitive domain populated by world views, and transforming domain that establishes links to a surface or behavioural. It is in the latter domain that social structures are maintained. We normally refer to transformation as transmogrification - which according to the 1975 issue of the concise Oxford English dictionary is a transformation that may be subject to surprises.

This model acts as the basis for viable systems theory as developed by Yolles (1999), who in doing so called on the work of Stafford Beer and Eric Schwarz. It also connects with the work that has appeared in artificial intelligence and language theory. Chomsky (1975), in his attempts to develop a theory of transformational grammar of language, distinguished between the semantics of a message and its syntax. Semantics occurs at a "deep" or cognitive domain of knowledge that carries meaning. Syntax is a manifestation of semantics that is created through the "surface" that has structure and from which we make utterances. A structurally similar model is used in the field of artificial intelligence (Clancy and Letsinger, 1981) that distinguishes between deep and surface knowledge.

4. ARGUING TO EXTENDING THE *FUNDAMENTAL* PARADIGM

In considering the *fundamentalist* argument we must first direct our attention to methodology since it is here that their arguments are applied. The coordination of methodological behaviour is concerned with the idea that different systems methodologies intended to be used to intervene in complex problem situations each have attributes that can be used for benefit in different situations. It recognises that they may each operate out of different paradigms, and have different rationalities stemming from alternative theoretical positions which they reflect. Each methodology will generate a view of reality and intervention strategy that is itself connected to the penchant of the methodology that spawns it. The different paradigms can operate in ways which are complementary to one another, each finding strength of examination and evaluation that others might not have in respect of different classes of situation.

We have said that arguments against complementarism by *fundamentalists* are that methodologies derive from different paradigms that are incommensurable. This means they cannot be compared or used in a coordinated way. Now, paradigms create the cognitive basis that will be manifested as behaviour in a viable system, and within this context both corporate and inquiry organisations are different species of the same genus of purposeful adaptive activity systems. Enterprises operate out of different paradigms and cooperate conditionally as joint alliances for perceived benefit in the same way that structured inquiry is envisaged to occur through methodological complementarism.

Since corporate and inquiry organisations are of the same genus, we propose to use the generic term joint ventures to include joint alliances and methodological complementarism. In fact one might well just use the term complementarism, with the implicit understanding that we mean the complementary joint ventures of viable organisations. By viability we mean that the organisations have duration and adaptability, are purposeful, and have both a cognitive and behavioural domain.

From this perspective it is more than curious that *fundamentalists* are concerned with complementarism in the management systems field, but not in that of strategic management. Indeed, from the perspective of purposeful adaptive activity systems it seems paradoxical that it can apply to one field and not the other. We suggest that the reason that the arguments of *fundamentalists* only hold true is because their own paradigm is conceptually bounded - that is they do not possess enough conceptual extensions to enable them to explain the processes of complementarism.

Paradoxes can be seen as fundamental contradictions of a paradigm. Attempting to solve a fundamental contradiction from *within* the paradigm is not possible. Rather, a new paradigm must appear beyond that of the *fundamentalists*, with extra conceptual extension(s) that are able to deal with it (as illustrated by Zeno's paradox (Gale, 1968)). We know that different organisations operate in joint ventures even though they operate out of different paradigms defined in part by their cultures.

There is an argument that methodological complementarism *can* be undertaken legitimately. It extends the *fundamentalist* paradigm by introducing a new analytically and empirically independent conceptual extension that derives from the work of Habermas in his theory of human interests. Our own approach lies in a similar vein, but enables the paradigm to be extended. Within the context of the application of methodology to a given complex situation, the new conceptual extensions can exist in a single frame of reference and be thought of as *orthogonalities*. Two of them: *cognitive interest*, and *cognitive purpose*, will be explored below.

5. COMPLEMENTARISM BY EXTENDING THE *FUNDAMENTAL* PARADIGM

We are familiar with the weltanschauung principle that tells us that no view of reality can be complete, that each view will contain some information about reality, but that the views will never be completely reconcilable. The principle of finding a more representative picture of reality by involving as many weltanschauungen as possible generates variety through opening up more possibilities in the way situations can be seen. Those who adhere to this principle during an inquiry consequently regard weltanschauung pluralism as desirable.

We know that a plurality of weltanschauungen can form a shared weltanschauung, and that when this becomes formalised a paradigm appears. It is reasonable to consider then, that there should also be a paradigm principle that might be expressed as follows. A paradigm defines a truth system that results in a logical process that determines behaviour. The truth system is also responsible for recognising and producing what its viewholders consider to be knowledge about reality. Since different paradigms have different truth systems, knowledge across paradigms will never be completely reconcilable. Formal models of reality are built from paradigms, and each model will contain some knowledge that guides behaviour.

Paradigms are created by groups of people, and a paradigm principle should be analogous to the weltanschauung principle. Thus, no formal model of reality can be complete, and finding a more representative picture of a given reality by involving a plurality of formal models generates variety through opening up more possibilities in the way situations can be addressed through action. To have paradigm pluralism, paradigm incommensurability must be addressed.

Several approaches to methodological pluralism [Jackson, 1993, pp201-202] occur through the selection of paradigms that are based on ideas within Habermas' theory of human interests [Habermas, 1970]. It classifies for human beings two basic cognitive interests in acquiring knowledge: a *technical* interest relate to the human endeavour referred to as work, and a *practical* interest for interaction. Another cognitive interest is *critical deconstraining* that results in the human endeavour emancipation, seen to be subordinate to work and interaction because it results from exploitation and distorted communication. Corresponding to these three classifications of human endeavour, are three types of knowledge that can facilitate "ideal" qualities of human situations.

Systems methodologies may be validly used in a complementary way when viewed in terms of Habermas' classifications [Jackson, 1993, p290-291]. To do this, we should see Habermas' classifications of cognitive interest as providing distinctions between knowledge and technical/practical behaviour. Now we are aware that (a) given knowledges are generated within a given cognitive domain populated by paradigms (or more generally world views), and (b) inquiry behaviour is part of method/methodology. Thus, we see that inquiry behaviour and cognitive/paradigmatic processes are analytically distinct. This leads to the argument that while paradigms guide knowledge production and therefore determine knowledge type, systems methodologies should be seen to serve cognitive interests. Various approaches would seem to adopt this distinct classification.

The relationship between cognitive interests and cognitive purposes is defined by Yolles (1999), in which a viable system model was constructed that distinguished between and links three domains together that are analytically and empirically independent. A cognitive domain is linked to a behavioural domain through one of transformation that occurs through organising processes. The behavioural domain is argued to be the place in which cognitive interests are manifested, and in the transforming domain we see cognitive purposes being manifested.

The notion of cognitive purpose comes from the idea that paradigms have associated with them purpose, and that cognitive purpose exists and has an autonomous status in a similar way to cognitive interest. It enables the creation of frames of reference that are cognitive purpose related. There are three types of cognitive purposes that effectively correspond to Habermas' types of cognitive interest. The relationship between Habermas' cognitive interests and cognitive purposes are provided in table 1. Here, *cybernetic cognitive purpose* is concerned with intention and is a precursor for technical cognitive interest; *rational cognitive purpose* is concerned with logico-relational constructions, and determined the ability for practical behavioural matters to be dealt with; *ideological cognitive purposes* is connected with the manner of thinking, and is a precursor for the conception of critical deconstraining.

As a result of this approach, it is argued that different methodologies have different independent cognitive purposes. The plurality of methodologies can now be established within a frame of reference that connects them together through their cognitive purpose. Thus, it is possible to see each methodology of a complementary plurality as an orthogonality in an n-space of inquiry.

A similar argument can be used to explain how corporate organisations manage to develop and maintain joint alliances, despite the fact that each of the plurality of organisations involved operates out of its own paradigm. Thus, in any joint alliance, we can see the plurality of organisations defined within the context of a cognitive purpose defined frame of reference. Each of the organisations can now be seen as an orthogonality that contributes to the joint venture in a way peculiar to itself.

Table 1: Relationship between human cognitive interests and purposes

	Technical	Practical	Critical deconstraining
Behavioural domain Cognitive interests	Work. This enables people to achieve goals and generate material well-being. It involves technical ability to undertake action in the environment, and the ability to make prediction and establish control.	Interaction. This requires that people as individuals and groups in a social system gain and develop the possibilities of an understanding of each others subjective views. It is consistent with a practical interest in mutual understanding that can address disagreements, which can be a threat to the social form of life	Emancipation. This enables people to (i) liberate themselves from the constraints imposed by power structures (ii) learn through precipitation in social and political processes to control their own destinies.
	Cybernetical	Rational	Ideological
Transmogrific domain Cognitive purpose	Intention. This is through the creation and strategic pursuit of goals and aims that may change over time, enables people through control and communications processes to redirect their futures.	Logico-relational. Enables missions, goals, and aims to be defined, and approached through planning. It involves logical, relational, and rational abilities to organise thought and action and thus to define sets of possible systemic and behaviour possibilities.	Manner of thinking. An intellectual framework through which policy makers observe and interpret reality that has a politically correct ethical and moral orientation, provides an image of the future that enables action through politically correct strategic policy, and gives a politically correct view of stages of historical development in respect of interaction with the external environment.

6. CONCLUSION

There are parallels between the arguments that are made about the creation of joint alliances between organisations, and those that come into being under the banner of methodological complementarism. A common platform that enables them to be considered as different species of the same genus is to consider them to both be purposeful adaptive activity systems. This enables them to be considered in terms of three domains: cognitive, transformational, and behavioural. The possible importance of this is that it may provide the potential for further developing the theory associated with joint alliances, beyond that explained in the current literature. It may also more closely link in the theory of methodological complementarism, enabling us to formulate a more general viable systems theory of joint ventures.

7. REFERENCES

Beer,S., 1985. *Diagnosing the System for Organisations.* Wiley
Berger, P., Luckman, T., 1966. *The Social Construction of Reality.* Penguin
Bergquist, W., Betwee, J., Meuel, D., 1995, Building Strategic Relationships: How to extend your organisation's reach through partnerships and joint ventures. Jossey Bass, San Fransisco.
Burrell, G., Morgan, G., 1979, *Sociological Paradigms and Organisational Analysis.* Heinemann, London
Checkland, P.B. Scholes,J., 1990, *Soft Systems Methodology in Action.* John Wiley & Son, Chichester
Chomsky, N., 1975, *Problems of Knowledge and Freedom.* Pantheon, New York.

Churchman, C.W. 1979, *The Systems Approach*, 2nd ed. Dell, New York

Fedor, K.J., Werther Jr., W.B.,1996, The Fourth Dimension: Creating Cultural Responsive International Alliances. *Organisational Dynamics*, Autumn, pp39-52.

Flood, R.L., 1995, *Solving Problem Solving*. Wiley, Chichester

Flood, R.L., Romm, N.R.A., 1995, Diversity Management: Theory in Action. *Systems Practice*, **8**(4)469-482.

Gale, M., 1968, *The Philosophy of Time*. Macmillan, London.

Habermas, J., 1970, Knowledge and interest. *Sociological Theory and Philosophical Analysis*, pp36-54, (Emmet, D., MacIntyre, A., eds), MacMillan, London

Jackson, M.C., 1992, *Systems Methodologies for the Management Sciences*. Plenum, New York

Jackson, M.C., 1993, Don't bite my finger: Haridimos Tsoukas' critical evaluation of Total Systems Intervention. *Systems Practice*, 6, 289-294.

Kelly, A., Parker, N., 1997, *Management Directions: Joint Alliences*. Institute of Management Foundation.

Kuhn, S.T., 1970, *The Structure of Scientific Revolutions*.University of Chicago Press, Chicago.

Yolles, M.I., 1997 (Aug.). From Viable Systems to Viable Inqury Systems. *Systemist*, 19(3)154,173.

Yolles, M.I., 1998, A Cybernetic Exploration of Methodological Pluralism. *Kybernetes*, **27**(4 and 5), 527,542.

Yolles, M.I., 1998a (Sept. 8-10), Exploring the Practice of Mixing Methods. UK Operational Research Society conference (OR40), Lancaster Universty

Yolles, M.I., 1999, *Management Systems: Viability in a Complex World*. Financial Times Management, London.

METAPHORS FOR SYSTEMIC INTERVENTION

Rosalind Armson

Systems Discipline
The Open University
Walton Hall
Milton Keynes
MK7 6AA

INTRODUCTION

This paper arises from the author's practice as a systemic, and reflective, process consultant and facilitator. Practice and the desire to practise responsibly, raise a number of questions about the practitioner's role in practice and the practitioner's role in the intervention. In this paper, the role of the practitioner is explicitly addressed.

METAPHORS FOR PRACTICE: THE CLIENT, THE SITUATION AND THE PRACTITIONER

Attention has already been drawn to the complexity that the systemic practitioner brings to the problem situation and to the potential mismatch between the way that the practitioner constructs the encounter and the ways the clients construct the encounter (Armson, 1997).

The consequences of this mismatch, which may not be visible initially, are manifold and complex, adding to the complexity of the problem situation. For example, the client may assume that the consultant is able to 'make it better' by applying her expertise, handing the improved situation back to the client at the end of the intervention. The client may, on the other hand, wish to work in a client-centred way, facilitating the emergence of improvements from the bedrock of the clients' own expertises. The mismatch may go beyond misunderstandings. The client may see their problem situation in terms of a need for simple adjustments while the consultant sees the situation in terms of multiply-faceted soft-complexity. Discussions of who is 'right' in such contexts is meaningless. Trying to impose an interpretation of what the intervention could be, should be, or ought to be, risks losing important insights into the client's construction of the problem, as well as being ethically suspect in a client-centred intervention. The responsible practitioner will seek to negotiate sufficient commonality in her own and the client's understandings of the situation to allow for a responsible intervention to occur.

Synergy Matters: Working with Systems in the 21st Century,
Edited by Castell *et al.*, Kluwer Academic / Plenum Publishers, New York, 1999.

Negotiating a common understanding for a systemic intervention is made more difficult if the practitioner is unaware of the basis on which she is constructing her own interpretation of the situation and the relationship between herself and the client. Indeed, it may be very rare for the practitioner to be fully aware of all the presuppositions she is bringing to the encounter. This imposes on the practitioner the obligation for self-reflection as well as rational reconstruction as part of her reflective practice. Reflective practice which focuses on self-knowledge and brings her unconscious agendas to consciousness is "rich in consequences" (Habermas, 1972).

This self-awareness is realised in the aspiration to become *continuously aware*, as well as reflective. Approaches which facilitate the emergence of this awareness are explored elsewhere (Armson, 1998).

Lakoff and Johnson (1980) demonstrate that "... *metaphor is pervasive in everyday life, not just in language but in thought and action. Our ordinary conceptual system, in terms of which we think and act, is fundamentally metaphorical in nature.*" Metaphors structure understandings, including those of the systemic practitioner: "... *metaphors have entailments through which they highlight and make coherent certain aspects of our experience*" (*ibid.*) It has been claimed that metaphors create realities when acted upon (Krippendorf, 1993). McClintock (1996), following Hausman (1989), follows the idea that exploration of the underlying metaphors that structure how experience is described allows access to the epistemological basis of the understandings of that experience. If metaphors can be said to project upon their subject a "*set of associated implications*" (Black, 1979), then these associated metaphors "*select, emphasise, suppress and organise the primary subject*" (McClintock, 1996). The workshops described below were designed to allow exploration of how practitioners understand their practice and were based on this theoretical framework.

WORKSHOPS

The author has been using Moustakas' (1990) heuristic methodology to explore the nature and experience of the consultancy encounter. As part of this process she has engaged with other consultants and systemic practitioners in conversations and workshops which explore the nature of the encounter and how it is experienced and reflected upon. A workshop experience was designed and run on three separate occasions in which participants were facilitated in bringing forth their metaphors for intervention. The first of these was run with a group of systems practitioners, mostly consultants, attending the 1997 UKSS Conference. The second was run with a group of student, graduate and tutor members of the Open University Systems Society attending the Society's 1997 AGM. The last was run for a diverse group of facilitators with whom the author runs training events. Between six and 25 people participated at each workshop event.

The workshop was designed to elicit metaphors by a number of means; to explore what might be revealed and concealed (Lakoff and Johnson, 1980) by them; and to facilitate conversations around the experience of systemic intervention. The workshop design embodies a certain pragmatism. Using conversation to elicit metaphors is relatively easy to organise and most of the participants at the three workshops were familiar with, and happy to use, rich pictures. Alternative approaches, such as cognitive mapping, psychodrama (Moreno, 1964) and cognitive sculpting (Sims and Doyle, 1995) would be possible in other circumstances.

Workshop design

The workshop was designed as a half-day activity. With small variations between events, the workshop took the form described below.

After initial introductions, including the negotiation of a confidentiality contract, participants formed pairs. For five minutes one of the pair, A, described to their partner, B, a recent systemic intervention they had made. The listener, B, was asked not to interrupt or to comment but to support A by attentive listening and by facilitative prompting only if necessary. At the end of this period, the pairs were asked to identify individually any visual or other imagery that had been used to describe the intervention. The formulation 'visual or other imagery' was used deliberately to discourage any formal metaphorical analysis and to retain an informal and exploratory light-heartedness. A and B in each pair then compared notes. The roles of A and B were next reversed. Since, in this case, the speaker B was more alert to the use of imagery, A was asked to be more searching in their questioning. Following feedback as before, the group discussed briefly the relative richness of metaphorical language in their mutual presentations.

Each participant was then invited to draw a rich picture around the theme of the image, or images, they had just identified. They were asked not to pay too much attention to matching the image with the experience of practice that had generated it but to explore features of the image and its story. Participants were then invited to consider whether their image revealed anything about their practice and whether the image was likely to conceal anything. This was intended as an entirely open question. There was no preconception that images used by consultants in describing their practice actually relate strongly to the practice, or to how that practice is constructed. The intention was purely to trigger conversations around the theme.

The whole group then discussed the images that had been elicited, with participants who were willing explaining features of their rich picture and the image that lay behind it.

Metaphors elicited

In all three workshops surprise was expressed at just how rich the spontaneous imagery was. The listener in each pair generally found it easier to identify visual imagery used by the speaker and typically generated a much richer list of images. This suggests that the consultant's use of imagery was not fully conscious or, at least, was more accessible when the recognition was facilitated by a listener.

Many of the images elicited in the exercise fell into one of four categories. These categories were animals (wily fox, snake, tiger), other professionals (physician, psychotherapist, plumber, repairer, 'thief in the night', 'counsellor for organisations'), fictional characters (Christopher Robin, Alice in Wonderland) and relationship (big sister, 'shoulder to cry on').

Participants often expressed excitement and interest at the images elicited and expressed surprise at the density of connection between the image-metaphor and the way they thought about their practice. Few of the participants had been consciously aware of the image before the workshop although the few that were aware had consciously constructed their practice around a particular interpretation of their role.

METAPHORS FOR PRACTICE: EXPLORING THE IMPLICATIONS

The elicitation of the metaphors for practice generated further insights into how the consultant's practice might be construed by clients and triggered further discussion about the effect of the metaphor on the clients' perception of both consultant and the consultancy intervention. Two examples illustrate the kinds of insights and 'questions for exploration' that emerged from these conversations.

One consultant found herself using some of the following terms: *sliding under, wiggling around, insinuating myself, beguiling the client, unearthing the goodies*. The metaphor that emerged was that of a snake. This metaphor highlighted an approach that the consultant consciously took with clients which was to do with getting into the depths of the organisation through routes which the clients were not necessarily offering directly. She consciously sought out the voices in the organisation which were 'unofficial' and found that to be an effective way of accessing information and understandings about the organisations with which she worked. She was also aware that there was an element of 'beguiling' in her practice which was a deliberate management of the impression she was creating and served to distract the contact client from the unofficial information-seeking she was engaging in.

The emergence of the snake metaphor prompted the consultant to consider whether her practice had become more deceitful than she wished. Although there was not, and never had been, a wish to deceive or to be dishonest, she now felt that there was a real danger that she was beginning to enjoy the stealth of what she did and that 'management of impressions' had actually become an exciting form of illusionism. She was also concerned that clients' image of her might be one of uncertainty. It was possible, she thought, that they perceived her as a disturbing and uneasy presence in the organisation - much like the unease that the presence of a snake might produce. She wondered if the snake image had, in fact, beguiled her and seduced her into habits of practice that she herself now found unattractive. She reported an urgent need to think through what she had been doing and to find alternative modes of operating that, while respecting her skill at accessing rich understandings of the organisations she worked with, were more open to scrutiny by clients.

A second consultant was already aware that she described her practice role as one of 'naive questioner'. Her clients, in answering her naive questions were, she believed, ordering their own appreciation of their situation-of-interest in ways that generated new insights for them. Exploring this metaphor in the workshop allowed her to recognise that this construction of her role enabled her to maintain her own low valuation of her skills. She expressed concern that she habitually limited herself through lack of self confidence but had felt unable to get past the problem. She explicitly espoused an ethic which was client-centred and in which the client retained ownership of the problem. Following the workshop however, she felt that the clients probably found it difficult to discern what her role was and how it should be understood. In the role of naive questioner, she brought little to the engagement except questions, effective though these were in eliciting strategies for improvement.

Following discussion with the group, this consultant decided to explore the meaning of 'discussant'. She understood this role to be one in which she was engaged to discuss the situation with the client. This would allow her to continue her effective critical questioning of the client's interpretation of the situation but also liberated her to propose alternative interpretations and to open up new possibilities.

At the time of writing, no formal follow-up to the original workshops has been undertaken. The first consultant reports herself as unsure about whether the new insights have taken root sufficiently to change her practice although she is more conscious of the need for transparency. The second consultant reports herself more confident and thinks her clients may be more comfortable about her role. She is nor sure whether this is simply because of her increased confidence or whether the role of discussant is better aligned with what the client expects or is able to understand.

CLIENT'S METAPHORS

The workshops described above were not designed to elicit clients' metaphors, either for the consultant's role or for the consultancy intervention. A shared concern for understanding clients' expectations and interpretations emerged from the discussions however. Many consultants experienced difficulties in arriving at an understanding of what the client expected from them, in terms of behaviour, role and output. Despite careful negotiation of contract, many experienced a residual misunderstanding of their role by the client. This was manifest through surprise at the direction the intervention took, at the areas explored, and at what the consultant deemed to be relevant. Typically, clients were reported as expressing surprise at the scope of the intervention. This may be a manifestation of the wider boundaries drawn by systemic practitioners but it may also indicate underlying and unexplored metaphors of what is appropriate and relevant to a consultancy intervention.

The richness of the insights generated by consultants' own exploration of their metaphors for their roles and interventions suggests that parallel explorations of what metaphors clients hold would also be fruitful.

ADDITIONAL COMPLEXITY

The workshop design outlined above leaves open the question of the extent to which the metaphors elicited actually influence behaviour in the consultancy situation. The insights generated as a result of eliciting the metaphors do suggest however that espoused or explicit metaphors may be at variance with metaphors-in-use, whether conscious or not. Eliciting metaphors allowed consultants to question their roles and the ways that these might be perceived by clients.

Without exploring the extent of the presence of the metaphors in the actual consultancy engagement it is hard to evaluate whether unconsciously held metaphors bring additional and unaccounted for complexity to the consultancy intervention. It is clear nonetheless that a closer alignment between espoused metaphors for both role and intervention are likely to reduce misunderstanding. It is important, however, not to assume that unconsciously held metaphors for the consulancy role and for the consultant's intervention are necessarily dysfunctional.

ACKNOWLEDGEMENTS

The author gratefully acknowledges her debt to participants in the workshops described and to those who have been willing to explore their metaphors for practice in other conversations. The work of her former colleague, David McClintock, continues to be a stimulus in exploring the use of metaphors to elicit new and shared understandings.

REFERENCES

Armson, R., 1997, The invisible practitioner or the holistic practitioner?, in: "Systems for Sustainability: People, Organisations, Environments," F. A. Stowell et al., eds., Plenum, New York.

Armson, R., 1998, Becoming aware: a personal reflection on trying to be more wholly systemic, in: "Creative Systems Practice: Proceedings of the Fourth Annual Australia New Zealand Systems Conference", University of Western Sydney, Hawkesbury, New South Wales, October 1998.

Black, M., 1979, More about metaphor, in: "Metaphor and Thought", R. Ortony, ed., Cambridge University Press, Cambridge

Hausman, C. R., 1989, "Metaphor and Art: Interactionism and Reference in the Non-Verbal Arts", Cambridge University Press

Lakoff, G., Johnson, M., 1980, "Metaphors We Live By," University of Chicago Press, Chicago

Krippendorf, K., 1993, Major metaphors of communication and some constructivist reflections on their use, *Cybernetics and Human Knowing*, 2:3-25

McClintock, D., 1996, "Metaphors that Inspire 'Researching with People': UK Farming, Countrysides and Diverse Stakeholder Contexts", PhD thesis, The Open University, UK

Moreno, J. L., 1964, "Psychodrama", Beacon, New York (first published in 1946)

Moustakas, C., 1990, "Heuristic Research: Design Methodology and Applications," Sage, London.

Sims, D. B. P., Doyle, J. R., 1995, Cognitive sculpting as a means of working with managers' metaphors, *Omega, Int J. Mgmt. Sci.*, 23:117-124

MATURANA'S NOTION OF PHENOMENIC REDUCTIONISM AND ITS IMPLICATIONS FOR SSM

John Brocklesby and John Mingers

University of Warwick
Coventry CV47AL

INTRODUCTION

The question of what might be extracted out of the work of Humberto Maturana in relation to the process of understanding and intervening in complex systems has once again come to the fore. Partly this is a result of publications such as Capra (1996) and Mingers (1996) which have brought ideas emanating from the biology of cognition to the attention of a wider audience than was the case previously. Partly it is due to some successful application of Maturana's ideas in domains such as family therapy and in psychotherapy.

Against this background, the following paper aims to juxtapose one important aspect of Maturana's theory of human cognition – the notion of *phenomenic reductionism* – against a more conventional view of cognition that is enshrined in Peter Checkland's SSM. This short paper is extracted from a larger project that aims to examine the relevance of Maturana's ideas to systems practice across different research paradigms.

THE BIOLOGICAL AND RELATIONAL DOMAINS

Maturana's epistemology (see Maturana 1988 for a comprehensive description) rejects the conventional account of human cognition that regards neural information processing as the core activity. Instead it asserts that there are two broad non-intersecting phenomenal domains involved in the process which Maturana refers to as 'observing'. The first of these is biology/anatomy and internal dynamics; the second, social relations and interactions. The term observing describes processes such as making distinctions, uttering cognitive statements and explaining experiences.

The origins of Maturana's views about the domain of biology/internal dynamics go back to experimental work done in the 1960's that led him and Varela to seriously question the prevailing view that the nervous system is open to environmental inputs. Instead they claim that it operates, ". . . as a *closed* network of interactions in which every change in the interactive relations between certain components always results in a change of the interactive relations of the same or other components*" (Maturana and Varela 1987:22 emphasis added). From this basic insight arises the idea that composite systems, including human beings, are *structure determined*, i.e." . . . *everything that happens in them happens as a structural change determined . . . either in the course of their own internal dynamics or triggered but not specified by the circumstances of their interactions."* (Maturana, 1990:13).

On this view human perception is the outcome of dynamic relations within a closed circular nervous system. Because human beings are structure determined, perception cannot involve building representations that mirror an external reality. Neither can external stimuli determine how they are experienced. We can correlate our naming of objects with states of neuronal activity, but not with the

Synergy Matters: Working with Systems in the 21st Century,
Edited by Castell *et al.*, Kluwer Academic / Plenum Publishers, New York, 1999.

403

stimulus that triggered the experience in the first place. According to Maturana and Varela this applies to all perceptual modalities.

Extending these ideas of circularity and closure of the nervous system. Maturana and Varela (1987:22) propose that all living systems are *"networks of molecular production such that the molecules produced, through their interaction, generate the network that produced them and specify its extension"*. The term autopoiesis is employed to describe this dynamic, or - as Maturana often puts it – this peculiar 'manner of relating' of the molecules. Because this dynamic is invariant across all living systems, the term autopoiesis captures what is referred to as the *organisation* of living systems. The term *structure* is employed to describe how the same organisation might be realised in particular cases. Whereas organisation is invariant, structure is subject to change, and it is the distinction between these concepts that allows us to see how stability and change – including learning - can exist side by side.

Maturana and Varela use another important term - *structural coupling* – to refer to the congruence or 'fit' that exists between the living system and medium. When there are recurrent interactions between say an organism and what an observer would regard as its environment (or between one organism and another), structurally determined changes occur in both; i.e. the two structures change *congruently* according to their respective structure determinism. Through this process the structure of the organism at any point in time becomes a record of previous interactions. While these changes are taking place there is a conservation of autopoiesis and adaptation.

In his explanation of the development of observing in human beings, Maturana (1988) extends the notion of a living system structurally coupled to a medium to encompass the structural coupling between two or more living systems. He describes the various actions that go on there as pertaining to the so-called *relational* domain. This is where *languaging* takes place. Now in popular discourse language usually refers to some system of symbolic communication in which there is a transfer of meaning from one entity to another. Maturana disputes this logic because of the organisational closure of the nervous system, and on the grounds that it violates the principle of structure determinism. Instead, languag*ing* is rooted in behavioural relationships. In its minimal form this simply involves one entity doing something on the consequences of the behaviour of another. However this becomes infinitely more complex for human beings as various recursions of behavioural coordinations occur. These take place in networks of structural coupling which provide the context and opportunity for people to agree on using linguistic tokens for specific behavioural coordinations. Through such a process objects and situations 'arise'. The bases of these are rooted in behavioural coordinations. However, with consistent and prolonged usage, they effectively become 'entities in-themselves' and the behavioural coordination becomes obscured. Once they have been 'brought forth' in the manner just described, objects and situations can be referred to. This foreshadows the development of more abstract entities such as height, weight, speed etc. Moreover it allows people to refer to themselves thereby becoming self-aware and self-conscious.

According to Maturana, this continual weaving of linguistic networks through social interaction as a distinctive *'manner of living'* constitutes the human being as the class of living system that it is. It involves a continual bringing forth of new descriptions, new explanations, and new realities - with others - through an on-going social process. On this view human beings do not *use* language. Instead the species is constituted as the living system it is through this distinctive capability.

THE LOGICAL ERROR: COLLAPSING TWO PHENOMENAL DOMAINS INTO ONE

Let us now turn to the *relationship* between the two phenomenal domains that are involved in observing. On this matter, Maturana claims that if we look to the origins of observing (the conditions that generate it) it is apparent that there are these two domains and that each one has a specific role to play. He believes however, that we are inclined to regard observing as an endowment or a property of the human condition. When this happens; when we do not – as he puts it - *'ask the question of the observer'*, we tend collapse the two domains into one. Maturana regards this as a serious logical mistake for although the two domains interact, phenomena belonging to one should not be explained in terms that are more appropriate to the other. Thus phenomena such as thinking, explaining and describing arise in the relational domain and must be explained through what goes on there. However, because there is a reciprocal relationship between the two domains, the structure of the nervous system must still be taken into account. Obviously organisms without an advanced nervous system cannot observe in the manner of human beings. The structure of the nervous system makes observing possible, and, to the extent that structure - at any point in time - is a record of previous historical interactions, it determines what is possible and what is not. At the same time, what goes on in the relational domain can trigger structural changes in the nervous system, i.e. what some - Winograd and Flores (1987) for example - would regard as learning.

THE PHENOMENAL LOCATION OF *WELTANSCHAUUNG* IN SSM

Focussing now on SSM, it is clear that Checkland acknowledges the role played by relational circumstances in the formation of peoples' thoughts and descriptions. For example, Checkland (1981:219, 1986:111), claims that the *Weltanschauungen* that participants bring to bear on a problem situation depend upon their backgrounds and their historical interactions. Yet the SSM literature does not pursue this social dynamic in any great detail. It has very little to say about how *Weltanschauungen* are derived. More seriously, Checkland does appear to commit the logical error that Maturana speaks of. In his writings there is clear evidence that he takes it to be axiomatic that there is some direct *Weltanschauung*-related physiological or anatomical structure hard-wired into the human brain. We have, he says, " . . . *in our heads stocks of ideas by means of which we interpret the world outside ourselves"* (1990:19 emphasis added, see also 20, 217 and 1981:215).

This observation is not mere pedantry. Because *Weltanschauungen* are taken to reside 'inside the head', and – correspondingly - because language is taken to be inextricably tied to the operation of the nervous system, it immediately narrows the scope of the inquiry down to the people involved in the debate. Yet technically, on Maturana's view, even if various thoughts and descriptions about a problem situation are the main focus of debate, these do not – as phenomena - logically pertain to the people who might express them. As Kay (1997:77) puts it, *"an individual's worldview is an emergent property or distinction placed upon them by someone's description of their behaviour – it does not exist as part of the individual's cognitive process . . . from an autopoietic perspective an individual's worldview is just as much a function of the observer as it is a function of the individual's nervous system".*

Technically Kay is correct. However in the context of SSM, it would be ludicrous to suggest that the various descriptions (and *Weltanschauungen*) that emerge during the debate do not - nominally at least - 'belong' to the people who provide them. A softer line might be to argue that descriptions are realised *through* people but *exist* independently of them.This echoes a debate of some years ago (Jackson 1982, Mingers 1984), in which Checkland was chastised for refusing to give credence to the idea that there are external forces that govern peoples' thoughts and descriptions. Now Maturana refuses to be drawn on the question of the existence of such structures on the grounds that even if they were to exist human beings are biologically incapable of gaining access to them. However, Maturana (1988) does speak about how human beings are constrained by the coherences of their experiences; about how these 'just happen'; and, about how – for the most part - they appear to the observer as regular and predictable. From this it is not unreasonable to speculate that such constraints and regularities are due to independently existing forces and structures. Another possible explanation is that the relational circumstances in which observing takes place imposes limits on what thoughts and descriptions are possible. Language, for example, is the mechanism through which human beings experience and explain their worlds, and language, as we have seen, is by its very nature an inter-subjective phenomenon. In this sense we are always-already constrained by social forces even during our most individually reflective and most singularly introspective moments. Relatedly, the structure of the nervous system, which allows us to have experiences, and which (in conjunction with the stimulus that triggered it) determines what the experience is, is itself partly a product of our previous interactions in various relational domains.

This leads us to suggest that there is a contradiction in SSM. On the one hand Checkland acknowledges that changing our views of the world is not unproblematic," . . . *it is characteristic of us that we cling tenaciously to the models which make what we observe meaningful." (Checkland 1980:216).* Moreover he acknowledges that we are constrained by our backgrounds and past experiences, *"our previous experience gives us in-built readinesses to notice or not to notice certain features of a complex problem situation as significant, and to judge them by criteria developed experientially"* (Checkland and Scholes 1990:192). On the other hand because he commits the logical error that Maturana speaks of, and because he ties *Weltanschauung* to the individual, he invokes a sense of creativity and possibility that in many cases may simply not be sustainable. While Checkland is correct to point out that our thoughts and descriptions are free intellectual constructions in the sense that there is no independently existing external point of reference through which they may be judged 'true' or 'false', this does not mean that an observer can construct whatever interpretation he or she likes. That observing belongs to and takes place in the relational domain not only suggests that observing phenomena such as *Weltanschauungen* exist independently of the observer, but that they are constrained by what goes on there, and by what has gone on there in the past.

In this regard, the value of Maturana's work to a user of SSM is that while SSM purports to bring about changes in *Weltanschauung* but says little about the extent to which this is possible and how it might be achieved, Maturana provides an explicit description of this dynamic. There is insufficient space here to elaborate on the practical implications of this; however the process itself is quite straightforward. Through recurrent interactions between the structure of a living system and its medium, the structure of both the system and the medium change each according to its own structure determinism. Thereafter, the structure of the living system, which – at any moment in time - is a record of previous interactions, places broad constraints on the range of possible actions that the system can perform. In human beings the range of descriptions, explanations, perceptions, ideas, and behaviour that are possible depend upon the observer's structure.

Maturana invites us to dispense with the idea that language is merely a means of representation, and with the notion that interpreting the world is a mentalistic information-processing activity that takes place in the brain. Instead he proposes that languaging (and therefore observing) is – by its very nature - a social phenomenon that arises in and belongs to the relational domain. To the extent that language is the mechanism through which people have experiences, and to the extent that language is necessary to explain an experience, then it follows that our experiences, our explanations (and the behaviours that arise out of these) are inherently social. They cannot be explained – as Checkland would have it – solely through reference to the individual concerned, specifically through reference to a *Weltanschauung* that he appears to regard as being carried around in the head. The only 'thing' that one might say is 'carried around', is a *plastic* structure that makes observing possible, that circumscribes or delimits the range of thoughts that can emerge out of the relational circumstances, and which – importantly - is altered (according to its structure determinism) by what goes on there. In simple terms it can be said that thoughts and descriptions do not reside in, or emanate from the head, or, as we normally think about it – from an embodied mind. To a large extent the nature of the relational circumstances determines the description. Maturana reminds us that a description is a *reformulation* of an experience that takes place *in* the relational domain; the description does not replace the experience. A description of an experience arises in the moment according to the nature of the relational circumstances. Although the structure of the observer is important in explaining why a particular description emerges, the latter does not, as Checkland seems to suggest, reflect a cognitive filtering device that is hard-wired in the brain.

ON SSM AS A LEARNING SYSTEM

Making this distinction between the two phenomenal domains that are involved in observing and pondering on it, leads us to wonder about the extent to which SSM can be realistically expected to deliver on all of what its literature promises. Accordingly, couching *Weltanschauung* – a relational phenomenon – in terms that are more appropriate to the domain of anatomy and biology, has important consequences for our understanding of the sorts of things that are likely to occur during and as a result of the debate. When, for instance, we regard peoples' descriptions as reflecting the operation of some filtering and information processing structure in the brain, we are almost inevitably compelled to explain changes in descriptions through reference to changes that are occurring to the individual concerned. This being the case, it is hardly surprising that SSM is commonly referred to as a learning system. SSM, claims Checkland (1989:78), is "... *a learning system*"; it "... *embodies a paradigm of learning* ..." (Checkland 1981:287); "... *the outcome* ... *is* ... *a learning which leads to a decision to take certain actions* ..." (Checkland 1981:213)

Clearly it would be absurd to deny the possibility that SSM can promote learning. As we have seen, what happens in the relational domain can trigger changes in the structure of the observer. What happens during the SSM debate can therefore trigger changes in the structure of the participants. To the extent that this happens, one can say that learning has taken place. On the other hand, it would be incorrect to automatically associate shifts in peoples' thoughts and descriptions with learning. An equally plausible explanation is that such shifts merely reflect changes in the relational circumstances.

406

Because Checkland commits the logical error he cannot account for changes in thoughts and descriptions other than to assume that there has been learning.

Once again, we would reject any suggestion that this does not matter. It matters because the traditional individually-focused and learning-oriented image of SSM conveys an impoverished sense of the complex dynamics that are involved when a group of people convene to debate a matter that is of some concern to them. Maturana's more precise clarification of the nature of the two phenomenal domains that are involved in observing, and his clarification of how each one relates to the other, opens up the theoretical space for a much more comprehensive rendition of what these complex dynamics involve. Now in order to show what the implications of this are for SSM, it is necessary to look more carefully at what Maturana has to say about what goes on in the relational domain. With this knowledge it should be possible to look at SSM in a rather different light to that which has been possible hitherto.

THE RELATIONAL DYNAMIC THAT UNDERPINS OBSERVING

Maturana's term *conversation* is pivotal in understanding the relational dynamic that underpins observing. It circumscribes two interrelated processes: *languaging* - which has already been discussed – and *emotioning*. Emotioning describes that which an observer might distinguish when he/she observes a body disposition that specifies a domain of possible actions. Unseen, but always-already present, these predispositions determine the actions that are possible and those that are not. Thus, as we flow from one emotional state to another, changes take place in the kinds of things that we will and will not do. We behave differently, we see differently, and we describe and interpret things differently according to the emotion in which we are doing these things. And like languaging, although emotional predispositions become embodied in the structure of the observer, they are relational phenomena that arise in the context of social networks. We learn our emotioning with other people, and, through recurrent interactions, structural patterns of emotioning become conserved. Finally, languaging and emotioning are braided; each process effects the other.

Maturana's claim that we participate in not one, but in an infinitely large number of conversations, further complicates the picture. At any moment a person's *bodyhood* is a node at the intersection of many different conversations, each one of which has its own braided flow of distinctions and emotions that have been learned through recurrent interactions over time. This means that just as descriptions within a single conversation are subject to change depending upon the flow of the conversation, they can also alter as the observer shifts from one conversation to another.

SSM AS MULTIPLE SIMULTANEOUS FLOWS OF CONVERSATION

What, on this view, can be said about SSM? In simple terms it can be said that when someone provides a description about a problem situation it arises in the moment depending upon the specific conversation in which he/she is participating, and where he/she is in its flow. As the participant shifts from one conversation to, and/or as the flow of the conversation changes, so too will his/her descriptions, thoughts, explanations, rationalities, and actions. The situation is one of constant flux and transformation within a range of possibilities that are dictated by historical interactions embodied in the structure of the individual.

This presents a very different understanding to that which a neophyte user of SSM might extract from its own literature. Reading this literature one gets the impression that when someone participates in *the* SSM debate, his/her contributions reflect a specific *Weltanschauung* that arises out of previous experiences and group memberships that is hard-wired in the brain. Maturana's more precise rendition of what is involved in explaining experiences leads to a different understanding. Firstly, it suggests that when someone enters the debate he or she will do so having a predisposition to provide a number of *different* descriptions of the situation under investigation. The specific description that arises depends upon what happens as the conversation proceeds, i.e. through its flow of languaging and emotioning. Secondly, the description depends upon the specific conversation in which the participant is involved. This might be with the other members of the group or it could be a conversation that someone is having with him or herself. In this regard it would be wrong to assume that all of those present are participating in the same conservation, or if they are, that the dimensionality of the conversation is the same for them all. Moreover, it would be wrong to assume that participants do not switch from one conversation to another as the debate proceeds, or that they cannot be involved in a number of different conversations concurrently.

In daily life we know that some of our thoughts and descriptions are stable over time while others are more changeable. Maturana provides a theoretical explanation for this. His various propositions suggest that while it may be difficult (although not impossible) to get someone to take on board a *new* set of thoughts and descriptions (learning), people will often oscillate from one viewpoint to another depending on the prevailing relational circumstances. In this regard we can re-conceptualise SSM as involving not one but potentially many different concurrent flows of languaging and emotioning. As a conversation's languaging and emotioning flows different descriptions are possible, and as one conversation replaces another, again different descriptions are possible. It would be unwise therefore to assume that a description provided at a particular point in times will automatically replace or nullify earlier descriptions, or to conclude – as Checkland does - that learning has taken place. To the extent that what happens in the debate triggers changes in the structure of the participant then we can speak of learning. Once the structure has been changed the observer participates differently in that and in other conversations. However it is clearly incorrect to assume that when it is possible to observe one viewpoint superseding another then learning has taken place.

CONCLUSION

This paper presents some preliminary thoughts about the implications for SSM that arise out of a careful demarcation of the phenomenal domains that are involved in observing, and consideration of the relationship between them. There are numerous practical implications that follow this analysis for which space limitations preclude discussion here. These implications draw attention to the extent to which intervention methodologies such as SSM can deliver on their promise, and they provide plausible explanations as to why the organisational outcomes achieved do not always meet the expectations of sponsors. On a more positive note Maturana's work offers insights that may assist practitioners in using SSM with greater finesse and skill. These insights provide a better understanding of the basic pre-conditions that are necessary for meaningful conversations to occur, and a clearer picture of the dynamics that must be present to maximise the chances that there will be an emergent shared understanding of the problem situation among the various participants. Finally clarification of the matter under discussion here can promote a better idea of what can be done to maximise the chances of bringing about enduring change not only in the people involved in the debate but also in the organisational situations that the intervention is designed to 'improve'. These issues will be discussed in a follow-up paper.

REFERENCES

Checkland, P., 1981, "Systems Thinking, Systems Practice," John Wiley & Sons, Chichester.

Checkland, P., 1989, "Soft Systems Methodology", in *"Rational Analysis for a ProblematicWorld,"* J. Rosenhead, ed., John Wiley & Sons, Chichester.

Checkland, P. and Scholes, J., 1990, "Soft Systems Methodology in Action," John Wiley & Sons, Chichester.

Checkland, P., and Davies, L., 1986, "The use of the term weltanschauung in soft systems methodology." *Journal of Applied Systems Analysis* 13: 109 - 115.

Jackson, M. C., 1982, "The nature of soft systems thinking, the work of Churchman, Ackoff, and Checkland." *Journal of Applied Systems Analysis,* 9: 17 - 28.

Kay, R., 1997, "Applying autopoiesis to the facilitation of worldview change." Australian and New Zealand Systems Conference, Brisbane.

Maturana, H., 1988, "Reality: the search for objectivity or the quest for a compelling argument." *Irish Journal of Psychology,* 9: 25-82.

Maturana, H., and Varela, F., 1987, "The Tree of Knowledge - The Biological Roots of Human Understanding", Shambhala, Boston.

Mingers, J., 1984, "Subjectivism and soft systems methodology - a critique." *Journal of Applied Systems Analysis,* 11: 85 - 103.

Winograd, T., and F. Flores, 1987, "Understanding Computers and Cognition," Addison-Wesley, Reading MA.

RETAINING AND MAINTAINING SOFT SYSTEM MODELS

David W. Bustard,[1] Raymond Oakes,[2] and Desmond D. Vincent[2]

[1]School of Information & Software Engineering
University of Ulster
Coleraine, BT52 1SA, Northern Ireland
E-mail: dw.bustard@ulst.ac.uk
[2]CITUNI
Northern Ireland Civil Service
Craigantlet Buildings, Stoney Road
Belfast, BT4 3SX, Northern Ireland

INTRODUCTION

Soft Systems Methodology (SSM) (Checkland, 1981; Wilson, 1990; Checkland and Scholes, 1990) facilitates a broad approach to the investigation of any situation considered problematic. As such it is beneficial in many professional areas (Mingers, 1992; Stowell, 1995). For engineers, seeking technical solutions to business problems, it provides a sound way of understanding the context in which the problems exist so that they are better able to develop effective solutions.

Engineers are fundamentally concerned with handling systems that change over time. It is somewhat disappointing, therefore, to discover that topics like 'maintenance' and 'evolution' are scarcely mentioned in standard SSM texts. In particular, there is no suggestion that SSM models, once created, might be retained and modified as necessary to match changes in the systems they describe. Indeed, it is commonly assumed that the models should be discarded once recommendations for improvement have been developed because the models are part of the learning process rather than a product of that process. The models do, however, seem to have value to people other than the analyst and so might reasonably be included as part of the standard documentation for system change.

The purpose of this paper is to consider the potential advantages and resulting implications of retaining and maintaining SSM models. This work is being undertaken as part of a research project, RIPPLE (Bustard, 1998), concerned with managing the co-evolution of a business and its computing support, described through the matching evolution of associated models. SSM is used for the business modelling part of the work. Computing models, such as those developed through object-oriented analysis, are linked to the SSM models (Bustard, 1996), and maintained through the standard engineering technique of 'configuration management' (Tichy, 1995). This provides motivation for maintaining SSM models.

Synergy Matters: Working with Systems in the 21st Century,
Edited by Castell *et al.*, Kluwer Academic / Plenum Publishers, New York, 1999.

409

The first section of the paper summarises the basic SSM process and discusses its interpretation. This is followed by a 'debate' putting forward the case for not retaining models, point by point, and providing matching counter arguments. The concluding section draws together the two sides of the debate.

INTERPRETING THE SSM PROCESS

Figure 1 is Checkland's preferred way of describing the "basic shape of SSM", with numbering added to facilitate the discussion that follows. SSM helps identify and model *relevant systems of purposeful activity* for a *real-world situation of concern*. Comparing the models with the perceived real situation identifies differences that suggest improvements to the situation. Implementing those improvements changes the real-world situation, which can then be investigated again to identify further improvements. This process can be repeated indefinitely.

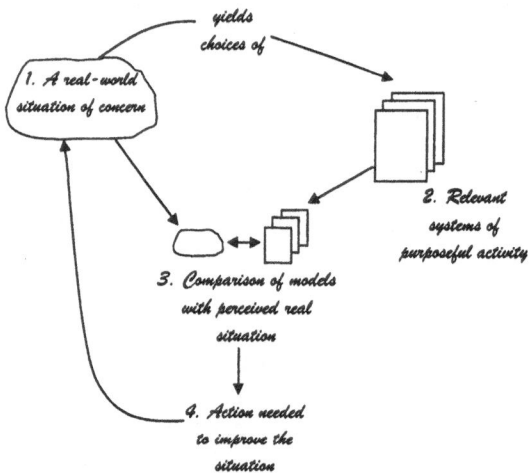

Figure 1. Checkland's "basic shape of SSM"

The process summarised in Figure 1, and the accompanying explanation provided by Checkland (1990) implies that the original cycle is repeated after each change is implemented. This in turn means that a fresh set of models are created on each cycle. The suitability of this process will depend to a large extent on the time taken to complete a cycle. Certainly if a system is revisited after five years it would be prudent to repeat a full analysis. The decision is not so obvious, however, if the period is, say, five months or five weeks, which can be the case if the implementation is incremental.

After the completion of one such phase, some form of validation is desirable to assess the impact of the change. This may lead to a refinement of the planned phases that follow. Such validation is not explicit in Checkland's process but can either be handled as part of step 4 or covered by the next instance of the cycle if it is performed immediately. Unfortunately, this is another aspect of the SSM process that is not examined in detail in the standard texts. It seems reasonable to suggest that validation could usefully be performed by comparing the models of relevant systems, developed in step 2, with the modified real-world situation. However, this implies that the models are still available at that stage, which is often not the case. The models are typically used to develop recommendations for change, but need not be included with those recommendations. Consequently change is often implemented against the recommendations and validation

410

either performed in terms of perceived benefit or through a repeated SSM analysis.

For incremental change, the SSM models are unlikely to become invalid during the implementation of one phase so it seems cost-effective to make use of them in the validation process. This argument led to the design of the evolutionary development cycle shown in Figure 2 (Bustard, 1998). A system description is created initially, followed by a cycle of incremental change, in which each increment is planned, implemented and reviewed. Exceptionally, following a review, a completely new evolutionary plan may be required if the system is to change substantially.

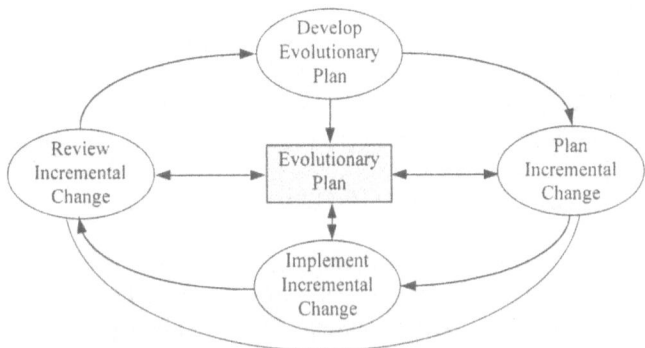

Figure 2. RIPPLE Evolutionary Development Process

The RIPPLE approach to handling change is based on Gilb's notion of Evolutionary Delivery (1988). The intention is to define incremental system changes that bring benefit in relatively small steps, reviewing objectives after each step. The Evolutionary Plan is based on a sequence of SSM models and relies on retaining these models as an essential element of the description of a system and how it is expected to evolve. Unfortunately this is not consistent with the commonly held views on how SSM models should be used so a detailed justification is required. The next section provides that justification in response to the likely criticisms of the approach.

JUSTIFYING THE RETENTION AND MAINTENANCE OF SSM MODELS

There are at least seven arguments against retaining and maintaining SSM models. These are presented and debated in the sub-sections that follow.

Inconsistent with Established Practice

At least some if not the majority of SSM analysts will have difficulty accepting the notion that models should be retained and maintained. This is due to the way they perceive and practice the methodology.

Practitioners in most fields are prepared to change their way of working if it provides a clear cost-benefit gain overall. SSM analysts who have become used to making recommendations for improvement without having to be involved in the system changes concerned will receive no benefit from providing models. They will also incur a significant cost in preparing the models for presentation and in ensuring that these models are understood. It is highly unlikely, therefore, that those analysts will chose to provide models unless they are a required component of the analysis report, even though they may well be of value to the client.

Retaining models is highly desirable, however, in cases where models have been prepared with the clients or where models are linked with other forms of documentation to provide an integrated system description.

Elevating the Status of SSM Models

Many do not see SSM models as a product in the same way that, for instance, SSADM analysts view their models. Rather SSM models are perceived as intermediate intellectual tools that help the analyst arrive at the recommendations for change.

Providing recommendations without models is much like providing an answer to a mathematical problem without showing the intermediate work. While the answer may be 'right', the availability of a step by step explanation is also beneficial. Providing models puts the recommendations into a context that enables them to be more easily understood by the client. In particular, the recommendations can be traced to desirable system behaviour, which will help avoid misunderstandings and may detect analysis errors.

Inappropriate Focus on Primary Task Models

The idea of retaining and maintaining models is only of relevance to *Primary Task* models (Checkland, 1990), which are closely related to business processes. This emphasis may deter the use of *Issue-Based* modelling, which can be equally or more important in a given situation.

SSM, being a methodology rather than a method, already allows for considerable variation in its use. Certainly the focus here is on building and maintaining Primary Task models but all other models and descriptions produced during analysis will contribute to the collective documentation for the system. All such material may be valuable in explaining how recommendations were deduced; not all of it need be maintained. Issue-Based modelling is just another technique in this context, much like rich picture diagramming (Lewis, 1992), and can be included or excluded according to the preferences of the analyst.

Enforced Standards for Model Presentation

The main SSM texts give little guidance on model construction and certainly prescriptive methods for drawing models are not widely advocated. Each SSM practitioner has therefore formulated their own method for drawing models based on a mixture of logic, creativity, and personal presentation preferences. Thus it is by no means certain that different practitioners will take the same meaning from the same model. This is exasperated by the vagaries of the English language.

Standard SSM texts show significant variations in the way that models are presented. For an experienced analyst this may be a strength but for novices or tool builders it seems to present an unnecessary obstacle to the effective use of SSM. The SSM models are relatively simple and conventions would not be difficult to develop. The most significant problem will be finding a form that is acceptable to the main SSM protagonists who have developed their own styles and established them through extensive publication. There is really no incentive for them to contribute to a common notation that would invalidate their own presentation style.

SSM Models are Incomplete

The act of drawing SSM models is as important as the models themselves in developing the thinking of the practitioner. A model on its own does not capture the understanding gained and the ideas provoked by the drawing process.

This is true of model building in general, for architects and in all branches of engineering. The SSM analyst seeks to capture the essence of a system in a model but if that does not reflect his or her full understanding, then additional descriptive material should be prepared as necessary.

Opportunities for Improvement are Missed

Maintaining models to keep them in line with system changes may mean that significant movements in the nature of the system and opportunities for radical improvement are missed. Models are a representation of a range of viewpoints, which will change over time. Thus the emphasis should be on understanding viewpoint changes rather than defining models of the changing system.

While it is true that developing fresh models in each improvement cycle will yield more opportunities for improvement, there will typically be many more ideas generated than can be pursued immediately. As there is no shortage of ideas, missing some possibilities is not a significant disadvantage.

Opportunities for Creativity are Missed

Maintained models will suffer from a lack of creativity, which is inherent in building new models.

The creative thinking approach underpinning SSM generates many possibilities for change that need to be prioritised and implemented in a practical order, as funds permit. As mentioned against the previous point, some loss of ideas is of no consequence overall. Starting afresh on each improvement cycle is expensive and is also potentially disruptive. Even De Bono (1971) warns against the "dangers of creativity", when applied too frequently.

CONCLUSIONS

This paper has made a case for retaining and maintaining SSM models. This should not be interpreted as a mandatory requirement for every use of SSM but there certainly seem to be advantages to the client in all circumstances. Where the models are explicit, perhaps developed collaboratively with the client, the case for retention is particularly strong.

To be able to retain models effectively some recommendations should be made for an SSM presentation model standard. This would be used in published examples of SSM models and would encourage tool support to help in the management of model development. It would also be of educational benefit, as novices prefer clear guidelines when learning a new notation or technique.

Retaining and maintaining SSM models reduces scope for creativity and the identification of beneficial change but the approach is already so rich that some loss is tolerable.

Overall, the impression is that if SSM models are worth building they are worth retaining. There are resulting constraints to be accepted but these have little significant effect and may well help promote SSM in other problem solving areas, such as engineering.

ACKNOWLEDGEMENTS

The work described in this paper has been undertaken as part of the RIPPLE project (Retaining Integrity in Process Products over their Long-term Evolution) funded by EPSRC, GR/L60906, under the SEBPC (Systems Engineering for Business Process Change) Programme. The project is in collaboration with the Northern Ireland Civil Service and British Telecom, who are helping to evaluate the proposed approach.

REFERENCES

Bustard, D.W., Dobbin, T.J., and Carey, B., 1996, Integrating soft systems and object oriented analysis, *in*: "Proc. IEEE International Conference on Requirements Engineering", Colorado Springs, USA, 52-59, IEEE Computer Society Press, Los Alimitos.

Bustard, D. W., and He, Z., 1998, A framework for the revolutionary planning and evolutionary implementation of a business process and its computing support, *in*: "Proc 3rd International Conference on ISO 9000 and TQM", 111-116, 3ICIT, Hong Kong.

Checkland, P., 1981, "Systems Thinking, Systems Practice", John Wiley & Sons, Chichester.

Checkland, P., and Scholes, J., 1990, "Soft Systems Methodology in Action", John Wiley & Sons, Chichester.

De Bono, E., 1971, "Lateral Thinking for Management", Penguin Books, London.

Gilb, T., 1988, "Principles of Software Engineering Management", Addison-Wesley, Wokingham.

Lewis, P.J., 1992, Rich picture building in the Soft Systems Methodology, *Eur. J. Inf. Systs.* 1(5), 351-360.

Mingers, J. and Taylor, S., 1992, The use of Soft Systems Methodology in practice, *Journal of the Operational Research Society*, 43(4), 321-332.

Stowell, F.A. (ed.), 1995, "Information Systems Provision: The Contributions of SSM", McGraw-Hill, London.

Tichy, W.F. (ed.), 1994, "Trends in software: configuration management", John Wiley & Sons, New York.

Wilson, B., 1990, "Systems: Concepts, Methodologies, and Applications", 2nd Edition, John Wiley & Sons, Chichester.

IMPLEMENTATION CONCERNS IN SOFT-SYSTEMS INTERVENTIONS: SOME LESSONS FROM INTERACTIVE MANAGEMENT

A. R. Cardenas[1], F. R. Janes[2] and G. Otalora[1]

[1] ITESM, Campus Monterrey, Suc. Correos "J", Monterrey, N. L. 64 849, MEXICO

[2] City University Business School, Northampton Square, London EC1V 0HB, UK

INTRODUCTION

Soft systems interventions are often characterized by an important emphasis on finding out ways to promote effective communication among groups of stakeholders and on helping them to understand and integrate their diverse interests and beliefs regarding complex situations. From the point of view of the outcomes of these interventions it is interesting to note that most available literature concentrates on reporting the processes through which the communicative efforts and the accommodation of interests take place. However, there are not many reports regarding the actual implementation of the agreed decisions or designed solutions arrived at with these interventions.

Considering that the Interactive Management (IM) process can be conceptualized as a soft systems approach, this paper discusses an application of IM in Mexico in which particular attention is given to the implementation of the decisions arrived at by a group of stakeholders who were involved in the application. The study of this case points out some important issues that need to be considered if soft systems approaches are meant to provide good basis for implementation.

SOFT SYSTEMS APPROACHES AS SYSTEMIC INTERVENTIONS

The contributions of systems methodologies in addressing problematic situations have been analyzed by Jackson (1992), among other things, in terms of how they serve the technical, the practical and the emancipatory interests of human knowledge identified by Habermas (1971). Within this context Jackson (1992) relates soft systems thinking to the practical interest:

> "The main value of soft systems thinking, in terms of Habermas's schema, lies in the support it offers to the practical interest in promoting intersubjective understanding. All the methodologies considered offer effective means of securing and expanding the possibility of mutual understanding among individuals in social systems - whether through dialectical debate, focusing attention on an idealized design, or engaging in a cyclic learning process... " (Jackson, 1992:168).

In accordance with this emphasis on the practical interest, soft systems approaches promote the engagement of stakeholders in different types of communication processes aimed at developing mutual understanding and at reaching agreements on how to tackle the situations under study. Under this perspective problematic situations are approached through intellectual, inquiring, processes based on human interaction. This focus on mutual understanding and learning seems to imply that the changes in perceptions, attitudes and/or behaviours that may derive from appropriate interactive processes are the major changes needed to improve a problematic situation.

However, considering that an important outcome of these processes is that of the agreements reached on how to tackle the problematic situations under study, the question remains open as to whether or not soft systems approaches provide a good basis for the implementation of the decisions made by the stakeholders. A possible way to explore this question is to look at soft systems methods from the point of view of systemic interventions in general.

Systemic interventions are conceptualized here as temporary organized and planned efforts which promote changes that significantly improve the situations under study according to the perspectives of the people involved. Systemic interventions are seen as temporary efforts, i.e., they are marked by a beginning and an end, because otherwise they would turn into processes of continuous improvement where a possible closure could not be reached. Also, considering that systemic interventions can be viewed as a special type of problem-solving effort, the most general procedural components of systemic interventions are: inquiry, action and evaluation (Ulrich, 1977):

a) The inquiry component is related to the development of satisfactory understandings of the situations of concern and the design of appropriate plans or decisions to deal with those situations. If an inquiring effort is carried out on the basis of thorough communication processes that promote learning and mutual understanding, as in the case of soft systems approaches, the kind of changes that might be expected from this endeavour refer to changes in perspectives, attitudes and/or behaviours of the people involved in the inquiring process.

b) The action component refers to the implementation of the designed plans that derive from an inquiring effort, and it implies the involvement of people and resources in making changes of a varied nature such as technical, structural, procedural, and so on:

"*Action* .... includes activities other than the purely cognitive, and it cannot be expected to derive automatically from decisions, or choices, or problem-solving activities." (Brunsson, 1985:7)

According to Brunsson (1985), a new organizational action requires not only planning and co-ordination, but also calls for a set of three socio-psychological conditions: a) positive expectations regarding the accomplishment of the required actions, i.e., the people involved should perceive that the plans are going to be put into effect; b) motivation, through which a strong positive connotation is attributed to the planned action(s); and c) commitment, which means "that a person is regarded by other people as being clearly tied to a specific action. The ties must be perfectly evident to others, explicit and 'strong', if they are to persist through the effort and energy required to overcome the foreseen and unforeseen difficulties that are typical of change actions." (Brunsson, 1985:52). Within this perspective there are at least two important factors that may inhibit the accomplishment of a new action effort: uncertainty which diminishes motivation and positive expectations, and conflict which operates against commitment.

c)   The evaluation component of systemic interventions involves a holistic assessment of the processes, products, and consequences of an intervention. Evaluation can be conceptualized as a continuous endeavour during the intervention effort, but it is also a final task which brings to a closure the whole process in order to learn from both the processes involved and the situations being addressed.

Within this perspective on systemic interventions, it is clear that soft systems approaches are mostly concentrated on the inquiring component of an intervention. In soft systems literature the emphasis is generally put on describing the inquiring processes that are proposed; the application cases that have been published hardly mention the ultimate consequences of these methods in implementation terms (see for example Checkland and Scholes, 1990). This lack of explicit attention to the action component of an intervention suggests that it is supposed that implementation will naturally derive from the characteristics and outcomes of the inquiring processes that are proposed.

Among the characteristics typically found in soft systems methods that could be related to implementation purposes are the involvement of stakeholders, the definition of action plans (or their equivalent in terms of basic agreements or decisions on what changes should be implemented), and the possibilities of generating positive expectations, motivation and commitment through the social interaction processes that are promoted. Thus, the question raised is to what extent these characteristics actually contribute to put into effect the action component (implementation) of a soft systems intervention.

In order to explore this issue, the next section will briefly describe an actual application case of Interactive Management (Warfield and Cárdenas, 1994) emphasizing implementation concerns. IM is considered as a soft systems approach since it portrays the basic characteristics attributed to soft systems thinking by Jackson (1992): it is a process aimed at dealing with complex situations by engaging groups of stakeholders in a learning process, and it works with the multiple perceptions of reality of the stakeholders instead of focusing on the direct study of the real world.

AN APPLICATION OF INTERACTIVE MANAGEMENT IN A NON-PROFIT MEXICAN ORGANIZATION

The application case of IM is presented in this section in the following terms: a) the organization in which it took place, b) a brief description of the IM project and its outcomes, and c) a discussion of the implementation of the outcomes.

a)   The Organization.

CM is a non-profit charity organization located in the city of Monterrey, Mexico. Within the range of services offered by CM are: medical services, nurseries, shelters for homeless people, and food distribution. These and a wide variety of other services are managed through a central office and a number of specific-service branches of CM spread across the metropolitan area of Monterrey, as well as through more than 70 representative offices of CM located in the parish churches of Monterrey. All the representative offices are managed by volunteers; part-time and full-time employees work basically in the central office and in the specific-service branches. The main incomes of CM are obtained by means of charitable donations offered by some major sponsors and by the community at large. The operations of CM are co-ordinated by a general manager who is in charge of the central office. There is a Board of Trustees who are in charge of policy and oversight functions; the general manager reports directly to this board.

b)   The Project and its Outcomes

By the end of 1992 a project was undertaken in CM with the purpose of finding out ways to improve the quality of the services offered at the central office. The project was carried out with the participation of two external IM consultants.

From the point of view of the IM consultants the activities involved in carrying out the project were: a set of interviews and meetings with the people in charge of the sections that formed the central office (section co-ordinators), regular meetings with the general manager, an IM workshop which lasted three days, and several discussions with the general manager througout 1993 relating to the implementation of the actions decided as a result of the project. According to the plan, the central activity of the project was the IM workshop; the design of the workshop was elaborated by the general manager and the consultants, and reviewed by the section co-ordinators. The three-day IM workshop involved the direct participation of the section co-ordinators (10 people); the general manager and the president of the board of trustees were invited as observers.

During the first part of the workshop, the participants exchanged their points of view concerning the main obstacles that the organization was facing in offering a high quality service and jointly developed a structural model representing the interactions among those obstacles. As a result, they found that the main issues involved in improving the quality of the services were related to the motivation of the personnel and to some specific organizational problems. The second part of the workshop was devoted to defining and establishing relationships among a set of actions aimed at eliminating the obstacles, and to developing a plan to implement those actions.

Besides the specific outcomes just mentioned (the issues identified and the actions defined), the immediate results of the project indicated that the thorough processes of reflection and dialogue that were promoted through the use of IM helped the participants to understand better their mutual roles and challenges within the organization, and provided the basis for improving the communication among themselves and with the general manager. These results were attested to by the individual responses to the whole project. At the end of 1993 the IM consultants received a letter from the general manager indicating that most of the agreed actions had been implemented.

c)   Implementation of the Outcomes.

In order to assess the final results of this project in terms of the implementation of the actions that were agreed, during the summer of 1997 a set of individual interviews was held with five of the ten participants in the IM workshop and with the general manager; also some additional inquiries were made in the institution.

Four of the six people interviewed considered that most of the actions defined were not implemented and that the project did not have a successful follow-up; they said that the project was a great opportunity to share their concerns and to improve communication but it raised expectations that were not satisfied. In sharp contrast with these perspectives, two people, the general manager and the co-ordinator of management services, considered that most of the actions were implemented and that the project was very successful.

In analyzing these differences it was found that most of the actions were actually implemented, or at least they were analyzed in detail and partially carried out. However, the results of these efforts were not communicated to the people directly involved in the project. Furthermore, even though many of the actions defined implied the involvement of the Board of Trustees and the rest of the employees of the organization, these groups were not clearly engaged in the implementation process.

The study of the results of this project in implementation terms suggested that the factors that hampered the success of the implementation can be classified in two groups: i) weaknesses in the design of the project and in the way it was carried out, and ii) lack of an explicit consideration of the action system required to implement the agreed changes.

Regarding the weaknesses in the design and in carrying out the project, it was concluded that the most important single factor that impinged upon the success of the implementation was the fact that some key actors of the implementation process were not directly involved in the project. However, even though some of these people could have been involved, there was no possibility of directly involving all the people concerned with the implementation, mostly because of time and organizational constraints. Thus, even if the project could have been improved by directly engaging more people, there would still be relevant people not participating in the main inquiring process.

The lack of an explicit consideration of the action system required to implement the agreed changes refers mostly to the need to co-ordinate and supporte the social processes that are put in place when an organizational change is promoted. In our case study, the lack of conceptualization of this action system manifested itself in the piecemeal approach to implementation that was adopted by the general manager who was left by himself with the follow-up task. From the point of view of the three socio-psychological conditions for action identified by Brunsson (1985), it was considered that positive expectations, motivation and commitment were highly developed during the inquiring phase, but the lack of communication during the implementation resulted in deterioration of the levels of motivation and commitment of the participants and did not promote these attitudes in the non-participants. Two major conclusions were derived from the study of this case, one concerning the conceptualization of the action component of a systemic intervention, and the other one referromg to the relationship between inquiry and action.

Regarding the conceptualization of the action component, it is not just a question of establishing detailed implementation plans, it is a matter of enabling the development of a new action system and of appropriately responding to its needs as time unfolds. Also, a new type of learning is required, learning from action, as opposed to the type of learning that is promoted during the inquiring process. The only reference to this issue that was found in the soft systems literature is in the work of Ackoff: "...if an organization's ability to improve its performance continuously is to be developed, the implementation of plans should be undertaken experimentally. Experimentation is controlled experience; it enables us to learn much more rapidly and effectively than we can from ordinary experience and trial and error." (Ackoff, 1981:237).

From the point of view of the relationship between inquiry and action, the perspectives of the participants illustrated that a project of this type should be conceptualized as an entire systemic intervention. Inquiry and action cannot be approached independently from each other if an effective process to bring about relevant changes is to be effected.

CONCLUSIONS

At the beginning of this paper the question was raised regarding whether or not the involvement of stakeholders, the definition of action plans, and the possibilities of generating positive expectations, motivation and commitment through the use of soft systems approaches provided a good basis for implementation. The application case of IM presented here has been helpful in illustrating that these elements play an important role in an implementation effort. However, the results of this study also indicate that in order to provide adequate methodological support for implementation, the design of an inquiring process is not enough. If the idea of looking at soft systems approaches as holistic systemic interventions is taken seriously, then much more research is needed on the action component of these types of approaches.

REFERENCES

Ackoff, R., 1981, "Creating the Corporate Future: Plan or be planned for", Wiley, New York.
Brunsson, N., 1985, "The Irrational Organization: Irrationality as a Basis for Organizational Action and Change", Wiley, Great Britain (reprinted 1996).
Checkland, P. B., 1981, "Systems Thinking, Systems Practice", Wiley, Chichester.
Checkland, P. B. and Scholes, J., 1990, "Soft Systems Methodology in Action", Wiley, Chichester.
Churchman, C.W., 1971, "The Design of Inquiring Systems", Basic Books, Inc., USA.
Crozier, M. and Friedberg, E., 1977, "L'Acteur et le Système", Editions du Seuil, Paris.
Habermas, J., 1971, "Knowledge and Human Interest", Beacon Press, USA.
Jackson, M.C., 1992, "Systems Methodologies for the Management Sciences", Plenum Press, New York.
Ulrich, W., 1977, The design of problem-solving systems, *Management Science*, 23:1099-1108
Warfield, J. N. and Cárdenas, A. R., 1994, "Handbook of Interactive Management", Iowa State University Press, 2nd. Edition, Ames, Iowa.

OPERATIONAL SYNERGY AND META-METHODOLOGY IN THE PRISON SERVICE

Mr. Jack A. Castle and Mr. Steve Crago

Senior Lecturer, School of Operations Management,
The University of the West of England (Castle).
Principal Pharmacist, HM Prison Bristol (Crago).

INTRODUCTION

CVAM is a new systems based methodology which has evolved through an action research programme from an interpretive method to become a multi-paradigm approach in its own right. The triple flux model has also been employed as a meta-methodology to order approaches using a range of imported methodology. This paper describes a meta-methodology application of CVAM in hard flux form to improve the operation of the pharmacy service of HM Prison Bristol by Crago. It offers new insights into the practical achievement of operational focus and synergy. Perhaps most importantly, the approach has been widely acclaimed in the prison service for its utility, results and the clarity with which difficult operational concepts may now be communicated.

CVAM AS A META-METHODOLOGICAL FRAMEWORK

CVAM was initially developed as an interpretive methodology to enable complex strategic and operations problems to be addressed (Castle 1995, Castle and Spurrell 1997a). It soon became apparent that a multiple paradigm process was essential to address the range of issues surfaced by the action research programme (Castle 1998a). This reframing of CVAM addressed the need to enable users to relate the approach to their lived experience. Castle (1998b) has questioned the ontological complicity of other methodology. CVAM is now ontologically rooted in a life flux model based on three entwining objective, subjective and strategic /radical fluxes. Castle has noted its similarity to aspects of Bourdieu's constructivist approach to social theory (Bourdieu and Wacquant 1992). Each of these fluxes is itself constituted as a rope spiral composed of CVAM'ic capital. These constituents include uncontrolled elements described as circumstances (C's). Within the 'system' viewpoints, values (V's): activities (A's): and the means (M's) of their prosecution complete the capital . Thus each rope is described as CVAM'ic (see-vamic). The model uses systems theory, envisaging the system boundary as drawn around 'within the system', V's and A's and M's, which can be directly manipulated by the system actors. The full procedure requires use of a framework incorporating sub-methodology which engages all three paradigm ropes in a five stage process designed to avoid pitfalls or deceptions in improvement activity. The methodology is defined as:

"... an attenuated and attenuating guide, to enable users to avoid deceptions in the achievement of improvement action through the application of practical reason."
(Castle 1998b:4).

Synergy Matters: Working with Systems in the 21st Century,
Edited by Castell *et al.*, Kluwer Academic / Plenum Publishers, New York, 1999.

These deceptions include those associated with variety imbalance and complexity: deceptions of reductionist thinking: time based deception: communicative deceptions: deceptions of boundary judgement: isolationist deception: and methodological deception. Since any methodology is, 'an attenuated and potentially misleading view of the constellations of theory available' it may deceive users. To overcome this the process framework has been designed to guide analysts who may wish to import alternative methodology to that specifically designed for CVAM. The process framework involves five formal stages and three paradigms of inquiry in a mapping approach inspired by Mingers (Mingers and Gill 1997). The stages differ somewhat from those proposed and address systemic discovery, attenuation, validation, analysis, and improvement. For example an isolationist pass along the objective pathway might proceed as follows in a chemical research programme associated with a complex chemical process. The framework would require systemic hard data capture to deal with complexity. Here instead of premature attenuation of variables the use of partial factorial designs would enable wide screening, allowing the most likely to be selected (attenuation). Experimental statistical validation of the main effects would then lead to perhaps full factorial analysis and path of steepest ascent optimisation at the latter stages (Davies 1956). Consideration of the experiments using the hard CVAM framework would alert the researcher to consider the environment in terms of hard C's that might impact on the experiment. The formalised procedures and safety instructions needed as hard V's. The chemical procedures and processes involved as A's, and finally the materials, utilities, equipment, and skills needed (M's) to vary those conditions amenable to the control of the experimenter. Here the framework is used to guide the experimenter to import scientific experimental design methodology in an ordered way. Such a single flux pass would not avoid isolationist, boundary judgement, communicative or time based deceptions since it uses only functionalist approaches.

UNDERSTANDING OPERATIONAL THEORY USING FLUX IDEAS

These ideas have been used to reframe functionalist operations theory using the CVAM'ic hard flux model. The work of Skinner (1971), Slack in Wild (1989), Hill (1985), and Bozarth (1992) were seminal in this endeavour. The models were then used to structure and teach complex operations theory to the MBA programme at Bristol Business School with very encouraging results (Castle 1997b). Operations theory, particularly that concerning issues of focus, flexibility, and trade offs is conventionally described using hard systems models (Slack et al. 1995). The ideas of focus is particularly interesting in that it requires synergy between operating systems elements in order to achieve distinctive performance. Existing methods often use simple profiles with little guidance for the practising manager. The fact that CVAM models operational life as a time based flux was vital for clearer modelling of ideas like flexibility and focus. The intrinsic claim here is that any purposeful system can be modelled in terms of the flux capital. In this context external circumstances in terms of measurable demands on the operational system are clearly hard C's. Hard V's are the organisationally legitimised and published operational strategies for products, processes, capacity and flexibility. Processes are modelled as chains of activities. The means of pursuing these activities (M's) clearly include human and non-human resources, information, structures, utilities and facilities. The fundamental mapping is shown in its most basic form in figure 1, which is shown on the next page. Mappings have now been developed covering operational capability, volume, capacity, logistics, and flexibility etc. in more detail.

A REAL WORLD APPLICATION IN THE BRISTOL PRISON SERVICE

Having completed an MBA assignment using an early version of CVAM to analyse the pharmacy service at HM Prison Bristol, Crago independently and unaided used these mappings as a framework for the re-design of the service to excellent effect. Crago was concerned to try to implement a new pharmacy operating system whilst working within prison operational constraints. To ensure that the pharmacy operation provided an integrated and quality pharmaceutical service to its customers. Of particular interest is that Crago's use of these ideas was accomplished completely independently without any formal consultancy or other support. If methodology is to be used operationally it must not be dependent on experts who can never match the variety of user demand.

422

The south western service was facing severe market forces. It and other services had been clustered around Bristol, which would now serve a number of satellite customers at Leyhill, Gloucester, Erlestoke, Eastwood Park and Shepton Mallet. Crago was concerned that the size of this manufacturing and delivery challenge had not been fully appreciated. This involved the construction of a practical operating system that was capable of meeting the variety of a process involving the delivery of thousands of drugs in different forms and strengths.

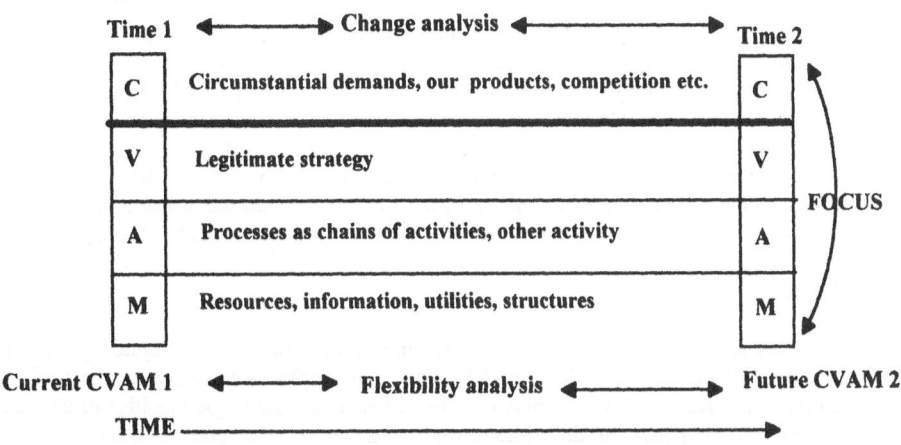

Figure 1: Using CVAM'ic hard flux ideas to map operational theory.

The CVAM process was used as a meta framework to inform the import of a variety of hard approaches. These included environmental variables in a hard C analysis, flow charting of activities and process evolution models as hard 'A' methods, and Job design approaches in hard 'M' analysis. The initial flux analysis identified new competitive circumstances, an internal established professional and ethical value system, but unbalanced processes, capacity, and human skills/job design. The initial CVAM cross correlations showed an out of focus position, with the pharmacy service trying to serve too wide a range of customers with an inadequate service bundle which lacked clinical input from the pharmacist, neglecting implicit intangibles (Sasser, Olsen and Wyckoff 1978). The process flows within the pharmacy were found to be inconsistent. Comparison of customer demands with the capacity of the process, particularly the loading on the staff showed the capacity of the operating system was inadequate. There was a huge variety of demand from the satellites and when that was cross correlated with an essentially jobbing process the system was seen to be both ineffective and inefficient. Detailed capacity calculations enabled Crago to demonstrate that the capacity of the present system was a critical variable. This led to recommendations that the jumbled process be replaced by parallel process flows, one for Bristol, the other for the satellites, with a product based orientation. It was found that this allowed greater flexibility as demands changed. It facilitated improved job designs, empowerment, and motivation of staff. This also allowed a dedicated service to the satellites with improved contact and a better service bundle. Crago saw this as a move towards a more batch orientated higher volume process flow. To sustain service levels inventory holdings were increased, but the introduction of a drugs formulary system helped reduce the overall variety of demand. Delivery time was identified as a key service attribute and the logistics were redesigned to, 'the needs of the customer and not the other way round'. Hence the negative correlations between external C's and internal V's, A's and M's were identified as clearly dysfunctional (as CVAM 1), and the system was redesigned to provide synergistic relationships in terms of the new structure (CVAM 2).

Having formed his solution Crago used the CVAM model in a presentation to the Prison Governor. The flux based models provided an effective way of communicating the proposals. It enabled a discussion of operational synergy to be achieved in a holistic way.

"... CVAM exposed the gaps and highlighted the strategic issues that required addressing. The Governor was particularly impressed with the way CVAM combined internal and external analysis and interlinked them across an imaginary flux... for too long we had been insular.. this provided a quality improvement of the existing system by taking the circumstances outside the system into consideration, rather than approach the problem piecemeal as was usually the case..." (Crago 1998)

Crago was now required to present these ideas to the Governors and Medical Officers of the satellites at a prison service conference in the South West. A series of overheads were used showing the strengths and weaknesses of CVAM 1 the present position. The negative and positive correlations contrasted clearly with the synergy and focus of CVAM 2, Crago's proposed new flux position. He noted that the immediate reaction from the floor was that CVAM 1, 'is where we are' and CVAM 2 is,' where we want to be'.

"This was not seen as a managerial miracle, what became obvious to all was that the holistic approach taken by the methodology, exposed a range of issues inside and more importantly outside of the existing system that needed to be addressed immediately..." (Crago 1998).

He noted that:

" ..it became evident to the audience that the model not only highlighted gaps but the existing internal strengths that could be built upon. After grappling with the concept of correlating across the CVAM spectrum, the audience seemed to gain an idea of how all the events that happened in the system were interrelated. The present system was value dominated and clearly unbalanced... CVAM 2 addressed that balance across the flux idea that I had built up in the audiences mind. It had the property of being understood... in the minds of the audience" (Crago 1998).

The model enabled him to demonstrate that an alternative solution spreading resources over five satellite pharmacies would not be capable of operating a legal service and would unfocus operations.

CONCLUSIONS

Crago used hard flux models under the guidance of CVAM as a meta-method to redesign a cluster pharmacy in a way that would work. CVAM exposed a host of issues that needed addressing both inside and beyond the system. The formulary was introduced, the process flow through the pharmacy was changed to a consistent set of tasks, job descriptions then evolved and capacity was tailored to the manufacturing and service task required. Crago concluded that he was able to link pharmacy operations with the overall prison corporate strategy of providing legal and safe healthcare. Operational experience over the last six months has demonstrated a significant reduction in health care costs, and more importantly, a much improved quality of service.

This study appears to offer some evidence that the ontological form of CVAM, modelling life as a flux, does enable untrained users to relate it to their lived experience. Life is not a metaphor or a linear equation but is an objective/subjective lived experience across time. The reaction of MBA students to this re-mapping of operations management theory has encouraged further effort in this direction. There is clearly a need for such technology. For example Wickam Skinner has powerfully argued for new methodology that can meet the variety of real life operational challenges:

" ..We say that manufacturing policy decisions must be made such that the system designed meets the manufacturing task. But the manager is left to figure out how to get from task to structure without much specific guidance...what is missing is the probable impact on system performance of each menu listing. That in brief is the gap." (Voss 1995:22)

In full form it is hoped that CVAM can contribute to the resolution of his challenge by providing managers and others with a useful methodology for operational synthesis.

REFERENCES

Bourdieu, P., and Wacquant, J.D., 1992, " An Invitation to Reflexive Sociology", Polity Press, Cambridge.

Bozarth , C.C ., 1993, A conceptual model of manufacturing focus, *International Journal of Operations and Production Management*, Vol.2, No 1: 81-92

Castle, J.A., 1995, The Development of an integrating methodology for strategic and operations management., *Proceedings of the British Academy of Management conference*, Sheffield University.

Castle, J.A., and Spurrell S., 1997a, CVAM a new systems methodology, and systemic sustainability, an Engineering Company, *in:* "Systems for Sustainability," F.A. Stowell, R.L. Ison, R. Armson., J. Holloway , S. Jackson, and S. McRobb, ed., Plenum, New York.

Castle, J.A.,1997b, "A Guide to the use of CVAM hard in operations strategy", *MBA course notes UWE*, Unpublished.

Castle, J.A., 1998a, New methodologies for integrated quality management. *The TQM Magazine.*Vol.10, 2:83-88. MCB University Press.

Castle, J.A., 1998b, CVAM as a meta-methodology, theoretical and practical progress, *A paper presented at the Operational Research Annual conference*, OR 40. Lancaster University 8-10th September 1998.

Crago, S., 1998, Pharmacy clustering, *Internal report*, HM Prison Bristol.

Davies, O.L., 1956, " The Design and Analysis of Industrial Experiments", Oliver and Boyd, Edinburgh.

Hill, T.J., 1985, "Manufacturing Strategy - The Strategic Management of the Manufacturing Function", Macmillan, Basingstoke.

Mingers, J., and Gill, A., 1997, "Multi methodology," Wiley, Chichester.

Sasser, W., Olsen, R., and Wyckoff, D., 1978, " The Management of Service Operations", Allyn and Bacon, Boston .

Skinner, W., 1971, The Anachronistic factory, *Harvard Business review*, January-February, Harvard University, Boston.

Slack, N., Chambers, S., Harland, C., Harrison, A., and Johnston, R.,1995, " Operations Management," Pitman, London.

Voss, C., ed. 1995, *"Manufacturing Strategy Process and Content"*, Chapman Hall, London.

Slack, N.D.C.,1989,Focus on flexibility, *in:* "International Handbook of Production and Operations Management", p 50, R.Wild, ed., Cassell Educational Ltd., London.

PURPOSEFULNESS AND COMPLEXITY-GRAPPLING CAPABILITY: SOME PRELIMINARY PROPOSITIONS

Indranil Chakrabarti,[1] Sheila R. Chakrabarti [2]

[1]Associate Professor
[2]Honorary Research Fellow
 Xavier Institute of Management
 Bhubaneswar - 751 013, Orissa, India
 email: indranil@ximb.stpbh.soft.net

The diversity and relative complexity of the behavioural patterns that have emerged even in these primitive [one-celled] organisms suggests that such developments must have taken place fairly early in evolutionary history. This is not to say that such organisms possess anything resembling the properties of thought or conscious experience we recognise in ourselves as humans. But nonetheless, the links between paramoecium avoiding obstacles and the activities of writing, or reading, this chapter, are evolutionary and capable of being mapped step by step.
Steven Rose (1976) The Conscious Brain

INTRODUCTION

'Purposefulness' is an important concept in systems thinking (ST) as in all of management science. The attribute enables the setting of objectives (Ackoff, 1994); it has also been shown to be contributing to the high complexity of management problems or 'messes' (Flood & Jackson, 1991).

While individual human beings as also social systems are seen as purposeful, purely technical systems, such as a robot or a fully automated plant, are not. That is, though the latter fulfil significant purposes, they cannot of their own accord set objectives or goals.

Implicit in all the discussion on purposefulness is the notion that it is an attribute mainly of humans (at most of the higher animals), more specifically of their brain-mind; the purposefulness of social systems is also derived from the purposefulness of humans. Huxley (1963) notes that: "In man alone do we find experiences and activities purposefully pursued". Echoing the prevailing view, Ackoff (1994) distinguishes purposeful entities from others, thus: "Purposeless things - for example, inanimate objects, plants, and lower-level animals". Could this attribute be seen as common to all living beings - indeed a defining characteristic of them? If that were so, it may in many ways enrich our understanding of purposefulness and related phenomena.

Monod (1972), among the foremost biologists of this century, who has also worked on the wider implications of the findings of modern biology, conceptualises a special attribute unique to living beings, which he calls 'teleonomy'. He defines it (1972), rather categorically, as: "one of the fundamental characteristics common to all living beings without exception: that of being *objects endowed with a purpose or project* ... which we shall call *teleonomy*" (italics in the original). How well does teleonomy correspond to purposefulness, as conceptualised in ST?

We explore this and argue mainly the following: One, that teleonomy actually is a bundle of attributes which if analysed and unbundled may be seen to be of two broad kinds: we call these autonomic teleonomy and conscious teleonomy. Two, conscious teleonomy when further unbundled may be seen to include purposefulness, as the attribute has been understood in ST, and another attribute which we call complexity-grappling capability. Three, the attributes purposefulness and complexity-grappling capability, though distinct, are closely and directly related.

Three propositions are advanced and some of the implications of these tentative observations for ST and management science are briefly noted. The paper ends with a statement on the further research programme.

PURPOSEFULNESS IN AN AMOEBA?

Nature is now commonly understood as ever dynamic. Rivers which once had a particular course, later shift to quite another. Mountains which today stand tall on top of the world were submerged below the sea at an earlier period. A quiet hill erupts into a volcano. Yet not one of these processes is supposed to involve any purposefulness. For instance, we do not deem that the river had purposefully steered itself to a new course.

Even the highly regular, periodic processes in inanimate nature, such as the spinning of the earth on its own axis or its revolving around the Sun, are not deemed by us to have involved any purposefulness. Nor for instance, is the spinning of Venus in a direction opposite to that of its motion around the Sun, alone of all the planets, seen as owing to purposefulness.

Contrast these processes with the case of a telecom company veering itself to a new course in a hypercompetitive milieu. Or with the saleswoman guiding her cycle-van through a crowded, free-for-all traffic junction in a small Indian town. We would say both these examples involve purposefulness.

Yet what about the other living beings - the very many varied animals and plants that populate the earth? Consider the following examples from the other extreme, the most primitive of organisms, possessing no nervous system, leave alone brain-mind.

Euglena, a one-celled plant prototype, when placed in an unevenly lit environment, tends to move to where the light, needed for its photosynthesis, is brightest (Rose, 1976).

Paramoecium, a one-celled animal prototype, moves to the region where the supply of its food material (bacteria) is the richest. It also tends to reverse away from varied hurdles in its path, such as fixed obstacles, extremities of temperature, or even irritating chemicals (Rose, 1976).

Dictyostelium Discoideum, a wandering, bacteria-eating single-celled amoeba, has a dual existence. When food is plentiful, these move about singly, ignoring one other. However, when food supply runs short, large number (around 100,000) of these organize themselves through a set of extremely intricate processes lasting several hours into a single multicellular organism called Plasmodium. "This organism has rudimentary locomotion organs and starts moving in search of food (performing this faster than an individual amoeba would). Having located a food store, the plasmodium 'breaks down' into free amoebas which resume their individual existence" (Ivanitsky, et al, 1987).

Is purposefulness an attribute common to even these primitive beings?

428

MONOD'S TELEONOMY

The Nobel laureate biologist Jacques Monod - whose main research area, molecular biology, interfaces the living and the non-living - has enquired into the characteristics by which the two may be distinguished. Monod (1972) identifies three such attributes which he calls teleonomy, invariance, and autonomous morphogenesis. All three are of interest to ST; for instance, there is a likely close correspondence between Monod's concept of autonomous morphogenesis and Maturana's and Varela's concept of autopoiesis (Mingers, 1995). But in this paper we would confine only to the idea of teleonomy.

"Rather than reject this idea (as certain biologists have tried to do) it must be recognized as essential to the very definition of living beings. We shall maintain that the latter are distinct from all other structures or systems present in the universe by this characteristic property", emphasises Monod (1972). He, however, alerts that "the notion most immediately and plainly inspired by the examination of the structures and performances of living beings, that of teleonomy, ... when analysed nevertheless appears to be a profoundly ambiguous concept since it implies the subjective idea of 'project'" (Monod, 1972).

Shorn of the technicalities of molecular biology, Monod's views on teleonomy, for our present purpose, may be summarized, simply, as follows: 1)The attribute implies *goal-seeking capability*. 2)It is "objectively discernible in all living beings" be they primitive bacteria or cyber-age humans. 3)It helps distinguish living beings from all other natural entities as also artifacts. 4)While living beings are capable of and show a great variety of such teleonomic activities, more so in the relatively evolved beings, all of these "may be seen as so many aspects or fragments of a unique primary project, which is the preservation and multiplication of the species" (Monod, 1972).

Monod (1972) stresses that "not only the activities directly linked with reproduction itself, but all those that contribute - *however indirectly* - to the species' survival and multiplication" (emphasis added) are part of teleonomy.

More recently, Corning (1995) refers to an "increasing acceptance" of the view that "biological (and social) systems are distinctive in being goal-orientated (or teleonomic)" (parentheses in the original). He however does not mention Monod; in fact he attributes the coinage of the term 'teleonomy' to Colin Pittendrigh.

AUTONOMIC AND CONSCIOUS TELEONOMY

The following two examples from our day to day life would both be readily recognizable as teleonomic activity. One, cooking a meal to feed oneself, and two, digesting the food towards providing energy and rebuilding body cells. Yet the kinds of goal-seeking in the two cases are perhaps very different.

The former example involves much more conscious goal-seeking. There is a wider exercise of choice, one may cook this food or the other, follow one recipe or another, and so on; there is also deliberate grappling with the complexity of cooking, during each run and over successive such runs. The latter, on the other hand, involves much more autonomic goal-seeking, as in a variety of other processes, such as "in the organism's system of defence through antibodies" (Monod, 1972), or in its attaining puberty. There is little exercise of choice by the organism in such cases, nor deliberate grappling with complexity. We shall tentatively call the two attributes *conscious teleonomy* and *autonomic teleonomy*, respectively.

Would such a distinction be manifest in the less evolved organisms? Let us again consider the single-celled paramoecium which eats bacteria. Moving to the bacteria-rich zone, reaching out for them and then apprehending them for consumption are quite likely to be seen as part of conscious teleonomy. While the digestion of the same, "a project demanding the synthesis of several hundred different organic constituents" (Monod, 1972), among other things, is likely to be seen as part of autonomic teleonomy. Not making such a distinction of the two categories of teleonomy - *as Monod fails to do* - would imply several readily refutable notions such as, the paramoecium's, or the human's, awareness of the food it eats is at the same level as its awareness of the highly complex chemical processes within itself.

We thus propose:
Proposition One. The unique attribute teleonomy (goal-seeking capability) which, according to Monod, distinguishes living beings from all other natural entities as also artifacts, may actually be unbundled into two distinct attributes: we tentatively call one, 'autonomic teleonomy', being mainly involuntary, and the other, 'conscious teleonomy' being more deliberate.

PURPOSEFULNESS AND COMPLEXITY-GRAPPLING CAPABILITY

Monod (1972) at one place paraphrases teleonomy as the capability to "decide on and pursue a purpose" (1972). It may be seen that there are actually two capabilities involved - one, for deciding on a purpose, or goal setting, and two, for pursuing a purpose, or grappling with the complexities involved towards attaining the goal. The distinction between the two may not be appropriate in the case of autonomic teleonomy. However, such a distinction is perhaps appropriate and also of importance in the case of conscious teleonomy. This, besides justifying our first extension of Monod's concept of teleonomy - unbundling it into two kinds, also leads to the next extension - unbundling conscious teleonomy.

Goal setting capability of an entity is of course what in ST has been called purposefulness. We shall call the other attribute, 'complexity-grappling capability'. A working definition of this attribute would be, quite literally, the capability of a system to grapple with a complex mess guided by one or more objective(s) towards fulfilling the same. Note that the guiding objective(s) need not be generated by the entity itself.

While conscious teleonomy comprises both the attributes, that they are distinct is readily seen. For instance, we may have only complexity-grappling capability with no purposefulness, such as in a robot, a computer, etc. We may also have purposefulness with little or no complexity-grappling capability, such as the many longings of a person with some kind of physical difficulty or impairment.

We thus propose:
Proposition Two. The attribute conscious teleonomy may in turn be unbundled into two distinct attributes: one pertains to goal setting capability, which we deem as identical to purposefulness; and the other, to grappling with the complexities involved in pursuing the goal, which we tentatively call complexity-grappling capability.

The attributes purposefulness and complexity-grappling capability, though distinct, are very closely and directly related in living beings. The enormous literature on learning - "defined as a change in behaviour ... that is due to experience or reinforcement" (Patterson, 1997) - across species, including in even the most primitive ones such as the bacteria (Corning, 1995), points to this. By 'close' relationship we mean a strong correlation between the two attributes, and by 'direct', simply, we mean the opposite of 'inverse'.

It is interesting to note that based on commonly accepted observations, different cultures in very diverse milieu have indigenously developed proverbs, such as 'Where there is a will, there is a way'. While distinguishing between purposefulness (to generate a will) and complexity-grappling capability (to generate a way), these adages also proverbialise their close and direct relationship.

We thus propose:
Proposition Three. The attributes, purposefulness and complexity-grappling capability, though distinct, are closely and directly related in living beings.

While the relationship between purposefulness and complexity has been widely explored in ST literature (Flood & Jackson, 1991), we would contend that the obverse, the relationship between purposefulness and complexity-grappling capability is as important, if not more. We hope to delve further into this in some of our subsequent papers.

CONCLUSION

Herbert Simon (1985) has remarked, "Nothing is more fundamental in setting our research agenda and informing our research methods than our view of the nature of the human beings whose behaviour we are studying". That actually practising this - in critiquing the idealistic assumption of hyperrationality for humans and proposing a more realistic 'bounded rationality' instead - has led him to pathbreaking works besides a Nobel Prize, would encourage one to take his precept rather seriously.

Purposefulness and complexity-grappling capability may be seen as attributes not only important to ST, but also underlying assumptions behind much of management science. Williamson (1996), while discussing 'transaction cost economics', an emerging, widely debated discipline influencing many areas of management science, observes that "all interesting problems of complex economic organization would vanish were it not for the twin conditions of bounded rationality and opportunism". It would be interesting to note the close correspondence between opportunism and purposefulness on one hand, and between bounded rationality and complexity-grappling capability on the other.

Based on a significant extension of Monod's work on teleonomy, a central contention of this paper has been that, contrary to the prevailing notion, purposefulness is perhaps not unique to only humans (or the higher animals) but is common to all living beings without exception. It is of course not our claim that the level of purposefulness of we humans is in any way comparable to that of say, the amoebas - infact, it is qualitatively apart by far from even that of the higher animals. Indeed, the paper is expected to provide the bases for a proposed deeper probing into this qualitatively different level of human purposefulness in at least two ways: insights from examining its evolution, and learnings from comparison and contrast with other purposeful entities.

These bases for deeper probing also hold for the other attribute derived from teleonomy, and identified in this paper - complexity-grappling capability. Its significance for the research on complexity as also problem solving, both very important to ST, as all of management science, is readily seen.

As we humans move on to the 21st century, it may be seen that most of our staggering problems - global or local - are related to the attributes purposefulness and complexity-grappling capability. Paradoxically, the two are also related to humanity's continuous quest to overcome problems, and for ever higher and truer synergy. We shall explore both aspects in some of our subsequent papers.

ACKNOWLEDGEMENT

The authors are grateful to Dr. Subhasis Mukhopadhyay for re-introducing them to Monod's "Chance and Necessity"; to Dr. J.L. Gnanarethinam, s.j., Director, XIM, who has known the late Monod as a friend, for his helpful comments on an earlier draft; and to the participants of the elective 'Quest for Unbounded Effectiveness with Systems Thinking (QUEST)' for providing the opportunity to explore some of these tentative ideas and their several startling implications for management practice.

REFERENCES

Ackoff, R.L., 1994, "The Democratic Corporation," OUP, New York.

Corning, P.A., 1995, Synergy and self-organization in the evolution of complex systems, *Systems Research.* 12(2):89-121.

Flood, R.L., and Jackson, M.C., 1991, "Creative Problem Solving: Total Systems Intervention," Wiley, Chichester.

Huxley, J., 1963, "Evolution in Action," Penguin, Harmondsworth.

Ivanitsky, G.R., Krinsky, V.I., and Mornev, O.A., 1987, Autowaves: an interdisciplinary finding, *in*: "Cybernetics of Living Matter: Nature, Man, Information," I.M.Makarov, ed., Mir, Moscow.

Mingers, J., 1995, "Self-Producing Systems: Implications and Applications of Autopoiesis," Plenum, New York.

Monod, J., 1972, "Chance and Necessity: An Essay on the Natural Philosophy of Modern Biology," translated by A. Wainhouse, Collins, Glasgow.

Patterson, M.M., 1997, Learning mechanism, *in*: "McGraw-Hill Encyclopaedia of Science and Technology, Vol 9," S. P. Parker et al, ed., McGraw-Hill, New York.

Rose, S., 1976, "The Conscious Brain," Penguin, Harmondsworth.

Simon, H., 1985, Human nature in politics: the dialogue of psychology with political science, *American Political Science Review.* 79:303.

Williamson, O.E., 1996, "The Mechanisms of Governance," OUP, New York.

MONITORING, EVALUATION AND DISSEMINATION MANAGEMENT IDEA, WULI-SHILI-RENLI SYSTEMS APPROACH AND THEIR APPLICATION

J.F Gu, F. Gao

Institute of Systems Science, CAS
Beijing, 100080, China

1 BACKGROUND OF THE WULI-SHILI-RENLI SYSTEMS APPROACH

The embryonic form of the Wuli-Shili-Renli (WSR) systems approach was formed at the center for systems studies, University of Hull, Hull, UK, in late 1994, this approach was put forward by J.F. Gu and Z.C. Zhu [1]. It is developed within the oriental concept and it is a method of system of the systems methodologies.

Compared with western philosophy based on science and democracy, Chinese philosophy has been characterized by its belief and intention towards integration, harmony and holism. It believes that human beings simultaneously understand and create the world, therefore the life world of Human conception, intention, and action cannot be properly investigated and researched as if separated from surroundings. The philosophy suggests that the Tao (moral), Zhi (knowledge or understanding), and Xing (action) of human beings are systemically related to the conditioning and support of one another, therefore cannot be 'artificially' isolated from one another. It focuses more on ethics and morality and has profoundly influenced Chinese society in politics, economy, academy, and daily life.

The traditional Chinese philosophy believes that everything in the world has a close relationship and it is complex, changeable and should be studied from a holism, complex and changing view, however the function of this philosophy focuses not on increasing active knowledge (active knowledge here means practical information or practical knowledge), but on raising mentality —reaching a state of super this life and getting upper value of moral value. While time goes, it is realized and agreed that additionally, the cultures are colorful and the values are pluralistic, with the development of science, technology, economy and information technique, the world is becoming more and more complex and full of competition, and all of them are under developing and changing. Under such condition a sharp, profound and unavoidable question faces to everyone who likes to investigate that is what point of view human beings should be taken when they try to understand and make practice in this complex, changing and competitive world. In order to answer this question, considering the advantages and disadvantages of traditional Chinese philosophy and western philosophy, the wuli-shili-renli systems approach mixes the oriental philosophy and western philosophy together and takes the great advantages of each of them to form a new point of view to achieve great effectiveness and efficiency of practical activities of reforming subjective world and society.

2 CONTENT, PROCESS, CHARACTERISTICS, AND APPLYING PRINCIPLES OF WSR SYSTEMS APPROACH

2.1 Content of WSR Systems Approach

This method, as its name suggests, is made up of three elements and two kinds of methodologies. The three elements are Wuli, Shili and Renli. Wuli is the laws of nature and objective existence; Shili means the

Synergy Matters: Working with Systems in the 21st *Century,*
Edited by Castell *et al.*, Kluwer Academic / Plenum Publishers, New York, 1999.

433

mechanisms in the universe or the methods of doing things efficiently and it includes two kinds of methodologies(tactical methodologies and strategic methodologies)﹔ Renli means the way of gathering people together to work effectively with the aid of psychology, sociology, behavioral science, combining with culture, legend, value, ethics, Wuli and Shili[2, 3, 4]. In this method, the substance of Renli varied with the circumstance, time, place and people, and is of most importance.

All rational activities of human beings, as they were described by Li[8], are either knowing or understanding activities (knowledge or understanding) or practical activities (practice). The methods base of WSR systems methodology is classified into two kinds of methods - the strategic and tactical methods base shown in table 1 (the reasons and principles of this classification are also provided by Gao and Gu[4]).

Strategic Methods	Tactical Methods
Delphi Method﹑ Brainstorming﹑ Interactive Planning Strategy Assumption Surfacing and Testing Strategy Choice﹑ Social Systems Design﹑ Social Choice﹑ Management Cybernetics﹑ Organization Cybernetics Strategy Options Development Analysis﹑ Contingency Theory Qualitative System Dynamics﹑ Soft Systems Methodology Problem Structure Method﹑ Meta-Synthesis Methods﹑ Shinayakana System Approach﹑ Critical Systems Thinking Total System Intervention etc.	Taylor's Scientific Management﹑ Operations Research System Analysis﹑ System Dynamics﹑ Management Cybernetics﹑ Organization Cybernetics Social Tech.-Systems Design﹑ Organization Design Viable System Diagnosis﹑ Problem Structure Method Meta-Synthesis Methods﹑ Total System Intervention System Engineering﹑ Shinayakana System Approach Interactive Planning﹑ Total Quality Management etc.

Table 1. The Classification of Methods in the Method Base of WSR Systems Approach

The relationships of the basic activities of human beings or a system (an organization or a corporation), Wuli, Shili, Renli and the two kinds of methods base, as it is described with the input-transformation-output schema with the aid of cybernetic terms [5], provide a basic theoretical framework and allow people easily to know how Wuli, Shili and Renli work in WSR systems approach and clearly and profoundly understand the process of the WSR systems methods. The schema describes the characteristics of the method and enables practitioners and academics to apply this method to copy with those problems arising from the practice of management easily.

2.2 Process of WSR Systems Approach

The process of WSR systems approach can be described as follows and easily understood.

Understanding the desire and investigation: By using the methods chosen from strategic level in methods base of Shili, combining with Wuli and Renli, a type of knowledge or understanding can be formed.

Define objectives or goals: Based on the obtained knowledge, some objectives or goals of an organization or a corporation (regarded as a system) can be defined.

Planning: After a kind of knowledge and objectives being gained, some inputs such as capitals, materials, machines, personnel and other environment factors will be sent into the transformation process (i.e. an organization or a corporation). Some suitable tactical methods chosen from the methods base of WSR systems approach, combining with Renli, will be applied to work out some detailed plans.

Implementation: Some concrete practice or activities will be carried out upon the plans. These activities are briefly regarded as transformation process. At last, some desirable and planned output, i.e. the objectives or aims of the system, will be gotten from this process.

Comparison: The output, as well as some evaluation of the performance of the organization or corporation, will be feedback to the high level of management and will be compared with the planned objectives, and the advantages and disadvantages of the system performance will be found

Review: Review is imperative. Then new knowledge or understanding about the system and adapted aims will be obtained

Back to Investigation or Fine-tune the goals: Based on the comparison and review, the strategy of the system will be fine-tuned and some more suitable goals of the system will be gotten. As some more suitable methods chosen from tactical methods base again, some adapted plans will be achieved. Those will be implemented in tranformation process.

The world is increasingly complex, rapidly changing and full of competition, The process of WSR systems approach can not come easily to an end, it will be iterated. It must be pointed out that there is iteration in the transformation process and there are many techniques to keep the process steadily and work efficiently.

2.3 Characteristics and Applying Principles of the WSR Systems Approach

The characteristics and applying principles of the WSR systems approach are as follows:

- **Meta-Synthesis:** WSR systems methodology is a meta-synthesis of science, technology and sociology. Based on the combination of oriental philosophy and western philosophy, the approach takes great advantage of modern science theory and technical means, utilizing computer as tools and experts as the medium to form a super intelligent, open system which enables users to achieve great success of the reforming subjective world and society in more effectively and efficiently way.
- **Computer as core tool:** This method uses computer as its main tool to set up database, model base, knowledge base, and methods base.
- **Co-operation and participation:** The method needs a group of experts from various backgrounds to co-ordinate their work to achieve the advantage of systems science, that is, the sum of parts is greater than the total. It emphasizes the stakeholders' participation at all stages.
- **Method of systems methodology:** This method makes all other methods as its potential chosen methods. It includes soft methods and hard methods [9,10, and 11], as well as other suitable science and technical approach. It provides a framework of dealing with the problems arising from practice and understanding.
- **Human-oriented method:** Drawn from traditional Chinese philosophy, this method focuses on ethics, moral, and value outlook. It delivers benefits to all stakeholders and focuses on the results of implementation.
- **Iteration:** As the world is very complex and changing at all time, iteration is an absolutely necessary way to reflect the changing situation.
- **Flexibility:** It is people who fine-tune strategy promptly upon judging actual circumstances. As there are more than one way to reach the goals, there is a broad range of choices.

3 MONITORING, EVALUATION AND DISSEMINATION (MED) MANAGEMENT IDEA

3.1 Background of Monitoring and Evaluation (ME)

Monitoring and evaluation theories have very long history in management field. The former has close relationship with Audit (a theory of making examination, adjustment, and certificate of the accounts of a business), which, at some degree and some aspects, is of monitoring function. As for evaluation, even in daily life, people already used to evaluate the things they face in their life, far beyond that it has become a popular management method in almost all fields. Although ME is considered as a general management idea, in order to explain them clearly and easily, what we talk about them is mainly from the view of project management.

Monitoring and evaluation project management approach is developed in the United States and other industrial nations and widely proposed and disseminated by the World Bank. Monitoring is a continuous internal management activity whose purpose is to ensure that the program achieves its defined objectives within a prescribed time-frame and budget, it involves the provision of regular feedback on the progress of program implementation, and the problems faced during implementation, and it consists of operational and administrative activities that track resource acquisition and allocation, production or the delivery of services, and cost records. Evaluation is an internal or external management activity to assess the appropriateness of a program's design and implementation methods in achieving both specified objectives and more general development objectives; and to assess a program's results, both intended and unintended and to assess the factors affecting the level and distribution of benefits produced [7].

3.2 Reasons of Implementation ME

It is customary to refer to them together (as in the term ME), many practitioners and academics treat them as distinct activities, and most of the U.S. evaluation literature assumes ME to be closely related, and frequently the term "program evaluation" is taken to mean both monitoring and evaluation. The function, importance and necessity of ME can be seen from the definition of them. ME studies can also help a country improve its method of identifying and selecting projects and programs by ensuring that these endeavors are consistent with national development objectives, that they will have a good chance of succeeding, and that they are using the most cost-effective strategy for achieving the intended objectives; it can determine whether the project is being implemented efficiently, is responsive to the concerns of the intended beneficiaries, and will have its potential problems detected and corrected as quickly as possible; ME measure whether projects and programs that are under way are achieving their intended economic and social objectives, as well as contributing to sectoral and national development objectives; evaluation studies can be used to assess the impact of projects on wider developmental objectives [7]. In one word, the purpose of ME is to improve the efficiency and effectiveness of human practical activities and get experience from practice.

3.3 Current Status of Monitoring and Evaluation

Now more and more emphasis is being placed on monitoring and evaluating the performance of implementation to ensure the planned objectives being achieved efficiently and timely, to assess the efficiency of implementation, to know the extent of the intended impacts of the program, to learn as much as possible from past experience that will enable government, organization or donor agencies to identify what kinds of delivery systems most likely to succeed and the factors most likely to contribute to success. Since the early 1970s, monitoring and evaluating most federal and state-financed project has become standard practice in the United States. The theory and practice of ME is also emerging as a separate social science and management discipline in the U.S.[7]. It has also become popular in many developing countries. Although a little has been done in monitoring and evaluating on social and economical projects in our country, lots of work has been done only on engineering technical projects. However, the most important thing is that now our national and sectoral leaders and other decision-makers own this idea and pay more and more attention on it. What need academics to do is to provide some suitable approaches for practice and make more people use them in management. In order to do this some methods are provided here.

3.4 Some Methods of Monitoring and Evaluation

There are many methods being used for monitoring and evaluation. The monitoring methods can briefly divided into two kinds of monitoring, one is input monitoring-that is monitoring whether resources being mobilized as planned; the other is output monitoring i.g monitoring whether services or products being delivered on schedule, monitoring can also be implemented during process. As to evaluation methods, there are various kinds of approaches and each kind make use of different ways as its tools and focus on different fields and situation [6]. Some methods used in monitoring and evaluation are provided in table 2.

Monitoring Methods	*Evaluation Methods*
Gantt Charts Follow-up Diagnostic Studies (Participant Observation Rapid Survey Direct Observation Interviews with Key Informants) Program Evaluation Review Technique Logical Framework Analysis Comparisons Analysis Network-based Systems for Physical and Financial Monitoring etc.	Comprehensive Scoring Approach Restraint Approach Visible Diagram Approach Order Number Approach Experts Systems Linear Allocation Approach Simulation Artificial Neural Network Analytical Hierarchy Process Logical Framework Analysis Quasi-experimental Design Total Systems Intervention Input-output Analysis Causal Networks Clustering Analysis Process Modeling Data Envelopment Analysis Path Analysis Systems Analysis Cost-effective Analysis Value Engineering approach Cost-benefits Analysis Social Analysis System Dynamics Utility Function Approach WSR Systems Approach etc.

Table 2: Methods of Monitoring and Evaluation

3.5 The Relationship of Monitoring, Evaluation and Dissemination

Monitoring, evaluation and dissemination is seen as a view of management here, and they have very close relationship. Besides to monitoring and evaluating human's practical activities to improve the quality of human understanding and practical activities, the more valuable thing is to find out good examples and experience which are best suitable to sustainable development of economy, society, environment, and human being, and let more people draw benefits from them.

To any activity, monitoring is implemented first. Based on the monitoring, evaluation can be made. Then some example and experiences that meet the fad of society and decision-makers can be drawn from the evaluation and be disseminated. Their relationships can be easily seen from Figure 1.

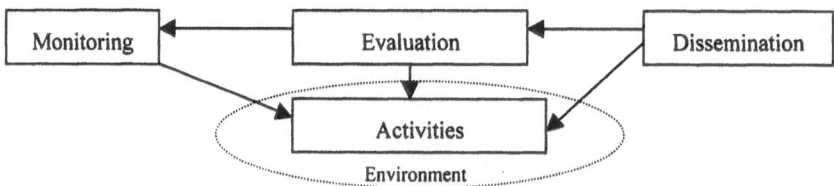

Figure 1: The relationship of monitoring, evaluation, dissemination and Human activities

4   THE RELATIONSHIP OF WSR SYSTEMS APPROACH AND MED MANAGEMENT IDEA

After introducing WSR systems approach and MED management idea, in order to explain them clearly and easily the relationship of them is described in Figure 2.

	Wuli	Shili	Renli
Monitoring Evaluation Dissemination	*Facts*	*Efficiency*	*Effectiveness*

Figure2: *The relationship of WSR systems approach and MED*

From Figure 2, it can be seen that monitoring corresponds with Wuli and examines the implementation condition of plans and objectives from facts. The efficiency of human's practical activities is evaluated with the aid of Shili. Dissemination mainly depends on effectiveness. As different people have different standards of effectiveness, which is based on their various backgrounds of culture, legend, ethics, and value, what being used as disseminated things are determined by people. So effectiveness is determined and decided by human being. Up to now, it can be learn that MED management idea is very useful and helpful. To apply this management idea into practice to achieve success, WSR systems approach is a wonderful and most suitable method. An example of its application is given as follows.

5   A CASE OF APPLYING MED MANAGEMENT IDEA AND WSR SYSTEMS APPROACH

5.1   China Labor Market Development Project and China Labor Market Monitoring, Evaluation and Dissemination Subproject

Since the late 1970s, China has pursued a policy of gradual reform, open its economy to the outside world, accelerating domestic and export production, and absorbing foreign investment and advanced technology so as to modernize and develop its economy more effectively. With economy reform, China has achieved great success in economy development and other field reform. However, a number of problems have risen. One of the important problems is the surplus workers in State-owned enterprises, and the big amount of unemployed and rural-to-urban migrant labor. In order to help Chinese government cope with this difficult problem and support China economy reform, the World Bank decided lend amount of money on some pilot municipality cities to implement China Labor Market Development (CLMD) project. In order to achieve the objectives of the project and extract lessons form CLMD project, a subproject—China Labor Market Monitoring, Evaluation and Dissemination (CLMMED) project is established to monitor and evaluate the implementation of the CLMD project and to disseminate good experience nationally.

5.2   Working Process of WSR System Approach on CLMMED Subproject of CLMD Project

The working process of WSR systems approach on CLMMED subproject of CLMD project is:
Understanding and investigation: Firstly, by using some strategic methods chosen from the strategic methods base of WSR systems approach, such as Brainstorming, Interactive Planning, combining with the facts, experts, decision-makers, officers and other stakeholders from the World Bank, Ministry of Labor, pilot cities, JP international consultant company and some other units reached agreement on some ideas about the CLMD and CLMMED projects.
Define objectives or goals or Fine-tune objectives or goals: Then, the objectives of CLMMED subproject were formed. That is to monitor and evaluate the performance of the implementation of CLMD project from four aspects (policy and legal reform, labor market service, in-service training, institutional development) with three types of indicators (input, output, and outcome indicators). Based on these monitoring and evaluation, combined with current society fads and leaders value standards, some good experience of implementation of CLMD can be draw and disseminated nationally
Planning: Once the objectives of MED subproject of CLMD being determined, some tactical methods such as Gantt chart, Table comparative analysis, Quasi-experiment (QE) designs, Statistics analysis, Logical framework analysis, Systems analysis, CATWOE analysis, WSR systems approach were applied to make the monitoring and evaluation designs of CLMMED subproject and formed some proper models.
Implementation: Fourthly, those plans and models were put into implementation.
Comparison: Fifthly, the results of the implementation of monitoring and evaluation feedback to the up stage and compared with the objectives of CLMD project and found out the good and bad impacts of CLMD project.
Review: Finally, a review must be made and some good lessons can be drawn from CLMD. With the development of the society and economy, new problems may be risen and adapted objectives, plans and

models have to be made and put into implementation again.

This process is iterations and it needn't implemented step by step. During these processes, lots of works have to be done to co-ordinate the works among the stakeholders. This project can also be expressed from the view of WSR systems in table 3.

	Wuli	Shili	Renli
Monitoring	*Inputs, output and outcomes Labor market information system Some service centers*		
Evaluation		*Gantt chart, Table comparative analysis, Quasi-experiment (QE) designs, Statistics analysis, Logical framework analysis, Systems analysis, CATWOE analysis, WSR systems approach*	
Dissemination			*Co-ordination Transformation of Value-standard, Culture, Idea*

Table 3. To see CLMMED subproject from the view of WSR systems approach

6 CONCLUSION

According to the above analysis, it can be seen that to manage large scale project MED management idea is a very helpful and useful way and WSR systems approach is a good method to implement it. With the economy and society reform, more large projects of both engineering technology and society will be involved. It is necessary to develop a suitable management idea to manage them. We hope that this discussion can "cast a brick to attract jade" and call more experts' attention to join in this work.

REFERENCES

1. J.F.Gu and Z.C.Zhu, The Wu-li Shi-li Ren-li Approach(WSR): an Oriental Methodology, in *Systems Methodology: Possibilities for cross-cultural Learning and Integration(G.Midgley and J.Wilby eds.)*, The University of Hull Press, United Kingdom, 1995, pp31-40
2. J.F.Gu, Wuli-Shili-Renli Systems Approach Methodology, Transportation System Engineering and Information, 1995, Vol.3 pp25-28(in Chinese)
3. J.F.Gu and F.Gao, See Wuli-Shili-Renli Systems Approach from the View of Management Science, Systems Engineering—Theory and Practice, 1998, 18(8), pp1-5 (in Chinese)
4. F.Gao and J.F.Gu , The Methodology Base of Wuli-Shili-Renli Systems Approach, Systems Engineering—Theory and Practice, 1998, 18(9), pp37-41 (in Chinese)
5. F.Gao and J.F.Gu , See Wuli-Shili-Renli Systems Approach From A New View, Proceedings of ICSSSE'98, Beijing, China, pp157-161
6. J.F.Gu and L.Y.Zhao, An investigation and research report about national evaluation methods and their application, Working Report, 1996. (in Chinese)
7. J. Valadez & M. Bamberger, ed., Monitoring and Evaluating Social Programs in Developing Countries: A Handbook for Policymakers, Managers, and Researchers, EDI DEVELOPMENT STUDIES, November 1994
8. Z.C.Li, ed., Methodology Volumes: Nature Science Methodology, Nanjing University Publishing House, 1995
9. R. L Flood, & M. C Jackson, Creative Problem Solving: Total Systems Intervention, Wiley, Chichester, UK. 1991
10. M. C Jackson, Contemporary Systems Thinking: Systems Methodology for the Management Sciences, Plenum Press, New York, 1991
11. P.Checkland, Systems Thinking, Systems Practice, Wiley, Chichester, UK. 1981
12. Z.Sun, Translated by J.B.Pan, etc., The Art of War, Military Science Publishing House, 1993
13. John B. Khu, etc., The Confucian Bible, Granhill Corporation, Philippines, 1991

CHOICE OF MODEL OR MODEL OF CHOICE ? - SYSTEMS THINKING AND THE PHENOMENOLOGY OF DECISION MAKING

John Hassall

University of Wolverhampton
Management Research Centre
Shropshire Campus
Shifnal Road, Priorslee
TELFORD TF2 9NT
UK

INTRODUCTION

This paper examines decision and "choice making" based upon three groupings of ideas which are then examined from a systems perspective. The three groupings are, firstly, the philosophy of choice with reference to classicism, modernism and post-modernism. Secondly, the works of Soren Kierkegaard, sometimes regarded as the first existentialist philosopher. Thirdly, the psychotherapeutic model developed by Eric Berne referred to as Transactional Analysis.

In the paper the phenomenological aspects of choice and decision making are considered most important and the ideas of Kierkegaard are considered to be of significance to an understanding of what it is to be human and to choose. An important feature of the paper is to link models and perspectives for choice and decision making within a systems framework so as to bring the philosophical ideas in to contact with practical concerns.

PHILOSOPHICAL BASES AND CHOICE

Three broad philosophical perspectives within which choice may be considered are modernism, post-modernism and classicism. Modernism is a way of dealing with the world based upon an "enlightenment" philosophy whereby a rational process of thought and action is assumed to be appropriate in all matters of truth and value. By contrast in the era of classicism, thought and action was prescribed by a number of authorities, the King or State, the Church and God. Finally, post-modernism rejects the "grand narratives" of both modernism and classicism (Lyotard, 1979); which attempt to explain behaviour in terms of underlying, abstract, but largely comprehensible forces and causes. So from the post-modernist perspective the possibility of a consistent model based rationale for choice is denied.

Modernist thought sits at the nexus of two competing perspectives, the classical world with its certainties derived from authority and post-modernism with its denial of certainty in any attempt to assign value. Any process, system or technology which aims to help us to make a choice is essentially a modernist enterprise and, moreover, as argued in earlier work (Hassall 1998) it may prove impossible to justify the choice of any model without reference to some form of "authority".

KIERKEGAARD AND CHOICE

Soren Kierkegaard (1813 to 1855) has detractors, in part because of his Christian religious beliefs which are perceived as classicist. However, a reading of Kierkegaard is possible from a phenomenological perspective reflecting the view that he was the first existentialist, (believing that value arises from action rather than being a guide to it).

In *"Fear and Trembling"* (Kierkegaard, 1843) the author examines what it is to decide by means of an exploration of the story of Abraham. Abraham was asked to sacrifice his only son to God and Kierkegaard argues that the choice by Abraham to carry out Gods request is based upon faith in an authority beyond himself and is to be admired above decisions made on other, implicitly lower, bases. He recognises that a choice can also be made on an ethical basis, dependent upon a consensus amongst members of a society of what is correct action (external to the chooser). Or, a choice may be made on the basis of personal taste , for Kierkegaard an aesthetic basis.

Kierkegaard makes two further points in relation to choice based upon faith. Firstly, such choices are not fatalistic. The basis of choice provides certainty of the best possible decision for the individual (and by extension the group to which she or he belongs). Secondly, that failure to choose based upon faith, is to suffer through lack of certainty, *"The Sickness Unto Death"* (Kierkegaard, 1849).

Kierkegaards arguments are often subtle and phrased in religious and poetical language, however his essential points apply to the phenomenology of choice, our inner experience of deciding. Every choice may contain elements of arbitrariness/aesthetic basis (taste..), ethical/rational basis (science..) and finally faith/authority (revealed truth..).

TRANSACTIONAL ANALYSIS AND CHOICE

Now widely used in the therapeutic community and beyond, transactional analysis (TA) in psychotherapy was developed initially by Eric Berne (Berne, 1961, 1964). TA uses a phenomenologically based (modernist) model of consciousness which sees both our internal mental states and our interactions with others as arising from 3 separable "ego states". These correspond to a "parent" component which is based upon our own parents, a "child" component evolving from our early emotional experiences and an "adult" component , a rational, reasoning component. The acronym PAC is often applied to this model.

PAC may be used as a framework for understanding both normal and pathological interactions between people, and there is a very extensive academic and popular literature on this (see for example Berne 1961, 1964, Harris 1967, Fensterheim and Baer 1976). Presently, we can consider the three "ego states" in terms of their possible involvement in choice and decision making.

Once familiar with PAC it is possible to identify the contribution of particular ego states to an interaction in terms of the words and language adopted. For example the influence of the parent ego state is observed in the use of such terms as "ought to" and "should"; the child by contrast says "I want" and "I feel like". Proponents of TA suggest that, in making a choice these "voices" will be present in our minds and influence how the decision process turns out. Moreover, the calm and logical voice of the adult ego state will play an important role.

In making a choice the parent appears as an authority which cannot be questioned. The child by contrast is characterised by a capricious approach to choice. The adult will exhibit a structured, logical and data based approach.

SUMMARY

Philosophical Bases **CLASSICISM**	Kierkegaard **FAITH**	TA and PAC **PARENT**
Characterised by:	Characterised by:	Characterised by:
Authority decides.	Faith in God,	"taught concept"
Choice:	Choice:	Choice:
Depends upon what the relevant authority has determined.	Depends upon what the transcendent entity requires.	Depends upon what our parents told us we should do.
MODERNISM	**ETHICS**	**ADULT**
Characterised by:	Characterised by:	Characterised by:
Rationality , logic.	Using the "universal" as a guide.	"thought concept"
Choice:	Choice:	Choice:
Depends upon a rational process.	Depends upon a process external to ourselves and shared with others.	Depends upon our own analysis. Influenced by "adult" input from others.
POST-MODERNISM	**AESTHETICS**	**CHILD**
Characterised by:	Characterised by:	Characterised by:
Rationality suspended No "grand narratives" (Lyotard, 1979)	The feelings of the moment dominate.	Feelings.
Choice:	Choice:	Choice:
An arbitrary matter. No basis for preference.	Depends upon our feelings at the time.	What makes us feel good.

A possible synthetic statement arising from the above summary is that the different philosophical bases have their root in phenomenology. Or, alternatively, that all three sets of frameworks reflect some underlying structure.

SYSTEMS PERSPECTIVE

Choice and action being inexplicably linked, a systems explanation of choices and behaviours seen in practical situations may be sought. Situated action may appear highly contingent (Hassall, 1997), (Dudley and Hassall 1995) because the chooser is 'embedded' within an organisational system which exercises a variety of constraints upon behaviour. In seeking a description of behaviours for an individual or system making a choice consider a recursive systems model as in (e.g.) (Beer 1979 and 1985).

In the first diagram below, a number of co-existing systems are embedded within the same meta-system. Their choices and behaviours may be affected by three main influences. Firstly, Authority based (external) influences which all feel (A). Secondly, influences from their partners in existence (O) and, finally, self determined (internal) influences (S). (A)uthority influences may be explicit but are often felt as normative constraints upon possibilities for choice (Dudley and Hassall 1997). (O)ther influences may be expressed within formal procedures, sometimes documented, or informal negotiations within a shared rationale.

Consider the view of a 'chooser' system. How does the system view its choosing? Is it aware of the influence of A (both explicit and normative)? or does it recognise an ongoing O process in partnership with other chooser systems. In practice this can be seen to present problems in any situation where change depends upon choice.

Influences upon systemic behaviour

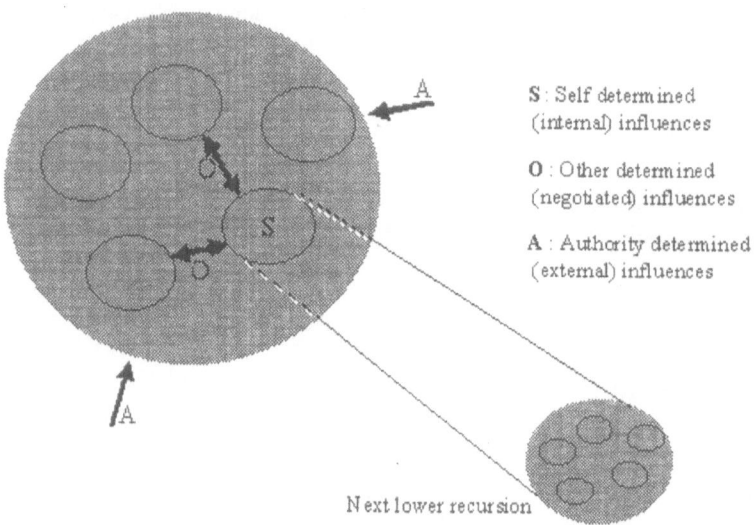

S : Self determined (internal) influences

O : Other determined (negotiated) influences

A : Authority determined (external) influences

Next lower recursion

The distinction drawn between A and O is crucial in determining whether the chooser system will engage with the choice as a rational exercise or merely act under some assumed authority. This issue of the boundary between A and O is central to the issue of empowerment for individuals who may be prone to overestimate the power of the organisation and its norms to control their choices and actions. In practice this my be observed by the personalisation of the organisation as an all-pervading system with control over individual choice ("The Company .....", "The University ....", or even "They...").

Influences upon systemic behaviour - Perspectives

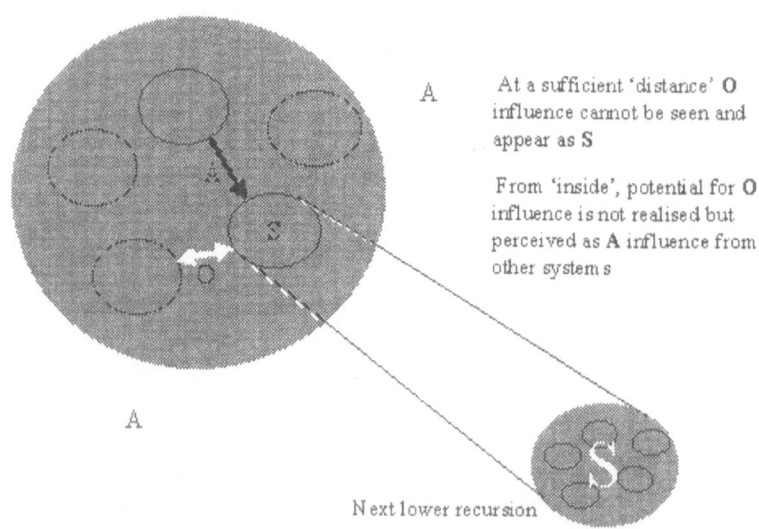

A

At a sufficient 'distance' O influence cannot be seen and appear as S

From 'inside', potential for O influence is not realised but perceived as A influence from other systems

A

Next lower recursion

442

If sufficient 'distance' is put between the observer and the chooser system the focus on O influences upon choice may be lost. The chooser system in focus is seen to act in a way that appears arbitrary because the interactions between internal and co-existing chooser systems are not immediately visible. (Perhaps looking at an ant colony or a flock of birds from a distance is a useful metaphor; the movements of the mass of individuals appears arbitrary, but in fact, its gross movement is a result of them interacting and exchanging information with their fellows all the time.)

CONCLUSION

Using the language of systems a model of choice has been derived which offers a meta-framework in relation to the earlier perspectives discussed. It appears from this that possibilities for judgement and choice may be related to how 'chooser' systems view the boundary interactions between their own and others systems; further, that understanding systemic choices and behaviour depends upon achieving the correct level resolution view so that interactions between partner systems can become visible.

REFERENCES

Beer, S., (1979), *The Heart of Enterprise*, Wiley, Chichester, UK.

Beer, S., (1985), *Diagnosing The System For Organisations*, Wiley, Chichester, UK.

Berne, E., (1961), *Transactional Analysis in Psychotherapy*, Grove Press, New York.

Berne, E., (1964), *Games People Play*, Grove Press, New York, USA.

Dudley, P, Hassall, J C, (1995), Systemic Refocusing Strategy: An Emancipatory Approach to Intervention, *Critical Issues in Systems Theory and Practice*, Ellis K, Gregory A, Mears-Young B R, Ragsdell G, Plenum Press, New York, pp 465 - 478, ISBN 0-306-45100-X.

Dudley, P, Hassall, J C, (1997), Systemic Refocusing - A Strategy for Sustainability, *Systems for Sustainability - People, Organisations and Environments*, Stowell F, Ison R, Armson R, Holloway J, Jackson S, McRobb S, Plenum Press, NewYork, pp 455 - 458, ISBN 0-306-45615-X.

Fensterheim, H, Baer, J, E., (1976), *Don't Say Yes When You Want To Say No*, Macdonald & Co (Publishers) Ltd, London.

Harris, T A, (1967), *I'm OK You're OK*, Avon Books, New York, USA.

Hassall, J C, (1997), Systemic Approaches and Contingent Actions in Consultancy Intervention, *Systems for Sustainability - People, Organisations and Environments*, Stowell F, Ison R, Armson R, Holloway J, Jackson S, McRobb S, Plenum Press, New York, pp 137 - 140, ISBN 0-306-45615-X.

Hassall, J C, (1998), Choice in Practice - Can Information Systems Really Help Managers and Consultants Decide?, *PAIS II, Second Symposium and Workshop on Philosophical Aspects of Information Systems: Methodology, Theory, Practice and Critique*, University of the West of England, Bristol, Chapter 12.

Kierkegaard, S, (1843), *Fear and Trembling*, Penguin Books, London: Translation from Danish by Hannay, A.

Kierkegaard, S, (1849), *The Sickness Unto Death*, Penguin Books, London: Translation from Danish by Hannay, A.

Lyotard, J, (1979), *The Postmodern Condition: A Report on Knowledge*, Manchester University Press, Manchester: Trans from French by Bennington G and Massumi B.

BOTTOM UP THINKING

Bill Hutchinson

Edith Cowan University
Churchlands
Western Australia 6018

INTRODUCTION

There are two approaches to evaluating hypotheses – deduction and induction (Kam, 1998, pp. 122-123). Deduction reasons from general rules, or theories to particular cases, whilst induction generalises from observation. Whilst there is not a complete correlation between these approaches and top down and bottom up thinking, the latter makes use of inductive thinking. Many systems approaches come from some preconceived model, for instance the Viable Systems Model (Beer, 1985), or the feedback loops of System Dynamics (Wolstenholme, 1981). To make sense of the world humans create boundaries around elements to produce concepts. Each boundary is necessarily incomplete to focus on certain elements in the world. The problem solver's worldview and the problem situation form the boundary (Checkland and Scholes, 1991; Wilson, 1990). However, because a boundary is drawn, paradoxes are created in the discipline known as systems thinking. Firstly, is the necessity to use reductionist techniques to examine the conceptual system. Often systems texts claim an *holistic* approach is being taken, but in practice, these approaches use reductionist techniques, which are just much more **inclusive** than traditional *hard* scientific methods. However, they are not the antithesis of reductionism as claimed (Klir, 1991, p.24-27). Although, for instance, Checkland (1981, p.162) describes his Soft Systems Methodology as the intermediate point between a philosophy and technique, the actual use of the methodology necessitates a *top down*, reductionist style. Systems approaches may expand the boundary but do not change the top down nature of the problem solving.

Another anomaly with system thinking is the concept of *emergent properties*. This is often explained by the statement that "a whole is greater than the sum of its parts" (Flood and Carson, 1993, p.17). Bertalanffy (1951, p.148) provides more detail: "You cannot sum up the behaviour of the whole from the isolated parts, and you have to take into account the relations between the various subordinated systems and the systems which are super-ordinated to them to understand the behaviour of the

Synergy Matters: Working with Systems in the 21st *Century*,
Edited by Castell *et al.*, Kluwer Academic / Plenum Publishers, New York, 1999.

445

parts". What does this mean? It implies a type of magical creation of properties when system elements are put together in a specific way. It is almost spiritual in nature, but explains very little. For instance, Capra (1996) claims that systems thinking provides a breakout from the reductionist Cartesian mindset, yet offers very little for practitioners to make use of this 'new' view. Even ardent, contemporary advocates of the idea of emergence such as Bossomer and Green (1998, p.8-9) admit they can offer no formal definition. This implies emergent properties can be observed, but not analysed in a strictly scientific sense. Hence, empirical data rather than theoretical frameworks are used to investigate emergent phenomena.

This paper will attempt to use a different systemic approach to attempt to alleviate these problems. For the want of better words, the term *bottom up thinking* will be used. It is not the opposite of top down approach, as the questions asked by the problem investigator, and the outcomes obtained are radically different. This approach is not claimed to be a substitute for top down thinking but a useful adjunct to it. However, it is claimed that this approach is becoming essential as the nature of contemporary organisational structures fundamentally change.

DEFINITIONS

In this paper, the term 'top down' is used to describe an approach to problem solving where the problem space is defined first. The worldviews of the participants are used to conceptualise the desired state of the proposed system. Once this is achieved, the system is developed within this boundary. The components, or subsystems are derived within the context of the predefined desired state.

The term 'bottom up' is used to describe an approach where no assumptions are made about the boundary of the problem space. The behaviour of component parts are rationally observed to determine the properties they have. Management decisions, or system designs are then based on the observed behaviour of components. It is accepted there are a number of paradoxes and limitations to this approach. The first is the nature of 'rational observation'. The term 'objective observation' was deliberately not use in recognition of the subjective nature of any observation. The idea of starting with no assumptions about the system is also rather idealistic, and because of the nature of human thought, impossible in practice. However, this paper is attempting to propose a practical and pragmatic way to deliver good management practice and design. It is the philosophy of the approach that is being described rather than the purity of its practical application.

In actual management practice, both approaches are necessary for quality decisions. For instance, a manager has the responsibility to map out organisational goals onto decision making. Thus, if a task is necessary, it must be assumed that it will be carried out in the manner specified by the manager. However, it is naïve to think that the actual situation will be the desired one just because a manager wants it. The **actual** interactions of people, technology, and so on need to be considered to evaluate what will really happen.

THE BOTTOM UP VIEW OF AN ORGANISATION

The organisation can be viewed as a conceptual whole, or as a set of semi-autonomous objects (Tsoukas, 1993). Here, it is assumed that the organisation has an overt, stated purpose. This view, **which is still using top down thinking**, sees the organisation as consisting of elements (usually individuals or groups) which are all carrying out their own activities. If the purposes of these elements are beneficial to the overall purpose of the whole system, the organisation will remain healthy (in its own terms). Within the system, elements have different purposes, which may contradict the main system purpose. The organisation thus can have a myriad of 'purposes'. If these sub-purposes do not dominate, the organisation will still carry out its main purpose. However, a situation could occur where the sub-purposes can dominate. Hence, the actions of the system elements will not serve the overt system purpose. The priorities of system purposes will change in fact, if not officially. This model has been included to give some idea of the chaotic character of organisations. Top down management thinking often assumes all the arrows on the system elements are all pointing toward the main purpose of the company. Experience tells us this is just not reality. People and groups of people have different agendas and motivations; assuming they correlate with the official organisational objectives is extremely naive. As Tsoukas (1993, p.514) says, "While social organisations are inevitably human artifacts, they are not necessarily the product of human design".

When an organisation is viewed in bottom up mode, each component has a set of attributes and potential. How each element behaves, will be determined by that element's internal state and the behaviour of other elements affecting it. Hence, people, groups, technology, and other resources determine a system's behaviour. It is these elements, and how they interact that will determine how the system works. Therefore, examining these will give a realistic image of what is **really** happening, and any achievable potential in the system. A criticism of managers using top down styles is that they do not know what 'really goes on'. Managers view staff behaviour with some preconceived notion of what **should** be happening. This can cause a discrepancy between ideas and practice. It might be valid to view management problems this way to get things achieved but can cause problems if the ideal and practice are very divergent. Bottom up views look at the actual practice of system elements, and develop ideas from them.

What is this the real state of affairs in the modern organisation? Increasingly, organisations possess flattened management structures. The move toward part-time and non-permanent work, and the practice of outsourcing are increasing. Surely, in this environment, the main aim of people will be self interest (not that of the organisation). Staff mobility will increase (hence, any loyalties will be transitory). Management will have to accomplish its task with much less authority. Has management practice come to grips with this change?

Whilst top down management approaches must still be used, a bottom up view might be more appropriate to predict outcomes, where the organisation consists of disparate, and loosely coupled elements. An environment where the individual element is supreme (in management practice terms), and not the organisation itself.

THE NEGATIVE SIDE OF TOP DOWN THINKING

In describing bottom up thinking, it is worth noting some of the real limitations of top down thinking. One major drawback of drawing a boundary around a 'concept' is that people believe it to be true, and then try to fit other people's behaviour into the context of their viewpoint. These concepts then allow managers, politicians, and ourselves to blame some abstraction rather than any decisions or actions we might have made taken. If things go wrong, or we wish to do something which is ethically unjustified, we can claim to be victims of trends, or of 'normal' business practice. Anything that will prove that people are powerless, and victims of the context. The abstraction is really to blame. Of course, who creates these abstractions is a mystery.

Differences in top down and bottom up views in organisations can manifest themselves in conflicts between upper management and operational staff. For instance, in a police service, the realities of the upper ranks, who have to report to political masters and the media, can be in stark contrast to the realities of the street, where the consequences of individual behaviours have to be handled. As Deacon (1988, pp. 302) says: "As a wartime officer who on at least two occasions had to ignore the orders of my superior officers, I can vouch for the risks of relying too much on orders when common sense tells me they won't work…In today's terms the tendency is for the powers-that-be to rely too much on the computer and not enough of the judgement of the man on the spot".

DESIGNING SYSTEMS USING A BOTTOM UP METHODOLOGY

When viewing problems from a bottom up perspective,' it is useful to have guidelines to aid in your approach. No methodology is perfect. In fact, methodologies when followed without reflection are no more than security blankets for the unthinking, something to blame if things go wrong. Having stated that, certain specified procedures can help in problem understanding and solution.

A bottom up methodology should allow the user to view the problem from the perspective of the system elements. Boundaries should not be assumed, but built up as the investigation progresses. The observed behaviours of system elements as they interact, in terms of their processes, and information and material flows, should be recorded to demonstrate the potential for different combinations of these elements. Thus, the system or problem investigation allows a solution to **emerge**.

A suggested bottom up methodology, which meets the above requirements, is offered below. It is not perfect, but may aid investigators. It consists of ten stages, they are:

1. Establish initial elements to be investigated.

   This stage is similar to the process of producing a Rich Picture in the Soft Systems Methodology (Checkland and Scholes, 1991; Checkland, 1981; Wilson, 1990). As many views as possible are obtained to find the elements which are to be considered initially. This process also determines the context of those elements. This is the starting point of the problem solving and, as the whole series of stages are iterative, other elements may be added later.

2. Establish internal attributes and states of elements.

   The individual characteristics of each element are determined to provide the limitations of their behaviour. If an element can have more than one state, the prevailing state should be recorded.

3. Evaluate needs (environment) of the elements.

   The general needs of each element are determined, as well as the different needs for each state. These general needs can vary from temperature to psychological condition. A general need is anything that will affect the state or existence of a system element.

4. Establish effects of other elements on each element.

   This is similar to the previous stage, except it is the effects of the other overtly stated elements on other elements (in terms of their state, or 'health') that is determined.

5. Determine how elements communicate.

   This stage determines the information needs of each element, and the information supplied by each element. The media for information transmission is determined, as well as the data format needed.

6. By observation, establish emergent properties of combinations of system elements.
   From the information gathered in the previous stages, the behaviours of groupings of elements in different states (and the environment needed) are explored. Any control mechanisms should be noted.
7. Set up feasible element combinations, and required environmental factors for desired emergent properties.
   At this stage, the desired behaviours are determined, and the elements and their required needs (including control systems) are set up to produce those outputs.
8. Set up contingencies for detrimental emergent properties of potential element combinations.
   Any undesired outputs from the element combinations set up in stage 8 are noted, and their probability and impact resolved. Contingency plans are drawn up to cope with them. This stage might involve the inclusion of more system elements. Hence, the investigator would to return to stage 1.
9. Determine if any other elements are implemented in your problem space. If there are, repeat stages with new elements. (These tasks should be considered at all previous stages, and if changes are made, process should return to stage 1).
   This stage decides if the boundary, which has been created by the previous stages, is suitable for the problem at hand. This is resolved by reflective pragmatism. This stage is important, as there can be a danger of constant iteration through the previous stages until the boundary becomes so wide it is unmanageable in the actual problem context.
10. When designed system in place, monitor any emergent properties and repeat when necessary. Iteration is important.
    This stage is performed after the implementation of a system. It reflects the iterative nature of problem solving, the dynamism of the real world, and the incomplete knowledge of all practitioners.

There are a number of assumptions made in the above. The first is that it is possible to isolate system elements, and determine their characteristics. Of course, this positivist approach should be tempered with the knowledge that complex interactions may not be predictable. Also, it does not overcome the problem of who determines what are **desirable** emergent properties. Therefore, stage 7 is problematic, and produces the same problem associated with top down approaches. That is, **who** decides **what** is desirable. (Added to that, is who decides **who is to decide**, *ad infinitum*). This problem will not be covered further in this paper, except to say that management decisions are made in a context of a power structure. Regardless of the overtly moral and ethical issues, it is this power structure which determines the outcome. The power structure might include pressure groups, physical laws, the media, or financial institutions. Whatever the powerful elements are, the real outcomes are resolved in this context.

The objective of this methodology is to start with the elements and proceed to build the system with no preconceived ideas of how the system elements will react. No emergent properties are assumed, and no boundary drawn. There is a paradox here. It can be argued that stage 1 automatically draws a boundary by the inclusion or exclusion of elements. However, these elements are intended as a **starting** point, not an end point. In real world problem solving, a decision has to be made. The intention here is to use these elements to build a system, and to add more as appropriate. Nevertheless, it has to be admitted that the choice of elements will be profoundly affected by the worldviews of the participants. Therefore, as in many system methodologies, it is extremely important to be as inclusive as possible with the choice of participants (see Hutchinson, 1998).

Stage 9 is an important one, and demonstrates the iterative nature of the methodology. Whilst it is place as the penultimate stage, it can be executed at any point during the methodology.

CONCLUSION

Views of a problem from a top down perspective are always incomplete. Making sense of the world inevitably means generalisation, and the exclusion of elements considered irrelevant to the 'big picture'. Whilst this approach has been successful in tightly coupled organisational structures, their limitations are exposed in loosely coupled, networked relationships. The Bottom up view is an aid in overcoming these limitations. In fact, it is valuable in providing a different perspective even in highly structured systems.

The bottom up design approach implies that control mechanisms are at the lowest levels, and responsibilities for actions are at the system element level. This is contrary to conventional management practice, where responsibilities for actions and consequences tend to be placed at a high level. For example, investigating a system failure in bottom up mode would look for causes[1] at the element level, or more precisely, at their interactions. The top down mode would tend to look at the overall system, and would put 'blame' on the system itself and how it functions, rather than emergent properties of system element interactions.

Finally, the methodology offered for consideration in this paper is an attempt to formalise an approach using this viewpoint. Constructive criticisms are invited about it.

REFERENCES
Beer, S. "Diagnosing the System for Organisations". John Wiley, Chichester. (1985).
Bertlanffy, L.von. An Outline of General System Theory. *British Journal for the Philosophy of Science*, vol.1, p.134-165. (1951).
Bossomer, T., Green, D. "Patterns in the Sand: Computers, Complexity and Life". Allen & Unwin: Sydney. (1998).
Capra, F. "The Web of Life". Harper Collins: London. (1996).
Checkland, P. "Systems Thinking, Systems Practice". John Wiley: Chichester. (1981).
Checkland, P.B., Scholes, J. "Soft Systems Methodology in Action". John Wiley: Chichester. (1991).
Deacon, R. "The Silent War". Collins Publishing: London. (1988).
Flood, R.L., Carson, E.R. "Dealing with Complexity – second edition". Plenum Press: New York. (1993).
Hutchinson, W. Scoping: designing information systems to meet people's needs, in: Matching Technology to Organisational Needs" D. Avison, D.Edgar Nevill. McGraw Hill: UK. (1998).
Kam, E. "Surprise Attack". Harvard University Press: Cambridge, Mass. (1988).
Leveson, N.G. "Safeware". Addison-Wesley: New York. (1995).
Tsoukas, H. Organizations as Soap Bubbles: An Evolutionary Perspective on Organization Design. *Sys. Pract.*, 6, 5, 501-515. (1993).
Wilson, B. "Systems: Concepts, Methodologies and applications 2nd edition". John Wiley: New York. (1990),

---

[1] It is accepted that it is simplistic to try to isolate a 'cause'. As Leveson (1995, p.43-51) illustrates, there can be a multitude of causes to any system failure. The particular set chosen will be dependent on the viewpoint of the investigator, and the context chosen. From the bottom-up perspective, it is preferable to investigate **outcomes** of element interactions rather than **causes** *per se*.

A SYSTEMS THINKING MODEL FOR ECONOMIC MANAGEMENT - THE CASE OF TANZANIA

Diodorus Kamala

Department of Administrative Studies
IDM Mzumbe
P.O.BOX 50
Mzumbe
Tanzania

INTRODUCTION

This paper presents a Systems Thinking Model for Economic Management (STMEM). The author argues that in the next millennium developing countries should adopt systems thinking ideas in the process of managing their economies so as to realise the benefits of globalisation. As the World economies converge towards a single market, developing countries must restructure their economies and create viable economies in order to withstand dynamic and emerging complexities.

STMEM is applied to analyse Tanzania's economy and used to promote a viable economy which will enable Tanzania and other developing countries to benefit from the globalisation process.

SYSTEMS THINKING

According to Checkland (1981) "systems thinking is the way of thinking about the World which takes into account the four pillars, namely emergency, hierarchy, control and communication". Emergent property is the behaviour of the system which emerges after the interaction of elements within a system. Hierarchy implies different levels of subsystems within a system and communication and control maintain the interrelationships of elements within a system.

In terms of the economy, we have variables such as full employment, economic growth, stable prices and balance of payment equilibrium to mention a few. These variables when combined produce a given type of the economy with different

Synergy Matters: Working with Systems in the 21st *Century*,
Edited by Castell *et al.*, Kluwer Academic / Plenum Publishers, New York, 1999.

451

characteristics. However, neo-classical economists neglect the impact of institutions to the modern market. Hodgson (1988:176) points out that "The mistakes of neo-classical economics are to conceive of rationality as unbounded in its scope, to assume the rationality is straight forward maximisation, and to assume that all action results from reason and calculation."

Systems Perspectives of Globalisation Process

OECD (1994:222) suggests that "In effect, over the past twenty years, we have lived through a fundamental transformation in the nature on international economy. We have seen the beginning of change from exchanges among set of inter-linked national economies to exchange within an integrated global economy" This process of globalisation has resulted in the emergence of new world economy.

Helleiner (1990:32) argues that "And, politically, we must condition ourselves to think more globally, both in terms of longer time horizons and in terms of fields wider of vision." The key issue to understand is that there is a new global economy emerging. STMEM has been designed to enhance the ability to think globally about emerging economies.

SYSTEMS THINKING MODEL FOR ECONOMIC MANAGEMENT (STMEM)

This section presents STMEM. The model employs the concept of the philosophy which is attributed to the emergent behaviour of global economy.

The Philosophy of the Model

- Communication and Control One is the determinant of the relationships between households and firms which is a price[1] of both factors of production and prices of goods and services.
- Communication and Control Two is monetary and fiscal policies that deals with the control of the economy within the economy at macro level.
- Communication and Control Three is the political power exercised by international organisations, economic blocks and multinational institutions which influence the World economy.
- In this model there are levels of hierarchies namely, International, National and Firms and Households that reflect the status of holistic economy at any level.
- All variables within the economy cannot be held constant and instead must be systemically studied over time.
- The interrelationships of macroeconomic variables are more important than the behaviour of a single variable within the economy.
- National economy is imaginary, what really exists is global economy.
- The model employs three types of economic analysis, micro, macro and international.

---

[1]The prices reflects not only the forces of supply and demand but also institutional forces arising from values, taboos, cultural activities, customary laws and many others.

Having established the philosophy of the model, it is now presented diagramatically below.

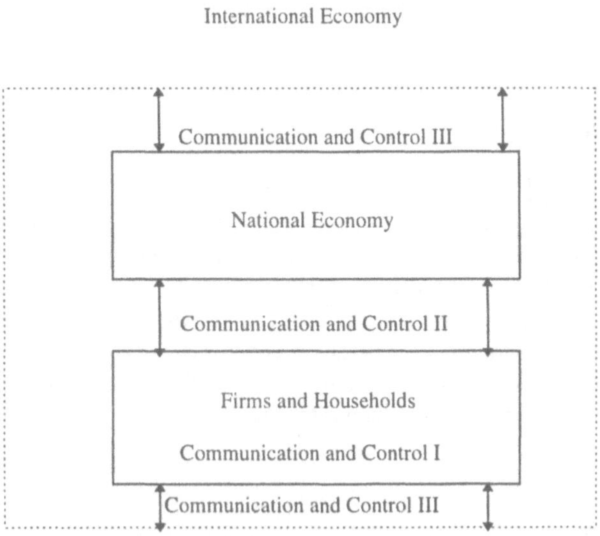

International Economy

Figure 1: Systems Thinking Model For Economic Management.

THE MODEL IN ACTION: THE CASE OF TANZANIA

At the beginning of 1980's Tanzania faced economic crisis. According to Havnevik et al (1988:145) "In 1973, the volume of export started to decline and from 1978 onwards Tanzania experienced a rapid deterioration in economic growth." Such issues have been well documented by writers such as Hyden (1980), Ellis (1982), Yeager (1982), Kahama (1997) and Havnevik (1993). STMEM is applied to the analysis and understanding of the Tanzanian economy in the following paragraphs.

Microeconomic Analysis

At this level we investigate the relationships of firms and households with the aid of Communication and Control I which is the price of factors of production and exchange of goods and services. At the same time we assess the impact of Communication and Control I to the households, firms, national economy and international economy.

In the year 1967 the ruling party at that time, TANU[2] issued the Arusha declaration. The declaration emphasised the philosophy of socialism and self reliance. According to Hodd (1988:52-55), the Arusha declaration committed Tanzania to a socialist development path. Inter alia, this shift in policy led to the formation of the Price Commission. Kahama et al (1986:113) argue that "In 1973 a National Price Commission (NPC) was established to control prices at several levels." This resulted to the disturbance of Communication and Control I and undermined the impact of interrelationships of households and firms.

---

[2]Tanganyika Africa National Union.

Nyerere (1969:154) had earlier suggested that "If we simply lay down hard and fixed rules for everything, we may finish up with the farmer being unable to buy things he wants at his convenient" Despite Nyerere's assertion, his government enacted the NPC which created problems in the economy. For example in Kagera Region, peasants decided to sell their cash crops to Uganda in return of better prices. Furthermore, most of the peasants abandoned coffee production at the expense of other crops which they could exchange among themselves.

Macroeconomic Analysis

At this level we assess the interrelationships of macroeconomic variables. Communication and Control II is a key tool in this analysis by consciously tracing the impact of monetary and fiscal policies to the national economy.

In 1980's Tanzania continued to pursue anti-market policies. The exchange rate was determined by the government through Central Bank of Tanzania. Free Money and Labour markets were not in the Tanzanian economic vocabulary. Tanzania adopted the policy of protectionism. The argument of protectionism aimed at protecting infant industries which surprisingly remained infant forever.

International Analysis

At this level, we consider Communication and Control III. Communication and Control III embraces political power and economic power exercised by international organisations, economic blocks and multinational institutions that influence the nation's economy. Attention is drawn to how these forces influence micro and macro levels of the economy.

In 1980's IMF[3] officials advised Tanzania to restructure her economy by devaluing her currency. However, Nyerere's philosophy perceived the devaluation of the Tanzania currency as a loss of sovereignty. Indeed, Tanzania underestimated the impact of the world economic and political power and it was a decade later when Tanzania initiated the process of restructuring the economy.

Understanding the world economy and holistic interpretation of international relationships and its impact to the economy are key issues which policy makers must always seek to understand. STMEM will enable developing countries to build a viable economy.

VIABLE ECONOMY FOR TANZANIA IN THE NEXT MILLENNIUM

A viable economy[4] is an economy that is competitive, flexible and adaptable. It is strongly recommended that Tanzania should build a viable economy. A viable economy will enable Tanzania and other developing countries to turn the challenges of globalisation into opportunities.

Conceptual Meaning of A Viable Economy

The creation of a viable economy is necessary for developing countries to benefit from the globalisation process. Five systems are suggested for a viable economy. The

---

[3]IMF stands for International Monetary Funds

[4]The concept of a viable system is well presented by Beer, 1979, 1981, 1984 and 1985.

systems are implementation, co-ordination, government, think tank and policy formulator.

The implementation system stands for the production of goods and services. It comprises households, firms, local governments, parastatals and global firms. Global firms, are likely to provide benefits to other subsystems in the implementation system. The benefits include new technology, increased social capital, creation of competent civil society and good governance.

Co-ordination and government systems provide guidance to the implementation system. Communication and control loops enhance the government's ability to manage and control the economy.

A think tank system need to be developed. The function of a think tank is to undertake research and develop common purpose for adopting the benefits of the globalisation process.

The policy formulator system deals with policy formulation, interpretation of global economy and directing thinkers and governments to perform better in the process of managing the economy. Given the benefits of good governance, the presence of civil society and social capital will increase policy makers accountability.

CONCLUSION

STMEM is presented here as an effective way forward for understanding and managing the economy in the forthcoming millennium. Viable economies will enable developing countries to benefit from the process of globalisation. Indeed, the possession of viable economies will enable developing countries not only to benefit from globalisation but also to withstand the emerging complex and turbulent world economic system.

ACKNOWLEDGEMENTS

I am extending my sincere thanks to Professor Michael Jackson for his advice and comments on my first draft.

REFERENCES

Beer, S. (1979) *The Heart of Enterprises*, New York, Wiley.

Beer, S. (1981) *Brain of the Firm, 2nd edition*, New York, Wiley.

Beer, S. (1984) "The Viable System Model: Its Provenance, Development, Methodology and Pathology." Journal of Operations Research Society, 35, pp 7-25.

Beer, S. (1985). *Diagnosing The System For Organisations*, New York Wiley.

Checkland, P. (1981) *Systems Thinking, Systems Practice*, First Edition, Wiley and Sons, London.

Ellis, F. (1982) *Agricultural Price Policy In Tanzania*, World Development Report.

Havnevik, J. K., Kjaerby, F., Meena R., Skarsten R. and Vuorella.U. (1988), *Tanzania, Country Study and Norwegian Aid Review*, Centre For Development Studies, University of Bergen.

Havnevik,J.K. (1993) *Tanzania, The Limits of Development From Above*, Nordiska of Africa Institute, and Mkuki na Nyota Publishers, Sweden and Tanzania.

Helleiner,G.K.(1990) *The New Global Economy and Developing Countries : Essays in International Economics and Development,* Edward Elgar Publishing Ltd, England & USA.

Hodgson, G. (1988) *Economics and Institutions,* Polity Press, United Kingdom.

Hodd, M. (1988) *Tanzania After Nyerere*, Printer Publishers Ltd, United Kingdom.

Hyden, G. (1980) *Behond Ujamaa in Tanzania Underdevelopment and Uncaptured Peasantry*, Berkeley, University of California Press.

Kahama, C.G., Maliyamkono,T.L., and Wells,.S. (1986) *The Challenge For Tanzania's Economy*,Tanzania Publishing House, Tanzania.

Kahama, C.G.(1997) *Tanzania Into The 21st Century*, TEMA Publishers Company Ltd, Tanzania

Nyerere, K.J. (1968) *Ujamaa, Essays On Socialism*, Oxford University Press, London.

OECD (1994) Globalisation : *What Challenges and Opportunities For Governments ?*, Working Paper Volume 28, Number IV, OECD, Paris.

OECD (1995) *New Dimensions of Market Access in A Globalising World Economy*, OECD Documents, OECD, Paris.

Yeager, R. (1982) *Tanzania: An African Experiment*, Gower Publishing Company Ltd, England.

MEASURES OF PERFORMANCE: THE THREE Es OF SOFT SYSTEMS METHODOLOGY AND THE INDICES OF PERFORMANCE OF THE VIABLE SYSTEM MODEL.

Alberto Paucar-Caceres

Senior Lecturer
Department of Business Studies,
Manchester Metropolitan University,
Aytoun Building, Aytoun Street,
Manchester, M1 3GH, England

E-MAIL: a.paucar@mmu.ac.uk

INTRODUCTION

This paper explores the concepts of Efficiency and Effectiveness as managerial measures of a system's performance. Problems involving the measuring the performance of a social system are outlined and two approaches are considered and discussed: (a) Checkland's systems ideas of 'managing and controlling' a system throughout a set of three Measures of Performance: Efficacy, Efficiency and Effectiveness; and (b) Beer's claims that the performance of a system has to be quantifiable and resumed on 'pure' numbers which should reflect the survivability of the firm. A parallel is drawn between the two approaches concluding that although the paradigms underpinning them are in some way different, the practicalities of the these approaches to control, measure and improve the performance of a system are very similar. A case involving the provision and improvement of a Business Studies course is used to illustrate the approaches.

MEASURING THE PERFORMANCE OF A SYSTEM

The issue of measuring the Performance of a System's behaviour is linked to the problem of improving systems. A cursory revision of the Systems literature reveals that the problems on these issues have only been studied by few people; to my present knowledge

Synergy Matters: Working with Systems in the 21st Century,
Edited by Castell *et al.*, Kluwer Academic / Plenum Publishers, New York, 1999.

457

only Checkland, Beer and Ackoff have studied this theme and elaborated some theory around it. The aim of this paper is to describe Checkland's and Beer's view with the objective of drawing a comparison.

SOFT SYSTEMS METHODOLOGY: MANAGING AND CONTROLLING THE TRANSFORMATION PROCESS.

The issue of measuring the performance of a systems is first mentioned by Checkland with relation to the attributes of the Formal System Model (Checkland, 1981). The Formal System Model, 'S' is described as a devise that can be used as a way of validating the model of the system currently being studied: "S is a formal construct aimed at helping the building of conceptual models which are themselves formal" (Checkland 1981, pp. 174). Checkland lists a number of conditions for 'S' to be formal, one of these conditions is that 'S' should be measured; 'S' is said to have a measure of performance capable "to signal progress or regress in pursuing purposes or trying to achieve objectives" (Checkland 1981, pp 174). Later, Checkland accepts the fact that the use of the Formal System Model as a validation devise has been almost abandoned; in the late 80s and 90s most of the SSM practitioners seem to prefer to use the CATWOE analysis when validating a conceptual model (Checkland, 1990). CATWOE is the mnemonic of the six crucial characteristics which should be included in a well-formulated root definition. **T**, the transformation process, is one of these characteristic and is the element on which Checkland elaborates the notion of measuring the performance of the system. According to Checkland, at the most fundamental level, any purposeful activity may be expressed through a Transformation Process which "changes or transforms some input into some output". In other words, for the Transformation to be relevant, Inputs are present in Outputs but in changed state. Then if the Input is abstract (e.g. 'need for nursing services') then the Output must also be abstract (e.g. 'need met'). If the Input is concrete (e.g. 'a patient') then the Output must be concrete (e.g. 'a treated patient'). This distinction is important because it helps to differentiate between the resources and the inputs of the system. SSM also stresses the fact that there are many ways of expressing a purposeful activity: more ways of expressing the activity in terms of Input-Transformation-Output will enrich the thinking.

THE THREE Es OF SOFT SYSTEMS METHODOLOGY: EFFICACY, EFFICIENCY AND EFFECTIVENESS.

According to SSM, when we try to 'manage' purposeful 'systems', it is useful to think of this situation in terms of:
(a) a purposeful system arranged as a set of activities which we may call the "operational system" (a set of linked activities to do 'x');
(b) a set of activities which will inspect the performance of the operational system an eventually will take action to bring it into line with aims, and expectations; this is the "monitoring and control" system which monitors and controls the doing of 'x' and shown by the **inner** subsystem in Figure 1.
(c) the system can be thought as part of a wider system which decides to do 'x' (the 'what') or decide the way (the 'how') in which 'x' is carried out; these decisions are carried out by its own "monitoring and control" system which monitors and controls the long term objective of the system located on an upper level. This is shown by the **outer** system of Figure 1. The criteria by which the Transformation can be judged gives the elements by which we can measure the performance of the system. If we think of the two levels

expressed in Fig. 1., we should ask the question: **How can the Transformation fail?**. For controlling purposes and ultimately for 'managing' this activity, the following reflections and possible answers are useful.

(a) The way chosen to do **T** might not work, therefore, we manage **T** by asking: Does the means selected work? The answer measures the *Efficacy* of **T**, measured by the monitoring and controlling activities at the 'operational system' level.

(b) **T** might not be being done with minimum resources (including time as a resource). We manage **T** by asking: Is **T** being done with minimum resources?. The answer measures the *Efficiency* of **T**, measured by the monitoring and controlling activities at the 'operational system' level.

(c) **T** could be the wrong activity to be doing. We manage **T** by asking: Is **T** the right thing to be doing? The answer measures the *Effectiveness* of the System, measured by the monitoring and controlling activities at the 'planning system' level.

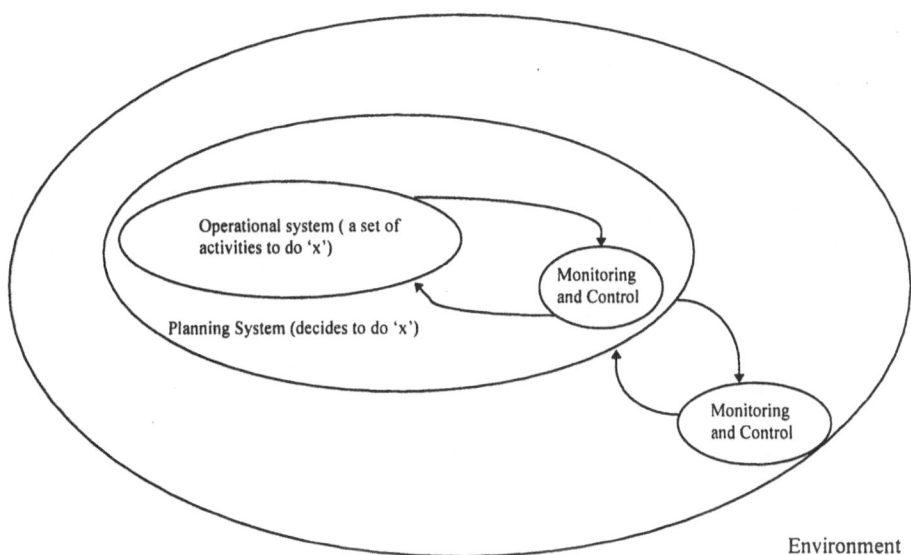

Figure 1. Monitoring and Controlling a System (From Checkland, 1986)

THE VIABLE SYSTEM MODEL: MEASURING INDICES OF PERFORMANCE.

Measures of Performance in Beer's account are part of the Viable System Model (VSM) epistemology and the requirement of its survivability in the long term. Beer argues that the dynamic structure of VSM will depend on the quantification of its performance. Information flowing throughout the different channels (The Formal Model has five Systems) will show how the organisations as a whole is performing in relation to its aims and objectives. Beer argues that in most organisations this is measured by the cost-accounting function and dismisses this profit maximisation and cost minimisation criteria because it does not take into account the long term view of the firm for survival (to show that the organisation has potential for survival is one of the main aims of the Viable System Model); instead Beer proposes to see the organisation performance in terms of three levels of achievement that he calls: *Actuality, Capability and Potentiality*. A combination of these three levels produces three indices: *Productivity, Latency and Performance*; these indices expressed in the form of 'pure' numbers can be used as measures of Performance throughout the firm (Beer, 1981, pp 162-166).

Levels of Achievement: Beer defines the following terms as useful concepts to understand the level of achievement. **(a) Actuality**: This is what we are managing to do now, with existing resources, under existing constraints; **(b) Capability**: This is what we could be doing (still right now) with existing resources, under existing constraints, if we really worked at it; and **(c) Potentiality**: This is what we ought to be doing by developing our resources and removing our constraints, although still operating within bounds of what is already known to be feasible.

From these concepts, three kinds of plans can be developed that can use these three levels of achievement: **(1) Programming**:- This is planning on the basis of actuality: Tactical or operational level; **(2) Planning by objectives**:- This is done when new objectives are set up and the firm tries to achieve them: Strategic Planning; and **(3) Normative Planning**:- This is the type of planning which sets potentiality as its target.

According to Beer, in planning what really matters is to relate *capability* and *potentiality* to whatever may be actual (*actuality*). So, these three indices are useful in relating the three levels of achievement:

(a) Productivity: ratio between *actuality* and *capability*;

(b) Latency: ratio between *capability* and *potentiality;*

(c) Performance: both ratio between *actuality* and *potentiality* and also product of *latency* and *productivity. F*igure 2. Relates the ideas behind Beer's Measures of achievement.

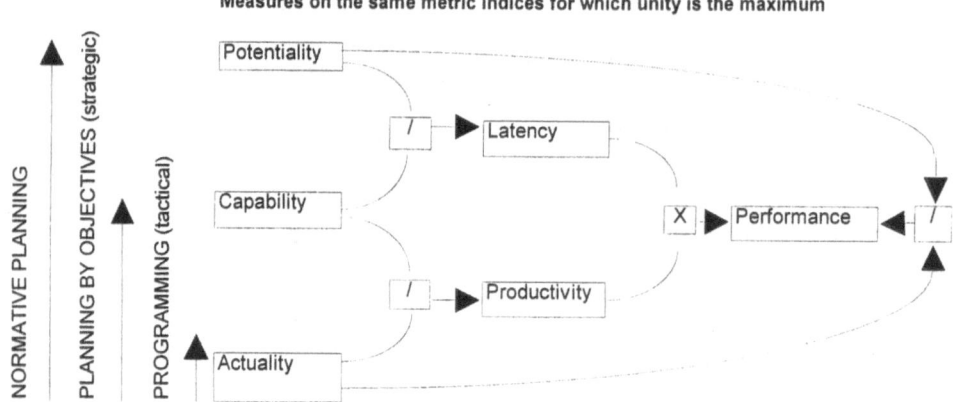

Figure 2. Three measures of capacity generating three measures of achievement. From Beer, 1981.

MEASURING THE PERFORMANCE OF A SYSTEM: COMPARING SSM AND VSM.

The three Es of SSM: Efficacy, Efficiency and Effectiveness can be related to the three different levels of Achievement of the VSM: Actuality, Productivity and Potentiality respectively. Table 1 shows a comparison of the approaches to the issue of measuring the performance of a system at both operational and planning level.

MEASURING THE PERFORMANCE OF A SYSTEM: THE FOREIGN EXCHANGE ELEMENT OF A BUSINESS STUDIES COURSE.

The assessment of a course is here used as a case-study to illustrate and compare the two approaches. The Business Studies Department at Manchester Metropolitan University runs a BA/BSc Business sandwich degree. This course has a Foreign Exchange (FX) element. Students can choose to go to European Universities during the second semester in their second year. Over the last five years the number of students taking this option has declined dramatically and the general view of the Department is that this issue needs to be re-examined with view of either re-structuring or dropping it completely.

SSM AND THE TRANSFORMATION PROCESS

The concept and the different views around the FX issue were described using Checkland's Transformation Process: The following is a partial list of how FX is viewed by different members of the Department:

- A way of exercising undergraduate English students business skills in different managerial cultures.
- A way of exercising and improving language skills.
- A reason for having a nice holiday in the middle of the course.
- A way of attracting potential students by offering a FX element in the Business Studies Degree.

If we use the Checkland's model on the first of the above views, we can apply the following control and monitoring concepts to this situation:

Input		*Output*
Students business skills	**T**	Students business skills having been exercised.

To manage successfully this process we should ask the following questions:
- Does the means selected to do **T** work? In other words does the FX allow students to exercise their skills when in Europe?- This answer will be a measure of **Efficacy** of **T**.
- Is **T** being done with minimum resources (including time)?- This is relatively easy to be measured (lecturers and administrative staff time involved, travelling cost for supervision, etc.). This answer measures the **Efficiency** of **T**.
- Is **T** the right thing to be doing? Here we need to question the FX element itself. This question can be asked only if we move to the upper level that is to the planning system which may decide to 'exercise students business skills' by other means: i.e. sending students to other non-European, English-speaking countries. The answer will be a measure of the **Effectiveness** of the **system**.

VSM AND THE LEVELS OF ACHIEVEMENTS OF THE SYSTEM

In terms of the VSM, the three levels of achievement can be defined as follows:
- Actuality; This is where we are now. FX is declining in numbers and students are not motivated and perceive FX as an interruption of their four years study. Focusing on Actuality will lead to question why the means to implement FX are not working properly (Efficacy).
- Capability: What we could be doing. The department has built some strengths in this area and it is 'capable' of structuring and running a FX element that can be better perceived by students and staff. Focusing on Capability will lead to improvements on Productivity of the System (Efficiency).
- Potentiality: What we ought be doing. Here the Department needs to question the FX element and ask if their commitment should be in this area; this is related to the long term survival of the system. Focusing on Potentiality will address the Effectiveness of the system.

461

Using these definitions we can derive the following measures of achievement:

(a) Productivity: actuality/capability. The department can 'quantify' its productivity. The number of students sent abroad and potential numbers can be gathered;

(b) Latency: capability/potentiality.;

(c) Performance: either Latency times Productivity or Potentiality divided by Actuality. It can be seen here that in order to have useful measures of the system's performance, the VSM of Beer relies on obtaining numerical values for the above levels of achievement.

Table 1. Measuring the Performance of a System: Comparing SSM and VSM.

THE THREE 'Es' OF SOFT SYSTEMS METHODOLOGY-Checkland	MEASURES OF ACHIEVEMENT OF THE VIABLE SYSTEM MODEL-Beer
Efficacy of the Transformation Process. Measured by the monitoring and control activities of the **operational** system (the doing of 'x') **Question**: Does the means to achieve T work?	**Actuality** of the firm (What are we doing now) Planning on the basis of Actuality: **Programming**
Efficiency of the Transformation Process. Measured by the monitoring and control activities of the **operational** level (the doing of 'x') **Question**: Is T being achieved by minimum resources?	**Productivity:** What we are doing now/ what we could be doing. : actuality/capability Planning on the basis of Capability: **Planning** by objectives.
Effectiveness of the System. Measured by the monitoring and control activities of the **Planning** System (the systems that decides to do 'x' instead of 'y') **Question**: Is T the right thing to do?	**Potentiality** of the firm (what we ought to be doing) Measured by Performance of the firm: Actuality/Potentiality Planning on the basis of Potentiality: Normative **Planning.**

CONCLUSIONS

This paper has discussed the problems of measuring the performance of a system. The Three Es of SSM have been compared with the levels of achievement of Beer's VSM. There seems to be a correspondence between the two first Es of SSM (Efficacy and Efficiency) and the Actuality and Productivity levels proposed by Beer. The third E of SSM, Effectiveness (is 'x' the right thing to be doing?) corresponds to the Beer's level of Potentiality (what we ought to be doing). The case of improving the Foreign Exchange element on a Business Studies course has been used to illustrate the two approaches. A general conclusion is that SSM is more easy to apply in this case mainly because numerical data seems not to be available; VSM levels of achievement will give more insight if factual data is available. Both approaches are useful to shed light into situations whose performance needs to be measured. More conceptual work is needed in this area and perhaps reference to the work of Ackoff would be productive (Ackoff, 1995). Also, further research on the application of these systems concepts are necessary with the view of comparing the use and practicalities of them.

REFERENCES

Ackoff, R., *'Whole-Ing' The Parts and Righting the Wrongs*, Systems Research, Vol. 12, N. 1, 1995.
Checkland, P., *Systems Thinking, Systems Practice*, Wiley, 1981.
Checkland and Scholes, *Soft systems in Action*, Wiley, 1990.
Checkland, *Some Basic ideas of Monitoring and Control for Managers*, Lancaster University, 1986.
Beer, S., *The Heart of the Enterprise*, Wiley, 1981.

NEW LIGHT THROUGH OLD WINDOWS - BUT IS IT THE RIGHT KIND OF LIGHT?

Geoff Peters and Joyce Fortune

Centre for Complexity and Change
The Open University
Walton Hall
Milton Keynes
MK7 6AA

INTRODUCTION

This paper can best be described as part of an ongoing reflection by the authors of their work to develop systemic methods for the identification, analysis and prevention of failures. In part it is a progress report, but more importantly, it is a vehicle for exploring allied thinking and drawing in further stimuli to support the enhancement and improvement of the approach.

In an earlier paper (Peters and Fortune 1992) the authors considered work that had been conducted over a twenty year period and traced the progress of the application of systems-related ideas to the analysis and modelling of failures. Since then they have worked with others to develop the Systems Failures Method further and by practical application extended its use to allow the exploration of the potential for failure to be incorporated into the design phase of new ventures. A major consolidation of this work (Fortune and Peters 1995) has signalled the direction of future developments and triggered a wider critique of candidate topics for incorporation into the Method. As one commentator on the work has observed, "The idea that methodology must change as a result of its application and reflection on its use is fundamental to the status of systems thinking and practice as a respectable academic discipline" (Mansell, 1996). This paper first considers some of the suggestions for consideration as candidates for incorporation into the Method and then examines some of the underlying issues which these deliberations raise.

CONTEXT

At the heart of the Systems Failures Method is the preparation of a description of the context which has given rise to a failure or potential failure in a format which

Synergy Matters: Working with Systems in the 21st Century,
Edited by Castell *et al.*, Kluwer Academic / Plenum Publishers, New York, 1999.

463

allows it to be tested against a model of a robust system that is capable of purposeful activity without failure. Comparison with the Formal System Model yields both an agenda for further investigation of the situation and selection of other relevant and more specifically -targeted abstract models. Use of these secondary models can in turn further enhance the insights gained form comparison with the primary Formal System Model.

The Method has been designed to evolve with use and experience. The intention being to achieve what Burgess (1997) has described as "a reasonably loose framework that provides opportunity for the deployment of other systems-related analytical methods". The objective of improving the Method's effectiveness by increasing its breadth and scope can be achieved in a number of ways. The most straightforward is the addition of further secondary models for comparison which can in turn provide new insights and an enhanced systemic description. Alternatively, the idealised system model could be reformed or replaced, with for example, the Viable Systems Model (Beer, 1979) being used instead. A more far reaching reconstruction of the format, ordering and techniques of the method, as was reported in Fortune and Peters (1995), could even be carried out to make it a part of a wider, more general, systems approach.

CANDIDATE FOR INCLUSION 1 - CULTURE

When a specific accident or failure is examined there are usually firmly held views as to its causes although the same views may not be shared by all the examiners. Sometimes the views may be at a level of generality which is not particularly illuminating, and not even very helpful (for example, "human error", "incompetence" or "Act of God"); at others they may be so specific as to have little bearing upon cases other than the one being considered on that occasion. However, sometimes new concepts emerge from individual systemic analyses which can be tested and reapplied elsewhere. One such example is the notion of vulnerability (Horlick Jones, Peters and Fortune 1991).

When the Method is reviewed by colleagues or when systems practitioners apply it, a wealth of personal experience and knowledge of other systemic approaches and other disciplines is often brought into play to test its validity. Sometimes this leads to suggestions of concepts, techniques or models as candidates for future incorporation into the Method. One of the recurring themes in many such suggestions is the inclusion of additional mechanisms for applying concepts such as culture, cultural perspective, conflict and metaphors of conflict. (See, for example, Mansell, 1996).

The case for including culture seems at first sight to be a powerful one. The cultures of the various organisations associated with the failure and the cultural background of the individuals and society(ies) in which it occurs would both seem relevant factors. This view is supported by other work in allied areas. Holmes and Poulymenakou (1995), for example, in their studies of the failure of information systems argue that although such failures show some generic attributes, failure of an information system is treated as a situationally (sic) sensitive issue contingent upon the organisation, its employees and culture as well as the external markets and environments in which the organisation operates.

Others support the idea that organisations have cultures of their own and Schein (1985) has argued that every organisation is concerned about the extent to which people fit into it, with newcomers rapidly learning the culture or else running the risk of feeling alienation. He characterises the elements that influence the degree of internalisation of the culture using the following set of parameters:

- Commonality of language and conceptual categories;
- Group boundaries and criteria for inclusion and exclusion;
- Power and status both structure and rules for change;
- Rules for intimacy, friendship, and love;
- Rewards and punishments appropriate to particular forms of behaviour;
- Ideology and "religion" which includes managing the unmanageable and unexplainable.

Although the terms which Schein uses are very general it would seem reasonable to assume that within his framework some organisations could have a "culture" which made it more able to respond to or anticipate potential failures whilst other organisations might be disadvantaged in this regard.

Indeed, later, Schein (1993) advocates a learning culture for organisations and describes culture as being about shared mental models, shared ways of perceiving the world, and common mental categories for sorting it out. He sees culture as being about shared tacit ways of being, "it reflects the deeper and more pervasive elements of our group life, and it operates out of our awareness so we are often quite ignorant of the degree to which our culture influences us ...." Schein also makes a strong link between the concept of culture and a willingness to undertake what is variously known as "generative learning" (Senge, 1990), "double loop learning" (Argyris and Schoen, 1985) or "learning how to learn" (Michael, 1985 and Bateson, 1993).

Schein then seems to make a very clear link between the internal culture and the ability of an organisation to learn, and he sees this learning as systemic in nature. However, although this linkage strengthens the case for including the consideration of culture in the exploration of particular failures there are some potential traps for the analyst. First Schein in effect argues that there are cultures that support organizational learning and presumably make them more successful. Meanwhile Fortune and Peters (1995) propose a model of a learning organisation which is willing to learn from its mistakes and they advocate the adoption of the failures method as a means by which organisations will be better able to take advantage of the opportunities which failures present. It would not be a large step from these two positions to the argument that organisations which had a learning culture were ones which regularly applied the failures method or that evidence of the application of the failures method was an indicator of a learning culture.

A second reason for caution concerns the fact that culture can be characterised in a variety of ways. For example O'Mahony and Dampney (1997) have linked the culture of an organisation with its success and failure with Information Systems. They surveyed 100 Australian secondary schools to map them on to a dichotomy between bureaucratic and organic cultures, they then found that schools which showed tensions between bureaucratic and organic cultural forms displayed different levels of satisfaction with the IS function.

CANDIDATE FOR INCLUSION 2 - CONFLICT

A similar picture to that above emerges when the role of conflict is examined in relation to failures. Conflict within or between organisations may well be an important aspect of failure but like culture it is difficult to characterise. However, having said this, a lengthy investigation of the crisis which surrounded UNESCO between 1978 and 1987 has yielded empirical evidence of a link between a model of conflict and foreign policy behaviour which may have more general applicability.

The severe UNESCO crisis resulted in the unprecedented withdrawal of both the USA (in 1983) and the UK (in 1984) from a UN family organisation. It is now generally held that the crisis had three underlying factors: first, the assertion by developing countries, supported by the USSR and others, of the need to restructure the global flow of information away from what they characterised as a one way flow from North to South; secondly accusations of weak organisational efficiency and mismanagement which were focused on Director-General M'Bow; and finally disagreement about the extent to which UNESCO could be used for overtly political activity such as that seen in a number of anti-colonial, anti-racist and peace campaigns.

Kittel, Rittberger and Schimmelfennig (1995) reviewed the position of the different countries during the period from 1978 to 1987 and mapped how they changed. They then looked at these changes using a conflict model which was based upon a simple three factor typology of dissatisfaction with the performance of an organisation which Hirschman (1970) called exit, voice and loyalty, coupled with later work by Keohane (1984).

Their work used a five point scale which ranged from co-operation to conflict:

1 Loyalty - the adaptation of one's policy to the behaviour decided and demanded by the organisation;
2 Constructive Voice - ready to negotiate, to compromise and to make prior concessions unconditionally and independently of the other party's behaviour;
3 Conditional Voice - ready to negotiate or to make concessions dependent on the behaviour of the other party;
4 Confrontational Voice - not ready to negotiate or to compromise, instead ignores or rejects the offers of the other party;
5 Exit - from the organisation.

In the same way as Kittel et al, were able to map this model onto the behaviour of nations in the UNESCO crisis it may well be possible to use it to test for evidence of conflict and co-operation within and between organisations involved in other forms of failure or potential failure.

CRITERIA FOR INCLUSION

There are strong parallels between the examination of previous accidents, disasters and failures and the historical analysis and methods of historians. Indeed Trotsky (1930 has opined that "The entire historical process is a refraction of historical law through the accidental. In the language of biology we might say that the historical law is realised through the natural selection of accidents". The historian too is faced with the task of sifting through the evidence and piecing together an explanation. In both approaches the aim is to attribute causes. Carr (1961) quotes the views of both Herodotus who saw the purpose of history as being to preserve a memory of the deeds of the Greeks and barbarians "and in particular, beyond everything else, to give the cause of their fighting one another", and more recently, in the eighteenth century, Montesquieu in *Considerations of the causes of the Greatness of the Romans and of their Rise and Decline* wrote "there are general causes, moral or physical, which operate in every monarchy, raise it, maintain it, or overthrow it. All that occurs is subject to these causes".

However, identifying causes may not be sufficient even in history. At the time at which Carr was writing the notion of an accidental view of history led him to construct an illustration of a pedestrian being killed by an intoxicated driver on a tight bend when the pedestrian endeavoured to cross the road to buy cigarettes. Whilst it may well be the case that the pedestrian would not have suffered the accident if he or she had been a non-smoker, Carr argues this accidental cause (being a cigarette smoker who had run out of cigarettes) cannot be generalised whereas the link between alcohol and accidents can be. As in the application of the Systems Failures Method it is the hunt for explanations which are both adequate and significant which is needed.

Both historians and systemists are looking for explanations which are adequate, significant, are potentially applicable to other situations and so lead to fruitful generalisations. The criteria for modifying a systems method, particularly one which has as one of its objectives trying to understand aspects of the past so as to throw light on the future, will at least include the extent to which they support adequacy, significance and wider applicability. As for fruitful generalisations, systemists may strive for generalisation but few can expect to achieve many.

CONCLUSION

The process of reviewing the armoury of systems-related models that can be deployed within the Systems Failures Method amounts to a re-appraisal of the boundaries of the subject itself.

If the Method and its application is to be pragmatic, and issue related, then the inclusion, or not, of any particular concept or technique can simply be validated against its effectiveness in aiding understanding and, where appropriate, in facilitating action. However, since the Systems Failures Method is most usually applied retrospectively, such a scrutiny would seem to bring little which would distinguish a systems failures analysis from an historical one.

The authors' conclusion is that, since the application of the method does itself occur within a contemporary context in which other analysis and approaches are well known, it beholds the systems analyst to pay particular attention to incorporating those valuable approaches which are also implicitly systemic or which might otherwise be ignored because they are not currently fashionable. So if as Carr (op cit.) claims history is a process of selection in terms of historical significance then a systems failures analysis can involve selection in terms of systemic significance.

Nevertheless, the comparison with history is instructive in many ways, not least because it lends support to the case for examining those things which have not worked in the past alongside those which do work now. However, in both subjects the art is in being able to move beyond the explanation which is entirely bound in the particulars of the situation to one which can be expressed in a more general form. As Montesquieu said two hundred years ago: "If a particular cause like the accidental result of a battle has ruined a state there was a general cause which made the downfall of this state ensue from a single battle".

REFERENCES

Argyris, C. and Schoen, D., 1985, "Organizational Learning", Addison-Wesley, Reading, MA.

Bateson, G., 1993, "Towards an Ecology of the Mind", Palladin, St. Albans.

Beard, A.N., 1995, Book review, *Fire Safety*, 25:365-368.

Beer, S., 1979, "The Heart of the Enterprise", Wiley, Chichester.

Burgess, T.F., 1997, Book review, *J. of OR Soc*. 48.

Carr, E.H., 1961, "What is History?" Macmillan, London.

Fortune, J. and Peters, G. 1995, " Learning from Failure", Wiley, Chichester.

Hirschman, A.O., 1970, "Exit, Voice, and Loyalty. Responses to Decline in Firms, Organisations, and States", Harvard University Press, Cambridge, MA.

Holmes, A. and Poulymenakou, A.1995, Towards a Conceptual Framework For Investigating Failure, in Proceedings of the 3rd European Conference on Information Systems, Athens, Greece.

Horlick-Jones, T., Peters G., and Fortune J., 1991, Measuring Disaster Trends Part Two: Statistics and Underlying Processes, *Disaster Management*, 4:1:41-46.

Keohane, R. O. 1984, "After Hegemony. Co-operation and Discord in the World Political Economy", Princeton University Press, Princeton, N.J.

Kittel, G., Rittberger, V., and Schimmelfennig F. Tübingen 1995, Between Loyalty And Exit: Explaining the Foreign Policies of Industrialized Countries in the UNESCO Crisis (1978-87),Tübinger Arbeitspapiere zur internationalen politik und friedensforschung, Nr. 24, Tübingen, Germany, ISBN 3-927604-17-8.

Mansell, G.,1996, Book review, *Systems Research*, 501-904.

Michael, D. N.,1985, "On Learning to Plan and Planning to Learn", Jossey Bass, San Francisco.

Montesquieu, [1965], "Considerations of the Causes of the Greatness of the Romans and of their Rise and Decline", translated by David Lowenthal, Free Press, New York.

O'Mahony, C.D. and Dampney, C.N.G., 1997, Linking Organisation Culture to Strategy: A Model and Interim Results, paper presented to Pacific Asia Conference on Information Systems.

Peters, G. and Fortune, J., 1992, Systemic Methods for the Analysis of Failure, *Systems Practice*, 5:5:529-42.

Schein, E.H., 1985, "Organizational Culture and Leadership", Jossey Bass, San Francisco.

Schein, E.H.,1993, How Can Organisations Learn Faster? *Sloan Man Review,* 34:85-92.

Senge, P. M., 1990, "The Fifth Discipline", Doubleday Currency, New York.

Trotsky, L., 1930, "My Life", Charles Schribner & Sons, New York.

INTERSUBJECTIVITY IN INTERPRETIVE RESEARCH: THE DIALOGICAL CASE

Robert Stephens

Faculty of Computer Studies and Mathematics
University of the West of England
Bristol, BS16 1QY, UK

ABSTRACT

In this paper the concept of intersubjectivity as it is employed in Information Systems research is reviewed from the perspective of dialogism developed in the work of Mikhail Bakhtin. A form of communicative understanding is presupposed in most IS analytic activities; intersubjectivity is the general criteria for interpretive studies that are acclaimed by the attempt to understand phenomena according to the meanings that people assign to them. Typically, it is said to be achieved empathetically by the generation of shared interpretations of the actions, practices and objects within the cultural systems that give them significance. The quality of engagement between participants is increasingly regarded as a key element in successful IT development, for example Participant Design, Joint Application Development, cooperative and interpretive methodologies, all recommend that system development is enhanced by user participation, albeit in varying degrees of involvement, and authenticity of analytic categories is assured in common understanding.

In Bakhtin's dialogic concept, this engagement, though welcome, is not an option but an epistemological requirement. However, the convergence of understanding in shared alterity (otherness) is radically resisted, and there is an insistence that difference must always exist for meaning to have any authenticity. This implies a rejection of both the foundationalism of the Other's viewpoint, as well as the notion of synthesis found in many reports of intersubjectivity. Instead of privileging the understanding of one or another person, whereby one agent (the informant) possesses knowledge or truth and the other (the researcher) acquires it, a focus on the interaction between the two is recommended. Interpretivism and hermeneutics are understood as a response to the other, a 'giving-meaning-to' the alterity presented by an other culture or person's

Synergy Matters: Working with Systems in the 21st Century,
Edited by Castell *et al.*, Kluwer Academic / Plenum Publishers, New York, 1999.

469

understanding and action rather than an overcoming of the difference presented by the Other.

INTRODUCTION

In this paper the concept of intersubjectivity as it is employed in Information Systems research is reviewed from the perspective of dialogism developed in the work of Mikhail Bakhtin (1981). In a recent survey of information systems research, Klein (1996) sees researchers taking an increasing 'linguistic turn', largely inspired by social constructionist theory, hermeneutics and phenomenology. The consequence of this departure is to direct attention at social context, which could be IS modelling, planning, organizational action, etc., and how this is created and sustained through social interaction. A key challenge to the IS traditions of positivism and empiricism is the dismissal of objective language, and a willingness to accommodate multiple perspectives within the ambit the inquiry. The convergence of various persons' understandings in intersubjective accounts, common or shared languages and the meeting of minds are positive injunctions associated with this departure. From a dialogic perspective, intersubjectivity has a dynamic that renders shared interpretation in consensual meanings suspect because in dialogism the very idea of meanings implies difference, perspective, and distance from the other.

Information systems researchers outside of the objectivist tradition are generally keenly aware of the sensitivity of meaning systems to technology (e.g., Checkland and Scholes, 1990; Mingers, 1995; Stamper, 1996), not the least because, as Bainbridge (1983), Zuboff (1988) and others have made clear, the more sophisticated the technical system, the more crucial and demanding the contribution from the human. The increased sophistication of IT and its extensive penetration into novel fields of human activity, the evolution of organizational members from *users* to *utilizers* of IT and contemporary concerns for knowledge work means IS scholarship needs to attend to the assimilation computational technology within a meaningful social context (Stamper, 1996). The linguistic turn identified by Klein (1996) is associated with theories that attempt to approach such meanings as social constructs that are linguistically mediated, and available in linguistic form (in all its different guises). In contrast Bakhtin's linguistic philosophy locates meaning in the use of language in action and communication, and that these meanings are generated and heard as social voices anticipating and answering one another.

These two pillars of Bakhtin's linguistic ontology, that meanings are founded upon difference or alterity and that language is unreducibly responsive, are used to explore the interpretivist research yardsticks and development milestones of shared and common understandings, languages and meanings. Although largely unacknowledged, linguistic ontology exerts a significant influence in IS (Stephens, 1998), from the appropriate role for IS and computer artifacts (Boland and Tenkasi, 1995) to regimes of articulation and research programmes (Button, et. al., 1995). How the socio-technical environment is constructed through language is a legitimate research concern which will also interest systems developers confronted with both organizational and technical complexity. This paper reports a dialogic perspective of the issues involved in investigating meaning, understanding and interpretation.

COMMUNICATION IN IS INQUIRY

It is almost axiomatic within the Information Systems community that the success of a system is proportional to the degree of 'user' involvement in design and development (Hirschheim and Newman, 1991). The general form is the working group of participants, typically representing a number of different backgrounds and interests, who engage in various group activities with a view to eliciting and refining ideas. An overwhelming need is to develop a common working culture and shared language that allows participants to cope with practical problems.

In fact, a form of communicative understanding is presupposed in most IS analytic activities. Conventional methods usually support written communication based on formalized languages, but do not cope well with a diversity or plurality of viewpoints on information (e.g. Davenport, 1997). Interpretive studies (Walsham, 1995) on the other hand, are generally distinguished by the attempt to cater for such variety by addressing phenomena according to the meanings that people assign to them, rather than relying on the imposition of a narrow technical semantics. Interpretive methods in IS are 'aimed at producing an understanding of the *context* of the information system, and the *process* whereby the information system influences and is influenced by the context' (Walsham, 1993: 4-5). Actors own interpretations are accepted as a potent factor in contexts, not just in the hermeneutic sense (e.g., Boland, 1987, Stowell, 1993), but also in the way that these situations themselves are the product of the joint actions of intelligent and knowledgeable agents, the so-called 'double-hermeneutic' (Giddens, 1984).

IS inquiries from the interpretive perspective therefore, must proceed with an investigation of the significance of these situations to the actors concerned. These may not be immediately transparent to the observer, so there is a particular resistance to substituting informants' discourse with that of the researcher's. Intersubjectivity is the general criteria for interpretive studies. Typically, it is said to be achieved empathetically by the generation of shared interpretations of the actions, practices and objects within the cultural systems that give them significance. The distinctive concern of interpretivism is primarily, not so much what the phenomenon *is*, but what people's *understanding* of it is. Efforts to explain understanding or meaning almost universally rely on the construct 'linguisticality' or 'language'. Gadamer puts it:"above all [understanding] takes place by way of language and the partnership of conversation."

Language and the partnership of conversation, participation in a linguistic community (c.f. Wittgenstein) is then, the site for intersubjective and intercultural access to meaning and the understanding of the Other's viewpoint. Thus a significant challenge is how does one escape the particularity of one's own subject position to align oneself with *another's* social viewpoint (Shields, 1996). This move is mandatory for such approaches, but the nature of intersubjectivity is rarely theorized; instead it is assumed because the sociality of language is taken to imply shared meanings and a common referential background. Further, understanding itself is *a priori*, teleologically motivated by an epistemology rooted in an unstratified symbolic order (c.f. Bourdieu, 1998); it is always achieved - there are never radical or insurmountable differences (Shields, 1996).

DIALOGIC UNDERSTANDING

Interpretive research is about comprehending human behaviour, products and relationships solely in terms of reconstructing the self-understandings of those engaged in creating or performing them (Fay, 1996:113). It is therefore a project of inquiry rather than theory. Moreover, interpretive method is not a special process, totally different from everyday human understanding; it is just one example of an everyday (hermeneutic) process through which persons make sense of their world. According to the classical hermeneutic approach, field work as well as frequent and regular exchange between participants offers the opportunity for an iterative alignment of researchers' implicit models with those of the subject(s). In IS the idea is to arrive at a shared description that includes users' interpretations and visions as well as the developer's own ideas on how to support organizational information practices.

The interpretive project is methodologically focused on a pathology of difference: difference is suspect, in need of resolution, or in some way problematized. In stark contrast the relation of difference is an epistemological necessity for dialogical understanding. Rather than being a barrier to be overcome, difference is the *sine qua non* for meaning, and therefore understanding. Bakhtin writes:

> "In the realm of culture, outsideness is a most powerful factor in understanding. It is only in the eyes of *another* culture that foreign culture reveals itself fully and profoundly .... A meaning only reveals its depths once it has encountered and come into contact with another, foreign meaning: they engage in a kind of dialogue, which surmounts the closedness and one-sideness of these particular meanings, these cultures. We raise new questions for a foreign culture, ones that it did not raise for itself; we seek answers to our own questions in it; and the foreign culture responds to us by revealing to us its new aspects and new semantic depths. Without *one's own* questions one cannot creatively understand anything other .... Such a dialogic encounter of two cultures does not result in merging or mixing. Each retains its own unity and *open* totality, but they are mutually enriched." (Bakhtin, 1981)

In the dialogic model, language is *sui generis* responsive, and its symbolic and meaningful quality is derivative only of the concrete situations of linguistic actions between addresser and addressee, which in turn belong to wider chains or networks of utterances. The term 'dialogism' means 'double-voicedness', rather than 'relating to dialogue', and refers to the presence of two distinct voices in one utterance. Bakhtin extends this sense into a property of all language, so that dialogism is characterized by 'the mixing of intentions of speaker and listener', the creation of meaning out of past utterance, and the constant need for utterances to position themselves in relation to one another. The interest in utterances is not as artifacts of language but as instances of a generative process, as a particular articulation of an ongoing process in which a person 'comes to see in a new way'. The utterance in dialogue contains within itself diverse, discriminating, often contradictory 'talking' components, and the more it is used, the more contexts it accumulates and the more its meanings proliferate. By their very nature, utterances resist unity and homogenization: there is no monological tendency to a shared univocal 'horizon of values' at any level.

Intersubjectivity is presupposed in Bakhtin's social philosophy, but it is based on the responsiveness of the person (there is 'no alibi' for existence), rather than shared meaning, intention or understanding. Indeed, a leitmotif of dialogism is the distinctiveness of the other. It is the very 'otherness' of the other, the fact that the other speaks from a different and unique horizon, that constitutes the enabling condition for the productivity of dialogue, and Bakhtin explicitly rejects the notion of a shared apperceptive mass as either ground or goal of communication:

> "In what way would it enrich the event if I merged with the other, and instead of two there would be now only one? And what would I myself gain by the other's merging with me? If he did, he would see and know no more than what I see and know myself; he would merely repeat in himself that want of any issue out of itself which characterizes my own life. Let his rather remain outside of me, for in that position he can see and know what I myself do not see and do not know from my own place, and he can essentially enrich the event of my own life." (Bakhtin, 1981).

This is a major departure from consensual approaches to intersubjectivity, because in the Bakhtinian version, the distance and difference of the other, and the struggle with misunderstanding, is not only always retained but deemed essential for the productivity of dialogue and thought.

CONCLUSION

In dialogism, self-other differences, rather than impeding understanding are its condition, and motivate and generate communication. Moreover, differences, which include conflict, domination and contradiction, are built into the dialogical model of language itself: there are no 'neutral' words. From a Bakhtinian perspective, intersubjectivity as shared meaning or consensus in language could not be more inappropriate.

Shields (1996) reports that contemporary interpretivism is constructed on a particular reading of Verstehen that motivates synthesis as a research goal (arising from the Hegelian notion of the dialectical resolution of antithetical elements). However, such synthesis is neither inevitable or singular for every proper inquiry context is dilemmatic; that is, a context structured by the possibility of the activities within it allowing criticism and/or requiring justification, and in which people must adopt a rhetorical or argumentative position in what is always at least a two-sided controversy (Billig, 1987). Rather than seeking to overcome the difference presented by the other, effectively silencing the other, Shields calls for a return to the original sense of Verstehen as a response to the other, a 'giving-meaning-to' the alterity presented by an other culture or person's understanding and action.

Dialogism focuses on the interaction between researcher and informant thereby increasing flexibility and widening analytical options:

> "Instead of trying desperately to defend the notion that individual utterances, or texts, have a fixed, original meaning which it is the business of criticism to recover, we can locate meaning in the dialogic process of interaction between speaking subjects, between texts and readers, between texts themselves." (Lodge, 1990: 86)

From this perspective intersubjectivity identifies with 'dialogized heteroglossia', the combative relations different languages enter into when they come into contact. These are the dialogically-interrelated speech practices operative in a given society, culture, organization or group at a given moment, wherein the idioms of different classes, races, genders, professions, locales, perspectives and attitudes compete for ascendancy.

REFERENCES

Bainbridge, L., 1983, Ironies of automation. *Automatica*, 19: 775-779.

Bakhtin, M., 1981, "The Dialogic Imagination: Four Essays by M.M. Bakhtin", ed. Micheal Holquist, *trans*. C. Emerson and M. Holquist, University of Texas Press, Austin.

Billig, M., 1987, "Arguing and thinking: a rhetorical approach to social psychology", CUP, Cambridge.

Boland, R.J., 1987, The in-formation of information systems, *in*: "Critical Issues in Information Systems Research",. R.J. Boland and R.A. Hirschheim (Eds.), John Wiley and Sons, Chichester.

Boland, R.J and Tenkasi, R.V., 1995, Perspective making and perspective taking in communities of knowing, *Org. Science*, 6(4): 350-372.

Bourdieu, P., 1998, "Practical Reason: On the Theory of Action", Polity Press, Cambridge.

Button, G., Coulter, J., Lee, R.E. and Sharrock, W., 1995, "Computers, Minds and Conduct", Cambridge: Polity Press.

Checkland, P. and Scholes, J., 1990, "Soft Systems Methodology in Action", Wiley, Chichester.

Davenport, D.H., 1997, "Information Ecology: mastering the information and knowledge environment", OUP, Oxford.

Fay, B., 1996, Contemporary Philosophy of Social Science: A Multicultural Approach, Blackwell Publishers, Cambridge, MA.

Giddens, A., 1984, "The Constitution of Society". Polity Press. Cambridge.

Klein, H.K. (1996), Preface: The potential contribution of semiotics and systems the to the continuing evolution of information systems research, *in*: "Signs of Work: Semiosis and Information Processing in Organisations", B. Holqvist, P. Bogh Andersen, H. Klein, , and R. Posner, eds., Walter de Gruyter and Co, Berlin.

Hirschheim, R.A. and Newman, M., 1991, Symbolism and information systems development: Myth, metaphor and magic, *Info. Sys. Research*, 2(1): 29-62.

Lodge, D., 1990, "After Bakhtin", Routledge, London.

Mingers, J., 1995, Information and meaning: foundations for an intersubjective account, *Info. Sys. Jnl.*, 5: 285-306.

Shields, R., 1996, Meeting or mis-meeting? The dialogical challenge to Verstehen, *Br. Jnl. of Sociology*, 47(2): 275-294.

Stamper, R., 1996, An information systems profession to meet the challenge of the 2000s, *Systems Practice*, 9 (3): 211-230.

Stephens, R.A., 1998, The meta-linguistic model in information systems research, *Proc. Fourth Americas Conference on Information Systems*. Baltimore, August 14-18.

Stowell, F. A., 1993, Hermeneutics and Organisational Inquiry, *The Systemist*, 15(2): 87-103.

Walsham, G., 1993, Reading the organization: metaphors and information management, *Jnl. Info Sys.*, 3: 33-46.

Walsham, G., 1995, Interpretive case studies in IS research: nature and method, *European Jnl. Info. Sys.*, 4: 74-81.

Walsham, G., 1995, The emergence of interpretivism in IS research, *Info. Sys. Research*. 6(4): 376-394.

Zuboff, S., 1989, "In the Age of the Smart Machine: The Future of Work and Power", Heinemann, Oxford.

TOWARDS SYNERGY IN THE SEARCH FOR
MULTI-PERSPECTIVE SYSTEMS APPROACHES

Zhichang Zhu

Lincoln School of Management
Lincoln, LN6 7TS, UK

INTRODUCTION

This paper presents a comparative study of two multiperspective systems approaches to management: the American TOP (T, O and P perspectives) and the Chinese WSR (*wuli-shili-renli*), analysing their commonality and differences, investigating their cultural roots, and reporting on a mutual learning process and its outcomes (for the theories and applications of TOP and WSR see Linstone 1984, 1985, 1989, Linstone *et al.* 1981, 1987, Linstone and Mitroff 1994, Mitroff and Linstone 1993, Gu and Zhu 1995, 1996, 1997, 1998, Zhu 1996a, b, c, d, 1997, 1998a, b, c, d, e).

COMMONALITIES

Systems means interconnectedness. The quintessence of systems thinking is the concept of interconnectedness. Every problem is to be found within every other problem; every model presupposes every other model; every science is to be found within every other. Technical and human aspects, thus, must be consciously brought to bear on our problems and managed as a whole. Separation of, or one-sided emphasis on, technical or human aspects is suitable only for tackling structured, bound problems, not for dealing with complexity in the 21st century in which multiple views are not a luxury but a fundamental necessity. 'Systems' to TOP and WSR is *not* confined merely to systematic procedures, *nor* is it equivalent with interpretivism. The either/or mentality is seen as inherently incompatible with systems thinking.

Adopting a circular, rather than a hierarchical or bifurcated, archetype of sciences. In TOP, the traditional hierarchical ordering of sciences and professions – as well as the pejorative bifurcation of the sciences into 'hard' versus 'soft' – is replaced by a circular concept of a relationship between them. Every one of the sciences and professions is considered fundamental; none is superior to or better than any other. In a similar way, WSR does not perceive *wuli*, *shili* and *renli* as forming a hierarchy, whereby a particular *li* is more basic, fundamental, primary, central, higher, reliable, certain, neutral or scientific than others. Rather, WSR sees *wuli, shili* and *renli* as interwoven in a dynamic network within which different *li*s condition, imply, inform and transform each other. There can be no 'real' facts or 'neutral' data upon which, alone, contending parties can arrive at 'objective' description, 'rational' judgement, or 'scientific' decisions. Any investigation, modelling and management of today's complexity are seen by WSR as but a localised and historically situated process of the continual and dynamic interplay of *wuli, shili* and *renli*.

Objectivity, if it exists, is a collective property of multi- perspectives or lis. No human can ever be said to be purely objective or unbiased in his or her mental process; every perspective or *li* is hence inherently limited. Thus, if it exists, objectivity is a systemic property of the systems of sciences taken collectively. Objectivity results, if at all, only from a process of weeding out differing perspectives or *li*s, of intense debates. TOP and WSR hold that multiple combination of models and observations are more likely to lead to 'truth' than any single model or a single set of observations.

Striving not for a better single truth but for a provisional model for action. 'Truth' in TOP and WSR does not mean reducing all issues in the ultimate desire to produce a 'single best' or optimal solution to a particular problem. Any 'truth' is always a provisional or a 'working' one. To achieve a better single model of a problem is to miss the whole point and spirit of multiperspective management. The central interest of managers, practitioners, and hence TOP and WSR is not Truth *per se*, but models or positions that decision-makers and problem-solvers can adopt for urgent actions in historically situated contexts.

*Multi-perspective/*li *approach as a system looking at systems.* TOP declares explicitly that it is a meta-IS (inquiry system), that is, it includes all the other ISs. As to WSR, it is presented as an approach for dealing with 'relations among relations', or, 'relations of relations'. Referring to Han-fei's teaching (?-233 BC, XX/VI/VII) that 'All things/affairs have each their own different *li*, whereas Tao brings the *li*s of all things/affairs into an embracing conception', the multi-*li* WSR is happy to be presented as a Tao-approach.

Not assuming fixed order, but learning to live with uncertainty. For TOP, real problems and hence various issues only unfold over time. Rarely do we have a clear definition of the problem before we begin working on it. Both organisations and individuals change their perspectives over time – 'beware of thinking statically in dynamic environments'. WSR explicitly presents itself as a 'bubble-*li* management' and a 'spiral-learning' process. W, S and R are metaphorically seen as bubbles, of which the motions do not follow routine paths. The patterns of their bubbling are uncertain. What is certain is that W, S and R will all bubble up, together or otherwise, here and/or there, at one time and/or another. We must prepare to manage W-S-R bubbles at every moment during the whole problem-solving process.

*Multi-perspectives/*lis *are not* the *solution.* One cannot pose an *a priori* argument that multiple perspectives/*li*s are invariably superior to a single (T) perspective or a single (*wu*) *li*. Obviously, three poorly done perspectives or *li*s will be inferior to one excellent T perspective or *wuli*. The answer lies in the quality of the multiple perspectives or *li*s, that is, their integrity and appropriateness, as well as the willingness, openness, skills and experiences of participants. We can only suggest in principle that, on a *ceteris paribus* basis, multiple perspectives and multiple *li*s should be preferred to single perspectives or *li*s.

P-perspective and shili*: a part or the whole?* If the P perspective is defined as the individual's view of the world, and if *shili* is defined as the ways individuals choose to see, to think and to act, do they not encompass whatever mix of T, O and P, or W, S and R which resides in the mind of that person? Admitting such a dilemma, TOP and WSR postulate that a qualitative discussion of an individual's personal balance or bias in perspective types and in various *li*s can be meaningful and useful. In real world problem-solving, we do, in fact, often distinguish what is best objectively, what is best for our organisation, and what is best for us personally; we do often consider such issues: do we know the *wuli*, are we concerned with our *shili*, have we cared for *renli*?

Refusing to be formalised. TOP and WSR consistently distance themselves from the obsession and temptation of formalisation, such as, in its most extreme form, being programmed into a marketable, computerised 'decision support system'. In our view, such a kind of formalisation will immediately shrink multiperspective approaches back into merely T perspective or *wuli* - the way in which we are looking will inescapably become once again one-dimensional. The ultimate purpose of TOP and WSR projects is not to establish T-O-P or W-S-R as *the* correct or true model of the world, but to stimulate multiperspective management, for promoting creativity, for nurturing an unbounded mind, not for encouraging conformity. To formalise TOP or WSR as *the* model of the world or *the* way we should follow, would be directly opposite to the very intention of TOP and WSR.

Living with post-modern contingency. TOP and WSR hold that there can be no single combination of perspectives or *lis* that enables us to say with absolute finality, 'This is it!'. Every picture we construct of the world is destined to be superseded by some other; therefore, it is not absolute but contingent. On the other hand, TOP and WSR believe that together multiple perspectives and *lis* may form a common language or grammar which makes human learning possible, a learning to see the world and our problems as a dynamically interconnected whole, rather than as purely contingent, unrelated parts. While perspectives/*lis* are not absolute, they are contingent 'enough' only to the extent that together they are still able to permit individual actions to make differences, and to allow individual visions to transform into collective ones in the face of the 'totality' of imbalanced relations in power, information, expertise and resource distribution.

Being idealist. Since there is no final, absolute picture of the world that is open to human beings, the decision to pick up a particular picture of the interplay of multiple perspectives or *lis* on which to base one's actions is ultimately a heroic act and not a 'logical' one. TOP and WSR take sociotechnical systems as the major domain of their application. And, seen through the O-P perspectives as well as *renli*, sociotechnical systems design and management are inherently political. While the contemporary intellectual community turns away from ethical issues, TOP and WSR explicitly draw the ethical and idealist issues back in, as central and critical features of management/systems approaches.

DIFFERENCES

Differing Meanings in the Trinitarian Relation Inquiry System

There are three dimensions in both the TOP and WSR frameworks, the first of which, 'relations with the world', is called the T perspective and *wuli* respectively. TOP confines it mainly to technology, a man-made artefact. In WSR, this dimension covers a much broader range of 'things', i.e., from material objects to data and the educational structure of the workforce, even formal reporting and accounting systems. In other words, *wuli* denotes both physical existence and man-made artefacts.

Regarding the second dimension, 'relations with the self', the 'self' in TOP denotes the Personal perspectives, which is unmistakably confined to each individual, whereas in WSR it is almost always a 'bigger' one, i.e., 'our' modelling, 'our' assumptions, 'our' ways of seeing and doing. We suggest that this difference may stem from the differing cultures: for the Americans, self-interests, individual rights and characteristics are unshakeable values and principles, but for the Chinese the 'self' has proper meaning only when it is located in various social networks.

Another difference along this second dimension is that, in TOP, P is one of the three perspectives, all of which are really 'perspectives', i.e., ways of looking problems, whereas, in WSR, *shili* is the central, if not the sole, dimension that addresses the psycho-cognitive aspect of 'doing projects', with *wuli* concerning 'objective existence', and *renli* caring for political relations and interactions.

As to the third dimension, 'relations with others', the O perspective in TOP denotes both cultural and political considerations, while in WSR, *renli* deals with interactionist behaviour among actors, with 'groupthink' and organisational cultures being part of *shili*.

Differing Mappings of TOP and WSR

TOP and WSR map their relationship differently, as illustrated bellow.

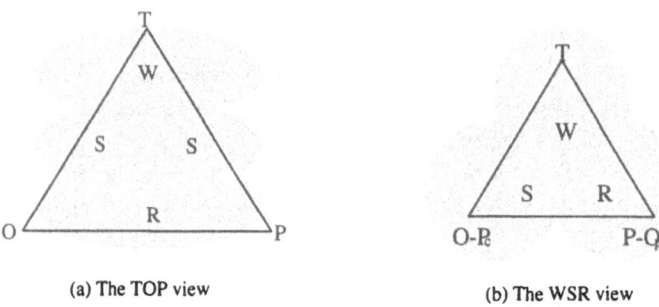

(a) The TOP view   (b) The WSR view

Figure 1. Mapping the relationship between TOP and WSR

The different views of TOP and WSR on the relationship between themselves can be explained by the differences in their methods of differentiation. Both WSR and TOP take as their starting-point the conceptualization of sociotechnical systems into two spheres: technical-material and human. The difference begins when the two approaches further differentiate the human sphere.

To the American TOP, individual and organisation dimensions are equally important; one cannot hide behind the other. To say 'our' ways, 'our' assumptions or 'our' modelling is pointless. In this view, *shili* cannot be conceptualised as 'a messy lump-sum'. Only when it denotes the differentiation-interaction between individuals and organisations, can *shili* be a useful concept. This is why TOP 'splits' and 'maps' *shili* along the two TOP dimensions, O and P. This also explains why TOP puts *renli* at the interaction area between O and P.

As WSR, the Confucian trinitarian teaching of 'virtuous action' is deep-seated in their mind; 'virtuous' in the sense of *rationality* for studying nature, *sincerity* for rectifying and cultivating one's mind, *benevolence* for enhancing social relations. Viewed by such a mind, both O and P have two dimensions, a cognitive-cultural one towards sincerity and a socio-political one towards benevolence. Thus WSR is happy to 'regroup' the cognitive-cultural components in O and P under *shili*, and the socio-political ones under *renli*.

Differences in Methodological Treatments

To TOP, 'The multiple perspective concept is not simply another methodology to add to the analyst's tool kit. There is no six-step procedure, no formula to weight perspectives' (Linstone 1989: 326); whereas WSR, although it, too, refuses to be formalised, does have a six-step, or seven-activity-element, a 'methodological face'. How profound is this difference and where does it come from?

We will suggest that the different treatments are shaped by the differing experiences and observations of the two groups of researchers, as well as by the situations where the two approaches apply.

The main experiences and observations of the TOP advocates are in the area of high level decision-making. Further, in the United States, the education levels of and among those involved are relatively high and equal. All these factors indicate that the major focus of TOP is primarily to improve participants' strategic thinking.

In contrast, in contemporary China, the education level of the workforce is relatively low and the education levels of participants are relatively unequal. Consultants on projects have to go through from A to Z, i.e., from strategic thinking to detailed operations. In such circumstances, it is not a bad idea to introduce a methodological guideline as a starting point for the involved to learn, to discuss, to compare, to select, to combine, and eventually to create their own methods in accordance with their own situations and abilities.

Actually, WSR is not a methodology but a multi- or meta-methodology. It does not have a particular set of techniques or tools of its own. It merely provides method-matching guidelines.

MUTUAL LEARNING

Example 1. In its early form, WSR was basically a 'doing approach' with an obvious ontological emphasis in its conceptual framework. Having been informed by TOP, WSR advocates recognise that imbalanced emphasis on ontological vision without conscious epistemological considerations will result in another 'better single *li*'. This recognition has opened a new ground for enhancing WSR. Interestingly, the enhancement of WSR is 'done' not through an explicitly introduction of the 'foreign' Singerian IS. Rather, WSR encourages indigenous users to move beyond the 'Virtue without Truth' mind-set in a way 'natural' to them, using their familiar 'doing' language.

The advocates suggest that WSR can be used in one, or in a combination, of the following ways: as an ontological model, an epistemological co-ordinator, a methodological guideline, a method-matching device, and an educational vehicle. All these usage can be meaningful and legitimate, so long as users consciously reflect on their usage, asking two questions: what are we doing (in what way(s) are we using WSR), and why are we doing so (why do we use WSR in such a way rather than other way(s))?

Example 2. TOP has benefited from becoming more culturally sensitive in a 'better way'. From its early applications, TOP found it not equally easy to put the P and the O perspectives across to local users. TOP suggested that local people, say, the native Chinese, should be fully involved in developing the O and P perspectives.

Through communication with WSR, TOP recognises that the multiple-view ideal can be materialized in a form(s) more natural to the indigenous culture(s). To promote multiperspective management, TOP may be most suitable in situation A, WSR can be more appropriate to location B, an explicit comparative presentation of TOP and WSR together might stimulate more creative applications in context C, etc. What is crucial is the multiperspective ideal, rather than a particular format or language.

CONCLUSION

It has been maintained for decades that it is a good idea to promote synergy between Western and Eastern approaches to sociotechnical systems management. But so far, mutual learning and synergy, other than simply 'importing' and 'inserting', have proved to be difficult, however desirable. The case reported in this paper can be seen as our attempt to take part in such a promising yet challenging endeavour, to make a real difference, however small or tentative. We have reasons to expect significant development and synergy in multiperspective approaches in the new millennium.

REFERENCES

Gu, J. and Zhu, Z. (1996) Tasks and methods in the WSR process, in *Systems Methodology II: Possibilities for Cross-Cultural Learning and Integration*, ed., Wilby, J., Centre for Systems Studies, England, pp. 15-22.

Gu, J. and Zhu, Z. (1997) Evaluation through the WSR Approach: the China case, in *Systems Methodology III: Possibilities for Cross-Cultural Learning and Integration*, eds., Wilby, J. and Zhu, Z., Centre for Systems Studies, England, in press.

Gu, J. and Zhu, Z. (1998) Knowing *Wuli*, sensing *Shili*, caring *Renli*: methodology of the WSR approach, *Systems Practice and Action Research*, under review.

Linstone, H. (1984) *Multiple Perspectives for Decision Making*, North-Holland, New York.

Linstone, H. (1985) Multiple perspectives: overcoming the weaknesses of MS/OR, *Interfaces*, **15**, 77-85.

Linstone, H. (1989) Multiple perspectives: concept, applications, and user guidelines" *Systems Practice*, **2**, 307-331.

Linstone, H. (1996) Personal communication with Zhu, Z., fax materials, 28th January and 8th February.

Linstone, H. A., *et al.* (1981) The multiple perspective concept, *Technology Forecasting and Social Change*, **20**, 275-325.

Linstone, H. A., Fried, J., Wang, Y., and Shu, H. (1987) Multiple Perspectives in Cross-cultural-cultural Systems Analysis: the China Case, Systems Science Ph.D. Program, Portland State University, Portland, Ore.

Linstone, H. and Mitroff, I. I. (1994) *The Challenge of the 21st Century: Managing Technology and Ourselves in a Shrinking World*, State University of New York Press, New York.

Mitroff, I. I. and Linstone, H. (1993) *The Unbounded Mind*, Oxford University Press, Boston.

Zhu, Z. (1996a) Systems approaches: Where the East meets the West?, in *Sustainable Peace in the World System, and the Next Evolution of Human Consciousness, Proceedings of the 40th Annual Meeting of the ISSS*, ed., Hall, M., Club of Budapest, Budapest, pp. 413-430.

Zhu, Z. (1996b) The WSR concepts for systems practice, in *Sustainable Peace in the World System, and the Next Evolution of Human Consciousness, Proceedings of the 40th Annual Meeting of the ISSS*, ed., Wilby, J., Spalding University, Louisville, pp. 703-707.

Zhu, Z. (1996c) Method choice in action research and systems practice: context-driven or process-oriented?, in *Forum One: Action Research and Critical Thinking*, ed., Wilby, J., Centre for Systems Studies, England, pp. 107-112.

Zhu, Z. (1996d) International conversation (1995) on the WSR approach: edited e-mails and letters, presented at *The 2nd China-Japan-UK International Workshop on Systems Methodologies*, Kyoto, 8-10 May (available on request from the author).

Zhu, Z. (1997) The naked emperor's clothes: A reply to comments on WSR, in *Systems Methodology III: Possibilities for Cross-Cultural Learning and Integration*, eds., Wilby, J. and Z. Zhu, Centre for Systems Studies, England, pp. 105-122.

Zhu, Z. (1998a) Confucianism in action: new developments in Oriental systems methodology, *Systems Research and Behavioural Sciences*, **15**, 111-130.

Zhu, Z. (1998b) WSR: A systems approach for information systems development, *Systems Research and Behavioural Sciences*, under review.

Zhu, Z. (1998c) A Trinitarian relation inquiry system in systems/management approaches? – More findings, in *Sustainable Technology and Complex ecological and Social Systems*, proceedings of the 42nd Annual Conference of the International Society for the Systems Sciences, July 19-24, Atlanta, Georgia, ISBN 0-9664183-0-1, eds., Allen, J. and Wilby, J., the last chapter.

Zhu, Z. (1998d) Cultural imprints in systems methodologies: the WSR case, presented at *The 3rd International Conference on Systems Science and Systems Engineering*, August 25-28, Beijing in proceedings, ed., J. Gu, pp. 401-408.

Zhu, Z. (1998e) Dealing with a differentiated whole: philosophy of the WSR approach, *Systems Practice and Action Research*, under review.

PEERING INTO OURSELVES: CRITICAL REFLECTIONS ON PEER LEARNING

Dr Susan J Byrne[1] and Dr Judith McMorland[2]

[1]Dept of Management Science and Information Systems
[2]Dept of Management and Employment Relations
University of Auckland
Private Bag 92019
Auckland
New Zealand

ABSTRACT

We have been engaged, with two others, for the last four years in a peer learning partnership. We are exploring the question of how we learn to learn together. In this paper, we reflect on the process in which we have been engaged and discuss questions that have arisen for us, as we have critiqued our own practice. Specifically we inquire into the systemic nature of our peer learning process. We discuss insights we have discerned from our experiences together and the implications, as a work-in-progress, for further critical subjective inquiry, within a critical reflective systems paradigm.

INTRODUCTION

This paper reports and reflects on an extended period of subjective experience. We present and discuss our experience of four years of peer inquiry, centred on the meta question *How do we learn to learn together*? We started from the tacit premises that we can learn *how* to learn (metalearning), that this is a human capability and skill developed through intentional social interaction, and that as individuals we *need* to learn how to **learn together** because within our own cultural milieu, collective learning is not yet a well developed attribute of social engagement. We take seriously Peter Senge's dictum that development of the capability of collective learning is important for human evolution (Senge, 1990). Can such a subjective endeavour be deemed serious research? We believe so, and in the pages below we identify the ways in which what we have been doing together can be deemed both critical and reflective inquiry in a serious practice domain, albeit at a micro level. We have aspired to bring critical systemic thinking both to our reflection-in-action, and our reflection-on-reflection.

Synergy Matters: Working with Systems in the 21st Century,
Edited by Castell *et al.*, Kluwer Academic / Plenum Publishers, New York, 1999.

Flood and Ulrich (1989) suggest that being critical is having the quality of "remaining self-reflective with respect to particular and all positions and approaches". In Jackson's (1991) view, critical systems thinking embraces five major commitments: critical awareness, social awareness, dedication to human emancipation, complementary and informed use of systems methodologies in practice, complementary and informed development of all varieties of systems approach.

We explore the extent to which these aspects can be applied to our Peer Partnership inquiry, but first we describe the essential components of our four year learning journey.

THE PEER PARTNERSHIP EXPERIENCE (1995-MID 1998)

Our Peer Partnership was started as an initiative of the New Zealand Organisation Learning Foundation to deepen understanding of collective learning endeavours. Our paper documents the experience of the four members who have been in the group throughout. Though the salient characteristics of membership are hard to define, in brief, we are two women and two men, spanning an age range of twenty years (mid-30s to mid-50s) from *pakeha* (non-indigenous) educated backgrounds. We share commitment to spiritual exploration and to the humanistic, emancipatory values base of the 'learning organisation'. We have each undertaken extensive personal development. We all have professional skills in group work and organisational change and development.

We knew we were not consciously skilled in the domain of collective learning, hence our central organising question. The challenge was to strip away the usual reliance on received knowledge, to use ourselves as the instruments of inquiry by researching deeply our own interactive practice. This meant letting go of the 'expert' authority we were each used to in our own fields of practice (in the academic and commercial realms) and opening ourselves up to exploration - both personal and collective.

PEER PARTNERSHIP PROCESS

We have met regularly once a month for three-four hour sessions, usually on a Friday afternoon, since early 1995. Context has been important. We met first in Mark's office (three years) and now work in Susan's home. We started the group as strangers to each other. Though we had met severally in situations outside the group, those meetings had not been under the rubric of the Learning Organisation inquiry.

Typical session process: Each session is an action-reflection process. For the first couple of years, we followed a ritual process of *check-in*, *dialogue* on emergent themes, *reflection* on what had occurred, and finally *check out*. More recently, we have woven reflection more deeply into on-going conversation, avoiding ritualisation of this form. The meanings we have developed for the terms *check-in* and *dialogue*, whilst related to Senge's interpretations (Senge 1990, Senge et al 1994), are distinctive. They are discussed in detail below.

Check-in: To begin well, we needed to tell each other who we were. We chose the process of *check-in* as our initial starting point, crafting it to our own purposes, beginning an iterative process which we have sustained and developed across the four years. *Check-in* is about declaring who we are in the here and now. There is paradox here - we meet each other afresh at each meeting yet we acknowledge that we are each shaped by our encounters from earlier meetings. Now we are deeply known to each other, yet when we meet we are open to surprise. Part of checking in seems to be rediscovering ourselves within as much of our lives as we are able to embrace within the present moment of our *check-in*.

482

Dialogue: We experimented with sessions of different styles. We told personal stories and explored the resonance of symbol and meaning in several folk tales. We used elements of psychodrama, concretisation and other action methods to expand understanding of personal situations. We crafted extended metaphors, conceptualised our experience in diagrams and played with ways of finding the appropriate language to express our insights. Throughout we conceived the peer partnership as a deep discipline of connection and exploration.

Check-out: Sessions typically concluded with a brief period of *check-out,* when each person identified their own learning and the value they had taken from the meeting. We reiterated the key themes that had been the substance of our conversation, noting the weaving together of our separate contributions and the ways in which dialogue had emerged, giving substance to our collective development.

STAGES OF GROUP DEVELOPMENT

We experienced the *check-in* process as a spiralling warm-up to shared collective consciousness. We recognised early in our reflections on the process of check-in that, whilst we all started on 'individual mountain peaks' in positions of relative isolation and difference - personal mental models (see figure 1) - we became increasingly aware of the themes of the group and of the reality of a 'group consciousness'. Our individual 'mountains' stood within a deeper river of connection. The river was, for us, a strong interpretive metaphor in the initial stage of the group.

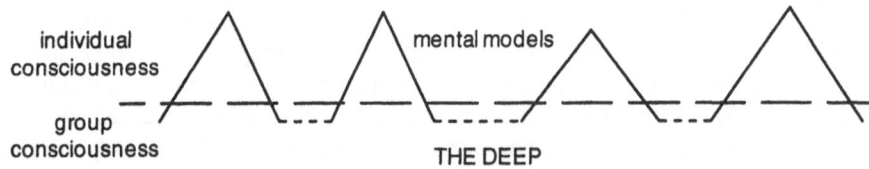

Figure 1. Individual and group awareness

The 'work' of the group has been to deliberate on, and be liberated by, our prime question. Contrary to some views of experiential learning, we rejected the need for a focusing 'task'. We believed that the *experience* dimension of experiential learning was grounded in **being** with each other, as well as **doing**. We sought not to *talk about* our experience, but to experience deeply our talk - our interactive practice.

Reflection is an essential part of this experiencing - it is the process whereby we understand (or construct) intention and meaning. Through this we have become more aware of the extent to which our language and imagery draws on our respective cultural and spiritual backgrounds and philosophical reading. We have been challenged to note the patterning processes of storying and self-presentation and the stages of change in the group itself. Figure 2 identifies four different stages in the evolution of our peer partnership to date:

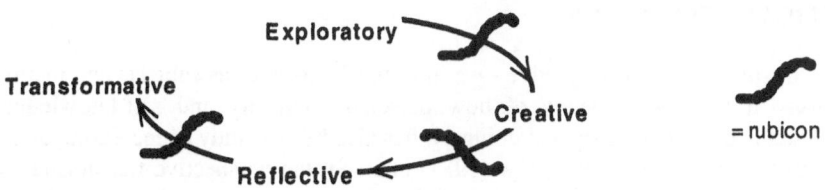

Figure 2. Stages of peer partnership evolution

This figure captures our reflection on the iterative nature of our learning and, building on Juch (1984), the 'rubicon' crossings - points of no return - that we encountered. From feeling swept along by change, we have gained confidence in our collective endeavours and recognise that provided there is an adequate container (by which we mean deep commitment to each other) then transformation can occur within it. In this respect we have experienced for ourselves the phases of dialogue identified by Isaacs(1994).

Out of the experience of transition, a new metaphor emerged in our conversations: that of spinning a fleece into strong strands of wool. In this image we are both the spinners and the spun. Spinning is a transformative process whereby the weaknesses of separate strands can be overcome. In this shift of metaphor we recognise a shift in our collective awareness of relationships and agency within the partnership. It expresses the ways in which we are implicated together, and from which we can individually draw strength when acting alone. The image takes us out into the world in a way that the river did not capture.

LEARNING FOR COLLECTIVE AND INDIVIDUAL ACTION IN THE WORLD

Deep learning is without meaning unless it gets translated into action. Our conversations take place against the backdrop of societal change, and our collective desire to be part of the shaping process of such change. Our meeting together is essentially purposeful - to create new experience from which action might be taken, within a world of turbulent change, at the intersection of new paradigms of thought, science and belief. In this respect our study meets a central criterion of 'participatory action research' - that of informing (emancipatory) practice (McTaggart 1998). Though we find it hard to identify exactly the ways in which specific explicit conversations have shaped our actions, we each report that our practice with colleagues, clients and student groups has been both subtly and profoundly deepened. Having had experience of meeting deeply, so we call this out of others we meet. The skills we have developed are those of sustaining focused attention on others; maintaining a conscious, but non-intrusive, awareness of self; creatively developing and weaving multi-threaded relationships within and beyond the group; and engaging in the art of meaningful conversation and inquiry.

We recognise the deep need we have for intimacy, and the under-development of our skills to create it. By intimacy, we are not referring to physical or intellectual attraction. Rather, it is that which leads to profound relationships of trust, where we can be deeply open and deeply appreciative of the revelation of self to others. We discovered for ourselves the importance of 'appreciative systems' of engagement (Checkland, 1985, citing Vickers, 1970). In this group we have found intimacy flourishes when status-defining behaviours and sexual politics are absent. Whilst the group came together in the spirit of equality - as a peer partnership - the meaning of that equality, of our essential common value as human beings has emerged as we have engaged each other. Herein lies emancipation, and the deep appreciation of personal integrity to which Bawden (1995) refers.

CRITICAL REFLECTION

Our study has been subjective - we have used ourselves as subjects and instruments of inquiry. It has been a study of how personal interiority and self-knowledge can be expressed in collective being and doing. It has also been a study of increasing awareness of collective consciousness. Mezirow (1981) claimed that perspective transformation occurs through the process of 'becoming aware of our awareness' and critiquing it. He identified seven levels of critical consciousness which could be brought to this task: *simple, affective, discriminant, judgmental, conceptual, psychic,* and *theoretical reflectivity.* Theoretical reflectivity he defined as 'the act of self-reflection by which one becomes aware that the conceptual understanding and the use of judgements or assessments are founded on a set of taken-for-granted cultural or psychological assumptions to explain personal experience'.

By deepening the skills of different levels of critical reflection within the group (and we make no claims for the fullness of these) we have sought to make a shift from collectively unconscious to collectively conscious learning. We have focused critical attention on our *experience* of separation and joining as well as on our meta-understanding of these processes (collective learning). Axiomatically, collective consciousness can only be known from subjective exploration of shared experience. The phenomena of our inquiry are the multi-stranded relationships which we develop through interaction : both being and doing together, both belonging (being joined) and being separate (individuated) in and through the processes of self-expression. Our relationships have been characterised by equality, intimacy and authenticity.

Subjectively, we identify that our relationships and reflectivity have had a character distinctive from those experienced in other groups. (We each have multiple experiences of group situations on which to make this comparison.) We are neither evaluative nor judgmental. Instead of being personally 'closed down' by judgement (evoking emotional responses of shame, fear, guilt or anxiety) we have experienced expansion, joyfulness, acceptance, love. For collective learning to occur we discern we need to value and develop deep skills of unconditional loving! Can such discernment be tested? Are these skills that can be 'taught' or does each collective of people who wants to learn together have to find for themselves ways of relating profoundly?

At this stage of work, we are caught in the 'reflexive turn' (Altheide and Johnson, 1994) of making explicit what we are experiencing and presenting it to others. If we are serious about this activity as a piece of critical systemic inquiry (or any other appropriate mode of qualitative inquiry) our research endeavour would seem to need to:

- identify what might count as collective experience (phenomenology)
- provide adequate account of and theoretical explanation of the experiences (epistemology)
- make explicit the grounds of our critique (ontology).

The problematic nature of all these dimensions is compounded when researchers attempt to research their own 'groupness' and the processes and outcomes of collective as well as individual learning. These are big challenges. It was the group's intention from the outset that we should document and critique our work but none of us has felt able to write about the process of the group until now. We were too intensely involved with the sensing to make sense of it. Now two of us, working collaboratively, have attempted the task (McMorland and Byrne 1998). We are drawn on by the excitement of exploration and the opening up of new areas of research and inquiry. Some of the research questions that emerge at the level of group experience are:

- How do we remain systemically open as a group?
- How do we maintain authenticity and avoid self-indulgence?
- How do we develop a meta-process for involving all four of us in a radical critique of our practice?

Part of the answer to the last question would seem to lie in delving more deeply into the praxis of critical learning systems, without violating our commitment to avoid being trapped into 'expert' knowledge systems, or of setting ourselves up as experts as a result of what we are discovering from within. Psychoanalytic and narrative methodologies may offer ways forward, as may the genre of spiritual discipline. These methodologies however refer more specifically to individual experience. As we see them, some of the tasks that lie ahead are honing the skills of making relational phenomena explicit; developing a richer vocabulary of collective learning, and collective experience; and generating and accounting for spontaneity and emergent creativity in group process (McMorland and Piggot-Irvine, 1997, 1998). To accomplish these we will need all the skills we have developed through the peer partnership, as well as a wider circle of companions and fellow explorers committed to collective learning.

AUTHORSHIP

We wish to acknowledge fully the other members of the peer partnership; although they are not direct authors of this paper, we are all co-authors of the inquiry. Mark Feenstra is director of Strategic Learning Systems. Tony Silvester-Clark is self-employed as a management consultant and social researcher. Judith and Susan also have their own consultancies.

REFERENCES

Altheide, D.L. and Johnson J.M. (1994). "Criteria for assessing interpretive validity in qualitative research" *in*: Denzin, N.K. and Lincoln, Y.S. (1994)

Bawden, R. (1995). "Systemic Development: a Learning Approach to Change" Faculty of Agriculture and Horticulture, University of Western Sydney, Hawkesbury

Checkland, P. (1985). From optimizing to learning: a development of systems thinking for the 1990s *Journal of the Operational Research Society* 36(9):757-767

Denzin, N.K. and Lincoln, Y.S. (1994). "Handbook of Qualitative Research" Sage, California

Flood, R.L.and Jackson, M.C. (ed.) (1991). "Critical Systems Thinking: Directed Readings" Wiley, Chichester

Flood, R.L. and Ulrich, W. (1989). Testament to conversations on critical systems thinking between two systems practitioners *Systems Practice* 3:7-29

Isaacs, W.(1994). Dialogue *in:* Senge et al pp357-364

Jackson, M.C. (1991). Social systems theory and practice: the need for a critical approach *in*: Flood and Jackson (1991)

Juch, B. (1984). "Personal Development" John Wiley & Sons, Chichester

McMorland, J. & Byrne, S. (1998). Diving deep: reflections on an extended exploration of peer learning Paper presented at ANZSys Conf., UWS Hawkesbury, Oct

McMorland, J. & Piggot-Irvine, E. (1997). Group learning: convergence and challenge at the edge of experience Presented at World Congresses 4th on Action Research, Action Learning and Process Management, 8th on Participatory Action Research, Cartagena, Colombia June

McMorland, J. and Piggot-Irvine, E. (1998). Facilitation as midwifery Unpublished conference paper, HERDSA Conf, Auckland University, New Zealand, July

McTaggart, R. (1998, forthcoming). Is validity really an issue for action research? *Studies in Cultures, Organisations and Society*, 5(1)

Mezirow, J. (1981). A critical theory of adult learning and education *Adult Education* 32(1):3-24

Senge, P.M. (1990). "The Fifth Discipline: the Art and Practice of the Learning Organisation" Doubleday, New York,

Senge, P.M., Roberts, C., Ross, R.B., Smith, B.J., Kleiner, A. (1994). "The Fifth Discipline Fieldbook: Strategies and Tools for Building a Learning Organisation" Nicholas Brealey Ltd. London

Vickers, G. (1970). "Freedom in a Rocking Boat" Allen Lane, London

GETTING TO GRIPS WITH GENDER AND SYSTEMS

Gender Issues in Systems Thinking Group

The Open University in Yorkshire
2, Trevelyan Square
Boar Lane
LEEDS LS1 6ED
UK

INTRODUCTION

The Gender Issues in Systems Thinking Group, which will be referred to as the *'group'* for the rest of this paper, has as its members tutors in the Yorkshire region who teach, or have taught on, the Open University (OU) systems courses: T245 Managing in Organisations, T247 Working with Systems and T301 Complexity, Management and Change: Applying a Systems Approach. The first two courses introduce systems ideas generally and in the context of organisations; T301 explores systems concepts and systems methodologies in some depth. Other actively involved members of the group include OU regional academic staff from the Faculty of Technology and an OU graduate – a former student on systems and women's studies courses. The group currently has ten members, and meets regularly. Two former members of the group have moved to new pastures.

As is characteristic of OU systems tutors, members of the group are applying and developing systems thinking in their work elsewhere which includes training childcare professionals, voluntary work, running small businesses and a charity, and teaching in other local universities. The contexts of these dual roles led them to consider some conjunctions of gender theory and equal opportunities practice with systems thinking and practice, for example, in information technology, within organisations, in organisational learning, and in supporting effective learning styles for students. This diversity fostered a rich and varied discussion of gender issues from which a number of paths of enquiry have crystallised. This paper describes the development of the group and maps the main avenues of thought it has explored in the few years of its existence. Figure 1 sets out the range of the group's interests, this paper explores a few of these in depth, work is progressing in all of these areas.

Synergy Matters: Working with Systems in the 21st Century,
Edited by Castell *et al.*, Kluwer Academic / Plenum Publishers, New York, 1999.

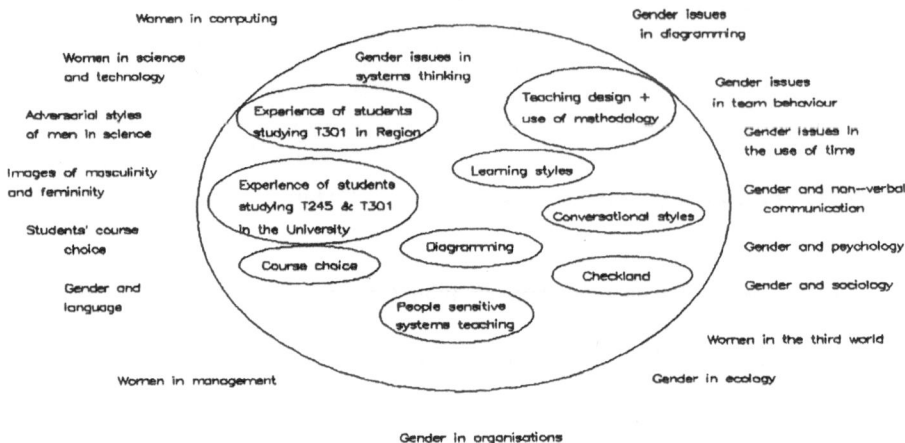

Figure 1. Systems map: areas of interest explored by the Gender Issues in Systems Thinking Group.

BACKGROUND

The formation of the group came about after two OU regional staff development days at the beginning of 1996: the first was in Manchester for systems tutors from a number of regions, this was shortly followed by one in Leeds for local tutors from all Technology Faculty courses. Regional and central academic staff from the Technology Faculty also attended these events. The timing of these two meetings was a significant factor in the group's formation; the meetings brought together Yorkshire Region systems tutors (and faculty academics) in a unique way since most of the time OU tutors are working in isolation at a distance and opportunities to meet each other are rare.

During these meetings two particular issues emerged and were discussed informally between some of the Yorkshire systems tutors and central academics. The first issue was the decline in the percentage of female students studying systems and related courses in the Technology Faculty. The second issue concerned the lack of gender specific context apparent in the then current systems courses offered by the OU. This opened up an ongoing dialogue with central academics which has recently resulted in an invitation for members of the group to join the course development team and contribute to the writing of the current revision of the level 2 systems courses.

Coincidentally during 1996, the Technology staff tutors had organised a series of seminars, in the Yorkshire region, under the general title 'Technology in Society' and two members of the group had already agreed to present papers at these sessions. One demonstrated the use of the OU 'hard' systems methodology in a staff development exercise with four women managers and another presented the results of her work on teaching technology to women students. Though these seminars were not formally part of the work of the group, they inevitably contributed to the development of thinking within the group.

ISSUES CONSIDERED BY THE GROUP

Students' Experience on Systems Courses

OU data presented at the 1996 staff development meeting showed that the percentage of women on Technology Faculty courses had declined from a peak of 28% to 22% (Ashby, 1996). This reflects a national pattern that despite a large expansion in overall student numbers, the proportion of women on computing/IT courses has fallen (UCAS, 1997) and begs the question: why?

One member of the group took a quantitative approach to investigating gender differences in the experiences of students on the systems courses in the Yorkshire region. The results of a sample of students who had done the T301 course were examined. It was found that there were no significant differences between the academic performance, submission rates of tutor marked assignments or project choice of women and men, though women were slightly less likely to favour the 'hard' systems approach taught on the course. This finding is echoed in similar work by Cifuentes and Lockwood (1996) which concluded that there is no significant difference between males and females in their ability to use a CASE tool in a learning environment.

These findings suggest that gender is not a significant factor once women have chosen to take a particular course but does not explain why so few women chose the course in the first place. It also relates to the wider question of why there are so few women in systems explored by Ruth Carter (1990). She points out that women are nurtured as systems thinkers and rely, routinely, on their holistic understanding of complex relationships to resolve dilemmas.

Members of the group are currently working with more detailed survey data provided by the OU's Institute of Educational Technology relating to the University's T245 and T301 courses which might enable the group to explore further some of the issues relating to recruitment to systems courses, retention and success. Questions being asked are for example: what do women enjoy on the courses? In which areas of the courses do women do better? What are students getting out of the courses? The results of this work should be available soon.

Gender Issues in Teaching Systems Thinking

Being collectively aware of over twenty years of research in, among other areas: linguistics, psychology, interpersonal behaviour, sociology, organisation studies, management, science and technology, which suggested some of the gender issues that might be worth exploring, the group began by posing a number of questions based on members' own experience. Questions such as: does being of a particular gender predispose people to systems thinking? Is it helpful to talk about 'masculine' and 'feminine' characteristics? From this discussion the group identified a number of areas as worthy of further investigation and members undertook to work, individually or together, in areas of interest to them.

Choice of Methodology

For her contribution to the 'Technology in Society' seminar one member drew together, through a literature search and her own experience, a number of issues relating to how people design an information technology application. Work by Sherry Turkle and Seymour Papert, identifying different epistemologies in computing, particularly drew her attention (Turkle and Papert, 1990). It resonated with her experience in teaching systems courses that women are less happy when a structure for thinking or doing is imposed. She had observed that women's interest was strong when she was introducing systems principles but fell off when the course moved on to specific techniques or methodologies. They were happy to see interactions or connectivity but not with the imposition of a method. She argued that methodologies influence design, application and creativity and their advocates (usually male) often fiercely defend them as if, without the methodology, there would be no output. Women students felt comfortable with their understanding and flexible approaches to supply outputs.

Preferences for particular methodologies by men and women were investigated among T301 students. This course explores the use of three different methodologies: Failures (a retrospective, understanding what went wrong analysis); an inhouse hard methodology (a synthesis of the features of most well known hard methodologies) and a soft methodology (Checkland's Soft Systems Approach). Analysis of feedback collected by two group members at the end of a final T301 tutorial found that women and men were equally satisfied with the course. However, small differences occurred in women's preference for the 'failures' methodology as one they would use again and men for the 'hard' methodology. In their answer to the question which methodology they thought most useful in the real world - men said 'hard', while women said 'soft' even though they did not prefer 'soft' as a project methodology to use again.

Approaches to Teaching Systems Design:

Studies of interaction within laboratories and workshops have shown that there are different communication styles and different ways of thinking behind those styles (Hall & Sandler, 1986; Pearl et al, 1990; Widnall, 1988) Men tend to use adversarial styles in science and technology and refutational practices from which women withdraw. Women are more likely to ask 'Have you thought of ...?' and use generally more inclusive styles. (Guzzetti & Williams, 1996) However, studies of participation that suggest that some women do not participate, may be criticised on the grounds that they do not consider active listening as participation. The group explored these ideas in relation to their own approaches to teaching and following on from this developed a workshop as a vehicle for raising tutors' awareness of these issues. With an observed mixed gender group exercise based around responses to a futuristic case study, it enabled tutors to look directly at how they teach systems thinking and to become more self-aware in what they were doing. The 'People Sensitive Systems Teaching' workshop had its first trial with a group of systems tutors and students in June 1997 at the Yorkshire Regional Centre.

Another group member followed his interest in extending Belbin's work on why teams fail or succeed to include a gender dimension (Belbin, 1981). The group discussed various approaches starting from Belbin's own but recognised that the scale of his work was a big barrier to replication. An alternative approach explored was Cognitive Mapping; the idea was to find if this could facilitate identification of gender differences in perception. During this enquiry the prominence of 'cooperation' became apparent and the link was made with Axelrod's work covered in T247 (Axelrod, 1990). He identified the Systems Game, used in the OU Systems summer school as a means to investigate this further. The intention was to run the game with large numbers of gendered groups, where the participants would not be aware of this active configuring. There would be some groups all female, all male and mixtures. Theory would suggest that female characteristic groups would tend to emerge as winners. Unfortunately the gender ratios in summers schools and the intrusiveness of the experiment in an already concentrated week have precluded carrying it through yet. However there are plans to pilot the approach in a large comprehensive school in the near future.

Checkland ungendered

The group has twice explored aspects of soft systems analysis under the title 'Checkland ungendered', prompted by a comment that Checkland's line between the 'real world' and the 'world of systems thinking' is an example of masculine 'either/or' thinking rather than the more inclusive style of feminine thinking. However, this starting point was not considered to be as worthy of exploration as other aspects of soft systems analysis (SSA). Though Checkland identifies the social and political streams of analysis (Checkland & Scholes 1990), these do not seem to go far enough to encompass the emotional aspects of 'human activity systems'. It would not seem possible to use to use SSA to investigate, for example, the processes that followed the death of Diana, Princess of Wales.

At present SSA is intended to investigate the 'human activity systems' related to particular 'tasks' and 'issues' and lead to action. This genders SSA in that Deborah Tannen (1991) has argued that men typically respond to any situation by action rather than by expression. It seems worth exploring the use, and possible extension of SSA in relation to human systems whose 'transformation' is no more than the maintenance of themselves. This might take us down the line tentatively explored by Checkland & Casar (1986).

SSA also assumes that you can create a 'root definition' that describes a single transformation; this can be seen as an extension of the assumptions of linear causality which underlie Western science (Adam 1989, Capra 1981). It makes the analysis of the multiple transformations that typically go on in a wide range of human systems impossible without reducing them to a series of separate, or dependent, transformations.

Though the original title for this series of explorations may no longer be appropriate, the group is continuing to explore the themes of emotion, expression and multiple transformations in human systems.

OTHER FACTORS IN THE DEVELOPMENT OF THE GROUP

Why did the group start, grow and develop? Firstly, the coincidence of people coming together to discuss related issues cannot be over-emphasised; secondly, the role of Ruth Carter in facilitating the group throughout its life has been significant - in arranging the space and facilities, in putting the group in touch with key academics in the University, in advising group members on the most efficacious ways of taking work forward, including advice on how to access staff development funds, and in keeping the group to task; thirdly, the group is unlike many faculty groups in traditional universities in including members with academic backgrounds in a wide range of the Arts, Sciences and Social Sciences, many of whom continue to practice within their original discipline as well as being systems tutors and who bring awareness of developments in a range of disciplines to the group.

In addition, dining out has been a not uncommon end to a group meeting suggesting that, as with an Oxbridge High Table, interdisciplinary discussion is promoted as much by gatherings which take place outside formal teaching or research.

GROUP MEMBERSHIP

Former members of the group: Beth Bibby and Katrina Lancaster.

Current members of the group: Gerry Broad, Alan Burke, Maria Burniston, Sue Butler, Ruth Carter, Wendy Fisher, John Hudson, Sue Palmer, Ley Robinson, Deborah Trayhurn

REFERENCES

Adam, B. 1989, Feminist Social Theory Needs Time. Reflections on the Relation between Feminist Thought, Social Theory and Time as an Important Parameter in Social Analysis. *Sociological Review*, 37(3): 471

Ashby, A., 1996, "Analysis of Applicants for Undergraduate Study with the Open University for 1996 Entry", [IET, Student Research Centre Report No: 104], Open University, Milton Keynes.

Axelrod, R.M. 1990, "The evolution of Cooperation", Penguin, London.

Belbin, R.M., 1981, "Management Teams: why they succeed or fail", Butterworth-Heinemann, Amsterdam

Capra, F., 1981, "The Turning Point", Wildwood House, London.

Carter, R., 1990, Women and systems. *Systems Practice* 3(6):561

Checkland, P. & Casar, A., 1986, Vickers' concept of an appreciative system: a systemic account, *Journal of Applied Systems Analysis*, 13:3

Checkland, P. & Scholes, J., 1990, "Social Systems Methodology in Action", Wiley, Chichester.

Cifuentes, C. and Lockwood, C., 1996, Introduction of a case tool to teach structured analysis. *Computers Educ.* 27(3/4):197

Guzzetti, B. and Williams, W., 1996, Gender text and discussion: examining intellectual safety in the science classroom, *Journal of Research in Science Teaching* 33(1):5

Hall, R.M. and Sandler, B.R., 1986, "The Classroom Revisited: chilly for women, faculty, administrators and graduate students". Project on the status and education of women. Association of Women Colleges, Washington DC

Pearl, A. et al, 1990, Becoming a computer scientist: a report by the ACM committee on the status of women in computing science. *Communications of the ACM*, 33(11):47

Tannen, D., 1991, "You Just Don't Understand", Virago, London.

Turkle, S. and Papert, S., 1990, Epistemological pluralism: styles and voices within the computer culture. *Signs Journal of Women in Culture and Society* 16(1):128

Widnall, S.E., 1988, AAAS Presidential Lecture: voices from the pipeline, *Science* 241:1740

UCAS, 1997, "Annual Report 1996 Entry", UCAS, Chelternham.

USING SSM IN DESIGNING A NEW NURSING INFORMATICS CURRICULUM

Peter Kokol

Faculty of Electrical Engineering and Computer Science
University of Maribor
Smetanova 17, 2000 Maribor, Slovenia

INTRODUCTION

Better health care for all and an overall health care restructuring are two among many processes introduced by the health care reform in Slovenia. These two processes have, in addition to many existent assignments, exposed many new tasks and responsibilities for the nursing personnel. Each of these responsibilities entails information as input and also generates information as the output. Concerning it has been recognized that the role of nursing informatics is and will be of the greatest importance in Slovenia. But the concept of nursing informatics is new in Slovenia and nurses are not used to employ the information technology in their every day work practice. Thereafter it is of the greatest importance to properly educate the nurses in nursing informatics and to design an appropriate nursing informatics curriculum. The aim of this paper is to show one of such curriculums and how it has been developed during the NICE project [3, 8].

NURSING INFORMATICS

Nursing informatics can in its simplest form be defined as "a point" where caring and advanced information technology meet. In radiology, laboratory, hospitals, health centres, outpatient centres etc., nursing, informatics is the application of information technology to enhance the quality of care - in other words: better health - care at a lower cost. Other definitions of the nursing informatics are collected in [9, 10] for example:
- The discipline of applying computer science to nursing processes;
- The application of the principles of information sciences and theory to the study, scientific analysis, and management of nursing information for purposes of establishing a body of nursing knowledge;
- The combination of nursing science, information science and computer science to manage and process nursing data, information and knowledge to facilitate the delivery of health care.

Synergy Matters: Working with Systems in the 21st *Century,*
Edited by Castell *et al.*, Kluwer Academic / Plenum Publishers, New York, 1999.

NURSING INFORMATICS EDUCATION

The review in this chapter is based on the invited talk of Hovenga at HTE98 [9]. One of the earliest training opportunities in health informatics was provided by the Laboratory of Computer Science at Massachusetts General Hospital in the mid 1960s where students worked on one or more on-going research projects focused primarily on hospital information system development. A training program emphasising clinical decision analysis was initiated during the 1980s at Tufts-New England Medical Centre and by the Harvard School of Public Health with support from the National Library of Medicine. In Europe, one of the first medical informatics educational program began in 1972 at the University of Heidelberg in Germany.

Nurses constitute the largest group of health professionals in any health system, yet this group has not been well served regarding health informatics education. The Nightingale project represents the first concerted effort to educate many nurses in the discipline. By 1997 only 27% of undergraduate nursing programs had actually incorporated informatics in some way and of those only 19 schools in the entire USA offered specific courses in informatics. However a fairly heavy use of computer technologies to support the education process and its administration was reported at that time. It was noted that the development of nursing classification systems in patient care, the evolution and research on nursing taxonomies and vocabulary, and the use of computers to store both patient and nursing data were hardly covered in all programs. The similar situation can be found also in Australia and Europe.

THE CURRICULUM DEVELOPMENT

To prepare a broad and general curriculum including many different viewpoints the development team consists of various processional profiles like nurses, engineers, computer scientists, medical doctors, administrators, health care managers, WHO representatives, government representatives, teachers and students from different countries. As anticipated these groups had very divergent background, views, opinions and beliefs regarding the nursing, nursing informatics, telemedical information society and especially the nursing informatics curriculum and have it should be taught. To find the consensus, that means an appropriate curriculum for all parties we decided to use the Checkland's Soft System methodology [1, 2], a methodology successfully used in such settings [4 -7].

According to their Weltanschaunng (W) five main groups have been identified:

- Nurses (W: the information technology can help by the documentation of care, making the care and nursing work more visible – and that's all),
- Physicians (W: the information technology can help by the documentation of health care, making the nurses more useful in the overall health care process, and at the same time reducing the routine part of the work of physicians, but a special concern should be placed on ethics, security and the division of responsibilities between nursing personal and medical personal),
- Informatics experts (W: the information technology can help nurses in all aspects of their work including education; not only the general knowledge about information

technology, but also the knowledge about computer supported medical instrumentation is needed);

- Teachers at the college of nursing: (W: the information technology can help by the education of nurses producing better educated nurses)
- Students (W: the information technology is absolutely necessary in both education of care and performing the care).

A root definition and a simple conceptual model describing the nursing informatics teaching process for each W have been constructed. Each model clearly identified the desired outputs (i. e. the required "computer" knowledge divided into topics needed to justify the weltanschauung of a group) and required inputs and activities needed to generate the outputs. The model was first discussed with each group individually and then when all groups was satisfied with their individual models, the "paired" discussions have been initiated (each group with each other group). At the end a workshop facing all groups have been organized resulting in a common root definition presented bellow:

Teaching nursing informatics is a process in which students obtain knowledge from basics of informatics and computer science, informatics in health care (i.e. medical informatics), informatics in care, midwifery and nursery (i.e. nursing informatics), computer communications (especially Internet), medical instrumentation and medical simulations via modern computer supported teaching methodologies with the aim to support real world nursing care process, making nursing work more visible and enjoyable; and finally and most important provide and enable better health for all at lower cost. In addition the students learn how to use computers in learning the conventional nursing subjects.

Table 1: CATWOE Elements of a Root Definition

Customer	Lecturers, students, patients…
Agent	Project management group
Transformation	Nurse → better educated nurse in basic nursing skills knowing also the basics of computer technology
Weltanschaunng	The information technology taught in an appropriate manner can help nurses in all aspects of their work resulting in better health for all at lower cost
Owner	EU, contractor, coordinator, dean of the college of nursing
Environment	Ethics, Law, moral considerations, technology…

At the same workshop we used the common root definition and all the individual conceptual models to construct a common model. Like individual models the common conceptual model clearly identified the desired knowledge which should be taught in the course, together with required inputs and activities needed to generate the outputs. Activities were then grouped in the manner that groups represent the subjects of the curriculum.

The final form of the curriculum is presented in Tables 2. and 3. It consists of six subjects, each subject having 45 hours of lectures and 45 hours of exercises

Table 2. The Nursing Informatics Curriculum – first three subjects

I. Basics of Computer Technology:
Computer hardware (special units used in health care)
System software and computer programming (languages)
Storing of computer data -- file systems
Computer communications (data transfer) and computer networks (protocols, standards)
Software engineering (prototyping, software ergonomy, users' role in designing the applications)
Embedded systems in telematics for health care
II. Medical Informatics:
Informatics in health care
Data and databases (data organisation and access, data files, relations among data)
Data security and protection, legislation, ethics
Information systems (variety and contents, design and goals of information systems, use in the health-care activities)
Patient record (general integrated view, health-insurance connections, standardisation)
Imaging and the health-care process (medical images as parts of the information systems, computer-assisted handling and transfer of images)
III. Nursing Informatics:
Nursing versus medical informatics
Information systems in nursing (goals, contents, implementation)
Design, use, and maintenance of nursing information systems
Understanding and interpretation of computer-assisted information in continuous patient care
Survey of available nursing information systems (in Slovenia and world-wide)
On-going projects on nursing informatics and their outcomes

CONCLUSION

The aim of this paper was to present the new Nursing informatics curriculum and the process of its development which was based on Checkland's Soft System Methodology. It has been shown that SSM can be successfully used to resolve conflicts due to various divergent views (Weltanschaunng) also in the "educational" environments.

REFERENCES

1. Checkland P. "Systems Thinking, System Practice". John Wiley & Sons, Chichester, (1981).
2. Checkland P and Scholes J. "Soft Systems Methodology in Action". John Wiley &Sons, Chichester (1990).
3. Kokol P, Brumec V, The first report about the NICE Project -Nursing informatics and computer aided education in Nursing informatics / edited by U. Gerdin, M. Tallberg and P. Wainwright, IOS Press : Ohmsha, (1997).
4. Gregory W J, Designing Educational Systems: A Critical System Approach. *System Practice* Vol 6 No 2.

5. Maruyama G, Application and Transformation of Action Research in Educational Research and Practice. *System Practice* Vol 9 No 1.
6. Bathany B. H (Ed.), Transforming Education By Design. *System Practice* Vol 8 No 3, Special issue on education.
7. Jeffcutt P (Ed.): Management Education and Critical Practice. *System Practice* Vol 10 No 6. Special issue on education.
8. Kokol P, Zazula D, Brumec V, Kolenc L, Šlajmer Japelj M: New Nursing Informatics Curriculum – an Outcome from the NICE Project in Proceedings of THE 98 / edited by J. Mantas, University of Athens, (1998).
9. Hovenga E J S: Global Health Informatics Education, Proceedings of THE 98 / edited by J. Mantas, University of Athens, (1998).
10. Ball M. J. et al. "Nursing Informatics". Springer, New York, (1995).

Table 3. The Nursing Informatics Curriculum – next three subjects

IV. Computer-Aided Continuous Patient Care:
Computer technology in various stages of the health-care process
Computer-assisted health-care documenting (administration)
Common features of computer-based devices supporting the continuous patient care
Managing the computer-based devices in the nursing and health-care process
Use of computers to improve nursing interventions (nursing diagnoses, nursing activities, therapy, remote
Patient-care interventions simulated and verified by a computer
V. Computer Networks and Information Retrieval:
Computer networking - client-server principle
Basics of the database queries, their formulation and logic
Internet and its services Intranet versus Internet, searching tools and browsers
Publicly available services and catalogues (in Slovenia and
Multimedia approaches, support to the nursing and health-care activities
VI. Computer-Assisted learning Tools:
Computer as learning tool (simulation of real processes)
Conceptual learning for nursing and health care (experimenting in simulated environments and situations)
Simulators of physiological phenomena
Simulations of nursing interventions the nursing classroom
Simulations of physiological reactions
Teleconferencing - transfer of a real clinical environment to

TEACHING SYSTEMS AT THE OPEN UNIVERSITY: RECONCEPTUALISING THE CURRICULUM BY CREATING MEANINGFUL CONVERSATIONS

Andy Lane

Systems Discipline
Centre for Complexity and Change
Open University
Milton Keynes MK7 6AA UK

INTRODUCTION

Systems thinking has been applied to many domains, including higher education. Indeed there has been a UKSS workshop (see Systemist Volume 19 No 4) and an International Workshop (see Systems Research and Behavioural Science Volume 16 No 2) dedicated to systems thinking in education. This paper adds to that discussion by trying to show how systems thinking has been used to help re-design a systems thinking curriculum delivered at a distance through supported open learning. All the activities I discuss overlapped and influenced each other in planned and unplanned ways; the 'story' that I relate is much clearer and cleaner than appeared the case at the time.

The Open University (OU) was founded 30 years ago. Since then it has expanded greatly in terms of its curriculum and the number of students it teaches. The original Systems Department was founded as part of the Technology Faculty in 1971 through the vision of the founding Dean, Geoff Holister: *"I felt that a concern for and systematic study of the social and environmental aspects of technology was essential. Certainly environmental problems were approachable only by means of systemic and interdisciplinary methods and I felt convinced that any Faculty of technology that did not concern itself with such problems could not claim to be either modern or responsible, whether socially or academically" (Holister, 1969)*

Arising from that vision, and given the original emphasis on modular, unnamed undergraduate degrees within the OU, the Systems Department was responsible for producing and presenting 3 linked but independent courses (or modules) to teach systems thinking and practice. This became our main academic project for the next 25 years, and more than 20,000 students have participated in these undergraduate systems courses to date. A description of this curriculum has been provided by Paton (1995).

Looking at this academic project systemically, I begin with an influence diagram (Figure 1) where the primary purpose of the system described is the development and

Synergy Matters: Working with Systems in the 21st Century,
Edited by Castell *et al.*, Kluwer Academic / Plenum Publishers, New York, 1999.

presentation of a loosely coordinated set of systems-based courses for OU students to take as part of their unnamed degree. This was not the sole purpose, as we contributed to other undergraduate courses and undertook research and consultancy work, but the latter had limited influence on that primary purpose. Attempts to change this primary purpose over the years were hampered by the predominant OU Worldview, due to the dominance of the undergraduate programme, which is that of looking simply at the individual modular courses as the focus of curriculum development and not other component levels e.g. groups of courses (qualifications) or individual teaching materials (co-published books, readers, videos, software, etc.), except as incidental by-products.

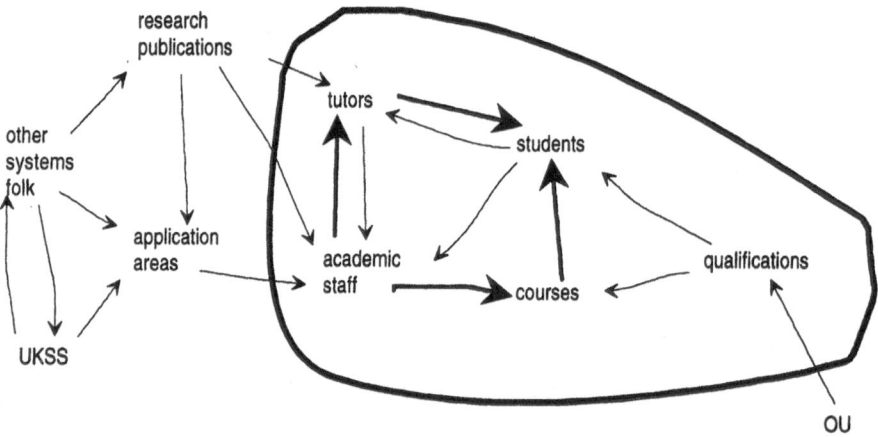

Figure 1 An influence diagram of a system for developing and presenting a loosely-coordinated set of systems-based courses from which students can select those that meet their needs

However, one significant development in recent years was the Department's expansion into producing *systems-related* courses at postgraduate level as part of named awards(dealing with specific subjects such as manufacturing, technology, and development management that included some systems material) rather than *systems-based* courses (that teach systems first and foremost in a variety of domains). This expansion of courses and subjects covered within the Department both challenged that predominant OU Worldview, due to its focus on groups of courses, but also led to intellectual turf wars in the Department over what is and isn't systems (e.g. just because an area **can** be looked at in a systemic manner is no *a priori* reason for using systems or claiming it **is** systems if the course authors do not want to do so). This resulted in our metaphorically likening the Department to a polo mint (Armson and Ison, 1995).

An important feature of Figure 1 is where I have placed the boundary and the connections of greatest influence. This reflects my view that we were too introspective and did not sufficiently acknowledge these new developments or listen to the views of tutors and students and consequently put too much focus on courses rather than the totality of our outputs. The main behaviour is one of influence flowing from ourselves to others through the courses with little going the other way.

Specific developments both within the Department and OU have given impetus to devising a new academic project. First, the size and variety of work within the Department led to organisational restructuringin late 1997, second the Technology

Faculty instituted a major review of its curriculum in the same year, third we increasingly have more rapid feedback from tutors and students through electronic communications, while fourthly the University bowed to student pressure and agreed to implement named, undergraduate degrees. It was within this rapidly changing context that we experienced 3 different types of conversations between participants in our system, the synergy of which helped us reconceptualise the structure and content of our curriculum.

CONVERSATIONS WITH STUDENTS AND TUTORS

A critical feature of reconceptualising the curriculum has been the gathering of views from students and tutors. Gathering the views of students is very important, not so much because they are the 'customer' but because they are mainly in full-time employment whilst studying and have to apply what they learn in the courses to their own domestic or work situation - they are a test of what works and does not work. In a similar vein tutors provide a test of credibility either because they teach systems at their own place of work in Further and Higher Education or because they are using it within their own work as an employee or freelance consultant. Interaction with students and tutors comes in two forms.

First, there is interaction at the individual course level. Until recently the distance teaching model has been characterised by the production and delivery of specially designed courseware to students' homes (an open loop system). This provides a poor environment for feedback and joint learning due to severe time delays (Bell and Lane, 1998). Although 'local' tutors are provided by the OU, geographical or social isolation leaves little room for contact between teacher and student or between student and student in a collegiate sense. The recent introduction of electronic communication provides a better environment for rapid feedback and support to a wider student body (a closed loop system). It has been used on some of our courses and does indeed provide a much richer and sustained discussion of academic issues than was previously possible (amongst the mass of more trivial communications).

A second development was the establishment of the OU Systems Society for (i) past and present students and (ii) any persons who have or have had an appropriate contractual link with the OU i.e. full-time staff, course tutors, residential school tutors, and consultants. This Society provides appropriate communications between the members on issues of systems thinking and practice, through the publication of a quarterly newsletter, an electronic conference, and one-day Annual Meetings.

A third approach is canvassing the views of students and tutors through surveys. Maiteny and Ison (1997) report on one such survey looking at how they benefit from systems thinking while more recently I have undertaken a survey seeking OU Systems Society members' views on how they encountered systems thinking, which concepts and techniques they most value and regularly use and what should be covered in a basic systems thinking course.

CONVERSATIONS WITHIN THE CENTRE

As part of the discussions at the time of the re-organisation of the old Department into a Centre there was a prolonged set of sometimes heated, but good natured, debates about what constituted systems thinking and practice conducted through an electronic conference. These conversations ran for some weeks but represented an intense and thought provoking conversation unmatched by any we had previously had at seminars (usually too focussed on one aspect of systems and not getting wide participation), Department meetings (having greater participation but too focussed on basic information sharing and administrative issues and prone to conflict) and course teams (focussed on systems thinking but restricted to a small subset of the participants). These unplanned and unguided electronic conversations covered many interconnected themes and allowedg time for reading and reflection before contributing.

Although spontaneous and representing thinking-in-progress of people with different perspectives these electronic conversations helped shape much of our subsequent thinking and cleared up many misunderstandings and aired many concerns that had previously remained hidden or suppressed. Such were the quality of these conversations that they have been edited down into a series of 6 informal papers which we have made available to members of the OU Systems Society. They have also sparked off ideas for other publications.

CONVERSATIONS WITHIN THE DISCIPLINE

As Head of the new, but smaller, Systems Discipline within the Centre for Complexity and Change, there were 3 aspects to the system described in Figure 1 that I thought needed redesigning to give us a new focus to our work. The first was to revise the main purpose, the second to change the system boundary and the third to widen the range of outcomes.

I proposed that our main purpose should be to encourage and foster the use of systems thinking and practice to as wide a range of individuals and institutions as possible through a coordinated set of teaching materials and other publications dealing with systems thinking and practice at all levels of higher education that can be assembled into courses and qualifications in a clearly defined set of broad application areas but do not necessarily have to be embedded in a specific course. This new system is described in a revised influence diagram (Figure 2). By widening the boundary to include other people more directly and by diversifying the ways in which we interact with them, we believe our overall influence and impact should be enhanced.

When considering the question as to what distinguishes the different levels and nature of our courses we started from where we were and then looked at what we wanted to achieve. If we ask the question "what, in essence, does the content of each of our *current* undergraduate courses enable a student to do?" we decided that:

1. Our contribution to the Faculty's first level course merely introduces the idea of systems thinking and simple practice in diagramming techniques.

2. The second level courses enable students to use systems concepts and techniques to transform information about, or experience of, real world situations into systemic descriptions of those situations.

3. The third level course subsumes the second level courses and enables students to transform systemic descriptions into plans for change, using the systems methodologies which are the core of our third level teaching.

4. A fourth level project course subsumes the previous systems courses and transforms plans for change into action.

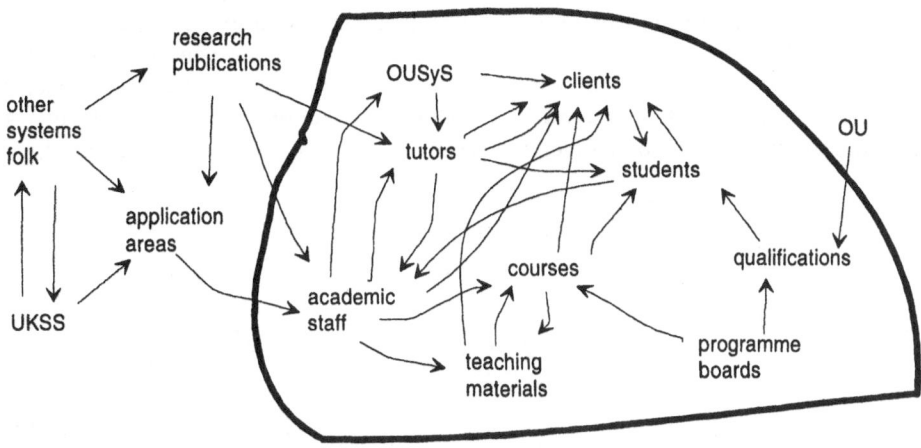

Figure 2 An influence diagram of a system for developing and offering a co-ordinated range of systems-based teaching materials, courses and publications from which students and other clients can select those that meet their needs

The content of each of these courses can therefore be seen as a system to enable students to transform something into something else. Thus, in crude terms, the outputs from the second level courses provide the inputs to the third level, while the outputs of the third level provide the inputs to the fourth, so each level can be seen as a transforming subsystem of the whole "working systemically" system, which transforms the real world through systemic action. All this can be shown on a simple input-transformation-output diagram (see Figure 3). What is missing from this diagram is any sense of feedback between the sub-systems, but these have been deliberately missed out in order to keep things simple.

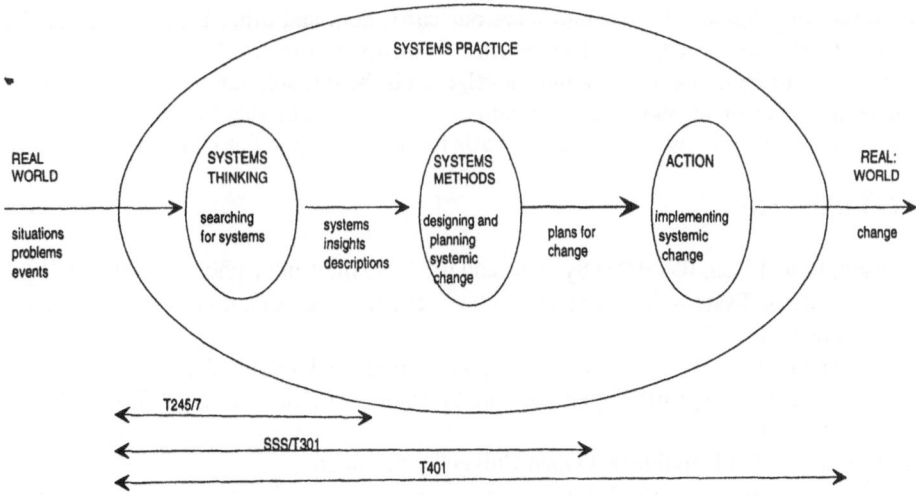

Figure 3 Defining levels of study in the current curriculum of courses

If we now look to the *future* and include new courses at level 1 and M (postgraduate) we can modify this model in 2 ways. First we can recognise that there are elements of theory and praxis at all levels (i.e. action is not just at one level) and that

503

there is feedback between them. Second, the progression of transformations moves from systems methods on to systems philosophy where there is more consideration of the epistemology as opposed to ontology of systems thinking and practice, with personal reflection on theoretical frameworks, methodological developments, etc. Figure 4 provides a revised model for defining levels and how planned courses fit to the model.

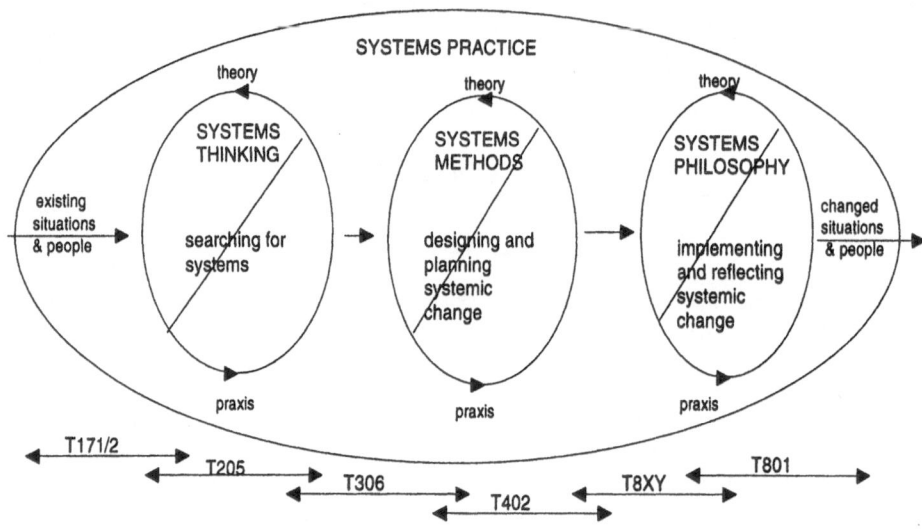

Figure 4 Defining the levels of study in the new systems curriculum of courses

POSTSCRIPT

This paper has described work in progress. Many of the conversations mentioned above are ongoing and further informing our curriculum and other areas of work. We have been able to use what we have learned so far to restructure the ways in which we deliver our curriculum at a programme level, even if there is still much detail to be sorted out at the level of an individual component or course. But the main message is the success in using systems thinking as the basis for deciding upon our own actions.

REFERENCES

Armson, R. and Ison, R. (1995) Systems and the 'polo mint' metaphor, in "Critical Issues in Systems Theory and Practice", eds Ellis, K. et al, pp 637 - 642, Plenum Press, New York.

Bell, S. and Lane, A. (1998) From teaching to learning: technological potential and sustainable supported open learning, *Systemic Practice and Action Research*, Vol 11 No 6.

Holister, G. (1969) Unpublished Open University document.

Maiteny, P and Ison, R. (1997) Learner-centered evaluation of systems, systems courses, and future needs in systems learning, in "Systems for Sustainability", eds Stowell et al, pp 257 - 262, Plenum press, New york

Paton, G. (1995) Opening up systems: a review of systems teaching in the Open University, in "Critical Issues in Systems Theory and Practice", eds Ellis, K. et al, pp 659 - 665, Plenum Press, New York.

USING MULTIMEDIA TO TEACH SYSTEMS THINKING

Sheila S. Stone

Technology and Manufacturing Management Discipline
The Open University
Walton Hall
Milton Keynes
MK7 7AA

INTRODUCTION

Having written several introductions to systems thinking for various courses I felt there was a need to provide a pack module that would require minimal additions to satisfy a course team. Each team wants a module that relates directly to its own contextual situation. However students taking more than one of these courses often found that the definitions of the concepts appeared to be different in each course. There is also a problem of perception on the student's part because this apparent anomaly occurs although the written definition appears to be the same or nearly so in two courses. The student may still complain that the definitions are different. It seems that the context within which it is used causes the problem.

As definitions are the basic building blocks of systems thinking and its communication to others it is important that students should always have the same definition and get to understand its fuller meaning by its application to different contexts. Not only are definitions of concepts and use of techniques influenced by the topic they are related to but also by the student's previous understanding of what the word means. The more mature the student the greater the problem can become. Checkland[1] and others have suggested using 'Holon' in place of 'system' to get over this problem. However, that raises other problems and, if adopted, would only deal with a single concept.

Part of the problem lies with the lecturers themselves who do not make explain the definition they are using. It is easy to highlight one aspect that is important to your application of a concept without recognizing that the definition being used is incomplete. Thus when the student encounters it again taught in another course, the definition could seem very different because the context in which it is applied is different. The student then has difficulty in relating their understanding of the two 'definitions' to each other. It is important that we should provide an environment in which the student builds on the previous knowledge and develops a more generalized definition.

As the Open University teaching is based on independent study and flexible learning of the type described by Entwhistle[2] and others to mature students, one cannot be sure what experience and knowledge each student is bringing to the course from their own

Synergy Matters: Working with Systems in the 21st Century,
Edited by Castell *et al.*, Kluwer Academic / Plenum Publishers, New York, 1999.

505

background and previously taught courses. This independence and flexibility thus creates a problem when several courses require some of their teaching to be based on a given set of knowledge but use it in different context. It is not surprising that each course teaches that knowledge in a way seen most suitable for that course. This negates mathemagenic activities[3], that is, the activities that give birth to learning. This type of confusion, in the environment in which they learn, causes the student to spend time on irrelevant areas in order to achieve the aims of the course. This often causes the student to become frustrated and not learn. In this case the affordances [3,4], what their learning environment provides with a student for good or ill, are not mathemagenic. At present at least six courses currently provide this initial introduction to systems thinking[5,6,7,8,9,10]. Though attempts have been made to ensure the basic teaching is similar problems with definitions still exist. Therefore there is a need to provide a preparatory pack. This multimedia demonstration is being developed to see whether using a CD-ROM or possibly the Internet could provide the teaching required.

DESIGN ISSUES

Problems with current teaching

I am the presentation chair of one of the System Programme Board's courses and have also tutored that course for many years. Analysing my experience together with comments from other tutors, monitoring assignments and students' comments, it is clear that there are problems with the current way of teaching the subject.

These are:
- Developing generalizable concepts and techniques;
- The students' ability to apply concepts and techniques in different contexts;
- Developing the art of analysis, especially at the understanding stage.

It appears that part of the problem is that the affordances of the current teaching environment are not all mathemagenic.

There are also some fundamental problems caused through the teaching being mainly done through text. Text provides a linear presentation of a non-linear subject. This linear representation loses the randomness of gaining information about a situation, which one would normally expect to occur in real life. Randomness can affect the subsequent analysis and cause the analyst at times to iterate in their analysis. Students get very little exposure to this type of problem when analysing a situation. It also dictates the learning environment within which students work but cannot as a rule create for themselves. This means that for many students some of these affordances will not be conducive to mathemagenic activities. The students need to be able to tailor this environment to their own style of learning more than they currently can.

Aims of project

The initial aims of the project are:
- To provide a package which can be tailored for different courses at different levels.
- To provide materials which can be equally usable for the first and subsequent times of use and provide the possibility of different learning environments.
- To create a prototype package to teach the basic concepts needed for systems thinking, incorporating the possibility of randomness in the collection of the information.

This paper concentrates on designing a package to satisfy the first two aims and indicates how the latter will be tackled.

Software for development

For a prototype of this kind I first needed a software package which would allow me to express my ideas without too many programming problems. A desirable package would allow me to translate my initial program for the CD-ROM into one for the Web so the course could be used in either form.

I considered Director, Authorware and Toolbook. Authorware proved unsuitable initially as it required the components to be already developed in another system. I began using Director 4, which provided me with very good control, dovetailing sound and text and animation. However the time required to develop this prototype proved excessive as I found that Director had a very steep learning curve that I did not have the time to scale. I decided to use Toolbook whose Instructor II would allow me to construct the course more quickly for either for CD-ROM or Web output, with possible translation between the two.

Toolbook also had the advantage of providing several ready made but adaptable items to insert in the package such as scrolling bars and navigation together with suggested scripts, not provided in Director 4. Animation, which had already been developed in Director, was imported into in Toolbook so work developed in Director was not wasted. That animation has since been improved using Shockwave. I have experimented with translating part of the course to a Web type course but have found that there is still quite a lot of work to be done in the new format. I therefore decided to continue with the CD-ROM format.

DEVELOPMENT OF THE TEACHING PACKAGE

Developing a New Teaching Package

The first design issue is how make the package relevant to its companion course. The approach here is to allow each course to integrate its own case studies into the package. The students will then see how the generalized teaching in the theoretical part of the package is applied to different contexts. By having an obviously common theoretical part with generalized definitions it is hoped that the application to case studies will provide the mathemagenic activities required for students to develop a greater understanding of the concepts. It will ensure that the same basic definitions are used in every course and that their application to a specific context is made explicit. If this is achieved it will then allow the students to successfully apply concepts in different contexts. It is hoped that this skill will be transferable to other subjects.

It is also important to give the student the freedom to create a learning situation more congenial to them. To do this the student is free to move wherever they wish to after a small section of teaching is done.

One of the important choices a person can make with this package is whether to have sound or text only or both media. I decided on providing this choice because at times when I feel I would prefer to hear rather than read or vice versa. I know from the reports on the audio tapes we use in our current teaching, some people get tired of hearing the tapes and would prefer to read the information and vice versa. The package allows the student change from sound to text or both at any point in the course. However this choice is not available when an imported video is played. The decision to provide the students with this choice has provided some problems. When text is included the screen becomes crowded, however there is also a danger that when sound only is used the screen becomes bare.

The screens were initially designed for text and sound together. This was done to ease the burden of developing a course using a medium I was unfamiliar with. However when moving on to allowing the student the choice of media I found that when either media was not used the screens needed to be adapted slightly. For instance, with sound only, the part of the screen devoted to text was a blank page so some picture or animation had to be added. In some cases this actually improved the teaching so was incorporated into the textual part of the course. This was an interesting example of using different perspectives in the development of the package so that it improves the package as a whole. When only text was used it was clear that there was a problem when animation was involved of how the student could control what was happening. When sound is used the student can watch and listen but with text this was not possible. It was clear that the student would need more control over the speed at which the information was given. The problem with text resulted in including continue buttons to allow the student to move on when they wished. This was also incorporated to improve the student's control when sound was used. It has the benefit that if the student started the wrong page they can move from it after a very short time. A further possibility is to allow the student to decide whether to run the sequences as one continuous one or not. This has not been investigated, as I am afraid that too much choice would be given.

The changes described above have improved the way the student can create their own environment, but it also provides a danger of flipping from one place to another without understanding anything. It is most important that a first time user does not do this.

First and recurrent exposure to the package

The package needs a structure for the student's first use while allowing freedom of choice for any subsequent use. It is important that the affordances between the learning environment and each student allow mathemagenic activities to occur. As these affordances differ between students this means that I have to try to cater for many approaches to learning. From students' comments it is clear that some would like to start with an example of what should be possible to achieve by the end of the course. Others wish to learn the theory and then how to apply it in easy stages. Using multimedia allows us to cater more for each type of learning environment in that the student is given the chance to construct, to some extent, their own environment. They will be free to make choices about how they tackle the subject. This is very important for subsequent users.

However this freedom has its own problems for a student who is unfamiliar to the navigation. Laurillard[2] found that students new to a package spend most of their time trying to decide what they can do next and not concentrating on the subject matter.

If the student is allowed freedom to navigate where they will at subsequent uses requires some guidance to be provided for first time users and others who have forgotten what is involved with the navigation. This will allow them to spend their time concentrating on learning the concepts and techniques and not the navigation. This guidance must be made clear from the start so that the student will soon gain confidence how to move about the package. The method chosen was the use of a deep orange colour on the suggested next button to click on. This colour only appeared when the student was required to move on. To get the new student used to making choices for themselves after the initial linear progression into the course, they are presented with choices. It is hoped that this will develop their independence.

For students encountering the package for the second and subsequent times, they can use the first time route until they remember the basics of navigation or go their own way. The support they are given is described next in the navigation design.

NAVIGATION DESIGN

Navigation design needs to allow you to have some idea of where you are, where to go next, how to get out of a mess, what you have done in the course, and how to find bits of the course you think are important. There are several aspects that need considering when developing a navigation design, especially when the user can go where they like.

- Identifying the type of teaching to be expected: Introduction, theory, practice, questions etc.
- The types of buttons to be used, standard Microsoft or course specific.
- Identifying the route through the work for a first time user.
- Providing menus to allow people to move quickly to where they wish.
- Providing information about the parts of the course the student has visited.
- Providing the ability to create a personal index.
- Providing the ability to construct a personal revision notebook.

So far the first four on the list have been at least partially implemented, as they are essential for the package. Providing information the last three will be done when the teaching part has been completed.

To help the student identify their current area of teaching it was decided to use background colour as an aid to location. Systems thinking requires one to start the analysis in the real world and move into thinking about the systems that might exist. As a system is a construct within a person's mind, systems thinking is a thought process that should be seen as such. It was decided to use green for the real world and the case studies (the practice part) and blue for systems thinking and its associated concepts and techniques. Initially, in the introduction of the subject, the screen was divided with green at the top and blue at the bottom to denote the movement from reality to thinking about it. All the additional learning questions had a darker blue background to indicate deeper thinking.

There is also, in parts of the package, textual information that indicates the area.

There is a debate as to whether it is better to use conventional Microsoft type navigation, which students might already be familiar with, on use some non-standard type. There seem to be good arguments on both sides but I decided to go down the non-standard route because I wanted to make the students recognize that they had to think in a different manner.

The navigation consists on a systems map which shows the structure of the teaching, see figure 1, and the use of buttons mostly at the foot of the page.

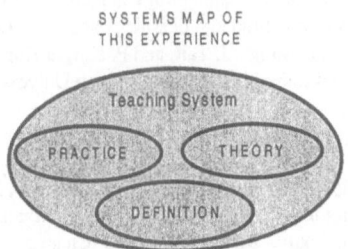

Figure 1 Systems map navigation

The basic navigation from theory to practice and accessing definitions of concepts is via the systems map. This reflects one type of diagram used in systems thinking and provides the students with the ability to move quickly from theory to practice and back again, or look up the definition of some concept. The student could also access these definitions by clicking on hotwords in to text. The use of theory and practice in the systems map allows for movement between the two often within the context of the work they are doing. This is especially true when they are working on a case study. The link takes the student to the relevant part of the theory they need.

Another essential button is QUIT to allow the students an escape route out of the package altogether. I personally have found that the lack of this button in a package has caused me to have to switch off the computer to escape from the situation I got myself into.

The other buttons that appear used at the foot of the page are continue, next, back and OK buttons so that the student always knows where to find them. Some other buttons appear elsewhere on the screen when there is a particular reason. It is always made clear to the student where these buttons are.

To allow the new user to find their way around the package the students are instructed to click on dark orange colour. Initially this colour is on a door through which they go to find out more about systems thinking. Clicking on buttons when they change to dark orange takes the student through the course. Initially these buttons provide no choice as the method of analysis is explained in general, but after that they can tailor the learning environment to their own preference. They can choose to explore a case study, and access theory as they need it, or concentrate on the theory. In the case study they could go through it without looking at the theory or use it as an introduction to the next part of the theory. The movement from a theoretical route to the case study is suggested at several points and it is hoped that by doing this the student will devise their own learning environment.

Because the screen is rather busy I have used a drop down menu to allow the student to jump to either a basic theory or a practice menu. The practice menu lists the case studies and the theory menu chapter headings in the course. Eventually there will be a large index and the students will be able to create their own bookmarks.

FUTURE WORK

The package has been demonstrated a few times and will shortly be ready for the first people to test it properly. This will most likely be before the book marking and identification of what one has already done has been implemented. The results of those tests will then be incorporated into the package and the final parts added.

REFERENCES

1. P. Checkland, and J. Scholes, "Soft Systems Methodology in Action", Wiley, Chichster (1990).
2. N. Entwhistle, "Recent Research on Student Learning and the Learning Environment" *in:* "The Management of Independent Learning", J. Tait, and P. Knight Kogan Page Ltd., London (1996).
3. D. Laurillard, "Studying learning and rethinking teaching", Open University: Professorial Inaugural Lecture (1997).
4. J. J. Gibson, "The Ecological approach to visual perception", Lawrence Erlbaum Associates Inc. New Jersey, (1986).
5. T247 course team, "T247 Working with Systems", The Open University, Milton Keynes, (1996).
6. T245 course team, "T245 Managing in Organisations", The Open University, Milton Keynes, (1995).
7. T301 course team, "T301 Complexity, Management and Change: a systems approach", The Open University, Milton Keynes (1993).
8. T836 course team, "International Operations Management ", The Open University, Milton Keynes (1996).
9. T840 course team, "Technology Management: An Integrated approach", The Open University, Milton Keynes (1994).
10. T841 course team, "The Strategic Management of Technology "The Open University, Milton Keynes (1994).

SYNERGY OF THE 'SOFT' AND 'HARD' SCIENCES FOR IMPROVING INFORMATION SYSTEMS

Joseph Akomode

School of Computing and Mathematical Sciences
Liverpool John Moores University
Byrom Street
Liverpool L3 3AF
England   United Kingdom
Tel: +44 151 231 2416   Fax: +44 151 207 4594
E-Mail: J.O.Akomode@livjm.ac.uk

INTRODUCTION

Various forms of Information Technology Systems (ITSs), more commonly known as Information Systems (ISs) are often employed by enterprise managers and other staff to support decision making and operational activities. The artefacts (i.e. ITSs) may largely result from technically biased systems-analysts/developers who have inadequate knowledge of the *unique* business needs of an enterprise. On the other hand a wealth of knowledge about the business(es) of an enterprise combined with little or no knowledge of the technical complexities associated with ITSs is unlikely to deliver a suitable IS to support managers and their staff effectively. Current research (e.g.: Avison and Wood-Harper, 1990; Flood and Jackson, 1991) indicates that to have a balanced knowledge-set for supporting the development of an effective ITS, in order to minimise business risks requires the combined knowledge of: (i) the business needs of an enterprise and (ii) the capabilities of Information Technology (IT). The implication of not having an appropriate combined knowledge-set being that the resulting ITS may be inappropriate to the situation that it is meant to resolve. A suitable relationship between the '*soft*' and '*hard*' sciences has the potential for a synergy of methodologies and their associated methods for providing an appropriate ITS (cf. Jackson, 1997). This paper discusses the following: (i) the issue of adopting a 'learning' process based on *Action Research* (AR) as a form of 'soft' science; (ii) the application of a *multi-criteria decision making* approach involving the *Analytic Hierarchy Process* (AHP) as a form of 'hard' science; (iii) the results of a synergistic ITS obtained from an industrially based research. An example of a framework for the development of an ITS is presented to support the potential for synergistic working, in order to cope with the challenge of obtaining a suitable IS for an enterprise.

EMPLOYING A LEARNING PROCESS BASED ON ACTION RESEARCH AS A FORM OF 'SOFT' SCIENCE

Managers and other staff in an enterprise may be viewed as 'owners' of the enterprise and its businesses. That ownership referred to implies that managers (and staff) are in a better position to be more conversant with the *strengths, weaknesses, opportunities and threats* associated with the enterprise. The expertise of the manager and staff often include 'deep' knowledge about key elements for management and operational activities, costs, possible suppliers, customers, markets and competitors. Consequently, they are better placed to know about the business needs concerned with the enterprise and its businesses. With regard to business risks and associated risk assessment/management for mitigating risks, managers often employ their human expertise which may be difficult to articulate let alone measure. The expertise of managers may involved heuristic, personal and judgmental components of risk. Developing an ITS to support such managers in decision making may become problematic due to the difficulties often encountered in Knowledge Elicitation (KE) (cf. Feigenbaum, 1984). Various KE techniques have been advanced by many experts and commentators and the techniques include: Repertory Grid based on Kelly's (1955) theory of personal constructs; Protocol analysis; Multidimensional scaling and Concept Sorting (Gammack and Young, 1985).

Also, Neale (1988) and Boose (1989) discuss various types of structured and unstructured interview techniques. In order to alleviate the inherent problem of KE the *Appreciative Inquiry Method* (AIM) has been adopted for iterative 'learning' about the domain of risk assessment and for obtaining key elements appropriate for developing a customised IT-based IS to support risk evaluation. The main aim for employing IS being to obtain more accurate and relevant information for supporting managers in effective decision making. The literature of AIM is available in Stowell, West and Fluck (1991); West (1995). The philosophy of the method is based on a 'systemic' approach and involves the strategy of AR (Lewin, 1946; Susman and Evered, 1978). More specifically, the epistemology of the method and its practical tools impact on Soft Systems Methodology (SSM) (Checkland, 1981; Checkland and Scholes, 1990); phenomenology and hermeneutics (Winograd and Flores, 1990; Burrell and Morgan, 1994); Vickers concept of appreciation (Vickers, 1965), all of which are located in the 'interpretive' paradigm for analysis of social theory (or organisation). Though the systems models employed in AIM are adopted from SSM, the objectives of the two are not the same. While SSM provides the process to bring about *change*, the purpose of AIM is mainly as a *means of finding out* what is considered to be the case in a given situation (West 1995, p.144). However, they both articulate AR and iterative 'learning' as a process of (organisation) inquiry and anlysis. Consequently, the key elements employed in Figure 1 below (especially the criteria and sub-criteria elements) were obtained by the use of AIM as an AR-based method and a form of 'soft' science.

APPLICATION OF THE ANALYTIC HIERARCHY PROCESS AS A FORM OF 'HARD' SCIENCE

Most (if not all) human endeavours require initiative and action to select one of several alternatives. Thus decision making is a practice we often get ourselves involved in from day to day. For the purpose of managing an enterprise effectively, appropriate and timely decision making may serve as a critical success factor. The *Analytical Hierarchy Process* (AHP) is a multi-criteria decision making method which has been widely employed in different types of decision making situations or problems, for example (Saaty, 1978; 1986; 1995; Saaty and Vargas, 1991; Zahedi, 1986; Yau and Davis, 1993).

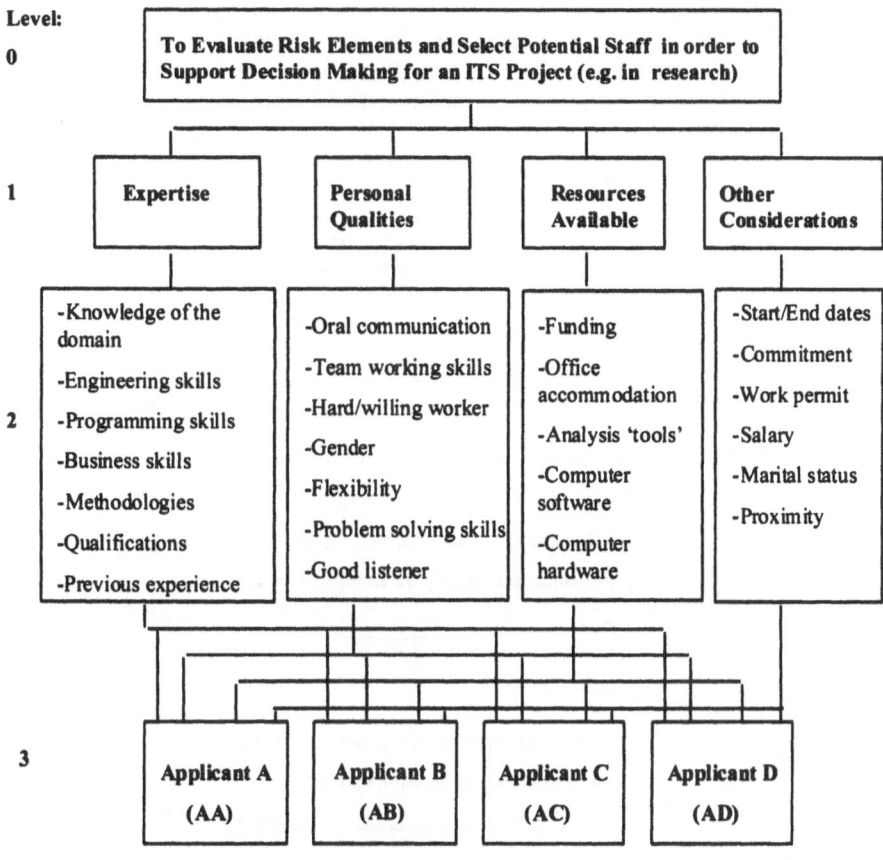

Key: 0 = Goal; 1 = Criteria elements; 2 = Sub-criteria elements; 3 = Alternatives

Figure 1. <u>Example of a Hierarchical Structure for Risk Evaluation of Potential Staff</u>

The AHP method has the capability of converting subjective assessments of relative importance into a set of linear priorities (or weights) which can be used to rank alternatives, in order to support decision making more effectively. The application of the method consists of a number of stages of which the main ones are:

(i) the development of a hierarchical format for attributes of the problem (e.g. Figure 1);

(ii) the identification of relative importance of attributes through pairwise comparison;

(iii) obtaining a set of *quantitative data* as a measure of the subjective judgments for the relative performance of each available alternative with regard to each element of the hierarchy (e.g. Figure 3).

In this case the problem to be addressed is that of evaluating risk elements associated with the appropriate selection of suitable staff for an ITS project. As an example consider the case of an ITS position where the number of the potential staff short-listed is four. The primary and associated key elements to be considered as shown in Figure 1 were obtained through a method based on AR. In order to employ the method of AHP, the problem and associated parameters including the four alternatives are arranged in a hierarchical format as shown in Figure 1. Based on that diagram (Figure 1) a *customised* IT-based IS was developed as an IT supported decision making artefact (i.e. the ITS model) for use by an evaluator (or a group of evaluators). The aim being to quantify subjective judgments and obtain more accurate information to support decision making. Figure 2 represents a framework for the development of the ITS. Detailed examples and results obtained from the application of AR-based investigation in an enterprise which further help to elucidate Phase-1 of Figure 2 are available in Akomode *et al* (1996; 1997a).

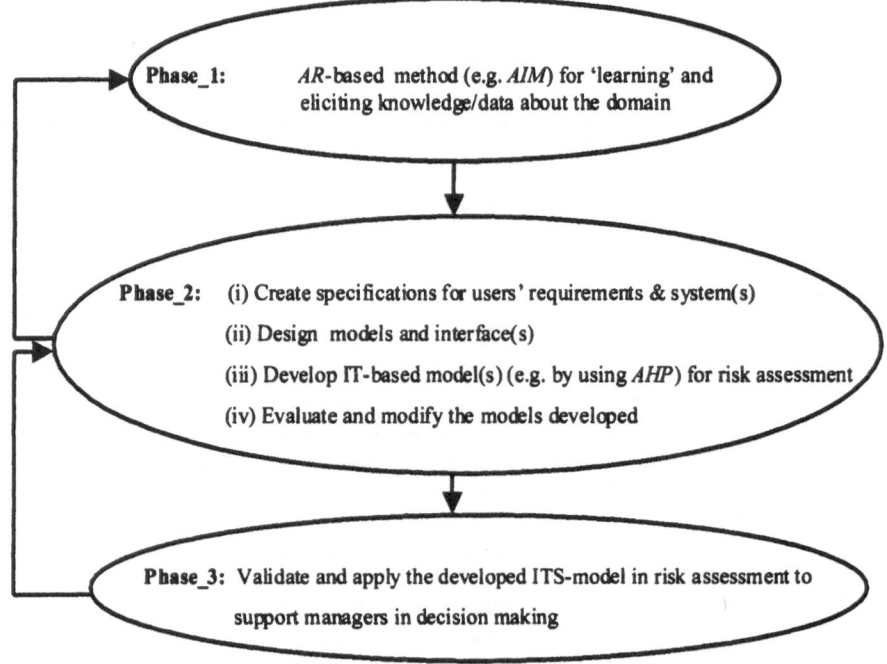

Figure 2. <u>Example of a Framework for the Development of an ITS for an Enterprise</u>

In using the developed ITS the evaluator typically has to give answers to two main questions, one in relation to the criteria elements and the other in relation to the sub-criteria elements in conjunction with the possible alternatives. For example: (i) *in terms of risk minimisation and quality improvement, which of these criteria elements is more important to us as an organisation and by how much?* (ii) *in risk minimisation and quality improvement with regard to the sub-criteria element (x), which of the four possible alternatives in Figure 1 best suits our purpose and by how much?* The variable (x) represents any of the sub-criteria elements at any point of the evaluation process. Based on an example from a similar evaluation process using the developed customised ITS, the resulting *quantitative* and *pictorial information* from a computer synthesis is given in Figure 3. From the result the 'optimum' choice is given to applicant C, hence AC = 31.7%; followed by applicant B, i.e. AB = 26.8%; AD = 22.1% while AA = 19.4% is the list preferred by the evaluator. Effective application of the ITS model requires the user to have a knowledge of the 'AHP scale for pairwise comparison', see Saaty (1978, p.152; 1995, p.83). Note that in using this model another evaluator with different opinion may obtain different results. The different results from two or more evaluators may be aggregated based on their agreement or used as a basis for discussion, in order to arrive at a final decision. Examples of technical details for developing an ITS based on the method of AHP are available in Akomode *et al* (1997b, 1998); Akomode (1998).

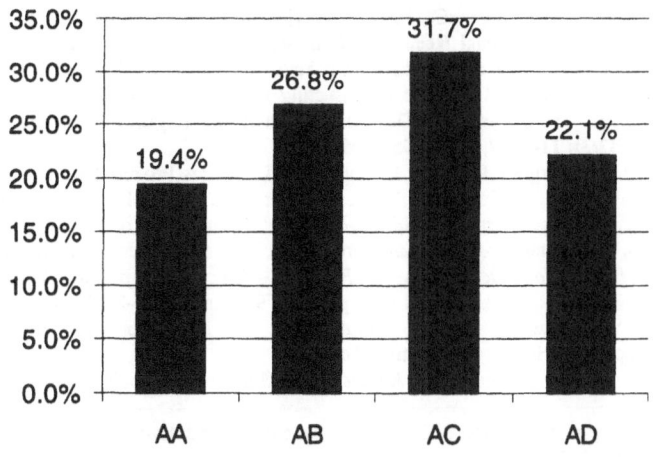

Figure 3. Risk Evaluation Results Obtained in the Selection of Potential Staff

CONCLUSION

To support the development of a suitable ITS for managers (or users) in an orgainsation an *Action Research* based method has been proposed as a form of 'soft' science and a precursor to the 'hard' system development phase. On the other hand a method based on a *multi-criteria decision making approach* (e.g. the AHP) has been employed as a form of 'hard' science in the development of the ITS for users. The main issue addressed here is that *a suitable combination of 'soft' and 'hard' sciences* has the potential to provide an effective customised IT-based information system for an organisation. The roles and advantages of the ITS presented include: (i) the opportunity to represent key elements associated with the selection of potential staff in a hierarchical format; (ii) the ability to optimise subjective judgments; (iii) the use of an alternative method for obtaining relevant information and explanation as to *how* such information was arrived at; (iv) the model may be used by one or a group of evaluators working together. This paper makes no claim to the effect that the approach presented will solve all the problems of risk minimisation and quality improvement associated with risk evaluation for the appropriate selection of suitable staff for an organisation. But the approach has the capability to help in obtaining useful and lucid information to support effective decision making.

ACKNOWLEDGEMENTS

Appreciation and gratitude are offered to Alex Paterson of Renfrewshire Enterprise, Paisley, Scotland and to the Manufacturing Director, Ian McNair, and other management personnel at Compaq Computer Manufacturing Limited, Bishopton, Scotland, for their support in the research programme.

REFERENCES

Akomode, O. J.; Lees, B. and Irgens, C., 1996, Eliciting risk assessment knowledge for decision support in manufacturing management, in: *British Computer Society, Information Systems Methodologies, Lessons learned from the use of Methodologies*, Jayaratna, N. and Fitzgerald, B., eds., University College Cork, Ireland, 12-14 Sept. 1996, 143-152, ISBN 0901865 87 7.

Akomode, O. J.; Lees, B. and Irgens, C., 1997a, Applying Information Technology to Minimise Risks in Satisfying Organisational Needs, in "Information Infrastructure Systems for Manufacturing", (eds), Goossenaerts, J. *et al.*, Chapman & Hall, 242 – 253, ISBN 0 412 78800 4.

Akomode, O. J.; Lees, B.; Irgens, C., 1997b, Constructing Customised Models and Providing Information to Support IT Outsourcing Decisions. Journal of: *Logistics Information Management*, 1998,11(2/3): 114-127, ISSN 0957-6053.

Akomode, O. J.; Lees, B.; Irgens, C., 1998, Evaluating Risks in New Product Development and Assessing the Satisfaction of Customers Through Information Technology, *International Journal of Production Planning and Control, 1998* (in print by Taylor & Francis Ltd).

Akomode, J., 1998, Quality and Risk Evaluation for Enhancing Performance in Enterprises, *in Proceedings of the British Computer Society, INSPIRE III – Process Improvement Through Training and Education,* (eds) Hawkins, C. *et al*, September '98, London, 69-81, ISBN 1 902505 03 4.

Avison, D. and Wood-Harper, A.,1990, "Multiview: An Exploration into Information Systems Development," Blackwell Scientific, Oxford.

Boose, J. H., 1989, A survey of Knowledge Acquisition Techniques and Tools, *Knowledge Acquisition*, 1: 3-37.

Burrell, G. and Morgan, G., 1994, Sociological Paradigms and Organisational Analysis, Ashgate Publishing Ltd.

Checkland, P. B., 1981, *Systems Thinking, Systems Practice.* Chichester; Wiley.

Checkland, P. B. and Scholes, J., 1990, *Soft Systems Methodology In Action.* Chichester; John Wiley and Sons.

Feigenbaum, E. A., 1984, The Applied Side, In: Hayes, J. E. and Michie, D. (Eds): Intelligent Systems-The Unprecedented Opportunity. Ellis Horwood, Chichester, 37-55.

Flood, R. L. and Jackson, M. C.,1991, "Creative Problem Solving: Total Systems Intervention," Wiley, Chichester.

Gammack, J. G. and Young, R. M., 1985, Psychological Techniques for Eliciting Expert Knowledge, in, Bramer, M. A. (ed.) *Research and Development in Expert Systems,* Cambridge University Press, 105-112.

Jackson, M. C., 1997, Towards Coherent Pluralism in Management Science, *Working Paper No. 16 1997, Lincoln School of Management, Lincoln, U.K.*

Kelly, G. A., 1955, "The Psychology of Personal Constructs" (two volumes), Norton, New York.

Lewin, K., 1946, Action Research and Minority Problems. *Journal of Social Issues*, 2:34-46.

Neale, I. M., 1988, First Generation Expert Systems: a review of knowledge acquisition methodologies, *The Knowledge Engineering Review*, 3(2):105-145.

Saaty, T. L. (1978). Modeling unstructured decision problems - the theory of Analytical Hierarchies. *Mathematics and Computers in Simulation*, 20(3):147 - 157.

Saaty, T. L., 1986, Exploring optimization through hierarchies and ratio scales. in, *Soci-Economic Planning Sciences - An International Journal,* (ed) Harker, P., 20(6): 355 - 360.

Saaty, T. L., 1995, Transport planning with multiple criteria: the Analytical Hierarchy Process applications and progress Review. *Journal of Advanced Transportation,* 29 (1):.81-126.

Saaty, T. L. and Vargas L. G.,1991, "Prediction, projection and forecasting - application of the Analytical Hierarchy Process in economics, finance, politics, games and sports", Kluwer Academic.

Stowell, F. A.; West, D. and Fluck, M., 1991, The Appreciative Inquiry Method: An Approach to Knowledge Elicitation as an Inquiry System, *Systemist*, 13(4):154-65.

Susman, G. I. and Evered, R. D., 1978, An Assessment of the Scientific Merits of Action Research, *Administrative Science Quarterly*, 23:582-603.

Vickers, G., 1965, "The Art of Judgement: A study of policy Making", London: Chapman and Hall.

West, D., 1995, The Appreciative Inquiry Method: A Systemic Approach to Information Systems Requirements Analysis, in: Stowell, F. A. (ed.) "Information Systems Provision: The Contribution of Soft Systems Methodology", pp.140-158.

Winograd, T. and Flores, F., 1990, "Understanding Computers and Cognition: A New Foundation for Design" Addison-Wesley.

Yau, C. and Davis, T., 1993, Using Analytic Hierarchy Process (AHP) to Prioritize Auditing Tasks for Large-Scale Software Systems, *Journal of Systems Management*, 44(11): 26-31.

Zahedi, F., 1986, The Analytical Hierarchy Process - a survey of the method and its applications, *Interfaces*, 16(4) July – August: 97-108.

WHAT IS THE POINT OF WORKING IN AN INDUSTRY WHERE THERE IS A BETTER THAN EVENS CHANCE OF YOUR WORK BEING SHELVED?

Steve Armstrong [1] and Aidan Ward [2]

[1] Computing Department, Faculty of Maths and Computing, The Open University, Milton Keynes, England MK7 6AA, tel: +44 1908 654056, fax: +44 1908 652140, email: s.armstrong@open.ac.uk
[2] Antelope Partnership, 9 Underhill Road, East Dulwich, London, SE22 0AH. tel: +44 181-299-1399, email: aidan_ward@antelopes.com

INTRODUCTION

The world of professional engineers of all kinds consists largely of business organizations who buy their services. Problems and opportunities are perceived by management and engineers are engaged to work on them. In particular, software engineers are contracted to do diverse jobs in large software projects. Often, a large organization will use software engineers from more than one consultancy or software house on a given project.

The ability of people within an organization to ask crucial questions about the purpose of their job is undermined by the number and variety of their allegiances. As a consequence, we find taboos that restrict what can be done for fear of what might be uncovered. Software engineers face a pressure to collude with the sponsoring organization about the nature of its problems so as to reduce or avoid their anxiety. The dynamic of avoidance is a driver towards authoritarian systems that turn their members into interchangeable and dispensable objects, which is how the accountants in their parent organization might treat them.

The technical matters involved in a large software project are part of a larger organizational and business process whose meanings and values are profoundly different from the objectivity that engineers pride themselves in. Software engineers have difficulty understanding the ways that their methods and techniques get caught up in this wider field of meaning. Projects can fail because of the impact of subjective issues that are not even acknowledged. The very basis of their self-respect can be undermined. Could it be that engineering is an inappropriate paradigm for software development?

This paper investigates the kind of skills that might improve the chances of your work being used. We are interested in promoting the chances of software engineers gaining work satisfaction and reducing the incidence of stress and grief.

Synergy Matters: Working with Systems in the 21st *Century,*
Edited by Castell *et al.*, Kluwer Academic / Plenum Publishers, New York, 1999.

CONTEXT

Some Links to Failure

There is no doubt that software projects can be late and over budget. But, the most significant problem relates to the suitability of product for the customer. In one survey of US federal projects by Jayaratna (1990), at least 76% of the developed software was not used by the customer.

The technical abilities of software engineers are necessary but not sufficient to guarantee improvements in organizational performance according to Stamper (1996). This is echoed by Sauer's recommendation (1993) that we need to acknowledge local politics and act upon them. Both authors agree that some improvement in human ability is required for successful software projects. For example, "...the practitioner must be competent in such a wide range of disciplines, both social and technical" (Stamper, 1996).

A Source of Large Software Projects

The offices and departments of the UK government sponsor large software projects to meet the moving legislative targets set by parliament. Such projects are open to competitive tender from computer companies, who may be constrained to a list of preferred suppliers. So, company A might win a tender for supply of an Information System (IS) in response to some policy change, such as a new set of rules for benefit payment. The project team might comprise people from both company A and the sponsoring department. As primary contractor, company A might identify the need for skills that it cannot provide at the required time, which leads to a search. Competitive contracts being what they are, the search begins at the cheapest sub-contracting supplier and this can lead to conflict. Furthermore, people from comapny B may not be acceptable to company A.

Over the lifetime of a large project, the membership of the team varies; people come and go. They may come from company A or B or even C or D, which leads to an ebb and flow of constant relocation that can lead to stress for the individual (Lu, 1996).

The Software Engineering Crisis

We now state our hypothesis as follows. The observed rate of failure on projects working from an engineering paradigm is due not to poor engineering but to engineering being the wrong paradigm for most of the underlying problems.

For individual engineers, who want to be more effective in their work, education must focus on what is missing in the engineering paradigm, not by increasing engineering skills. From the perspective of the software engineering profession, the same observations can be made. The reported failure rates result in some defensive dynamics. People attempt to draw boundaries between real software engineers and mere programmers. People talk about technical paradigms such as object-orientation (OO) and rule out success via the 'wrong' paradigm. People talk about the development capability of an organization or a development process. Experts work on standards and procedures and methodologies to bridge the gap.

Within organizations, the lack of success of projects generates political battles for the control that would give success, whether that be internal management control or out-sourced. Organizations are increasingly dependent on there IS, but almost never feel that they get business value from their investments. How many organizations develop lengthy business cases for their projects only to continue the projects long after the business case has been invalidated? What is the relationship between this behaviour and professional engineering?

The engineering crisis is embedded in a double business crisis:
* Investments in IS do not actually deliver enough return to justify the risk
* Large organizations have become completely and irreversibly dependent on their technical infrastructures

If projects fail for reasons other than lack of engineering professionalism, then increasing the professionalism will not have much affect on the project failure rate.

How can software engineering education be used to intervene in the software engineering crisis? The problem is equivalent to asking the distilling business how to deal with alcohol abuse. The distillers are not directly responsible for the abuse but they need to watch the image their products acquire. They may need to intervene in the abuse question to demonstrate a responsible restriction of their product to people not prone to addiction.

There are many perspectives on a business some insist that to measure success in a business it is necessary to have balanced information from a number of domains (e.g. Stacey, 1996). We should not talk of success if it is only financial success, only market penetration or only a potent research programme.

Software engineering is a service that businesses need in order to exert control over the deployment of their technical infrastructure. Equally they need to be able to manage the organizational change that is almost always attached to technical change (Armstrong, 1994). Equally they need to understand the uncertainty and risk in their business situation both to avoid disaster and to grasp chances as they come along (Stacey, 1996).

The addict thinks that the answer to his/her problem lies in the addictive substance. Businesses, in that sense, are addicted to an engineering perspective on their problems, either to propose technical solutions or to attribute technical blame.

Before we feel pleased with this analysis we need to remember that an addicts' family can rarely help with the addiction, being part of a system that produces the addictive symptoms in the sufferer. We focus in this paper on a putative software engineer, needing employment in the industry, and the actual choices open in gaining work satisfaction. Part of the solution is clearly to deal with the perspective problem. If a range of business disciplines is not brought to bear on business problems, and if a set of professional services supporting those disciplines are not maintained during a business change, then that business is not exerting the available control. Conversely, our software engineer can know that if the cultural, market, risk, financial, organizational aspects of change are all represented and all maintain services to the others, then the engineering job is probably feasible.

OUR RESPONSE

Software Engineering Education

One view of education is that it generates choice for the successful student. Since choices between engineering approaches to typical business problems tend to have similar outcomes, then these choices are only meaningful when the more pressing issues have been confronted. In order to generate any meaningful choice, software engineers must be able to distinguish engineering problems from other types of problem. They must be able to practice voluntary restraint from the wrong sort of problem. That is to say, they must not become addicts themselves.

Consider, for example, the arms industry where the solution being offered promotes more of the problem as well as providing a short term fix. The arms industry promises to reinforce your diplomatic initiative with a credible threat of force. The sting comes in being dependent on the force as your diplomacy withers.

We are now talking about the use of the products of engineering. Professionals help their clients to use their services effectively. This must include decommissioning systems, making systems smaller and more manageable, making them less prominent in people's working lives. If all that ever happens is that new systems are built in addition to and on top of old systems, then it is difficult to argue that this represents balanced professional advice. If users are put on a treadmill of mandatory upgrades, are they are being exploited?

One root of the problem for software engineers is the pressure to collude with the clients about the nature of organizational problems. It is this feature that distinguishes software engineering from other sorts of engineering, although this is a matter of degree. There are plenty of examples of engineering disasters brought about by collusion to avoid an accurate problem definition (e.g. Grabowski, 1996). The personal strength and the support systems to avoid being pulled into collusion are rare in software engineers in our experience. Engineers choose their profession to avoid the need for such confrontation: a technical job dealing with objective issues.

Perspectives on Problems

We perceive not data but structured, meaning-laden situations (Stamper, 1987). What we perceive is strongly conditioned by education and experience, by what we are looking for. Software engineering education needs to present a range of organizational problems to show where an IS plays a role in their solution and where it is likely to exacerbate the problem. The primary awareness can then be focused on contributing to the organization not on building the system that was asked for.

Every software engineer we have ever met has tales about the ludicrous requirements that they have encountered. Without taking away from the real world pressures to follow orders in business organizations, the ethical issues for educational programmes are glaring: what do you do when confronted with inconsistent or mistaken system requirements? Is following orders a good enough excuse? Who is in a position to influence the immense economic drain from failed IS projects?

Our work with business organizations always starts with an assessment of how individuals and groups perceive their world. For example, sub-contracted software engineers avoid contact with others of different allegiances. They might seek out colleagues from the parent company (who may be working on the same site for another project team), but otherwise they 'put their heads down and get on with the job'.

When they perceive a symptom of a problem do they assume that:

1   The problem is logical, requiring experience of similar problems, training in particular techniques, the application of particular tools for its solution?
2   The symptoms stem from a lack of vision about where the organization is headed and a viable strategy for getting there, from a lack of business understandings in the marketplace?
3   The root of the difficulties is bad faith, a lack of true listening, a deficit in the human trust necessary to do work?

We have found in many such assessments that engineers typically take the first perspective, and we surmise that there is a predisposition on the part of people drawn to engineering as a career to perceive problems as soluble by engineering effort.

Organizational problems respond to 'solutions' in many domains. All sorts of interventions from all sorts of perspectives are capable of generating change, both intended change and side effects. However, it is vitally important to recognise problem ownership. If you are angry or anxious, there is little we can do to 'solve' the 'problem' of anger or anxiety. We may be able to help but you have to solve the problem yourself. The members of an organization own problems in a similar way, hence the importance of perspective in understanding the problems that they can address for themselves and those they cannot.

Organizations consist of interacting groups of people. To do work these groups of people need sufficient anxiety to be motivated but not so much that it distorts their perception of the reality they must deal with. The dynamics produced by groups of people fall into fairly regular patterns so that generalisations can be made about behaviour;

"Organizations progressively dismantle the ability of their members to ask crucial questions" (Hinshelwood, 1994). That is, there comes to be an organizational reality that is not shared by those outside the organization, and this effect is more pronounced the more important that reality is to the organization.

We can now describe a common scenario where groups within an organization collude in a problem identification that they know is flawed in order to reduce their anxiety. They look at a problem that requires perspective 2 or 3 for its solution and describe a solution from perspective 1 because it allows them to avoid their own responsibility for the problem. Perspectives 1 and 2 imply, require personal change and perspective 3 does not. Who is likely to collude with whom? The ultimate problem owners are management teams who find a ready pool of expert engineers wanting to have their expertise valued and to be paid for doing work. If people have been treated fairly in the past, the more likely you can expect them to provide a positive contribution to include a software project. Effective planned change may only be possible in a climate of fairness (Novelli et al., 1995).

The maximum anxiety for organizations is often around failures with formal and technical systems. IS failures, critical audit reports, evidence of financial fraud; these things are hard to hide and to hide from. Where the anxiety is highest, organizations are least likely to be able to see what the problem is. Fuel is poured on the fire by attempting ever more and ever grander perspective 1 solutions. This is addictive behaviour at an organizational level.

Looking for the Key Skills

To understand the key skills you need to have a particular problem solving perspective. In negotiating settlements, especially when many parties are involved, it is necessary to understand:

* where the views that people have of their opponents are based on some mature perception of the behaviour of the other parties and
* where it is based on their own need to externalise their problems, to see them in others rather than in themselves.

In a mature organization, different parts of the organization and different external providers all have different views on the nature of the problems facing the organization. It is not necessary, indeed it is counter-productive, for one view to dominate the others. Even when a solution, such as a major systems development, is proposed and agreed its effectiveness, its side effects, its currency given the changing situation need to be kept under review from all perspectives.

A software engineer can determine an organization's ability to deal with a problem by asking a variety questions of those service providers, such as:

* "What is the cost benefit equation for this project and how is it changing?"
* "What market risks could affect the applicability of the system?"
* "Which users will have to change their work practices and how will management information change?"
* "In what ways do the implications of this system not fit the culture?"

It is not the answers to the questions that matter but establishing where an answer would come from and what it would be based on. If those questions can be answered independently of the project then there may well be a place for successful deployment of engineering. If they cannot be answered, or the answers are out of date, then the business is probably in a dangerous place for the engineers.

The first key skill is so foreign to most engineers that they cannot understand what it might be. It is the ability to distinguish me from not-me, my own psychological world from that of others around me and the groups I belong to (Boxer, 1991). This skill helps avoid the use of technical arguments in personal and group defences. It is a pre-requisite to the responsible use of software engineering.

The second is an ability to use systems thinking to understand organizational behaviour. Interventions of any kind in an organization produce change throughout the organization so that designing interventions and managing their impact requires a systems view. Note that the observer is always inside and playing a role in this system. This perspective is the main bulwark against 'trigger-happy' software engineering.

The third key skill is facilitation. The ability to catalyse the ability of others to work through their problems without projecting them into the technical domain is a pre-requisite to arriving at workable set of requirements where an IS does have a role to play. Software engineers must be able to promote debate and observe the group process inside and outside the debate to understand when a contribution can be valuable.

Real work takes place in mature relationships where a common goal is pursued by parties with different perspectives, skills and tools. Once the skills and tools are assumed to encompass a solution, the possibility of real work is lost. Good engineering in the wrong context leads to grief for all parties. Bad engineering muddies the water and is probably an evolved response allowing deferral of the problems to another day.

CONCLUSIONS

What is the point of working in an industry where there is a better than evens chance of your work being shelved? It is a matter of choice and that choice must not be avoided.

Our own evidence is that it is perfectly possible to make visible the set of underlying motivations around an organization's desire to possess a technical solution, so that the organizational decision process is strengthened. This has a far greater effect on the chances of success for the organization than strengthening the engineering process. Our observation is that it also improves the engineering because it tends to remove organizational stress from engineers.

If engineers are taught the skills to be able to recognise the context where engineering can contribute, and to be able to play an organisational role using those skills and tools then the chances of their work being used and being useful are greatly enhanced.

REFERENCES

Armstrong, S. and Ward, W. A., 1994, I've found the will. Now, what does it mean?, at : IEE Colloquium on Legacy Systems; barriers to business process re-engineering, London.

Boxer, P. and V. Kenny, V., 1991, The economy of discourses, *Human Systems Management,* **9**(4):205-224.

Grabowski, M. and Roberts, K. H., 1996, Human and organisational error in large scale systems, *IEEE Transactions on Systems, Man and Cybernetics,* 26(1): 2-16.

Hinshelwood, R. D., 1994, Attacks on the reflective space. Containing primitive emotional states, in: *Ring of Fire, primitive affects and object relations in group psychotherapy* (V. L. Schermer and M. Pines, eds.), pp86-106, Routledge, London.

Jayaratna, N., 1990, Systems analysis: the need for a better understanding, *Int. J. Inform. Manage.,* **10**:228-234.

Lu, L. and Cooper, C. L., 1995, The impact of job relocation: future research, in: "Trends in Organizational Behaviour", Volume 2, C. L. Cooper and D. M. Rousseau, eds., pp.51-65, Wiley, Chichester.

McWhinney, W., 1992, "Paths of Change", Sage Publications, Newbury Park.

Novelli, L., Kirkman, B. L. and Shapiro, D. L., 1995, Effective implementation of organizational change, in: "Trends in Organizational Behaviour", Volume 2, C. L. Cooper and D. M. Rousseau, eds., pp.15-36, Wiley, Chichester.

Sauer, C., 1993, "Why Information Systems Fail: A Case Study Approach", Alfred Waller Ltd., Henlry-on-Thames.

Stacey, R. D., 1996, *Strategic Management and Organizational Dynamics,* Pitman, London.

Stamper, R., 1987, Semantics, in: "Critical Issues in Information Systems Research", R. A. Boland and R.A. Hirschheim, eds., pp.43-78, Wiley, Chichester.

Stamper, R., 1996, An information systems profession to meet the challenge of the 2000s, *Systems Practice,* **9**(3): 211-229.

RAPID AND PARTICIPATORY IS ANALYSIS AND DESIGN: A MEANS TO DEFY THE 'ANATOMY OF CONFUSION'?

Dr. Simon Bell,

Systems Discipline,
Centre for Complexity and Change,
Open University,
Milton Keynes, MK7 6AA

1. CONFUSION IN IS?

Those working in the field of information systems (IS) development are well aware that the field is confusing and complex. It is confusing in that the basic rules and properties of the field are in a constant state of flux. It is complex in that the field is young and is continuing to diversify and extend into new domains. The potent mixture of confusion and complexity extends across all areas of the field. At any time it appears that the developers of new technologies are confused (e.g. see Collins and Bicknell 1997), that the providers of IS 'solutions' are confused (e.g. the description of the London Ambulance information system Bicknell 1993), that the developers of information systems (no matter how well supported and financed) are confused (e.g. City of London Taurus project Drummond 1996) and that the final users of all technology based information systems are in a constant state of confusion, occasionally relieved by plateaus of peace (e.g. the brief period when the 286 chip was dominant and the rate of change in applications software seemed to slow down for a while). Technologists, IS developers, retailers and users seem to regularly share in the confusion of the field.

Checkland and Holwell ascribe this confusion largely to the accelerating pace of technology and the inability of theory to keep up. As soon as the practitioners have tested and documented some new case of IT development and the theorists are pondering upon the meaning and impact of such technology (e.g. the advent of ISDN lines and groupware and implications for distance working) then the technology moves on again and the process of theorising is groping to keep up.

The processes by which IS are planned and designed come under the category of analysis and design. Of course there are numerous IS design approaches:
• The highly structured and documentational (e.g. see the various texts which deal with Structured Systems Analysis and Design Methodology (SSADM) such as Ashworth and Goodland 1990).
• The people centred (for example the approaches advocated by Mumford 1995; Checkland and Holwell 1998).
• The approaches which are essentially machine-based such as Rapid Applications Development - RAD - (see the description in Jones and King 1998).

These three categories are not exhaustive but are indicative of the range of approaches. Structured, machine-based and human activity focused approaches all have their values and related problems and all require different skills and training in their practitioners. However, even with the range of approaches available to the organisation for planning the development of IS - there continues to be confusion and copious examples of IS failure.

Nowhere is the process of confusion more evident than in the experience of the developing countries and transitional economies in their relationship with IT/IS. Theorists and practitioners in industrialised countries have severe problems with IT and the development of IS. These problems are compounded when issues of poverty, poor infrastructure and lack of training and education are evident.

2. DEVELOPING COUNTRIES/ TRANSITIONAL ECONOMIES AND IS

IS projects and Developing Country/ Transitional Economy (DC/TE) status combine to make very complex projects. Anecdotally the failure rate of such projects is very high and the negative impacts such projects can have on organisations in these contexts can also be high. Researchers and consultants moving into the area of IT/IS projects in DC/TEs are in a particularly vulnerable position - especially if these researchers and consultants are not indigenous people and have the further difficulty of not having a wide ranging understanding of the local context. They can see themselves as being between two highly risk-prone realities of confusion and complexity - development projects and IT projects.

Thinking about projects in DCs in particular and drawing from the work of Biggs (Biggs 1989) it can be argued that IT/IS projects are following along a well-worn path of technological assistance. Biggs describes the experiences of research scientists working on agricultural research stations in developing countries in the 1980s. Although they were developing agricultural technologies for local farming communities they were often isolated and isolating in their practice. Often remote from the community which they are expected to serve, there is a detailed history of research based on agricultural research stations being misdirected and misapplied. Biggs argues that this was because scientists based their work on professional assumptions about problems rather than on problems actually faced by farmers. In ascertaining why many projects based on research–station technology are not successful with local farmers, Biggs addresses two issues:

• Firstly, the process of technology adoption and questions relating to the value of experts' work when divorced from local needs and aspirations.

• Secondly, the capability and willingness of farmers to make use of 'solutions' not tested and improved by themselves.

More recently, Chambers (Chambers 1997 / page 205) has emphasised this point concerning the 'expert' and has criticised the traditional development project as being essentially non-participatory and working off an expert-view rather than the equally (if not more so) important view of local stakeholders. The conclusion to this is the startling observation that expert opinion grounded in local need and undertaken with local consent and participation has a greater chance of being relevant and sustainable than approaches which are characterised by remote, aloof, technology-driven fixes. Yet the fix or "solution" model would appear to be the model driving most IT projects.

If the development project side of the issue is confusing and complex, the IT element might be seen as being worse. The litany of disasters relating to IT are numerous. Both private and public sectors are prone to massive IT project failure (£80 million in the case of Taurus). It has become an issue beyond the computer press with main-line television programmes such as the UK BBC's "Money Programme" alluding to a "Computer Triangle" causing the failure. The "triangle" is an interesting phenomena. In this programme the problems of IT projects are related to:

• Experts. These are often more of a source of problems than an aid to problem solving.

• IT systems designed without reference to existing packages and tried and tested fixes.
• The ownership of the IT project - who is in charge?
These three issues can be summarised as relating to:
• The appropriateness of the expert,
• The appropriateness of the package and,
• The identification of the owner

Similar issues have been expressed by Collins (Collins and Bicknell 1997). The triangular model described above has a lot in common with the views of development project failure set out earlier in this section. Biggs identified the expert as remote from the problem context. Chambers notes the problem of inappropriate technological solutions and also the problem of ownership of projects in terms of participation and inclusion in decision making processes. Most recently Checkland has indicated that information systems projects are often driven by outdated theories of organisational processes (Checkland and Holwell 1998). The triangle would seem to have relevance to both IT and development project contexts. My concern as an academic and a practitioner is to address the difficulties in the domain of IS adoption in DC/ TEs and set out a process approach which might aid all those involved in avoiding the confusion.

3. RAPID INFORMATION SYSTEMS DEVELOPMENT (RISD) AS A HEURISTIC DEVISE IN IS PLANNING

An observation arising from preliminary research was that highly detailed and exhaustively documented approaches to IS (such as SSADM) were generally not applied (through lack of training or fear of the daunting 'manuals' which came with them maybe). Where such approaches are applied they are often applied badly and the only alternative appeared to be the imposition of technology without any form of planning at all. RISD/ Multiview arose as a suggestion for dealing with this confusion. The RISD/ Multiview idea is simple in form, being an adapted and simplified version of the Multiview approach originally devised by Avison and Wood-Harper (Avison and Wood-Harper 1990). Multiview, as originally devised is a five 'view' methodology for investigating and developing information systems.

This five-step approach advocated by Avison and Wood-Harper was further expanded in Bell and Wood-Harper (Bell and Wood-Harper 1998) to a sixth stage involving the development of hardware, software, training and implementation strategies. It is argued that the team involved in analysis and design is often best placed in the organisation to recommend on procurement and implementation strategy. This view was further reinforced by research/ experience which indicated that those who produce the procurement list often pay scant attention to the results of analysis and design.
In essence Multiview in its RISD form is intended to allow the problem context in which an IS is being proposed to be viewed from a number of perspectives, from a number of epistemological stances, in a number of sensible ways. The reasoning behind this approach relates to the value of gaining a range of perspectives - this value can be gauged in terms of:
• Triangulation - perspectives will re-inforce or contradict each other reducing the potential for single view errors (e.g. taking one standpoint and then seeing all problems in terms of that standpoint).
• Confirmation - the process of RISD/ Multiview increases the likelihood that presenting problems will be confirmed as actual problems.
• Reflective practice - the action research requirement of the RISD/ Multiview approach requires the analysis team to change role in the context and reflect upon their practice

The major intended strength of the RISD/ Multiview approach is the manner in which the approach can be applied rapidly by non-specialists in a cyclic and learning/ reflective practitioner manner (a theme later developed in Bell 1996; Bell 1997). The central themes of RISD as developed from Multiview was that information systems could be planned rapidly (or existing plans rapidly reviewed) by means of a non-specialist team, ideally with assistance initially anyway by a person with familiarity with the approach, with the approach being used as a template of good practice and as a means to develop the successive stages of the IS in a systemic and systematic manner. Further, the approach can be seen as a cycle of IS thinking with the team, or elements of it, constantly reviewing the IS in place (the implemented IS) and the IS in plan (the potential evolution of the IS). This cyclic form of adoption is set out in Figure 1. The figure demonstrates RISD as a cycle with the review and planning of IS always beginning with an adapted Soft Systems Methodology analysis (SSM) and resulting in the planning of the procurement components of the system and implementation strategy. To reiterate, the cycle indicates that the approach is on-going and that the themes for RISD are participation, evolution and continuous learning.

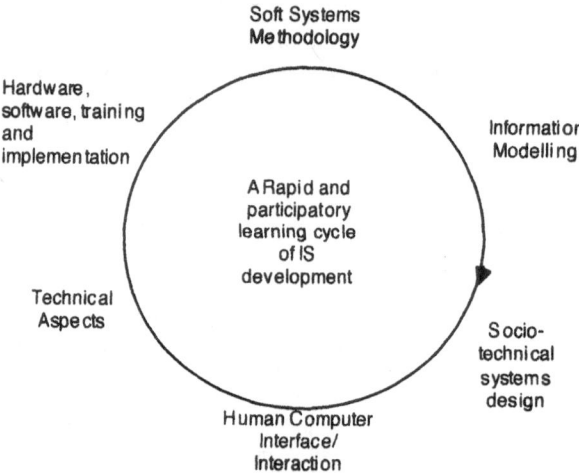

Figure 1. RISD/ Multiview as a Learning Cycle

4. AVOIDING CONFUSION?

One of the primary requirements of RISD is to provide a rapid and participatory approach to IS design which avoids the majority of the problems of IT and development projects set out in section 2 above and at the same time allows the planners of such systems to respond rapidly to changes in technology. Such systems will inevitably tend to be 'fairly quick and fairly clean' and, although far from perfect, are designed to provide sensible and useful information. Further, I argue that this approach, derived from work in participation in developing countries and transitional economies and eclectic methodology building in the UK has potential for use in many contexts in the industrialised countries where approaches to IS tend to exclude the majority of employees within organisations whilst appearing bureaucratic and over proceduralised.

In 1994 I was involved in the development of a Management Information System (MIS) in China (see Bell and Minghze 1995). This first phase of an MIS project was developed making use of both RISD/ Multiview analysis and design approach and the Logical Framework approach to project planning. The 1994 cycle of RISD/ Multiview activity undertaken in participation with Chinese colleagues produced a design which was to prove relevant and useful as well as practical. During Phase 1 I worked with Chinese colleagues both in China and the UK on the six stages of RISD, developing the plan of the system in four weeks and then assisting local staff in developing this as a software platform over the following six months. To some extent only time will tell how valuable the exercise has been but despite inevitable difficulties with implementing and developing the system in a rapid manner (e.g. the rate of change evident in the Chinese economy and the inevitable impacts which this had in terms of necessary modifications to project design), there were significant successes in terms of levels of training, competence in systems development and the ability to turn learning into new, evolutionary systems. Now in 1998 there is an opportunity for further development in the second phase of the project. This phase is also being planned in participation with local staff and is now based upon the learning cycle approach to analysis and design - building off the success and failures, strengths and weaknesses of the first phase (see Figure 2.). Phase 2 will begin with the new team reviewing the good and bad points of the first analysis and developing their new IS from the basis of this evaluation. RISD/ Multiview now contains explicit use of the Learning Organisation approach as developed by Senge (Senge, Ross et al. 1994) and this approach has now been adopted and adapted by the Chinese organisation in question (Linsheng 1998).

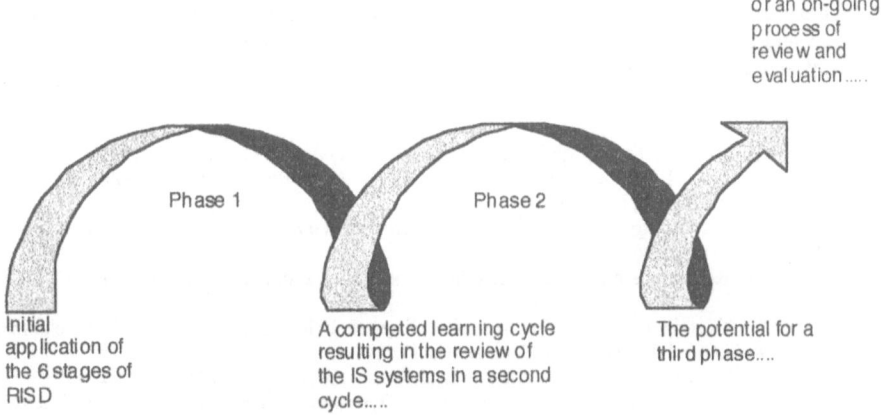

Figure 2. Learning Cycles in the use of RISD

The Learning Organisation (LO) five elements (systems thinking, personal mastery, mental models, team working and shared vision) are applied by means of a reflective exercise during each of the six stages of RISD/ Multiview. It is the intention of the approach that LO, when applied with systems analysis, should provide the practitioner team with the prompts to question the information that is being gathered and reviewed. RISD/ Multiview is intended as a learning approach undertaken as an action research project. The inclusion of reflection on the five learning organisation disciplines should result in learning (both positive and negative) being retained by the team and taken forward for future analysis.

All analysis and design produces mixed results. As indicated in the section above in discussing phase 1 of the current project and in section 2 of this paper, all approaches to analysis and design are capable of producing spectacular negative as well as positive results and this is in part a limitation of the approach and in part is due to unforeseen problems and changed contexts in the organisation in question. In my own use of RISD/ Multiview I have recorded mixed results (e.g. see Bell 1996; Bell 1996). My claim would be that the use of RISD when applied rapidly is low cost, practical and results in the useful outcome of the organisation benefiting from significant learning as well as significant potential for IS development. At the very least, if confusion still results from the implementation of the IS project then, following RISD procedures, the sources of confusion should be more clearly understood.

5. CONCLUSIONS

In reflection a number of points arise:
• First and most importantly - to avoid the confusion of the organisation - it is essential

for the analysis and design process to be owned and employed by the organisation in question. Each analysis and design methodology which builds up levels of complexity and unnecessary technocracy in the development of information systems effectively robs the organisation in question of the capacity to develop its own information system and share in the learning which this analysis produces.

• Secondly, the developing countries and transitional economies have been the recipients of many projects over the years which constitute impositions of remote technocrats over local needs. The recent experience of information systems design indicates that this phenomenon is developing into the relatively new area of IS. The primary intention of the RISD/ Multiview approach is to provide an approach which is useable by the wide range of new practitioners now emerging in the developing countries and transitional economies.

• Thirdly and finally, one of the hall-marks of analysis and design to date has been the stunning inability of authors to explicitly describe failure and problems. For confusion to be avoided and complexity to be understood and simplified it is essential that lessons are learned and practitioners share their experiences of both the science and art of the analysis and design process.

References

Ashworth, C. and M. Goodland (1990). *SSADM: a practical approach*. London, McGraw-Hill.

Avison, D. E. and A. T. Wood-Harper (1990). *Multiview: an exploration in information systems development*. Maidenhead, McGraw-Hill.

Bell, S. (1996). *Learning with Information Systems: learning cycles in information systems development*. London, Routledge.

Bell, S. (1996). "Reflections on Learning in Information Systems Practice." *The Systemist* 17(2): 54-63.

Bell, S. (1997). "Not in Isolation: The necessity of systemic heuristic devices in all development practice." *Public Administration and Development* 17: 449-452.

Bell, S. and L. Minghze (1995). MIS and systems analysis applications in China: a case study of the Research Institute for Standards and Norms. *Global Information Technology and Socio-Economic Development*. M. Odedra-Straub. Nashua, Ivy League: 153 - 160.

Bell, S. and A. T. Wood-Harper (1998). *Rapid Information Systems Development: systems analysis and systems design in an imperfect world: Second Edition*. London, McGraw Hill.

Bicknell, D. (1993). Any takers for a stretcher case? *Computer Weekly*. London: 14.

Biggs, S. D. (1989). The Role of Management Information Systems in Agricultural Research Policy, Planning and Management in the Indian Council of Agricultural Research. New Delhi.

Chambers, R. (1997). *Whose Reality Counts? Putting the first last*. London, Intermediate Technology Publications.

Checkland, P. and S. Holwell (1998). *Information, Systems and Information Systems: Making sense of the field*. Chichester, Wiley.

Collins, T. and D. Bicknell (1997). *Crash: ten easy ways to avoid a computer disaster*. London, Simon and Schuster.

Cyranek, G. and S. Bhatnagar, Eds. (1992). *Technology Transfer for Development: the prospects and limits of information technology*. New Delhi, Tata McGraw-Hill.

Drummond, H. (1996). *Escalation in Decision Making: The tragedy of Taurus*. Oxford, Oxford University Press.

Jones, T. and S. King (1998). "Flexible Systems for Changing Organizations: Implementing RAD." *European Journal of information Systems* 7: 61-73.

Linsheng, H. (1998). Learning Organisation and Computer Supported Management. Beijing, China International Engineering Consulting Corporation.

Mumford, E. (1995). *Effective Requirements Analysis and Systems Design: the ETHICS method*. Basingstoke, Macmillan.

Senge, P., R. Ross, et al. (1994). *The Fifth Discipline Fieldbook: Strategies and tools for building a learning organisation*. London, Nicholas Brealey.

CULTURE, IDEAS, IDIOSYNCRASY

Malcolm K. Crowe[1] and Sandy Kydd[2]

[1]University of Paisley,
 Paisley, UK, PA1 2BE
[2]DMC Ltd,
 3 La Belle Place,
 Glasgow, UK G3 7LH

INTRODUCTION

The paper views changes to business processes, specifically related to computer support, from the viewpoint of the culture of the organisation. It arises out of research on workflow systems (Kydd 1998), aimed at developing non-prescriptive, dynamic mechanisms to support information systems. Instead of demanding adherence to rigid procedures, users have the flexibility to make and respond to suggestions that alter the process flow and can contribute to process evolution. This allows procedures to begin to be used while parts such as exception-handling are as yet incompletely specified, or have been left open for participants to organise themselves.

This paper aims to consider the take-up of suggestions for changes to business processes, from the viewpoint of the culture of the organisation. At one extreme, ideas that completely conform to the culture will offer nothing new, while if they are completely out of line with the culture they risk being dismissed as cranky. To be accepted, ideas must contain aspects that take account of established values and meanings: to be valuable they must contain aspects that are new and original. In many everyday examples, no special analysis is required to evaluate a new idea from this point of view. However, in the information systems field (Ciborra and Lanzara 1994, Checkland and Holwell 1998), this provides a useful perspective on the failure of many computer-based innovations: both of products designed for a market and of projects for specific clients. Here a conscious analysis of the cultural aspects might reduce the numbers of failures in the future. Although the focus in this article is on information systems, it is helpful to see how general are the principles that need to be applied. We can widen the discussion to almost any kind of open human system associated with a culture, so that a business organisation in the ordinary sense is only one example. Culture here implies the sharing of some concepts and values.

For example, it may seem surprising to non-cycle enthusiasts that there are many different kinds of bicycle (e.g. Davidson and McGurn, 1998). For example, there are clever sorts of portable cycle that can be folded up into a small package for taking on the train or hiding beside an office desk. There are cycles where the rider's position is "recumbent", sitting, or lying with the legs out straight in front, with the feet sometimes well in front of the front wheel, and in some cases where the handlebars are positioned below the saddle. There are even examples, where the rider looks up from a prone position and the legs are

Synergy Matters: Working with Systems in the 21st Century,
Edited by Castell *et al.*, Kluwer Academic / Plenum Publishers, New York, 1999.

529

out the back Other human-powered vehicles are adapted for carrying loads, or have more than two wheels. Here the ideas, as original designs, and the notion of culture in which they appear, are easy enough to grasp. The associated concept of organisation has as its components cycle manufacturers, cycle magazines and other media, cycle clubs and shops and their members and customers. The acceptance of a new idea in that culture can mean commercial success for the small companies that design and build these cycles: based presumably on some successful appeal to notions such as utility, elegance and economy as understood in that culture.

In economics and management, there are people who create new followings for their ideas. In the economics field there have been Keynes and Friedman. In management, there have been Taylor and Ford. Here the culture is that of politics, business and industry, and the elements of organisation include banks, newspapers, political parties, and their members, shareholders, voters, employees. Acceptance of the ideas results in their incorporation into policies and projects affecting the lives of people and communities: based presumably on some successful appeal to notions such as benefits (over costs), transparency and efficacy as understood in that culture.

In business organisations, a new idea might be for a new product or new business process: it is successful in that context if it is adopted into the portfolio of products or processes, as something new to do, or a new way of doing it. If it is too much out of line with the culture of the business organisation it will fail, or be dismissed, as irrelevant or unworkable. This aspect of the ideas of this article applies particularly to the introduction of new computer-based support for the information system of the organisation (Keen 1981, Ciborra and Lanzara 1994), and we consider this aspect in the rest of this paper.

CULTURE AND INFORMATION SYSTEMS

This section will review in more detail the ways in which information systems evolve within organisations, by drawing together a number of ideas from the literature. It is a starting point that people may identify information systems in considering the routine processes that define their business organisation. There is of course a very wide spectrum of such processes, some allowing their participants great freedom of action and discretion, as in troubleshooting or creative applications, and some necessarily requiring strict adherence to precise scripts, as in some tele-sales or safety-related applications. Typically, information analysis of these processes is carried out when actions aimed at improving the business process are in prospect, such as the introduction of some computational support, but the reasons why such innovations are being considered are not the focus of this paper. Here it is taken as a starting point that ideas for changing the business process are being canvassed. At such times, it may seem that there is a chance to make major changes, to initiate radical changes in the processes that are normally carried out.

However, attempts at radical change are usually frustrated by the culture itself. People develop their own ways of making sense of the needs of their roles (as they perceive them), and ways of using or adapting the processes and tools to enable them to carry out their job (Suchman, 1987), (Ciborra, 1996, p.8), (Wynn, 1996, p.234). This is what humans are extremely good at; they are not at all good at being treated like cogs in a machine. Information should be seen as emerging from this sense-making (Checkland and Holwell 1998) and the information system should be seen as *self-organising* because of the purposeful activities of the people creating the information through making sense of the data they deal with. Ciborra (1996a, p. xii), calls this the "organisingness" of everyday life populated by real people and not abstract decision-makers; an "organisingness" that makes business processes work.

In large organisations, the very number of people whose sense-making has built the culture makes sudden change difficult. The new ideas must permeate a larger mass of people, not all of whom will be equally receptive to them. This slowness to change, as a negative capability, has its positive side, since if processes changed irrevocably in response to every environmental fluctuation there would soon be no culture and no organisation. Business organisations can only do business and stay in business to the extent that they mak a name for something that is valued in the culture that forms their market. The slowness to a adopt change in a large, well-known company is mirrored by the slowness to perceive change in the market it addresses. If it has invested in promoting itself as good for one product, it is bad business to discontinue that product on the basis of one week's bad figures.

These considerations apply also to information within the company and how it is handled. They also apply to information abou the company as it is perceived by the culture in the market it addresses.

STRUCTURAL COUPLING AND PROCESS EVOLUTION

From the Maturana viewpoint (Mingers, 1995, p.37), the organisingness of business processes is an example of *structural coupling*: that is, the way in which people accommodate to each others' working and adopt a socially-mediated consensus creates (for them) new composite unities. This ontogenic coupling applies to the organisation itself as well as to its process and its information system. Checkland and Holwell (1998) argue that this is a distinguishing characteristic of business organisations: both members and non-members of the organisation are ready for many purposes to treat the organisation as if it is a conscious, person-like entity capable of unitary purposeful action. Non-members identify the organisation as an entity from the ways they encounter it (its products, its advertising, its employees), just as members identify the organisation by abstracting its operations and business process from their day-to-day experience as employees. To someone from outside the culture in which the organisation is identified, this can be seen as just a linguistic convention, or simplifying model, enabling people to encapsulate some complex set of coordinated actions of individuals into a statement about something "done by the organisation". Furthermore, it can be seen that different observers may well conceptualise the organisation, and attribute purposes and motives to it, according to their own views, interests, and agendas.

Each individual retains to a greater or lesser extent their own way of working in the environment, with their own idiosyncrasies and personal trademarks. As a result of structural coupling, the overall information system can cope with small enough individual variations. A new computer-based support that is seen as more restrictive in this respect will be less likely to gain acceptance.

Moreover, because of structural coupling, working methods that are adopted into the culture come to be treated as separate from the individuals that have contributed to their development. They are like a familiar part of the furniture, and to this extent cease to be primary objects of attention. But every so often, a member of the group may exercise a part of the procedure that they do not normally encounter, and be struck that something has changed since the last time they used it. They would quite naturally consider that the "system had changed" even though this is merely a shorthand for a more complicated sequence of events. Thus working processes and any identified information systems are seen to evolve, and so can even be considered to organise themselves.

Since any identified information system has as its only context the unique, instantaneous organisational culture in which it is identified, it follows that as and to the extent that this culture appears to evolve, the information systems identified will become less and less relevant. These considerations become important if computer support is built for work processes. These aspects of information systems were of little general interest before the rise of information technology. But now the static nature of many computer "solutions", and the limited extent to which they participate in structural coupling compared to human assistants, has thrown into sharper focus the dynamic and cultural nature of information systems.

ADAPTIVE COMPUTATIONAL SYSTEMS

In this paper, therefore, an attempt is made to move towards computer-supported information systems that possess some capability of adapting to their users. It is to be hoped that this will enable the processes of structural coupling so that as the business process continues to evolve the computer-based support can evolve with it. The price to be paid by the process analysts is that they lose the privilege of having the last word. As Mingers has observed (Mingers, 1995a, p.46):

> [One reading of (Winograd and Flores, 1987)] suggests that systems should be developed that allow the users to create their own language and conversations rather than defining everything for people. This would mean that the same system might appear quite differently to different users and would contain different concepts and sets of information, each reflecting different shared views of reality.

In other words, by giving the users the ability to impose their own language usages and meanings on the process, we open the possibility that the same situation may bear different interpretations, and different processes would be seen by different users as appropriate. We rely only on the organisingness of business processes and the processes of structural coupling to ensure that shared meanings and process paradigms emerge.

We argue that in the past people lacked the patience to allow these processes to occur, or feared the disruption and turbulence while the pre-existing mechanisms were disrupted. The resulting over-specification of procedure served in many cases only to frustrate the whole enterprise of innovation, producing static and unadaptive mechanisms. However, it is obvious that much depends on the starting point for the innovation, that is, the existing culture of how things are done. If people are already working in extremely regimented ways then regimentation itself would probably not be an issue, but they might be the more ready to find fault with details of the script. But it is more usually the case that it is prescriptive aspects of proposed innovations that jar with current practice.

WORKFLOW MECHANISMS

In this section and the next, a prototype non-prescriptive workflow system (Kydd, 1998) is considered which, we claim, makes a positive contribution to the problematic aspects considered above.

The simplest and most pervasive workflow system is represented by the in-tray on an employee's desk. For paper-based transactions, each document in the in-tray represents a task to be done, which may be simple (e.g. check and sign) or complicated and time-consuming. As the proliferation of desktop computers continues, many documents arrive electronically, and the in-tray in the office e-mail application supplements the in-tray for paper-based documents. Workflow-based computer support seeks to develop this concept further, by taking account of the business processes of which the e-mail exchanges form a part. Workflow processes, as expressed in these office applications, model and support these business processes (Kydd and Crowe, 1994).

In current workflow systems, the workflow processes need to be designed by someone. In the design, all the possible routes for the flow of work through a series of activities need to be mapped out. In some systems the process definition can be changed, and subsequent instances of that process will follow the new template. Other systems even allow the workflow of a running process to be changed on the fly, but there is no connection between the two, since all the paths in the workflow have to be defined in both cases.

This approach is unrealistic and unworkable except for the very simplest business processes. Even the simplest procedures in everyday experience involve discussion and consultation over unusual aspects of individual cases, and including all such pathways in the workflow model will overburden it and make it hard to understand and use. In addition, the exceptional nature of some of these pathways may be hard to specify (as may the nature of the exception!). Many documented failures of such systems are in the literature (e.g. Ciborra and Suetens, 1996, p. 200).

It is much better to allow flexibility in use of the model, on the understanding that all steps, and especially all departures from routine, will be recorded in a database. If the workflow system is also allowed to suggest incorporation of some recorded departures from routine into the procedure itself, then a new responsive sort of workflow mechanism becomes possible, called dynamic workflow in this paper.

Current workflow systems over-specify tasks, denying the delegatee the opportunity of deciding how best to approach the task: such a restrictive approach is once again only suited for the simplest and most routine of tasks. In the current implementation of dynamic workflow, this aspect has yet to be addressed, but a notion of responsibility or delegation could be fairly easily built into the model. Further, in many processes, responsibility for tasks can also be dynamic rather than specified in advance, residing with whichever member of a designated group takes up the task.

DYNAMIC WORKFLOW

Dynamic workflow (Kydd 1998) is a concept where a workflow system does not need to be wholly prescriptive when designing a process (i.e. defining all possible routes before the workflow is enacted). The workflow software instead can use the options that are defined in conjunction with an analysis of previous instances of the process running to *suggest* to the user the possible routes which might be taken next at any stage in the workflow execution. With this mechanism, a much more flexible workflow system can be designed, where processes do not need to be periodically redesigned to take account of changes in working practice, but can evolve along with the work practices themselves.

The workflow process is dynamic - although elements in it are defined, they can be changed when a job is run for the process. Elements can be edited for individual jobs, and exception conditions handled. A logical extension of this model is to allow process definitions to evolve based on what is actually being done in jobs for the process, when users make changes in the jobs to the definitions initially supplied from the process. This leads to the idea of automatically suggesting changes to a process definition in a process review phase, based on analysis of the jobs which have been run for the process.

The concept of dynamic workflow provides two main advantages over current workflow models: suggested workflow paths and feedback into the process definition.

All paths between activities in a workflow process do not need to be defined at process definition time. In such cases, other constraints defined in the workflow can be used to provide a partial ordering of activities where necessary, and the user can be presented with a choice where there is no clear option for the next activity to execute.

Analysis of the running of previous instances of a process can be fed back into the calculation of possible paths in the current running of a process instance. This feedback can be used to alter the suggested next options after an activity completes, according to the predominant options that are normally taken in the running of the process at this point. This means that if one path proves more popular with users when running instances of a process, this will have a higher priority in the suggestion list. In effect the workflow in the process is evolving along with the predominance of certain work practices.

CONCLUSIONS

In evaluating proposed changes to computer support for the information system in an organisation, three main aspects need to be borne in mind.

To begin with, those who will use them should be able to recognise the proposed solutions as supportive of the organisation's processes or purposes. If the solutions seem to users irrelevant or counter-productive, their introduction will not yield any of the benefits expected by their proposers. As a first step, therefore, users should at least be able to see some familiar landmarks in the outline of the proposed changes.

Secondly, where the participants have already formed a view that the current procedures have some problems or need to be changed in some way, they will feel frustrated if the proposed new solution seems not to address these issues. It will improve the likelihood of success of the innovation if the participants can perceive some advantage in the new mechanisms. For example, computer-based support might allow faster transmission of a message, or make for a cleaner copy for record keeping purposes, free of erasures. If these aspects are seen as improving the process, participants will be the more ready to adapt to the new procedures.

Thirdly, however, users should also be able to identify ways in which the personal adaptations and adjustments that they have made to standard procedures (structural coupling) can also be made in the proposed new mechanisms. Furthermore the new mechanisms must be able to adapt as far as possible to future evolution, at least to kinds of incremental change that can reasonably be anticipated. For the participants, this judgement will be made at the time of introduction of the new support, whereas the organisation may well have evolved in various ways since the support was designed and the processes it supports were analysed.

REFERENCES

Checkland, P. B., Holwell, S. (1998) *Information, Systems and Information Systems: making sense of the field*, Wiley, Chichester.

Ciborra, C. U., Lanzara, G. F. (1994) Formative contexts and information technology: understanding the dynamics of innovation in organizations, *Accounting, Management and Information Technology*, **4**, p. 61-86.

Ciborra, C. U. (ed.) (1996) *Groupware and Teamwork: Invisible Aid or Technical Hindrance?*, Wiley, Chichester, 1996. Specifically, Ciborra, C. U.: Introduction: What does Groupware Mean for the Organizations Hosting it?, ibid., p. 1-19.

Ciborra, C. U. (1996a) *Teams, Markets and Systems: Business Innovation and Information Technology*, 2nd edition, Cambridge University Press, Cambridge.

Ciborra, C. U., Suetens, N. T. (1996) Groupware for an Emerging Virtual Organisation. In (Ciborra, 1996), p. 185-209.

Davidson A., McGurn, J. (1997): *Encycleopedia*, Open Road Ltd, PO Box 141, SK2 7BX. See also http://www.bikeculture.com/home.

Keen, P. G. W. (1981): Information Systems and Organizational Change, *Communications of the ACM*, **24**, p.24-33.

Kydd, S., and Crowe, M. (1994) Co-operative Working for End-Users with MMTCA. In Proceedings of the Conference on Information Systems Development. Bled, Slovenia, 1994, 667-674

Kydd, S. (1998) Sharing Data in Co-operative Workspaces, Ph.D. Thesis, University of Paisley (in preparation)

Mingers, J. (1995) *Self-Producing Systems: Implications and Applications of Autopoiesis*, Plenum Press, New York.

Mingers, J. (1995a) Using Soft Systems Methodology in the Design of Information Systems. In Stowell, F.A. (ed.): Information Systems Provision, McGraw-Hill, London, p.18-50.

Suchman, L. (1987) Plans and Situated Actions, Cambridge University Press, Cambridge.

Winograd, T., Flores, F. (1987) Understanding Computers and Cognition, Ablex, New York.

Wynn, E. (1996) Groupware in a Regional Health Insurer: Local Innovations and Formative Context in Tension. In (Ciborra, 1996), p. 211-236.

A KNOWLEDGE REPRESENTATION APPROACH TO COMPUTER-BASED SUPPORT FOR SOFT SYSTEMS ANALYSIS

Quan C. Dang

Department of Computing and Information Systems
University of Paisley
Paisley PA1 2BE
Scotland

INTRODUCTION

This paper describes an approach to supporting Soft Systems Methodology (SSM) (Checkland, 1981) using Knowledge Representation techniques (Barr and Feigenbaum, 1982). The approach attempts to provide SSM with a computer-based support, without reducing it to a set of rigid 'hard' techniques.

The paper begins with some background, identifying issues inherent to the computer-based support for SSM and common problems of the tools proposed in the literature. This is followed by a description of the knowledge representation approach in terms of its rationale, the representation of soft systems knowledge and a knowledge-based tool to support soft systems analysis. In the final section of the paper, an evaluation of the proposed approach is given and directions for further research are discussed.

COMPUTER-BASED SUPPORT FOR SOFT SYSTEMS

In principles, supporting SSM with the software tool is not straightforward. This is related to intrinsic characteristics of SSM. Firstly, SSM is an interpretivist, user-dependent methodology (Checkland, 1981; Kreher, 1993), i.e. SSM may be used in flexible manners, depending on a particular user and a particular situation. Secondly, the modelling language of SSM is informal and open-ended. Any appropriate symbols, pictures or words from the natural language may be used to express soft systems models (Checkland, 1981). Thus, the software tool designed to support SSM should support these features of SSM in order not to reduce SSM to a hard method.

The usefulness of the computer-based support for SSM have long been advocated in the literature, e.g. (Avison and Golder, 1991; Stowell et al., 1991). Main benefits of

Synergy Matters: Working with Systems in the 21st *Century,*
Edited by Castell *et al.*, Kluwer Academic / Plenum Publishers, New York, 1999.

535

supporting SSM with a software tool are: (1) tools would make the use of SSM easier, particularly for the analyst to cope with model intergrity and complexity, (2) the use of tools would narrow the gap between SSM and structured analysis and design methods in information systems analysis and design, hence it encourages information systems analysts to use SSM in information systems development, (3) tools would be helpful for novice users to learn SSM. For more detailed dicussions of this topic, see e.g. (Avison and Golder, 1991; Mingers, 1995).

The appreciation of benefits of computer support for SSM has been demonstrated in a number of SSM tools proposed in the literature (e.g. Stowell et al., 1991, Avison et al., 1991; Davenport and Ayers-Hunt, 1995; Zhang et al, 1997). For example, (Stowell et al., 1991) reports on a tool based on an expert system to assist teaching SSM with a fixed set of rules and guidelines in using SSM. (Avison et al., 1992) offers a tool for drawing SSM rich pictures with a pre-defined set of 'standard' icons. The SoftCase tool (Zhang et al., 1997) provides a set of tools, such *as Rich Picture Builder, Relevant Systems Namer, Root Definition Builder* and *Conceptual Model Builder* allowing the maintenance of textual and graphic representations of soft systems models.

A common feature of existing SSM tools is the provision of repositories of soft systems models. However, all of the tools are limited in that they can only provide a fixed (and finite) set of symbols and/or icons to represent the soft systems model or part of it. Consequently, the use of the tools reduces the richness of soft systems models. Besides, as the repositories are inactive, the existing tools can not any intelligent support for soft systems analysis, e.g. to generate an error message when a modelling rule being violated.

The indicated shortcomings are related to the requirement for a formalised representation of the soft systems knowledge, which is informal, so that they can be stored and handled by a software tool. Additionally, in order to support the activity of modelling, meta knowledge about the methodology should also be embedded in the supporting system. In the following section, an approach based on knowledge representation to satisfy this requirement is described.

THE KNOWLEDGE REPRESENTATION APPROACH

Rationale

The use of knowledge representation comes from two ideas. Firstly, knowledge representation does not inhibit the analyst from exploiting SSM for knowledge elicitation because the knowledge representation activity takes place after the knowledge elicitation activity (see Figure 1). Secondly, knowledge elicited by SSA may be represented in a formalised format using an appropriate knowledge representation language. This would enable the elicited knowledge to be stored in a knowledge base, which can be manipulated with a knowledge base management system.

Representation of Soft Systems Knowledge

The soft systems knowledge may be seen being composed of (1) the meta knowledge about SSM, i.e. soft systems modelling concepts and heuristics, and (2) the knowledge about the problem situation, which is obtained through a soft systems analysis.

The knowledge representation language *Telos* (Mylopoulos et al., 1990; Jarke et al. 1995) is employed for the representation purpose. Since Telos implements the *semantic network knowledge representation scheme* (Minsky, 1968), it can represent knowledge which is represented in natural language as in SSM models. The language provides facilities

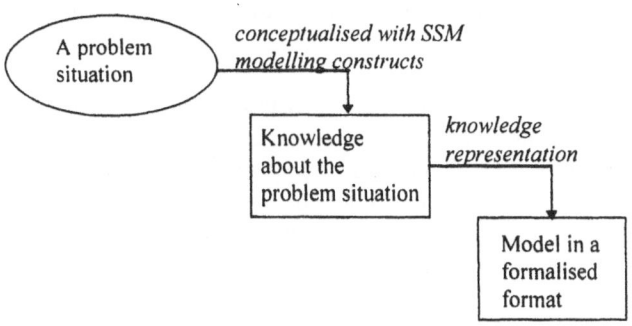

Figure 1. Conceptualisation and representation of knowledge about a problem situation.

for representing logical rules and constraints, which may be used to automatically enforce modelling rules by a knowledge-based tool. Besides, Telos is an object-oriented knowledge representation language. In Telos, concepts and things are represented as classes and objects that are instances of the classes. For example, the concept of conceptual model of a human activity system is represented as a Telos class called `ConceptualModels`. All conceptual models of realistic human activity systems are represented as instances of the `ConceptualModels` class. Using the knowledge representation language Telos, both bodies of the soft systems knowledge can be represented in a coherent fashion.

Representation of the SSM Meta Knowledge. All of the soft systems modelling concepts such as rich picture, human activity system and conceptual models are represented as classes in Telos. To illustrate the representation of SSM modelling concepts as Telos classes let us consider the specification of the concept of 'conceptual model'. The specification of the object `ConceptualModels` (see Figure 2) indicates that a conceptual model is built based on a root definition (viz. the attribute `systemDefinition`) and is composed of sub-activities linked by logical links (viz. the attributes `subActivity` and `logicalLink`). Moreover, the specification requires that at least two sub-activities must be specified in a conceptual model (viz. the constraint `containsTwoDifferent-SubActivities`). This constraint is an example of executable rules representing the SSM modelling heuristics. Non-executable heuristics such as definitions of SSM concepts

```
Individual ConceptualModels in ScoreObjects,MetaClass,Class isA SSMConstructs with
   attribute,necessary,single
      systemDefinition : RootDefinitions
   attribute,necessary
      subActivity : HumanActivitySystems;
      logicalLink : Links
   attribute
      monitorAndControl : HumanActivitySystems
   attribute,constraint
      containsTwoDifferentSubActivities : $ forall m/ConceptualModels
a1/HumanActivitySystems (m subActivity a1) ==> exists a2/HumanActivitySystems ((m
subActivity a2) and not (a1==a2)) $
   end
```

Figure 2: The specification of the *conceptual model* concepts

or modelling guidelines may also be embedded into the knowledge base as narrative text attributes of classes.

Relationships between the modelling constructs are represented in Telos using attributes of classes and supplementary classes. For example, the `Links` and `DirectedLinks` classes are defined for representation of links in rich pictures and conceptual models. The resultant collection of Telos classes that represent the meta knowledge about SSM forms a meta-model of the soft systems model. Details of the SSM meta-model can be found in (Dang, 1997).

Representation of the Problem Situation Knowledge. Soft systems models of real-world problem situations are represented by objects that are instances of the meta-model's classes. These objects inherit all the properties of the classes (i.e. the modelling concepts and their relationships) specified in the meta model. For example, a conceptual model of the human activity system "selecting applicants" in a university admission problem situation is represented as an instance of the class `Conceptual-Models` (see Figure 3).

Furthermore, the meta model can be extended with new concepts to capture specific knowledge of a particular problem situation. This feature is useful because it is not unusual for the knowledge about a problem situation gained from a soft systems analysis to be much more than what is captured in rich pictures, relevant human activity systems and conceptual models. The feature shows that the use of knowledge representation does not reduce SSM to a hard technique.

```
Individual CM_SelectApplicants in
ConceptualModels,SBU_AdmissionObjects with
   attribute,systemDefinition
      rootDef : RD_SelectApplicants
   attribute,subActivity
      a1 : HaveApplications;
      a2 : AssessApplicantQualifications;
      a3 : MakeDecisionAboutOffer;
      a4 : ConfirmOffers;
      a5 : InformApplicantsAboutOffer;
      a6 : ObtainReplyFromApplicants
   attribute,logicalLink
      l1 : LinSA_1;
      l2 : LinSA_2;
      l3 : LinSA_3;
      l4 : LinSA_4;
      l5 : LinSA_5;
      l6 : LinSA_6;
      l7 : LinSA_7
   attribute,monitorAndControl
      monitorAndControl : MonitorAndControlOfSelecting
end
```

Figure 3: Representation of a conceptual model of a particular human activity system of a problem situation.

A Knowledge-Based Tool to Support Soft Systems Analysis

A knowledge-based tool has been prototyped on top of the ConceptBase system (Jarke et al., 1995). ConceptBase is a knowledge-base management system that implements Telos. The system has a graphical user interface with facilities for storing knowledge in the form of Telos classes and objects in a knowledge base. The stored knowledge can be retrieved and queried in textual and graphical formats. Rules and integrity constraints specified in object specifications are automatically checked throughout the knowledge base when changes are made to the knowledge base.

The prototype tool includes three components that extend the ConceptBase to support soft systems analysis. The first component is the SSM meta-model's collection of Telos classes described in the previous section. The second component comprises of a collection of queries that are typical to the soft systems analysis. These queries are Telos *query classes*. For example, the query *"what are sub-activities defined in a given conceptual model?"* has been embedded in the system. The third component is a customised *graphical type model* that defines the graphical properties for displaying the SSM concepts' classes and their objects. In the example shown in Figure 4, human activity systems are displayed with ellipses and links are displayed with diamonds. The figure also illustrates the graphical visualisation of the textual specification given in Figure 3.

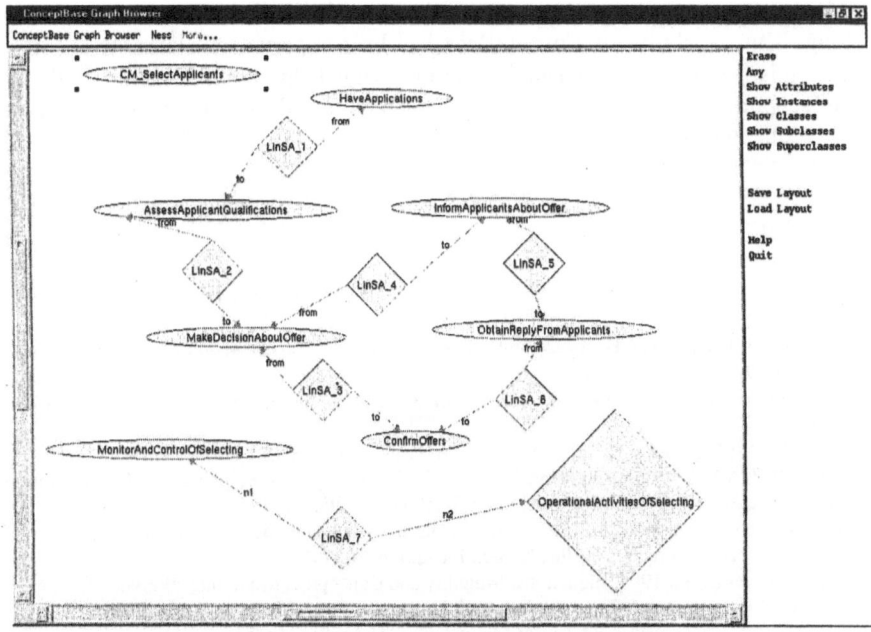

Figure 4. Visualisation of a conceptual model in ConceptBase.

By and large, the knowledge-based tool helps the analyst to maintain the knowledge base of Telos objects representing soft systems models of problem domain. Additionally, the systems helps the analyst, to an extent, with automatic enforcement of modelling rules (by the execution of constraints and rules embedded in the SSM meta-model), and with the none-executable modelling heuristics stored in the system.

CONCLUSIONS AND FUTHER RESEARCH

In this paper, a support for SSM based on knowledge representation of soft system models is proposed for the first time. In (Gregory and Merali, 1993), a method for representing SSM conceptual models using Prolog is described. However, the method has not addressed the general issue of the representation of the knowledge captured by soft systems analysis. To be specific, the knowledge captured by the rich picture is not included in the Gregory and Merali's model.

In comparison with other SSM tools, the knowledge-based tool presented in this paper has several advantageous features. Firstly, the knowledge representation approach does not restrict the analyst to carry out soft systems enquiry using SSM as such (see Figure 1). Secondly, the tool allows the analyst to maintain both the SSM meta knowledge and knowledge about modelled problem situations. Moreover, the knowledge base can be extended with new concepts if that is necessary. Finally, the tool provides, to an extent, an intelligent support for modelling by the execution of executable modelling rules embedded in the knowledge base.

A direction for future work is to enhance the tool with a SSM-specific user-interface and built-in features on top of ConceptBase, which guide and assist the analyst through the SSM 7-stage cycle (Checkland, 1981). Generic queries for soft systems model validation, analysis and exploitation are to be embedded into the tool. The other direction is to experiment the use of ConceptBase in multi-user and distributed mode for co-operative modelling. This capability of ConceptBase has been reported in (Jarke et al., 1995). If succeeded this feature is a further indication that Knowledge Representation could provide a type of support that encourages the SSM appreciation of different people's viewpoints on the same problem situation.

REFERENCES

Avison, D. E., and Golder, P.,1991, The need for tool support for soft systems, in: "Systems Thinking in Europe", M.C. Jackson, ed., Plenum Press, New York.

Avison, D., Golder, P. and Shah, H., 1992, Towards an SSM toolkit, *European Journal of Information Systems.* 1:397.

Barr, A., and Feigenbaum, E. A., 1982, The Handbook of Artificial Intelligence, Vol. 1, Addison-Wesley, London.

Checkland, P.B., 1981, Systems Thinking, Systems Practice, John Wiley & Sons, Chichester.

Dang, Q. C., 1997, A Soft-systems-conceived Model with Knowledge Representation for Information Systems in the Office Environment, PhD thesis, South Bank University, London.

Davenport, M.S., and Ayers-Hunt, J., 1995, Soft systems analysis and modelling Tool, in: "Critical Issues in Systems Theory and Practice", Ellis, K., ed., Plenum, New York.

Gregory, F., and Merali, Y., 1993, Inductions, Modality and Conceptual Modelling, Warwick Business School Research Bureau Research Paper 79, Warwick University, Warwick.

Jarke, M., ed., 1995, ConceptBase V4.1 User Manual, RWTH Aachen, Aachen.

Jarke, M., Gallersdörfer, R., Jeusfeld, M.A., Staudt, M., and Eherer, S., 1995, ConceptBase - a deductive object base for meta data management, in: *Journal of Intelligent Information Systems.* 4:167.

Kreher, H., 1993, Critique of two contributions to soft systems methodology, in: *European Journal of Information Systems.* 2:304.

Mingers, J., 1995, Using soft systems methodology in the design of information systems, in: "Information Systems Provision - the Contribution of Soft Systems Methodology", Stowell, F., ed., McGraw-Hill, London.

Mylopoulos, J., Borgida A., Jarke M., and Koubarakis M., 1990, Telos: a language for representing knowledge about information systems, in: *ACM Transactions on Information Systems.* 8:325.

Minsky, M., ed., 1968, Semantic Information Processing, MIT Press, Cambridge, Mass.

Stowell, F., West, D., and Stansfield, M., 1991, The application of an expert system shell to unstructured domain of expertise: using expert systems technology to teach SSM, in: *European Journal of Information Systems.* 1:281.

Zhang, J., Smith, R., and Watson, R.B., 1997, Towards computer support of the soft system methodology: an evaluation of the functionality and usability of a SSM toolkit, in: *European Journal of Information Systems.* 6:129.

AN ANALYSIS OF 'FAN TRAPS' IN AN EER SCHEMA BY USING A SET OF 'INFO CONCEPTS'

Junkang Feng

Department of Computing and Information Systems
University of Paisley
Paisley, UK PA1 2BE

INTRODUCTION

The conectedness of an EER (Enhanced Entity-Relationship) schema can be complex and problematic. One prominent connection problem is so called 'fan traps' (Howe 1983). Any EER schema will contain potential connection traps, as long as there is a path of length >1 in the EER. This problem appears to have not been well addressed in the literature of database design. Well-known experts and text authors, such as Elmasri and Navethe, (1994) and Date (1995) do not address these issues. Howe (1983) addresses it but not thorough enough, it seems.

In this paper we will tackle this problem from an information perspective and by using a mechanism for capturing information flows. These were developed through the integration of a number of ideas and frameworks, namely the ideas of soft systems thinking (Checkland and Scholes 1990, Checkland and Holwell 1998), the idea of a 'sense-making' system put forward by Lewis (1994, 1995), the framework of sign, information and meaning proposed by Mingers (1995), Barwise's situation theory (1997), and Devlin's information flow theory (1991).

The basic idea is this. Using the ideas of the soft systems thinking, a database can be viewed as a sign system that bears interpretations of some actor(s) in some human purposeful actitvities. The interpretations can be seen as the information that is represented by the data, which are a type of signs, in the database. The 'fan trap' problem is therefore a mistake that some user or designer of the database might make in terms of the relationship between some signs and the information that the signs bear. When the relationship is fully understood, some solution to this problem will emerge. To this end a mechanism would be required whereby the relationship can be expressed in a concrete fromat. The main constructs of the mechanism that will be made ùse of in the rest of the paper are the following two:

1) $S_1 = [s_1' \mid s_1' \mid \sigma]$

where σ is a set of information elements called infons (after Devlin 1991), s_1' is a parameter of situation, $s_1' \mid \sigma$ means that within s_1' σ is true. The above expression defines a situation type. For example, $S_1 = [s_1' \mid s_1' \mid$ tutors(lecturer', student', 1)] defines a situation type where a lecturer tutors a student.

2) $C = (S_1 \Rightarrow S_2)$

where S_1 is termed *1st position situation type*, and S_2 *2nd position situation type*. Through perception and cognition we find specific individuals for all the free occurrences of parameters in the structure of the 1st position situation type, then we obtain *infons* that are made true by the instance of the 2nd position situation type. We call this process the instantiation of a pair of situation types that are connected by an *info connection*. This is how information is created and flows.

In the rest of the paper, we will look at 'The meaning of a path of length >1', 'False and undesirable transitive connections', 'Coupling fans structutes and fan traps', and conclude with a summary.

THE MEANING OF A PATH OF LENGTH >1

Synergy Matters: Working with Systems in the 21st *Century,*
Edited by Castell *et al.*, Kluwer Academic / Plenum Publishers, New York, 1999.

The meaning of a path of length >1 can be divided into two categories. One is simply the sum of the individual relationships in the path, and nothing else, which we will call 'basic meaning'. The other is any meaning implied by the path other than the 'basic meaning', which we will call 'implied meaning.'

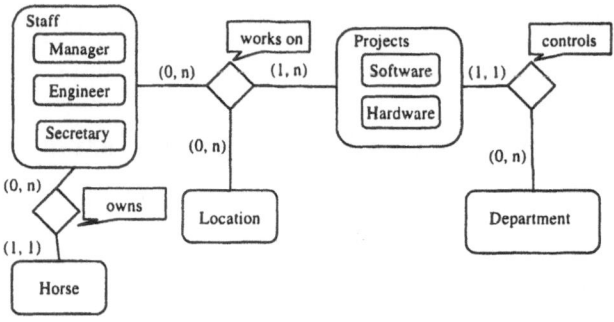

Figure 1. The meaning of a path of length >1

For example, the basic meaning of path(owns, works-on) in Figure 1 is just that some staff own horses, some staff own horses and also work on some projects at some locations, and some staff work on some projects at some locations and owns no horses. As far as the basic meaning is concerned, the two binary relationships, namely 'owns' and 'works on', are two quite separate binary relationships except that some staff may be involved in both relationships.

Two or more non-transitive relationships' connecting to the same entity implements the 'connections' of *infon*s. Such a connection is realized by occurrences of the same parameter in more than one *infon*. For example, we may have a situation type

$$S_2 = [s_2' \mid s_2' \mid \text{owns(staff', horse', l', t', 1)} \land \text{works-on(staff', project', location', t', 1)}].$$
The two occurrences of parameter staff' connects the two *infon*s.

So any instance connection between Horse and Project in the EER schema in Figure 1 is merely that a same member of staff happens to own a horse and work on a project. There can hardly be any implied meaning in path(owns, works-on).

Path(works on, controls) in Figure 1 could be a completely different case. The basic meaning of the path is that some staff works on projects at locations and departments control projects. An implied meaning of the path could be a binary relationship 'work-for' between some staff and some departments if the following is true, namely, 'If a member of staff works on a project and the project is controlled by a department, then the member of staff works for the department.'

The basic meaning of a path is unlikely to be misinterpreted. Any use of a path to represent information that is based solely on its 'basic meaning' will be unambiguous and always supported by the path.

But 'implied meaning' of a path is prone to misinterpretations, which gives rise to connection traps. In principle any construct in an ER schema can be misinterpreted. But certain types of constructs are more prone to a misinterpretation than others. In the section that follows, we will examine one types of these structures.

FALSE AND UNDESIRABLE TRANSITIVE CONNECTIONS

A false and undesirable transitive connection is that a path between two instances exists but it is not valid. We will use Figure 2 to illustrate this point. To understand the problem fully, we shall examine the meaning of path(belongs-to, belongs-to), which can be expressed as a situation type

$$S_3 = [s_3' \mid s_3' \mid \text{belongs-to(employee}_1', \text{department', 0)} \lor (\text{belongs-to(employee}_2', \text{department', 1)} \land \text{belongs-to(department', division', 1))]} \qquad \text{Exp 1}$$

Let $\sigma_1 = \text{belongs-to(employee}_1', \text{department', 0)}$, $\sigma_2 = \text{belongs-to(employee}_2', \text{department', 1)}$, and $\sigma_3 = \text{belongs-to(department', division', 1)}$, Exp 1 can be re-written as

$$S_3 = [s_3' \mid s_3' \mid \sigma_1 \lor (\sigma_2 \land \sigma_3)]$$

542

σ_1 is caused by the partial participation of Employee in 'belongs-to' with Department, and employee$_1$' and employee$_2$' are so named that they always refer to different individual employees. That is to say, an employee cannot both 'not belongs to' and 'belongs to' a department. In σ_2 and σ_3 though, the same parameter department' appears, which is caused by the total participation of Department in both 'belongs-to' relationships. That is, the two occurrences of parameter department' shall always refer to the same individual department whenever the situation type S_3 is instantiated. In S_3, ($\sigma_2 \wedge \sigma_3$) provides the only mechanism for an employee to be connected to a division.

Suppose that the rule 'If an employee belongs to a department, and the department belongs to a division, then the employee belongs to the division as well' does **not** exist. Then the path should not be interpreted as capturing a 'belongs to' relationship between (even) some employees and some divisions.

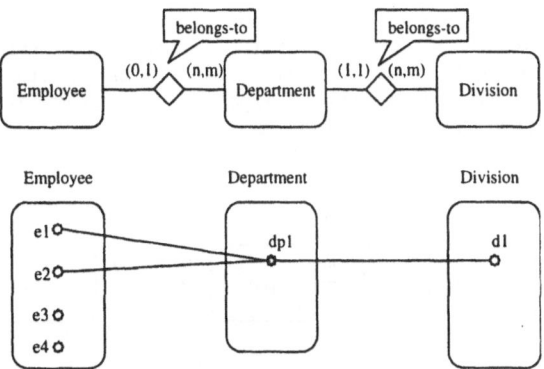

Figure 2. False and undesirable transitive connections

For example, suppose that employee e2 belongs to department dp1, and belongs to no division, and e1 belongs to both dp1 and d1. The ER schema in Figure 2 is not capable of telling the difference between e1 and e2 as far as their linkages with a department and a division are concerned. A transaction using 'natural join':

((Emplyoee[E-no,Dpt-no]join Department[Dpt-no,Div-no])join Division[Div-no]) [E-no, Div-no]

to navigate through the path (belongs-to, belongs-to) will treat e1 and e2 in the exactly same way and give the result in
Figure 3.

E-no	Div-no
e1	d1
e2	d1

Figure 3. A query result that may be misinterpreted due to 'false and undesirable transitive connections'

How to interpret the result is crucial. If we use only the basic meaning of the path, namely just the sum of the two independent relationships 'belongs-to' and 'belongs-to' connected through entity Department, then the result should be interpreted as both e1 and e2 belong to a department, which may or may not the same department, that belongs to division d1. This is a correct interpretation. But if we used the path to capture a 'belongs to' relationship between some employees and some divisions, and interpret the result as both e1 and e2 belong to division d1 accordingly, then the path would be misinterpreted.

We shall now state the false and undesirable transitive connection problem above as follows:
The EER schema is not capable of bearing information made true by the situation type S_4:

$$S_4 = [s_4' \mid s_4' \mid \text{belongs-to(employee', department', 1)} \land \text{belongs-to(department', division', 1)} \land \text{belongs-to(employee', division', 0)}]$$

In other words, it is not true that $S_3 \Rightarrow S_4$.

A false and undesirable transitive connection can arise in any path of length >1 as long as the 'total valid transitive connection' condition is not satisfied. The 'total valid transitive connection' condition will be dealt with shortly. One type of the structures that is most likely to give rise to a false and undesirable transitive connection is so called 'fan trap'.

'COUPLING FANS' STRUCTURES AND FAN TRAPS

Relationship fans

A relationship fan exists where two or more relationship instances of the same relationship type fan out from the same entity instance. In Figure 4, edges (s1, c1) and (s2, c1) form a relationship fan.

It is possible that two relationship fans connect to the same instance of an entity. When the structure of part of an ER schema is such that a situation like this is possible, the relationship types and the entity type in question are said to constitute a 'coupling fans' structure.

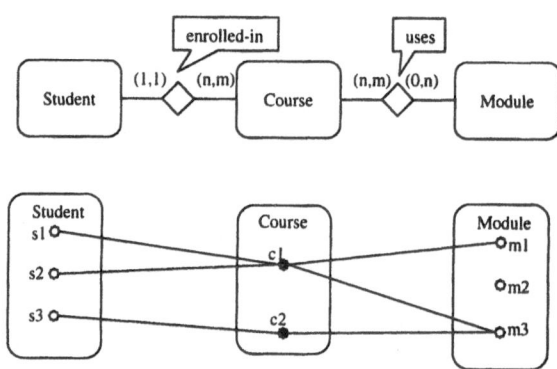

Figure 4. A 'coupling fans' structure

For example, in Figure 4, c1 has two relationship fans connected to it. So path (enrolled-in, uses) is of a 'coupling fans' structure. A 'coupling fans' structure is determined and indicated by the 'max' number in the 'structure constraint', namely (min, max), of a relationship being grater than 1 on both sides of the entity type in the middle of a path of length 2. In Figure 4, entity Course is such an entity.
We define the 'coupling fans' structure formally as follows:

'Coupling fans' structure
A 'coupling fans' structure exists if:
An entity is involved in two or more non-transitive relationships (whether with itself or other entities), and the "maximum" numbers in the *structural constraints*, namely (*min, max*) on its participation in any two or more of the relationships are **greater than one.**

A 'coupling fans' structure can be problematic under certain circumstances. We shall use the notion of **'total valid transitive connection'** to tackle this problem.

Let a path consist of three entities A, B, and C, and two binary relationships $r_1(A, B)$ and $r_2(B, C)$. Entity A has a total valid transitive connection with entity C if for any instance of A, say a_i, that is connected to an instance of B, say b_i, and if b_i is connected to instances of C, say $c_1, ..., c_n$, where $n \geq 1$, the transitive connections between a_i and all $c_1, ..., c_n$ are valid.

It can be seen that the reverse is also true. That is, if entity A has a total valid transitive connection with entity C, then entity C has a total valid transitive connection with entity A.

Proof

Let b_i be connected to $a_1, ..., a_m$ of A, and $c_1, ..., c_n$ of C.   (1)
Let also every of $a_1, ..., a_m$ be validly connected to all $c_1, ..., c_n$.   (2)
Assume that c_i is validly connected to a proper subset of $a_1, ..., a_m$, that is, there is at least one of $a_1, ..., a_m$, that is not validly transitively connected to c_i. This contradicts (2).

A 'coupling fans' structure with a total valid transitive connection is not a hazard as there can be no false and undesirable transitive connections when the transitive connections are used to represent some information. For example, in Figure 4 if all modules that are used by a course are compulsory, then all paths between instances of Student and Module involved in the 'coupling fans' structure are valid. That is, (s1, m1), (s1, m3), (s2, m1), (s2, m3) are all valid connections.

Now we shall explain the information-bearing capacity of a 'coupling fans' structure with a total valid transitive connection by means of our 'info' concepts. The basic information that the EER in Figure 4 is able to represent when no valid transitive connection is considered is the *infon*s made true by the situation type

$S_5 = [s5' \mid s5' \mid$ (enrolled-in(student', course', 1) $\wedge$ (uses(course', module$_1$', 1) $\vee$ uses(course', module$_2$', 0))]

where parameters module$_1$' and module$_2$' refer to different individual modules. The two non-transitive binary relationships 'belongs-to' are independent of each other in that either is determined by only two of the entities that are connected by a relationship.

When the following is true:

$$S_6 = [s6' \mid s6' \mid \text{ enrolled-in(student', course', 1)} \wedge \text{uses(course', module', 1)}]$$
$$S_7 = [s7' \mid s7' \mid \text{ takes(student', module', 1)}]$$

and $S_6 \Rightarrow S_7$, then path(enrolled-in, uses) has a total valid transitive connection. This is because the *info connection* $S_6 \Rightarrow S_7$ means that in any situation where a student is enrolled in a course and the course uses a module then the student takes the model. The *info connection* $S_6 \Rightarrow S_7$ is therefore the reason why all paths between the vertices are valid.

A 'coupling fans' structure without **'total valid transitive connection'** is likely to be misinterpreted and/or used to represent some information it is not capable of representing. For example, in Figure 4, if not all modules that are used by a course are compulsory, then entity Student and entity Module have a partial valid transitive connection. That is, not all connections between entity instances involved in a coupling fans structure are valid. For example, if s1 takes m1 and m3, and s2 takes m1 only, then (s1, m1), (s1, m3), (s2, m1) are true whereas (s2, m3) is false. An important point here is that the ER schema in Figure 4 is not capable of telling the difference between the above true and false connections. In other words, it is unable to bear the information regarding 'which student takes which module', which can be expressed as the following *infon*

takes(student', module', 1).

Fan traps

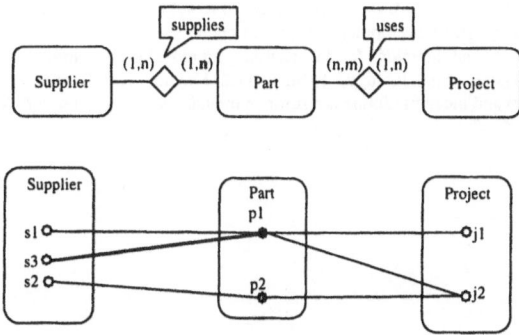

Figure 5. A path with 'wholeness' and a 'coupling-fans' structure

We will use Howe's term 'fan trap' (1983, Page 114) to indicate a 'coupling fans' structure without 'total valid transitive connection' in relation to a given set of true instance connections. We will re-define the term 'fan trap' formally as follows:

Fan traps

A fan trap is a 'coupling fans' structure without a 'total valid transitive connection' in relation to a given set of true instance connections. A fan trap is likely to arise when a path of length 2 comprises three entities and two relationships, and the path captures a kind of 'wholeness', for example, Supplier supplies Part to Project

Figure 5 shows that suppliers s1 and s3 both supply p1. Path(supplies, uses) is now incapable of capturing 'who supplies what parts to which projects' as s3 may supply p1 to either j1 or j2 or both (which means that there is no 'total valid transitive connection'), the path cannot tell which one. After you have written, say, (s1, p1) and (p1, j1) into the database, it is lost immediately as you would not be able to tell them from (s1, p1) and (p1, j2), which is false. In fact, what has happened shows that we cannot use (s1, p1) and (p1, j1) in path(supplies, uses) to represent supplier s1 supplies p1 to j1. That is to say, if we used the path to capture 'who supplies what parts to which projects,' we would fall into the fan trap.

SUMMARY

This paper tackles a confusing issue in EER modelling, namely **fan traps** by using a mechanism, which was developed on the basis of soft systems thinking and some ideas regarding signs and their meanings. The mechanism enables the problem to be looked at from both the topological perspective and semantic perspective. The former can be seen as being at the 'sign' level, the latter the 'information' level. An innovative concept presented here is the 'basic meaning' and 'implied meaning' of a path. The 'basic meaning' of a path is the set of true instance connections that the path's topological structure can always represent without referring to any organizational rules. Applying 'organizational rules' onto the basic meaning of a path of some certain structure arrives at some implied meaning of the path. An implied meaning of a path is either part or the whole of a set of true semantic connections (i.e., information) that the path is capable of representing under the 'organizational rules'. The essence of a fan trap is that 'false and undesirable semantic connections are not recognized', which is part of a larger problem of whether and how a piece of information is represented by the structure of an EER schema.

REFERENCES

Barwise, J. and Seligman, J., 1997, "Information Flow – the logic of distributed systems," Cambridge University Press, Cambridge.

Checkland, P., and J. Scholes, 1990, "Soft Systems Methodology in Action," John Wiley & Sons, Chichester, New York.

Checkland, P., and S. Holwell, 1998, "Information, Systems and Information Systems," John Wiley & Sons, Chichester, New York.

Date, C., 1995, "Introduction to Database Systems," 6th edition. Addison Wesley, Reading.

Devlin, K., 1991, "Logic and Information," Cambridge University Press, Cambridge.

Elmasri, R., and Navathe, S.B., 1994, "Fundamentals of Database Systems," 2nd edition. Benjamin/Cummings, Redwood City, Califonia.

Howe, D.R., 1983, "Data Analysis for Data Base Design," Edward Arnold, Maryland, USA.

Lewis, P., 1994, "Information-systems Development," Pitman, London.

Lewis, P., 1995, New Challenges and directions for data analysis and modelling, in "Information Systems Provision: The Contribution of Soft Systems Methodology," F. Stowell ed., McGraw-Hill, London.

Mingers, J., 1995, Information and meaning: foundations for an intersubjective account, *Info Systems J*, 5:185-306

THE INTRINSIC AND EXTRINSIC ASPECTS OF
INFORMATION SYSTEMS EVALUATION

Dr Misha Hebel[1] and Chris Davis[2]

[1]Business School
University of Greenwich
Woolwich Campus
Wellington Street
London SE18 8PF

[2]School of Information Systems
UWE, Bristol
Frenchay Campus
Bristol BS16 1QY

INTRODUCTION

It can be contended that Information Systems (IS) are simply the combination of information gathering methods with Information Technology (IT). Nevertheless, innovations in IS development methodologies the process of IS design and implementation still involves substantial trade-offs. These impose an increasingly reductionist design ethic particularly as the project nears completion. Our research shows that this reductionist ethic affects performance measures used to judge systems in operation. Our research activities have, from very different starting points, revealed a substantial dissonance between the espoused (conceptual) and actual (operational) criteria used to judge performance.

We suggest that the synergy of IS in their organisational context has largely been ignored particularly during evaluation and assessment. Common interest in the impact of human values on the design, implementation, use and assessment of management information systems has emerged from overlapping research activities which explore perceptions of value used to judge performance. From research carried out in varied organisational settings[1] using different approaches we have found that evaluation is grounded in human values. These adjust perceptions of value with each intervention and evaluation. As information systems proliferate and their impacts become more complex it is important to re-examine the validity of the evaluation process. Such re-examination is necessary to ensure that the values used to assess

---

[1] The settings included hospitals, schools, police, construction industry and Quaker meetings.

performance maintain a similar dynamic to the systems they purport to judge. The convergence of our research activities enables us to reveal and explore the dissonance between espoused and actual value systems through analysis of responses to aspects of management information systems (MIS). The first approach focuses on the use of performance measures, measurement process and underlying values in a range of organisational settings. The second approach explores interpretations of the impact of the same MIS when used for a range of purposes.

LOOKING IN FROM THE OUTSIDE

The starting point of the first approach was an observation that the introduction of aspects of performance measurement to organisations prompted many different responses other than the utilisation of information to aid management. Many useful and well-established methods of gathering information for performance measurement were identified (Lucey, 1991; Holloway et al. 1995). The means of accumulating and validating were established and on paper there seemed no reason for MIS to go astray (Armstrong, 1994; Robson, 1997; Collins & Bicknell, 1997). This appeared to occur in both the public services, businesses and other organised groups (Hebel, 1996).

From this observation four case studies were identified in order to compare and contrast measures used by each group. The analysis is based on the hypothesis that the problems are generic and not job specific and consequently that solutions may also be generic. A sample of organisations representing a range of organisational life spans and purposes (Hebel 1996) were needed to test this. Some were able to present a formal statement of goals and performance measures others were not. They were both office and site based, white-collar and blue collar. These groups were specifically a Metropolitan Police CID, a Quaker business meeting[2], a small-scale building contractor, Mark Richardson & Co. and a 400 year old Jesuit school. Both research and analysis were rooted in pragmatic and humanist ideas. Information on the case studies was gathered using questionnaire, interview, observation, literature review and content analysis.

Comparison of performance measurement reality with the models suggested in the literature (Armstrong, 1994; Holloway et al, 1995; Lucey, 1991; Robson, 1997) prompted a search for an underlying theme that might better explain the difficulties encountered. A promising area appeared to be the impact of human values on the design and response to performance measurement. A number of key characteristics of value systems were identified using this approach. Firstly, values are formed early in life and are consequently deeply entrenched. Secondly, continual reinforcement of those early values makes the value system both closed and complicated, where change only takes place over a long time or by trauma. Thirdly, different combinations of values are held for each situation encountered and fourthly measuring values encounters similar problems as measuring performance (Hebel-Holehouse 1998).

---

[2] Although not a conventional work group Quakers are included as a case study because in addition to worship, business meetings are also held on a regular basis at a local, district and national level (Hubbard 1985). All meetings are based on similar principles and are in some senses inextricable so both are included in this research.

An early finding using the first approach was that there were inherent difficulties in measurement especially using questionnaires due to the relativity of interpretation and changes over time. This was revealed through the emergence of a number of collectively 'right' answers. It appears that if the question emphasis could be interpreted as personal then 'good' organisational qualities would be emphasised. Identification of such 'ideals' (Frantz 1995) is a good way of sharing what we think is important but the action runs the risk of building an inappropriate picture of the environment: Farbey et al (1995) refer to this phenomenon as a false consensus. Working towards such a false consensus could in part explain why some change initiatives and in particular performance measurement schemes fail to be accepted as the devisors expect. Designers appear to work towards the espoused ideal, whilst those being measured work toward what is practical. Both do so within the limits of their own experience. Additionally, the continual reinforcement of personal values carries huge implications for the management of change in organisations.

From the case studies and reviews of the performance measurement and IS evaluation literature (Armstrong, 1994; Farbey et al, 1995; Holloway et al, 1995; Walsham, 1993) it became apparent that the differences of the world-view of the performance measurer and those of the people being measured often caused problems. Different backgrounds and world-views result in different priorities and vocalisation. Loyalty may be held to be important by all the members of a particular team but the way it manifests itself - in resisting management initiatives for instance - may cause more conflict than benefit.

A values survey, based on the assumption that successful implementation of performance measurement is dependent on value systems, was designed to demonstrate the diversity of values. This consequently provided further evidence of the variability and subtlety of value systems through the introduction of terminology used by case study participants. These were seen as indicative of world-views representing values in organisations. The survey intends to validate the earlier conclusions on value systems composition and manifestation. The questionnaire used consisted of a series of 54 statements which the respondents are asked to indicate a level of agreement on a four point scale. A total of 254 questionnaires were returned.

The most striking characteristic of the results is the varying extent of agreement. The responses cover a wide range: rarely were any of the statements supported so strongly that a confident prediction of commonly held views was possible, indicating the variety of values held within any organisation. Statements overwhelmingly agreed are mostly descriptions of ideals, such as *an organisation needs to adapt in order to stay viable; people who have been in an organisation for some time are a valuable resource; people work best if they feel secure; colleagues and peers should be loyal to each other* and *the purpose of management is to support and develop staff.* Statements indicating a high level of disagreement included *the older a person is, the better they will be at management; questionnaires only ask for personal details so critical comments can be traced back; people in authority know best by virtue of their position* and *people learn best if dropped in at the deep end.*

The results show a strong preference for learning over experience and confirm the dislike of change indicated by many people in the case study interviews. This ties in with the link of security to motivation and raises a divergence with the need for an organisation to remain viable. The characteristics of management are also not deemed to be simply a matter of age or status according to these results. Instead a more supportive role is indicated as valued. Consideration of the reinforcement, self-organisation and issues arising from implementation of performance measurement systems prompted reconsideration of the meaning of 'value system'. Values were initially assumed to be components interconnected for the purpose of providing some framework for an individual's value judgement process. The systemic nature of values inferred that emergence was likely but also that there was perhaps a hierarchical arrangement to their priorities. The second approach explores these issues further.

LOOKING OUT FROM THE INSIDE

The second approach emerged from a concern with the problematic nature of the evaluation of information systems. The reasons for persistence of the problem seem revolve around the increasing pace of innovation and diffusion of information systems: Kling (1996) points out that rather than events of greater or lesser magnitude, the impacts of information systems are emergent, in that they depend on a succession of responses throughout the process of implementation. Kling (1996) conceptualises the assessment of information systems impacts as a dynamic process rather than a singular event driven by empiric measures and the search for consensus: Walsham (1993) supports this approach, suggesting that such rational assessment events are incomplete since the consensus on which they seek to rely represents an imposed external reality, leading to the persistence of narrow goal achievement measures. This may be exacerbated by the dominance of a linear model of technological innovation and adoption (Scarbrough and Corbett, 1992, p7) over processual models, such as that proposed by Iivari (1992). This approach to evaluation fails to address the dynamism of IS innovation. This mismatch between the pace of the evolution of IS and the pace of the evolution of evaluation practice renders the premises on which these evaluations are based as incomplete: we might infer that the search for consensual measures by which to judge performance or benefit is futile since the condition referred to as alignment (Iivari, op cit), equilibrium (Friedman and Cornford, 1989) or maturity (Nolan, 1979) is never reached. Rather than an event, the assessment of impacts is thus reconceptualised here as a dialogue which explores those impacts as they emerge. A key issue is the need to ensure sufficient openness in the dialogue in order to capture as wide a range of these emergent, often serendipitous, impacts as possible.

The second approach addresses human values as the basis for interpretation and judgement. In contrast to the first approach, the goal is shared appreciation rather than consensus. The second approach is thus individualistic rather than systemic (Pidd, 1996). Repertory Grid Analysis (RGA) has been used to elaborate a process of negotiation, giving rise to a dialogue enabling participants to move towards a shared appreciation of IS and their impacts. The goal of such a dialogue is not consensus: such a concept might impose yet another false boundary on the processes surrounding the design and use of information systems. A key contribution of this research approach is to provide a forum for discussion and negotiation which is capable of raising awareness about the wide range of expectations among the participants. Although its use in this context is at any early stage, RGA has shown itself to be a useful technique by providing some evidence of the range and diversity of the expectations and underlying values surrounding the implementation and operational use of IS. RGA has thus been used to supplement observation and interview techniques in a way which provides a deeper and more broadly based interpretation of the particular research setting.

The National Automated Fingerprint Identification System (NAFIS) is currently the largest fingerprint recognition system in the world. Although initial research at one of the pilot sites revealed great enthusiasm among the community of fingerprint experts, it also immediately challenged the view that there was agreement about the impacts of NAFIS: this erroneous assumption of consensus became a major issue which the research sought to address. Conflicting expectations about what NAFIS would, and would not, do meant that there was substantial dissonance among those responsible for decision making: this meant that there was substantial potential for suboptimisation of NAFIS brought about by well intentioned, but ill-informed, management interventions as the system was implemented.

The data gathered through the interview and grid elicitation processes are analysed both quantitatively and qualitatively. The goal of developing a cogenerative dialogue is fulfilled by qualitative analysis, using the grid data as a "focus for dialogue" (Gammack and Stephens, 1994, p79). No judgements or inferences are made: participants continue to construe their own meaning during discussion of the results. Similarly, constructs identified as non-discriminatory and therefore redundant by the algorithm are not abandoned since later interviews may reveal utility not immediately exposed by the analysis: this is particularly important in the context of the assessment of emergent phenomena such as the impact of IS. This qualitative analysis of RGA data through 'talkback' interviews, where participants are asked to elaborate the meaning of both the terms used and their interrealtionship represents a movement away from consensus seeking, which is the goal of the more complex arithmetic analyses. This use of RGA allows the technique to contribute to the provision of insights into specific (application) contexts, thus enriching the dialogue surrounding their evaluation. The capacity of RGA to facilitate the construction of values not ordinarily associated with the assessment of information systems can be demonstrated by appraisal of data gathered to date. Collectively, the grids comprise 85 elements (activities or processes which associated with fingerprint identification) and 55 bipolar constructs (used to differentiate the elements). The constructs span a range of categories within which NAFIS could be assessed, which include skills required for task completion; quality, integrity and continuity of data; the capacity to support creativity and discovery; the balance between microscopic and macroscopic examination; the balance between intrinsic (fingerprint detail) and extrinsic ('intelligence') information.

REACHING A CONCLUSION

The argument that the pattern of values and value systems development in an individual is transferable to groups and organisations as well as those products or ideas developed by them is problematic in the context of MIS. From the values held arises a sense of value, worth or benefit. It may be value for money it may be information richness, academic or philosophical advancement, but without underlying values supporting it will be meaningless. It is this perception of value that drives the demand for and acceptability of MIS (among other things) and it impacts on the design specification of MIS, choice of MIS, application and evaluation. Accordingly value is nothing without recognising that human values are inherent to worth. The dilemma for MIS is that in some cases values underpinning their assessment will never be appropriate or useful. If structured measurement systems are not compatible with a value system they will not on their own be successful. A broader philosophy needs to be adopted where either total abandonment or continual adaptation is needed to fit MIS to the source of the information. Evaluation in this case is essential but must be continual, adaptable and aware of the impact it has on the process.

This paper has taken an exploratory viewpoint concerning practical consequences of human values philosophy in information systems. These values are taken to be core beliefs or ideals about the way things should be. For individuals and organisations they are fixed very early in life but their aspect changes over time according to circumstance and priorities. Although these ideas are transferable to organisations as they are composed of expectations and paradigms that share common ground, they are interpreted not in the light of facts but models of reality based on individual human values. The consequences for information systems lies in the suggestion that they are composed of and therefore represent values in the same way as organisations and individuals.

REFERENCES

Armstrong, M.,1994, "Performance Management" Kogan Page, London
Collins, T. & Bicknell, D., 1997, "Crash" Simon & Schuster, London
Farbey, B; Land, F and Targett, D., 1995, "Hard Money, Soft Outcomes" Alfred Waller, London
Frantz, T.G., 1995, "Imagine the Ideal, Make It Real: Bringing Shared Values to Life" *Systems Practice* 8(3):289-306
Friedman, A and Cornford, D., 1989, "Computer Systems Development" John Wiley, Chichester
Gammack, J and Stephens, R., 1994, Repertory Grid Technique in Constructive Interaction in
        Cassell, C and Symon, G. "Qualitative Methods in Organisational Research" Sage, London
Hebel, M., 1996, The Impact of Value Systems on Performance Measurement *Systemist* 17(2):64-78
Hebel-Holehouse, M., 1998, "Exploring the Impact of Value Systems and Worldviews on
        Performance Measurement" Unpublished PhD thesis, City University, London.
Holloway, J., Lewis, J., & Mallory, G., 1995, "Performance Measurement and Evaluation" Sage,
        London
Iivari, J., 1992, The Organisational Fit of Information Systems *Journal of Information Systems* 2:3-29
Kling, R., 1996, "Computerization and Controversy " (2nd Edition) Academic Press, San Diego
Lucey, T., 1991, "Management Information Systems" (6th edition) DP Publications, London
Nolan, R., 1979, Managing the Crisis in Data Processing *Harvard Business Review* 115-126
Pidd, M., 1996, "Tools for Thinking " John Wiley, Chichester
Robson, W., 1997, "Strategic Management & Information Systems" (2nd edition) Pitman, London
Scarbrough, H and Corbett, J., 1992, "Technology and Organisation" Routledge, London
Walsham, G., 1993, "Interpreting Information Systems in Organisations" John Wiley, Chichester.

"WE FOCUS ON THE PARTS AND IGNORE THE WHOLE": THE STORY OF INFORMATION SYSTEMS TEACHING AT OUR UNIVERSITIES.

Brian Hopkins

School of Design and Communication Systems
Anglia Polytechnic University
Chelmsford CM1 1LL
j.b.hopkins@anglia.ac.uk

BACKGROUND

The research project which provides the foundations for the findings and proposals contained in this paper was the product of a continuing and growing sense of disquiet on the part of the author concerning the consistent level of criticism over the years regarding the quality, as perceived by the clients, of the information systems delivered to them by professional practitioners.

This generalised concern became a more focused one as the author became acculturated into the IS/computing community within the UK higher education sector, having been for the previous fifteen years a practitioner in the field. This process of assimilation proved to be characterised by reservations and doubts about the validity of our pedagogic approaches and, increasingly, alterations in outlook and mindset as a result both of a process of reflection on experiences as a practitioner and of an intensive period of reading of the body of texts which espoused a more participative, human-centred approach. Examples were the work of Mumford (1979), Checkland (1981, 1991), Suchman (1987), Hirschheim and Klein (1989), Winograd and Flores (1988), Ehn (1988), Greenbaum and Kyng (1991) and Walsham (1993).

Also among the texts was Schon's *The Reflective Practitioner* (1983), a seminal book containing sets of concepts which provided persuasive ammunition against the primacy of technical rationality in a wide range of fields. In it he promoted the idea that:

> "From the perspective of Technical Rationality, professional practice is a process of problem *solving*. Problems of choice or decision are solved through the selection, from available means, of the one best suited to the established ends. But with this emphasis on problem solving we ignore problem *setting*." (Schon, 1983: italics in original).

Synergy Matters: Working with Systems in the 21st Century,
Edited by Castell *et al.*, Kluwer Academic / Plenum Publishers, New York, 1999.

553

In these words he encapsulated the essence of the philosophical, pedagogical and, indeed, practical stumbling blocks which seemed to be impeding an ameliorative process in the development and implementation of effective IS.

The author's mounting concern found expression in a variety of ways. For example, over many years continued attempts were made to introduce a more human-centred approach into our courses (Hopkins, 1997). There were occasional successes but the overall pattern was one of resistance to change and unargued maintenance of the *status quo*. The present research project was the consequence of frustration resulting from a perception that the traditionalists were unshakeable in their stance and their conviction, without really engaging with an analysis of the potential consequences of their adherence to the technical rationalist position.

The results for our students included a restricted and limited view of the nature of the problems which they were to encounter in their later careers. The education and training which they received served to encourage a mindset which saw problems in the IS field within organisations as being essentially concerned with debates about means on the assumption that the ends were easily defined and usually non-contentious. In other words, their expertise rested in the acquisition and then the effective implementation of the requisite tools, techniques and methods to enable technically efficient information systems to be constructed.

Such a worldview can be fairly summarised in the words of the title of this paper; we are indeed focusing on the parts (the techniques) and ignoring the whole (the rich and complex socio-organisational environment).

OUR PEDAGOGICAL SIGNALS

Analysis of a range of documents produced by a sample of the IS/computing departments of UK universities (and by the major professional body) provided most illuminating insights into the mindsets of the authors/designers. This section provides examples of these, using the techniques of discourse analysis (Potter and Wetherell, 1994), to illustrate and draw out the signals and messages which are, sometimes perhaps unconsciously, transmitted by our publicity material and internal documentation, and to attempt to gauge what these word and phrases actually **do** in the context of our teaching and our students' later practice.

The documents under review included university prospectuses, student handbooks (for courses and for modules), in-course assignments and examination papers and answer scripts. In addition, an analysis was conducted of the publicity and examination syllabus documentation of the British Computer Society (BCS). This was seen as relevant because of the influence - indirect in many cases, but present nonetheless - of this institution, our primary professional body in IS/computing, on our university syllabi and hence on our pedagogy.

The detail of the analysis is too extensive to include in this brief paper but has been written up (Hopkins, 1998) as part of the research project. For our immediate purposes it should suffice to provide a flavour of the findings and of the sources. As well as analysing the IS/computing entries in the prospectuses a comparative study was also made of the entries for "companion" disciplines i.e. those where "design" was a core component, such as architecture.

The prospectus entries in our discipline area provided a rich (sometimes contradictory) mixture of messages. In many computer science courses the prospective student was alerted to the need to cover "knowledge of formal techniques", "use of mathematical ideas to develop concepts with useful applications" (Hopkins, 1998), would be expected to acquire skills in "methods for designing and implementing computer applications", and were given the advice that "software systems need to be properly designed in much the same way that engineering structures need careful design" (ibid). One might argue (reasonably) that this is very much what one would expect in courses with those titles, but that in fields such as business information systems one would encounter a fuller, more inclusive approach. Indeed the prospectus entries for such courses did include phrases like "human and managerial issues as well as technical ones", "full understanding of the business context", "management of change and conflict" and "organisational politics" (ibid) but we also come across (often on the same pages as the above) the following "up-to-date techniques and tools", "the fundamentals of computing", "industry-standard methodologies", "design and construction of quality software" and "a variety of architectures ... databases, client-server networks" (ibid).

The signals being transmitted indicated not only a discipline suffering from uncertainty regarding its direction but also one which (perhaps because of that very uncertainty) resorted to what were seen as the "fundamentals". In this case those core concepts were the traditional, well-practised, widely understood, non-contentious principles and practices built up over the past four decades.

By comparison, the prospectus entries in, say, architecture make reference to "distinguishing architecture from mere building" - a particularly apposite and fortuitous comparative example - and to "giving students an awareness of the many factors which influence how we use and relate to buildings" (ibid). More than that the breadth of required knowledge and involvement is brought out in phrases like "the ability to collaborate through the design process becomes very important" and statements that architecture "embraces disciplines from the physical and social sciences to the arts and humanities". The contrast between the messages is surely stark.

When the focus was transferred to the internal documentation relating to courses in the IS area then the "fundamentalist" messages detected above were in fact found to be re-inforced. At this point is needs to be acknowledged that in many modules within our courses it is in fact mandatory to adhere to such rigid sets of rules and to instil respect for the order, structure and discipline implied thereby. It would clearly be fatuous to suggest that the design and construction of programs, databases and computer networks could be achieved in an effective manner without following the relevant guiding principles precisely. In fact we are well aware of the consequences of ignoring or flouting these fundamental principles **in these clearly-defined, and bounded, technical areas**.

However, we discover (not surprisingly) in our courses the replication of the philosophical position referred to earlier in which application of a design approach (based around tools and techniques) which has achieved uncontestable success within a tightly delineated problem environment is then transplanted, without significant amendment, into a very different (and arguably, alien) environment. The analogy with organ transplant surgery in human medecine, while certainly not perfect, provides us with an insight into the potential dangers of basing our actions on such seductive assumptions.

Of course it will be argued that within our courses there are modules which do address the human and social aspects of IS development within organisations. Indeed, such modules do exist and from their titles (for example, Systems Development) we could reasonably expect to find in their outline learning objectives and content a more human-centred approach. It was therefore surprising and disappointing to encounter such a module, the learning objectives of which were:

"* Review different approaches to systems development
* Understand the need for engineering principles to be applied to the construction
        of large software systems
* Demonstrate an understanding of the skills required to investigate, analyse and
        design an information system and also some of the techniques used in systems
        development
* Account for and illustrate techniques for project control
* Describe and use in a basic fashion CASE tools for the aid of systems
        specification and design" (Hopkins, 1998)

This was an introductory module and the impact of the words and phrases written above is clearly apparent. Reference to "engineering principles", "techniques" and "CASE tools" sends unambiguous messages about the perceived nature of this process. On the other hand, it could be argued that "understanding of the skills required to investigate, analyse and design an information system" would include the necessary socio/psychological understanding and skills. Sadly, further analysis of student performance in, for example, role play investigative interviews demonstrated that such hopeful expectations were not realised in practice.

A further influential (although to some, peripheral) factor is the role of the British Computer Society (BCS) in forming, instilling and monitoring the mindset which underpins our continued loyalty to technical rationalism as a guiding philosophy for our pedagogy and practice in IS development. The BCS as the leading professional body in the UK has a pervasive impact on our university curricula through its accreditation of courses. In spite of its ongoing struggle to gain ascendancy and recognition for itself in our field (comparable to other professional bodies in their fields) the Society nevertheless does exert a considerable and, probably, increasing influence. Socio-cultural shifts in the UK at present seem likely to promote the role of such professional bodies as we seek certified "verification" of the competence of our professional practitioners generally. Specifically in the case of the IS field, as computer systems occupy an increasingly crucial role in our daily lives so will society seek reassurance through perceived guarantees of quality and high standards as claimed by bodies like the BCS.

Against this background it was instructive to peruse some of the publicity material issued to intending members; in particular the examination syllabus leaflet provided invaluable insights into what is essentially a parallel stream of thinking and outlook to the ones identified within the university material. For example, in the introductory sections we read that:

" ... their knowledge should enable them to discuss technical matters with others who have
specialised *in different areas of computing from their own* and to learn from such discussions."
(BCS, 1997: italics added)

and that

" ... it should also enable them to *explain to a layman in simple language* the problems, dangers and difficulties inherent in the implementation of any computer system." (BCS, 1997: italics added)

The above quotations obviously are only a snapshot of the material in the examination syllabus document (for further detail and associated analysis, see Hopkins, 1998), but the wording could not be clearer in its message and impact. The reader - the putative IS practitioner or possibly a university teacher - is invited to become part of an **exclusive** group whose membership "discuss technical matters" amongst themselves and "explain to a layman in simple language" other knotty, problematic issues connected with IS development and implementation. The condescending tone of the second quotation is not only breathtaking in its arrogance but extremely disquieting in its vision of the nature of the IS implementation process. After some forty years of experiencing the "problems, dangers and difficulties" it is likely that the "layman" has a reasonable experiential understanding of these issues! More than this, this same "layman" surely has earned the right to be recognised as a full, contributing partner in the processes of setting the problems and working to devise ways of improving those problem situations?

In conclusion, we have discovered in this (unavoidably) brief coverage of the vital area of the education of future IS practitioners that our universities and our major professional body together perpetuate the idea that development and implementation of IS are essentially susceptible to technical methods based on tried and tested tools and that training in the use of this range of techniques is an adequate preparation and training for future practice. In spite of continuing expression of dis-satisfaction with the outcomes of such thinking and practice we are reluctant to fully explore alternatives, particularly those which may take us and our students into the "swamp" which characterises the organisational environment in which so many of our IS are destined to operate.

FUTURE PROSPECTS

It would be easy to conclude that, in the face of many years of entrenchment of the "traditional" worldview regarding the education of the future IS professionals(and its evident supremacy within many of the IS departments in our universities at present), there is little possibility of radical change in the foreseeable future.

Such a stance would be understandable but would also be unduly pessimistic. Although we should always be wary of the "panacea syndrome" which has been an all too frequent visitor to IS thinking over the decades, there does appear to be a set of novel and innovative influences (as far as the IS field is concerned at least) at work around the periphery of our discipline.

This hopeful development lies in the fusion of the burgeoning multimedia field and the traditional IS practices. Clearly there is significant overlap between the two; indeed, it is unclear just how the fusion is going to work out within university courses. However, the interesting aspect as far as this paper is concerned is the markedly differentiated design approaches being adopted within the multimedia field. These approaches have their roots in a creative, flexible, emergent and participative way of thinking more readily associated with the "artist" than with the "engineer".

Some indicators of these changes in attitude and approach were evident in the university prospectuses, viz.

" ... offers creative students of any discipline the opportunity to design and develop effective multimedia systems using computers; ... consideration ... given to the wider impact on our lives; ... the course will be wide-ranging and holistic, integrating design, technology, sociology, psychology and business." (Hopkins, 1998)

and sought to attain

" ... a broad coverage of the engineering, artistic and social aspects of multimedia systems and their application." (ibid.)

When compared with the traditional IS course publicity there is a qualitative difference displayed in the multimedia approach pointing to a markedly different philosophical starting point but drawing on the best of the IS tradition and extending it. Observation and recording of comparative work by multimedia students suports the contention that they represent a way of designing which is in contrast to our IS students and is more in sympathy with the holistic, inclusive methods which this paper is advocating as a path to more effective IS.

REFERENCES

British Computer Society (BCS), 1997, "Examination Syllabus," Swindon, UK

Checkland, P., 1981, "Systems Thinking, Systems Practice," Wiley, Chichester.

Checkland, P., and Scholes, J., 1991, "Soft Systems Methodology in Action," Wiley, Chichester

Ehn, P., 1988, "Work Oriented Design of Computer Artefacts," Arbetslivscentrum, Stockholm.

Greenbaum, J., and Kyng, M., 1991, "Design at Work: Co-operative Design of Computer Systems," Lawrence Erlbaum Associates, New Jersey.

Hirschheim, R,. and Klein, H.K., 1989, Four paradigms of information systems development, *Communications of the ACM*, 32, 1199-1216.

Hopkins, J.B., 1998, Unpublished papers from an ongoing research project, Open University, UK.

Hopkins, J.B., 1997, "Twenty five years before the class: a personal reflection on changing times," an unpublished paper from an ongoing doctoral project, Open University, UK.

Mumford, E., and Weir, M., 1979, "Computer Systems in Work Design: the ETHICS Method," Wiley, New York.

Potter, J., and Wetherell, M., 1994, "Discourse and Social Psychology: beyond Attitudes and Behaviour," Sage, London.

Schon, D.A., 1983, "The Reflective Practitioner: How Professionals think in Action," Basic Books.

Suchman, L.A., 1987, "Plans and Situated Actions: the problem of Human-Machine Communication," Cambridge University Press, UK.

Walsham, G., 1993, "Interpreting Information in Organisations," Wiley, Chichester.

Winograd, T., and Flores, F., 1988, "Understanding Computers and Cognition: a New Foundation for Design," Ablex Corporation, Norwood, NJ.

A SYSTEMIC MODEL FOR PARTICIPATORY CHANGE IN INFORMATION SYSTEMS MANAGEMENT.

Andy Hyde

Clinical R&D
Nycomed Amersham PLC
Nycoveien 1-2
PO Box 4220 Torshov
N-0401 Oslo
Norway

1. INTRODUCTION

Computers, whether we like it or not, are a growing part of our daily existence. As with any technological advance throughout history it brings with it great change. Many people fear change of any kind but where does this fear come from? In IS I believe it has something to do with a feeling of a lack of control. Computer users become more aware of the possibilities for each new generation leaving education and joining the workforce. The days when a system developer knew more than the user are fast disappearing. A new systems development methodology is required that puts the developer and user on the same level.

Participatory methods of development do exist but have been more a theory than a practice. I present here a development methodology based on the principle that people who have the possibility to change their own future and experience that change as positive when compared with their mental model of the world will willingly and enthusiastically participate in that change. For this, the process of change is more important than the outcome because if the user participation has been effective, whatever the outcome, it will be correct.

The methodology presented in this paper has a systems epistemology through the feeding back of positive experience and new knowledge into the user's mental model so that these can affect future decisions. I suggest the use of contextual evolutionary prototyping as a vehicle for providing the required feedback whilst developing the IS solution.

Synergy Matters: Working with Systems in the 21st Century,
Edited by Castell *et al.*, Kluwer Academic / Plenum Publishers, New York, 1999.

559

2. THREE GENERATIONS OF DEVELOPMENT METHODOLOGIES

Somebody once said "We do not fear change but we fear *being* changed". These words succinctly describe the situation that the computer revolution has created in modern society. The change that computers have brought has happened at a pace which for many has been too fast. Individuals as well as society as a whole needs time to absorb change, to understand it and become a part of it otherwise it can develop into fear.

We do not fear change if we are a part of that change. The rate of development in Information Technology and Information Systems has perhaps left more people behind and created more fear than many previous changes in society because of its penetration into so much of our daily existence. The change has been driven largely by 'experts'. The domain of computer science has been a closed domain to outsiders, but that is changing.

Benathy (1997) has discussed these changes in the context of social systems but the same can be seen in Information Systems based on computer technology. Benathy identifies three generations: Developing *for* the users, developing *with* the users and developing *as* a user.

The first generation of developing *for* the users in the IS context was the 'expert' deciding what the user needed, developing and delivering a completed system. This, in the early days of the computer where there were few educated in the new science was perhaps understandable.

The second generation of developing *with* the users started as, and to a large extent is still, more an impression than a reality. A short consultation may be the most interaction that takes place between a developer and the user before the system is developed and delivered. Methodologies for development based on Participatory Design (PD) have made a genuine attempt to involve the user. These methodologies however have lacked a teachable theoretical background and have therefore not become part of computer education.

The third generation is developing *as* a user. In the social systems discussed by Benathy we are all part of the system being designed. With Information Systems the technological part of the system is external, we interact with it but are not a part of it. The parallel though is that the developer of the system should become part of the domain in which the user exists and into which the new system will be placed. From this position the developer can understand how the user perceives the system from his or her own view of the world, not as is most often the case, from the developer's external view of the world. One way to achieve this is as Shackel (1990) suggests, that the developer becomes a user. This approach is similar to Ethnography, more commonly associated with sciences such as anthropology, but which is seen as having more relevance to the study of Human-Computer Interaction (Preece *et al*, 1994)

If the developer is developing the system from within the user's domain and not from outside, he or she is in a much better position to see the users as individuals and account for this in the change process.

3. DEVELOPING IN THE NEW GENERATION

Many IS projects still fail and much research has been done to identify the reasons for failure. Traditional reasons are given as time and cost overrun and failure of the system to perform according to specification, however, these can be seen as emergent from two other reasons for failure; a failure to manage the complexity (Kokol, 1997; White, 1997) and a failure to manage expectations (Bennetts and Wood-Harper, 1997). Failure to manage complexity will lead to time and cost overrun and failure to manage expectations will lead to the system not performing according to specification. The latter because the specification is built on an expectation of what the system will be able to do.

Methods of development in the early second generation are still used. Little or no contact with users is sought by the developer before detailed design and implementation begins. In traditional development methodologies a complete description of the required system, the system specification, should be created before any design work is done. Beeby *et al.* (1997) believe that this is not possible. The requirements will be generated over time as the development progresses. From this view it is clear that a running dialog between user and developer is required throughout the process.

A conceptual definition and understanding of what is being developed is required. An Information System can be viewed as many things to many people. Kammersgaard (1990) describes these views as perspectives. Two of these are especially pertinent to this discussion. The system perspective and tools perspective. The system perspective is that most often seen by the technologist. The new system being developed is a technological achievement. The system is comprised of computers, networks, software and interfaces. The tools perspective, however, looks at the development as an extension of the tools available in the user's domain, the tools necessary for employees to carry out their daily tasks. System developers need to realise this and see their role as the development of a tool for the users that will require constant refinement through iterative design and testing. Developers need to see the knowledge and skill that individual users posses as essential to that development.

Users can contribute constructively, contrary to the beliefs of many IS developers. Drucker (1998) succinctly captures this realisation in the context of worker involvement in management in saying that he and others were surprised to find that workers were neither "Dumb oxen nor immature nor maladjusted" as others had maintained earlier.

4. A NEW METHODOLOGY

The theoretical underpinning of the methodology is to provide constant feedback to enable change at an absorbable pace. For users to be able to absorb the change it is necessary to understand how the user perceives the world of technology and technological change. Each individual has a different perception based on their mental model of the world. This mental model is based on the individual's motivation, values, visions, knowledge, experience, and often assumptions and guesses. Oppenheim (1992) suggests that action is determined by attitudes which are only abstractions - "though they are real enough to the person who holds them". This point is central. Each user's attitude to another may appear unfounded but still the system developer must address the issue. It may not be possible to build a system that encompasses every individual's needs but each must have been addressed to the satisfaction of user who perceived it otherwise the user's mental model will be negatively reinforced. The mental model is used to filter information before action is taken and the results of the actions are also used to shape the mental model. Bad experiences can therefore be a block to action but good experiences can be a catalyst.

The methodology suggested in this paper has five elements. The elements each contribute to experiential learning by the constant use of feedback during the development allowing the users to fully understand the implications of the change and the way the change is being implemented, to influence the change and absorb the change.

The elements alone and in some combinations are not new, but as with Senge's five disciplines (Senge, 1993) it is the synergy achieved from the combination of all five that makes the methodology what it is.

The elements are:

1. Ethnographic Requirements Analysis
2. Participative development
3. Contextual development
4. Evolutionary prototyping
5. Systems Thinking

4.1 Ethnographic Requirements Analysis

In order for the system developer to obtain a full understanding of the domain into which the new system will be placed and an understanding of the users it is necessary for the developer to be within the user domain for more time than is traditionally the case. The developer needs to spend time looking at the environment, the conflicting tasks, the organisation and the individual users. The belief that the user will behave the same as the developer or even that one user will behave like another is described by Landauer (1990) as "naïve intuition fallacy". Davenport (1994) observes that technocrats are constantly caught off guard by the "irrational" behaviour of end users. Only by close and constant interaction can these misunderstandings be corrected.

4.2 Participative Development

PD is traditionally short for Participative Design but Chin *et al.* (1997) suggest there is a need to involve the users much earlier in the process. This view is shared by Mumford (1997) who has developed the QUICKethics methodology to involve users in the requirements analysis phase. Carmel *et al.* (1993) and Blomberg and Henderson (1990) found that effective participation is not just inviting users to a design meeting, the users need to be involved and need to perceive their input as having meaning and effect. The term Participatory Development is therefore used to emphasise this requirement as distinct from the detailed design phase where other participatory methodologies tend to focus.

PD has no fixed definition but Blomberg and Henderson (1990) have identified three tenets that guide practitioners: the goal is to improve the quality of working life, the orientation is collaborative and the process is iterative. These three tenets underpin the methodology described in this paper.

The methodology is inescapably participative because of the contextual evolutionary prototyping and the way that the requirements are developed iteratively from an ethnological interaction.

4.3 Contextual Development

The tradition in the design of computer systems for human use through Human-Computer Interaction techniques has been to use laboratory testing to design a usable system. However, designing in a laboratory does not provide for feedback based on real world conditions. Whiteside *et al.* (1988) highlights several failings of the laboratory approach based on environmental differences and task design and Landauer (1990) questions it on psychological grounds. Checkland and Scholes (1990) in describing action research point out that this requires involvement in the problem situation.

If one is to understand how the new development will interact with other aspects of the user's human activity systems, development must be done in the context of those systems. Systemic interactions of the technology cannot effectively be separated from other parts of the system.

4.4 Evolutionary Prototyping

Evolutionary prototyping will enable the iterative development of requirements and the iterative design and testing of the new tool whilst allowing the users to absorb the change.

Established methods of Participative Design (PD) described in Clement and Van den Besselaar (1993) and Blomberg and Henderson (1990) use passive methods of design such as paper based scenarios to create a design. However, Beeby *et al.* (1997) emphasise that the best way for users to understand the change is by experiencing it first hand.

The methodology's epistemology is based on the need for experiential learning. Everybody in the change process needs to learn from each other and from the experience in order to change their mental model of the world.

4.5 Systems Thinking

No system can be developed separate from its environment and organisation. Soft Systems Methodology (SSM) (Checkland and Scholes, 1990) clearly emphasises this in the SSM that evolved from experience of its use. The seven stage model was replaced by the model with a cultural stream of analysis and a logic based stream of analysis. The logical based stream of analysis in the outlined methodology of this paper is the contextual evolutionary prototyping. Mental based analysis is replaced with action research based experimentation. The implications of change resulting from the implementation of Information Systems is complex, just the type of complexity SSM sets out to address, perhaps too complex to be fully understood before it is experienced.

5. CONCLUSIONS

The methodology is developed out of the need to involve the users in the process of change and in the belief that the development of Information Systems which are the tools of our daily work can only be done effectively with the synergies resulting from a co-operation between users and developers.

The developer must develop a better understanding of the user's domain and the user must learn, at an absorbable pace, the capabilities of the technology used to build Information Systems. The effects of implementing technology based Information Systems are increasingly complex, especially when a systemic view of the world is held. It is therefore suggested that a new methodology with a systems epistemology is required.

The methodology places the developer in the user's domain to teach and to learn. The development project is incremental in order to avoid the generation of fear. The development must be in the context of the final implementation so that it will function in the real world when completed. Although, because the world keeps changing, the development is never complete until it is replaced by a successor.

The percentage of new technology projects that fail is alarmingly high, even after so many years of experience. Analyses of the reasons for these failures consistently point to problems of co-operation or communication between developer and user. If the user is alienated by the change process they are more likely to resist it or even sabotage it. The methodology suggested involves the user in a way where they can see that they have a real possibility to affect the change and therefore their future. By using positive reinforcement to emphasise this and replace bad experience with good and ignorance with knowledge it is hoped that users will want to become involved in the change process and will actively and constructively contribute.

REFERENCES

Beeby, R.B., Gammack, J.G., and Crowe, M.K., 1997, "Constructing End-User Design Environments: implementing client-led design", in *Proceedings of the 5th conference of the United Kingdom Systems Society*, Stowell *et al*. eds., Plenum Press, N.Y., USA. pp. 537-541

Benathy, B., 1997, "Designing Social Systems in a Changing World: a journey to create our future", *Systemist*, Vol.19, No.3, pp.187-216

Bennetts, P.D.C., and Wood-Harper, A.T., 1997, "Soft Systems Methodology: a metaphor for the process of data analysis", in *Proceedings of the 5th conference of the United Kingdom Systems Society*, Stowell *et al*. eds., Plenum Press, N.Y., USA., pp.543-547

Blomberg, J.L., and Henderson, A., 1990, "Reflections on Participatory Design: lessons from the trillium experience" in *CHI '90 Proceedings*, The Association for Computing Machinery, Inc., N.Y., USA, pp. 353-360

Carmel, E., Whitaker, R.D., and George, J.F., 1993, 'PD and Joint Application Design: a Transatlantic Comparison', *Communications of the ACM*, Vol.36, No.4, pp.40-47

Checkland, P., and Scholes, J., 1990, *Soft Systems Methodology in Action*, Wiley, Chichester, UK

Chin, G., Rosson, M.B., and Carroll, J.M., 1997, "Participatory Analysis: shared development of requirements from scenarios', in *Proceedings of CHI '97 Electronic Publications*, The Association for Computing Machinery, Inc., N.Y., USA.

Clement, A., and Van den Besselaar, P., 1993, "A Retrospective Look at PD Projects", *Communications of the ACM*, Vol. 36, No 4, pp 29-37

Davenport, T.H., 1994, "Saving IT's Soul: human-centered information management'. *Harvard Business Review*. 1994 (March-April). pp. 119-131

Drucker, P.F., 1998, *On The Profession of Management*, Harvard Business School publishing, Boston, MA., USA.

Kammersgaard, J., 1990, "Four Different Perspectives on Human-Computer Interaction" in *Human-Computer Interaction*, Preece, J. and Keller, L. eds., Prentice Hall, Hemel Hempstead, England. pp.42-63

Kokol, P., 1997, "Reasoning about software system design with SSM", in *Proceedings of the 5th conference of the United Kingdom Systems Society*, Stowell *et al*. eds., Plenum Press, N.Y., USA. pp.579-582

Landauer, T.K., 1990, "Relations Between Cognitive Psychology and Computer System Design", in, *Human-Computer Interaction*, Preece, J. and Keller, L. eds., Prentice Hall, Hemel Hempstead, England. pp.141-160

Mumford, E., 1997, "Requirements Analysis for Information Systems: The QUICKethics approach", in *Proceedings of the 5th conference of the United Kingdom Systems Society*, Stowell *et al*. eds., Plenum Press, N.Y., USA. pp.15-20

Oppenheim, A.N., 1992, *Questionnaire Design, Interviewing and Attitude Measurement*, Pinter, London, England

Preece, J., Rogers, Y., Sharp, H., Benyon, D., Holland, S., and Carey, T., (1994), *Human-Computer Interaction*, Addison-Wesley Publishing Co., Wokingham, England

Senge, P., 1993, *The Fifth Discipline*, Century Business, London, UK.

Shackel, B., 1990, "Human Factors and Usability", in *Human-Computer Interaction*, Preece, J. and Keller, L. eds., Prentice Hall, Hemel Hempstead, UK.

White, D., 1997, "Risk Management and project failure" in *Proceedings of the 5th conference of the United Kingdom Systems Society*, Stowell *et al*. eds., Plenum Press, N.Y., USA. pp.525-529

Whiteside, J., Bennett, J, and Holtzblatt, K., 1988, "Usability Engineering: our experience and evolution", in *Handbook of Human-Computer Interaction*, Helander, M. ed., North Holland, Amsterdam, Netherlands.

INFORMATION AND COMMUNICATION TECHNOLOGY:
TACIT KNOWLEDGE AND INNOVATION

Jon-Arild Johannessen,[1,2,3] Bjørn Olsen[1]

[1] Bodø Graduate School of Business
[2] Lillehammer College
[3] Agder Research Foundation

INTRODUCTION

Companies of today are facing an exponential advancements in technology, a frequent shifting in the nature of customer demand, and growing global competition, leading to increased turbulence and complexity in the business environment. D'Aveni (1994) categorizes the situation in its extreme form as hyper-competition. To meet these challenges, both the popular and the academic press are advising companies to focus their attention toward innovation in order to create and sustain competitive advantages. Jacobson (1992) argues that it is the continuous changes in the state of knowledge that produce new disequilibrium situations and, therefore, new profit opportunities. Hence, we have witnessed an increasing focus on knowledge as the most important resource for companies. Drucker (1993) postulates that knowledge as an input resource will in the future mean more than physical capital. This has also been underlined by Quinn et al. (1996), Thurow (1997), Sveiby (1997), Solow (1997) and Stewart (1997). In the wake of this development we have seen that rapid access to knowledge and information (Grant, 1996) is becoming paramount. Consequently, spending on information and communication technology (ICT) has surged during the last decade. However, as this technology is limited to the transfer of explicit (codifiable) knowledge (Antonelli, 1997), our concern is that this may relegate tacit knowledge (Polanyi, 1962; 1966) to the background, in spite of this knowledge being emphasized by the literature as an important strategic resource for most companies (Gøranzon and Florin, 1990; Gøranzon, 1993; Black and Boal, 1994; Nonaka and Takeuchi, 1995, Howells, 1996). Hence, leading to the mismanagement of knowledge.

KNOWLEDGE AND ICT- INVESTMENTS

Although we observe a general optimism, in the society as a whole, concerning ICT's potential for creating sustainable competitive advantages, recently, challenges to the earlier optimism have risen from the emerging empirical evidence which indicate a lack of support

Synergy Matters: Working with Systems in the 21st Century,
Edited by Castell *et al.*, Kluwer Academic / Plenum Publishers, New York, 1999.

565

for the positive economic impact of ICT investments (Strassman, 1990; Brynjolfsson, 1993; Wilson, 1993; Dos Santos et al., 1993; Loveman,1994; Powell and Dent-Micallef, 1997). This is denoted as the "productivity paradox of information technology". Sweeny (1996:6) argue that: "The expectation that investment in science and high technology would result in higher levels of economic prosperity has not been fulfilled. Something has gone wrong".

Various explanations to the lack of empirical findings have been proposed. Brynjolfsson (1993:73) grouped the explanations into four groups: 1) Mismeasurement of outputs and inputs, 2) Lags due to learning and adjustments, 3) Redistribution and dissipation of profits, and 4) Mismanagement of information and technology. Our main concern in the present paper is on mismanagement. However, when Brynjolffson (1993:76) discuss the issue, he argue that the problems of mismanagement stem from "....the lack of explicit measures of the value of information, which makes it particularly vulnerable to misallocation and over consumption by managers". We take a rather different approach by arguing that the mismanagement of ICT is found in the lack of understanding of tacit knowledge, and the relationship between tacit knowledge and ICT.

Knowledge can be categorized in two different categories: explicit and tacit knowledge (Nonaka & Takeuchi, 1995). Explicit knowledge can relatively easily be formulated by means of symbols and can be digitized. This knowledge can thus with relative ease be transferred to others by e.g. the use of ICT. Tacit knowledge (Polyani, 1962;1966) is entrained in action (practice) and is linked to concrete contexts (Schøn, 1987; Gøranzon, 1993; Molander, 1993; Rolf, 1995). This knowledge is difficult to communicate to others as information, and can at best be difficult to digitalize. Tacit knowledge is defined by Howell (1996: 92) as: "non-codified, disembodied know-how that is acquired via the informal take-up of learned behavior and procedures. Tacit knowledge does not involve the generation and acquisition of tangible products and processes, or the more formal element of intangible knowledge flows associated with specific research, technical or training programs". Polany (1966:4), who was the first one to introduce the concept tacit knowledge expresses the meaning of the concept in the following simple and precise way: "We can know more than we can tell". In the strategy literature the resource base perspective (e.g. Barney, 1991), the knowledge theory (see Grant, 1996), in addition to the dynamic capability approach (e.g. Teece et al., 1997), have addressed parts of this development, but still we know little about links between tacit knowledge, and ICT (see Howells, 1996).

What happen when enterprises unilaterally invest in ICT? The focus will easily be on the part of the knowledge base which can be formalized, i.e. which can easily be communicated to others as information. The tacit knowledge can then easily be de-emphasized. However, a lot of literature underscores the very fact that it is tacit knowledge which will determine to what extent companies will be competitive in a turbulent market, and a global economy (Nonaka and Takeuchi, 1995; Spender and Grant, 1996; Sweeny, 1996; Teece et al., 1997). If this is the case, i.e. that the tacit part of knowledge is important to generate sustainable competitive advantages for companies, companies will easily lose their competitive edge if they emphasize investments in, and use of ICT without taking tacit knowledge into consideration. To put it more directly, companies invest into a position where they lose, and do not improve competitive advantages, if they do not emphasize the entire knowledge base.

HOW TO BALANCE THE USE OF KNOWLEDGE?

The remaining question is: how could the use of tacit- and explicit knowledge be balanced to positively impact the outcome of a company? To answer the question, we need to know more about how tacit knowledge and ICT affect continuous improvements, innovations, performance, and subsequently, sustainable competitive advantages.

Organizational knowledge processes and organizational learning constitute an integrated process (see Sobol and Lei, 1994; Spender and Grant, 1996), and it is impossible to study one element without studying the other element too. This is clearly underlined Nonaka (1994). The tacit knowledge constitutes a "core capability" (Grant, 1996: 380) for companies, as it distinguishes companies from their competitors and promote their strategic advantages (Leonard-Barton, 1995). This type of core capability, according to Hamel and Prahalad (1994), is developed through collective learning processes in the company. Learning by doing, using experimenting and interacting is here seen as the processes constituting tacit knowledge. This is also seen by i.e. Dosi (1988); Dierickx and Cool (1989); Reed and de Fillippi (1990); Badaracco (1991).

But what about innovation? We argue that tacit knowledge, on its own, does not enhance innovation, only continuous improvements. Also, that tacit knowledge can be a key barrier to innovation. This is because tacit knowledge usually is part of a long term learning process in a specific context, being embodied in the structure of thinking, the way of thinking, and therefore functions as a conservative element in relation to innovation. Tacit knowledge, states Fleck (1996: 119):" is the most crucial in restricting the social distribution of knowledge, and has been widely identified as a major constraint on the diffusion of both science and technology". But, tacit knowledge also is a sort of organizational "immune"-system hindering imitation from other social systems. The function of tacit knowledge is then both conservative, i.e. stabilizing the system, and acting as an imitation "guard".

Continuous improvements, on the other hand, are enhanced by tacit knowledge. Solow (1997: 24) argues that: "The routine continuous improvement of products and processes is arguably the most important source of increased productivity in mature industries", i.e. the experience based part of the firms knowledge base. Young (1993:447) makes the following statement: "Following models of learning by doing, I assume that production experience generates new knowledge on how to produce good more efficiently.--- Experience in production increases the productivity of the new technology---". However, tacit knowledge is bounded by a negative feedback factor. This factor is found in that when no innovation occurs in the organization, continuous improvement increases performance, but only to a certain degree. Thus tacit knowledge promotes continuous improvement only to a certain level, and then declines. This is in line with Solow (1997: 25) who denotes this phenomenon as "bounded learning by doing". We assume that all tacit knowledge has this effect on continuous improvement.

We have argued that tacit knowledge "on its own" does not improve innovation. A major point is that the entire knowledge base for the individual company, the tacit part included, is developed in a social and cultural context, where the interaction between companies and among companies and external systems constitutes important elements for both development and transmittance of tacit knowledge (see Adeboye, 1997). Hence, we argue that it is only when tacit knowledge is linked to the explicit knowledge in the system, and the systems external knowledge base, that continuous innovation appears, increasing performance and promoting sustainable competitive advantages. Hence, linked to the external knowledge base, something new may occur and innovation is brought into the system. Both continuous improvement and continuous innovation are created by the interaction between the system-specific knowledge base, the link to other systems in the environment and organizational learning. That continuous innovations promote continuous improvement is shown by Fruin (1997) at Toshiba. Continuous innovations may be understood as punctuation's of continuous improvements (Fruin, 1997:27). When at a certain time innovation enters the organization, performance will increase to a higher level, which gives a positive bandwagon effect upon learning by- doing,- using , interacting and-experimenting, because something new brings the learning process to a higher level of achievement, even if the bounded negative feedback factor still is in operation. I.e. it is in operation, but at a new level of achievement.

We further argue that the use of ICT influence the use of the companies external knowledge base by effectively transferring explicit knowledge. Huber (1990) argued that the use of information technology leads to more quickly retrieved information. Kessler and Chakrabarti (1996) argued after an extensive literature review that there is a growing recognition that speed is important in the development of successful innovations. To enable firms to initiate innovations within a turbulent and complex environment, they need fast access to information. Hence, using ICT increases the speed on the availability of information, which in turn enhances innovations.

However, we argue that a prerequisite for the companies external knowledge base to promote continuous innovation, it need, in addition to the use of ICT (facilitating the transfer of explicit knowledge), to be linked to the company's tacit knowledge, accompanied by sensitivity to change and external meeting places (facilitating the transfer of tacit knowledge).

Lee (1994:143) argued that: "In the view of information richness theory, electronic mail filters out important cues such as body language and tone of voice and, unlike face-to-face meetings, is not conductive to immediate feedback. Hence, for continuous innovation to appear there is a need for external meeting places. External meeting places constitute an arena were tacit knowledge can be converted to explicit knowledge. This is also underlined by Nonaka and Takeuchi (1995), because the interaction between individuals at such meeting places is psychologically close and the information media is rich. Daft and Lengel (1986:560) argue that: "In a sense, richness pertains to the learning capacity of communication".

We further argue that sensitivity to change is important, as it represent is a prerequisite for seeking the kind of information that enable the firm to meet customers future demand through external meeting places and by using ICT.

CONCLUSION

Tacit knowledge is recognized as playing a key role in determining the extent to which companies are able to create and sustain competitive advantages. However, investments in ICT may lead to a de-emphasizing of tacit knowledge with devastating consequences. Emerging empirical evidence have indicated a lack of support for the positive economic impact of investments ICT. We have argued that this may be due to the mismanagement of knowledge. This mismanagement is first of all created by an imbalance between the emphasis on tacit- and explicit knowledge, often manifested trough firms ICT-strategies prior to investing in ICT. A central point in this is article the focus on establishing the right balance between tacit and explicit knowledge. Central in our argument is the creation of a learning loop whereby continuos innovation enhance a higher level of learning by doing-using-, experimenting-, and interacting, creating a positive spiral. To create such learning loop, there is a need to link the company's tacit knowledge to the company's external knowledge base by the use of ICT. However, this need to be accompanied by a sensitivity to change and the creation of external meeting places, in order to promote innovation, which in turn promotes performance and enhances a higher level of learning by doing-, using, experimenting - and interacting. The learning loop also improve the company's sustainable competitive advantages, by limiting competitors possibilities for imitation (through a higher level of tacit knowledge), and by increasing continuous improvements, innovation and performance.

REFERENCES

Adeboye, T. (1997). "Models of innovation and sub-Saharan Africa's development tragedy". *Technology Analysis & Strategic Management*, **9**(2). pp. 213-235.

Antonelli, C. (1997). "New information technology and the knowledge-based economy. The Italian evidence". *Review of Industrial Organization*, **12**(4), pp. 593-607.

Badaracco, J. (1991). *The knowledge link*. Harvard Business School Press, Cambridge, MASS.

Barney, J. B. (1991). "Firm resources and sustained competitive advantage". *Journal of Management*, **17**(1), pp. 99-120.

Black, J.A., and Boal, K.B. (1994)."Strategic Resources: Traits, configurations and paths to sustainable competitive advantage". *Strategic Management Journal*, **15**, pp.131-148.

Brynjolfsson, E (1993)."The productivity paradox of information technology". *Comm. ACM*, **35**, pp. 66-77.

Daft, R.L., and Lengel, R.H. (1986). Organizational information requirement, media richness and structural design, Management Science, **32**(5), pp. 554-571.

D'Aveni, R. (1994). *Hypercompetition: The dynamics of strategic maneuvering*, Basic Books, New York.

Dierickx, I., and Cool, K. (1989). "Asset stock accumulation and sustainability of competitive advantage". *Management Science*, **33**, pp. 1504-1511.

Dosi, G. (1988). "The Nature of the innovative process". In G, Dosi, et.al. *Technical Change and Economic Theory, pp.* 221-238, Pinter, London.

DosSantos, B.L., Peffers, K.G., and Mauer, D.C. (1993). "The impact of information technology investment announcements on the market value of the firm". *Information Systems Research*, **4**(1), pp. 1-23.

Drucker, P.F. (1993). *Post-capitalist Society*, Butterworth Heineman, New York.

Fleck, J. (1996). "Informal information flow and the nature of expertise in financial services". *International Journal of Technology Management*, **11**(1-2), pp. 104-128.

Fruin, W.M. (1997). *Knowledge Works: Managing intellectual capital at Toshiba*, Oxford University Press, Oxford.

Grant, R.M.(1996). "Prospering in dynamically-competitive environments: Organizational capability as knowledge integration". *Organizational Science*, **7**(4), pp. 375-387.

Gøranzon, B. (1993). *The practical intellect: Computers and skills*, Springer-Verlag, Heidelberg.

Gøranzon, B., and Florin, M. (1990). *Dialogue and technology. Art and technology*, Springer-Verlag, London.

Hamel, G., and Prahalad, C.K. (1994). *Competing for the future*. Harvard Business School press, Boston, MA.

Howells, J. (1996). "Tacit knowledge, innovation and technology transfer". *Technology Analysis & Strategic Management*, **8**(2), pp. 91-105.

Huber, G. (1990). "A theory of the Effects of Advanced Information Technologies on Organizational Design, Intelligence, and Decision Making". *Academy of Management Review*, **5**(1), pp. 47-91.

Jacobsen, R. (1992). "The Austrian School of Strategy". *Academy of Management Review*, 17(4), pp. 782-807.

Kessler, E.H., and Chadrabarti, A.K. (1996). "Innovation speed: A conceptual model of context, antecedents, and outcomes", *Academy of Management Review*, **21**(4), pp. 1143-1191.

Lee, A.S. (1994). "Electronic mail as a medium for rich communication: an empirical investigation using hermeneutic interpretation". *MIS Quarterly*, June, pp. 143-157.

Leonard-Barton, D.L. (1995). *Wellsprings of Knowledge: Building and Sustaining the sources of Innovation*, Harvard Business School Press, Boston, MASS.

Loveman, G.W. (1994). "An assessment of the productivity impact on information technologies". In T.J. Allen and M.S. Scott Morton (eds.). *Information technology and the corporation of the 1990's*: Research studies, MIT-Press, Cambridge, MA, pp. 84-111.

Molander, B. (1993). *Kunskap i handling*, Daidalos, Gøteborg.

Nonaka, I. (1994). "A dynamic theory of organizational knowledge creation". *Organizational Science*, **5**(1), pp. 14-37.

Nonaka, I., and Takeuchi, H. (1995). *The Knowledge Creating Company*. Oxford University Press, Oxford.

Polanyi, M. (1962). *Personal Knowledge*. Routledge & Kegan Paul, London.

Polanyi, M. (1966). *The tacit dimension*, Gloucester, Mass.

Powell, T.C., and Dent-Micallef, A. (1997). "Information technology as competitive advantage: The role of human, business and technology resources", *Strategic Management Journal,* **18**(5), pp. 375-405

Quinn, J.B., Anderson, P., and Finkelstein, S. (1996). "Leveraging intellect". *Academy of Management Executives,* 10(3), pp. 7-27.

Reed, R., and de Fillippi, R.J. (1990). "Causal ambiguity, barriers to imitation and sustainable competitive advantage". *Academy of Management Review,* 15(1), pp. 88-102.

Rolf, B. (1995). *Profession, tradition och tyst kunskap,* Nya Doxa, Nora, Sverige

Schøn, D. (1987). *Educating the reflective practioner.* Jossey-Bass, London.

Sobol, M.G., and Lei, D. (1994). "Environment, manufacturing technology and embedded knowledge". *International Journal of Human Factors in Manufacturing,* 4(2), pp. 167-189.

Solow, R.M. (1997). *Learning from learning by doing: Lessons for economic growth.* Stanford University Press. Stanford: California.

Spender J.C., and Grant, R.M. (1996). "Knowledge and the firm: Overview". *Strategic Management Journal,* 17, Winter, Special Issue, pp- 5-9.

Stewart, T.A. (1997). *Intellectual capital: The new wealth of organizations.* Doubleday, London.

Strassman, P.A. (1990). *Business value of computers, Information Economic Press.* New Canaar, C.T.

Sveiby, K.E. (1997). *The new organizational wealth: Managing & measuring knowledge-based assets.* Berrett-Koehler Publisher, San Francisco.

Sweeney, G. (1996). "Learning efficiency, technological change and economic progress". *International Journal of Technology Management,* **11**(1-2), pp. 5-27.

Teece, D.J., Pisano, G., and Schuen, A. (1997). "Dynamic capabilities and strategic management". *Strategic Management Journal,* **18**(7), pp. 509-533.

Thurow, L.C. (1997). *The future of capitalism.* Nicholas Breeley publishing, London.

Wilson, D. (1993). "Assessing the impact of information technology on organizational performance", In R. Barker., R. Kauffman and M.A. Mahmoad (eds.). *Strategic information and technology management,* Idea Group, Harrisburg, PA.

Young, A. (1993). "Invention and bounded learning by doing". *Journal of Political Economy,* **101**(3), pp. 443-472.

QUALITATIVE MODELLING OF INFORMATION SYSTEMS

J Korn

Visiting Academic
Middlesex University
Bounds Green Road
London, N11 2NQ

INTRODUCTION

In the late 19th century Maxwell introduced the idea of a demon which managed to separate fast from slow moving molecules of gas in a container[1]. This was a matter of concern since the action of demon led to a temperature difference apparently without any energetic input. In fact, the demon was using information operating as a part of a control system with energy and information carrying regions interfaced by an amplifier[2]. Later, investigators developed a quantitative theory of information leading to a logarithmic function connecting the amount of information to the size of an assembly[3,4,5]. An approach to the question of information with meaning was also evolved[6,7].

The term 'information system' appears to be applied to modelling human activity situations such as business, manufacturing and other organisations involving the use of computers[8,9,10].

It appears from the literature that the concept of information to a large extent has been treated without reference to that of information systems. Information systems are used extensively by living things : humans, animals and plants at the micro and macroscopic level and by the vast variety of their organisations. Information is prepared and applied by purposive systems including hardware based control, for creating a change of state, mental or physical, directly or indirectly, in much the same way as energy is utilised. There is not as yet an accepted, comprehensive method leading to a more in depth understanding, modelling and design of such systems.

Thus, the aim of this paper is to suggest a concept of information that carries meaning and quantity and fits into a method of representation of information systems called linguistic modelling, and to outline this method.

Synergy Matters: Working with Systems in the 21st *Century,*
Edited by Castell *et al.*, Kluwer Academic / Plenum Publishers, New York, 1999.

CONCEPT OF INFORMATION

The mind through the brain can form visual images as a collection of properties[11,12], which, when reproduced as informatic, lead to iconic models. The problem with such models is that they reflect parts of the world in their entirety. They cannot be used for : 1. expressing a view, asking a question and issuing an instruction, 2. manipulating images other than changing spatial position in the mind, reasoning and operating rules are not possible.

These mental feats can be accomplished by symbols[13]. Information as the term is understood here, is used for expressing a view and is implemented as a collection of symbols. This is done by separating one or more properties from an image rather than viewing it in its entirety. The symbolism is called subject-predicate. This symbolism is best known in natural language but all means intended to carry information such as signs, use it. For example, the triangular road sign refers to a section of the road and one of its features : that there is some form of danger. The subject-predicate form represents an object and a feature or an event and a feature as a noun phrase and an adjectival or verb phrase respectively.

Information when embodied in the subject-predicate form as a message which is understood and accepted by a receiver, is used by a sender for exerting specific influence so as to change the mental state of the receiver. Thus, the subject-predicate form is attached to special verbs representing influence[12] and capable of carrying messages as adjectival or verbal subordinate clause. For example, 'man *notifies* his friend *that he is ill*'.

A message usually contains more than a single subject-predicate form called here the unit of information. Total information, just like energy, to accomplish a specific task, has to be constructed from a specific number of meaningful units of information. A letter asking for a loan or a business proposition should carry such total information. Data are numerical form of or a supplement to predicate, often used in technical discourse.

CONCEPT OF INFORMATION SYSTEM

A theoretical object can be regarded as a conjunction of properties[12] one or more of which emerges depending on the situation in which the object finds itself. Having identified one or more objects whose mental state expressed in terms of properties such as happy, sad or instructed, is to be changed, an information system is an assembly of interacting objects with the task to accomplish the change. Total information delivered by an information system to an identified object called the 'receiver', is regarded as a product which has to be designed[14]. Thus, information carried in a letter, for example, is constructed so as to achieve the expected effect. A system - product – receiver chain has emerged[14]. When the product is a form of energy or an artifact, the term 'receiver' is replaced by that of 'user'.

Interaction in an information system is called influence which carries the appropriate information modelled as a verb with a subject-predicate form attached. Theoretically influence requires zero physical power. In practice, however, information is carried by a medium such as paper called a letter, a leaflet or an order, or, when exerted verbally, by air pressure. In such a case the medium itself may be treated a part of the structure of a situation. An instrument as an extension of a sense organ generating geometric, material, numerical or energetic property as informatic for carrying information[12] is also seen as medium. An appropriate kind of energy carried by physical power is a matter of the physics of a situation, the appropriate information attached to a medium or to an instrument is a matter of convention.

QUALITATIVE MODELLING OF INFORMATION SYSTEMS

Qualitative or linguistic modelling uses well defined empirical input leading into a symbolism capable of being manipulated and of the application of computing[12,13]. Usually the starting point is a narrative, or story, in natural language which is subjected to linguistic analysis. This is necessary : 1. to reduce a story to a collection of one- and two-place context-free sentences and to the modifiers of their constituents, 2. to identify clauses which model the structure of a situation and those which carry information.

One- and two-place sentences with dynamic verbs are considered as basic constituents of which a situation is constructed. They carry the initiating and affected objects and their modifiers as properties : driving (dp), enabling (ep) and calculating (cp), also the verbs, or interactions (in), and their adverbial modifiers[12,13]. As a result of assumptions, a basic constituent is expressed as a pair of logical conditionals which can be diagrammed[12,13].

The logic sequences generated from the diagram of a situation admit no uncertainty. For given input conditions the outcome will occur. Uncertainty is introduced through : 1. Personality profiles, 2. Graded adjectives, 3. Certainty factors, 4. Comparators, and 5. Explicit purposive systems[13].

The above can be captured by the symbolism :

$$zp(on, rp, lp, (prop_i(mod_j))).. \rightarrow ..in(vn, do, ao, (adv_k(mod_m))).. \qquad 1.$$

$$in(vn, do, ao, (adv_k(mod_m))).. \wedge wp(on, rp, lp, (prop_i(mod_j))).. \rightarrow ..ap(on, rp, lp, (prop_i(mod_j))).. \qquad 2.$$

where ap stands for acquired property[12], z – d and/or a, w – e and/or c and/or a, on – object name, rp,lp – reference, live positions specifying topology as numerals, prop - property name, mod – grade,personality factor, vn – verb name, do,ao – driving, affected objects' position, adv – adverbial name, $i = 1,2,..I$, for each 'i', $j = 1,2,..J$, similarly for k,m.

To make the topology of a situation more explicit, eqs.1,2 reduce to

$$zp(-,x) \rightarrow in(x,-) \qquad 3.$$

$$in(-,y) \wedge wp(-,y) \rightarrow ap(-,-) \qquad 4.$$

where x and y are numerals specifying topology.

AN EXAMPLE

A story reads as follows : 'Customer sends an order specifying the required items to manufacturer where it is received by a clerk who verifies it if it is complete. The clerk returns an incomplete order to customer. The clerk checks that all items are available at the local depot. If the answer is yes, the clerk generates a local delivery advice for the order. If it is no, he prepares a transfer delivery note. The store uses the local delivery advice to prepare delivery and delivery docket at the local depot and to send transport request to vehicle scheduler to arrange a vehicle to deliver the ordered items'.

Only one sentence '..clerk checks that...' satisfies the description of concept of information. The other sentences indicate the medium presumed to carry information.

The story yields nine context-free sentences the first four of which are : 'Customer sends order', 'Order is received by clerk', 'Clerk changes status of order' an additional sentence, 'Order is verified by clerk'. Context-free sentences enable the diagram of the situation to be drawn as shown in Fig.1 with properties and interactions indicated. For example :

dp(1,1) – customer : buying habits ; frequent, occasional,
standing ; long, recent,
in(1,2) – sends : sending ; quickly, slowly,
where : manufacturer,

showing graded adjectives and adverbs introducing uncertainty into the situation. Some of these have been added intentionally to the story.

From Fig.1 logic sequences as eq.3,4 can be derived[12] :

dp(1,2) → in(1,2)   5.

in(1,2) ∧ ep(2,2) → ap(3,3)   6.

.... and so on for 18 expressions.

There are two comparisons in the story : '..verifies it if it is complete..' and '..If the answer is yes..'. These are taken into account by 'cp' properties and logical 'OR' function and shown at objects 7 and 9 in Fig.1.

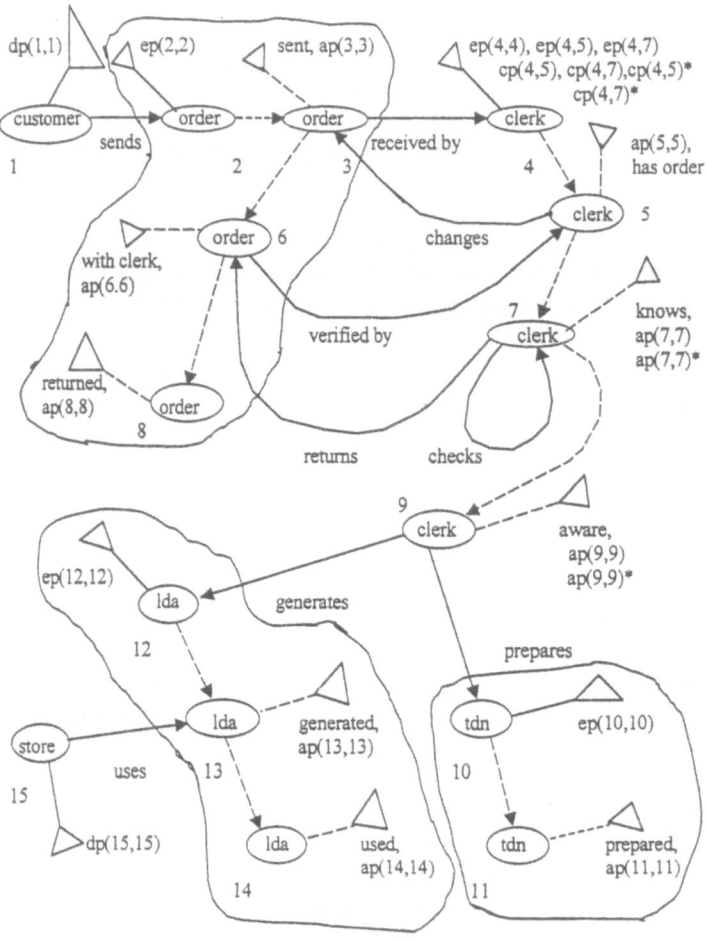

The contours indicate changes of state of products as a result of interaction.

Fig.1. Diagram of situation

Further to Fig.1, the situation with objects 1 to 8 is considered in detail. Using the format of eqs.1,2, in order to compute uncertainty, personality profiles, graded adjectives/adverbs and certainty factors are introduced. Thus, the first two terms of the logic sequence corresponding to eqs.1,2,

dp(customer,1,1(buying(frequent,90/.8,occasional,60/.7)),
        (standing(long,70/.8,recent.70/.6)))(..) →
(..)in(sends,customer,1,order,2(sending(quickly,slowly)),(where(manufacturer)))(..)   7.

in(sends,customer,1,order,2(sending(quickly,slowly)),(where(manufacturer)))(..) ∧
ep(order,2,2(dispatch(recorded,100/.3,ordinary,90/.7)),(items(required,100/.8)))(..) →
(..)ap(order,3,3(sending(sent)))(..)   8.

.... and so on for 10 expressions.

In general, the logic sequences a sample of which is given by eqs.7,8, contain the topology, objects, interactions, their modifiers and all the factors required to introduce uncertainty into a situation. With reference to eqs.1,2,7,8 the personality profile of an object consists of a number of property names each with its grades and personality factors. For example, the customer's personality profile has two components. A personality factor has two numbers : the number, 0-100, expresses the significance of a property and grade as far as the consequent, or outcome, is concerned, the number from -1 to $+1$ is intended as a measure of certainty that an object has a particular grade of a particular property.

From combining the grades the personality profile can be made variable and a certainty factor (cf) can be calculated for each variation. Such cf is then written in the brackets empty so far. For example, in eq.7 'long' and 'recent' can be combined with each of 'frequent' and 'occasional'. Thus, for instance, 'frequent,90/.8,long,70/.8' gives a cf = 0.8 which was obtained by dividing the sum of weighted whole numbers by their sum i.e. (90x0.8 + 70x0.8)/(90+70). Cf = 0.8 means that the 'customer is almost certainly long standing and a frequent buyer'[15].

Using the idea of variable personality profiles, eqs.1,2, through their particular cases such as eq.7,8 can be expanded and each variation is also combined through the logical operations of 'AND' and implication. Introducing numerals, eqs.7,8 can be written as

dp(1,1,1(1,2),2(1,2))(..) → (..)in(1,2,1(1,2),2(1,2))(..)   9.
giving 4x2 = 8 possibilities and

in(1,2,1(1,2),2(1,2))(..) ∧ ep(2,2,1(1,2),2(1))(..) → (..)ap(3,3,1(1))(..)   10.
giving 8x2 = 16 possibilities.

.... and so on for 10 expressions.

The number of possibilities expands very rapidly. Using all 10 expressions and with reference to Fig.1, in terms of the acquired properties the situation can occupy any of the following states : ap(3,3) = 16, ap(5,5) = 32, ap(6,6) = 512, ap(7,7) = 98304 and ap(8,8) = 100663296. The expansion of possibilities can be kept within limits by explicit, purposive systems which introduce acceptability criteria. For example, only possibilities with cf greater than a specified figure, are allowed to propagate further.

In the absence of other criteria, all states occur with equal probability if unchecked. For calculating particular values of cf, we choose the first state and inserting the cf for the rule after the implication sign, we can write:

$$dp(1,1,1(1),2(1))(.8) \rightarrow (.9)in(1,2,1(1),2(1))(.72) \qquad\qquad 11.$$

$$in(1,2,1(1),2(1))(.72) \wedge ep(2,2,1(1),2(1))(.55) \rightarrow (1)ap(3,3,1(1))(.55) \qquad\qquad 12.$$

.... and so on for 10 expressions.

The result is $ap(8,8,1(1)) = 0.55$ which says that 'the order maybe returned'. If the personality profile of the 'customer' in 'dp' is allowed to vary to 'occasional,recent', for instance, with all other conditions remaining the same, it turns out that the result remains unchanged.

CONCLUSIONS

A concept of information has been introduced based on the well known subject-predicate form which fits into linguistic modelling of information systems. Since the behaviour of living, especially human, beings is by and large unpredictable, repeatability cannot be taken for granted in constructing a formal model. Thus, the model presented here can only say that if objects possess the given features and behave in the manner specified then the computed outcomes will occur.

It is stipulated that the structure of a situation given in terms of objects and interactions remains invariant over a time span but their modifiers expressed as personality profiles, are allowed to fluctuate due to moods, human frailty, competence etc. The effect of fluctuations on the occurrence of outcomes possibly over time can be assessed.

It has been demonstrated that the number of states in which a situation can find itself due to fluctuation of modifiers, rapidly expands leading to some form of a chaotic operation. The expansion can be checked by introducing limiting criteria through certainty factors carried by calculating properties (cp) and explicit purposive systems.

The example considered shows information as understood here and as carried by medium. The latter can be eliminated by rewording the sentences : 'Customer communicates with the clerk that he requires certain items', for example.

REFERENCES

1. J.H. Jeans."Dynamical theory of gases," Cambridge UP, New York (1921.
2. J. Korn. Theory of spontaneous processes, *Structural Engineering Review.* v7, n1 (1995).
3. L. Brillouin. "Science and information theory," Academic Press, New York (1956).
4. R.V.L. Harltey. Transmission of information, *Bell System Technical J.* (1928).
6. C.E. Shannon and W. Weaver. "The mathematical theory of communication," University f Illionis Press, Urbana (1964).
6. Y. Bar-Hillel. "Language and information," Addison-Wesley Pub.Co., London (1964).
7. K. Devlin. "Logic and information," Cambridge UP, New York (1991).
8. I.T. Hawryszkiewycz. "Introduction to systems analysis and design," Prentice Hall, New York (1988).
9. F.A. Stowell et al. "Systems for sustainability," Plenum Press, New York (1997).
10. Anon. Information systems as a discipline, *Systemist.* v17, n1 (1995).
11. P.N. Johnson-Laird. "The computer and the mind," Fontana Press, London (1988).
12. J. Korn. Linguistic modelling of situations, *Systemist.* v18, n4 (1996).
13. J. Korn. "Linguistic modelling of information systems," Proc. of PAIS II Symposium, . of West of England, Bristol, UK, 27-29 July (1988).
14. J. Korn. Domain-independent design theory, *J. Engineering Design.* v7, n3 (1996).
15. J. Durkin. "Expert systems," Macmillan, New York (1994).

POLY-AGENT SYSTEMS THEORY: EVOLUTION MODEL AND ITS APPLICATIONS

Kyoichi   Kijima

Tokyo Institute of Technology
Tokyo 152-8552, Japan

Introduction

The purpose of this paper is to discuss static and dynamic characteristics of evolution of poly-agent systems. Poly-agent system is a model of societal system where intelligent and autonomous decision makers (or agents) with essentially diversified preference are related to and contest each other to realize their desire in the network of mutual interdependence[6]. The agents may perceive even the same situation they face differently and subjectively. Furthermore, the state of the agents themselves as well as of the interrelationship should continuously change by interaction.The present author proposed intelligent poly-agent learning model (I-PALM) as a typical operational model of poly-agent systems theory (PAST) [4][5].

We modify and elaborate conventional concepts such as ESS and replicator dynamics originally defined for two person non-cooperative games in such a way that they are applicable to I-PALM. We also apply the model to spoof-proof problem in virtual (or network) society [4] to demonstrate its validity. In virtual societies, the people involved may not have incentive to behave seriously particularly because there is no Substantial mechanism to impose sanctions to those who report the false. Spoof-proof problem of motivating the people to reveal their private information honestly is a critically important problem for successful realization of network society supported by information technology.

Conventional Model of Evolution

We first focus on relationship between two agents since two-agent decision situation is the most basic and provides the starting point of the research. Our analysis starts with a non-cooperative game [1]. Let p and q be agents.

Definition 1 A two person non-cooperative game with agents p and q is a quadruple $G = (S_p, S_q, >_p, >_q)$.
We assume that S_p and S_q are finite while $>_p$ and $>_q$ are linear orderings and can be represented by some ordinal utility functions. $s^* = (s_p{^*}, s_q{^*})$ in S_p x S_q is called Nash equilibrium iff there is no incentive for either of the agents to change their strategy as long as the other does not change its strategy.

Synergy Matters: Working with Systems in the 21st Century,
Edited by Castell *et al.*, Kluwer Academic / Plenum Publishers, New York, 1999.

Evolutionary Stable Composition

Maynard Smith and Price [6] analyzed a number of animal conflict games in which several pure strategies are available to each player. In order to consider which type of the strategies can prevent others from invasion, they proposed a concept of ESS (Evolutionary Stable Strategy) as "a strategy with the property that if most members of a large population adopt it, then no mutant strategy can invade the population"[3]. It is intended to reflect a stationary situation in the evolutionary process. It is well known that ESS is an refinement of Nash equilibrium [8].

However, in this paper, according to Colman [3], we employ concept of Evolutionary Stable Composition (ESC) rather than that of ESS. He claims that ESC is more appropriate than ESS to describe a systems condition because it refers to the structure of the system, not only to strategies. Suppose a proportion k of the population adopt strategy A and the reminder adopt strategy B. Let us write the expected cardinal payoff to each individual adopting A by $E(A, k)$ while let the expected cardinal payoff to each individual adopting B be represented by $E(B, k)$. By definition, it is clear that A is ESS iff $k = 1$ is ESC while B is ESS iff $k = 0$ is ESC.

Definition 2 *Suppose that k increases by an arbitrary small amount t or decreases by an arbitrary small amount $-t$. Then, k is called an ESC if we have either $E(A, k + t) \leq E(B, k + t)$, or $E(B, k - t) \leq E(A, k - t)$.*

Replicator Dynamics

Now assume that the set of possible phenotypes is finite and they coincide with the set of pure strategies; $S = \{s_1, s_2, ..., s_m\}$. In this paper we present the evolutionary dynamics in a discrete-time framework so that let time be measured by $t = 1, 2, \ldots$. Denote the population profile over pure strategies prevailing at t by $v(t) \equiv (v_i(t))_{i=1,2,...,m}$, where $\Sigma v(\cdot) \equiv 1$. For simplicity, each member of the population is assumed to live only one period, leaving some offspring which inherits the same phenotype as the parent. Reproduction, therefore, is assumed asexual, with each member of every new generation having only one parent.

Naturally, the number of offspring left by each member of the population is taken to depend on the pay-off earned during his (one-period) lifetime. Specifically, we shall adopt a strictly biological approach here and identify the "pay-off" of an individual with the expected number of viable offspring he is able to produce. Thus, if an individual of "generation" t plays strategy s_i against a population profile $v(t)$, the expected number of offspring he is assumed to produce (with the same phenotype) exactly equals its pay-off $\pi(s_i, v(t)) \equiv \pi_i(v(t))$.

With the previous interpretation for pay-offs, the phenotypical dynamics induced becomes a matter of sheer definition. Normalize the size of the population to one. Then, if the population profile at t is $v(t)$ the size of the population at $t + 1$ is given by $\Sigma_i v_i(t)\pi_i(v(t))$, where recall that each $\pi_i(v(t))$ has been identified with the number of offspring produced by each individual which plays strategy s_i. Thus, $v_i(t + 1)$, the fraction of the population which plays strategy s_i at $t + 1$, is obtained by

$$v_i(t + 1) = v_i(t) \frac{\pi_i(v(t))}{\Sigma_{j=1}^m v_j(t)\pi_j(v(t))}, \quad \text{where } i = 1, 2, \ldots m$$

or, denoting the average pay-off $\bar{\pi}(v(t)) \equiv \Sigma_{j=1}^m v_j(t)\pi_j(v(t))$, we have for $i = 1, 2, \ldots m$

$$\frac{\Delta v_i(t + 1)}{v_i(t)} \equiv \frac{v_i(t + 1) - v_i(t)}{v_i(t)} = \frac{\pi_i(v(t)) - \bar{\pi}(v(t))}{\bar{\pi}(v(t))}.$$

It intuitively shows that the share of the population which plays any given strategy changes in proportion to its relative pay-off (i.e. in proportion to its deviation, positive or negative, from the average pay-off).

Theorem 1 *ESS is an asymptotically stable equilibrium of replicator dynamics [8].*

Intelligent Poly-agent Systems Theory

Let N be a set of agents. PAST assumes each agent $i \in N$ is intelligent so that $i \in N$ is associated with an internal model M_i and a decision criterion or rule R_i. M_i represents i's subjective understanding of its environmental surrounding, which is usually constructed through interactions with the environment including other agents. The decision criterion R_i tells agent i what alternatives are reasonable or rational for the perceived environment. In I-PALM, a basic relationship is formalized by a simple hypergame:

Definition 3 *A simple hypergame of agents p and q is a pair of (G_p, G_q), where $G_p = (S_p, S_{qp}, \geq_p, \geq_{qp})$ is a game that p believes both sides perceive while $G_q = (S_{pq}, S_q, \geq_{pq}, \geq_q)$ is a game that q believes both sides perceive.*

In this case, G_p and G_q represent p's and q's internal models, respectively. One of the most natural ways to describe rationality of p and q is to consider Nash equilibrium in G_p and G_q independently.

As the time goes on, a simple hypergame situation may change to symbiotic hypergame, whose formal definition is given by:

Definition 4 *A symbiotic hypergame with players p and q is a pair $((G_p, f), (G_q, g))$, where we have $G_p = (S_p, S_{qp}, \geq_p, \geq_{qp})$ and $f: S_q \rightarrow S_{qp}$, while $G_q = (S_{pq}, S_q, \geq_{pq}, \geq_q)$ and $g: S_p \rightarrow S_{pq}$ hold.*

In $((G_p, f), (G_q, g))$, G_p and G_q are simple hypergames while the function f represents how p interprets S_q of strategies of q. Similarly, g formulates how q interprets S_p of strategies of p. We refer to f and g as interpretation functions of p and q, respectively. $((G_p, f)$ and $(G_q, g))$ represent p's and q's internal model, respectively.

The following is a natural and straightforward way to define rationality S for a symbiotic hypergame.

Definition 5 *Let $((G_p, f), (G_q, g))$ be a symbiotic hypergame where $G_p = (S_p, S_{qp}, \geq_p, \geq_{qp})$ and $G_q = (S_{pq}, S_q, \geq_{pq}, \geq_q)$, while $f: S_q \rightarrow S_{qp}$ and $g: S_p \rightarrow S_{pq}$. $(s_p^*, s_q^*) \in S_p \times S_q$ is called symbiotic Nash equilibrium of $((G_p, f), (G_q, g))$ iff $(s_p^*, f(s_q^*))$ is Nash equilibrium of G_p and $(g(s_p^*), s_q^*)$ is Nash equilibrium of G_q.*

Spoof-proof Problem

Static Analysis

Now we will adopt these concepts to an example of spoof-proof problem. We assume a virtual society consists of a lot of message senders and receivers. We denote a sender and a receiver by p and q, respectively, and represent the interactive activity by a simple hypergame (G_q, G_q) (Refer to Tables 1 and 2). We distinguish the people in the society into two types; the serious and spoofing.

In G_p, we assume p possesses three strategies, namely, T, S and M. T means a strategy to send truth while S indicates a strategy to send message containing some false but p has no malice and just wants to play tricks on the receiver. M means sending message filled with malice. Furthermore, we presume that p believes q has two alternatives B and NB, where B denotes a strategy to believe message while NB does a strategy not to believe message.

It may be legitimate to call the agent serious if (1) he/she believes the counterpart is also serious and his/her preference as a sender is shown by (G_{p1}, f_1) and that as a receiver by (G_{q1}, g_1). (G_{p1}, f_1) indicates that the agent as a sender basically prefers T most and S next while M is least preferred. (G_{q1}, g_1) implies that he/she as a receiver prefers normal communication most. On the other hand, G_q shows that q believes that p's message is either acceptable (A) or unacceptable (UA). G_q also implies q has two choices, *i.e.*, to respond to message (R) or to ignore it (I).

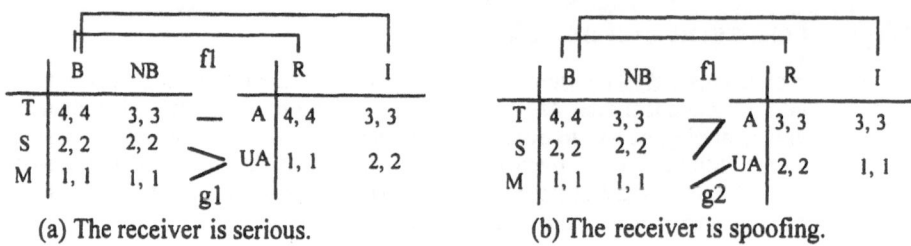

(a) The receiver is serious.   (b) The receiver is spoofing.

Fig.1 When the sender is serious.

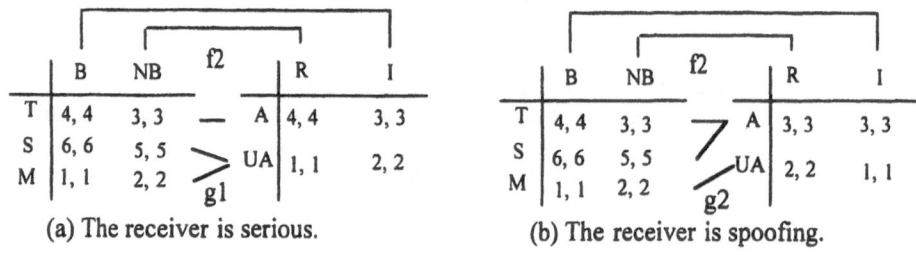

(a) The receiver is serious.   (b) The receiver is spoofing.

Fig.2 When the sender is spoofing.

We call the agent spoofing if (1) he/she believes the counterpart is also spoofing and (2) his/her preference as a sender is shown by (G_{p2}, f_2) and that as a receiver by (G_{q2}, g_2). (G_{p2}, f_2) indicates that he/she basically prefers S most and T next while M is least preferred. (G_{q2}, g_2) shows he/she as a receiver prefers normal communication most but does not care to ignoring the messages. Then, we can easily find out that Nash equilibria of G_{p1} are $\{(T, B)\}$ while those of G_{p2} are $\{(S, B)\}$. We have that Nash equilibria of G_{q1} and G_{q2} are $\{(A, R)\}$ and $\{(A, R), (A, I)\}$, respectively.

In our symbiotic hypergame it is reasonable to suppose a serious sender is associated with f_1 such that $f_1(R) = f_1(I) = B$ and a spoofing sender is with f_2 where $f_2(R) = NB, f_2(I) = B$. A serious receiver is supposed to employ g_1 such that $g_1(T) = A, g_1(S) = g_1(M) = UA$ while a spoofing receiver is to be with g_2 where $g_2(T) = g_2(S) = A, g_2(M) = UA$. In summary, we have characterized a serious and spoofing sender by (G_{p1}, f_1) and (G_{p2}, f_2), respectively, while we have featured a serious and spoofing receiver by (G_{q1}, g_1) and (G_{q2}, g_2), respectively.

By definition of symbiotic Nash equilibrium, we can find that, for example, in the symbiotic hypergame $((G_{p1}, f_1), (G_{q1}, g_1))$ there is a symbiotic Nash equilibrium (T, R). By applying similar arguments to other three cases, we can obtain symbiotic Nash equilibria for four cases, which are summarized in Table 3. It also shows cardinal utility for each symbiotic Nash equilibrium. α denotes the reward to behave properly while β shows the temptation to spoof. γ represents the cost induced by improper behavior. We can easily find $((G_{p1}, f_1), (G_{q1}, g_1))$ and $((G_{p2}, f_2), (G_{q2}, g_2))$ are both Nash equilibria.

Evolutionary Analysis

Our evolution problem is primarily concerned with what sort of situation happens after a certain time of the contest between the serious and spoofing. Is there an agent of a certain type who is able to survive and dominate the society? Or, could agents of both types find their niche to coexist together?

Let k of the society is occupied by the serious, that is, let k be probability with which a person in the society comes across a serious receiver or sender when he/she sends or receives some message. Furthermore, we assume that he/she becomes a sender with probability θ and

Receiver / Sender	Serious		Spoofing	
Serious	SNE	(T, R)	SNE	(T, R), (T, I)
		(α, α)		$(\alpha \cdot \gamma, \alpha \cdot \gamma)$
Spoofing	SNE	void	SNE	(S, I)
		(0, 0)		$(\alpha + \beta, \alpha + \beta)$

Fig. 3 Symbiotic Nash Equilibria (SNE) and their Payoff

becomes a receiver with probability $1 - \theta$, where we have $0 < \theta < 1$. Then, the expected utility for a serious person in this situation $E(\text{serious}, k)$ is obtained by:

$$E(\text{serious}, k) = \theta(\text{Expected utility as a sender}) + (1 - \theta)(\text{Expected utility as a receiver})$$
$$= \theta\alpha + (1 - \theta)\alpha k.$$

Similarly, the expected utility for a spoofing person $E(\text{spoofing}, k)$ is calculated by:

$$E(\text{spoofing}, k) = \theta(1 - k)(\alpha + \beta) + (1 - \theta)\{k(\alpha - \gamma) + (1 - k)(\alpha + \beta)\}$$
$$= \theta(\alpha + \beta) + \theta(\alpha + \beta)k + (1 - \theta)(\alpha + \beta) - (1 - \theta)(\beta + \gamma)k$$
$$= (\alpha + \beta) + \{\theta\alpha + \beta + (1 - \theta)\gamma)\}k.$$

Then we see that the two lines necessarily intersects at $k^* = \frac{(1-\theta)\alpha+\beta}{\alpha+\beta+(1-\theta)\gamma}$, where we have $0 < k^* < 1$. Simple calculations imply the following.

Proposition 2 *(1) Any k except $k = k^*$ is ESC, i.e., almost any composition could happen. (2) The composition k^* is too unstable to continue in the society. (3) If θ increases, then so does k^*, that is, if chances for the serious to send messages increase, then the unstable composition ratio k^* goes up. (4) In particular, if we set $k^* = 1$ by adjusting the parameters, then the domination of the society by the spoofing should become unstable.*

Finally we discuss dynamic evolutionary trajectory. Let us define $\pi(\text{serious}, (k, 1-k)) = E(\text{serious}, k)$ and $\pi(\text{spoofing}, (k, 1 - k)) = E(\text{spoofing}, k)$. Then by using the replicator dynamics discussed in Section 2.3 we can simulate several cases. For example, if we set $\alpha = 5$, $\beta = 3$, $\gamma = 2$ and $\theta = 0.3$, then we have $k^* = 0.691$. We can see $k^* = 0.691$ is actually unstable by examining simulations where $k = 0.69$ and $k = 0.70$, respectively, each of which is given by Figs. 4 and 5. The horizontal axis represents the generation t while the vertical axis indicates the proportion of both type; $v1(t)$ and $v2(t)$ show the proportion of the serious and the spoofing at time t, respectively. The fugures al so illustrates both $k = 0$ and $k = 1$ can become asymtotically stable equilibria of replicator dynamics depending on the inisital state due to *Theorem 1*, since both $k = 0$ and $k = 1$ are ESC and, hence, ESS.

Conclusion

This paper possesses at least the following three original contributions. (1) Each agent is characterized by its internal model (*i.e.*, a pair of its subjective pay-off matrix and interpretation function). (2) To discuss the dynamic evolution process we introduce what we call a two-step approach, *i.e.*, first calculating HNE for symbiotic hypergames and, then applying ESC to the result. (3) We clarify that HNE is concept defined on ordinal payoff while ESC is essentially calculated with cardinal payoff. Furthermore, we explained these ideas through an interesting example.

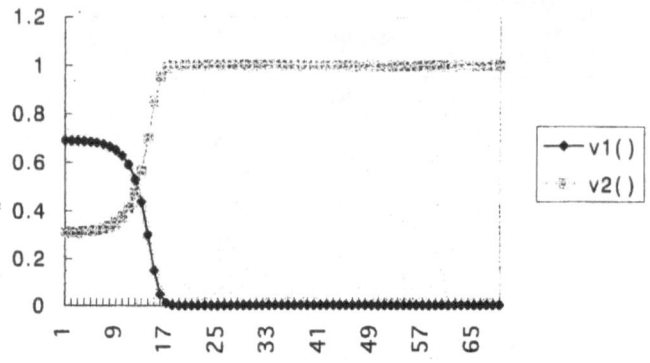

Fig. 4   A Simulation Result for *k* = *0.69*

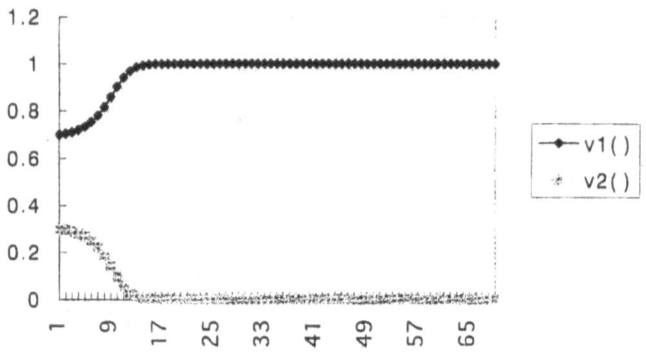

Fig. 5   A Simulation Result for *k* = *0.70*

References

[1] Bennett, P.G., Hypergames: Developing a Model of Conflict, *Futures*, 12(6)(1980) 489-507

[2] Bennett, P.G., Cropper, S. and Huxham, C., Modelling Interactive Decisions: The Hypergame Focus, in *Rational Analysis for a Problematic World*, John Wiley and Sons, Chichester, 1989

[3] Colman, A.M., *Game theory and its Applications*, Butterworth, 1995

[4] Kijima, K., Intelligent Poly-agent Learning Model and its Application, *Information and Systems Engineering*, 2, 47-61(1996)

[5] Kijima. K., Truth-telling Problems in Virtual Society: Higher Order Hypergame Approach, *Proceedings of Pan-Pacific Asia Conference on Information Systems*, Brisbane, Australia, 1997

[6] Maynard Smith, J. and Price, G.R., The Logic of Animal Conflict, *Nature*, 246, 15-18, 1973

[7] Takagi et al., *Human and Society in Multi-media Age*, Nikkagiren Shuppan, 1995 (Japanese)

[8] Vera-Redondo, F., *Evolution, Games, and Economic Behaviour*, Oxford Univ. Press, 1996

WHAT DRIVES THE DEVELOPMENT OF AN INTERNATIONAL INFORMATION SYSTEM? A GROUNDED THEORY INVESTIGATION

Hans Lehmann

Department of Management Science and Information Systems
University of Auckland
New Zealand
h.lehmann@auckland.ac.nz

INTRODUCTION

Until recently, international information systems (IIS) technology as a field has been "..sometimes ignored altogether" (King & Sethi, 1993) and academic research is sparse (Cash, McFarlan & McKenney, 1992). This paper investigates the driving forces for an IIS architecture. Any such theory can then contribute to an IIS development framework, aimed at reducing their complexity and the risk inherent in building them.

LITERATURE REVIEW

As set out in more detail in literature reviews elsewhere (Hamelink ,1984, Sethi and Olson, 1993, Lehmann, 1996a,b), past research into international information systems is sporadic and spread over a wide array of topics. Only in the last few years have researchers begun to direct their attention onto the design and development of IIS. Some of this recent research focuses on the structure and architecture of IIS.

Many researchers of IIS architectures use a framework for the classification of enterprises operating in more than one country which was developed by Bartlett and Ghoshal in 1989. Their model is centred around the level and intensity of global control versus local autonomy:

- 'Global' means high global control while 'Multinationals' have high local control;
- 'Internationals' are an interim state, transiting towards a balance of local and global;
- 'Transnational' organisations balance tight global control whilst vigorously fostering local autonomy. This strategy of "think global and act local" is considered optimal for many international operations;

Butler Cox (1991) put a developmental perspective on the Bartlett-Goshal framework. They show that companies become active internationally first by applying a *'Global'* business strategy. Increased activity in any one location encourages autonomy for local operations, taking on the role of *'Multinational'*. In the next phase this degree as an *'International'* firm. Finally, as global operations mature, firms transit through an *'International'* phase to where local autonomy is counterbalanced by global control in the *'Transnational'* strategy.

Such a developmental perspective on global business strategies puts a strong requirement of flexibility to any systems architecture for international systems.

A number of researchers developed models of IIS with a direct, one-to-one relationship between Bartlett and Ghoshal's global business strategies and these systems architectures. Table 1 contains an overview of the main architecture frameworks:

Table 1: Comparison of major architecture styles/configurations identified in the literature

Bartlett & Ghoshal	Butler Cox (1991)	Kosynski et al (1993)	Sankar et al 1993)	Ives & Jarvenpaa (1994)
Global	Centralised	Centralisation	Centralised	Headquarters-driven
Multinational	Autonomous	Decentralisation	Decentralised	Independent
International	Replicated	Inter-organisational	(undefined)	Intellectual Synergy
Transnational	Integrated	Integrated	Integrated	Integrated

Whilst the centralised and decentralised structures have been researched over a number of years and are by now well understood, the nature of the 'integrated' structure/architectures has rarely been an object of empirical study. This research sets out to shed some light on them.

RESEARCH METHODOLOGY

The dearth of IIS research makes qualitative, theory building methods an appropriate choice. Such methods are well established in organisational research and are becoming accepted in information systems research too (Benbasat et al, 1987, Galliers et al, 1987, Yin, 1989, Lee, 1989, Orlikowski et al, 1991, Zinatelli et al 1994). In particular, Eisenhardt (1989) describes the process of building theory, focusing especially on its inductive nature. In Sociology, Glaser and Strauss (1967) had already developed a specific inductive method which they termed the 'Grounded Theory' (GT) approach, where theory is left to 'emerge' from the data - in which it is 'grounded'. Turner (1983) was one of the first to apply the GT approach to management studies. Since 1984, GT had been used in a number of business studies. Orlikowski (1993, 1995) has pioneered GT in Information System (IS) Research. Yoong (1996) and Atkinson (1997) are recent studies. Glaser and Strauss (1970) have also set out how to use grounded theory with cases. This approach was selected for this research project.

The following section gives a - highly abridged - description of the case history in question.

CASE HISTORY: AUSTRALASIAN FOOD PRODUCERS' CO-OP

Background

The marketing authorities for land-based industries (such as fruit growers, meat producers, dairy farmers, forestry, etc.) are often large companies with strong international presence. The Australasian Food Producers' Co-op (later also referred to as the 'Co-op') with some $4.5bn[1] revenue is one of the largest of those. Like the others, the Co-op is a 'statutory monopoly', as there is legislation which prohibits any other organisation from trading their produce in international markets. With about a quarter from raw materials and manufacturing outside Australasia, the Co-op is a mature transnational operator. Structured into nine regional holding companies, in 1997 it has a presence in 135 offices in 40 countries.

The 15,000 primary producers are organised into 18 co-operative 'Production Companies' (ProdCos), in which the farmers own in shares proportional to their production. The ProdCos collectively own the Co-op. This tight vertical integration is seen as a big advantage. It allows the Co-op to act as one cohesive enterprise and to develop a critical mass needed in most of its major markets.

Business background

Prior to the mid-1970s Australasia exported the vast majority of its produce to the United Kingdom, who, under Commonwealth rules, used to accept it all. Once the UK had joined the European Union, however, they had to give free access to all other EU members, and cut the Co-op's quota severely. Australasia had to develop new markets. A number of subsidiary offices was set up rapidly and agencies were nominated in the US and Canada.

This policy of local autonomy was successful. Within a decade the Co-op had built a presence in more than thirty countries and had managed, throughout, to secure a satisfactory return for the all their primary producers.

At the onset of the 90s, however, competition for the Co-op had become increasingly global. With the emergence of global brands (such as Coca Cola, McDonalds, etc.); the Co-op needed to

---

[1] All names within the enterprise have been changed. All figures are in US Dollars

develop global brands themselves and had to have sufficient command (and control) to mount synchronised international marketing and logistics operations. With the arrival of a new Chief Executive Officer in 1992 the Co-op began a concerted campaign to shift authority and control back to head-office. Part of this new policy was a critical look at the role of information systems throughout the Co-op's operations.

During the 'global' phase, the Co-op had built up a sizeable IS department with a mainframe operation at the head-office, linking up with all the main subsidiary offices and ProdCos throughout the country. Foreign activities were few and hardly needed computer support. The forced expansion drive in the 80's, however, lead to an increased need by local operations to be supported with information systems. By 1992 a number of regional offices had bought computers and software to suit their own, individual requirements

The Global Information Systems Project

Against this background of a proliferation of uncoordinated local systems on the one hand and a declared policy of more control from the Co-op's centre on the other, the Co-op's IS Department, in April 1992 took the initiative to establish a "Framework for Information Systems" as the basis for globally common technology, communications, data/information and application software standards, effective for all of the Co-op's 135 offices in 35 countries. Subsequently, late in 1992, the 'Food Information Systems Technology' (FIST) project was created by the IS Department to implement the 'Framework's' in three stages:

1. Development of a 'prototype' system with a representative site;
2. Implementation of the prototype in a small number of 'pilot' sites;
3. Synchronised 'roll-out' of the 'global system' into all the regions and offices.

Estimated completion dates were late 1993, early 1995 and mid 1996 respectively.

In 1992 the North America region (NA) had started to embark on a review of its ageing IBM S/34. At the same time, Singapore was also looking to upgrade their fragmented PC-based installation to cope with the rapid growth in the region. Both sites thus became candidates for the development of the prototype and as pilot sites for further implementation.

The Pilot Project(s)

For NA, the FIST team agreed to have selected technology, completed the prototype, tested and modified it as a pilot and to have gone live with the new system (which would at the same time be the first global system) by June 1994 - nine months hence. The requirements for NA were set out as a 'benchmark' for all other sites. However, Singapore were quite concerned when the FIST team restricted itself to comparing the 'benchmark model' with the South East Asia region and found a "90 - 95% match". Subsequently, Singapore opted out of the Pilot because they felt that as North America's predominant business is in the industrial produce market, this would not at all fit South East Asia ("nor Europe, for that matter"), as their scope of business mainly covers the consumer and food manufacturing markets. He was also very critical of what he called the "top-down-approach" taken by FIST. With very little participation by the regions, he feared the systems would be missing the actual requirements of the local business - "just like the other past failures of the Computer Centre". By the end of 1993, North America was therefore the only pilot site.

As the North America pilot missed the June 1994 deadline, a Request for Proposal (RFP), asking for firm quotes for software, hardware and communications technology to be used internationally in the Co-op's 130 offices in 35 countries, was issued to ORACLE, IBM and UNISYS. After a rapid evaluation by the FIST team with some North America input, ORACLE was chosen as the main provider for data base middleware and, together with DATALOGIX, for applications software. Hardware and communications technology was not selected. At the same time, the 'benchmark model' was now compared with Europe and another "90% to 95%" match was experienced. However, as the regional manager Europe remarked: "These models are so general, they'd make Disney look like us." The European region subsequently opted out of the FIST programme.

At this stage, to counter the mounting resistance to one global, standard system for every subsidiary office, the FIST team began to look at what applications should be the same throughout the Group and which could be different for local subsidiaries.

The outcome of these definitions were a re-formulation of the 'standard' global system which has the main business operation ETC. (the "Enquiry To Cash") as the framework for the 'Core' information systems and leaves manufacturing and marketing operations as the 'Local" applications to be selected by each office individually.

Thereafter, the FIST team began with the implementation of the software in North America in September 1994 - and immediately encountered serious problems: The manufacturing and distribution modules would not conform with the business processes they were selected to support. The changes were estimated to cost $1,8m. However, Oracle were negotiating with Datalogix about absorbing the Datalogix Distribution modules into their own ones. For the duration of these negotiations no work on the software was done. By mid 1995 North America reached an agreement to abandon the pilot efforts and to alter its software so that it reflected their local requirements.

In early 1995, the Co-op decided to open a new office in the Middle East region, in Dubai and by mid 1995, there were 12 people in the office. To replace North America, the FIST team selected Dubai as the new pilot site to test out the common global system for the Co-op. The first installation was going to be the 'standard' Oracle Financials together with business procedures defined around the system. The first target date for completion was September 1995. However, for want of adequate local support, the systems could not be developed on site - it was therefore decided to develop the first prototype at head office. In November 1996 the standard Oracle Financials were handed over to Dubai as a working system.

Developments concerning FIST at the head office

The major difficulties with the FIST project, especially the missed deadlines, the significant costs (by 1995 approx. $ 8m) without any noticeable results and the refusal by a major region to accept the FIST system began to attract the attention of the CEO. In mid 1996 he commissioned Ernst & Young to evaluate the FIST projects. Their report was critical of FIST as being overly ambitious and not achievable within the time frame or the existing project set-up. This proved to be a turning point: The CEO re-aligned the IT portfolio - and with it FIST- into the Finance department, whose General Manager had been an open critic of the project for a long time. As early as March 1995 he had called for a critical review of the "real" reasons for wanting to spend $21m and had advocated that business reasons should drive the project, not technology.

LESSONS LEARNED FROM THE CASE

The building blocks of a grounded theory are the 'categories' of facts distilled from the data and their 'properties', i.e. the various aspects, manifestations, etc. of the category so described. The main theoretical categories derived and comments, outlining the main 'lessons learned' from the Co-op's case are summarised in Table 2 below.

Table 2: Theoretical Categories and Comments

Categories	Lessons and Commentary
Co-op migration through the global strategies: *From a 'global' through a 'multinational' stance and now moving towards 'transnational'.*	The Co-op's transitions bear out the linkage postulated in the literature on IIS architectures and the migration path presumed by Butler Cox (1991).
Mis-alignment of FIST with the Co-ops global strategic intent: *FIST attempted to install a globally standardised, functionally fixed, common system across the Co-ops regions.*	This was seen by regional management as a move backwards - re-instating an extended, replicated version of the mainframe systems architecture of the 1970s, whereas the Co-op's executive management was seen to move forwards to a new level of maturity.
Definition of Common and Local Systems: *the inability to find a framework for the definition of 'common' and 'local' processes/systems despite four attempts.*	This seems to indicate that the problem is deeper than just lack of skill or biased standpoints: regional management interprets 'core' systems as synonymous with central control and subsequently lowers their level of co-operation.
Domestic skill/mind-set: *The shortcomings of only knowing local information systems manifest themselves in a number of ways:* *1. Wrongly estimating activities;* *2. Assuming that activities in the same business areas are the same across different countries;* *Ignorance of important legislative and/or business culture restrictions in foreign countries;*	The main lesson is that what works locally, for a domestic system very often does not work for an IIS. Two conclusions follow: 1. Support is given to the notion that domestic systems are different from IIS 2. Local participation is a *sine qua non* for IIS development and also important for political reasons

CONCLUSION

Failure of the Computer Centre to recognise the pattern of the Co-op's strategy migration led them to attempt an IIS regime that was diametrically opposed to the business strategy. The Computer Centre's use of biased, political 'justifications' for their actions lost them further creditability. These political problems were exacerbated by the Computer Centres inability to define the IIS's common and local modules with traditional methods, despite numerous tries. Inexperience with the business realities outside the domestic arena, coupled with inadequate adherence to professional 'good practice' furthered the eventual failure of the project.

For the formulation of a theory - in terms of the Grounded Theory approach - it is now necessary to carry out more analysis with additional cases to obtain theoretical saturation in one or more categories.

REFERENCES

Atkinson, G. J. 1996. *A Study of Perception of Individual Participants of a Client Group Undertaking a Series of Meetings Supported by a Group Support System.* Doctoral Thesis at the Curtin University of Technology, Perth, W.A., Australia.

Bartlett, C.A., Ghoshall, S. (1989). *Managing Across Borders: The Transnational Solution.* Boston. Harvard Business School Press

Benbasat, I., Goldstein, D.K. and Mead., 1987. The Case Research Strategy in Studies of Information systems, *MIS Quarterly*, September, 369-386

Butler Cox plc. (1991). Globalisation: The Information Technology Challenge. *Amdahl Executive Institute Research Report.* London.

Cash, J.I. Jr., McFarlan, W.F. and McKenney, J.L. (1992). *Corporate Information Systems - The Issues Facing Senior Executives.* Homewood. Irwin.

Eisenhardt, K.M., 1989. Building Theories from Case Study Research. *Academy of Management Review,* 14,4, pp532-550

Galliers, R.D. and Land, F.F., 1987. Choosing Appropriate Information Systems Research Methodologies. *Communications of the ACM,* 30,11,pp.900-902

Glaser, B.G., and Strauss, A.L., (1967). *The discovery of grounded theory.* Aldine Publishing Company, Hawthorne, New York.

Glaser, B. G., Strauss, A. L.. 1970. *Anguish.* , Ch. 11: *Case Histories and Case Studies;* Sociology Press, Mill Valley, California

Hamelink, C.J. (1984).Transnational data flows in the information age. Lund, Sweden: Student-litteratur AB

Jarvenpaa, S.L., Ives, B. 1994. Organisational Fit and Flexibility: IT Design Principles for a Globally Competing Firm. *Research in Strategic Management and Information Technology,* Vol 1, pp1-39

King, W. R. and Sethi, V. (1993). Developing Transnational Information Systems: A Case Study. *OMEGA International Journal of Management Science,* 21, 1, 53-59.

Konsynski, B. R. and Karimi, J. (1993). On the Design of Global Information Systems. *Globalization, Technology and Competition.* Bradley, S.P., Hausman, J.A. and Nolan, R.L. (Ed's). Boston. Harvard Business Press.

Lee, A., 1989. A Scientific Methodology for MIS Case Studies, *MIS Quarterly,* 13,1,pp32-50

Lehmann, H.P. 1996a A Research Note on the methodology and domain of international information systems research. *Working Paper of the Department of Management Science and Information Systems, University of Auckland,* May 1996

Lehmann, H.P. 1996b Towards a common architecture paradigm for the global application of information systems. *Proceedings Of The IFIP WG 8.4 Working Conference On The International Office Of The Future,* Tucson, Arizona, April 1996.

Orlikowski, W.J., 1993. CASE tools as organisational change: Investigating incremental and radical changes in systems development. *MIS Quarterly,* Sept 1993, p309-337

Orlikowski, W.J., and Baroudi, J.J., 1991. Studying information technology in organisations. *Information Systems Research* (2:1), March 1991, p 1-28

Orlikowski, W.J.,(1995). *Organisational change around Groupware.* Working Paper CCS 186, Massachusetts Institute of Technology.

Sankar, C., Apte, U. and Palvia, P.(1993). Global Information Architectures: Alternatives and Trade-offs. *International Journal of Information Management,* (1993), 13, 84-93.

Sethi, V. and Olson, J.E. (1993) An integrating framework for information technology issues in a transnational environment. In *Global Issues in Information Technology,* Idea Publishers, Harrisburg

Turner, B.A., 1983. The use of grounded theory for the qualitative analysis of organisational behaviour. *Journal of Management Studies* (20:3) July 1983, p333-348

Yin, R.K. 1989. *Case Study Research: Design and Methods*. Sage Publications, Newbury Park, Ca.

Yoong, S. P. 1996. *A Grounded Theory of Reflective Facilitation: Making the Transition from Traditional to GSS Facilitation*. Doctoral Thesis at the Victoria University of Wellington, New Zealand.

ELECTRONIC COMMERCE: MORE HYPE THAN SUBSTANCE?

Peter Marshall and Judy McKay

School of Management Information Systems
Edith Cowan University
Churchlands, Western Australia 6018

INTRODUCTION

We live at a time when we are being told that Electronic Commerce (EC) is fundamentally changing our world. We are exhorted to embrace the electronic revolution or risk being swept aside by this virtual juggernaut. "Cyberspace or Cyberia"[1] is the catch cry of the moment. Cyberia is not known as being a very comfortable place for academics to reside. Dissidents (i.e. those who express views countervailing to those put forward by the cyberprofs) are banished there for failing to jump onto the latest academic bandwagon, and an inability to get research grants, or to get material published at international conferences because it is insufficiently trendy are amongst the punishments metered out.

In his keynote address to the European Conference on Information Systems in 1995, Land (1995) argued that business people generally did not have the same regard for, nor place the same emphasis on, IT as did IT professionals. With all the excitement surrounding EC, could a similar effect be occurring? Is the entire business community abuzz with e-commerce (literally e-mmerced), or are the catchcrys of transformation and revolution largely falling from the mouths of IT people, those with vested interests in the uptake of e-commerce, and the media (always keen for something new)?

In the rush to develop new degree offerings in EC to become part of the trend, or to change existing ones to reflect the imperatives of internet-based business or whatever, it seems timely to pause and reflect for a moment on what this new phenomenon is, and what changes are required to better equip graduates entering the workforce at the dawning of a new millennium. This paper does not put forward any great philosophical insight, but does aim to generate debate amongst colleagues.

---

[1] This was the title of Professor Janice Burn's excellent inaugural lecture at Edith Cowan University, April 1998.

Synergy Matters: Working with Systems in the 21st Century,
Edited by Castell *et al.*, Kluwer Academic / Plenum Publishers, New York, 1999.

589

Another important aim is to try to establish some degree of equilibrium in the great debate on EC, and to urge a little more caution and empirical evidence to support the type of techno-diatribe with which we are currently being assaulted.

HOW DIFFERENT IS ELECTRONIC COMMERCE FROM INFORMATION SYSTEMS?

Let us start the dialogue with a discussion on the differences between EC and Information Systems (IS). It would appear from the marketing of 'new' courses and media hype that these two things are quite distinct entities. However, do these differences hold up when subject to careful scrutiny and reflection?

An IS is traditionally viewed as having a number of components: for example, there would be little disagreement if we were to characterise an IS as being made up of people, computer hardware and software, telecommunications, data / information and procedures (Hutchinson & Sawyer 1996). Students of IS typically undertake courses covering topics and skills such as programming, systems analysis and design, databases, networks and telecommunications, business problem solving and information systems management. The assumptions inherent in these courses were that IS/IT would improve organisational efficiencies and effectiveness, as well as business competitiveness, and would facilitate business transactions and relationships. While much of the focus may have been on the study of IS within one organisation, these courses have also increasingly made mention of IS that spanned organisational boundaries and facilitated business transactions and relationships. These interorganisational systems which enabled the development of networks both within and between organisations, and supporting technologies such as Electronic Data Interchange (EDI), are not really new, and could scarcely be considered products of the 1990s. Strategic IS such as American Airline's SABRE system, and American Hospital Supply's ASAP system were frequently cited examples, dating back to the late 1960s, of IS designed to bring the customer into a closer relationship with its supplier (Applegate et al. 1996). While undoubtedly being seen as offering the consumer improved service from the supplier, these systems also dramatically increased the switching costs for the consumer, thus making it much more difficult for them to consider doing business with any of their supplier's competitors. These systems were widely lauded in the IS world, and were oft cited examples of the potential of interorganisational systems and networks to transform contemporary business practices and processes (Callon 1996). It thus seems fairly logical that a further development would involve the linking of network to network, from which was spawned the concepts we now know of as the Internet, e-commerce, and the electronic marketplace (Applegate et al. 1996).

So what then, is EC, and in what ways does it differ from IS? EC has been defined in a variety of ways. To many, EC is defined as the buying and selling of products and services over the Internet. However, often definitions imply that EC is viewed as being a lot more than the handling of purchase transactions and fund transfers over the internet. Kosiur (1997) defines EC as a system that includes not only those transactions that centre on buying and selling goods and services to directly generate revenue, but also those transactions that support revenue generation, such as generating demand for those goods and services, offering sales support and customer service, or facilitating communication between business partners. Minoli and Minoli (1998) incorporate any electronic media, such as fax, TV and on-line networks in their definition of electronic commerce. Perrot (1997) extends the definition to argue that electronic commerce is any organisational strategy or effort related to the transmission of electronic information or signals, activated for the ultimate purpose of transacting commerce.

Given the definitions of IS and EC, one can surely be forgiven for thinking that there are a lot more similarities than there are differences between the two. Indeed, a systematic comparison of these two areas suggests that the differences between EC and IS have a lot more to do with the hype and cant surrounding EC than anything of actual substance. Take Perrot's definition for example. What would the major components of EC thus defined? We would suggest that a cogent argument could be developed to support the claim that EC thus defined would be seen as consisting of people, computer hardware and software, telecommunications capabilities, data / information, and procedures. Also implied of course would be access to the Internet. It all sounds rather familiar.

What we are left with then, is the belief that, given a certain generosity, the only 'new' aspect of EC that stands up to any scrutiny is the Internet itself, and even this was argued to be far from a new concept in the field of IS. Networks themselves are not new, and fall safely within the domain of IS, and particularly interorganisational systems. What may be new, and hence different, is the notion of linking network to network, and the pervasiveness and the potentiality of this internetworking in the 1990s. Of interest then is how radical a change this linking of network to network really is.

HOW RADICAL IS THE ELECTRONIC COMMERCE REVOLUTION?

Cyberprofs and their supporters in the media write and speak as though through, and because of, EC, we are about to witness the complete remaking of the world. Take, for example, the recent enthusiasm of Mack (1998: 28):

With the Internet set to cause the next revolution in business, e-commerce is gaining ground...the Internet revolution will be just as dramatic as the industrial revolution and have the same effect of changing the world order with globalisation of business...Japan is lagging behind the US in IT infrastructure...and one only has to look at these economies to see which is ahead. To survive, countries will have to make the transition to the new world of the Internet.

Not content with mere businesses, the very survival of countries is now being attributed to the Internet, and the national economic health apparently dependent only on the state of its IT infrastructure. While not wishing to underestimate the role of IT as a driver in the economy, this is surely simplistic to the extreme, and can do nothing but damage the credibility of the IT field in the eyes of the pragmatic business leader. In a similar vein, Mougayar (1998) writes that:

...electronic commerce is really allowing organizations to recast their market power and corporate profits in an entirely new universe...electronic commerce is BIG

and getting BIGGER...the marriage of the Internet with electronic commerce represents the cutting edge in business today...

A slightly more sagacious look at empirical evidence suggests a somewhat different picture. The uptake of EC with all its supposed advantages is not happening at anything like the speed at which some pundits are suggesting. Nor are the more moderate long term forecasts suggesting the same overwhelmingly electronic future (Markham 1998).

There appear to be sound reasons as to why the EC revolution will never occur as some writers portray. To develop this argument further, let us consider the retailing industry for example, as this is the one so often excitedly touted as being the showcase of EC wizardry. Protagonists of the EC revolution imply that the Internet will spawn literally thousands of new businesses, hence increasing competition, each benefiting from the substantially reduced costs of electronic transactions and thus being able to offer extremely competitive prices to the consumer (Phillips 1998). Furthermore, it is asserted that because these new "virtual" organisations have none of the traditional capital outlay of warehouses, shop fronts and equipment, for example, they should thus be relatively easy to establish. Indeed, there are writers who predict the end of the need for a physical shop presence for many retailers (Davidow and Malone 1992). The reality reported by other writers differs markedly, however. Mayoh (1996, cited in Markham 1998) suggests that to be successful on the Internet, retailers need an established street presence (This idea will be further developed in subsequent paragraphs). Mayoh goes on to write that *"significant costs are associated with building up a presence, and that retailers already benefit from the support and reassurance that high street visibility gives to customers"*. The fact that Levy Jeans can trade successfully on the Web is perhaps testimony to the fact that they have over many years, invested large amounts of money in building a brand name, and in always maintaining a highly visible store presence. Claims that the major fashion houses could do without retail outlets are also of concern (Davidow and Malone 1992), as it could be argued that their very success can in large part be attributed to their boutique presence in shopping malls, and so on. *Marks & Spencers* trading on the Web, for example, would hold for the customer promise of familiarity, quality and reliability, that a Web site for *Marshall and McKay* just might lack.

Rather than perpetually presenting trading on the Internet as signalling the end of the High Street store, a more moderate (and we would argue realistic) view is that trading over the Internet will prove to be a useful supplement to, and will co-exist with current retail practice (i.e. the physical store), as it will also with catalogue shopping (Markham 1998). Their viewpoint would be supported by the notion of co-opetition, previously described by Brandenburger and Nalebuff (1996). Co-opetition is the term coined to indicate that business represents a balance of co-operation and competition simultaneously. When it comes to creating a market, much can be achieved through co-operation, but competition will dictate how the market is divided up between various players. So, for example, the early manufacturers of cars such as Ford and General Motors, benefited enormously when Goodyear developed a successful tyre business, and similarly with the advent of a headlight manufacturer. These companies were all existing as complements to one another. This was further solidified when they together set up a Highway Association to lobby for and catalyse development of a coast-to-coast highway in the United States, an action which ultimately lead to increased demand for all their products. The actions of each of these businesses served to increase business for each of them (Brandenburger and Nalebuff 1996). A similar pattern emerges with the advent of video rentals. Many saw the move to release movies on video as signalling the death of the local cinema. This did not occur, and in fact, videos seemed to spark a renaissance in cinemas and in the movie industry (Brandenburger and Nalebuff 1996). Televised soccer matches in Europe have not had the effect of reducing match attendances. Far from it: attendances have markedly increased since the increased availability of television broadcasts (Markham 1998). Could it be that the two events act as complements for one another? The question then is whether this logic could also apply to Internet trading. We would argue that it could. If it does, then retailing over the Internet will not cause the demise of traditional nor catalogue shopping. Rather, it may well act as a complement to them both, serving to increase the overall size of the pie, rather than serving to benefit itself at the expense of other forms of retailing.

There are other issues surrounding EC which need to be addressed. For example, what proportion of the world population is sufficiently computer literate to contemplate trading over the Internet, and what proportion of those actually have adequate access to the computer technology to make EC a reality? We would suggest that in myopically considering the USA, Europe, parts of Asia (including Australia) as the leaders of the revolution, we are indeed looking at a very small section of the total world population and market. While Internet commerce will doubtless expand the options for educated and relatively wealthy people in developed nations, it will be unlikely to improve conditions for less fortunate members of the world community.

Amazon.com has received a lot of media exposure and all the plaudits in the world for showing the way in terms of e-commerce. But, at the risk of asking politically incorrect questions, has it ever made a profit? ("Ah, but that's just the creative fiddling of accountants" counter the cyberprofs.) Indeed, there are writers that question whether Amazon.com will ever make a profit: *"For all its on-line fame, Amazon has yet to prove that there is money to be made selling books on the Internet. Bezos lost $27.6 million...in 1997, and he will not say when he expects to make a profit"* (Willis 1998). IBM's World Avenue opened in Autumn, 1996, with the promise of a shopping mall for 20 retailers. That was subsequently scaled back to 6 shops, and then zero, as IBM pulled the plug on the whole venture in Autumn, 1997. ("Ah, but that IBM for you" retort our cyberprofs.) After excitedly, but erroneously, forecasting that 20% of its trade would be done over the Internet back in 1991, Winn-Dixie have likewise pulled the plug on a direct Internet selling presence in 1997, preferring instead to maintain an order facility via a toll-free telephone line (Markham 1998). As is so often the case in the IS world, successes, however premature they may be, get plenty of exposure, while failures are concealed so no one can learn from the events and contributing factors. This phenomena was recognised by the likes of Abdel-Hamid and Madnick (1990) and reported long before Internet successes and failures came to the fore.

IS IT ALL SOMETHING TO DO WITH ECONOMIC IMPERATIVE?

Those most vociferous in their claims about the future of EC often seem to be those who will benefit the most from its uptake. Is it cynical to suggest that part of

Bill Gates' vision for the future may be coloured somewhat by the extent to which Microsoft may gain if the world embraces and adopts his vision? The IT industry would not be the first where powerful economic interests were seen as promulgating a future that was desirable from their limited, economic perspective.

CONCLUSION

If only...computer people and promoters of electronic futures would discipline themselves to live in the real world of normal people, with human responses and psychologies, they would be able to judge possible effects in this world instead of appearing , at times, to be on another planet." (Markham 1998). We will let the prophetic words of Markham speak for themselves.

If EC is an attractive, highly marketable label, only the foolhardy in modern universities would resist its use. However, there is the potential for even more confusion in the discipline of IS if we pretend that EC is something substantially new and different. Let us knowingly and wisely adopt and use the term EC in a meaningful manner, and resist the temptation of being swept away by hype.

BIBLIOGRAPHY

Abdel-Hamid, T.K. and Madnick, S.E., 1990, The elusive silver lining: how we fail to learn from software development failures, *Sloan Management Review*, 32 (1) Fall (1990): 39-48.

Applegate, L.M., McFarlan, F.W. and McKenney, J.L., 1996, "Corporate Information Systems Management: Text and Cases," Irwin, Chicago.

Brandenburger, A.M. and Nalebuff, B.J., 1996, "Co-opetition," Currency Doubleday, New York.

Callon, J.D., 1996, "Competitive Advantage through Information Technology," McGraw-Hill, New York.

Davidow, W.H. and Malone, M.S., 1992, "The Virtual Corporation," Harper, New York.

Hutchinson, S.E. and Sawyer, S.C., 1996, "Computers and information systems, 5th ed," Irwin, Chicago.

Kosiur, D., 1997, "Understanding Electronic Commerce," Microsoft Press, Redmond.

Land, F., 1995, The new alchemist: or how to transmute base organizations into corporations of gleaming gold in "Proceedings of the Third European Conference on Information Systems," G. Doukidis et al., eds., Athens University of Economics and Business.

Mack, M., 1998, E-commerce is ready for fast takeoff, *Computerworld*, 21:28, 30.

Markham, 1998, "The Future of Shopping," Macmillan, London.

Minoli, D. and Minoli, E., 1998, "Web Commerce Technology Handbook," McGraw Hill, New York.

Mougayar, W., 1998, "Opening Digital Markets," McGraw Hill, New York.

Perrot, B., 1997, "Electronic Commerce: What is the Future?," School of Marketing, University of Technology Sydney, Sydney.

Phillips, M., 1998, "Successful E-Commerce," Bookman, Melbourne.

Willis, C., 1998, E-trade tests the strength of Amazon, *Business Review Weekly*, 20(15):120-121.

WORKFLOW FOR INTER-ORGANISATIONAL ENVIRONMENTS

Roger M Tagg

Department of Information Systems
Massey University
Palmerston North, NZ

INTRODUCTION

The phenomenon of Workflow owes its arrival largely to the early 1990s vogue for Business Process Reengineering (BPR)[1]. Economic conditions at that time placed, as they still do today, a surge of pressure on organisations to operate more efficently. There was a revival of interest in processes, looking at the whole value chain between inputs and customers, and including administrative activity which had grown into an unsustainable overhead.

The computing field has seen an analogous swing back from a possibly excessive concern with shared databases to a balance in which process aspects take a greater role. Workflow has been the spearhead of this re-balancing, and Workflow Management systems (WfMS) occupy a position somewhat analogous to Database Management systems (DBMS).

Much current practice and research in Workflow addresses processes which exist within a single organisation. However, today's administrative work is often not confined within a single set of organisational walls. The increasing number of joint ventures, cooperatives and inter-organisational working groups is evidence of this. Modern business activity is characterised by a growing number of organisational boundaries, across which management must negotiate supply or service contracts - and also exchange information.

This paper discusses current business trends and their consequent patterns of cooperative work. It then highlights some key needs for workflow management support in inter-organisational situations. A range of IT solutions are then introduced, including a prototype developed as part of our current research. Finally, some areas for further work are suggested.

Definitions and Scope

Workflow is defined by the Workflow Management Coalition[2] as "the computerised facilitation or automation of a business process, in whole or part". The same source also defines a Workflow Management system as "a system that completely defines, manages and executes workflow processes through the execution of software whose order of execution is driven by a computer representation of the workflow process logic".

Workflow Management is related to, but can be distinguished from, both Groupware and Document Management. Groupware, unlike Workflow, does not necessarily model a defined administrative process. Document Management refers to the system which controls a distributed database of documents in order to enable users in a group to refer to them in a consistent manner. Commercial software for Workflow Management is often integrated with Groupware and Document Management.

Commercial Practice

There are over 50 commercial Workflow Management software products, of which around a dozen can be regarded as major players in the market. Examples include Action Workflow, FileNet Ensemble, IBM FlowMark, InConcert, JetForm, Keyflow, SAP Business Workflow, Staffware, Worklogik and WorkMAN. Many recent versions of these products also offer a user interface based on Web forms.

Synergy Matters: Working with Systems in the 21st *Century*,
Edited by Castell *et al.*, Kluwer Academic / Plenum Publishers, New York, 1999.

Some of the leading Groupware products also offer some Workflow Management functions, in particular Lotus Notes (now acquired by IBM) and Microsoft Exchange. These products also allow for fuller-function WfMS to be integrated as components. Some Groupware products also include Document Management facilities, although there is also a market for specialised Document Management Systems. Example products are Documentum, Filenet, Keyfile and Saros.

Process Modelling tools are a key preliminary to workflow design. Many products have arisen from the need to support BPR exercises. Some of the products come in the form of simulation toolkits which allow potential workflows to be evaluated. Around 20 products are described in surveys, eg[3].

Standardisation

Many commercial product vendors are involved in the Workflow Management Coalition (WfMC), which is a group of over 100 software suppliers, users and consultants who are attempting to agree terminology, architecture and standards for workflow software and related tools. Within its Reference Model[2], the WfMC has defined interfaces for process definition tools, client applications, invoked server applications, administration/monitoring tools and WfMS-to-WfMS interoperability, the last named being of particular relevance to inter-organisational environments.

Current research directions

The current literature, eg[4], indicates that a wide range of workflow research topics is being followed up. Some of these topics are listed below:

- Complex transactions
- Exception handling
- Handling disconnected clients
- Use of CORBA in workflow
- Dynamic modification of rules
- Workflow design tools
- Design of workflow client GUIs

- Triggering models
- Availability and fault tolerance
- Workflow Message APIs
- Security and authentication
- Version handling
- Knowledge elicitation for workflow

Several leading workflow research groups are working on the topic of Transactional Workflow, eg the University of Georgia. This topic covers relatively close-coupled systems which are required to take responsibility for the consistency of databases accessed by the WfMS. The so-called ACID properties (Atomicity, Consistency, Isolation, Durabilty) have to be maintained as if the WfMS was a Transaction Processing system.

A general comment is that many of the existing research themes listed above is that they lean towards tight coupling, or high levels of integration, in the workflow management system. Also, much workflow research and development is currently aimed at larger organisations where more standardisation of infrastructure is possible. Such organisations are better able to see - and pay for - the value of improving the management of their workflows, than are loose confederations or joint ventures.

BUSINESS TRENDS

In both commercial business and public enterprise, the trend has been almost universally towards decentralisation and devolution into separate autonomous business units. At the same time, many previously independent units find themselves increasingly having to get involved in cooperative activitites.

Joint Ventures

Joint Ventures are an increasingly common mode of operation in many business sectors. With the internationalisation of business, and the influence of trans-national groupings such as the European Union, it is often advantageous in bidding situations to put in tenders as a consortium of companies from more than one country; this also helps spread the risk on large projects.

Cooperatives, Trading Associations and Working Groups

There is a whole range of cooperatives, trading associations and voluntary working groups, which are involved in activities involving individuals from different organisations or self-employed persons. Such groups may require support of both commercial and administrative information interchange. Trading associations have been quick to adopt technology such as EDI (Electronic Data Interchange) for inter-organisational business transactions.

Devolved Responsibilities and Purchaser/Provider Relationships

Many large multinationals and government departments have divested themselves of non-core businesses, even though they continue to be heavy users of the services these units provide. It is very desirable, in such environments, to avoid wasted effort by streamlining the processes of contract negotiations and monitoring of performance.

No Organisation is an Island

All of the above situations can be regarded as examples of Virtual Organisations. Just as with single organisations, there is a payoff from automated prompting of the execution of processes across a Virtual Organisation. However, even large monolithic organisations are rarely able to operate their processes in isolation, and will typically have to engage in processes involving external organisations, such as customers, suppliers, contractors, tax authorities and regulatory bodies.

COOPERATIVE WORKING PATTERNS

The patterns listed below are examples of what are seen as the most common types of workflow in inter-organisational environments.

Document Preparation
Some work on such documents as joint tenders, contract documents, sales literature, capability brochures, journals of development, newsletters and magazines may be ad-hoc and thus more the domain of Groupware. However other document development is often subject to formal procedures and committee work which approves changes to the documents.

Negotiation and Trading
This type of pattern includes multiple-round bidding on prices and contracted service terms; auditable procedures for contract tendering and evaluation, takeovers and mergers; and contract performance management.

Customer Contact Management
This covers customer complaints and claims, salesperson visit planning and tracking. With inter-organisational workflow, the customer can be brought into the process first-hand.

Committee Work
This involves managing the business of geographically dispersed working committees (forming motions, amendments, voting, approving etc).

Regulatory Compliance
This includes the tracking of laws, regulations and guidelines; processing of questionnaires, checklists, inspections; formal reporting; and training/awareness courses. Interchange of information often involves multiple stages rather than a simple one-off transfer of data.

INFRASTRUCTURE NEEDS

Supporting the Essence of Inter-Organisational Activity

In the absence of any unifying chain of command which binds the participants to follow enforceable procedures, one has to rely on participants being motivated to follow a jointly-agreed process framework for their own business benefit. Furthermore, participants may want to introduce "variations" to the processes. However responsibility for design and operational control of each process still has to be allocated, normally to one of the participants involved.

Client Technology Tolerance

No one authority has control over the participating users' hardware, operating system or personal computing tools. Nor can it be assumed that all participants have access to a WfMC-compliant Workflow Management System! Users who work for an organisation will often have access to a local workgroup server, which may be able to store shared files and operate a Document Management system. Short of using WfMS software that has clients for every possible workstation type, use of a Web browser is currently the only realistic way of achieving this goal.

Universal Transport and File Reference Mechanisms

By the same token, the very wide disparity in network facilities can only be cost-effectively addressed by the Internet. It is not generally necessarily to transport documents and other data physically between server and participant workstation (or vice versa). The simplest method is to pass to the recipient only the address of the document, page or file which the recipient is to process - assuming of course that the recipient can then access that data over the network. An originator of data can hold that data locally on a shared part of his/her own local disk storage, or can put it into a shared area on his/her local workgroup server. The natural mechanism for such reference is the URL system used with the Internet.

Matching of Semantics

Participants in different organisations may use different data names and "ontologies" to describe what is essentially the same process. This has already been recognised within the EDI (Electronic Data Interchange) community[5], as the demand for more "open" forms of Electronic Commerce becomes stronger. An ideal system would support the concept of "matching" of data names between different ontologies.

This problem is compounded where the participants include people who speak different languages. In such cases the system needs to act as an "interpreter" as well!

Mobile and Disconnectable Clients

This phenomenon occurs in single organisations, but is even more critical in inter-organisational environments. The workflow system must cater for participant not being logged on, mobile computers being unconnected, computers being switched off, and network connections or servers being down.

INFRASTRUCTURE SOLUTIONS

Summary of Approaches

The appropriateness of the world-wide web (WWW) as the means of delivering inter-organisational workflow is now quite well recognised. Table 1 shows one way in which current approaches can be classified. The vertical dimension represents the level of software autonomy in the distributed system, ranging from fully autonomous, peer-to-peer distribution at the top to centralised with thin clients at the bottom. The horizontal dimension distinguishes three generic workflow models, namely process-oriented, document-oriented and transportable agents; this is discussed further below.

Table 1: Some Approaches to Inter-Organisational Workflow

	Process-oriented models	Document-oriented models	Transportable agents
Distributed servers, no central site	Exotica FMQM OrbLink (U. Georgia)	Wing, Liu & Colomb	
Distributed servers + some central functions			DartFlow (Cai et al)
Workflow clients + nominated central server	Exotica FMDC Most commercial WfMS	A few commercial WfMS	
Only web browser on clients, rest on nominated central server	ActionWorks Metro WebWork (U. Georgia) Lightweight		

Process Oriented Models

In this, the most common approach, business processes are represented as pre-defined workflow patterns, using a Petri Net or similar method. The approach assumes that the details of the process are known and can be defined. Uncertain outcomes can be allowed for through conditional branching. Variations to processes can be supported by new workflow patterns which inherit, but partly override, the original. This approach may be less suitable for some types of inter-organisational processes, where the workflow rules are evolved by negotiating from the bottom up.

IBM's FlowMark is a typical example of a process-oriented WfMS. Like most WfMS products on the market, it is oriented to a central server with workflow client software on user workstations. Researchers at IBM Almaden have proposed FMDC[6] as a solution to the problem of clients which may be disconnected from the server at times. Another development of FlowMark, FMQM[7], addresses the situation in which several autonomous servers collaborate to manage workflows through persistent messages. Several approaches have been proposed as part of the University of Georgia METEOR project[8]. OrbWork is a peer-to-peer CORBA-based system, while WebWork uses a server-based Web Browser-CGI-Database model. Similar to the latter is ActionWorks Metro, a thin-web-client version of Action Workflow.

The author's own research[8] has taken a similar approach to these last-mentioned systems. The primary target is distributed work groups who have many inter-organisational processes that are pre-definable, but where cooperation is voluntary and significant investment per user cannot be justified. We use the term "lightweight" to describe a server, operating an existing relational DBMS which incorporates a Rule or Trigger system, which acts only as a simple "prompter" of action to be taken by participants. No built-in facilities for data or document management are envisaged. Any transaction control would be either manually controlled by human participants, or handled by the TP facilities of auxiliary applications. We also envisage a transition stage in which some participants only have email, and the range of clients may include old PCs, Network Computers, PDAs and Pager Devices.

Document Oriented Models

A few commercial WfMS adopt a document-oriented model in which each document carries its current routing instructions as it is transported from user to user. Workflow patterns do not exist independently of documents; instead documents may inherit instructions from other documents in the same document class or "dossier".

Wing, Liu and Colomb[10] describe an document-oriented approach to distributed workflow where relationships between participants are at arms length, and cooperation rules are evolving rather than being predefined - trading groups are a primary target environment. Like FMQM above, the system uses message queues to control the transmission of EDI-type messages which trigger actions at participating nodes.

Transportable Agents

Cai, Gloor and Nog[11] have proposed DartFlow, a system based on transportable agents which are software components each representing the logic of one process. Agents are replicated to whichever site has data which is subject to the process. Generally, there is less transmission involved in moving code than in moving data. The system is essentially peer-to-peer as above, and hence suitable for inter-organisational environments, but it requires a central server to coordinate roles, worklists and status tracking,. The user interface is again web-based.

FURTHER WORK

Modelling of the Process of Cooperation Itself

Choice of the best approach for inter-organisational workflow depends on a good understanding of how patterns of cooperation are built up. The approaches in the previous section are based on very different assumptions, and we have started to look into this aspect. An interesting approach is suggested by Hawryszkiewycz[12] who has applied a Soft-Systems-like model to government-fostered collaboration between SMEs (Small to Medium Enterprises).

Multi-lingual and Multi-ontology Situations

Particular needs for multi-lingual support arise in Europe and S.E.Asia; there would seem to be an opportunity for the use of Translation Agent software provided as a service on the network. The multi-ontology situation is somewhat different; a standard such as the Basic

Semantic Repository for EDI[5] might be gradually adopted by SMEs, while larger organisations would require translation to and from the standard.

The Potential for Workflow Agent Services

There would appear to be an interesting, though probably small, commercial opportunity in the provision of a Workflow Agent as a third party service to a group which may not find it convenient to use the facilities of one of its participants.

Measures of Effectiveness

We are also starting to consider what measures of performance are important when judging one solution against another. Wing et al[10] emphasize the need for no bottlenecks and minimum downtime. However use of multiprocessor "non-stop" servers may be as effective as distributing all the functions onto multiple nodes. Pure processing speed, as long as it is not painfully slow, is probably less critical. Important in our thinking is the ability to work with existing client computer systems with a minimum of add-ons. We also rate cheapness (capital cost, maintenance, clerical time, training) as highly important.

Experimentation with a Prototype

Our current plan is to install a test-bed system and offer it to a selected user group as an experiment, preferably in place of an existing paper or ad hoc messaging system. Reactions would be observed and measures of effectiveness logged.

Since workflow systems are an attempt to provide better assistance to humans in their administrative tasks, there is a good opportunity to investigate qualitative outcomes in addition to measurable administrative efficiency. One possible example is the effects of different system styles on the motivation of the separate organisational units.

CONCLUSION

This paper does not so much represent a report of research done, but rather an exploration of a business need and an attempt to provide directions for further work. The paper has reviewed trends in inter-organisational business activity, and has considered various approaches for proving automated workflow support. It has identified a number of research directions, somewhat different from the mainstream of workflow research, where effort might most valuably be spent. Hopefully, it is the start of an interesting and challenging journey.

REFERENCES

1. M. Hammer and J. Champy. "Re-engineering the Corporation", Harper/Collins, New York (1993).
2. D. Hollingsworth. "Workflow Management Coalition: The Workflow Reference Model", Workflow Management Coalition, Brussels, Belgium (1994).
3. L. Jennison. "Some tools for business re-engineering" in "Software Assistance for Business Re-Engineering",. K. Spurr, P. Layzell, L. Jennison and N. Richards, eds, John Wiley, Chichester, UK (1993).
4. A. Sheth and K. Kochut. "Workflow applications to research agenda: scalable and dynamic work coordination and collaboration systems" in Proceedings of NATO ASI on Workflow Management Systems and Interoperability, Istanbul, Turkey (1997).
5. K. Steel. "The Basic Semantic Repository" http://www.cs.mu.oz.au/research/icaris/bsr.html
6. G. Alonso, R. Gunthör , M. Kamath, D. Agrawal, A. El Abbadi and C. Mohan. "Exotica/FMDC: handling disconnected clients in a workflow management system", in Proceedings of 3rd International Conference on Cooperative Information Systems, Vienna, Austria (1995).
7. G. Alonso, C. Mohan, R. Gunthör, D. Agrawal, A. El Abbadi and M. Kamath. "Exotica/FMQM: A persistent message-based architecture for distributed workflow management", in Proceedings of IFIP Working Conference on Information Systems for Decentralized Organisations, Trondheim, Norway (1995).
8. J. Miller, A. Sheth, K. Kochut and D. Palaniswami. "The future of web-based workflows" in Proceedings of International Workshop on Research Directions in Process Technology, Nancy, France (1997).
9. R.M. Tagg, W. Lelatanavit and S.S. Reddy, "Preliminary design of a lightweight workflow server", in Proceedings of Australasian Conference on Information Systems, Adelaide, Australia (1997).
10. H. Wing, C. Liu and R. Colomb "A bottom-up approach to distributed workflow", in Proceedings of Pacific Asia Conference on Information Systems (PACIS), Brisbane, Australia (1997).
11. T.Cai, P. Gloor and S. Nog. "DartFlow: a workflow management system on the web using transportable agents" Dartmouth University Technical Report PCS-TR96-283 (1996).
12. I. Hawryczkiewycz. "A design method for choosing services for large distributed teams" in Proceedings of Conference on Design of Cooperative Systems, Juan-les-Pins, France (1996).

RESEARCH AND REALITY: CO-EXIST OR CO-INHABIT?

Professor Sam Waters, George Bakehouse, Kevin Doyle
School of Information Systems
Faculty of Computer Studies and Mathematics
The University of the West of England (Bristol)
Coldharbour Lane, Frenchay
Bristol, BSI6 1QY

INTRODUCTION

The aim of this paper is to stimulate a debate centred around the relationship between research and reality. Central to this debate is the notion of two separate and sometimes competing approaches to solve real life problems, theory and practise. The last three or so decades has witnessed the development of numerous methodologies which vary across a wide spectrum from the "very hard" to the "very soft" most claiming to have practical benefits in the real world. An area of Systems Science that has grown quickly amongst all the confusion is that of Information Systems, a new and highly dynamic subject area where academics and practitioners often fail to agree at any level about things as fundamental as the meanings of 'information' and 'system'. There is ample evidence to show that in the world of Business Information Systems, technologists do not understand the world of business and vice-versa. This paper asserts that a similar gulf exists between Systems theorists and Information Systems practitioners.

Research and Reality: Different Worlds?

Tsagdis (1997) went as far as describing two different worlds. World one (W1) the real world and world two (W2) the world of research, emphasising the requirement for discourse between the two. Kline (1995) tells us that " the two worlds should be viewed as actional contexts that mutually constrain each other." Undeniably there is a close relationship between the two paradigms, the question being, which is the dominant partner? The authors response is to emphasise the merits of action research which takes place in W1 using theories from W2, in the intersection of the two worlds. We all spend some time in W2 but undeniably we spend all of our time in W1.

As Becker (1965) so eloquently stated: "..... there is more to doing research than is dreamt of in philosophies of science, and texts in methodology.... the best laid research plans run up against unforeseen contingencies in the collection and analysis of data; the data one collects may prove to have little to do with the hypotheses one sets out to test; unexpected findings inspire new ideas. No matter how carefully one plans in advance, research is designed in the course of its execution....."

Land (1995) in his discussion of the new alchemists warned that we must be aware of the dangers of jumping onto every new bandwagon that appeared to be the panacea. The solution to all IT/IS problems, the ultimate methodology. He stated "if management is to be regarded as a science it has reached the maturity that physics and chemistry had reached in the middle ages" he continued "..... we look to wise men or gurus to provide panaceas and wise men (consultants) come forward and proclaim that they, and only they have the knowledge which will attain corporate success, if only corporate mangers followed their prescriptions. We seek, and they offer remedies, which sometimes resemble the notions of the medieval alchemists - the holy grail and the philosophers stone."

Historically some of the holy grails have included Scientific Management (Taylor, 1910) Cybernetics and General Systems Theory (Beer 1994, Von Bertalanffy, 1969, 1973), Total Quality Management (Demming, 1992, Juran, 1989), SSADM (SSADM, 1990), SSM (Checkland, 1981) and Business Process Re-engineering (Hammer, 1990) . The sequence is something like this; here is the panacea, try it, after a period of time when the results are not as expected try this new one. Often the new panacea is no more than a revamped version of its predecessor or a mixture of several which have been repackaged and are apparently new. Huczynski (1987) reported that during this century theorists have provided us with in excess of three hundred relevant theories and methods all aimed at improving organisational performance, yet none has withstood the challenge of time.

ACTION RESEARCH IN THE REAL WORLD OF PRACTICE

In order to think about, understand and explain the world about us, it is necessary to develop models or abstractions of the world and ways of using them to think about it. These abstractions and approaches then become the epistemological constructs which form the basis of our reasoning, communication and discussion about the world. As disciplines have emerged, each has developed its own ways of modelling and reasoning about the world. Many providing different descriptions of what is essentially the same set of phenomena. Religion and Philosophy, Politics and Economics, Sociology and Anthropology, all have their own brand of models and approaches, rules and evidence. The differences between them often lead us to the mistaken belief that the phenomena which they study are not the same.

The problem of fragmentation exists because of the differences in the epistemological constructs used to describe the world and its phenomena (observable or otherwise). It is true that to move from one paradigm to another constitutes a fundamental shift and that a lack of isomorphy means that the models of each may not be easily mapped or transformed into an equivalent form in the others. Through the medium of action research, different models may, be considered to be complementary because despite their differences in emphasis, focus and use, they have a common point of contact; the real world of social action. Action research also provides the means by which theory may be developed and tested in the real world.

The notion of improving theory through practice is anything but new. The action research approach, exemplified by Checkland's Soft Systems Methodology, is well established. (Checkland, 1981; Checkland et al, 1990). Conventional research approaches following the paradigm of science (reductionism, experimentation, refutation and repeatability), seek to form and then verify theories. Action research promotes the emergence of (academically defensible) theories through practice and the speculation over how such practice might be bettered in some way. In this respect action research is both descriptive/interpretative and subjective/argumentative. (Galliers et al, 1987).

A clear point of contact for theory and practice and for their different tools and techniques, frameworks and approaches, is the world of information systems. Information Systems is an ideal area to consider this unification, as IS development and use is a natural point of contact for the 'natural' and the 'artificial', the 'hard' and the 'soft', the 'concrete' and the 'abstract' the 'physical' and the 'social'. This undertaking cannot come from abstract academic thought alone. Such a framework needs to be hammered out in the real world of human activity, through the medium of action research (Doyle, 1995).

"The modern specialist field-worker soon recognises that in order to see the facts of savage life, it is necessary to understand the nature of the cultural process. Description cannot be separated from explanation, since in the words of a great physicist, 'explanation is nothing but condensed description.' Every observer should ruthlessly banish from his work conjecture, preconceived assumptions and hypothetical schemes, but not theory". (Malinowski, 1936)

A series of action research projects at the University of the West of England (formerly Bristol Polytechnic) spanning over a decade, has seen the emergence of an approach to embedding the tools and techniques of systems engineering in an action research framework (Bakehouse et al,

1995, 1997, Doyle, 1994, Waters et al, 1994). These projects have involved strategic, tactical and operational systems in education, health care, construction, banking and other areas of the private sector. The research team has worked with a wide range of organisations, at a number of different levels of involvement in an attempt to define a general purpose framework of open utility.

The research began by taking and adapting a recognised framework for the development of IS strategy (Waters, 1987 ; Waters, 1988) and interleaving it with SSM. This strategic framework made use of established strategic tools such as failures theory (Waters, 1986), SWOT analysis (Waters, 1989), and PEST analysis (Johnson & Scholes, 1988). The interleaving process added: Rich pictures, Conceptual Models, a focus on cultural feasibility as well as organisational desirability and a loop back to the beginning of the framework to take account of the learning cycle. The dominant aspect of the research programme has been to help organisations to improve the way that they manage their information resource.

INFORMATION, ITS DIMENSIONS AND QUALITY

The Industrial Revolution gave us the '4 M's' of men, money, machines and materials, the Green Revolution gave us the environment (the natural world that we have inherited, that we briefly inhabit and that we must conserve for our future generations) and the Computer Revolution gave us information; today, we would re-phrase 'men' as 'people' and 'machines' as 'technology'. Information is widely regarded as the intangible resource. Seminal works (by Kent, Stamper and others) warn us of the dangers of dealing with this abstraction whilst in the practical world of commerce, industry and administration the President of the Confederation of British Industry maintains that 'managing information is the greatest challenge facing all organisations today'. The continuing action research program addresses the question of how can we interpret this theoretical abstraction of the information resource into the practical reality of 'helping people to get better with information'.

Given that the main objective of IS is to deliver the right information to the right person to support the right activities at the right time in the right place at the right cost with the right quality in the right presentation and with the right availability (in the same sense of Drucker's definition of improving organisational effectiveness and efficiency as 'doing the right thing right'). In practice, if people understand and improve upon these dimensions of information then they will 'get better with information'.

Right information

We mean by this that information should be "the truth, the whole truth and nothing but the truth" and have found this to be generally understood by people in practice. Our information dimensions are accuracy (and consistency), completeness and non-redundancy. Wang (1995) identifies these same dimensions as accuracy, consistency, completeness and relevance.

Right person

We mean by this that information should only be delivered to people who are authorised to access and/or change it; Wang (1995) identifies this same dimension as privilege and other common synonyms include privacy, security and confidentially within the overall umbrella of an information control system. It is important for all of us to recognise that no system is infallible therefore there is no guarantee that the right person has in fact been identified (whether it be by password, signature, iris, fingerprint, voice, lip, or whatever recognition method); unfortunately, many influential people mistakenly believe or knowingly propagate the myth of "IT infallibility".

Right activities

We mean by this that there are a myriad of activities that a person can perform when using authorised information; however, not all of these activities may be authorised for that person by the organisation, professional ethics or the laws of the land. Examples of information mis-use and abuse include fraud, "insider-dealing", profiteering, blackmail and extortion therefore monitoring and control meta-systems should be designed and operated to counteract these violations of "privilege", to use Wang's terminology.

Right time

We mean by this that information should be delivered to a person when and whenever it is required (subject to any business constraints imposed through the information system, e.g.:- access to customer credit files is only permitted during normal business hours). Business trends such as globalisation and homeworking are extending this time dimension to "around-the-clock" (24 hours a day, 7 days a week, 52 weeks a year). Wang (1995) does not specifically identify this dimension but uses "timely" to describe how up-to-date any information is.

Right place

We mean by this that information should be delivered to a person where and wherever it is required (subject to any business constraints imposed through the information system, e.g.:- access to personnel files is only permitted on company premises). Business trends and IT ubiquity are extending this space dimension to global, "around-the-world". Wang (1995) identifies this dimension as the means by which the user can "get the data", implying access to a terminal local to the user.

Right cost

Almost a century ago, Elbourne (1914) observed that "it is quite possible for management to collect more information than it can use to advantage or which is more costly than the information is worth". This notion that the value of the information delivered to a person should exceed its cost of delivery would seem to be just as relevant, if not more so, today (with its stringent economy drives). However, we have yet to witness people or information systems that monitor this cost dimension at, for example, even the level of a database query (although some DBMS do pre-advise the expected volumes of a content-retrieval response); in some case, the costs of delivering information can be vast as evidenced by the hordes of UK civil servants and research assistants scurrying around to answer a question posed by a Member of Parliament from the floor of the House to a Minister of the Realm. It is only very recently that the Parliamentary tradition of answering any such question has been relaxed by the dismissive "such information can only be provided at disproportional costs". Wang (1995) does not specifically identify this cost dimension (but then neither does the world at large?).

Right quality

We mean by this that information should meet other quality criteria over and above those identified above; in particular, information must be sufficiently up-to-date to meet its users' decision-making requirements. Wang (1995) identifies this same dimension as currency ("when the data item was stored") and volatility ("how long the item remains valid") but uses "quality" to describe the entire gamut of information dimensions.

Right presentation

We mean by this that information should be in a form that fits a person's requirements of understandability; this involves all aspects of HCI (Human Computer Interface). Wang (1995) identifies this same information dimension as interpretability meaning "understandable in terms of syntax and semantics". Although many people might not be familiar with such terms as "syntax" and "semantics" (apart from such professionals as linguists, lawyers, neurosurgeons, etc.), they certainly know what they cannot understand.

Right availability

We mean by this that "on a good day" all the above information dimensions will be perfectly delivered; however, on other (hopefully, few) occasions some of these requirements will not be satisfied due to IT, IS and/or IM failures. This holistic view of the availability dimension extends Engineering's MTBF and MTTR from "machine" faults to embrace any information failure that impairs the delivery of the expected service to a person. A key issue of (lack of) availability is the design and provision of interim services using emergency, fall-back, graceful degradation, back-up, recovery and other support sub-systems. Wang (1995) does not identify this dimension but uses "available" as a synonym for "accessible"

Using this taxonomy and definition of information dimensions as a framework, the quality of information in practical real-world settings can be measured in terms of the occurrences of defects. Thus, ethnographic field research methods (particularly observation) can be applied to identify information failures and to verify, classify and quantify their occurrences; ultimately, this helps people to prioritise their information problems in order to propose and implement solutions.

CONCLUSIONS

Our ongoing empirical research compares leading technological organisations in four sectors of the UK economy; these are Banking (Citicorp), Construction (Kvaerner - Trafalgar House), Health (Frenchay NHS Healthcare Trust) and Transportation (LEX). This comparison identifies their stages of IS development, their relative timescales and costs (measured in terms of IS investment per employee per annum) and their information quality (indicated by the average number of defects suffered by each employee each day). A goal is to improve information quality control by back-tracking the causes of defects and evaluating their effects by forward-tracking, where possible.

The underlying focus is to help organisations manage their information better. No attempt has been made to suggest "come listen I have found the holy grail ". The approach adopted focuses on the belief that it is possible to work within the bounds of theory and practise simultaneously, if successful, both areas will benefit . The authors have previously worked in practise for many years and fully appreciate the freedom and opportunity that action research offers as opposed to working within the bounds of a prescribed methodology. The freedom to experiment, adapt, adopt and develop theories with the added bonus that the results may make a difference in the real world (W2).

Finally, our observational field research forces us to wade through mud, blood, grease and boardrooms wearing hard-hats, surgical greens, blue boiler-suits and city slickers so that we may try to understand the practical realities of information mismanagement. As one eminent IS Professor told us "...somebody has to do this work; I am glad it is not me!".

To answer the question in the title: Within the IS discipline there is scope for both coexistence and co-inhabitance, the authors believe the best results will be gained by a fully committed marriage between research and reality.

REFERENCES

Bakehouse, G, Davis, C, Doyle, K and Waters S.J, 1995, "Putting Systems Theory into Practice, The role of Observation in Analysing the Real World", In proceedings UKSS 4th International Conference, Systems Theory and Practice.

Bakehouse, G & Davis, C, & Doyle, K. G, & Waters, S. J, 1997, "Anthropological Reflections on Systems Engineering: Seeing is Believing." in Philosophical Aspects of Information Systems, Winder R. L, Probert S. K, & Beeson I. A (Eds.): Taylor & Francis. pp 181 - 200

Becker, H, 1965, "Review of Sociologists at Work: American Sociological Review"

Beer, S, 1994, "Diagnosing the System for Organisations" Wiley, Chichester.

Bertalanffy, L. Von, 1969 "The Theory of Open Systems in Physics and Biology" in Emery, 1969

Bertalanffy, L.Von, 1973, "General Systems Theory": Penguin University Books

Checkland, P. B, 1981 "Systems Thinking Systems Practice":John Wiley, Chichester.

Checkland, P. & Scholes, J, 1990, "Soft Systems Methodology in Action": John Wiley, Chichester

Deming, E.W, 1992, Comments made during a teleconference on Total Quality broadcast by George Washington University in Spring 1992.

Doyle, K.G & Wood, J,R,G & Wood-Harper, A,T, 1993, "Soft systems and systems engineering: on the use of conceptual models in information systems development":   Journal of Information Systems, 1993, 3, pp 187- 198

Doyle, KG, 1994 "Growing computer based information systems organically: The integrated clinical workstation project": in "New Systems Thinking and Action for a New Century": Proceedings of the 38th International Conference of the ISSS, Asilomar, Pacific Grove, California. June pp 527 - 534

Doyle, K. G, 1995, Uniting Systems Theory With Practice in *Critical Issues in Systems Theory and Practice* Proceedings of the 4th international conference of the UKSS, Hull. July. Ellis, Gregory, Mears-Young & Ragsdell (Eds.): Plenum. pp 297 - 302

Ellbourne, E.T. 1914, "Factory Administering and Accounts", The Library Press

Galliers, R, D & Land, F, F, 1987, "Choosing appropriate information systems research methodologies": Communications of the ACM, vol.30, no.11, pp900 - 902

Hammer, M, 1990, "Re-engineering work: Don't automate, Obliterate", Harvard Business review.

Huczynski, A, 1987, "Encyclopaedia of Organisational Change Methods", Gower, Aldershot.

Johnson, G. & Scholes, K. 1988, "Exploring Corporate Strategy": Prentice Hall.

Juran, J.M, 1989, "Juran on Leadership for Quality", New York, The Free Press.

Kline, S.J, 1995, "Conceptual Foundations for Multidisciplinary Thinking", Stanford University Press, Stanford.

Land, F, 1995, "Conference Preface", In proceedings, 3rd Annual European Conference on Information Systems, Greece.

Malinowski, 1936, "Anthropology" : in The Encyclopaedia Britannica : (first supplementary vol)

SSADM, 1990, CCTA. SSADM Version 4. Reference Manuals, Vol 1,2,3,4: NCC, Blackwell.

Taylor, F.W, 1910, "Shop Management", Harper.

Tsagdis, D, 1997, "Business Process Re-engineering and Sustainable Superior Organisational Performance: The Contribution of Research", Proceedings 14th International conference of WACRA. Sustainable Development.

Wang, R.Y., Reddy, P. R and Kon, H.B. 1995,"Toward Quality Data : An attribute-based approach", Journal of Decision Support Systems, Vol.13, Elsevier.

Waters, S. J, 1986, "Three C's of Successful IT Projects", Annual International Banking Conference, Institute of Banking, Kuwait.

Waters, S.J, 1987, "An approach to planning an IT strategy", in Annual Top Executive Forum, NCC, Exeter

Waters, S.J, 1988, "Managing information technology by objectives", International CIS Journal, command and Control, Communications and Informations Systems vol.2 no.1

Waters, S.J, 1989, "SWOT analysis in IT projects", International CIS Journal, Command and Control,Communications and Informations Systems vol.3   no.1

Waters, S.J & Bakehouse, G & Davis, C & Doyle, K, 1994, "Integrated clinical workstation: user requirements for a neurosciences directorate.": NHS Executive, IMG

AN EMPIRICAL STUDY OF EIS SATISFACTION - SOME PRELIMINARY FINDINGS[1]

Xianzhong M. Xu

Luton Business School
University of Luton
Park Square
Luton, UK, LU1 3JU

INTRODUCTION

The application of computer-based information systems within business organisations has followed an upwards movements in the corporate managerial level (Edwards and Peppard, 1993). This is driven by a): the increasing needs to cope with the dynamic and uncertain business environments at corporate level; b): the rapid development of more sophisticated and affordable computing technology. The upwards movement can be seen by the emerging and the application of Executive Information Systems - EISs (Rockart and De Long, 1988; Watson, Rainer, and Koh, 1991; Holtham, 1992; Gray, 1994; Watson, *et al.* 1996). Although the definition of EIS is rather confusing in practice[2], most researchers like Nord and Nord (1995); Rainer and Watson (1995); Elam and Leidner (1995) tend to refer EIS to "a computer-based information system designed to provide senior managers access to internal and external information that is relevant to his or her management activities and decision making". The function of EIS has evolved from tracking Critical Success Factors and Key Performance Indicators (Rockart and Treacy, 1982), to electronic "briefing books", executive decision support, personalised news alert, and e-mail, e-diary. Different types of EISs have been seen in practice, e.g. conglomerate EISs, control and monitoring EISs, competitive and intelligence EISs, communication EISs (Edwards and Peppard, 1993).

Rockart and De Long (1988) suggest that the primary purpose of implementing EISs is to improve strategic information available to the top management team. However, empirical studies (Nord and Nord, 1995; Watson, *et al.* 1995, Walstrom and Walson, 1997a) reveal

---

[1] The findings reported in this paper is a part of the Ph.D. research project undertaken by the author at the Open University Business School. The author would like to thank Professor G. R. Kaye for his supervision of this project.

[2] Edwards and Peppard (1993) observed that some organisations give their middle management team EIS, and some vendors use EIS to repackage their existing MIS, and Marketing Information Systems.

that most EISs focus on the feature of "interface design", the function of "easy data access" and "information communication", but not sufficiently on strategic information provision. It seems that the direction of EISs is driven by professional EIS developers. It is not yet clear whether executives agree with the design of the EISs, and satisfied with the systems. There is a large body of literature documenting the key factors to EISs success (Rainer and Watson, 1995a, Overton, Frolick and Wilkes, 1996; Bajwa, Rai, and Brennan, 1998), however, EISs have been reported failures in many cases (Watson and Glover, 1989; Rainer and Watson, 1995b; Walstrom and Walson, 1997b). The EIS application has barely reached beyond middle management, and has not penetrated the top management team to any significant extent (Wheeler, Chang and Thomas, 1993, Frolick, 1994). There is a need to investigate whether an emphasis on EISs "user interface design" and "improved data access" will lead to an acceptable EIS, and also to what extent executives are satisfied with their EISs. Ascertaining the reasons underpinning the satisfaction or dissatisfaction could influence the future design of EISs.

METHODOLOGY

This study uses questionnaire survey to examine Executives' EIS satisfaction. The survey[3] was conducted in UK industries - computers, electronics, food, chemicals, and transport. 1518 questionnaires were sent to named senior managers selected from the FAME database. All the selected companies have over 100 employees, this is to ensure that the companies participated in this study is not too small to engage in strategic activities. The sample hence excludes small entrepreneurial firms. The questionnaire was exposed internally and externally to a pilot survey to ensure content validity of all items used and to gain constructive criticism. Posting the questionnaire was conducted in two stages which resulted in a total of 242 returned questionnaires. The valid responses were 155, representing a response rate of 10.2%. It is worthy of noting that the concern over this relative low response rate should take the following particular factors into account: a) this is a large size, detailed questionnaire on not only EIS satisfaction, but also strategic information acquisition; b) the subject of the survey requires top level managers' knowledge and commitment; c) the sensitive subject; d) the uncontrollable factors across multi-industry sectors, e.g. company policy of not participating in any surveys. However, the respondents profile is very encouraging, as 78.1 percent of the respondents are Managing Directors, CEO, Directors, and company Chairman. Table 1 shows the details:

Table 1. Respondents Profile

Respondent's position	Number of Respondent	Percent %
Chairman	7	4.5
Managing Director / CEO	46	29.7
Director*	68	43.9
Company secretary	3	1.9
Division / Branch Manager	23	14.8
Other♦	8	5.1
Total	**155**	**100**

* Directors include financial directors, technical directors, commercial directors, etc.

♦ Other respondents include strategic planning manager, Group controller, Logistical manager, etc.

---

[3] This survey is conducted for a project examining senior managers' behaviour in strategic information acquisition. The EIS satisfaction is a part of the questionnaire.

It can be assumed that the data generated from this study reflects UK executives' view on the current practice of EISs. In order to examine the possible association between EIS satisfaction and company's performance, all the sample companies are classified into leaders or laggards by comparing their companies' performance (ROCE) against their industry average. A follow-up case study was carried out to explore the reasons underpinning executives' dissatisfaction with the EIS.

RESULTS AND DISCUSSION

The Survey

By asking executives whether they are satisfied with their EISs for providing strategic information, the survey reveals that overall, 41.7 per cent of the respondents expressed their dissatisfaction with the EISs, while other 41.7 per cent of the respondents indicate that they are satisfied with their EISs. Further investigation shows that there is no significant difference between leaders and laggards in terms of EIS satisfaction. Table 2 shows the result.

Table 2. Executives' Perceptions of EIS Satisfaction

	Leaders % (n = 63)	Laggards % (n = 76)	Total % (n = 139)
Satisfied	44.4	39.5	41.7
Not satisfied	42.9	40.8	41.7
No EIS & Under development	12.7	19.7	16.5

The reasons why executives are dissatisfied with the EISs are sought and summarised as follows:

- A narrow focus on internal data, this is described as "no external, future prospects available to enable better strategic / long term planning", "not yet able to handle / analyse / present external threats well", "not readily available for external sources".
- Poor information quality which is described as "limited scope of data subjects", "too often out of date and inaccurate", and "data is not selected on a consistent basis".
- Lack of effective data filtering and refining facilities to diminish overwhelmed operational data. Executives expressed that "the system produces too much data, too little information", there are "too much unrelated information", the information delivered from the EIS is "too detail - not summarised in a friendly manner" and is "too generalised, not interpreted".
- Poor interface and limited flexibility which is evident by those comments: "systems are not user friendly, it does not allow executive (i.e. non technical) use to be performed easily", "poor presentation and format". "The information on the database is often inflexible and can not anticipate other ad hoc information shared between individuals", "limited selectivity - it is either lots of detail or nothing".
- No experience / expectation in using EIS as a strategic information source. One executive claims that "our EIS is not designed to produce strategic information, nor do we want it to", "EIS concentrates on operational data, not strategic information".

It can be argued that executives' dissatisfaction with the EISs mainly caused by the lack of external, strategic, meaning information. Mintzberg (1980) reports that senior managers demonstrate a thirst for external-oriented information to make strategic decisions, but they have limited capability in obtaining all the needed strategic information (Kiesler and Sproull, 1982; Martinsons, 1994). The dissatisfaction thus can be explained by the fact that, firstly, current EISs often contain large quantity of data from operational systems and/or the MIS, which is primarily internal-historical data in nature. Because there is no systematic external information being scanned in, the EISs produce much internal and static information which is not relevant to executives' needs. This is what is called a "static picture-show approach to EIS" (Frolick, 1994). Secondly, emphasising on "data access" and "interface design" may not effectively improve EIS satisfaction. This is because improving executives' access to mass data without appropriate filtering and refining can create a problem of "data deluge", which implies that identifying critical strategic changes becomes a daunting task, i.e. the daily flow of EIS-generated data can be too much, making it hard for executives to spot trends, patterns, and exceptions in detailed data. Finding a problem from mass data becomes the real problem. "Data deluge" and "information meaningless" runs the risk of compromising the advances of colourful, graphic design of an EIS. Without data refining, increased data access through EISs to executives could make data overload problem even worse.

The Case Alpha

Alpha is a medium sized company that sells intellectual products such as books, computer software to the UK and international market. Through observation and structured interview, it found that the primary purpose for the company's management to use computers is for "sales control and management", e.g. examining exceptions, conducting comparisons. "communication with others" is ranked second. This is followed in turn by "support decision making", "explore business opportunity" and "knowledge enhancement". The company's management attributes their satisfaction with the systems to the effective information support provided by the information specialists, which reveals an important consideration of EIS satisfaction. It was observed that the company uses a commercial assistant to retrieve, prepare weekly, and monthly synthesised information for senior managers. The commercial assistant is the immediate user of the computer-based systems. There is also an IT facilitator, but the role is in technical support. The senior managers are information users. The "Information Support" role at Alpha is depicted in Figure 1.

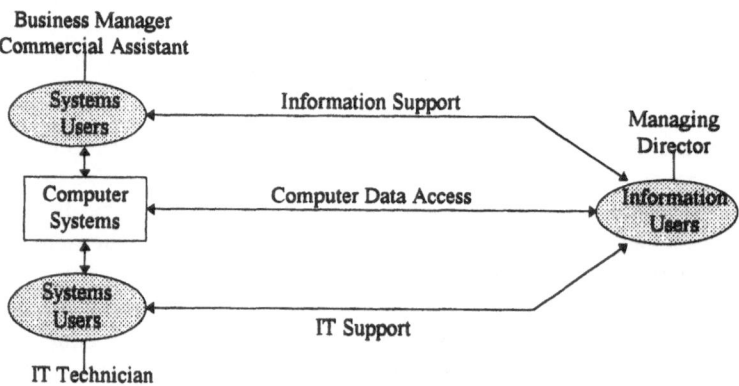

Figure 1. The Role of Information Support

It can be argued that senior managers prefer useful information to sophisticated computer systems. Computer-based information systems are formal and over-rigid in the sense of processing and delivering information, which is difficult to fit managers' intuitive and informal process of strategic information acquisition (Cottrell and Rapley, 1991; Wilson and McDonald, 1994). Even with IT support, executives may not be satisfied with their EISs due to the fact that some senior executives may have no interest, no time or capability to use computers (Schmitz, Armstrong and Little, 1992), but they certainly desire for meaningful information. IT professionals are helpful in operating the EISs, but they are incapable in scanning, analysing, and presenting information in an effective way to executives. This, however, can be done by an information specialist who acts as a buffer between the computer systems and the senior managers. The information specialist can help executives gather information from the environments, and keep them instantly informed about the most significant changes. It can be argued that "Information Support" could be more effective to improve executives' satisfaction than "IT support". This is akin to Frolick's (1994) argument that executives need information specialists to support them using the EISs, and EISs are increasingly being designed to be used by most, if not all, knowledge workers.

In summary, this limited study suggest that current EISs focus on the features of easy of use, data access, but failed to address information scanning and information provision. Thus, the EISs failed to meet the primary purpose of implementing EISs, that is to improve strategic information available to the top management team. It reveals that some EIS tools are inward looking in that they are more concerned with displaying information from internal databases and less concerned with what strategic information ought to be produced, and how. In addition, lack of flexibility of EISs also contributes to dissatisfaction, due to the fact that executives' information requirements could not be known beforehand and so could not have been specified during the EIS development. The EIS data structure could be too rigid to match executives' changing information requirements.

CONCLUSION

Current EISs failed to systematically scan, process and provide meaningful information to senior managers engaged in strategic decision making, which has been seen the main cause leading to EIS dissatisfaction. The design of current EISs concentrates a great deal on technical issues, i.e. ease of use, graphics capability, data access and drill-down features, this may not be attractive enough to get executives' hands on the EISs. Rather, through EISs "to improve strategic information available to the top management team" appears essential to increase executives' satisfaction. The EISs hence should be able to scan both external and internal information of relevance to executives (Heley and Watson, 1996; Vandenbosch and Huff, 1997; Frolick, et al. 1997), in particular, should be able to cater for soft, non-quantitative and highly aggregated information (Watson, et al. 1996). This can be implemented by embedding the EISs with information scanning facilities to scan both factual and soft information from the external environment, and tools for filtering and refining the received information. In addition, the "Information Support" role needs to be created as opposed to traditional IT support, to enhance senior executives', as well as an organisation's information scanning and processing capability.

REFERENCES

Bajwa, D., Rai, A. and Brennan, I. (1988) Key antecedents of Executive Information System success: A path analytic approach. *Decision Support Systems (DSS)* **22** (1), P 31-34

Cottrell, N. and Rapley, K. (1991) Factors critical to the success of executive information systems in British Airways *European Journal of Information systems* 1(1), 65-71

Edwards, C. and Peppard, J. (1993) A taxonomy of executive information systems: Let the 4 Cs penetrate the fog. *Information Management & Computer Security* **1** (2) P 4-10.

Elam, J. and Leidner, D. (1995) EIS adoption, use, and impact: the executive perspective *Decision Support Systems* **14**, P 89-103

Frolick, M., Parzinger, M., Rainer, R. Jr, Ramarapu, N. (1997) Using EISs for environmental scanning *Information Systems Management* **14** (1), P 35-40

Frolick, M. (1994) Management support systems and their evolution from executive information systems *Information Strategy: The Executive's Journal* Spring, P 31-38

Gray, P. (ed.) (1994) *Decision Support and Executive Information Systems* Englewood Cliffs, NJ, Prentice-Hall

Holtham, C. (ed.) (1992) *Executive Information System And Decision Support* Chapman & Hall, London

Kiesler, S. and Sproull, L. (1982) Managerial response to changing environments: Perspectives on problem sensing from social cognition *Administrative Science Quarterly*, **27**, P 548-570

Martinsons, M. (1994) A strategic vision for managing business intelligence *Information Strategy: The Executive's Journal* Spring, P 17-30

Mintzberg, H. (1980 edition) *The Nature of Managerial Work*. Prentice-Hall, Inc., Englewood Cliffs, N.J.

Nord, J. and Nord, G. (1995) Executive information systems: A study and comparative analysis *Information & Management* **29**, P 95-106

Overton, K., Frolick, M., and Wilkes, R. (1996) Politics of implementing EISs *Information Systems Management* 13 (3), P50-57

Rainer, R. and Watson, H. (1995a) What does it take for successful information systems? *Decision Support Systems* **14**, P 147-156

Rainer, R. and Watson, H. (1995b) The keys to Executive Information System success. *Journal of Management Information Systems*, Fall, **12** (2), P 83-98.

Rockart, J & Treacy, M. (1982) The CEO goes on-line. *Harvard Business Review*, **60**(1), P 84-88.

Rockart, J. and De Long, D. (1988*) Executive Support Systems: The emergence of top management computer use*. Dow Jones-Irwin, Homewood, Illinois.

Schmitz, J. Armstrong, G. and Little, J. (1992) CoverStory - automated news finding in marketing In Holtham, C. *Executive Information Systems and Decision Support* Chapman & Hall, London, P 227-238.

Vandenbosch, B. and Huff, S. (1997) Searching and scanning: How executives obtain information from executive information systems *MIS Quarterly* **21** (1), P 81-107

Walstrom, K. and Wilson, R. (1997a) An examination of executive information systems (EIS) users *Information & Management* **32** (2), P 75-83

Walstrom, K. and Wilson, R. (1997b) Gaining user acceptance of an EIS *Information Systems Management* **14** (1), P 54-59

Watson, H. O Hara, M., Harp, C., Kelly, G. (1996) Including soft information in EISs *Information Systems Management* **13** (3), P 66-77

Watson, H. and Glover, O. (1989) Common and avoidable causes of EIS failure *Computerworld*, December 4, P 90-91.

Watson, H., Watson, R., Singh, S. and Holmes, D. (1995) Development practices for executive information systems: findings of a field study *Decision Support Systems* **14**, P 171-184

Watson, L. Rainer, R. and Koh, C. (1991) Executive information systems: a framework for development and a survey of current practices. *MIS Quarterly*, 15(1), P 13-30

Wheeter, F., Chang, S. and Thomas, R. (1993) Moving from an Executive Information System to Everyone's Information System: lessons from a case study *Journal of Information Technology* **8**, P 177-183

Wilson, H. and McDonald (1994) Critical problems in marketing planning: the potential of decision support systems *Journal of Strategic Marketing* **2**, 249-269

SELF-SPACE AND VISION

Louis Jacques Filion

HEC, the University of Montreal Business School
3000 Côte-Sainte-Catherine
Montreal (Québec) H3T 2A7 Canada

INTRODUCTION

The first part of this paper examines self-concept and learning. It demonstrates that self-concept results as much from the sociological components of social surroundings as from the immediate context in which the individual acts. Self-concept enables the individual to identify areas of interest, and then to define intentions, images and visions. This determines the learning needs and then the learning choices that will gradually shape the individual's self-awareness and know-how and eventually prepare the individual for professional activity.

The second part of the paper deals with the questions of self-space, self-concept, learning and vision. It demonstrates that the individual moves from a received self-space to a constructed self-space, allowing the self-concept to express itself through the development and implementation of visions whose scope will depend on the extent of the constructed self-space. The learning needed to perform the activities required for the implementation of the vision is then determined. The relations system plays a key role here.

Several elements influence the constructed self-space, including the six examined and explained here: (1) the possibilities for survival in a given environment; (2) the ways in which power is exercised; (3) the historical approach to freedom and individual rights within the society concerned; (4) the social maturity and educational level of the surroundings; (5) the social conventions governing respect between individuals; and (6) the consensus established on the standards governing social and interpersonal relationships.

BACKGROUND

There is a growing trend to associate the concepts of organizational actor and learning. The soft systems methodology developed by Checkland (1981) gave this approach its initial impetus, since it was among the first to suggest an articulated approach to the management

Synergy Matters: Working with Systems in the 21st Century,
Edited by Castell *et al.*, Kluwer Academic / Plenum Publishers, New York, 1999.

of self-learning. In the field of entrepreneurship, the pioneering work was done by Collins and Moore (1970) in their empirical field study of entrepreneurs. The University of Lancaster, in Great Britain, established a department of organizational learning in the 1970s. Senge (1990), like other researchers before and since, proposed models to improve the articulation of learning in general, and organizational learning in particular. Our interest in the phenomenon has arisen mainly through empirical studies of various categories of entrepreneurial actors (Filion, 1991a and b; 1996). This research led to the identification of the concept of vision as a structural basis for the design and organization of activity systems. It also allows the learning needed to perform the projected activities to be specified, but is unable to explain why certain actors are more successful than others. Various other notions, including self-concept and self-space, can help to explain the evolution of a visionary system and the learning that makes it possible.

SELF-CONCEPT

In the following sections, we will examine what this learning involves. Without analyzing learning as such, we will discuss certain elements that trigger and support the learning mechanism. Figure 1 below sets out some of the elements involved in the process.

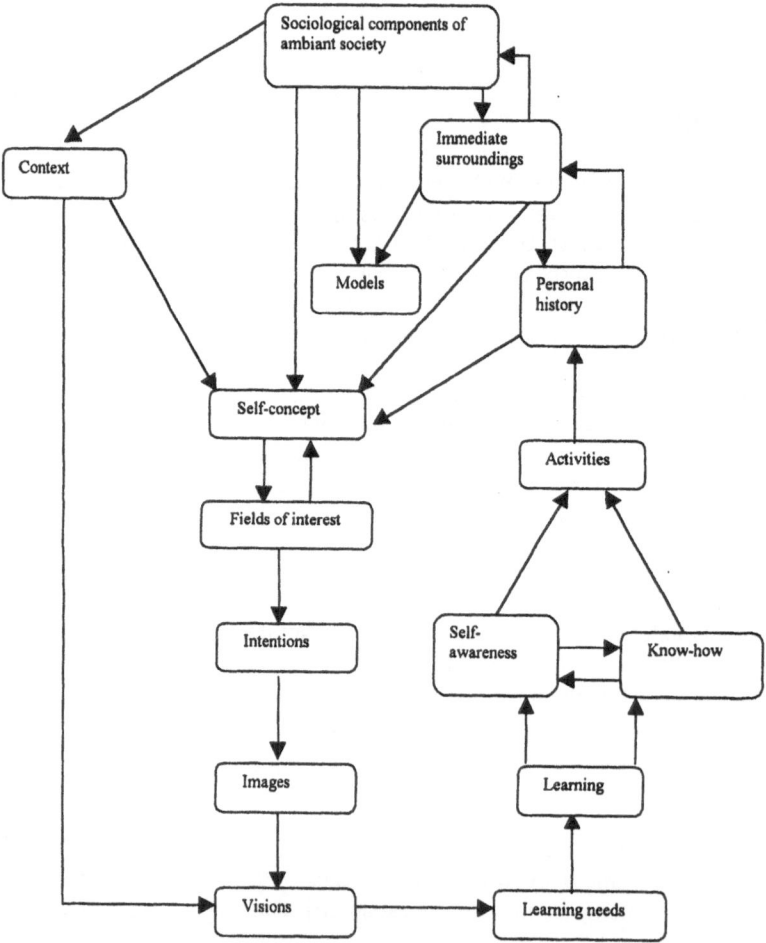

Figure 1. Self-concept and learning

A central, even capital element at the heart of the process is known as "self-concept" (Burns, 1979; Coopersmith, 1967; Cross and Madson, 1997; Hamachek, 1995; L'Écuyer, 1978; Martinek and Zaichkowsky, 1977; Marsella, Devos and Hsu, 1985; Shavelson and Bolus, 1982; Taylor, 1998). It refers to the way in which individuals perceive themselves, and their self-esteem (Maslow, 1970), and to the understanding individuals have of their own abilities that constitute the underlying foundation on which knowledge and the visionary process are based. The self-concept is conditioned by certain other elements: the economic and social context in which the actor operates, the immediate surroundings and the models encountered by the actor, and the actor's personal history, including his or her education and life experience. All these elements are influenced by the sociological components of the social surroundings, which contain values, standards, working methods and ways of directing energy that influence self-awareness, behaviour and actions. Certain societies, surroundings and families provide models that focus on certain activities, tending to produce self-concepts that also focus on those activities. Obviously, the nature of a given individual, and his or her characteristics, physical abilities and mental capacity will also condition the perception of what is possible or accessible.

The way in which the self-concept is organized seems to be strongly linked to the interests that guide the individual's intentions, engendering a selection process for the perception and retention of the images that initiate the visionary process. It is only when the individual's intentions have crystallized into visions, in other words genuine projects for the future, that learning needs can be defined. This is the trigger that spurs the future actor to learn. However, it is the self-concept that conditions the scope of the developing vision and of the learning process that the individual will be willing to undertake. The skills and proficiencies developed will enable the individual to achieve the levels of self-awareness, know-how, managerial capacity and self-direction needed to perform the activities relating to the selected organizational role. Performance of the activities will, in turn, influence the individual process adopted by each individual actor, and consequently the overall social process as well. In short, it can be advanced that the self-concept constitutes a central concept from which the visionary process emerges, and onto which the learning process is grafted. Several elements provide support for the construction of a self-concept, including self-space, which is discussed in the next section.

3. SELF-SPACE

In every society, self-space develops under the influence of the society's history, social class structure, sociological characteristics, level of development and population density. The number of self-space alternatives will be directly proportional to the social, ethnic and religious diversity of the society itself, and to its educational levels.

Self-space can perhaps best be defined as the place where the self is found, the psychological space occupied by each individual; it corresponds to the space in which the self-concept develops and operates. The self-space is the spatial, systemic configuration surrounding the self-concept. Its boundaries determine the space available to the self-concept for development and deployment.

Very few researchers have focused on the concept of self-space (Filion, 1993; Latane and Liu, 1996; Hall, 1959; 1966; 1976). To understand the concept of self-space, it is necessary to refer to various concepts including freedom and the extension of freedom. Freedom requires the existence of an area where mutual respect will prevent the anarchy that leads to dictatorship and a curtailment of freedom. Conventions, laws and rules are established, and must be complied with to guarantee the continued exercise of freedom. For example, motorists are free to drive their cars where they like as long as they follow the rules and, for example, obey traffic signals. Traffic lights impose a restriction on the freedom to drive a car, but allow that freedom to be extended to a greater number of people. Custom and convention also have a role to play, for instance when it is stated that references to the masculine gender include the feminine. Many women would probably agree that this convention does not provide them with enough self-space to support the full development of the feminine self-concept with its specific attributes.

Self-space implies that a psychological distance both separates us from, and joins us to, other people. It reflects the space that usage and custom have established as being reserved for each individual; it results from the social compromises and the set of formal and informal conventions that govern relations between human beings in a given society. For instance, two grammatical forms, "vous" and "tu", are used in French as in many other languages to express the distance existing between individuals and, by extension, to define the self-space we keep for ourselves and the self-space we assign to others in a relationship. The boundary of our self-space corresponds to the outer edge of our aura.

The notion of psychological space is an element that has received very little attention in studies of management, but one that we have observed attentively in our empirical studies of organizational actors. Figure 2 provides an outline, in model form, of the process by which an individual psychological space, or self-space, is established.

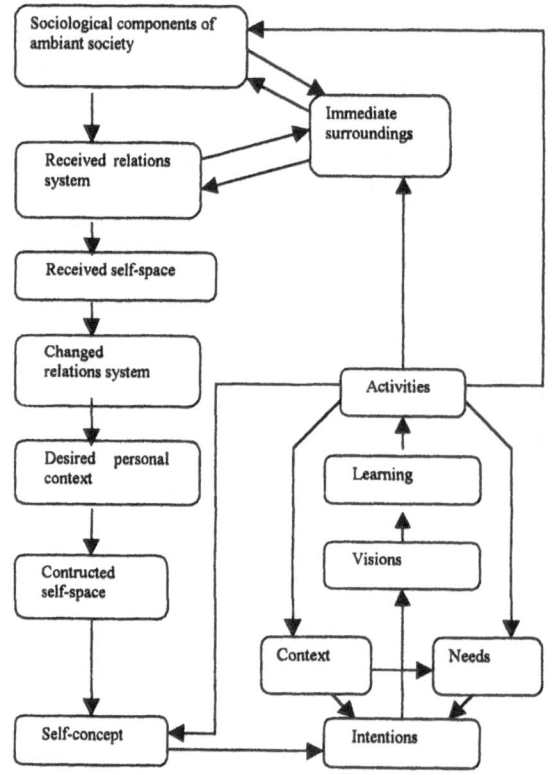

Figure 2. Self-space, self-concept, vision and learning.

It can be seen that each individual develops within a received self-space, based on the characteristics of the sociological components of the social surroundings. However, it is especially influenced by the relations system and by the immediate surroundings in which the person lives: culture, ethnic group, religion. Next, the individual psychological space is gradually transformed to reflect the individual's personal desires. Many people leave behind their families, move to a new region, or even emigrate in order to establish a self-space that will allow them to develop on their own terms. This is extremely important in explaining how individuals go about conquering a space in which the self-concept can develop. For example, future leaders need a minimum space in which to grow, and if the space is not available in the immediate surroundings, they will seek it elsewhere. In most societies, people from other cultures seem to be granted more space than individuals from within the society concerned. The eventual scope of the self-concept will be determined by the self-space constructed in this way; the individual's intentions, dependent on need and context, will condition possible visions and the learning required to take action. The activities performed will, in turn, influence the process by which the self-space and self-concept are constructed and their future development.

A reverse correlation seems to exist between the geographical space available and the self-space established in a given society. Many long-standing cultures, mostly in Asia and Europe, seem to provide for a better defined, broader self-space than in other cultures. The higher the population density, the greater the degree of respect shown for the psychological space of each individual which is, in fact, governed by a large number of standards and rules. Self-space also seems to vary according to social class. Clear rules impose courtesy and respect for others, for the elderly, and for those in a certain social rank or position. Signs of deference, and even reverence, can be detected, and the degree of respect found in interpersonal relations is striking. This is clearly observable in organizations, if only on the basis of the physical attitudes between various individuals, and is especially true in the case of social superiors and the elderly. Table 1 shows some of the determinants for the extension of self-space within a given society.

Table 1. Determinants of self-space

- The possibilities for survival in a given environment
- The ways in which power is exercised
- The historical approach to freedom and individual rights within the society concerned
- The social maturity and educational level of the surroundings
- The social conventions governing respect between individuals
- The consensus established on the standards governing social and interpersonal relationships

CONCLUSION

Every human activity seems to involve a certain number of prerequisite conditions. Learning is closely linked to self-concept, which is expressed within the psychological space of each individual, in other words the self-space that is first received and then constructed. This is also the place where intentions can be expressed, where the image individuals have of themselves, for both the present and the future, is reflected. All these elements nourish the underlying basis of the actor's system, expressed in the form of visions of varying degrees of ambitiousness. The way in which the self-space is constructed will often be a key factor not only for the initial and on-going development of the self-concept, but also for self-fulfilment.

REFERENCES

Burns, R.B., 1979, "The Self-Concept: In Theory, Measurement, Development and Behaviour", New York, Longman.

Checkland, P., 1981, "Systems Thinking, Systems Practice", Chichester, Wiley.

Collins, O., Moore, D.G., 1970, "The Organization Makers, A Behavioral Study of Independent Entrepreneurs", New York, Appleton-Century-Crofts.

Cross, S.E., Madson, L., 1997, "Models of the self: Self-construals and gender", *Psychological Bulletin*, 122:1, p. 5-37.

Coopersmith, S., 1967, "The Antecedents of Self-Esteem", San Francisco, Freeman

Filion, L.J. , 1991a, Vision and relations: Elements for an entrepreneurial metamodel, *International Small Business Journal*, 9:2, p. 26-40.

Filion, L. J., 1991b, "Vision et relations: clefs du succès de l'entrepreneur", Montreal, Éditions de l'entrepreneur.

Filion, L.J., 1993, Entrepreneur, organisation et apprentissage: nécessité de s'aménager un espace de soi. Partie 1: l'entrepreneur et l'apprentissage, *Revue ORGANISATION*, 2:2, p.59-69.

Filion, L.J., 1996, Différences dans les systèmes de gestion des propriétaires-dirigeants, entrepreneurs et opérateurs de PME, *Canadian Journal of Administrative Sciences.*, 13:4, p. 306-320.

Hall,E.T., 1959, "The Silent Language", New York, Doubleday.

Hall, E.T., 1966, "The Hidden Dimension", New York, Doubleday.

Hall, E.T., 1976, "Beyond Culture", New York, Doubleday.

Hamachek, D., 1995, Self-Concept and School Achievement: Interaction Dynamics and a Tool for Assessing the Self-concept Component, *Journal of Counseling and Development*, 73:4, p. 419.

Latane, B., Liu, J.H., 1996, The intersubjective geometry of social space, *Journal of Communication*, 46:4, p. 26-34.

L'Écuyer, R., 1978, "Le concept de soi", Paris, Presses Universitaires de France.

Marsella, A.J., Devos, G., Hsu, F.L.K. (Eds.), 1985, "Culture and Self: Asian and Western Perspective", New York and London, Tavistock.

Martinek, T. J., Zaichkowsky, L.D., 1977, "Manual for the Martinek-Zaichkowsky Self-Concept Scale for Children", Jacksonville, Illinois, Psychologists and Educators.

Maslow, A.H., 1970, "Motivation and Personality", New York, Harper & Row.

Senge, P., 1990, "The Fifth Discipline", New York, Doubleday.

Shavelson, R.J., Bolus, R.,, 1982, Self-concept: The interplay of theory and methods, *Journal of Educational Psychology*, 74:1, p. 3-17.

Taylor, C., 1998, "Les sources du moi: la formation de l'identité moderne", Montreal, Boréal.

ORGANIZATIONAL LEARNING AND IMPROVEMENT BASED ON THE ART OF BUSINESS COACHING.

M. Adrian Flores

The Monterrey Institute of Technology University System (ITESM)
Campus San Luis Potosi, Mexico

INTRODUCTION

Many tools have been presented to solve and improve problems in organisations, but they almost always miss talking about how to manage the human factor effectively. Organisational learning depends on individual learning. Individuals are the ones that can change organisational structures. By becoming a different observer, a second order observer, individual change and learning are more effective, and as a consequence, organisational change and learning can be achieved. This paper aims to present how organisational learning and process improvement can be accomplish through expanding our effective action and developing core competencies in business processes, considering the interaction and conversational networks between people in organisations.

Organisational learning is now an issue of discussion, seen as a dilemma and controversy that is now been implemented in many businesses. Our current reality shows that the world has been changing and will continue doing this for a longer while (Flores, 1998a). Learning becomes a strategic tool to capitalise effective actions in organisations. Now we can observe how the world is connected through a network of information, which may alter most country activities, some related to the economy, social behaviour, politics and business administration. I can observe two choices in today environments: do what you are doing now or transform your organisations through the learning process based on effective actions. In some countries, like Latin American ones, 80 % of the firms want to change their organisations, but 60 % of them declare to have no idea of how to do it (Gomez, 1998). An interpretation, about it, is that organisations feel the environment is changing them and this is affecting their processes, and they may not have a clear strategy to effectively cope with these changes. I also observe the continuous change is challenging the obsolete ways of doing business. The effectiveness of their actions is in doubt. Something that before was bringing success now can be the wrong step in this new changing environment and becomes a condition of failure.

The answer to these challenges is what we call **learning**. Learning is understood as the auto-transformation capability of an entity, which can be an individual or an organisation. Both, seeking the assurance of its viability and expanding its possibilities for success. An entity must be, at least, in condition of transforming accordingly to its environmental changes, that is, organisations need to develop the capacity to learn new ways to operate. They need to reconfigure themselves (Espejo, 1997).

Synergy Matters: Working with Systems in the 21st *Century,*
Edited by Castell *et al.*, Kluwer Academic / Plenum Publishers, New York, 1999.

619

INDIVIDUAL AND ORGANIZATIONAL LEARNING

Organisational learning requires individual learning, because organisational actions are produced by the actions of its members. But it is not enough to modify only individual actions to achieve the desired change in individual behaviour, and as a consequence, to produce individual learning (Flores, 1998b and 1998c). In order to do this a structural transformation of the system is required. A modification of the system's structure is an important condition to promote individual learning that will finally be translated into organisational learning.

According to Echeverria (1998a), individual learning depends on the kind of observer an individual is. The kind of observer that each individual is determines individual's behaviour. Ckeckland (1990) argues that we perceive the world through a filter of internal ideas to interpret the world. Becoming a different observer is the way to open the windows to individual learning.

At first level (superficial), the capability of individual learning is conditioned to a set of motivational elements. If an individual resist learning, it will be very difficult to achieve it. Strebel (1996) proposes the used of reciprocal obligations and mutual commitments as agreements called "Personal Compacts". Those are propose to manage employee's resistance to change. But, there is another factor that blocks individual learning. The kind of observer, an individual is, determines individual behaviour. Every action is a result of the kind of observer we are. The way we make sense of a certain situation, before we intervene on it, depends on the way we observe. The kind of observer we are, determines the way we act (Echeverria, 1996) and, in the same way, strains our learning possibilities. For instance, a cost reduction policy can work for a short term, but in the long may affect futures sales and produce unmotivated employees.

As mentioned before, organisational actions are represented by individual actions. An organisation acts through its members. However, its members not always observe how their actions can affect the organisation. That is why it is important to transform the kind of observer we are into a systemic observer.

In a certain manner, organisational actions are not a result of individual actions. Organisations don't act by themselves, they act through their members' actions, and they represent the way individuals express their being. Therefore, if we want to produce organisational learning, this has to be translated in terms of effective member's actions, in other words organisational learning requires to be translated into individual learning (Echeverria, 1998a).

If individuals don't modify the way they act, organisations wont modify their acting either. Argyris (1996) describes that organisational learning is when the individuals within an organisation experience a problematic situation and inquire into it on the organisation's behalf. The term 'problematic situation' used by Argyris, is not necessarily a condition for individual learning. Individual learning can also be produced when an organisation needs a change as a consequence of its vision.

As individuals, we learn when we detect and correct errors. Learning is manifested by our effective action in a situation rather than by our accumulated information about it.
Organisational learning is creating, acquiring, and transferring distinctions and practices in the organisation. Organisational learning implies behaviour modification, including changes in relationships, in order to create the conditions for creating, acquiring and transferring distinctions and practices. An organisation is learning when people in it succeed in overcoming defensive practices (Espejo, 1997 and Argyris, 1994b).

As an individual, we need to learn to learn in order to have organisations capable to increase their productivity through a continuos process dissolving attractors (e.g. mental models) that inhibit possibilities of action, and by generating attractors that open possibilities for effective action.

BUSINESS COACHING (ontological approach)

The American approach for business coaching is feeble compared with the one I will present here, proposed by Rafael Echeverria.

Through several years of work, Echeverria (1996) has developed, based on the work of Maturana (1996, 1997), Flores (1989) and others, an approach called Ontology of Language. It aims to interpret the meaning of being human in order to expand our possibilities of action (e.g. in executives, managers and people who want to help others). During his most recent program, "The Art of Business Coaching", Echeverria has shown that business coaching represents a tremendous tool for organisational and individual learning and transformation.

Business coaching is based on conversational competencies and observing human action. Regarding conversational competencies, the next issues are considered as important elements: language, emotionality, and conversations.

Language is divided for better understanding in listening and speaking. Listening refers to the way we hear what an individual says, and his intentions or expectations when he or she is speaking. That is, listening public or private conversations. A Public conversation is what an individual says during a conversation. A private conversation is what an individual thinks during a conversation but he or she doesn't say it. In the other hand, speaking is divided for better understanding in speaking modalities and basic linguistic acts. Speaking modalities refer to purposing and inquiring. Basic linguistic acts refer to assertions, statements, assessments and promises. What I have described is very important for an effective communication in organisational conversations.

Emotionality is divided for better understanding in moods and emotions. Moods are the expressions of our corporeality that preconditions our actions and as a consequence, our learning. Emotions are produced by something that alters the pace of your living (e.g. wining a game, losing a game, a car accident, etc.) and leave us in a certain state of being or mood. Emotionality is a key element for learning and teamwork. Marcial Losada (1998) has presented, in a recent research, that teams with a high proactive emotionality (called nexi level) are more productive and effective in organisations.

The power of conversations is amazing when we want to break up a problematic situation. Many times individuals don't spend enough time conversing and they lose the opportunity to improve relationships. Relationships between individuals can be more effective through conversational actions. A conversation to coordinate actions and conversation for assessments and explanations are samples that help us improve relationships in organisations. Conversations are an essential element in learning organisations.

Becoming an observer of human action promotes second order learning (see figure 1). Echeverria (1998b) refers to second order observer as a requisite for individual and organisational learning. Observing human action should be in a conscious way, that is, being aware of what is happening or experiencing individuals and organisations. Self-insight or self-observation is the first step to becoming an effective observer.

A second order observer expands the possibilities for action both in individuals and organisations. This can be accomplished by being aware of our actions and observing the results from those actions. In manufacturing companies this is presented when production management is focus on a pull approach instead of pushing production due to excess capacity (first order learning). We have learnt the consequences by producing without observing the whole system.

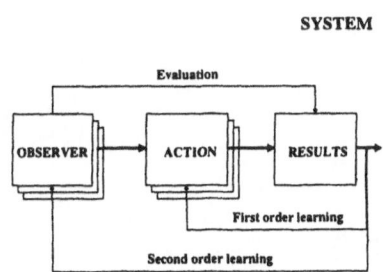

Figure 1. Observer of human action (Echeverria, 1998b)

BUSINESS COACHING AS A TOOL FOR ORGANISATIONAL LEARNING

Learning is not as simple as we though, there are factors beyond our capabilities. As an example, consider the situation when two individuals don't match their expectations (e.g. sales manager and production manager). They don't want to live this situation. The problem is that sometimes they are not capable of observing the consequences of their own behaviour. They know it, however, they don't know what to do. They don't know that is possible to learn and change what they recognise doesn't work. It doesn't mean that we don't have a solution. With these conditions "the ontological coaching" appears as a possibility when we confront the assessment that we have a problem without solution. Is not in discussion availability of social knowledge to solve a problematic situation and act effectively. The ontological coaching looks for a different observer, a second order observer for double-loop learning. The practice of ontological coaching is directed to overcome the learning horizon boundaries of an individual. Individuals have learning boundaries, and they are normally represented by the kind of observer we are. What ontological coaching seeks is to expand the individual boundaries, identifying and dissolving those barriers that strains his observing limits as presented in figure 2. Organisational learning requires an effective structure that promotes individual and collective learning.

New organisational challenges require people with new skills for positions as chief executive learning. Business coaching arises as a new role in organisations to identify and dissolve barriers for a better effective action. Michael Hammer (1997) argues that the new executive role requires coaching skills to facilitate and guide development on individuals and teams. Hammer adds that in future organisations we will have three kind of people. We will have a big amount of people adding value to processes, a group of coaches and a few executive-coaches (leaders).

To becoming a business coach, the following characteristics should be acquired:
- Second order observer (double-learning loop)
- An individual focus on actions
- Systemic thinking
- To be a support to others, in order to help them to overcome defensive attitudes and be successful
- Different observer

ITESM CASE

At the ITESM we are working in a new project called Redesign of the teaching-learning process, which aims to put the student in the centre of learning process.

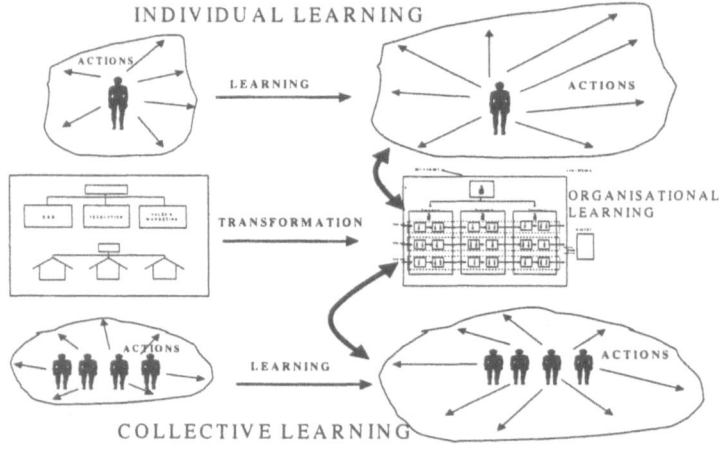

Figure 2. Individual, collective and organisational transformation

The model implies that the teacher should become a facilitator of that process. The teacher is now focused on the learning process and not only in the teaching process. It doesn't mean that the teacher's workload is now reduced. In addition the teacher has to focus on the students' learning process using high technology to support it (called learning space from Lotus Notes Co.). This project is an answer to Mexico's community needs stated in the ITESM's mission.

In others words, what we are experiencing is a new learning process. Mental models are our big barriers to change the new learning process. The ITESM strategy is to implement 100 % of courses using the new redesigned model for year 2002. The ITESM has 27 campuses throughout Mexico where the new model will be implemented. In campus San Luis Potosi, we have developed a structure and processes to support teacher and students change[1]. Figure 3 shows the structure to support the redesign project. This structure aims to facilitate the learning process for the two main individuals: students and teachers. The ITESM Campus San Luis Potosi has trained four ontological business coaches in order to support and lead this important educational project. Their main role is to promote and enhance individual, collective and organisational learning. One of the most important roles of an executive or manager is to convert individual actions into organisational actions.

Effective structures are a fundamental requirement to manage complexity. They allow organisations to create opportunities for themselves, as well as, to respond to disturbances and change (Espejo, 1997 pp.15). By structure we mean the set of arrangements by which the resources of an organisation, human and others, are connected through relationships.

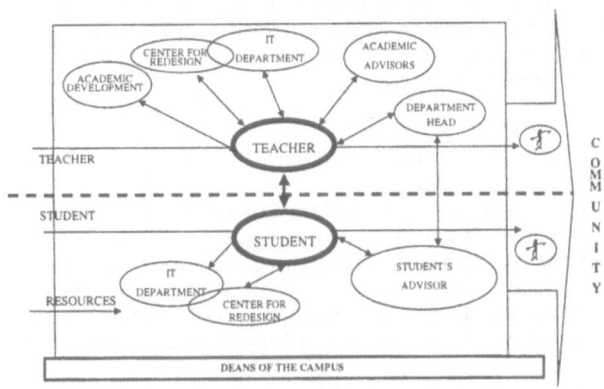

Figure 3. Structure for support new teaching-learning process

CONCLUSIONS

Learning lets individuals and organisations grow and change in order to be able to confront the challenges that this new era of fast paced societies brings about every moment. People today need to broaden their paradigms and give themselves the opportunity to observe from a different perspective the organisational phenomenon. One of the principles of the ontology of language state that we are as we act, and we act as we are. Actions generate being, if we want to have successful organisations we need to act up to the new challenges. Business coaching is becoming an important tool needed to manage and overcome human beings in organisations that want to learn through effective actions. Many approaches of individual and organisation learning are being introduced but none propose a serious way of closing the gap between fads and effective actions.

---

[1] The structure to support the new teaching-learning model was developed by a group of colleagues that includes mainly the job of the department heads and of the San Luis Potosi campus.

REFERENCES

1. Argyris C. and Schön D. A. (1996). Organizational Learning. Theory, Method, and Practice. Addison-Wesley. USA
2. Argyris Chris. (1994a). Good Communication that Blocks Learning. Harvard Business Review, July-August. USA
3. Argyris Chris. (1994b). Overcoming Organizational Defenses. Facilitating Organizational learning. Boston: Allyn and Bacon. USA
4. Brown David. (1996). The Essences of Fifth Discipline: Or Where Does Senge Stand to view the World? Systems Research vol. 13 No. 2. UK
5. Checkland P. and Scholes. (1990). Soft Systems Methodology in Action. John Wiley and Sons. England
6. Echeverria Rafael. (1996). Ontologia del Lenguaje (Ontology of Language). Dolmen Estudio. Chile
7. Echeverria Rafael. (1998a). The Business Coaching: a Tool for organizational Learning. Newflied Consulting. Venezuela
8. Echeverria Rafael. (1998b). El Caracter del Coach Ontologico. Newflied Consulting. Venezuela
9. Espejo R. , Schuhmann W., Schwaninger M. and Bilello U. (1997). Organizational Transformation and Learning. A Cybernetic Approach to Management. John Wiley & Sons. England.
10. Flores C. Fernando. (1989). Inventando la Empresa del Siglo XXI. HACHETTE. Chile
11. Flores M. Adrian. (1998a). An Integrated Model for Operations Management. Aided Production Engineering. Poland
12. Flores M. Adrian. (1998b). Information Systems in Complex Manufacturing. Elsevier Science. Switzerland
13. Flores M. Adrian. (1998c). Learning Organisations: The New Challenge of Business Process Reengineering. International Congress of Industrial Engineering and Sysytems. Cihuahua, Mexico
14. Goleman D. (1995). Emotional Intelligence. Bantam Books. USA
15. Gomez M. Magdalena. (1998). A Reengineering Methodology for Mexican Enterprises. MSc Thesis, ITESM San Luis Potosi. Mexico
16. Hammer Michael. (1997). Más allá de la Gestión Empresarial. Rowan Gibson, Preparando el Futuro. Ediciones Gestión 2000 S.A. Barcelona, Spain
17. Jackson Mike C. (1995). Beyond the Fads: Systems Thinking for Managers. Systems Research vol. 12 No.1. UK
18. Lane D. and Jackson Mike C. (1995). Only Connect! An Annotated Bibliography Reflecting the Breadth and Diversity of Systems Thinking. Systems Research vol. 12 No. 3. UK
19. Losada Marcial. (1998). Emotionality in high performance teams. The Art of Business Coaching. Second conference. Madrid, Spain.
20. Maturana Humberto. (1996). El sentido de lo Humano. DOLMEN, octava edicion. Chile
21. Maturana Humberto. (1997). Emociones y Lenguaje en educación y Política. DOLMEN, novena edicion. Chile
22. Rummler G. A. and Brache A. P. (1996). Improving Performance. Jossey-Bass. USA
23. Senge P. (1990). The Fifth Discipline. The Art and practice of the Learning Organizations. Doubleday. USA
24. Strebel Paul. (1996). Why do Employees Resist to Change? Harvard Business Review, May-June. USA

EDUCATING THE KNOWLEDGE WORKER

James G. Howell

Department of Computing and Information Systems
University of Paisley
Paisley
Scotland
UK.
e-mail: Howe-ci0@cs.Paisley.ac.uk

INTRODUCTION

It is predicted that a new type of worker, née the Knowledge Worker, is required to suit the new Information Age (Despres and Hiltrop, 1995). The combination of economic and technical advance (Bell, 1980, 87) has brought about increased information handling activities within business and the increased development of information rich products. Generally, the Knowledge Worker is associated with this changing ethos of information rich business environments and the networked organisation (Coulson-Thomas, 1991). This individual should be able to use communicating information technologies for remote groupworking: technologies that aid the formation and dissolution of working groups as "projects" demand. The Knowledge Worker utilises intellective skills in dealing with information; information that is mainly computer mediated. From a technical perspective it is oft inferred that given access to a repository of data, the knowledge worker can mine a relevant chunk of data and make sense of that data in a simple transformational way that creates knowledge. It is wrongly inferred that data. information and knowledge is a scalar quantity. The transformational process from data to knowledge is not easily understood. From an IS perspective, the basic concepts of "data", "information" and "knowledge" are confused - as well as the relation between them (Checkland and Holwell, 1997). Nor is it enough to focus on the individual alone, other classical human themes are relevant to the understanding of this knowledge worker, for example, working effectively in a team (Coulson-Thomas, 1991).

The phenomena of this new type of worker and the type of work they engage in is poorly understood. (Despres and Hiltrop, 1995). Research often focusses on understanding this individual by trying to compare their type of work according to extant but ageing classifications. Such research may prove to be potentially fruitless if we accept that this individual is emerging within new societal systems and we might not yet know their appposite work patterns and skill sets. In relation to extant disciplines. they are likely to be transdisciplinary in nature. The Knowledge Worker label may simply be a manifestation of another management fad or fancy but it is evident that some of the deep changes associated with the Knowledge worker may endure. It is of singular importance to learn about this phenomenon.

This paper aims to provide some discussion and analysis of conceptual and pragmatic issues related to understanding appropriate development programmes for this new breed of worker. There are some deep problems with any such analysis: knowledge does not have an ontological status. It may be simply a state of mind, but the fact is that the expression 'I know' commonly has a performative rather than a descriptive use (Ayers, 1974). This paper will adopt a view that knowledge can be categorised as *performative knowledge* and *representational knowledge*. This paper takes the position that the Knowledge Worker is expected to be able to create information relevant to a group of individuals, for a period.

Synergy Matters: Working with Systems in the 21st Century,
Edited by Castell *et al.*, Kluwer Academic / Plenum Publishers, New York, 1999.

625

THE CONTEXT – SOCIETY AND TECHNOLOGY

Within the UK, there has been a quantum change in our industrial base; from a predominantly manufacturing base towards a dominant service sector. There is a temporal association - the devlopment of the Knowledge Worker and the growth of the service sector. Kumar (1981) leaves us in little doubt that no assocjation can be inferred.

Paralleling the decline of the manufacturing sector has been the decrease of government initiatives to sustain the "skill systems" as evidenced by the large Training Boards of the 70s and 80s, Skillcentres and large F.E. skill courses (Howell and Gammack, 1997). There has been no large-scale initiative to create an Information Technology Training Board (ITIB), albeit some operational level training is provided through the ITECs. New Universities and colleges have filled some of the vacuum in IT skill development within courses of study and in creating information technology rich modules. These courses of study are now being increasingly financed by the students and their parents. Constraints act not just upon realistic design of programmes of vocational and educational development but also upon the successful delivery of such programmes. It has become all too easy for programme providers to maintain a focus on technological developments and the creation of modules of study and development to provide for the latest set of technological "skills" without consideration of the context within which skills can be developed.

As applications become more accessible and usable, and technologies become more transparent, many more people will become engaged in forms of work which do not require detailed knowledge of the technologies. Instead of requiring detailed knowledge, classic human and business themes are likely to dominate specific forms of activity. One model of this presented by Peter Senge (Senge et al, 1995) shows the ideas and fads of the day which shape mundane activity and technologies as transient and ephemeral compared to the domain of deep change where meaningful work and development is realised.

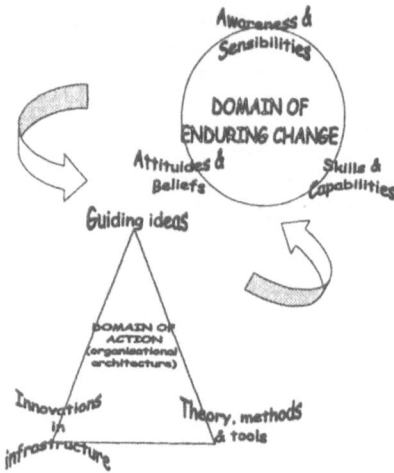

Figure 1. The Learning Organisation

As new skills and awareness increase, reflective practitioners can engage in prposeful activity without having to be concerned more than minimally with the technological media of the day. Whilst recognising the undeniable potential for technology to make a difference at this level of organisational activity, to understand more deeply the impact of the mechanistic variations, we require to provide a basis for the reflective practice of systems thinking. The merely mechanistic is then contextualised through Systems Thinking. Infusing our courses with this awareness is likely to be of immense future benefit to our young and mature students in whichever walk of life they find themselves. The vocational aspects of our teaching relate to the wider problem of changes in industry structure, working and technology. Much of the material in the teaching of computing science relates to the technology of the day and has, therefore, a short sell-by date stamped on it. The move towards flatter structures in organisations, in particular, creates a demand for effective teamworking and communications, and these aspects are increasingly designed into IS courses, both as incidental learning and as subject matter. Effective teamworking is a human theme of central importance to the Knowledge Worker.

THE KNOWLEDGE WORKER – THE INTELLECTIVE SKILLS ?

IT has permeated the workplace. Changes in the nature of work and the patterns of work have ensued. We have moved from more physical skills towards symbolic skills in manipulating "data" and traversing computer software interfaces. We have moved further away from a written culture to that of an electronic culture. Deep changes in the way we work as individuals and in groups is taking place. The obvious intellective skills that are being required include "abstraction, generalisation, induction, analogous reasoning and other forms of thought that allow one to grasp the essence of a situation" (Work, 1997).

More and more businesses are moving towards competency-based recruitment and appraisal. Inspection of the general competencies often reveals a mix of these intellective and behavioural "skills". For example,

Table 1: The Competence Framework.

Competencies	Behaviour/Intellective Skills.
Getting the job done	handles conflicting priorities without compromising effectiveness
Teamwork and working with others	builds and uses an effective network of contacts to the benefit of projects and colleagues
Communicating	conveys ideas and facts clearly, concisely and with persuasion
Thinking and Analysing	quickly grasps intricacies of new work, projects or instructions
Managing people	gains commitment from, and motivates staff to perform well understands and implements policies and Equal Opportunities objectives
Managing financial and other resources	manages change to achieve results with minimum disruption

Businesses are becoming more agile in adaptation to the new Information Age. They are changing by, amongst other tactics, recruiting the type of worker popularly described as the Knowledge Worker. The idea of a job, born with the Machine Age, is changing beyond all recognition (Angell, 1997). Growth will be created from the intellect of the Knowledge Worker (Drucker, 1992), an individual able to*"manipulate and orchestrate symbols and concepts, identify more strongly with their peers and professions than their organisations, have more rapid skill obsolescence and are more critical to the long-term success of the organisation"* (Despres and Hiltrop, 1995). In order to respond to this rapidity of change to individuals' skills sets, the Knowledge Worker will his/herself require to accept an increasing degree of responsibility for updating his/her skills, in particular IT skills.

THE KNOWLEDGE WORKER - A FORM OF NEW MANAGER ?

Some forms of behaviour have devolved from the traditional concept of the supervisory manager to the empowered worker, and hence are relevant to the role of the Knowledge Worker. The empowered knowledge worker is responsible for managing resources - time, tasks and money. Furthermore they are often responsible for developing and maintaining working relationships within an extended supplier chain. They are responsible for managing information. In short, they are a manager. It is instructive therefore to consider research undertaken in the modern manager sphere. A learning needs assessment process for identifying managerial learning needs may draw on experiential learning theory (Kolb, 1984) . Kolb provides one approach "for mapping the terrain of managerial competence based on what is called the 'competency circle'". The two dimensional map (shown below) shows the specialised adaptive competencies of managerial knowledge around the experiential learning cycle based on their association with the four basic modes of the experiential learning process,

> affective competencies (e.g. being sensitive to people's feelings)
> symbolic competencies (e.g. building conceptual models)
> perceptual competencies (e.g. gathering information)
> behavioural competencies (e.g. making decisions)

The experiential learning cycle can eventually be expected to lead to a formation of competencies which might be described as integrative. This does not simply imply a particular set of abilities that can be

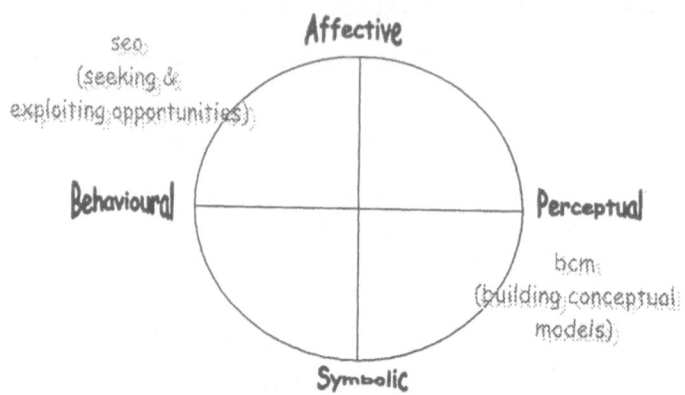

Figure 2. The Competency Circle

directly trained and added to the other four sets of competencies, but confers an ability to move fluidly among particular modes (or arcs of the competency circle) as required in a given situation.

Although the competency circle is attractively holistic as a tool, it is not easily adaptable to display integration of competencies. Profiles of competencies can be measured and plotted for different professions at different levels, resulting in spatial or 'star maps' of managerial competence. Integrative competence can be seen as the highest level of development, achievable through time and able to be shown on any candidate model. Programmes of development should consider methods of development for these integrative competencies, rather than expect these competencies to be an emergent property of any programme of development.

THE KNOWLEDGE WORKER, TECHNOLOGICAL CHANGE AND IT SKILLS

In the use of IT, technological change occurs within a timescale that often renders specific education within a technological skill domain obsolete within a few years. It then becomes unrealistic for anyone to keep conversant with and professional in all developments. If we do not wish the Knowledge Worker to embark on a labour of Sisyphus then we must be mindful of constructing appropriate programmes of development and updating. It is possible to identify and develop within a technological niche which allows for some persistence in IT skills, and some stability to be engendered within the particular work system for a period. If not, retraining every few years is one possibility, or having a high turnover of young IT-literate staff is another less expensive (in a cursory analysis) one. Identifying the core, transferable skills required is surely the basis for vocational education policy. Pedagogic concerns require that we should avoid teaching technical skills which may be out-of-date within the term of an undergraduate education in particular. It is a nonsense to focus on providing specific technical training on "systems" or software packages which enjoy only a short shelf life. In modelling the learning organisation, Senge(1995) explains the differences between the *triangle* of organisational architecture, and the *circle* of the deep learning cycle. The apex of the triangle locates the tools and methods currently in vogue within organisations (see figure1.), and changes made in this area can be short lived. The reflective practice of systems thinking encourages us to critically question the assumptions upon which these activities are based. We are left in no doubt that it is circle of deeper learning that informs and creates enduring change, and changes at this level are what matter. IT is properly located in the triangle of mundane operational tools and methods, and whilst recognising the undeniable potential to make a difference at this level of organisational activity, it is the reflective practice and adaptive employment of these skills that business requires and education must help develop.

We must concentrate on persistent and transferable skills, the knowledgeable and professional employment of these skills and in developing the understanding of the context within which they are employed, for example, the business context. Educationally, this is very much within the domain of information systems.

Within the technological domain, "IT reskilling has historically been viewed as a one-dimensional problem - that of giving people a very specific set of skills confined to products, methodologies and techniques........ reskilling should be thought of at three fundamentally different levels: applied skills, behavioural skills and cognitive skills." (Hayday, 1996).

Applied skills are the easiest to learn and are the IT industry norm. They are product- or technique-based, or pertain to a more general category of technology, such as Object Orientation. Behavioural skills, which relate to how people act, together with the impact and effect of these actions, are learned and can therefore be taught. Development activities can be used to bring about these changes, and supplemented by practice in an appropriate environment. "But cognitive skills will be the most important in shaping tomorrow's IT department..... people should be employed on desired cognitive skills and then taught the applied and behavioural skills they need" (Hayday, ibid). "It is no longer felt to be enough for the designer of a computer-based information system to be technically proficient; they must have knowledge of finance and economics to appreciate the business consequences of the information-system, the management sciences to satisfy the needs for management and control, the behavioural sciences to manage the intervention and achieve successful change. All this whilst still being competent enough in the computer sciences to properly understand the possibilities and limitations of the technology" (Lewis, 1994).

To develop this mix of skills, at Paisley we have located business IT courses within the field of information systems. Developing the students experiential knowledge base is achieved through the provision of a placement year, using appropriate business-oriented case study material and through other measures. Prime amongst these other measures is in locating students within business on an ad hoc basis in order to produce integrative work, for example a Strategic Information Systems (IS) plan for an SME or a consultancy report for the acquisition of a suitable IT-based business information system.

To provide a unifying approach, information systems is taught through each year of the Business IT courses. Although definitions vary, IS can be seen as a distinct subject area which embodies the philosophical context of systems science, with Von Bertalanffy's General Systems Theory providing many of the unifying and foundational concepts of the field. Along with Boulding's (1956) classic hierarchy of systems and Checkland's system types (1981) the relative places of the mechanistic and humanistic or social aspects can be seen. Systems thinking, (also elaborated in Peter Senge's "The Fifth Discipline") is, we believe, a good foundation for many courses. In so doing we are achieving a balance between offering an amount of continuosly update technical skills within a more stable educational context.

THE MULTI-DISCIPLINARY INDIVIDUAL AND IS

Historically the classification of new "types" of worker has been closely associated with the emergent raw material employed by skilful artisans. Inferentially, the Knowledge worker employs knowledge as a raw material. The source of this raw material is data, usually accessed through communicating technologies and IT. A scalar relationship between data, information and knowledge is logically implied. No such scalar relationship exists. Individuals are, using their technical and cognitive abilities, able to access data/information and interpret this in a way that is useful and of value. Value is associated with a chunk of information; a capital product that may be peddled in a free marketplace (Despres and Hiltrop, 1995), and value which can be translated into helping the organisation or individual to achieve some temporary form of competitive advantage. To achieve a more sustainable competitive advantage the new business or knowledge creating company will require realise that Knowledge is the new competitive resource (Nonaka and Tageuchi, 1995) and that the value of most products and servioces will depend primarily on how "knowledge-based intangibles can be developed" (Quinn, 1992). The knowledge worker does not, following historical patterns, use knowledge as a raw material. From a historical perspective, it is difficult to view the Knowledge Worker as the modern artisan.

The nascent Knowledge Worker will require to employ intellective skills to complement their technical skills. These skills must be developed distinctly and must not be a hoped-for consequence of increasing exposure to higher level technical skills. It is clear that businesses are placing more emphasis on the value of behavioural skills to complement intellective and technical skills. In order to develop these in the graduate or postgraduate students, at Paisley, we are trying to enhance practical links with businesses.

CONCLUSIONS

Human skill is emerging as the critical success factor and a limiting constraint (Coulson-Thomas and Brown, 1990) to the success of the modern business or "learning organisation". New types of work are now being performed by new types of workers, and both of these phenomena are poorly understood. Patterns of work will emerge that will help us better understand how to develop programmes suitable for the Knowledge Worker. We should endeavour to learn as much as possible about this emerging phenomenom.

Whether or not we view the Knowledge Worker as a faddish label loosely associated with information technology and a tele-worked organisation, deep change in the working environment is already taking place.

The Knowledge Worker is sometimes viewed as an individual able to mine data and transform this into knowledge but there is no scalar relationship between data, information and, in turn, knowledge. Knowledge does not hold an ontological status. From a historical perspective, the Knowledge Worker is not the modern artisan utilising knowledge as a resource.

The Knowledge Worker accepts greater responsibility for his/her own personal development and updating, managing their own time and tasks and, inductively, through the extended supplier-chain, the work of others. They are utilising some traditional management skills and responsibilities. In short they are a new form of manager. Like the more traditional manager, integrative and adaptive competencies are likely to be required and designing for these more holistic qualities within programmes of development will be challenging within our extant structures.

ACKNOWLEDGEMENTS

Acknowledgement is gratefully given to Dr. John Gammack for his contribution in developing earlier versions of this paper and Dr. R. B. Beeby for engaging in helpful and entertaining discussions which have progressed this work.

REFERENCES

Angell, I., 1997, "Welcome to The Brave New World" *in* "Information Systems: An Emerging Discipline", Mingers J. and Stowell F.A. (eds.), McGraw-Hill, London.

Ayers, A. J., 1974, The Problem of Knowledge, Penguin, Suffolk, UK.

Bell, D., 1980, "Sociological Journeys: Essays 1960-1980, Heinemann Education, New York.

Bell, D., 1987, "The Post-Industrial Society: A Conceptual Schema" *in* "Evolution of an Information Society", Cawkell A E (ed), AdLib, London.

Boulding, K.E., 1956, "General Systems Theory - the skeleton of Science", *Management Science* 2, 197-208.

Checkland, P.B., 1981, "Systems Thinking: Systems Practice", Wiley, Chichester.

Checkland, P.B. and Holwell, S., 1997, "Information, Systems and Information Systems: Making Sense of the Field", Wiley, Chichester.

Coulson-Thomas, C., 1991, "IT and New Forms of Organisation for Knowledge Workers: Opportunity and Implementation" *in* Employee Relations, Vol.13 no 4, MCB University Press.

Coulson-Thomas, C. and Brown, R., Beyond Quality, British Institute for Management, Corby, 1990.

Despres, C. and Hiltrop, J.M., 1995, "Human Resource management in the Information Age: current practice and perspectives on the Future". Employee Relations, Vol 17 No 1, pp.-23.

Drucker, P., 1992, "Post Capitalist Society", Butterworth, London.

Holtham, C., 1998, "Applied Knowledge Management: A Case Study in Knowledge management", ECIS

Hayday, G., 1996, "Shifting Hands" in Business and Technology, Sept.

Howell, J.G. and Gammack, J.G., 1997, "Systems, Crafts and Sustainability" in *Systems for Sustainability*, Proceedings of the 4th UKSS conference, Plenum, N.Y.

Kolb, D.A., 1984, "Experiential Learning: Experience as the Source of Learning and Development", Prentice-Hall, Englewood Cliffs, NJ.

Kumar, K., 1981, "Prophecy and Progress: The Sociology of Industrial and Post-Industrial Society", Pelican, Hammondsworth, UK.

Lewis, P., 1994, "Information Systems Development", Pitman, London.

Nonaka, I. and Takeuchi, H., 1995 "The Knowledge-Creating Company", Oxfrod University Press.

Quinn, J.B., 1992, "Intelligent Enterprise: A Knowledge and Service based Paradigm for Industry", The Free Press, New York.

Senge, P., Ross, R., Roberts, C., Smith, B. and Kleiner, A., 1995, "The Fifth Discipline Fieldbook", Nicholas Brierley Publishing Ltd., London, UK.

Work, B., 1997, "Is an Information Systems Education Relevant?", Presented Paper of the UKAIS Conference, UK. April.

MORE METAPHORS FOR ORGANISATIONAL TEAMS

Lawrence Reavill and Chris Brady

Department of Management Systems and Information
City University Business School
Northampton Square
London EC1V 0HB UK

INTRODUCTION

Earlier papers by the authors, both individually and jointly, have discussed the application of systems thinking and metaphor to the performance of teams in organisations. A paper by Reavill (1995) discussed the role of the individuals in high technology project teams. It concluded that the composition of the team might need to change during the course of the project to accommodate different specialisations as dictated by the phases of the project. Another paper in the UK Systems Society Conference of 1997 (Reavill 1997) looked more generally at how high performing teams could be developed within organisations, and in particular, what made the difference between a work-group and a team in the organisational context. The perception which appears to illuminate management writing on this subject is that there is considerable advantage in achieving "synergy" within the group, (the precise nature of this "synergy" will be discussed later), and performing "as a team". Handy (1990) for example extols the importance of teamwork, and assures us that "teams are here to stay". Although a group might be any collection of individuals brought together to perform a task, the perception is that a team is a highly coordinated entity in which all members perform with commitment to the group objective. The characteristic which Hilarie Owen calls "teamness" (Owen, 1996) could therefore generate improvement in working practices which in turn improve organisational performance.

While Lawrie Reavill concentrated on "management" teams, mostly in innovative technological environments, Chris Brady was researching the behaviour of teams of elite decision-makers in governmental situations. Brady and Catterall (1997) looked at management issues of cabinet governance, and Brady (1998) looked at the way the ultimate "Top Team" of the UK in 1956, Anthony Eden's "Egypt Committee", a group of government Ministers, dealt with the Suez Crisis.

The frequent use of sports based metaphors in current management literature ("maintaining a level field of play", "keeping an eye on the ball" etc.), and the authors' interest in various sports, prompted consideration of the extent to which the sports team could be used as a metaphor for the performance of decision-making teams in business and politics. A paper considering the possible contribution of the sports team metaphor was the first output of the new two-man team (Reavill and Brady, 1998).

Synergy Matters: Working with Systems in the 21st *Century,*
Edited by Castell *et al.*, Kluwer Academic / Plenum Publishers, New York, 1999.

SOME CONTRIBUTIONS FROM THE LITERATURE

Blake and Mouton (1964) first drew attention to the need for leaders of working groups to give attention to the task needs of the group and the personal needs of the group members. Adair (1979) added another dimension "team maintenance", the first indication of the need for a "teamness" element. Belbin (1981, 1993) examined the personal characteristics which were necessary in combination in the members of the team for a successful outcome. Owen (1996) examined the selection and training of the Royal Air Force "Red Arrows" aerobatic team, and introduced the concept of "teamness". Her work showed the high standards of team development and coordination which could be achieved under very specialised and tightly controlled conditions. She developed a model, the "Synergy Chain", which related the characteristics of the team members directly with those required by the team as an entity. This correlates well with the "Team Syntegrity" concept of Stafford Beer (Beer, 1994). This model shows that a group is converted to a team (or "infoset") by shared knowledge of the objective and shared information. An important pre-requisite of "Team Syntegity", which is evident from the title of Stafford Beer's book: *Beyond Dispute*, is that dispute can be eliminated by following the procedure he has devised, called the "syntegration game". Beer's model, "*collaborative* rather than competitive", and Owen's study based on the "Red Arrows", are limited in their application. Total commitment to the team objective is only obtained under very special conditions, though Owen's work shows the value of this element. The elimination of dispute within teams is difficult, but Beer shows the advantage of doing this to the greatest extent possible. The influence of individual objectives and individual competitiveness cannot be totally eliminated, and needs to be managed by the relevant leaders, be they directors (commercial) or Prime Ministers (political).

SPORTS TEAM METAPHORS

The common usage of sports metaphors in current management literature led the authors to consider the possible use of sports team metaphors as a means of illuminating the behaviour and characteristics of commercial and political teams. A team shows all the characteristics of a system, and a study of the operative mechanisms and personal interactions across a range of teams could generate managerial principles of a systemic and prescriptive nature which could be applied to business teams (Project Teams, Boards) or political teams (the Blair team, the Clinton team).

One author (Reavill) considered the simplest teams, such as the tennis doubles team and small rowing teams (coxless pairs; fours, eights, with or without cox), hoping to identify "teamness" in its most elementary form. This produced only a limited and obvious outcome, that performance was related to the combination of individual ability and the strength of the co-ordination between the players. A more complex team was needed. Some teams, for example squash teams and golf teams such as the Ryder Cup are no more than the sum of the parts. Others, such as teams of sailors in the Americas Cup have the same specialisation limitation noted in Owen's study, though it is rewarding to study the team building methods used by the New Zealand team who were spectacularly victorious in the most recent race. Rugby and cricket were considered, and the initial feeling was that they were too individualistic and compartmentalised to represent the business and political teams towards which the work was directed. Football (Association Football or Soccer) became the main focus of the work. Its team of 11 players plus 5 substitutes gives requisite variety, and the specialist background knowledge of the area retained by one of the authors (Brady) was helpful. However, the possibility that different types of sports team could be metaphors for different types of business or political team has not been discarded.

Support for the latter view comes from work by Keidel (1985), who examined the characteristics of the three major team sports in the USA: basketball; baseball; and American football. The latter is more akin to Rugby Football than to Soccer, though it does have eleven players like soccer. He argues that these team games can represent generic organisational forms common in business (and other sectors). Basketball is the most complex and fluid game, with high interconnection between team members. In basketball, the basic unit is the team. One of the authors (Reavill) is not entirely convinced of this, having observed a basketball coach during a "time-out" urging his players to concentrate on feeding the ball to a particular player who was "hot" at that time. Baseball concentrates highly on the individual, and thereby has some affinity to cricket. American football has functional sub-teams and rehearsed set-plays which give it a more structured style.

Soccer has elements of the characteristics of all three of the American major team sports. The contribution of the individual is important as in baseball. The fluidity of movement and team unity is critical as in basketball. The prior planning and set plays are part of the game plan as in American football. Another important aspect of soccer is its transitional ability. Players can be transformed immediately from attackers to defenders, and this reflects well the dynamism of current government and corporate behaviour. Thus soccer was selected as the sport for our initial investigations. The fact that one of the authors (Brady) had a background in football coaching might also have influenced the choice a little.

THE PERFORMING ARTS

Though some public perception of the "artist" may include the lonely painter starving in a garret while producing masterpieces, the truth can be far from this stereotape. The performing arts in particular are almost invariably team activities. A small team of individuals is needed to play jazz or string quartets. Teams of various sizes are needed to produce plays, from the pair required to perform the "two-hander" to the many required for the epic production with parts for leading actors and others for "spear-carrying" extras. Orchestras range from chamber to symphony. Dance performances range from experimental modern works for single performers or small groups to classic ballet ranging from an isolated *pas-de-deux* for two performers to major classic ballets with scores of dancers and a large orchestra. For opera, the size of the team varies from moderate to very large.

In the performing arts, an individual may have continuing prominence, such as the conductor of a symphony orchestra, a soloist in a concerto, or a ballerina in a classic ballet. The prominence may be temporary, such as a solo in an opera or an opportunity for improvisation in a jazz number. In an earlier publication, Reavill considered the example of the string quartet (Payne et. al., 1996). The most highly regarded string quartets are groups of excellent musicians who might not flourish at the international soloist level. Their success in this group activity may come from ability to adjust their individual styles to blend with the other musicians, the extra time to practise together, or a greater understanding of the needs and contribution of the other players. This may not avoid arguments on the interpretation of a particular piece, and the need for leadership and consensus within the group. The individualism required from the star soloist, either in an accompanied work or the combative relationship of the orchestral concerto may require characteristics for the musician which are not so compatible with a team performance, such as extroversion, a strong personality, and perhaps even a little arrogance. The instrumental concerto soloist needs to stand out from the accompanying orchestra, the members of the quartet need to blend together as one integrated unit.

The metaphor of the musician or the dancer parallels that of the tennis player or the oarsman. The single musician, dancer, tennis player or oarsman has complete freedom to perform as he or she thinks fit. As soon as two musicians play a duet, or two dancers perform a pas-de-deux, the quality of the outcome is dependant as much if not more on the coordination between the two performers, as on the quality of the individual performances. The analogy between the musicians, dancers, tennis players and oarsmen is strong.

At a systems science conference in 1998, Phillip Christian presented a paper which considered some aspects of the contribution insights from the performance arts could make to the understanding of collaboration (Christian, 1998). He argues that since the beginning of time to the present day, man has created collaboratively, ".... from cave paintings to Lennon and McCartney", and via Gilbert and Sullivan. He notes that in the performing arts, particularly in music, collaboration is a frequent occurrence, but feels that the literature does not deal adequately with the key elements that differentiate the success or failure of a collaboration. His hypothesis is that insights of how collaboration occurs in the arts might help us improve performance in business groups. His investigation was based on interviews with professional musicians, graduate students and business people. It is clear that Christian's work has parallels with that of the authors' and that there are principles and objectives in common.

It is not possible with the limited space available in this paper to address the detail of Philip Christian's paper, but his conclusions are supportive to the line the authors are following. He asks "why is it that certain teams just click, while others, well, don't? What is it about performers working together to bring forth collective excellence that could inform business process? Why do artists and performers seem to desire co-creation, while business people more often feel the need to succeed as individuals?" The point could be made that the answer to the last question might be that there can be material advantage to the businessman, probably financial, in individual success, and artistic advantage to the performer. This diversion does not lead us away from Christian's answer to these questions. He focuses on individual participation, personal values and expectations. Themes explored include responsibility, wholeness, individuality, and integration. He finds that the ideals and expectations expressed by performers are very different from those espoused by business people. However, honesty, trust, and the notion of being other-centric provide a framework for the emergence of collaboration. Coincidence of the values and beliefs of those involved provide a space for collaboration. He concludes that there is at the heart of every collaboration the desire to solve a problem, create, or discover something. The participants must share these goals if success is to be achieved. There must be a set of values that includes trust, honesty, appreciation of the contribution of others, openness and a desire for congruence. Performing artists come with an expectation of reciprocity. They are willing to set aside their individual goals and embrace the goal of betterment of the group. Elements which foster collaboration include good communication, intelligence, and leadership, but a common objective and shared values are the most important.

The authors note the correlation which it is possible to draw between the findings of Christian with those of Beer and Owen. Also, the contribution that the performing arts team metaphor can bring to the understanding of the requirements for successful team performance. It suggests that further study of the interactions which occur in the coming together of collaborators might be valuable in the creation of effective teams. However, there is a gut feel that artists might put a higher personal value on achieving an artistically satisfying result. To look back on the Maslow model of personal motivation (Maslow, 1987), it is possible that the need for "self-actualization" by the artist is much higher than that of the business-person or the footballer. The authors are thus encouraged to proceed further with their own investigations, particularly in the area of the football team metaphor, as the more diverse and less altruistic objectives of the footballer might mirror those displayed in the business environment.

STRATEGIC AND TACTICAL TEAMS

The authors were concerned that the word "team" was very general in its use and very broad in its application. A working definition of "team" was needed, and the authors settled for:

"a group of individuals committed to achieving specific objectives, explicitly or implicitly agreed, generated by a recognised rule-enclosed environment"

This covers a wide spectrum, including sports teams and cabinets, but the spectrum is too wide, and there are clear differences between sports teams and cabinets, the significance of the decisions taken within them, for example. The factors which appear to distinguish teams are: time; decision making process; task; and goals. This can be boiled down to two categories: strategic and tactical. Analysis proceeded with consideration of tactical and strategic teams in business, politics and sport.

Strategic teams are concerned with policy, and with the survival and prosperity of the organisation of which they are a part. They have long-term diffuse goals and engage in analytical decision-making, based on the identification and weighing of criteria, the evaluation of options, and the generation of the best option..

Tactical teams are responsible for the implementation of policy, and for the achievement of short-term defined goals. They perform "recognitional" or intuitive decision-making (Klein, 1998) based on previously experienced situations. The continuum on which teams can be located is illustrated in the diagram (Figure 1) below:

Figure 1.

Long-term	TIME	Immediate
Analytical	DECISION MAKING	Cybernetic
Policy	TASK	Implementation
Diffuse	GOAL	Single

STRATEGIC--TACTICAL

The Red Arrows aerobatic team (Owen, 1996) can be located near the extreme tactical end of the spectrum. There is remarkably high congruence between individual objectives and team objectives in this example. Footballers often change teams for personal advantage, but it would be difficult if not impossible for a Red Arrows pilot to do so. Team membership can change in project teams as the need for different skills and experience develops (Reavill, 1995). Cabinet teams change in what is known, often justifiably, as a "reshuffle". The first "real" cabinets of Attlee and Wilson were formed after a year in government, when they had assessed the situation, and had greater freedom of selection of ministers. The Labour party rules define the composition of the first cabinet on gaining power (Brady and Catterall, forthcoming).

The strategic end of the spectrum would be occupied by teams such as the governing body of a well established organisation with brand dominance in an unchanging market. The Roman Catholic Church, and the Politburo of the former Soviet Union furnish examples for this extreme end of the spectrum. Again their unique characteristics do not allow useful extrapolation. It therefore appears that further investigation should concentrate on the non-extreme areas. Organisations of all sizes can have teams in the strategic or tactical areas, but generally in the larger organisations, the more tactical decisions are taken by middle and lower management teams, and the strategic decisions at board level.

DISCUSSION AND CONCLUSION

A number of strands of meaning, or tentative conclusions, emerge from this "work in progress". The work considers teams, not necessarily successful teams, though the objective is to establish means of making them more successful. John Major's cabinet in the final stages of the life of his government could hardly be regarded as a successful team, but it was still technically a team despite doubts about this in some quarters. True synergy may not be possible with a team. It may only be possible to achieve the sum of the capability of the team members. Disputes and poor team maintenance may produce considerably less that this. However, some of the examples examined (tennis doubles, teams of musicians, pas-de-deux dancers) suggest that it may be possible for a team to produce more than the sum of its parts. "Teamness" is enhanced by shared values and the degree of commitment to the common objective. These provide the coordination strength. The skeleton of a three dimensional model is beginning to emerge, with strategic/tactical; shared values; and objective commitment as the dimensions.

REFERENCES

Adair, J., 1979 *Action Centred Leadership*, Gower, London.
Beer, S., 1994 *Beyond Dispute: The Invention of Team Syntegrity*, Wiley, Chichester.
Belbin, R.M., 1981 *Management Teams: Why they Succeed or Fail*, Butterworth, London.
Belbin, R.M., 1993 *Team Roles at Work*, Butterworth-Heineman, London.
Blake, R.B. and Mouton, J.S. 1964 *The Managerial Grid*, Gulf.
Brady, C., 1997 "Cabinet Government and the Management of the Crisis During Suez", *Contemporary British History*, Autumn.
Brady, C., and Catterall, P., 1997 "Managing the Core Executive", *Public Administration*, Autumn.
Brady, C., and Catterall, P., (forthcoming 1999) *Government by Committee: The Development of Cabinet Committees in Britain*, MacMillan, London.
Christian, P., "Insights from Performance Arts on Collaboration in Business", *Proceedings of the International Society for System Sciences*, Atlanta, July 1998.
Handy, C., 1990 *Inside Organisations*, p. 123, Plenum, London.
Keidel, R., 1985 *Game Plans*, Dutton, New York.
Klein, G.A., 1998 *Sources of Power: How People Make Decisions*, MIT Press, London.
Maslow, A.H., 1987 *Motivation and Personality*, Third Edition, Harper and Rowe, London.
Owen, H., 1996 *Creating Top Flight Teams*, Kogan Page, London.
Payne, A.C., Chelsom, J.V. and Reavill, L.R.P., 1996 *Management for Engineers*, p. 185, Wiley, Chichester.
Reavill, L.R.P., 1995 "Team Management for High Technology Projects", *Management of Technology Conference*, Aston University, April 1995, pp. 535-537.
Reavill, L.R.P., 1997 "High Performance Teams and their Development", *Systems for Sustainability*, Stowell et al (Eds), pp. 715-719, Plenum Press, New York.
Reavill, L.R.P., and Brady, C., 1998 "Football, Business and Government: Can Studies of High Level Teams Across Disciplines Produce Generic Principles for Management?", *Proceedings of the International Society for System Sciences*, Atlanta, July 1998.

SYNERGY IN A COMPLEX AND POST-MODERN WORLD

R.G.(Bob) Saunders

Centre for Complexity and Change
Open University
Walton Hall
Milton Keynes
MK7 6AA

INTRODUCTION

Synergy is an output from a system, an output which gives a result which is 'greater than the sum of the inputs'. Synergy is the concept that a team of people working together can produce more work in quantity and at a higher quality, with more creativity, than can individuals.

We know that we live in a complex world, a world which seems to become more and more complex as time goes by. This complexity has meant that the world becomes a more and more difficult place in which to live, and to understand. This, I submit has caused the human effects of that philosophical condition we know as 'post-modernism' to have more and more effect on the people, on all of us. Its consumerism, the availability of infinite information and almost infinite products, its relativity, individualisation, the very possibility of people being able to withdraw into a private world, the unwillingness to judge others, the distrust of logic and rationality, and of 'progress', the avoidance of hard decisions, and the replacement of discussion by babble. This spells out 'post-modernism and there is a real feeling that the enlightenment project is at an end, and that the western world - and the people in it - is therefore changing in its very essence.

In this paper I hope to explore the above, asking the question whether a 'team' as we know it can ever happen in a post-modern world, how we might deal with this, and hopefully I hope to be able to place before the community evidence as to whether the changes I mention above are having their effects even now. Let us start with the main conference topic:-

SYNERGY

The word comes from the Greek for 'co-operation' and has come to mean that the 'whole is greater than the sum of its parts'. It is also:-

Synergy Matters: Working with Systems in the 21st *Century,*
Edited by Castell *et al.*, Kluwer Academic / Plenum Publishers, New York, 1999.

- the state in which a team 'takes off', begins to work together.[1]
- The emergence of unexpected and interesting properties.[2]

The term is used in 'Systems', but has its place in organisation management theory. Like many such concepts it has a place at many hierarchical levels. These include such as Technology R&D and Marketing where technological and marketing synergy is defined as the ability to build on from existing resources/development and production.[3] Henry and Walker also suggest that synergy is a common thread that binds new business to old and is a key and significant factor to success. It is also used by Bennis et al in the same sort of way in regard to a whole economy.[4]

Bennis et al provide a detailed description of 'synergy' as it fits 'teams'. They quote from[5] M. Follet who said that there is no such thing as a self-contained unit or individual. That people delimit their identity by relating to other people. The unit of 'Society' in her estimation is the group- individual. There is no such thing as an independent (of outside sources) individual, no such thing as a separate ego or self-contained individual with separate independent existence of its own. She stated the people cannot operate outside 'society', or at the least to a specific group. It should also be noted that individuals affect others in the group. The dynamic is the same in all groups but where there is change of person in a group, the whole group changes and is affected. There is a number of different types of group, depending on the identity, size and form of organisation including those in organisations, work groups of various types with their own methodology, objectives and methods of operating. But these groups do depend on group-individuals, who affect and are affected by their groups. Synergy then is, by and large a desirable attribute of a team. Results from a group which can produce results which are 'greater than the sum of the parts' i.e. results which are greater in quantity and quality, with more creativity than can individuals working on their own, separately. To sum up, using Follets' work, synergy might be seen as reasonably easy to obtain, if people can work in this way. Unfortunately, it has never been that easy (and Follet, it must be said, didn't think it was easy either!) and is even less easy now.

TEAM WORKING

'Team working' therefore has become an important (not to say necessary) part of 'industrial' organisation and practice.[6] The author suggests, teamworking is an evolution of 'Fordism'. Fordism (from John Henry Ford), was the original 'line' operation (which, in effect is a sort of team operation, because people are dependant on others for a product on which to work) to I use here the word 'industry' in the broadest possible sense. 'Industry' for me ranges from hard engineering and manufacturing through software and IT to work in insurance companies and voluntary organisations. Anything which has a 'product' of some sort. Team working is used by all or most organisations which produce an output by the work of people and/or machines. As Morris[7] says, teams have always been around, but not particularly in the world of commerce and industry which until recent years have concentrated (not exclusively) on 'line' or departmental orientations. It has been the development of 'project' working which has developed with it many ideas of small groups and small group working. More latterly these have been cross-functional groupings, where people from many different skills and trades (and world-views) have come together to achieve common objectives. With this development the ideas and practice of team working within projects has grown up and these carry the considerable advantages of team operation - including the opportunity for synergy which has been described. This is true even in the type of matricized 'team' which involves multiplexing members of staff between projects. On this topic Cleland[8] quotes Reeser (Academy of Management Journal Volume 12

Dec.1979) 'project organisation has a built-in capacity for causing human problems of its own'. Insecurity about unemployment, career retardation and personal development is more keenly felt in project rather than functional organisations. There are also frustrations over ambiguity and conflict in the work environment, and over having more than one boss. This neglect of human needs can hinder the growth and development of the staff. The very idea of multi-functionality which brings with it the ideas and differing 'language' and the thinking which this brings with it, is also disturbing and frustrating to those involved.

These sets of people working together to achieve a common objective, therefore require special treatment and management. They need to be welded together to form 'the team'. As project managers will affirm, team building forms a large part of a project managers reading - and much time has been spent and much work has been done in this area. In the last few years, the idea of 'team' has become 'multi-functional' to ensure that the whole process of producing the deliverable is under the control of one group. This enables the process to be 'seamless' and overcomes the problems of independent departments being responsible for the onward progress of a project. (which is always a dubious way to organise a 'project'!) Team approach is therefore integral to the operation of the organisation and to the delivery of its products. So, we have team working which - hopefully - will lead to synergetic working, and a product which the individuals could not have produced on their own. And yet in this post-modern age, this may be where a problem might lie as far as team working is concerned. Apart from anything else, I have discussed the idea that there is no such thing as an entirely separate individual. There is evidence of change in this area which may well lead to this total separation of individuals. Even in [9] which was published in 1967 Argyle was talking about 'deviant' and 'independent' people able to change group norms by their behaviour, reducing cohesiveness. There is also much there on training in social skills, 'On the job', role playing, even reading, lectures, case studies and films being the most common approaches quoted. Argyle goes on to mention that this sort of training is needed by all age groups, but goes on to say that 'adolescents and young adults are those who seem to have the greatest need'. He says elsewhere says that 'relations with others are the most important part of human life, and that most ...essential human characteristics cannot be manifested by a person in isolation...for some people their social behaviour is based on obtaining maximum rewards, often at the expense of others'. Other anecdotal evidence, apart from that mentioned above reveals project managers finding this inability of people to work together more and more common and difficult to deal with, especially in 'knowledge based' R&D and IT projects. This evidence was obtained from an initial set of questions aimed at an Association for Project Management network of R&D project managers.

(I observe that this whole section may well be a very negative and extreme view, but I am known in some circles for this sort of prophetic thinking. Certainly there is a fair amount of evidence in Argyle and elsewhere for the problem as stated)

POST-MODERNISM

The very essence of the problem seems to lie in the assumption that people will always be able to work together and *can* be welded into a team, with all that means. But, even in the past, people were to be found who did not respond to this treatment. The whole purpose of this paper is to suggest that in a post-modern environment there may be a greater number of people who do not respond to 'team work'. Enough to make a real, and negative difference to successful project outcomes.

The problem arises because there has been a slow but continuing breakdown of the social consensus and a fundamental change in the way people understand the world.[10] There are theories which say that the complexification of the world is leading to there being more choice for people, but less opportunity to refuse to make the choices There is also evidence of increasing individualisation and increasing cynicism about the ideas of 'progress' and planning so beloved by the western world over the last 300 or so years of the Enlightenment project. People no longer believe in an over-arching common understanding - whether of social values or even rationality. This trend is being extended by commodification of everything -including people. It is also leading to the fragmentation of society one can remember that famous remark in 1988 'There is no such thing as society!' People are:-

- more and more looking for personal experience,
- having an awareness of the plural nature of the world
- having a belief that perspective ('where we're coming from', or 'weltanschauung') is everything.

This is having an effect on people, not only the young who are being educated and brought up in this environment. Certainly I have anecdotal evidence that young people are not always socialised when they enter the work environment, and this has to happen before they will work with others.

The preceding paragraph finished with an introduction to the state in human development that has been labelled as 'post-modern' (indicating what comes after the 'modern' period, or the 'enlightenment' project.) But there is more to come:-

- Post-modernism is the state of suggestion that the project has been discredited and closed.
- The commodification of society has caused a sea change in the way that people see the world and :
    - People are unwilling to adhere to logic and rationality, will not face hard issues,
    - People no longer believe in history or 'modern' progress or in over-arching narratives of life,
    - People have come to have an individual micro-life, personalised if you like, and there are people who are now almost totally separated from the rest of the social world.

Homeworkers and shift workers spring readily to mind, but there are also those whose life style is to be solitary, we are told, an ever increasing number, those who shop with, are entertained by, who live and work with and by their computers. To quote from [11] 'The slow...forgetting of social skills....What used to be...kept together by individual skills...now by technologically produced tools purchasable in the market. In the absence of such tools...groups disintegrate...presence and resilience of teams...are market dependent.'

THE FUTURE

If we are not, as Western humanity, (and post-modernism is, by and large at present relevant only to 'western' or '1st world' culture) to lose the ability to work together as teams it will be necessary to come to some understanding of people in this area so that we can seek ways which may well help to overcome the lack of socialisation being found in younger people and the individualisation of the population as a whole. Certainly the situation exists, as has been discussed above. Certainly it may result in lost projects if people cannot be found to work and learn together.

MEETING THE NEEDS

There is a need to ensure that the team is, in itself, socialised. That it is 'built'. The need for 'team building' has long been recognised. Stokes, in [12] gives a set of team building exercises which are called, collectively 'Synergetic Exercises'. These facilitate group work and team building. There are many 'team building' exercises and I guess many of us have made use of them. There are many people in the world who are not socialised, more, it must be said, than in the past, and we must learn to work with them, as they must learn to work with us. The trend is for this number to grow larger. Argyle has provided many ways of providing social training and the reference is recommended for deeper reading on this topic. Synergetic exercises seem to offer something more than the old team building exercises. These particular exercises start with the individual making decisions then move on to group decisions on various project management specific questions. The objectives and benefits:-

* learning to appreciate group work
* developing individual behaviour, attitudes and motivation
* training individuals
* developing group behaviour, attitudes and motivation
* training the group
* developing the organisation

He goes on to detail 'Temperament Analysis' of various sorts. Suggesting that a temperament type checklist can help people and groups to determine their own behavioural and thought profiles.

CONCLUSIONS

I set out to explore 'Synergy' in teams and its place in the world of 'meeting common objectives' (in projects, that being where I come from) and the effects of a change in mind set in the population with a philosophical slant that calls itself 'post-modernism'. As a label, this is a useful hook on which to hang some ideas about thinking, and they have been summarised here. Post-modern thinking is not an isolated occurrence in the 1st.world. Many people - including you and I - think in this way and exhibit these characteristics.

Generally I conclude that project managers need to give some thought to both socialisation and lettering about team working. That there is the possibility that they will become more difficult to achieve and thus team working, and synergy, will also become more difficult to achieve.

An extended form of team building may well form one route to synergy and some idea on that topic have been put forward from various sources. Team building already happens, more thought needs to be given to it, to ensure that it encompasses socialisation as well as building, and produces people who are able to work together even in this post-modern age.

REFERENCES

1. Carter et al. 'Systems management and change'. Harper & Row, London (1984)
2. Flood & Carson 'Dealing with Complexity Plenum, New York,(1988)
3. Henry & Walker. 'Managing Innovation' Sage, London (1991)
4. Bennis et al, 'Beyond Leadership' Blackwell, Oxford (1994)
5. Follett M, 'The New State', Longman London (1920) (quoted in 4.)
6. Kumar, K 'Post-industrial to Post-modern Society' Blackwell, Oxford (1997)
7. Morris PWG 'The Management of Projects' Telford,London, (1994)
8. Cleland & King, 'Systems Analysis & Project Management' McGraw- Hill, USA (1975)
9. Argyle M 'The Psychology of Interpersonal Behaviour', Penguin, London,(1994)(1st.1967)
10. Craig Y et al 'Tomorrow is another Country Church House Publishing, London, (1996)
11.Bauman, Z 'Postmodernism and its discontents' Polity, Cambridge (1997)
12. Stokes I 'Cultural Synergetics' 'Project Management Today', Wokingham, (1992)

MANAGING LEARNING AND KNOWLEDGE DURING I.S. DEVELOPMENT IN AN AUTOPOIETIC ORGANISATION: IMPLICATIONS FROM THE LITERATURE.

Clive Savory

Department of Computer and Information Sciences
De Montfort University
Hammerwood Gate
Kents Hill
Milton Keynes
MK7 6HP
E-mail: csavory@dmu.ac.uk

INTRODUCTION

This paper uses the framework of an autopoietic organisation, as the basis for considering the implications for managing learning and knowledge within the information system development process. The concept of autopoiesis is a relatively new one and has only recently been applied to organisations (e.g. Morgan 1986, Von Krogh and Vicari 1993). The principle of autopoiesis is used to explain the nature of autonomy within a living system, it suggests that organisations are closed systems exhibiting self-referencing behaviour. The implications for organisations is that they are cognitive systems that use their previous knowledge as a reference point for interpreting new knowledge.

The area of I.S. development is a relatively young one, though in its short life there has been a dramatic alteration in the perspectives taken. Several authors (e.g. Bell and Wood-Harper, 1998 pp. 227) have charted the history IS development in depth, but one of the main themes has been a move from a software engineering approach to ones that take into account the complications produced by including humans within systems. There are now several approaches that have taken more human centred approaches e.g. Soft Systems Methodology (SSM)(Checkland and Scholes, 1990), ETHICS (Mumford, 1995), Multiview (Avison and Wood-Harper 1990).

The major issue that is currently gripping the business world is how to cope with rapid, unpredictable change. For example Stacey (1996) highlights the problems associated with dealing with open ended change. The I.S. development area has also had to come to terms with rapid change in organisations, as often systems development is unable to cope with the pace of change. In response much of the management literature has focused on the problems of rapid change and considered the need for learning in organisations and the management of knowledge. It is suggested that only by the careful management of learning and knowledge that the pace of change in the business environment may be matched.

This paper looks at the scope for applying the concepts developed within the organisational learning and knowledge management literature to I.S. development context. Initially a review of the key concepts within organisational learning and knowledge management is given. The resulting discussion ends by introducing a rationale for the adoption of an autopoietic view of organisations. Though many of the ideas within this paper are rooted in a range of different philosophical areas, it seems useful where an idea helps us in understanding a situation to make use of it. This approach concurs with Dodgson (1993) who suggests the need for a multi-disciplinary approach to organisational learning.

Within the organisational learning literature much emphasis is put on the need for higher level learning (e.g. Argyris and Schon 1978, Senge 1991). IS development plays a central role in the structuration process (Giddens 1984) between the crystallisation of organisational routines and procedures, and the creation of new knowledge. Consideration of the learning mechanisms within an autopoietic system are therefore of interest

Synergy Matters: Working with Systems in the 21st Century,
Edited by Castell *et al.*, Kluwer Academic / Plenum Publishers, New York, 1999.

643

to the IS developer. By focusing on learning and knowledge management the process of IS development may be able to meet the challenge created by open ended change in the environment.

The paper concludes with a discussion of the issues that are raised by taking the perspectives of organisational learning, knowledge management and autopoiesis, particularly in the area of I.S. development.

CYBERNETICS

Before considering the very wide area of organisational learning literature, it is useful to summarise some of the concepts that underpin much of the work. The area of cybernetics provides fundamental ideas for understanding control and communication within an organisation. The means by which communication and control is used by organisations to cope with change is highly significant.

In taking a systems view of an organisation the primary problem is to cope with the complexity of system's environment. Complexity being defined "by the number of elements in the system, their attributes, the interactions among the elements, and the degree of organisation inherent in the system" (Schoderbek et al 1985 pp. 5). In order to cope with complexity the controller of a system must be able to exhibit an equal number of states or complexity. This idea has been encapsulated within Ashby's Law of Requisite Variety. In the context of I.S. development the control of complex projects must itself be done through a controller that exhibits an equal degree of complexity.

The second major concept of cybernetics is the concept of feedback systems. Schoderbek et al (1985 p.83) provides a useful description of 1st, 2nd and 3rd order feedback systems. The implications of considering the feedback loops within organisations are that in the case of 3rd order feedback we are given a model of feedback that encompasses learning through reflection on past decision making. Much of the organisational learning literature has built on the idea of feedback in building their models of learning. In terms of I.S. development this has the implication that to control I.S. development effectively, it is best done through 3rd order feedback and so the mechanism by which learning and reflection can be built into the development process must be understood.

One of the features of 3rd order feedback systems are the qualities of autonomy, self-awareness and self-organisation. These qualities have been expanded on in the work of Beer. In his viable systems model (VSM) of organisations he has provided a model to enable us to understand how these qualities come to exist in a system. Using the analogy of the human brain he describes a hierarchy of five sub-systems that each serve a different purpose in controlling the overall system. Through the delegation of activities to the next level down of the hierarchy the VSM model explains how a controller can cope with and diffuse variety. The implications of the VSM approach to organising I.S. development does not seem to have been explicitly addressed yet. Though Mumford does utilise VSM as a methodological framework for designing systems of work and hence the design of information systems (Mumford and Beekman 1994 p.83).

The area of cybernetics gives basis tools for understanding mechanisms for learning. At an organisational level though these ideas may seem rarefied. The organisational learning literature has filled the gap between conceptual level thinking and practice. The cybernetic principles described above though are still relevant and underpin many of the organisational learning approaches.

THE ORGANISATIONAL LEARNING LITERATURE

The initial question to be addressed by this paper then is to examine the nature organisational learning and highlight issues relevant to I.S. development. Definitions of organisational learning have been suggested from many authors. Initially the emphasis of definitions were based upon behavioural views of organisations and emphasised the change in behaviour aspects of learning (Cyert and March 1963). It is only later on that the emphasis of cognitive aspects of learning are introduced into definitions. Fiol and Lyles (1985) suggested that organisational learning was the process of improving actions through better understanding and knowledge. Clear cognitive approaches to organisational learning were identified by a range of authors that recognised that a process of conscious acquisition of knowledge or insight was at work (Argyris and Schon 1978, Hedberg 1981 and Huber 1991). Dodgson (1993) suggests that organisational learning definitions should include both behavioural and cognitive elements. This seems to be a sensible approach to the subject as underpinning ideas of cybernetics will show that to produce effective levels of control and communication both feedback loops from 1st to 3rd levels are beneficial. Behavioural approaches to learning emphasise the relationships between stimulus and responses are examples of 1st order feedback and it is only through the consideration of cognitive abilities can the goal direction of 2nd and 3rd order feedback be achieved. This highlights the problems in the study of organisational learning using reductionist approaches, characterised by behaviourist models of learning that ignore many of the emergent properties of learning in organisations. Holistic approaches to understanding organisational learning seem to have more value. For the purposes of thus paper the definition of organisational learning given by DiBella et al (1996) seems to be most appropriate (this definition builds upon the work of Huber 1990): "We define organisational learning as the capacity (or processes) within an organisation to maintain or improve performance based on experience. This activity involves knowledge acquisition (the development or creation of skills, insights, relationships), knowledge sharing (the dissemination to others of what has been acquired by some), and knowledge utilisation (integration of the learning so that it is assimilated, broadly available, and can also be generalised to new situations)."

Having defined organisational learning, an important set of questions is to identify where learning resides within an organisation. Several perspectives have been taken on this. Initially the view that learning is based on individuals was considered. The role of the individual as the main element in both learning and storing the results of learning has been the initial focus of organisational learning research. Dodgson (1993) has summarised the thinking suggesting that "individuals are the primary learning entity in firms". This reflects the work and approaches adopted by writers including Simon (1957), Argyris and Schon (1978) and Hedberg (1981). The use of individual learning theory to organisations though has been challenged. Schein (1985), Von Krogh and Roos (1995) suggest that learning lies within the domain of culture and not simply the individual. Other writers suggest that learning may reside within the organisations such as within organisational routines (Cyert and March 1963) or other aspects of organisational memory (Huber 1991). It is only through considering the role of culture and organisation as a repository for learning that the explanation of how learning transcends individuals can be explained.

The next question is to assess what entities are responsible for learning. The role of the individual as the primary learning entity seems a good starting point (Argyris 1991 and Dodgson 1993). However at the organisational level the process of learning may itself be defined. Nelson and Winter (1982) suggest that organisational routines may be used to systemise the process of learning and innovation. Levitt and March (1988) also discuss the influence of beliefs, paradigms and culture on organisational learning. In their discussion of competence traps they highlight how organisational learning is history dependent and relies on the interpretation of experiences.

The implications of this discussion of learning highlight important issues for I.S. development. In order to capture the results of organisational learning relevant to an information system, any approach must operate at three levels: the individual, the organisation and the culture of the organisation.

When considering the learning processes that make up organisational learning it is useful to start by looking at models of individual learning. Already this paper has made use of behavioural and cognitive models of learning. Straight application of these models to an organisational context seems to be problematic. The use of a behaviourist approach rests on the existence of stimulus-response (S-R) links. As discussed by Von Krogh and Roos (1995 p. 38-39) this assumes organisations to be input-output processing devices with the outputs being directly related to the inputs. They suggest this is not the case as the outputs have a more complex relationship with the inputs. In the case of cognitive views of organisational learning much of the theory rests upon the transfer of cognitive view of learning in individuals. The learning processes within organisations are therefore very complex. It is perhaps useful to see the development of learning in organisations as a interaction between behaviour and knowledge. The two being so closely linked that each affect the other. This can be seen a structuration process (Giddens 1984).

At the organisational level there seems to be a consensus that learning is closely linked with a process of reality creation (Argyris 1992 p.7). This view of learning also links with Vickers' description of an appreciative system (Checkland and Casar 1986) that assumes that learning is based on the interpretation of a flux of events and ideas. For example Levitt and March (1988) suggest that the development of organisational routines is history dependent, interpretation of an organisation's experience is through the combination of its commonly held beliefs, values, paradigms and culture. Hedberg (1981) uses this to develop the concept of unlearning where previously learned knowledge is discarded before being replaced by new knowledge. The ability of the organisation to unlearn is therefore as important as its ability to learn. Simon's theory of bounded rationality (Simon 1957) also highlights the issue of the problems of constructing knowledge without "perfect" knowledge. March et al (1991) identify a further issue when considering how organisation's learn from single "critical incidents". The construction of reality based on these infrequent events cannot guarantee the creation of valid knowledge. Senge (1990) suggests the use of systems as a strategy for coping with the tendency to react in a simplistic way to single events. One of the techniques that he suggests is to recognise patterns in the organisation's communication and control processes using system archetypes e.g. patterns of behaviour that reinforce poor performance.

Miller (1996) has attempted to make sense of the problems in understanding the process of organisational learning. Using two axes to describe the range of processes involved in organisational learning. The first axis describes the range of processes in terms of degree of determinism. This ranges from a highly deterministic process of learning such as within a learning ecology, to a voluntaristic process where actors within organisations are deemed as having high levels of autonomy and intelligence. The second axis uses the mode of thought and action in the learning process, this ranges from the methodical to the emergent. Using these two axes Miller suggests six possible learning modes: analytic, synthetic, experimental, interactive, structural and institutional. Using these modes of learning Miller proposes that different modes will be used depending on the level of complexity of the organisation. In particular the context defined by level of goal conflict or goal uncertainty will also make different modes of learning more or less effective. He suggests that viability of the organisation may well be effected by the mode of learning in operation. For IS this underlines the need to consider the learning mode that is provoked by a particular development approach.

The outcomes of organisational learning have also been discussed in the literature. At a relatively low level the organisational routine is a useful construct (Cyert and March 1963). By transforming learning into a new more appropriate way of operating in the form of an organisational routine the learning will be communicated to other individuals in the organisation. In forming organisational routines the knowledge created will be held tacitly within an informal routine or more explicitly in terms of a new formal procedure. The outcomes of learning therefore may be held in the formal fabric of the organisation or less formally within the cultural fabric of the organisation.

Core competences are an outcome of organisational learning that has been highlighted recently. Proposed as the basis of competitive advantage (Nonaka 1991), core competences place organisational learning as central to organisational strategy and link learning and strategy making (Prahalad and Hamel 1990). As Lei et al (1996) points out: "organisational learning alone does not translate into a core competence; rather, the firm must utilise and convert learning into firm specific resources and skills". From this it can be seen that the rationale for understanding the organisational learning process is given greater importance as it can provide the basis of competitive advantage for a firm. In the context of the resource based view of the firm defined by Amit and Showmaker (1993)"For managers the challenge is to identify, develop, protect and deploy resources and capabilities in a way that provides the firm with a sustainable competitive advantage and thereby a superior return on capital". Organisational learning is a key process and it is through a triple loop learning process that core capabilities of a firm will be developed (Andreu and Ciborra 1996). In their triple loop learning model, core competencies are a result of routinisation, capability and strategic learning loops. These levels of learning provide a clear demonstration of 1st, 2nd and 3rd order feedback. An interesting point made by Andreu and Ciborra is that the capability and strategic learning loops are the result of structuration processes between the capability and the organisational context. In essence these processes operate in a closed system.

Overall the most important outcome of organisational learning is the creation of knowledge (Nonaka 1994). This then leads into an examination of the nature of knowledge in organisations. Polanyi (1966) identified the distinction between tacit and explicit knowledge. Various other authors have then tried to make further distinctions between different types of knowledge. Senge (1993) highlights the role of knowledge creation placing the emphasis on management to mould the new business visions emerging from all levels of the organisation into a shared vision through the processes of communication, assessment and an on-going process of building on the ideas of one another. Perhaps the major issue for I.S. development is to cope with this knowledge management process and in particular approach the problem of how tacit knowledge in the organisation can be transformed and communicated as explicit knowledge.

From definitions of organisational learning it would seem important to note that organisational learning takes place when both the behaviour of an organisation is transformed perhaps in terms of its procedures, also organisational learning is concerned with changes in the cognition of the organisation. The improvement in terms of levels of understanding, insight, paradigms and frames of reference all contribute to improved levels of knowledge in the organisation. From the literature we also see that the entities responsible for both doing the learning and holding the subsequent knowledge will include the individual, the organisation and the cultures permeating through the organisation. To facilitate or capture organisational learning then we need to consider all three levels. The nature of learning has also been related to a process of reality creation, within the I.S. development process it is necessary to recognise that a purely positivist approach will not fully take in the richness of the learning process. In terms of learning processes again the I.S. developer must be ready to recognise processes that range from the methodical to the evolutionary. At the same time the degree of chaos in the learning process will range from deterministic through to voluntaristic modes. Finally the outcomes of learning will range from relatively locally oriented organisational routines to core competences of the organisations that provide the basic competitive advantage for the organisation. Finally perhaps the most valuable outcome of learning is an increase in the knowledge base. The lesson then for I.S. development is that a major focus should be knowledge management.

VIABILITY AND AUTONOMY

To consider the process of I.S. development from a learning and knowledge perspective it is useful to consider the I.S. development process as a system. From the work of Beer it would seem useful to consider what is required to make the system viable. There is insufficient space in this paper to discuss the applicability of the VSM to I.S. development. However it is perhaps useful to draw on some of the conclusions of Beer on what makes a systems viable. Beer places much emphasis on the concept of autonomy and uses as part of the explanation of the mechanism by which each of the sub-systems neutralise variety for each of the higher level systems. In considering the viability of the system of I.S. development requires an understanding of autonomy.

The main focus of this paper is to consider the I.S. development process in autopoietic terms. Autopoiesis is a process defined by Maturana and Varella (1980). The purpose behind its definition was to attempt to define and explain the nature of autonomy. In doing this they suggested that cognitive systems are not open systems but are closed to their environment. Autopoietic systems act autonomously in that they do not behave or produce outputs directly in response to inputs. In particular they relate to their environment by constructing reality through making distinctions in observations. An important consequence of treating systems as autopoietic is that they do not receive information directly from their environment. The signals received from the environment do not convey any information, simply the system through self-reference to previous experience makes distinctions that give meaning to the signals, consequently the knowledge of its environment is constructed. Though it would be difficult to clearly and firmly describe an organisation as autopoietic (Mingers 1995) it would seem useful to use autopoiesis in a metaphorical sense to help understand the I.S. development system. Further Von Krogh and Vicari (1993) have suggested that taking an autopoietic view of organisations will help in understanding the nature of knowledge within firms. This view of the organisation meets some of the views suggested by Andreu and Ciborra. (1996) as interpreting learning at the higher level as being a process of building and modifying the frames of reference held by the organisation. The adoption of an autopoietic view will also help in the understanding of how learning occurs at the voluntaristic end of Miller's axes and hence the learning modes linked to low levels of determinism.

DISCUSSION AND CONCLUSIONS

From this discussion then what are the implications for I.S. development? First of all in the cybernetic literature the importance of the I.S. development system being a viable one seems an important premise. Up to now emphasis has been placed on the system being developed matching Beer's Viable System Model. Should not the system producing the system also be viable? An important element of the VSM model is the autonomy of elements in a system, autopoiesis theory provides a frame work for considering how autonomy is created. The introduction of views of organisation that assume a closed system that operate on the basis of self-referencing behaviour allows the grouping of several important phenomena. Reality creation provides the basis for understanding the development of knowledge, autopoiesis provides a coherent model of how distinctions are made within a system about its external environment. The difficulty found by organisations in converting tacit knowledge to explicit knowledge is also explained by an autopoietic view. Reducing the inhibitors to learning in organisations can also be examined using an autopoietic view as the self-referencing behaviour required for reflective, goal changing behaviour is again embodied in the view.

The research questions that lead from these implications are wide ranging. It should be recognised that some of the issues have already been partially addressed. Some of the I.S. development methodologies with a soft focus have successfully worked on the problems of coping with constructed realities in organisations and have taken "relationship maintaining"(Stowell and West 1994) views of organisation models. SSM has successfully focused on the problems of how the different frames of reference of the world held within an organisation may be surfaced.

In applying autopoietic views however the question of how autonomy can usefully be integrated into the I.S. development process seems a useful question. As it is unlikely that all tacit knowledge can be efficiently converted into explicit organisational knowledge (Von Krogh and Roos 1995, p.51) then the creation of autonomous groups in the IS development process seem the only way that the triple learning loops identified by Andreu and Ciborra can be mobilised. It is probably only through the creation of autonomy that the de-selection of organisation routines that have become obsolete will occur at a rate that will match the pace of change in the organisation. In relying on autonomous groupings the problems of reducing the inhibiting factors to learning will also have be investigated. Finally using the autopoietic view a model of organisation that can cope with high levels of variety and complexity needs to be considered. The V.S.M. suggested by Beer may well provide a suitable framework for building an I.S. development approach that can support many of these suggestions.

REFERENCES

Amit and Shoemaker (1993) cited in Andreu, R. and Ciborra, C. (1996)

Andreu, R. and Ciborra, C. (1996) "Organisational learning and core capabilities development: the role of IT", Journal of Strategic Information Systems, Vol. 5, Part 2, pp. 111-127

Argyris, C. (1993) "On organisational learning", Basil Blackwell

Argyris, C. and Schon,D. (1978) "Organisational learning", Mass.: Addison-Wesley

Ashby, W.R. (1956) "Introduction to Cybernetics", Chapman and Hall

Avison, D. and Wood-Harper, T. (1990) "Multiview: an exploration in information systems development" Oxford, Blackwell Scientific Publications

Beer S., "Brain of the Firm", Wiley, 1981

Bell, S. and Wood-Harper, T. (1998), "Rapid information systems development" McGraw Hill

Checkland, P. and Casar, A. (1986) "Vickers' concept of an appreciative system: a systemic account", Journal of Applied Systems Analysis, Vol. 13, pp. 3-17

Checkland, P. and Scholes,J. (1990), "Soft systems methodology in action", Chichester, Wiley

Cyert, R. and March, J. (1963) "A behavioural theory of the firm", New York: Prentice Hall

DiBella, A., Nevis, E. and Gould, J. (1996) "Understanding organisational learning capability", Journal of Management Studies , MAY, Vol. 33, Part 3, pp. 361-380

Dodgson, M. (1993) "Organisational learning: a review of some literatures"", Organisation Studies, Vol. 14, Part 3, pp. 375-394

Fiol, C. and Lyles, M. (1985) "Organisational learning", Academy of Management Review , Vol. 10, pp. 803-813

Georg von Krogh and Johan Roos (1995) "Organisational epistemology", Macmillan Press Ltd, p. 61

Giddens, A. (1984) "The constitution of society", Polity Press, Cambridge

Huber, G.P. (1990) "Organizational learning: the contributing processes and the literatures", Organization Science , Vol. 1, pp. 88-115

Lei, D, Hitt, M. and Bettis, R. (1996) "Dynamic core competences through meta-learning and strategic context", Journal of Management, Vol. 22, Part 4, pp. 549-569

Levitt, B. and March, J. (1988) "Organisational learning", Annual Review of Sociology , Vol. 14, pp.319-340

March, J. Sproull, L. and Tamuz, M. (1991) "Learning from samples of one or fewer", Organization Science , FEB Vol. 2, Part 1

Maturana, H. and Varela, F. (1980) "Autopoiesis and cognition: the realisation of the living", London: Reidl

Miller, D. (1996) "A preliminary typology of organisational learning: synthesising the literature", Journal of Management , Vol. 22, Part 3, pp. 485-505

Mingers, J. (1995) "A comparison of Maturana's autopoietic social theory and Giddens theory of structuration", Warwick Working Papers

Morgan, G. (1986) "Images of organisation", Sage

Mumford, E. (1995) "Effective systems design and requirements analysis: the ETHICS approach" Macmillan Press, 1995

Mumford, E. and Beekman G. (1994) "Tools for change and progress", CSG Publications, Netherlands

Nelson, R. and Winter, S. (1982) "An evolutionary theory of economic change", Belknap Press

Nonaka, I. (1991) "The Knowledge Creating Company", Harvard Business Review , NOV DEC, pp. 96-104

Nonaka, I. (1994) "A dynamic theory of organisational knowledge creation", Organization Science, FEB, Vol. 5, Part 1, pp. 14-37

Polanyi, M. (1966) "The tacit dimension", Routledge

Prahalad, C. and Hamel, G. (1990) "The core competences of the corporation", Harvard Business Review, MAY JUN, pp. 79-91

Schein (1985) "Organisational culture and leadership" San Francisco: Jossey-Bass

Schoderbek, P, Schoderbek, C. and Kefalas (1985) "Management Systems: Conceptual Considerations", Business Publications

Senge, P. (1990) "The fifth discipline: the art and practice of the learning organisation", Doubleday

Senge, P. (1993) "Transforming the practice of management", Human Resource Development Quarterly , Vol. 4, Part 1, pp. 5-32

Simon, H. (1957) "Models of man", Wiley

Stacey, R. (1996) "Strategic management and organisational dynamics", Pitman, p.23

Stowell, F. and West, D. (1994) "Client led design", McGraw Hill

Von Krogh, G. and Vicari, S. (1993) "An autopoiesis approach to experimental strategic learning" in Ed. Lorange, P, Chakravarthy, B, Roos, J.and Van de Ven, A. (1993) "Implementing Strategic processes: Change, Learning and Co-operation", Basil Blackwell

ON REDISTRIBUTING VALUES: A MANAGEMENT METHODOLOGY

Martha Vahl[1] and Gerard de Zeeuw[2]

[1] Centre for Corporate (&) Community Renewal (CORE)
Nijenrode University
3621 BG Breukelen
The Netherlands
[2] University of Lincolnshire and Humberside
LN6 7TS Lincoln
United Kingdom

INTRODUCTION

The experience of being able to achieve together what doing alone appears impossible is widely spread. Examples include building the pyramids, creating companies and founding families (Corning, 1995). One of the main lessons seems to be that viable and effective organisations allow their members sufficient 'space' for self-determination—so they can match the organisation's division of labour to their personal tastes and abilities. Insurrection and disruption usually signal a lack of such space.

What is not clear is why the emphasis is on 'lessons' instead of on what we 'know'. There is a big difference. 'Learning' seems to refer to becoming able to separate what is useful from what is not. 'Knowing' may apply when this separation is exhaustive, and its instances completely recognisable. The statement not only tells us therefore that people in organisations need their 'space'—which as a 'lesson' seems undeniable—but also that we still do not 'know' enough to avoid using relatively vague words like 'sufficient'.

This does not appear due to a lack of effort. Recent examples of attempts to summarise our experiences in (re-) organising the organised include Ideal Planning, Business Process Redesign, the Viable System Model, Organisational Learning, to name but a few (Clegg, Hardy and Nord, 1966; Jackson, 1991). Each offers a formidable and operationally clear 'lesson'—but none seems eager to claim 'knowledge', or to enjoy much popularity on this score.

The lack of knowledge also does not seem due to a lack of awareness that going from 'learning' to 'knowing' can be important. What is difficult is to explain why this does not happen more frequently. One possibility is that it is too difficult *in principle*. Another that 'knowing' is still too difficult *in practice*, the difficulty being to systematically separate knowing what 'is' from learning what we can use such knowing for: the 'oughts'.

A number of authors appear to favour the first possibility. They appear inclined to let 'learning' replace 'knowing' (Kompier and Marcelissen, 1990; Peters and Waterman, 1982). Others prefer to focus for the time being on 'learning', hoping that later it will approach 'knowing' to any desired degree (Argyris and Schön, 1996; Reason, 1994; Ulrich, 1989). Or they accept the way both are interwoven (Checkland and Scholes, 1990; Guba and Lincoln, 1989).

The present authors wish to join the debate on the relation between organisational 'learning' and 'knowing'. Their starting point is the suspicion that 'learning' was introduced to answer the question how organisations can be improved even when it proves difficult to follow the traditional convention of science. The particular difficulty being that one can not

Synergy Matters: Working with Systems in the 21st Century,
Edited by Castell *et al.*, Kluwer Academic / Plenum Publishers, New York, 1999.

separate the 'is' and the 'ought' in any direct way (Hanson, 1958; Nagel, 1986; Putnam, 1983; Kripke, 1980; Lakatos, 1977; Steier, 1991).

The authors aim to explore one of the possibilities that remain to achieve the desired separation. What seems required is that the members of an organisation succeed in developing *viewpoints* that are *locked* into each other in the sense that their observations 'fit' their actions to help the organisation function effectively. The important insight here seems to be that viewpoints *embody* 'oughts': they constrain observers when making observations.

The paper is constructed as follows. Two examples are introduced to help define and clarify the problem to be solved (analysis). The authors next aim to characterise acceptable solutions. They emphasise observation and the development of viewpoints and continue to discuss their approach in terms of aspects like induction and of its similarity to traditional forms of research (transfer). They finish with some concluding remarks (conclusion).

ANALYSIS

A project by the first author seems to usefully exemplify the difficulties (Vahl, 1993). She was commissioned to evaluate two teams, which had been initiated to provide extra, non-standard services for people with physical disabilities. The evaluation was needed to decide on continuation. The teams were chosen as the 'units of analysis' of the evaluation. Aspects to be considered included the aims with which the teams had been set up, their achievements, and various indicators of possible shortcomings.

The difference between the aims and the achievements proved to be relatively small, which boded well for the teams—although there had been some significant failures. Then other developments intervened. When the workers in the teams were informed of the results, they self-organised improvements of their own choice—as did the clients. This changed the relevance of the results considerably, even to the point where the commissioner no longer wished to (or could) use them.

This did not mean that nothing had been 'learnt'—although the project clearly did not satisfy the commissioner's original need to 'know', something the project leader was politely (but irrelevantly) blamed for. The question thus arose what approach should have been taken. One answer would have been to select a different 'unit of analysis'—the obvious one being a 'unit' in which one would have included the reactions of the team members, and possibly of their clients—next to those of the commissioner.

This kind of design was implemented in a second project, an evaluation of five experimental teams (Vahl, 1994). First members' interactions were described focussing on observations that would trigger members' actions— in the form of statements 'If observation X, do action A'. Next team members were taught to negotiate changes in X, or A, or both. Any such (individual) change would 'reverberate' among the team members in that each would have to adapt, and consider further changes.

Implementation of this design was laborious in that it required large numbers of statements, which also had to be mutually connected. Eventually however members of each team did learn to implement the three changes (X, A, or both), and in this sense to behave *as if* they were 'commissioners' themselves. This had a second effect. The changes the teams implemented made clear what could be achieved still. They thus facilitated the decision by the 'real' (that is paying) commissioner to continue the experiments.

There was a third effect. Clients became able also to change their observations of the services provided (X), as well as their use of such help (A), or both. This implied that the decision by the commissioner was not simply his anymore: it required an evaluation by clients as decision-makers as well. In other words, what the project had achieved was not something that could be used only by the commissioner (which would not have been different from 'learning' about the teams), but rather by anyone using the teams' services.

To focus on what got 'known', these effects can be reformulated as follows. Firstly, being able to systematically negotiate among them implied that each team member acquired a 'space' to use and adapt to disturbances, e.g. changes in requirements from 'higher up' or from clients. Secondly, the teams became *organisationally closed*: they did not need any other organisational structure to 'know' how to adapt, even to 'unknown' disturbances.

The project proved to be an instance of 'research' therefore. Having achieved 'organisational closure' implied *observational closure* in the sense of traditional science, on three levels. *Members* could recognise each other as instances of contributors to the teams. The *teams* became instances of processes, the observations from which could support activities such as decision-making. Their *internal structure* made it possible to see the teams as instances of a service organisation with a clear outward identity and independence.

An interesting aspect of such research is that its resulting 'knowledge' is 'embodied' in the teams. They start to function as *reservoirs*, or retainers, in the sense that they can serve anyone who becomes a member: given organisational closure, the teams' functioning is invariant under replacement. One quality of such knowledge is that it will minimise misuse, as taking advantage of one's role would imply that roles of team members can be changed for external reasons, and this would destroy the teams' closure.

An even more remarkable aspect of such 'knowledge' is that it is related not only to (improved) *observations* (which makes it part of traditional science), but also to (improved) *values*, as embodied in the viewpoints of the team members in the resulting division of labour (which makes it non-traditional in its implications). As indicated, such values also will be 'honest' expressions, or in conventional terms, achieving observational closure will reduce any irrelevant 'laden-ness' of the values (Hanson, 1958; Vahl, 1997a; De Zeeuw, 1998b).

With this notion of (improved) values we seem to have hit on at least one good reason why it is difficult to go from 'learning' to 'knowing'. 'Learning' will usually be a mixture of changes in values (the 'oughts') and in observations (the 'is's). But 'knowing' in the traditional sense does not include any 'ought' (Kant, 1960; Churchman, 1971). From this it only follows, as the literature does indicate (Guba and Lincoln, 1989), that the best one can do is allow organisation to 'learn', as they can not 'know'.

Alternatively, the above analysis suggests that it is possible for organisations to 'know' in the traditional sense. What this requires is dealing with the 'oughts' in a special way. They have to be *locked into* an organisation through negotiations or *communications* among its members—within constraints obtained when observational closure is achieved (expressed, in the example above, as modifications of statements like 'if observation X, do action A'). When this happens, members become recognisable by their (preferred) *viewpoints*.

TRANSFER

The second project presented above may be summarised in terms of two important processes. On the one hand there is the emphasis on members exchanging and negotiating observations to achieve observational closure on different levels. On the other hand there is the emphasis on the process of satisfying the criteria, which implies changing existing organisations to become more robust against disturbances, that is more viable. This implements what traditionally is considered research, which also emphasises the need to collect and improve on observations. The effectiveness of improvements is taken to depend on how observations are transferred, that is on the form of transfer. Its power comes from the fact that it does not allow organisations only to gain experience: such experience must be checked against the criterion of observational closure.

Together the two processes seem to constitute a method that stands out, firstly, for its emphasis on combining (reports of) *observations* as a way of realising changes such as a new division of labour. Secondly, for its emphasis on *observational closure* as a criterion for choosing some combination. In contrast, other methods seem to prefer to take (reports of) *experiences* as their 'input' (Argyris, 1992; Covey, 1992), and as their criterion the *effects* an organisation hopes to achieve by operating on these experiences (Checkland, 1981).

A widely preferred form of transfer is 'if observation X, then observation Y'. It implies that improvement results from reducing observational deviations (without this, X would not be mapped one-to-one on Y, as intended). This reduction may be achieved by identifying a *class* of *observations*, and by recognising deviating observations as deviating from that class. In traditional research this class is called the 'object of study'. It summarises a general *viewpoint*, or the constraints on observations that a scientific observer finds are needed (De Zeeuw, 1998a).

In the previous section the authors introduced another form: 'if observation X, do action A'. Improvement here implies observing a *class of viewpoints*, in this case to recognise deviating viewpoints rather than observations. Such a class implies that two types of viewpoints are distinguished: that of the scientific observer (who summarises 'the' object of study), and those from which the observations are made that support actions A (we may call them *modified* or non-traditional objects of study).

Modified objects combine into a non-modified object only when their viewpoints 'lock into' each other. In this case there is no need for further observations to support actions A: each viewpoint sufficiently helps to maintain the other viewpoints as elements in the class. This implies that the class of viewpoints is observationally closed, all viewpoints having been modified to create the space for co-ordinating the actions A, thereby to achieve strong collective actorship (Axelrod, 1984; Casti, 1979; Haken, 1988; Lewin, 1992).

This analysis of this form of transfer justifies the conclusion that it is sufficiently similar, given a proper stretching of the notion of an 'object of study'. That is to say that it can also be seen as a form of research, and as leading to 'knowing', even though it requires a special process of construction. One has to implement *a form of transfer* that ensures that a class of 'locked' viewpoints develops (see also transfer), or in embodied form: that those to be involved become observers in a collective that itself is observationally closed (Arthur, 1988).

This process of construction is an interesting one in that it appears to turn traditional research inside out, although it remains similar. Such research aims to *exclude* the users of its results from the process of acquiring them. That is to say, their 'oughts' are excluded and become invisible. In contrast, our approach aims to *include* users. It tries to do so by making at least some 'oughts' visible (as 'modified objects') by enclosing them in an 'is' (the collective)—in fact as many 'oughts' as possible, to achieve observational closure on the 'is' (De Zeeuw, 1997; 1998b).

In this way it becomes possible to achieve what traditional research can not do: it can 'know' when the latter can only suggest to 'learn'. There is a cost however. Creating observers in a collective requires effort, and even continuous effort (financially, as providing information can be expensive; organisationally, as 'locking into' may have to be 'umpired'). There is some compensation however: in our first project we failed to maintain the commissioner as an exclusive 'authority', so his funding was wasted.

It seems useful to consider avoiding such waste. This is not what seems to happen however. Research funded by single 'authority' commissioners still seems to increase (in the Netherlands for example estimated at Mfl 750 per year; In 't Veld, 1997; p. 89). Lacking the resources to maintain the 'knowledge' from such research, its use often leads to unpredictable and large side effects. It seems advisable to avoid such costs, for economic reasons, but also for ethical reasons as it decreases the *discretionary power* of those involved (Vahl, 1996; 1997b; Vahl and De Zeeuw, 1997).

CONCLUSION

The literature on changing, improving and innovating is large and still growing. There still is a remarkable gap however, in the sense that what is actually 'known' appears to be relatively limited. What the literature appears to report on seems to consist mainly of reports on various experiences, and on approaches that have been applied more than once, although not necessarily leading to the same results. One such approach emphasises the need for organisations to 'learn' rather than to 'know'.

This may be the best solution to the problems of change and innovation yet. Still the question remains why 'knowing' or even 'knowing to learn' are so difficult to achieve. In the paper the authors wished to contribute to the discussion on possible answers to this question. They presented and analysed some of their own projects, both successful and unsuccessful. They conclude that insufficient use has been made of a major achievement of Western thought, the *disciplinarisation of observation*.

What they propose therefore is to acquire knowledge on organisations by studying them in terms of the way people act as 'observers', or as 'embodied points of view'. This kind of study requires careful designs, as it will take continued effort to maintain the distribution of 'points of view' that represents knowledge. One of the more attractive designs appears to be to identify forms of transfer which allow for collectives of observers to develop, all of whom can 'test' their observations on each other, to better co-ordinate their actions.

REFERENCES

Argyris, C., 1992, "On Organizational Learning," Blackwell, Cambridge, MA.

Argyris, C., and Schön, D., 1996, "Organizational Learning II: Theory, Method, and Practice," Addison-Wesley, Reading, Massachusetts.

Arthur, W.B., 1988, Self-reinforcing mechanisms in economics, *in*: "The Economy as an Evolving Complex System," P.W. Anderson, et al., eds., Addison-Wesley, Reading, Massachusetts.

Axelrod, R., 1984, "The Evolution of Co-operation," Basic Books, New York.

Casti, J., 1979, "Connectivity, Complexity and Catastrophe in Large-Scale Systems," Wiley, New York.

Checkland, P.B., 1981, "Systems Thinking, Systems Practice," Wiley, Chichester.

Checkland, P., and Scholes, J., 1990, "Soft Systems Methodology in Action," Wiley, Chichester.

Churchman, W.C., 1971, "The Design of Inquiring Systems," Basic Books, New York.

Clegg, S.R., Hardy, C., and Nord, W.R., 1996, "Handbook of Organization Studies," Sage, London.

Corning, P.A., 1995, Synergy and self-organisation in the evolution of complex systems, *Systems Research* 12: 89-121.

Covey, S.R., 1992, "The Seven Habits of Highly Effective People: Powerful Lessons in Personal Change," Simon & Schuster, London.

Guba, E.G., and Lincoln, Y., 1989, "Fourth Generation Evaluation," Sage, London and Lincoln.

Haken, H., 1988, "Information and Self-Organization," Springer-Verlag, Berlin.

Hanson, N.R., 1958, "Patterns of Discovery," Cambridge University Press, Cambridge.

In 't Veld, R.J., 1997, "Noorderlicht. Over Scheiding en Samenballing," Vuga, Den Haag.

Jackson, M.C., 1991, "Systems Methodology for Management Sciences," Plenum, New York.

Kant, I., 1960, "Kritik der Reinen Vernunft," Meiner, Hamburg (original 1781).

Kompier, M.A.J., and Marcelissen, F.H.G., 1990, "Handboek Werkstress," NIA, Amsterdam.

Kripke, S.A., 1980, "Naming and Necessity," Basil Blackwell, Oxford (revised from 1972).

Lakatos, I., 1977, "Proofs and Refutations. The Logic of Mathematical Discovery," Cambridge University Press, Cambridge.

Lewin, R., 1992, "Complexity. Life at the Edge of Chaos," MacMillan, New York.

Nagel, T., 1986, "The View from Nowhere," Oxford University Press, Oxford.

Peters, T.J., and Waterman, R.H. Jr., 1982, "In Search of Excellence," Harper and Row, New York.

Putnam, H., 1983, "Realism and Reason: Philosophical Papers," Cambridge University Press, Cambridge.

Reason, P., 1994, Three approaches to participative inquiry, *in*: "Handbook of Qualitative Research," N.K. Denzin, and Y.S Lincoln, eds., Sage, Thousand Oaks, CA.

Steier, F., ed., 1991, "Research and Reflexivity," Sage, London.

Ulrich, W.,1989, Critical heuristics for social systems design, *in*: "Operational Research and the Social Sciences," M.C. Jackson, P. Keys, S.A. Cropper, eds., Plenum Press, New York.

Vahl. M., 1993, "The Community Resource Teams for People with Physical Disabilities," The University of Hull, Hull.

Vahl, M., 1994, "Improving Mental Health Services in Calderdale. An Evaluation of 5 Schemes Funded through the Mental Illness Specific Grant," The University of Hull, Hull.

Vahl, M., 1996, Managing stories: a community research perspective on water, *in*: "Cybernetics and Systems '96," R. Trappl, ed,. Austrian Society for Cybernetic Studies, Vienna.

Vahl, M., 1997a, Doing research in the social domain: concepts & criteria, *in*: "Systems for Sustainability: People, Organisations, and Environments," F. Stowell,.et.al., eds., Plenum, New York.

Vahl, M., 1997b, "Choice of an Area for a Third Community Access Centre," Centre for Systems Research, University of Lincolnshire and Humberside, Lincoln.

Vahl, M., and Zeeuw, G. de, 1997, What does a self-referential research methodology look like?, *Systemica* 11:357-374.

Zeeuw, G. de, 1998a, Improving on differences among viewpoints, *in*: "The Complexity of Relationships in Action Research," B. Boog, H. Coenen, L. Keune and R. Lammers, eds., Tilburg University Press, Tilburg.

Zeeuw, G. de, 1997, Second order organizational research, *in*: "Organizational Cybernetics," J. Achterbergh, R. Espejo, H. Regtering, M. Schwaninger, eds., Nijmegen Business School, Nijmegen.

Zeeuw, G. de, 1998b, Interaction of Actors Theory (accepted for publication in the *International Journal of Human-Computer Studies*).

INDEX